The Pneumococcus

The
Pneumococcus

Volume Editor

Elaine I. Tuomanen
St. Jude's Children's Research Hospital

Associate Editors

Timothy J. Mitchell
University of Glasgow

Donald A. Morrison
University of Illinois at Chicago

Brian G. Spratt
Imperial College

ASM
PRESS

Washington, D.C.

Copyright © 2004   ASM Press
American Society for Microbiology
1752 N St., N.W.
Washington, DC 20036-2904

Library of Congress Cataloging-in-Publication Data

The pneumococcus/volume editor, Elaine I. Tuomanen; associate editors, Timothy J. Mitchell, Donald Morrison, Brian G. Spratt.

   p. ; cm.
  Includes bibliographical references and index.
  ISBN 1-55581-297-X (alk. paper)
  1. Streptococcus pneumoniae.
  [DNLM: 1. Streptococcus pneumoniae—pathogenicity. 2. Streptococcus pneumoniae—physiology. 3. Pneumococcal Infections—physiopathology. 4. Pneumococcal Infections—prevention & control. WC 217 P7378 2004] I. Tuomanen, Elaine I.

QR82.S78P548 2004
579.3′5—dc22

                        2004003049

10 9 8 7 6 5 4 3 2 1

Address editorial correspondence to ASM Press, 1752 N St., N.W., Washington, DC 20036-2904, U.S.A.

Send orders to: ASM Press, P.O. Box 605, Herndon, VA 20172, U.S.A.
Phone: 800-546-2416; 703-661-1593
Fax: 703-661-1501
E-mail: books@asmusa.org
Online: www.asmpress.org

CONTENTS

CONTRIBUTORS

Karita Ambrose
Department of Medicine
Emory University School of Medicine
Atlanta, GA 30322

Carina Bergmann
Department of Microbiology
University of Kaiserslautern
Paul Ehrlich Strasse 23
D-67663 Kaiserslautern
Germany

David E. Briles
Departments of Microbiology and Pediatrics
University of Alabama at Birmingham
Birmingham, AL 35294

Angela B. Brueggemann
Interdepartmental Academic Unit of Microbiology and Infectious Disease
John Radcliffe Hospital
University of Oxford
Headington
Oxford OX3 9DU
United Kingdom

Jay C. Butler
Arctic Investigations Program
Centers for Disease Control and Prevention
Anchorage, AK 99508

Fang Chi
Department of Microbiology
University of Kaiserslautern
Paul Ehrlich Strasse 23
D-67663 Kaiserslautern
Germany

Jesus Colino
Department of Pathology
The Uniformed Services University of the Health Sciences
4301 Jones Bridge Road
Bethesda, MD 20814

Derrick W. Crook
Interdepartmental Academic Unit of Microbiology and Infectious Disease
John Radcliffe Hospital
University of Oxford
Headington
Oxford OX3 9DU
United Kingdom

Ron Dagan
Pediatric Infectious Diseases Unit
Soroka University Medical Center
Ben-Gurion University of the Negev
P.O. Box 151
Beer-Sheva 84101
Israel

Christopher G. Dowson
Biomedical Research Institute
Department of Biological Sciences
University of Warwick
Coventry CV4 7AL
United Kingdom

Kathryn M. Edwards
Division of Infectious Diseases
Department of Pediatrics
CCC-5323 Medical Center North
Vanderbilt University School of Medicine
Nashville, TN 37232

Ernesto García
Departamento de Microbiología Molecular
Centro de Investigaciones Biológicas, CSIC
E-28040 Madrid
Spain

José Luis García
Departamento de Microbiología Molecular
Centro de Investigaciones Biológicas, CSIC
E-28040 Madrid
Spain

Pedro García
Departamento de Microbiología Molecular
Centro de Investigaciones Biológicas, CSIC
E-28040 Madrid
Spain

Regine Hakenbeck
Department of Microbiology
University of Kaiserslautern
Paul Ehrlich Strasse 23
D-67663 Kaiserslautern
Germany

William P. Hanage
Department of Infectious Disease Epidemiology
Imperial College London
St. Mary's Hospital
London W2 1PG
United Kingdom

Susan K. Hollingshead
Department of Microbiology
University of Alabama at Birmingham
845 19th South Street
Birmingham, AL 35294

Margaret K. Hostetter
Department of Pediatrics
Yale University School of Medicine
333 Cedar Street, LMP 4085
P.O. Box 208064
New Haven, CT 06520-8064

Edward N. Janoff
Mucosal and Vaccine Research Center
Veterans Affairs Medical Center
University of Minnesota School of Medicine
Minneapolis, MN 55417

Helena Käyhty
Vaccine Immunology Laboratory
National Public Health Institute
00300 Helsinki
Finland

Abdul Q. Khan
Department of Pathology
The Uniformed Services University of the Health Sciences
4301 Jones Bridge Road
Bethesda, MD 20814

Keith P. Klugman
Division of Infectious Diseases
School of Medicine
Emory University
1518 Clifton Rd. NE
Room 764
Atlanta, GA 30322

Sanford A. Lacks
Biology Department
Brookhaven National Laboratory
Upton, NY 11973

Marc Lipsitch
Department of Epidemiology
Harvard School of Public Health
677 Huntington Avenue
Boston, MA 02115

Rubens López
Departamento de Microbiología Molecular
Centro de Investigaciones Biológicas, CSIC
E-28040 Madrid
Spain

Paul Anthony Majcherczyk
Department of Fundamental Microbiology
University of Lausanne, Biology Building
CH-1015 Lausanne
Switzerland

P. Helena Mäkelä
Department of Vaccines
National Public Health Institute
Mannerheiminitie 166
00300 Helsinki
Finland

Larry S. McDaniel
Departments of Microbiology, Surgery, and Medicine
University of Mississippi Medical Center
Jackson, MS 39216

Tim J. Mitchell
Division of Infection and Immunity
Institute of Biomedical and Life Sciences
Joseph Black Building
University of Glasgow
Glasgow, G12 8QQ
United Kingdom

Philippe Moreillon
Department of Fundamental Microbiology
University of Lausanne, Biology Building
CH-1015 Lausanne
Switzerland

Daniel M. Musher
Infectious Disease Section
Department of Medicine
Veterans Affairs Medical Center
Houston, TX 77030

James C. Paton
School of Molecular and Biomedical Science
University of Adelaide
Adelaide, S.A. 5005
Australia

Timothy E. A. Peto
Interdepartmental Academic Unit of Microbiology and Infectious Disease
John Radcliffe Hospital
University of Oxford
Headington, Oxford OX3 9DU
United Kingdom

Shwan Rachid
Department of Microbiology
University of Kaiserslautern
Paul Ehrlich Strasse 23
D-67663 Kaiserslautern
Germany

Jeffrey B. Rubins
Mucosal and Vaccine Research Center
Veterans Affairs Medical Center
University of Minnesota School of Medicine
Minneapolis, MN 55417

Karen L. Sleeman
Oxford Vaccine Group
University of Oxford
Centre for Clinical Vaccinology & Tropical Medicine
Churchill Hospital
Old Rd., Headington
Oxford OX3 7LJ
United Kingdom

Clifford M. Snapper
Department of Pathology
The Uniformed Services University of the Health Sciences
4301 Jones Bridge Road
Bethesda, MD 20814

Brian G. Spratt
Department of Infectious Disease Epidemiology
Imperial College London
St. Mary's Hospital
London W2 1PG
United Kingdom

David S. Stephens
Department of Microbiology and Immunology
Emory University School of Medicine
Atlanta, GA 30322

Edwin Swiatlo
Departments of Medicine and Microbiology
University of Mississippi Medical Center
VA Medical Center Research Service
Jackson, MS 39216

Hervé Tettelin
Department of Microbial Genomics
The Institute for Genomic Research
9712 Medical Center Drive
Rockville, MD 20850

Elaine I. Tuomanen
Department of Infectious Diseases
St. Jude Children's Research Hospital
Memphis, TN 38105

Joerg R. Weber
Department of Neurology
Charité, Universitätsmedizin Berlin
D-10098 Berlin
Germany

Jeffrey N. Weiser
Department of Microbiology and Pediatrics
University of Pennsylvania
402A Johnson Pavilion
Philadelphia, PA 19104-6076

Zheng Qi Wu
Autoimmunity Branch
NIAMS
National Institutes of Health
9000 Rockville Pike
Bethesda, MD 20892

Janet Yother
Department of Microbiology
University of Alabama at Birmingham
Birmingham, AL 35294

FOREWORD

The pneumococcus, *Streptococcus pneumoniae*, a normal component of the microflora of the human upper respiratory tract and an opportunist pathogen, has been an object of scrutiny for almost a century and a quarter. Its initial isolations have been described succinctly by Welch (93):

> The micrococcus lanceolatus was discovered by Sternberg in September, 1880, by inoculation of rabbits with his own saliva. It was next found by Pasteur in December, 1880, by inoculation of rabbits with the saliva of a child dead of hydrophobia. Pasteur's observations were the first to be made public, being announced at a meeting of the Académie de Médecine in Paris on January 18, 1881. . . . Sternberg's first publication on this subject appeared on April 30, 1881.

Both scientists were able to grow the organism in vitro, and their descriptions of the morphology of their isolates were similar (76, 82), each recognizing what are now designated its diplococcal form and its capsule, attributes visible in the first microphotograph of the pneumococcus made by Sternberg (83). Sternberg was able also to recover pneumococci from the saliva of several carriers.

Since the initial observations of Sternberg and of Pasteur, much has been learned about the ecology of the pneumococcus in health and in disease. Studies of the pneumococcal carrier state have revealed that colonization of the human upper respiratory tract may occur on the day of birth (36) and that children may be colonized simultaneously with as many as 4 different capsular types (37) and, over a period of several years and at different times, with 1 or more of 12 different capsular types (55). Carriage of a single type over a period of 3 years in an adult in the absence of illness has also been observed (6, 91). Noteworthy are the findings demonstrating convincingly in an epidemiological study of the pneumococcal carrier state that half those colonized would not have been identified had mice not been inoculated with their upper respiratory secretions (47).

The mammalian lower respiratory tract, in contrast, is highly resistant to pneumococcal infection. Normal mice exposed to an aerosol containing pneumococci suspended in particles small enough to reach the pulmonary alveoli clear the organism rapidly. Mice infected with influenza A virus do not develop pneumococcal pneumonia when similarly exposed at a time when the viral titer is at its height but in the absence of histologic changes in the lung. Only when there is visible evidence of pulmonary injury or edema does pneumococcal infection develop following such exposure (39). These findings are emblematic of events that, in many instances, find their counterparts in humans. It is highly likely also that viral infection of the upper respiratory tract plays a similar antecedent role in pneumococcal infections of the middle ear.

Although it is probable that pneumococci were visualized in pulmonary secretions by Klebs as early as 1875 and in histologic sections of the lung described by Carl Friedländer in 1882, recognition of the pneumococcus as the principal cause of human lobar pneumonia was not to be clarified for another four years, when Anton Weichselbaum resolved the bitter dispute between Friedländer and Albert Fraenkel as to its etiology (4). It is of historic interest that Christian Gram, working in Friedländer's laboratory, developed the stain that now bears his name not to distinguish among bacteria but rather to facilitate their visualization in histologic sections of the lung (34).

The 1880s saw rapid extension of the association of the pneumococcus with extrapulmonary foci of infection that are now well recognized. Pneumococci were isolated from blood, from the lung by direct puncture during life, from pleural, cerebrospinal, and joint fluid, and from the middle ear (5). In the same decade, Netter described pneumococcal endocarditis and produced the disease experimentally by damaging a rabbit's aortic valve and injecting the animal subsequently with pneumococci (67).

Knowledge of the host's defenses against pneumococcal infection evolved slowly over the final 15 years of the 19th century. In 1885, Fraenkel recorded that a rabbit which recovered from a pneumococcal infection of the ear was refractory to reinfection with the same pneumococcal strain (31). Gamaléia, in 1888, was perhaps the first to suggest the role of phagocytic cells in defense against pneumococcal infection of the lung (33). In 1891, Metchnikoff described agglutination of pneumococci by immune serum (62), and two years later, Issaef reported that immune serum, although lacking both antitoxic and antibacterial properties, promoted phagocytosis (49). In the early 1890s, the Klemperers observed that the offspring of immunized rabbits were immune to infection with the same pneumococcal strain used to immunize the mother, that the serum of an immunized rabbit would protect an unimmunized rabbit against challenge with the same organism, and that the serum of a patient after crisis could protect a rabbit against infection. They also injected several patients subcutaneously with small amounts of rabbit serum (50). The rational development of serotherapy, however, had to await the clear definition of pneumococcal types, intimations of which are discernible in the observations of Bezançon and Griffin (14). In the same period, loss of virulence on passage of pneumococci in vitro and its recrudescence on subsequent passage in vivo were also described (53).

To Fred Neufeld must be credited major contributions to knowledge of the pneumococcus. In 1900, he described the bile solubility of pneumococcus, an observation which greatly facilitated its identification (68); in 1902, he described the Quellung reaction (69); and in 1904, with Rimpau, he showed that antipneumococcal antibody combines with pneumococci and not with the polymorphonuclear leukocytes that phagocytize them (73). Of equal, if not greater, importance was his definition with Händel of pneumococcal capsular types 1 and 2, making possible the introduction of rational serotherapy of pneumococcal infection (71). Interestingly, neither Neufeld nor others concerned with pneumococcal infection were to use the Quellung reaction in their studies for a quarter century following its original description, the identification of the first 30 pneumococcal capsular types having been achieved with other techniques (22). The term "Quellung" is probably a misnomer; for, as shown by Schiemann and Casper, a combination of a solution of capsular polysaccharide with homologous antibody results in the formation of a liquid gel which differs in no way in its refractile properties from the pneumococcal capsule when exposed to homologous antibody, giving rise to what might be termed more appropriately a capsular precipitin reaction (79). The Quellung technique was introduced by Neufeld and Etinger-Tulczynska as the preferred method for typing pneumococci in 1932 (70).

Both the capsule and the mode of cellular separation after division influence the morphological patterns of the pneumococcus in liquid and on solid media and have given rise historically to several descriptive terminologies at both the microscopic and macroscopic levels (2). The patterns of pneumococcal growth appear to be determined largely by its environment, the organism adapting to conditions optimizing its nutrition and its defense against phagocytosis. In liquid media, pneumococci, whether capsulated or noncapsulated, grow preferentially as diplococci distributed throughout the medium. Continued cultivation of capsulated strains in liquid medium in the presence or absence of antibody is associated with loss of capsulation, no longer essential to survival. Cultivated on solid media for prolonged periods as dispersed colonies, pneumococci, capsulated or noncapsulated, give rise to variants growing in long chains, with pseudomotility providing nutritional advantage over diplococcal forms. By contrast, in liquid media, the filamentous variants are at a nutritional disadvantage, being autoagglutinable, and revert to the diplococcal form. An additional type of morphological variation has been described recently by Weiser et al., again related to environmental circumstances and reflected in the transparence or lack thereof of colonies on transparent solid media (92). Pneumococci isolated from the nasopharynx show a higher ratio of cell wall polysaccharide to capsular polysaccharide and form transparent colonies, in contrast to similar organisms isolated from blood in which the augmented amount of the capsular polymer is associated with an opaque colonial form.

The virulence of a given pneumococcal type, of which 90 have been defined to date (46), is determined both by the chemical composition of the capsular polymer and the amount synthesized (52). Of the two, the former is the more important. In addition to capsular polysaccharides, the pneumococcus produces an additional polymer in this category, the so-called C or cell wall polysaccharide composed in part of a teichoic acid. Discovered during an investigation of the cellular constituents of the pneumococcal cell by Tillett and Francis in 1931 (87), this carbohydrate is common to pneumococci of all capsular types. Noncapsulated pneumococci grown in a suitable environment can give rise to mutants producing a capsule of C-like polysaccharide, the genes for which occupy a locus separate from those determining the synthesis of conventional capsules (16). Pneumococcal C polysaccharide is of additional interest for its ability to combine with an acute-phase protein of human hepatic origin. Unlike antibody, C-reactive protein appears at the height of the inflammatory response, be it of pneumococcal or unrelated origin, and disappears from the circulation with resolution of the inflammatory process (1).

The pneumococcus has played a major role in the development of immunology, and some of its beginnings in the 19th century have been cited above. Prior to the second decade of the 20th century, it was widely believed that all antigens were proteins. In 1917, Dochez and Avery published two landmark papers titled "Soluble Substance of Pneumococcal Origin in the Blood and Urine during Lobar Pneumonia" (25) and "The Elaboration of Specific Soluble Substance by Pneumococcus during Growth" (26). Thought initially to have some characteristics in common with proteins, the specific soluble substances of pneumococci of several capsular types and that of a Friedländer bacillus (*Klebsiella*) were shown by Heidelberger and Avery all to be polysaccharides and the substances comprising the capsules of both species (43). Subsequent studies by Heidelberger and Kendall marked the beginning of quantitative immunology (44).

Recognition by Stryker (85) that loss of capsulation by pneumococci resulted in loss of virulence led Avery to a new potential therapeutic approach, namely, that of enzymatic depolymerization of the pneumococcal capsule in vivo. Avery and Dubos isolated a bacillus from a cranberry bog which produced an enzyme which depolymerized the capsular polysaccharide of pneumococcus type 3 when the latter was provided as its principal source of energy (10). Although the enzyme had no effect on the viability of type 3 pneumococci in vitro or in vivo, it was capable, when injected into infected laboratory animals, of destroying their capsules, rendering them susceptible to phagocytosis and permitting survival of the host (11). Like anticapsular antibody, the enzyme depolymerizing the pneumococcal capsule had no effect on the viability of the organism, however; and in the absence of sufficient enzyme to permit elimination of all viable organisms, capsular regeneration would lead ultimately to the demise of the infected animal. Other potential drawbacks to enzyme therapy were the need for a specific enzyme for each capsular polysaccharide and the effects of antibodies that develop against the enzyme on its activity. Because neither anticapsular antibody nor capsular depolymerizing enzyme has any effect on the viability of infecting pneumococci, their protective effect being to facilitate phagocytosis of the invading organisms, the observation of Rich and McKee that noncapsulated pneumococci caused progressive infection in agranulocytic animals is noteworthy (77). Study of several proteins of the pneumococcal soma, notably pneumolysin, shows them to play a role in its pathogenic properties (17).

Treatment of pneumococcal infections has followed two somewhat parallel courses: immunotherapy and chemotherapy. Indication of the potential utility of immunotherapy had its inception in the observations of the Klemperers, cited above, but its practical application to humans had to await the clear definition of pneumococcal types 1 and 2 by Neufeld and Händel in 1910. A program for the treatment of type 1 pneumococcal pneumonia with type-specific equine antiserum was initiated at the Rockefeller Institute for Medical Research in 1913 (9). Over the ensuing years, efforts were made to minimize untoward reactions to the antisera by elimination of proteins other than antibodies and to quantify their potency. A unit of antiserum was defined by Felton (28) as "that amount of antibody that will protect at least 50 percent of a series of inoculated standard white mice against 1,000,000 fatal doses of a standard pneumococcal culture of the same type." The results of treatment of type 1 and type 2 pneumococcal pneumonia with equine antisera show an approximately 50% reduction in mortality among cases of diverse origins collated by Heffron (42). Recognition of differences in the antipneumococcal antibodies of the horse and the rabbit, including the smaller size of the rabbit antibody, led to the introduction in the mid 1930s of rabbit antisera for the treatment of pneumococcal pneumonia caused by a number of capsular types (48). Although type-specific rabbit antisera appeared to have significant advantages over equine antisera in terms of dosage and untoward reactions, their use was to be short-lived, the introduction of sulfonamides and, shortly thereafter, of penicillin leading to their abandonment.

Contemporaneously with the development of serotherapy, the initial investigations of pharmacotherapy were undertaken. In 1911, Morgenroth and Levi reported the protective effect of ethylhydrocupreine (optochin), a derivative of quinine, in mice infected experimentally with pneumococci (66): The following year, in what may have been the first published account of the development in vivo of bacterial resistance to a chemotherapeutic agent, Morgenroth and Kaufmann described the isolation of optochin-resistant pneumococci from animals similarly infected (65), an observation confirmed by the in vitro observations of Tugendreich and Russo in 1913 (88). Optochin was used for a brief period to treat pneumococcal infection in humans; a 20-fold increase in resistance to the drug was observed in patients undergoing therapy (63, 64). Its use was abandoned because of its toxicity to the eye and its limited efficacy. Its only use today is in a laboratory test for the presumptive identification of pneumococci (18).

The introduction of the sulfonamides in the 1930s made available for the first time relatively safe and effective antibacterial drugs. Although sulfanilamide proved ineffective in treating pneumococcal infections, sulfapyridine and a number of successive congeners were moderately successful in the management of pneumococcal pneumonia. Their use marked one of the first times, if not the first time, that levels of drug in the blood were used to regulate dosage (58), and although earlier experience with optochin was not cited, they became other examples of the pneumococcal capacity to develop resistance to antimicrobial agents (56).

The advent of penicillin treatment of both nonbacteremic and bacteremic pneumococcal pneumonia, reducing the overall case fatality rate to 5 to 8%, had a profound effect on professional attitudes toward these infections. Because therapy was effective irrespective of the infecting pneumococcal type, capsular serotyping of the organism was largely abandoned and its recognition declined significantly. Only when typing was reintroduced was it recognized that attack rates of pneumococcal infection had changed little, if at all (8). Although pneumococcal mutants showing incremental resistance to penicillin had been isolated from experimentally infected mice in 1933 (80), another three decades were to elapse before such mutants were recovered from humans (38), initiating a therapeutic problem of ever-increasing magnitude. Penicillin resistance in the pneumococcus is a result of mutation in one or more of the bacterium's penicillin-binding proteins (96).

The ability of pneumococci to adjust to exposure to other antimicrobial agents appears subject to few, if any, limitations. Mutants resistant to cephalosporins, tetracyclines, chloramphenicol, macrolides, quinolones, and aminoglycosides have all been recognized (51), and mutants tolerant of vancomycin have also been described (74).

Because of the persisting morbidity and mortality worldwide of pneumococcal infections and the increasing incidence of antimicrobial resistance, a program to reintroduce a polyvalent pneumococcal polysaccharide vaccine was initiated in the 1960s. Attempts to prevent pneumococcal infection date from 1911. Although Sternberg had reported in 1882 that injection of a rabbit with his saliva after its treatment with an antiseptic had rendered the animal immune to what would otherwise have been a subsequent lethal injection of his untreated saliva (84), no one seems to have recognized the prophylactic implications of this experiment.

The stimulus to develop a pneumococcal vaccine arose in the gold mines of South Africa, where pneumonia was epidemic in the early years of the 20th century. Sir Almroth Wright, credited with the development of typhoid vaccine, carried out with his collaborators several trials of pneumococcal vaccines of unknown serotypic constitution involving 50,000 miners (95). Although Wright, who had an antipathy to biostatistics, thought the vaccine to be effective, subsequent analysis failed to support this conclusion. Wright's protégé, F. Spencer Lister, continued studies in South Africa, described independently the serologic diversity of pneumococcal strains causing infection, and demonstrated that one could inject humans intravenously with 2 to 4 billion heat-killed pneumococci without untoward effect (54). Thirty years later, Heidelberger and his associates were to show that 10 billion pneumococci of types 1 and 2 yielded approximately 30 to 40 μg of capsular polysaccharide (45). Lister carried out several trials of polyvalent vaccines of heat-killed pneumococci and might have shown their efficacy in the second decade of the 20th century had it not been for his choice of controls. Because attack rates of pneumonia were not the same at different mining compounds, the decision to vaccinate all miners at one compound and to compare the attack rate of pneumonia with that in unvaccinated miners at another compound was faulted by the vaccine's detractors, and the vaccine's efficacy remained moot (3).

In World War I, several trials of vaccines of whole pneumococcal cells were carried out elsewhere in military and other populations. Reviewed by Heffron, who estimated that a million subjects may have been included in the several trials, the efficacy of vaccines of killed whole pneumococcal cells remained uncertain (41).

Antigenicity of the capsular polysaccharide of pneumococcus type 3 in the mouse was recognized in the late 1920s (79), and in 1931, Francis and Tillett, injecting intradermally small amounts of capsular polysaccharides to determine the adequacy of serum therapy in patients with pneumococcal pneumonia, discovered that polysaccharide of a type differing from that causing pneumonia stimulated the formation of antibodies, demonstrating thereby the immunogenicity of capsular polysaccharides in humans (32), an observation confirmed by Finland and others (29, 30). These findings led to several inconclusive trials of capsular antigens as vaccines (27), but conclusive evidence of the efficacy of a tetravalent vaccine in a military population experiencing epidemic pneumococcal disease was reported by MacLeod et al. (57). It was shown also in the same trial that if an individual was a carrier of a pneumococcal type represented in the vaccine, vaccination would not eliminate the carrier state, but if the individual was not a carrier, vaccination would reduce by half the likelihood of his becoming one. As a sequel to this study, two hexavalent vaccines were licensed in the 1940s, a time when interest in pneumococcal infection was at its nadir; after several years, they were withdrawn by their manufacturer for lack of use.

Study a decade later revealed that there had been little if any change in the epidemiology of pneumococcal infection despite the introduction of active therapeutic agents and that a significant number of deaths still occurred (8). The findings led to renewed interest in prophylaxis to protect those identifiably at high risk. Accordingly, trials of polyvalent capsular polysaccharide vaccines were carried out in South Africa involving 12,000 young adult males with a high attack rate of disease (7). The *aggregate* efficacy of a tridecavalent formulation in preventing disease caused by types represented in the vaccine was 78.5% ($P < 0.0001$), and its ability to prevent radiologically confirmed pneumonia irrespective of cause in a population in which pneumococcal infection predominated was greater than 50% ($P < 0.0001$). On the basis of the foregoing data, a 14-valent vaccine was licensed in 1978, and in 1983, the formulation was expanded to include 23 capsular polysaccharides, the most complex vaccine administered to humans, designed to prevent 23 immunologically distinct infections. A case-controlled study in the United States among immunocompetent persons over 40 years of age with infection of a normally sterile bodily site showed the *aggregate* efficacy of the vaccine to be 61% ($P < 0.0001$), with a more rapid decline in protection among those 65 years of age and older (81).

Although capsular polysaccharides are suitable antigens for adults, they are not immunogenic in the human infant, a fact recognized as early as 1937 (23). The rabbit, like the human infant, is similarly unresponsive to polysaccharide antigens, but as shown by Avery and Goebel (12), if the polysaccharide is linked chemically to a protein, antibodies to the polysaccharide will be made. Seeking to provide an effective vaccine against infection with *Haemophilus influenzae* type b, the most common cause of bacterial meningitis in children under 2 years of age, Robbins et al. coupled the type b capsular polysaccharide to each of several well-characterized proteins and succeeded in producing a safe and effective vaccine (78). These findings led to the production of analogous pneumococcal polysaccharide-protein conjugates which have proved similarly effective in preventing systemic infection with the capsular types predominating as causes of infection in infancy (15).

Several protein antigens of the pneumococcus have been studied for their prophylactic effects in laboratory animals with the aim of developing vaccines of lesser complexity than those of capsular polysaccharides. To date none has been the subject of field trials in humans, and their prophylactic potential remains to be determined (19).

No discovery arising from the study of bacteria has had a greater impact on biology than the one revealing genes are composed of DNA, the outcome of investigations of the pneumococcus. As noted earlier, noncapsulated pneumococci can often be recovered when capsulated organisms are grown in liquid medium containing homologous anticapsular antibody. In the early 1920s, the British epidemiologist Fred Griffith undertook a study to define conditions under which noncapsulated variants of pneumococcus type 2 would regain capsulation and virulence in vivo (35). His initial protocol entailed injecting mice subcutaneously with living noncapsulated pneumococci derived from a strain of capsular type 2 together with a large amount of vaccine of heat-killed capsulated type 2 pneumococci. From mice dying following inoculation with the foregoing mixture, Griffith recovered virulent capsulated type 2 pneumococci. Extending his study, he repeated the experiment, substituting on this occasion a heat-killed vaccine of type 1 pneumococci, and recovered pneumococci of capsular type 1. Griffith's controls were meticulous. As noted by Hayes 38 years later in an appreciation of Griffith's work, ". . . . it is strange that Griffith does not draw attention to, much less stress, what now seems to us the most striking feature of transformation, namely, that it results in an *hereditary* alteration of character wherein, of course, lies its real biological significance" (40).

Griffith's findings, published in 1928, were confirmed the same year by Neufeld and Levinthal in Germany (72) and by Dawson at the Rockefeller Institute (24). Further studies led to the successful transformation of pneumococci in vitro (60) and culminated in 1944 in the publication by Avery, MacLeod and McCarty entitled "Studies on the Chemical Nature of the Substance Inducing Transformation of Pneumococcal Types. Induction of transformation by a deoxyribonucleic acid fraction from pneumococcus type 3" (13). The report, the first to attribute biological activity to a nucleic acid, was greeted initially with skepticism by some, and acceptance of the role of DNA was not immediate. Purification of DNase initiated by McCarty (59), leading to the ultimate crystallization of an enzyme inactivating the transforming principle, and the transformation of additional pneumococcal attributes brought ultimate acceptance of the genetic properties of DNA (61). The work of Chargaff showing that the adenine-thymine:guanine-cytosine ratios of DNAs from different species differed provided evidence that all DNAs were not alike (20) and played a role in the development by Watson and Crick of their model of the DNA molecule that was to revolutionize genetics (89). It is beyond the scope of this foreword to describe the development of this discipline in the past half century, much of which can be found in the volume entitled *Recombinant DNA* by Watson et al. (90). It is fitting to note, however, that the complete genome of a capsulated pneumococcus was recorded in 2001 (86).

Problems remain. Can currently available polysaccharide vaccines be used more effectively, and will any other antigen(s) of the pneumococcus prove more effective? Will it prove possible to discover antipneumococcal agents to which the organism is itself unable to give rise to resistant mutants? Perhaps no problem is more important than unraveling the mystery of how pneumococcal infection leads ultimately to the patient's demise. The problem was stated tellingly by Osler (75) in 1897:

> Very large areas of the breathing surface may be cut off without disturbing the cardiovascular mechanism. In no way is this more strikingly shown than by the condition of the patient after the crisis. On one day with a lung consolidated from apex to base, the respirations from 60 to 65 and the temperature between 104° and 105°, the patient may seem in truly desperate condition, and it would appear rational to attribute the urgent dyspnoea and the slight cyanosis to the mechanical interference with the exchange of gases in the lungs. But on the following day, dyspnoea and cyanosis may have disappeared, the temperature is normal and the pulse rate is greatly lessened and yet the physical condition of the lungs remains unchanged. We witness no more striking phenomenon than this in the whole range of clinical work and its lesson is of prime importance in this very question showing that the fever and the toxins rather than the solid exudate are the essential agents in causing cardio-respiratory symptoms. The toxemia outweighs all other elements in the prognosis of pneumonia; to it . . . . is due in great part the terrible mortality of this common disease and unhappily against it we have as yet no reliable measures at our disposal.

Although the advances in antimicrobial therapy of the last century have bought at least temporary respite from the mortality of pneumococcal infection, the interactions of the pneumococcus with its mammalian host still remain, for the most part, to be unraveled. Advances in understanding of the minutiae of sepsis are revealing the enormous complexity of both host and parasite and of their interactions (21). Much remains to be learned from continued study of the pneumococcus and of its interactions with humans. The note of optimism sounded by Benjamin White in his monograph *The Biology of Pneumococcus,* published in 1938, remains relevant today:

> The study of the members of this small group of microorganisms in a subordinate branch of biology is bringing light into some of the obscure realms of the related sciences. The peculiarities of Pneumococcus are yielding a generous return to the investors who have cast in their resources with its lot, resulting in the accumulation of a store of solid bullion for the scientist and for mankind. (94)

REFERENCES

1. **Abernathy, T. J., and O. T. Avery.** 1941. The occurrence of a protein not normally present in the blood. I. Distribution of the reactive protein in patients' sera and the effect of calcium on the flocculation reaction with C polysaccharide. *J. Exp. Med.* **73:**173–182.
2. **Austrian, R.** 1953. Morphologic variation in pneumococcus. I. An analysis of the bases for morphologic variation in pneumococcus and description of a hitherto undefined morphologic variant. *J. Exp. Med.* **98:**21–34.
3. **Austrian, R.** 1977. Of gold and pneumococci. *Trans. Am. Clin. Climatol. Assoc.* **89:** 141–161.
4. **Austrian, R.** 1960. The Gram stain and the etiology of lobar pneumonia, an historical note. *Bacteriol. Rev.* **24:**261–265.
5. **Austrian, R.** 1981. Pneumococcus: the first one hundred years. *Rev. Infect. Dis.* **3:**183–189.
6. **Austrian, R.** 1986. Some aspects of the pneumococcal carrier state. *J. Antimicrob. Chemother.* **18**(Suppl. A)**:**35–45.
7. **Austrian, R., R. M. Douglas, G. Schiffman, A. M. Coetzee, H. J. Koornhof, S. Hayden-Smith, and R. D. W. Reid.** 1976. Prevention of pneumococcal pneu-

monia by vaccination. *Trans. Assoc. Am. Phys.* **89**:184–194.

8. **Austrian, R., and J. Gold.** 1964. Pneumococcal bacteremia with a special reference to bacteremic pneumococcal pneumonia. *Ann. Intern. Med.* **60**:759–776.

9. **Avery, O. T., H. T. Chickering, R. Cole, and A. R. Dochez.** 1917. Acute lobar pneumonia. Prevention and treatment. Rockefeller Institute for Medical Research monograph 7, p. 1–100. The Rockefeller Institute for Medical Research, New York, N.Y.

10. **Avery, O. T., and R. Dubos.** 1930. The specific action of a bacterial enzyme on pneumococci of type III. *Science* **72**:151–152.

11. **Avery, O. T., and R. Dubos.** 1931. The protective action of a specific enzyme against type III pneumococcus infection in mice. *J. Exp. Med.* **54**:73–89.

12. **Avery, O. T., and W. F. Goebel.** 1931. Chemo-immunological studies on conjugated carbohydrate proteins. V. The immunological specificity of an antigen prepared by combining the capsular polysaccharide of type III pneumococcus with a foreign protein. *J. Exp. Med.* **54**:437–447.

13. **Avery, O. T., C. M. MacLeod, and M. McCarty.** 1944. Studies on the chemical nature of the substance inducing transformation of pneumococcal types. Induction of transformation by a deoxyribonucleic acid fraction from pneumococcus type 3. *J. Exp. Med.* **79**:137–158.

14. **Bezançon, F., and V. Griffin.** 1897. Pouvoir agglutinatif du serum dans les infections expérimentales et humaines, à pneumocoques. *C. R. Soc. Biol.* **48**:579–581.

15. **Black, S., H. Shinefield, B. Firman, E. Lewis, P. Ray, J. R. Hansen, L. Elvin, K. M. Ensor, J. Hackell, G. Siber, F. Malinoski, D. Madore, I. Chang, R. Kohberger, W. Watson, R. Austrian, K. Edwards, and the Northern California Permanente Vaccine Group.** 2000. Efficacy, safety and immunogenicity of heptavalent pneumococcal conjugate vaccine in children. *Pediatr. Infect. Dis. J.* **19**:187–195.

16. **Bornstein, D. L., G. Schiffman, H. P. Bernheimer, and R. Austrian.** 1968. Capsulation of pneumococci with soluble C-like (C_s) polysaccharide. *J. Exp. Med.* **128**:1385–1400.

17. **Boulnois, G. J.** 1992. Pneumococcal proteins and the pathogenesis of disease caused by *Streptococcus pneumoniae. J. Gen. Microbiol.* **138**:249–259.

18. **Bowers, E. F., and L. R. Jeffries.** 1955. Optochin in the identification of *Str. pneumoniae. J. Clin. Pathol.* **8**:58–60.

19. **Briles, D. E., R. C. Tart, E. Swiatlo, J. P. Dillard, P. Smith, K. A. Benton, B. A. Ralph, A. Brooks-Walter, M. J. Crain, S. K. Hollingshead, and L. S. McDaniel.** 1998. Pneumococcal diversity: considerations for new vaccine strategies with emphasis on pneumococcal surface protein A (PspA). *Clin. Microbiol. Rev.* **11**:645–657.

20. **Chargaff, E.** 1951. Structure and function of nucleic acids as cell constituents. *Fed. Proc.* **10**:654–659.

21. **Cohen, J.** 2002. The immunopathogenesis of sepsis. *Nature* **420**:885–891.

22. **Cooper, G., C. Rosenstein, A. Walter, and L. Peizer.** 1932. The further separation of types among the types of pneumococci hitherto included in Group IV and the development of therapeutic antisera for these types. *J. Exp. Med.* **55**:531–554.

23. **Davies, J. A. V.** 1937. The response of infants to inoculation with type 1 pneumococcus carbohydrate. *J. Immunol.* **33**:1–7.

24. **Dawson, M. H.** 1928. Interconvertability of "R" and "S" forms of pneumococcus. *J. Exp. Med.* **47**:577–591.

25. **Dochez, A. R., and O. T. Avery.** 1917. Soluble substance of pneumococcal origin in the blood and urine during lobar pneumonia. *Proc. Soc. Exp. Biol. Med.* **14**:126–127.

26. **Dochez, A. R., and O. T. Avery.** 1917. The elaboration of specific soluble substance by pneumococcus during growth. *J. Exp. Med.* **26**:477–493.

27. **Ekwurzel, G. M., J. S. Simmons, L. I. Dublin, and L. D. Felton.** 1938. Studies on immunizing substances in pneumococci. VIII. Report on field tests to determine the prophylactic value of a pneumococcus antigen. *Public Health Rep.* **53**:1877–1893.

28. **Felton, L. D.** 1928. The units of protective antibody in antipneumococcus serum and antibody solution. *J. Infect. Dis.* **43**:531–542.

29. **Finland, M., and H. F. Dowling.** 1935. Cutaneous reactions and antibody response

of human subjects to intracutaneous injections of pneumococcus polysaccharides. *J. Immunol.* **29**:285–299.

30. **Finland, M., and W. D. Sutliff.** 1932. Specific cutaneous reactions and circulating antibodies in the course of lobar pneumonia. I. Cases receiving no serum therapy. *J. Exp. Med.* **54**:637–652.

31. **Fraenkel, A.** 1885. Bakteriologische Mittheilungen. *Z. Klin. Med.* **10**:401–461.

32. **Francis, T. J., Jr., and W. S. Tillett.** 1930. Cutaneous reactions in pneumonia. The development of antibodies following the intradermal injection of type-specific polysaccharides. *J. Exp. Med.* **52**:573–585.

33. **Gamaléia, N.** 1888. Sur l'étiologie de la pneumonie fibrineuse chez l'homme. *Ann. Inst. Pasteur (Paris)* **2**:440–459.

34. **Gram, C.** 1884. Ueber die isolierte Färbung der Schizomyceten in Schnitt- und Trokenpräparaten. *Fortschr. Med.* **2**:185–189.

35. **Griffith, F.** 1928. The significance of pneumococcal types. *J. Hyg.* **27**:113–159.

36. **Gundel, M., and F. K. T. Schwarz.** 1932. Studien über die Bakterienflora der obern Atmungswege Neugeborner (im Vergleich mit der Mundhöhlenflora der Mutter und des Pflegepersonals) unter besonderer Berüchsichtigung ihrer Bedeutung für das Pneumonieproblem. *Z. Hyg. Infektionskr.* **113**:411–436.

37. **Gundel, M., and G. Okura.** 1933. Untersuchungen über das gleichzeitige Vorkommen mehrerer Pneumokokkentypen bei Gesunden und ihrer Bedeutung für die Epidemiologie. *Z. Hyg. Infektionskr.* **114**:678–704.

38. **Hansman, D., H. Glasgow, J. Sturt, L. Devitt, and R. Douglas.** 1971. Increased resistance to penicillin of pneumococci isolated from man. *N. Engl. J. Med.* **284**:175–177.

39. **Harford, C. G., and M. Hara.** 1950. Pulmonary edema in influenzal pneumonia of the mouse and the relation of fluid in the lung to the inception of pneumococcal pneumonia. *J. Exp. Med.* **91**:245–259.

40. **Hayes, W.** 1966. The discovery of pneumococcal type transformation: an appreciation. *J. Hyg.* **64**:177–184.

41. **Heffron, R.** 1939. *Pneumonia with Special Reference to Pneumococcus Lobar Pneumonia*, p. 446–475. The Commonwealth Fund, New York, N.Y.

42. **Heffron, R.** 1939. *Pneumonia with Special Reference to Pneumococcus Lobar Pneumonia*, p. 888–903. The Commonwealth Fund, New York, N.Y.

43. **Heidelberger, M., and O. T. Avery.** 1923. The soluble specific substance of the pneumococcus. *J. Exp. Med.* **38**:73–79.

44. **Heidelberger, M., and F. E. Kendall.** 1929. A quantitative study of the precipitin reaction between type III pneumococcus polysaccharide and purified homologous antibody. *J. Exp. Med.* **50**:809–823.

45. **Heidelberger, M., C. M. MacLeod, S. J. Kaiser, and B. Robinson.** 1946. Antibody formation in volunteers following injection of pneumococci or their type-specific polysaccharides. *J. Exp. Med.* **83**:303–320.

46. **Henrichsen, J.** 1995. Six newly recognized types of *Streptococcus pneumoniae*. *J. Clin. Microbiol.* **33**:2759–2762.

47. **Hodges, R. G., C. M. MacLeod, and W. G. Bernhard.** 1946. Epidemic pneumococcal pneumonia. III. Pneumococcal carrier studies. *Am. J. Hyg.* **44**:207–230.

48. **Horsefall, F. L., K. Goodner, and C. M. MacLeod.** 1938. Antipneumococcus rabbit serum as a therapeutic agent in lobar pneumonia. II. Additional observations in pneumococcus pneumonia of nine different types. *N. Y. State J. Med.* **38**:245–255.

49. **Issaeff, B.** 1893. Contribution à l'étude de l'immunité acquise contre le pneumocoque. *Ann. Inst. Pasteur (Paris)* **7**:260–279.

50. **Klemperer, G., and F. Klemperer.** 1891. Versuche über Immunisierung und Heilung bei der Pneumokokkeninfection. *Berl. Klin. Wochenschr.* **28**:833–835, 869–875.

51. **Klugman, K. P.** 1990. Pneumococcal resistance to antibiotics. *Clin. Microbiol. Rev.* **3**:171–196.

52. **Knecht, J. C., G. Schiffman, and R. Austrian.** 1970. Some biological properties of pneumococcus type 37 and the chemistry of its capsular polysaccharide. *J. Exp. Med.* **132**:475–487.

53. **Kruse, W., and S. Pansini.** 1892. Untersuchungen über den Diplococcus pneumoniae und verwandte Streptokokken. *Z. Hyg. Infektionskr.* **11:**279–380.

54. **Lister, F. S.** 1916. An experimental study of prophylactic inoculation against pneumococcal infection in the rabbit and in man. *S. Afr. Inst. Med. Res. No.* **8:**231–287.

55. **Loda, F. A., A. M. Collier, W. P. Glezen, K. Strangert, W. A. Clyde, Jr., and F. W. Denny.** 1975. Occurrence of *Diplococcus pneumoniae* in the upper respiratory tract of children. *J. Pediatr.* **87:**1087–1093.

56. **MacLean, I. H., K. B. Rogers, and A. Fleming.** 1939. M & B 693 and pneumococci. *Lancet* **i:**562–568.

57. **MacLeod, C. M., R. G. Hodges, M. Heidelberger, and W. G. Bernhard.** 1945. Prevention of pneumococcal pneumonia by immunization with specific capsular polysaccharides. *J. Exp. Med.* **82:**445–465.

58. **Marshall, E. K., Jr., K. Emerson, Jr., and W. C. Cutting.** 1937. Para-aminobenzene sulfonamide; absorption and excretion; method of determination in urine and blood. *JAMA* **108:**953–957.

59. **McCarty, M.** 1946. Purification and properties of desoxyribonuclease isolated from beef pancreas. *J. Gen. Physiol.* **29:**123–139.

60. **McCarty, M.** 1985. *The Transforming Principle. Discovering That Genes Are Made of DNA,* p. 73–88. W. W. Norton & Co., New York, N.Y.

61. **McCarty, M.** 1985. *The Transforming Principle. Discovering That Genes Are Made of DNA,* p. 213–235. W. W. Norton & Co., New York, N.Y.

62. **Metchnikoff, E.** 1891. Etudes sur l'immunité. IV. L'immunité des cobayes vaccinées contre le Vibrio Metchnikowii. *Ann. Inst. Pasteur (Paris)* **5:**465–478.

63. **Moore, H. F., and A. M. Chesney.** 1917. A study of ethylhydrocuprein (optochin) in the treatment of acute lobar pneumonia. *Arch. Intern. Med.* **19:**611–682.

64. **Moore, H. F., and A. M. Chesney.** 1918. A further study of ethylhydrocuprein (optochin) in the treatment of acute lobar pneumonia. *Arch. Intern. Med.* **21:**659–681.

65. **Morgenroth, J., and M. Kaufmann.** 1912. Arzneifestigkeit bei Bakterien (Pneumokokken). *Z. Immunitaetsforsch. Exp. Ther.* **15:**610–624.

66. **Morgenroth, J., and R. Levy.** 1911. Chemotherapie der Pneumokokkeninfektion. *Berl. Klin. Wochenschr.* **48:**1560–1561, 1979–1983.

67. **Netter.** 1886. De l'endocardite végétante-ulcéreuse d'origine pneumonique. *Arch. Physiol. Norm. Pathol.* **8**[Ser. 3]**:**106–161.

68. **Neufeld, F.** 1900. Ueber eine specifische bacteriolytische Wirkung der Galle. *Z. Hyg. Infektionskr.* **34:**454–464.

69. **Neufeld, F.** 1902. Ueber die Agglutination der Pneumokokken und über die Theorieen der Agglutination. *Z. Hyg. Infektionskr.* **40:**54–72.

70. **Neufeld, F., and R. Etinger-Tulczynska.** 1932. Nasale Pneumokokkeninfectionen and Pneumokokkenkeimtrager im Tierversuch. *Z. Hyg. Infektionskr.* **112:**492–526.

71. **Neufeld, F., and Händel.** 1910. Weitere Untersuchungen über Pneumokokken Heilsera. III. Mitteilung. Über Vorkommenen und Bedeutung atypischer Varietäten des Pneumokokkus. *Arb. Kaiserlichen Gesundheitsamte* **34:**293–304.

72. **Neufeld, F., and W. Levinthal.** 1928. Beiträge zur Variabilität der Pneumokokken. *Z. Immunitaetsforsch. Exp. Ther.* **55:**324–340.

73. **Neufeld, F., and W. Rimpau.** 1904. Ueber die Antikörper des Streptokokken und Pneumokokken—Immunserums. *Dtsch. Med. Wochenschr.* **30:**1458–1460.

74. **Novak, R., B. Henriques, E. Charpentier, S. Normark, and E. Tuomanen.** 1999. Emergence of vancomycin tolerance in *Streptococcus pneumoniae. Nature* **399:**590–593.

75. **Osler, W.** 1897. On certain features in the prognosis of pneumonia. *Am. J. Med. Sci.* **113:**1–10.

76. **Pasteur.** 1881. Note sur la maladie nouvelle provoquée par la salive d'un enfant mort de la rage. *Bull. Acad. Med. (Paris)* **10:**94–103.

77. **Rich, A. R., and C. M. McKee.** 1939. The pathogenicity of avirulent pneumo-

cocci for animals deprived of leukocytes. *Bull. Johns Hopkins Hosp.* **64:**434–446.

78. **Robbins, J. B., R. Schneerson, P. Anderson, and D. H. Smith.** 1996. Prevention of systemic infections, especially meningitis, caused by *Haemophilus influenzae* type b. *JAMA* **276:**1181–1185.

79. **Schiemann, O., and W. Casper.** 1927. Sind die spezifisch präcipitablen Subtanzen der 3 Pneumokokkentypen Haptene? *Z. Hyg. Infektionskr.* **108:**220–257.

80. **Schmidt, L. H., and C. L. Sessler.** 1943. Development of resistance to penicillin by pneumococci. *Proc. Soc. Exp. Biol. Med.* **52:**353–357.

81. **Shapiro, E. D., A. T. Berg, R. Austrian, D. Schroeder, V. Parcells, A. Margolis, R. K. Adair, and J. D. Clemens.** 1991. The protective efficacy of pneumococcal polysaccharide vaccine. *N. Engl. J. Med.* **325:**1453–1460.

82. **Sternberg, G. M.** 1881. A fatal form of septicaemia in the rabbit, produced by subcutaneous injection of human saliva. An experimental research. *Natl. Board Health Bull.* **2:**781–783.

83. **Sternberg, G. M.** 1882. A fatal form of septicaemia in the rabbit, produced by the subcutaneous injection of human saliva. *Studies Biol. Lab. Johns Hopkins Univ.* **2:**183–200.

84. **Sternberg, G. M.** 1882. Induced septicaemia in the rabbit. *Am. J. Med. Sci.* **84:**69–76.

85. **Stryker, L. M.** 1916. Variations in the pneumococcus induced by growth in immune serum. *J. Exp. Med.* **24:**49–68.

86. **Tettelin, H., K. E. Nelson, S. Peterson, J. A. Eisen, T. D. Read, R. T. DeBoy, D. H. Haft, R. J. Dodson, A. Scott Durkin, M. Gwinn, J. F. Kolonay, W. C. Nelson, J. D. Peterson, L. A. Umayam, O. White, S. L. Salzberg, M. R. Lewis, D. Radune, E. Holtzapple, H. Khouri, A. M. Wolf, T. R. Utterback, S. Angiuoli, T. Dickinson, E. K. Hickey, I. E. Holt, B. J. Loftus, F. Yang, H. O. Smith, J. C. Venter, B. A. Dougherty, D. A. Morrison, S. Hillingshead, and C. M. Fraser.** 2001. Complete genome sequence of a virulent isolate of *Streptococcus pneumoniae*. *Science* **293:**498–506.

87. **Tillett, W. S., and T. Francis, Jr.** 1930. Serological reactions in pneumonia with a non-protein somatic fraction of pneumococcus. *J. Exp. Med.* **52:**561–571.

88. **Tugendreich, J., and C. Russo.** 1913. Ueber die Wirkung von Chinaalkaloiden auf Pneumokokkenkulturen. *Z. Immunitaetsforsch. Exp. Ther.* **19:**156–171.

89. **Watson, J. D., and F. H. C. Crick.** 1953. The structure of DNA. *Cold Spring Harbor Symp. Quant. Biol.* **18:**123–131.

90. **Watson, J. D., M. Gilman, J. Witkowski, and M. Zoller.** 1992. *Recombinant DNA,* 2nd ed. Scientific American Books, New York, N.Y.

91. **Webster, L. T., and T. P. Hughes.** 1931. The epidemiology of pneumococcus infection. *J. Exp. Med.* **53:**535–552.

92. **Weiser, J. N., R. Austrian, P. K. Sreenviasan, and H. R. Masure.** 1994. Phase variation in pneumococcal opacity: relationship between colonial morphology and nasal colonization. *Infect. Immun.* **62:**2582–2589.

93. **Welch, W. H.** 1892. The micrococcus lanceolatus, with a special reference to the etiology of acute lobar pneumonia. *Bull. Johns Hopkins Hosp.* **3:**125–139.

94. **White, B.** 1938. *The Biology of Pneumococcus,* p. XVI–XVII. The Commonwealth Fund, New York, N.Y.

95. **Wright, A. E., W. Parry Morgan, L. Colebrook, and R. W. Dodgson.** 1914. Observations on prophylactic inoculation against pneumococcus infections, and on the results which have been achieved by it. *Lancet* **i:**1–10, 87–95.

96. **Zighelboim, S., and A. Tomasz.** 1980. Penicillin-binding proteins of multiply antibiotic resistant South African strains of *Streptococcus pneumoniae*. *Antimicrob. Agents Chemother.* **17:**434–442.

ROBERT AUSTRIAN
Department of Research Medicine
The University of Pennsylvania School of Medicine

PREFACE

The pneumococcus represents both a vast burden of human disease and an enormous opportunity for medical and basic science discovery. This nemesis has a very long track record as a severe, invasive pathogen despite over a hundred years of medical progress. Current events indicate that the situation is only getting worse. The pneumococcus has pushed biological science to the discovery of DNA, of polysaccharide-based vaccines, of quorum sensing, of peptide-based bacterial communication, and many other basic tenets. It is the quintessential gram-positive pathogen and leads the path of discovery of the pathogenesis of this class of organisms, a group that now outnumbers the gram-negatives. Each time we think we are in control, the pneumococcus teaches us that it can beat any defense. Antibiotic resistance has become rampant and involves multiple drugs to the point that there are pneumococci that severely tax our capability to kill them. As time progresses, this naturally transformable pathogen will continue to harvest hazardous weapons while browsing in the DNA gift shop of the human nasopharynx. Good vaccines have been developed but, unfortunately, have been poorly used. Even when they are used appropriately, the bacterium switches capsular type to evade elimination. We have learned a great deal, but there is a long life ahead for investigators in this field.

The scientific database centered around the pneumococcus has been very rich and has become codified in a series of meetings of investigators in the field, always drawing an international crowd. The group was small in California in 1981 and expanded in Portugal in 1996, a meeting resulting in the publication of a book that is a milestone in the field. Since then, pneumococcal investigators, M.D.'s and Ph.D.'s alike, have met every 2 years just to try to keep up with the massive amount of new data and new problems presented by this pathogen. This book comes only 7 years since the last, but it is virtually entirely new. A major catalyst to this increased information is the publication of sequences of several pneumococcal genomes. We will mine these data for years to come. Taking stock of where the field is now is meant to provide a framework of where the biggest needs are for future investigation and, importantly, to show where the basic science can translate into changes in clinical care. Thus, this book emphasizes both the details of the bacterium and all it can do as well as the host response and mechanisms of disease. The pace of discovery is intense and richly rewarding, but seasoned investigators know from experience that the pressure is always on because this pathogen will stand still for no one.

It has been immensely rewarding and challenging to track the developments of the pneumococcus, and there are many to thank for making this book a reality. First, I thank my coeditors, who very actively interacted with the authors to expand and interconnect the components of this story. They demanded excellence and got it. My deepest gratitude goes to the authors. Everyone that I asked to write for this project accepted with enthusiasm, indicating that there was a pent-up desire to report and review a reservoir of new knowledge at this time for this field. Authors contacted colleagues and created liaisons as coauthors to bring in a broad representation of experts in the field. As word of the book spread, some investigators even volunteered as authors. Now that is enthusiasm and collegiality! Helpful comments on the scope of the book were provided by Liise-anne Pirofski, Albert Einstein College of Medicine; Michael Caparon, Washington University School of Medicine; Peter W. Andrew, University of Leicester; Larry S. McDaniel, University of Mississippi; and Russell W. Steele, Children's Hospital, New Orleans. The guidance behind the project came with the experienced and steady hand of Greg Payne at the American Society for Microbiology (ASM). He kept us true to time and worked through expansion of the book as more and more information came from more and more authors. ASM worked tirelessly even over the December holidays to assemble the final book. Finally, I would like to acknowledge that my energy to continue to read, write, and edit was renewed every day by my very industrious lab people, by the faculty in Infectious Diseases at St. Jude Children's Research Hospital, and by my crackerjack children, Laura and Stefan.

My hope is that this book will serve as a milestone in the race against this pathogen and invigorate young investigators in the field. Those of us in the "senior investigator group" all remember key people, now giants in biomedical science, who encouraged us early on in our careers. I hope that when new investigators of gram-positive microbiology and pathogenesis come to us for advice, we will reach up to the bookshelf and take down this book, enthusiastically saying, "This is a great place to start."

ELAINE TUOMANEN
January 2004

THE BACTERIA

WHAT IS A PNEUMOCOCCUS?

Christopher G. Dowson

I

BACKGROUND

The pneumococcus (*Streptococcus pneumoniae* or diplococcus) has been profoundly important in our understanding of the human response to infectious disease, the nature of genetic material, and natural transformation as a means of genetic exchange, as well as in the recognition of bacterial resistance to drugs (1). The past decade has witnessed more than a doubling of the research output as the pneumococcus has acquired multiple antibiotic resistance (16) and the scientific and medical communities have been gearing up to develop, introduce, and validate novel vaccines to counter this problem (8, 39, 46, 60, 76, 87). Yet there is still much to be discovered even about the natural biology of this organism.

The pneumococcus is related to commensal members of the oral streptococci, it is primarily carried asymptomatically, and it benignly inhabits a range of niches within the upper respiratory tract, although the better-known face of this organism, the one most commonly presented, is that associated with widespread morbidity and mortality (57). The rather paradoxical nature of this organism, illustrated above in its characteristics as both a commensal and a pathogen, continues. On the one hand the core genome of the pneumococcus is highly conserved with little genetic diversity (14, 84), while on the other there is significant diversity in the distribution of loci between genomes (58, 59); some strains, for example, possess a raft of ABC and PTS transporters that others do not. There are also a staggering array of diverse pneumococcal capsular types (27) and genetic malleability in the ability to acquire novel allelic variants of loci by horizontal gene transfer (42, 50, 78). It is therefore not surprising that because of these characteristics there are still problems in designing and implementing the means and methods of long-term control of an organism which in the main is a harmless commensal but at the same time is the leading cause of bacterial infection worldwide (57).

Developing strategies for effective control of pneumococci clearly requires an ability to detect and identify the organism during carriage and disease. There is also a need to study the population biology of the organism. Doing so will determine whether potential targets are found throughout the population or only among a subset of strains and whether these targets are conserved or highly diverse, are uniformly or differentially regulated, and express common or altered phenotypes. Addi-

Christopher G. Dowson, Biomedical Research Institute, Department of Biological Sciences, University of Warwick, Coventry CV4 7AL, United Kingdom.

The Pneumococcus. Volume Editor, Elaine I. Tuomanen,

tionally, it is important to understand how stable such targets might be if they become subject to the new selective pressure of widespread vaccination or chemotherapy. Finally, understanding the evolutionary origins of pathogens opens up the possibilities of using comparative genomics to help identify the potential raft of alterations and acquired virulence determinants that have resulted in the creation of the "Janus-like" two-headed diplococcus.

DETECTION AND IDENTIFICATION

Classically, microbiological characterization has relied upon different methods for the isolation of organisms from blood culture or sputum samples. More recently, a significant effort has gone into the standardization of procedures required for detecting upper respiratory tract carriage (57). For samples where there is likely to be a mixed flora it is important that transportation maintains a viable pneumococcal flora, as related streptococci tend to be more robust, outliving pneumococci and giving a skewed result (56).

Identification of pneumococci has relied heavily upon colony morphology and hemolytic activity on blood agar plates during the initial selection of organisms for subsequent typing (7, 47). In addition to this, optochin sensitivity, bile solubility, and agglutination with antipneumococcal polysaccharide capsule antibodies are the most frequently used means of identifying pneumococci. Although the colony morphology of related streptococci can be very similar, these other members of the oropharynx are typically resistant to optochin and the action of bile and do not react with antipolysaccharide antibodies. Such conventional methods allow identification of the majority of isolates, but there are many observations of pneumococci producing one or more atypical reactions to the standard tests and other streptococci giving positive reactions leading to difficulties in identification, including optochin-resistant pneumococci (22, 37, 52, 62), optochin-susceptible oral streptococci (44), bile-insoluble pneumococci (55), and bile-sol-

uble organisms that are not pneumococci (84). However, it may well be that a combination of optochin and bile susceptibilities may be at least as reliable as serologic methods (81).

Nontypeable pneumococci, i.e., unencapsulated pneumococci that do not react with typing sera, have been reported to comprise 2% of pneumococci isolated from normally sterile sites (4). Acapsulate pneumococci are frequently associated with conjunctivitis (23), and among these organisms infecting eyes there is at least one stable prevalent clone which has recently been identified (43). Pneumococci are able to reversibly switch off capsule production by the generation of direct tandem duplications (78, 79), to down-regulate the amount of capsule produced during opacity phase variation changes associated with changes in teichoic acid content of pneumococcal cell walls (36), or lose the ability to produce capsule as a result of other genetic alterations involving point mutational changes (40). Furthermore, there have been reported cross-reactions between α-hemolytic streptococci and polyvalent pneumococcal omniserum (29) resulting in the generation of false-positive results. Despite these problems, most diagnostic laboratories still frequently use conventional serologic identification techniques alongside optochin and bile solubility (35).

PHYLOGENETIC RELATIONSHIPS AND HORIZONTAL GENE TRANSFER

On the basis of 16S rRNA sequences S. mitis and S. oralis share over 99% sequence identity to S. pneumoniae and represent the closest members of the mitis group which continues to expand. This encompasses S. cristatus, S. gordonii, S. infantis sp. nov., S. oligofermentans sp. nov., S. parasanguinis, S. peroris sp. nov., and S. sanguis (32, 73). The use of additional single loci, including the superoxide dismutase gene (sodA) (44) or the groESL genes (72), also positions pneumococci closest to S. mitis and S. oralis.

What is interesting is that these two organisms have been previously implicated as donors of chromosomal DNA to pneumococci in the evolution of mosaic penicillin-binding protein (PBP) genes involved in penicillin resistance (13, 71). The horizontal transfer of homologous *pbp* genes by transformation among these naturally competent organisms is well recorded and involves the transfer of these genes not only to pneumococci but also back out into other members of the mitis group (15). This flow of genetic information is not restricted to resistance genes per se but extends also to those located nearby. The D-alanine:D-alanine ligase gene (*ddl*), which is located next to the gene for a PBP (PBP2B), clearly evolves by recombination, and presents allelic variants in pneumococci that are linked to resistance (18). Given the prevalence of PBP-mediated resistance among the oral streptococci (65), such targets are less than optimal for use to reconstruct phylogenetic relationships among this group of organisms (33).

Allelic replacement by horizontal transfer is not restricted to antibiotic resistance genes. There is clear evidence that some mosaic pneumococcal genes involved in virulence (*lytA, nanA, pspA, iga, pspC*) (12, 28, 30, 48, 61, 83) and competence (*com* loci) (82) are the result of recombination with homologs from related streptococci or bacteriophage (11, 83), again showing that the genetic pool and allelic diversity are not restricted simply to the named species. This has implications for the selection of both diagnostic targets and vaccine candidates.

POPULATION GENETICS

In addition to studies identifying resistance or virulence loci that have evolved by recombination, other data from population genetic studies using multilocus enzyme electrophoresis (66), multilocus restriction typing (50), and multilocus sequence typing (19) have shown that housekeeping loci around the chromosome, which are not subject to a positive selective pressure, are at linkage equilibrium; i.e., they are not linked, and allelic variants of different loci freely shuffle through the pneumococcal population. More recent analysis of

these data has quantified the relative contribution of recombination and mutation and shown that the generation of novel alleles of these housekeeping loci in pneumococci is driven predominantly (10 times more frequently) by recombination rather than point mutation (21).

By selecting loci that could be amplified across a range of oral streptococci, Whatmore et al. (84) were able to identify and understand the fine genetic relationships between pneumococci and their closest relatives. Sequence information from fragments of three housekeeping genes from each of these organisms, *hexB* (DNA mismatch repair protein), *recP* (transketolase), and *xpt* (xanthine phosphoribosyltransferase), was concatenated and used to construct a neighbor-joining dendrogram (Fig. 1). Interspecies genetic exchange and horizontal transfer between different strains is clearly an important evolutionary mechanism in pneumococci. Therefore, understanding the evolutionary processes impinging upon different loci is clearly a key issue when deciding upon the relative value of these as diagnostic markers for the species per se or of different strains within the species.

MOLECULAR DIAGNOSTICS

There is a range of molecular genetic identification tools either available commercially or described in research papers that have been reported to be highly sensitive and specific for pneumococci. These include a probe based upon 16S ribosomal DNA sequence (9, 51, 69) or digestion of the amplified 16S rRNA gene (68); probes for the major pneumococcal autolysin (63); and PCR amplification of the autolysin gene (*lytA*) (25, 74), the pneumolysin gene (*ply*) (45, 75), *psaA* (49, 70), *pit2* (5), and *pbp* genes (17, 88).

However, just as with the classical approach to the identification of pneumococci, where several diagnostic tests are applied to an isolate, there are repeated reports of organisms for which the results are ambiguous, being positive for one molecular diagnostic test and negative for another (31, 77, 84). Such ambiguities are

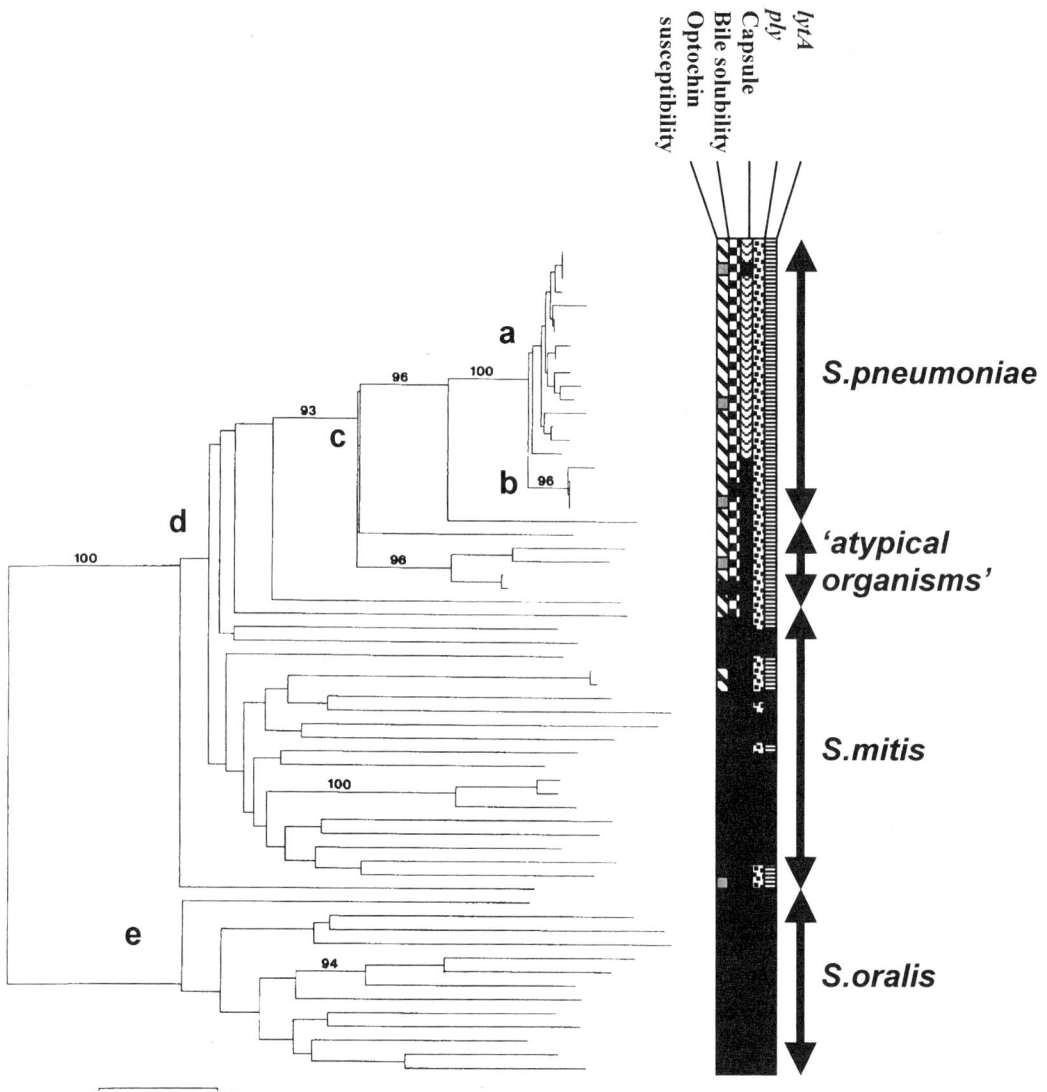

FIGURE 1 Dendrogram of genetic relationships between streptococcal isolates described by Whatmore et al. (84) and the presence or absence of characteristic pneumococcal diagnostic features. The dendrogram was constructed from housekeeping gene sequence data (*xpt, recP, hexB*) by using the neighbor-joining method. Only bootstrap confidence values exceeding 90% are shown. The scale represents the number of nucleotide substitutions per site. Overall identities of strains clustering together either as *S. pneumoniae*, atypical organisms, *S. mitis*, and *S. oralis* are indicated by arrows. Features include optochin susceptibility (▨), optochin intermediate resistance (□), bile solubility (▩), presence of an identified pneumococcal capsule (▨), presence of the pneumolysin gene *ply* (▩), and presence of the major autolysin gene *lytA* (▤). Strains deficient in these characteristics are highlighted black. Strains, identified here as *S. pneumoniae* using three loci and in a previous study examining nine different loci (16), arise from a common node with a bootstrap value of 100. Within this group two clusters were identified as a and b. Cluster b isolates, though clearly pneumococci, were all acapsulate, with one isolate exhibiting bile insolubility and another exhibiting intermediate optochin resistance. One acapsulate isolate was identified within cluster a. Atypical organisms, arising at node c, represent a rather diffuse collection of strains but are clearly separated from typical strains of *S. pneumoniae* and from strains identified as *S. mitis* and *S. oralis*. Strains identified as *S. mitis* form an ill-defined group the bulk of which arise at node d. Strains identified as *S. oralis* form a clearly distinguishable group arising at node e. What is interesting is the genetic conservation within pneumococci, compared to the diversity found among *S. oralis* and even greater diversity among isolates of *S. mitis*.

perhaps not surprising given the diversity that exists within organisms that are closely related to pneumococci and evidence that these organisms can harbor "pneumococcal" virulence determinants (84). To develop effective diagnostic tools for the pneumococcus, we need to understand more about its nearest neighbors and to include the breadth of diversity of these organisms in strain panels used to validate diagnostic tools—unfortunately, this tends to be the exception rather than the rule.

PNEUMOLYSIN AND AUTOLYSIN

The pneumolysin (80) and autolysin (10) genes have been flagged as potentially useful targets for molecular diagnostics for the past decade or so (25, 67), and reports throughout this time have always found typical human clinical pneumococci to possess *ply* and *lytA*. Recently it was discovered that pneumococci isolated from the respiratory tracts of horses have undergone a deletion event resulting in the loss of the 5' end of *ply* and the 3' end of *lytA*. This event was the result of a recombinational event between 13-bp regions of *lytA* and *ply* which possess ~92% identity (85). Clearly this event, which resulted in the loss of both pneumolysin and autolysin function, could occur in human isolates. However, the fact that such deletions are not found within the human population suggests that there is a strong selective advantage in maintaining the function of these proteins.

Amino acid alterations associated with changes in hemolytic activity have been reported for pneumolysin (41), although the underlying sequence diversity associated with these alleles is low. Greater diversity has been reported to occur in *lytA* (24), although a systematic sequence analysis has shown that *lytA* displays restricted allelic variation despite localized recombination events with genes of pneumococcal bacteriophage encoding cell wall lytic enzymes (83). Given that the majority of clinical isolates of pneumococci are lysogenic for one or more bacteriophage (64) and that these phage carry homologous autolysins, note has to be taken of this during the selection of

PCR target sequences. A further aspect concerning pneumolysin and strain diversity, highlighting that some strains of pneumococci might present radically different phenotypes, is the observation that several strains, including WU2 and GB05 (often used in animal models), have been reported to release pneumolysin prior to generalized cell lysis and are *lytA* independent. This is opposed to the more typical *lytA*-dependent postlysis release of pneumolysin for other pneumococci (2). The reason for this difference in phenotype still remains unresolved.

Studies examining the clinical use of pneumolysin and autolysin as diagnostic targets in sputum, pleural fluid, throat swabs, urine, blood, and ear swabs present a mixed set of results (6, 20, 26, 31, 53, 67, 86). They do, however, highlight several issues concerning the strengths and weaknesses of molecular diagnostics for pneumococci and of PCR-based detection in general. Samples such as blood, pleural fluid or cerebrospinal fluid that are usually sterile represent the cleanest samples with which high-resolution techniques, including nested or fluorescent quantitative PCR, can derive real advantages in speed and sensitivity over classical culture, especially if samples are taken after antibiotic administration. Moreover, the organisms recovered from such samples are highly likely to be associated with the disease. Samples from nonsterile sites, such as nasopharyngeal swabs, throat swabs, or sputum samples, that may be contaminated with other members of the oral flora are more problematic. This is especially true because related organisms, as shown in Fig. 1, may possess targets such as *ply* and *lytA* that are notionally diagnostic for pneumococci. Some of these related organisms possess both targets, others only one. However, it is not clear how important such confounding organisms may be in the clinical setting.

So, straightforward amplification of *ply* by PCR, or in combination with *lytA,* is unable to resolve all strains with equivocal reactions for serotype, optochin, or bile solubility (31,

84), as the nucleotide sequences of pneumolysin genes from typical pneumococci and members of the related oral streptococci are sufficiently similar to enable amplification (34). Nevertheless, it might be possible to design internal primers, particularly for *lytA*, that maximize the sequence diversity of this locus, enabling discrimination in amplification with different primers or by examining differences in dissociation characteristics of the different *lytA* amplicons.

This critique is designed not to discount the use of *ply* and *lytA* as valuable diagnostic targets, but rather to highlight the observation that occasionally organisms which are not pneumococci are found to possess these loci. Although more work is required to understand the associated pathogenicity of such organisms, each organism isolated to date was from the sputum of patients suffering from acute respiratory tract infections (84), and identification of such organisms carrying *ply* may well be of value to inform treatment, if less so for pneumococcal vaccination efficacy screening.

ALTERNATIVE APPROACHES TO IDENTIFICATION

Having begun to establish a phylogenetic framework for pneumococci, identifying their close relatives and understanding something of the distribution of some key virulence determinants among these organisms, it is possible to speculate about the ancestral origins of pneumococci. Data underlying Fig. 1 would suggest an evolutionary origin of the pneumococcus lying within the morass of organisms allied to *S. mitis*, presumably an organism already possessing *lytA* and *ply*. Serial rounds of genetic acquisition may then have enabled these ancestral organisms to acquire advantageous attributes enabling niche dominance, niche diversity, or host-to-host spread. Evidence for such acquisitions has been presented by Brown et al., who examined the role of pneumococcal iron transport proteins (Pit) (5). The *pit2* locus is contained within a 27-kb region of chromosomal DNA that has several features of pathogenicity islands of gram-negative bacteria. This proba-

ble pathogenicity island (PPI-1) represents one example of no doubt other events that will be revealed by comparative genomics, which has contributed to the evolution of the founding pneumococcus from which the extant population has arisen.

The identification of stable pathogenicity islands present in all pneumococci might contain useful diagnostic targets such as *pit2*. Although a preliminary screen would suggest that *pit2* is found only in pneumococci, the breadth and depth of isolates screened were restricted (5) and are currently being extended to help validate *pit2* as a useful locus by which pneumococci can be unambiguously identified (C. G. Dowson, unpublished data). The apparent uniformity of *pit2* and universal prevalence of *pit2* within pneumococci contrast with a pneumococcal metalloprotease, ZmpC, for which distribution is patchy and which appears to be predominantly associated with strains of pneumococci causing pneumonia (58).

It would certainly be very helpful for clinical diagnostic purposes to have a single locus that would categorically identify pneumococci. A single universal clinical diagnostic tool would be immensely valuable. The same may also be true for rapid identification of strains within the laboratory that may have an uncertain origin. For example, it has been reported that mutagenic inactivation of *zmpB* within a serotype 4 strain of *S. pneumoniae* (Norway 4) appeared to have ablated a raft of classical characteristics diagnostic for a pneumococcus and instead presented an organism showing extensive chain formation, defective lysis, and transformation deficiency (54). A hypothesis was constructed to account for these novel phenotypes and the role of *zmpB* in pneumococci. Subsequently, reports of essentially the same experiment, inactivating *zmpB*, using the well-characterized laboratory pneumococcal strain R6 did not result in chain formation, loss of lysis, or loss of transformation (3). Recent analysis of the original strain carrying the *zmpB* mutation (54), presented here for the first time (Fig. 2), reveals that the chain-forming, nonlytic, nontransformable organism does not fall

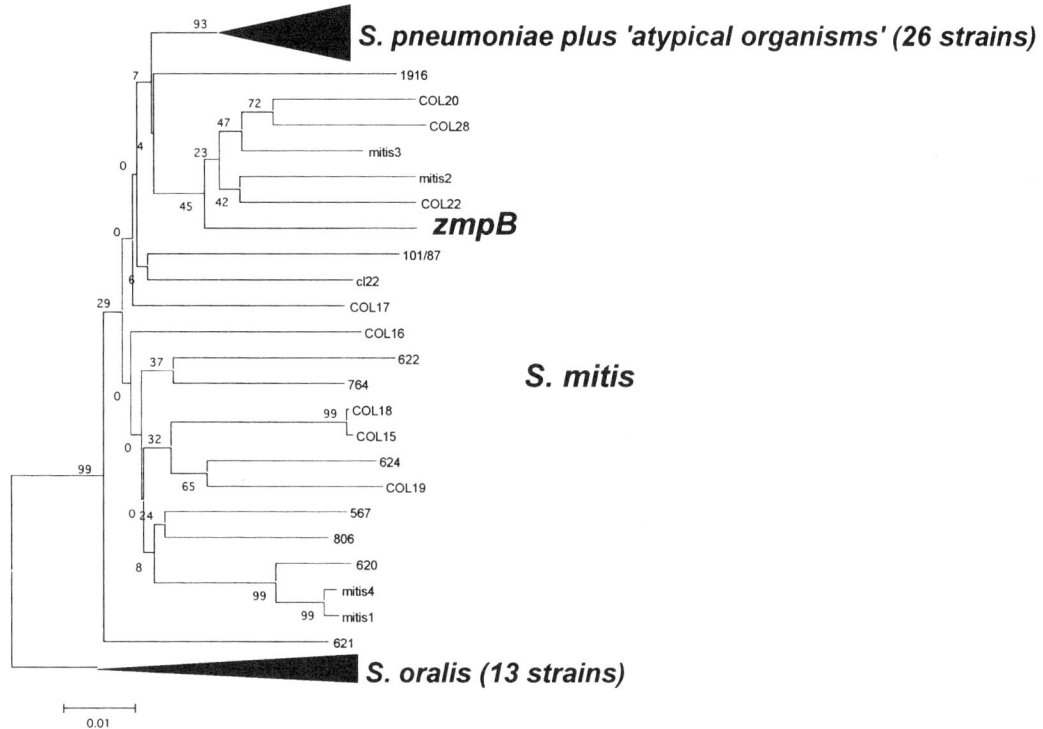

FIGURE 2 Condensed dendrogram of genetic relationships between the strain carrying the proposed Norway 4 *zmpB* mutation (in bold) described by Novak et al. (54) and a panel of genetically diverse well-characterized strains of *S. pneumoniae* (including strains Norway 4[54] and R6 [3]), atypical organisms, *S. mitis*, and *S. oralis* (84). The dendrogram was constructed by the neighbor-joining method using the program within MEGA version 2.1 (38). The scale represents the number of nucleotide substitutions per site.

within the group of strains clearly identifiable as *S. pneumoniae,* nor among the wider group encompassing "atypical organisms." These groups have bootstrap values of 100 and 93, respectively (Fig. 1 and 2). By contrast, the parental Norway 4 organism and strain R6 are unambiguously pneumococci, possessing quite different alleles of the three loci used for the analysis, *recP, hexB,* and *xpt,* to those found within the *zmpB* mutant (Fig. 2). Furthermore, the Norway 4 *zmpB* mutant is acapsular, and capsule-specific PCR and hybridization (C. G. Dowson et al., unpublished data) showed that it does not possess the type 4 cap locus. In fact, the genetic background in which the Novak et al. *zmpB* mutant is found, as proposed by Berge et al. (3), is not that of *S. pneumoniae* but clearly places it among the diverse

range of organisms currently identified as *S. mitis* (Fig. 2). Unambiguous identification of laboratory-derived mutants as pneumococci is clearly important if we are to understand results in the light of strain-to-strain or species-to-species variation.

However, the development of a diagnostic test based upon single-target identification might be an ambitious hope given the genetic plasticity of the pneumococcus and its naturally transformable relatives. Further work is required to validate the distribution of apparently pneumococcus-specific loci, such as *pit2* (5), located on pneumococcal virulence islands, especially as these islands will have arisen from another organism that would itself be *pit2* PCR positive. Combinations of two or three unique loci would therefore be needed to reduce the possibility of false positives. For a rapid diagnostic tool, a combination of primers designed against *pit2* plus *ply* or *lytA* would be a good start. However, for the time being, for absolute certainty of identification, sequence analysis of housekeeping genes (19, 84) is the best means of pigeonholing an organism. From this we can assign an organism to a species, clonal complex, or sequence type (19). Unfortunately, this will not necessarily correlate with total genomic identity. There are reports from different laboratories of the same (confirmed) lab strain possessing quite different phenotypes for virulence, presumably reflecting uncharacterized chromosomal variations.

The original signification of the two-headed god Janus meant vigilance and new beginnings, as in the word "January." So, recent descriptions of tandem duplications driving phase variation, and unpublished results of multiple chromosomal lesions (Dowson et al.), confirming previous comparative genomic analysis indicating this (59), all raise a note of vigilance before we launch headlong into interlab comparative transcriptomics and find that we are unable to reconcile the profusion of diverse results.

REFERENCES

1. **Austrian, R.** 1981. Pneumococcus: the first one hundred years. *Rev. Infect. Dis.* **3:**183–189.
2. **Balachandran, P., S. K. Hollingshead, J. C. Paton, and D. E. Briles.** 2001. The autolytic enzyme LytA of *Streptococcus pneumoniae* is not responsible for releasing pneumolysin. *J. Bacteriol.* **183:**3108–3116.
3. **Berge, M., P. Garcia, F. Iannelli, M. F. Prere, C. Granadel, A. Polissi, and J. P. Claverys.** 2001. The puzzle of zmpB and extensive chain formation, autolysis defect and non-translocation of choline-binding proteins in Streptococcus pneumoniae. *Mol. Microbiol.* **39:**1651–1660.
4. **Broome, C. V., and R. R. Facklam.** 1981. Epidemiology of clinically significant isolates of Streptococcus pneumoniae in the United States. *Rev. Infect. Dis.* **3:**277–281.
5. **Brown, J. S., S. M. Gilliland, and D. W. Holden.** 2001. A Streptococcus pneumoniae pathogenicity island encoding an ABC transporter involved in iron uptake and virulence. *Mol. Microbiol.* **40:**572–585.
6. **Butler, J. C., S. C. Bosshardt, M. Phelan, S. M. Moroney, M. L. Tondella, M. M. Farley, A. Schuchat, and B. S. Fields.** 2003. Classical and latent class analysis evaluation of sputum polymerase chain reaction and urine antigen testing for diagnosis of pneumococcal pneumonia in adults. *J. Infect. Dis.* **187:**1416–1423.
7. **Chandler, L. J., B. S. Reisner, G. L. Woods, and A. K. Jafri.** 2000. Comparison of four methods for identifying Streptococcus pneumoniae. *Diagn. Microbiol. Infect. Dis.* **37:**285–287.
8. **Darkes, M. J., and G. L. Plosker.** 2002. Pneumococcal conjugate vaccine (Prevnar; PNCRM7): a review of its use in the prevention of Streptococcus pneumoniae infection. *Paediatr. Drugs* **4:**609–630.
9. **Denys, G. A., and R. B. Carey.** 1992. Identification of *Streptococcus pneumoniae* with a DNA probe. *J. Clin. Microbiol.* **30:**2725–2727.
10. **Diaz, E., and J. L. Garcia.** 1990. Characterization of the transcription unit encoding the major pneumococcal autolysin. *Gene* **90:**157–162.
11. **Diaz, E., R. Lopez, and J. L. Garcia.** 1990. Chimeric phage-bacterial enzymes: a clue to the modular evolution of genes. *Proc. Natl. Acad. Sci. USA* **87:**8125–8129.
12. **Dowson, C. G., V. Barcus, S. King, P. Pickerill, A. Whatmore, and M. Yeo.** 1997. Horizontal gene transfer and the evolution of resistance and virulence determinants in Streptococcus. *Soc. Appl. Bacteriol. Symp. Ser.* **26:**42S-51S.
13. **Dowson, C. G., T. J. Coffey, C. Kell, and R. A. Whiley.** 1993. Evolution of penicillin resistance in Streptococcus pneumoniae; the role of Streptococcus mitis in the formation of a low affinity PBP2B in S. pneumoniae. *Mol. Microbiol.* **9:**635–643.
14. **Dowson, C. G., A. Hutchison, J. A. Brannigan, R. C. George, D. Hansman, J. Linares, A. Tomasz, J. M. Smith, and B. G. Spratt.** 1989. Horizontal transfer of penicillin-binding protein genes in penicillin-resistant clinical isolates of Streptococcus pneumoniae. *Proc. Natl. Acad. Sci. USA* **86:**8842–8846.
15. **Dowson, C. G., A. Hutchison, N. Woodford, A. P. Johnson, R. C. George, and B. G. Spratt.** 1990. Penicillin-resistant viridans streptococci have obtained altered penicillin-binding protein genes from penicillin-resistant strains of Streptococcus pneumoniae. *Proc. Natl. Acad. Sci. USA* **87:**5858–5862.
16. **Dowson, C. G., and K. Trzcinski (ed.).** 2001. *Evolution and Epidemiology of Antibiotic Resistant Pneumococci.* Marcel Dekker, Inc., New York, N.Y.
17. **du Plessis, M., A. M. Smith, and K. P. Klugman.** 1998. Rapid detection of penicillin-

resistant *Streptococcus pneumoniae* in cerebrospinal fluid by a seminested-PCR strategy. *J. Clin. Microbiol.* **36:**453–457.

18. **Enright, M. C., and B. G. Spratt.** 1999. Extensive variation in the ddl gene of penicillin-resistant Streptococcus pneumoniae results from a hitchhiking effect driven by the penicillin-binding protein 2b gene. *Mol. Biol. Evol.* **16:**1687–1695.

19. **Enright, M. C., and B. G. Spratt.** 1998. A multilocus sequence typing scheme for Streptococcus pneumoniae: identification of clones associated with serious invasive disease. *Microbiology* **144** (Pt. 11):3049–3060.

20. **Falguera, M., A. Lopez, A. Nogues, J. M. Porcel, and M. Rubio-Caballero.** 2002. Evaluation of the polymerase chain reaction method for detection of Streptococcus pneumoniae DNA in pleural fluid samples. *Chest* **122:**2212–2216.

21. **Feil, E. J., J. M. Smith, M. C. Enright, and B. G. Spratt.** 2000. Estimating recombinational parameters in Streptococcus pneumoniae from multilocus sequence typing data. *Genetics* **154:**1439–1450.

22. **Fenoll, A., R. Munoz, E. Garcia, and A. G. de la Campa.** 1994. Molecular basis of the optochin-sensitive phenotype of pneumococcus: characterization of the genes encoding the F0 complex of the Streptococcus pneumoniae and Streptococcus oralis H(+)-ATPases. *Mol. Microbiol.* **12:**587–598.

23. **Finland, M., and M. W. Barnes.** 1977. Changes in the occurrence of capsular serotypes of *Streptococcus pneumoniae* in Boston City Hospital during selected years between 1935 and 1974. *J. Clin. Microbiol.* **5:**154–166.

24. **Gillespie, S. H., T. D. McHugh, H. Ayres, A. Dickens, A. Efstratiou, and G. C. Whiting.** 1997. Allelic variation in *Streptococcus pneumoniae* autolysin (N-acetyl muramoyl-L-alanine amidase). *Infect. Immun.* **65:**3936–3938.

25. **Gillespie, S. H., C. Ullman, M. D. Smith, and V. Emery.** 1994. Detection of *Streptococcus pneumoniae* in sputum samples by PCR. *J. Clin. Microbiol.* **32:**1308–1311.

26. **Greiner, O., P. J. Day, P. P. Bosshard, F. Imeri, M. Altwegg, and D. Nadal.** 2001. Quantitative detection of *Streptococcus pneumoniae* in nasopharyngeal secretions by real-time PCR. *J. Clin. Microbiol.* **39:**3129–3134.

27. **Henrichsen, J.** 1999. Typing of Streptococcus pneumoniae: past, present, and future. *Am. J. Med.* **107:**50S–54S.

28. **Hollingshead, S. K., R. Becker, and D. E. Briles.** 2000. Diversity of PspA: mosaic genes and evidence for past recombination in Streptococcus pneumoniae. *Infect. Immun.* **68:**5889–5900.

29. **Holmberg, H., D. Danielsson, J. Hardie, A. Krook, and R. Whiley.** 1985. Cross-reactions between alpha-streptococci and Omniserum, a polyvalent pneumococcal serum, demonstrated by direct immunofluorescence, immunoelectroosmophoresis, and latex agglutination. *J. Clin. Microbiol.* **21:**745–748.

30. **Iannelli, F., M. R. Oggioni, and G. Pozzi.** 2002. Allelic variation in the highly polymorphic locus pspC of Streptococcus pneumoniae. *Gene* **284:**63–71.

31. **Kaijalainen, T., S. Rintamaki, E. Herva, and M. Leinonen.** 2002. Evaluation of gene-technological and conventional methods in the identification of Streptococcus pneumoniae. *J. Microbiol. Methods* **51:**111–118.

32. **Kawamura, Y., X. G. Hou, Y. Todome, F. Sultana, K. Hirose, S. E. Shu, T. Ezaki, and H. Ohkuni.** 1998. *Streptococcus peroris* sp. nov. and *Streptococcus infantis* sp. nov., new members of the *Streptococcus mitis* group, isolated from human clinical specimens. *Int. J. Syst. Bacteriol.* **48:**921–927.

33. **Kawamura, Y., R. A. Whiley, S. E. Shu, T. Ezaki, and J. M. Hardie.** 1999. Genetic approaches to the identification of the mitis group within the genus Streptococcus. *Microbiology* **145** (Pt. 9):2605–2613.

34. **Kearns, A. M., J. Wheeler, R. Freeman, P. R. Seiders, J. Perry, A. M. Whatmore, and C. G. Dowson.** 2000. Pneumolysin detection identifies atypical isolates of *Streptococcus pneumoniae*. *J. Clin. Microbiol.* **38:**1309–1310.

35. **Kellogg, J. A., D. A. Bankert, C. J. Elder, J. L. Gibbs, and M. C. Smith.** 2001. Identification of *Streptococcus pneumoniae* revisited. *J. Clin. Microbiol.* **39:**3373–3375.

36. **Kim, J. O., and J. N. Weiser.** 1998. Association of intrastrain phase variation in quantity of capsular polysaccharide and teichoic acid with the virulence of Streptococcus pneumoniae. *J. Infect. Dis.* **177:**368–377.

37. **Kontiainen, S., and A. Sivonen.** 1987. Optochin resistance in Streptococcus pneumoniae strains isolated from blood and middle ear fluid. *Eur. J. Clin. Microbiol.* **6:**422–424.

38. **Kumar, S., K. Tamura, I. B. Jakobsen, and M. Nei.** 2001. MEGA2: molecular evolutionary genetics analysis software. *Bioinformatics* **17:**1244–1245.

39. **Lee, C. J., S. D. Banks, and J. P. Li.** 1991. Virulence, immunity, and vaccine related to Streptococcus pneumoniae. *Crit. Rev. Microbiol.* **18:**89–114.

40. **Llull, D., R. Lopez, and E. Garcia.** 2000. Clonal origin of the type 37 Streptococcus pneumoniae. *Microb. Drug Resist.* **6:**269–275.

41. **Lock, R. A., Q. Y. Zhang, A. M. Berry, and J. C. Paton.** 1996. Sequence variation in the Streptococcus pneumoniae pneumolysin gene affecting haemolytic activity and electrophoretic mobility of the toxin. *Microb. Pathog.* **21:**71–83.

42. **Majewski, J., P. Zawadzki, P. Pickerill, F. M. Cohan, and C. G. Dowson.** 2000. Barriers to genetic exchange between bacterial species: *Streptococcus pneumoniae* transformation. *J. Bacteriol.* **182:**1016–1023.

43. **Martin, M., J. H. Turco, M. E. Zegans, R. R. Facklam, S. Sodha, J. A. Elliott, J. H. Pryor, B. Beall, D. D. Erdman, Y. Y. Baumgartner, P. A. Sanchez, J. D. Schwartzman, J. Montero, A. Schuchat, and C. G. Whitney.** 2003. An outbreak of conjunctivitis due to atypical Streptococcus pneumoniae. *N. Engl. J. Med.* **348:**1112–1121.

44. **Martín-Galiano, A. J., L. Balsalobre, A. Fenoll, and A. G. de la Campa.** 2003. Genetic characterization of optochin-susceptible viridans group streptococci. *Antimicrob. Agents Chemother.* **47:**3187–3194.

45. **McAvin, J. C., P. A. Reilly, R. M. Roudabush, W. J. Barnes, A. Salmen, G. W. Jackson, K. K. Beninga, A. Astorga, F. K. McCleskey, W. B. Huff, D. Niemeyer, and K. L. Lohman.** 2001. Sensitive and specific method for rapid identification of *Streptococcus pneumoniae* using real-time fluorescence PCR. *J. Clin. Microbiol.* **39:**3446–3451.

46. **McCool, T. L., T. R. Cate, E. I. Tuomanen, P. Adrian, T. J. Mitchell, and J. N. Weiser.** 2003. Serum immunoglobulin G response to candidate vaccine antigens during experimental human pneumococcal colonization. *Infect. Immun.* **71:**5724–5732.

47. **Meats, E., A. B. Brueggemann, M. C. Enright, K. Sleeman, D. T. Griffiths, D. W. Crook, and B. G. Spratt.** 2003. Stability of serotypes during nasopharyngeal carriage of *Streptococcus pneumoniae*. *J. Clin. Microbiol.* **41:**386–392.

48. **Morona, J. K., R. Morona, and J. C. Paton.** 1999. Analysis of the 5′ portion of the type 19A capsule locus identifies two classes of *cpsC*, *cpsD*, and *cpsE* genes in *Streptococcus pneumoniae*. *J. Bacteriol.* **181:**3599–3605.

49. **Morrison, K. E., D. Lake, J. Crook, G. M. Carlone, E. Ades, R. Facklam, and J. S. Sampson.** 2000. Confirmation of *psaA* in all 90 serotypes of *Streptococcus pneumoniae* by PCR and potential of this assay for identification and diagnosis. *J. Clin. Microbiol.* **38:**434–437.

50. **Muller-Graf, C. D., A. M. Whatmore, S. J. King, K. Trzcinski, A. P. Pickerill, N. Doherty, J. Paul, D. Griffiths, D. Crook, and C. G. Dowson.** 1999. Population biology of Streptococcus pneumoniae isolated from oropharyngeal carriage and invasive disease. *Microbiology* **145:**3283–3293.

51. **Mundy, L. S., E. N. Janoff, K. E. Schwebke, C. J. Shanholtzer, and K. E. Willard.** 1998. Ambiguity in the identification of Streptococcus pneumoniae. Optochin, bile solubility, quellung, and the AccuProbe DNA probe tests. *Am. J. Clin. Pathol.* **109:**55–61.

52. **Munoz, R., A. Fenoll, D. Vicioso, and J. Casal.** 1990. Optochin-resistant variants of Streptococcus pneumoniae. *Diagn. Microbiol. Infect. Dis.* **13:**63–66.

53. **Murdoch, D. R., T. P. Anderson, K. A. Beynon, A. Chua, A. M. Fleming, R. T. Laing, G. I. Town, G. D. Mills, S. T. Chambers, and L. C. Jennings.** 2003. Evaluation of a PCR assay for detection of *Streptococcus pneumoniae* in respiratory and nonrespiratory samples from adults with community-acquired pneumonia. *J. Clin. Microbiol.* **41:**63–66.

54. **Novak, R., E. Charpentier, J. S. Braun, E. Park, S. Murti, E. Tuomanen, and R. Masure.** 2000. Extracellular targeting of choline-binding proteins in Streptococcus pneumoniae by a zinc metalloprotease. *Mol. Microbiol.* **36:**366–376.

55. **Obregon, V., P. Garcia, E. Garcia, A. Fenoll, R. Lopez, and J. L. Garcia.** 2002. Molecular peculiarities of the *lytA* gene isolated from clinical pneumococcal strains that are bile insoluble. *J. Clin. Microbiol.* **40:**2545–2554.

56. **O'Brien, K. L., M. A. Bronsdon, R. Dagan, P. Yagupsky, J. Janco, J. Elliott, C. G. Whitney, Y. H. Yang, L. G. Robinson, B. Schwartz, and G. M. Carlone.** 2001. Evaluation of a medium (STGG) for transport and optimal recovery of Streptococcus pneumoniae from nasopharyngeal secretions collected during field studies. *J. Clin. Microbiol.* **39:**1021–1024.

57. **O'Brien, K. L., and H. Nohynek.** 2003. Report from a WHO working group: standard method for detecting upper respiratory carriage of Streptococcus pneumoniae. *Pediatr. Infect. Dis. J.* **22:**133–140.

58. **Oggioni, M. R., G. Memmi, T. Maggi, D. Chiavolini, F. Iannelli, and G. Pozzi.** 2003. Pneumococcal zinc metalloproteinase ZmpC cleaves human matrix metalloproteinase 9 and is a virulence factor in experimental pneumonia. *Mol. Microbiol.* **49:**795–805.

59. **Oggioni, M. R., and G. Pozzi.** 2001. Comparative genomics for identification of clone-specific sequence blocks in Streptococcus pneumoniae. *FEMS Microbiol. Lett.* **200:**137–143.

60. **Poland, G. A.** 1999. The burden of pneumococcal disease: the role of conjugate vaccines. *Vaccine* **17:**1674–1679.

61. Poulsen, K., J. Reinholdt, C. Jespersgaard, K. Boye, T. A. Brown, M. Hauge, and M. Kilian. 1998. A comprehensive genetic study of streptococcal immunoglobulin A1 proteases: evidence for recombination within and between species. *Infect. Immun.* **66:**181–190.

62. Powley, L., J. Meeson, and D. Greenwood. 1989. Tolerance to penicillin in streptococci of viridans group. *J. Clin. Pathol.* **42:**77–80.

63. Pozzi, G., M. R. Oggioni, and A. Tomasz. 1989. DNA probe for identification of *Streptococcus pneumoniae. J. Clin. Microbiol.* **27:**370–372.

64. Ramirez, M., E. Severina, and A. Tomasz. 1999. A high incidence of prophage carriage among natural isolates of *Streptococcus pneumoniae. J. Bacteriol.* **181:**3618–3625.

65. Reichmann, P., A. Konig, J. Linares, F. Alcaide, F. C. Tenover, L. McDougal, S. Swidsinski, and R. Hakenbeck. 1997. A global gene pool for high-level cephalosporin resistance in commensal Streptococcus species and Streptococcus pneumoniae. *J. Infect. Dis.* **176:**1001–1012.

66. Reichmann, P., E. Varon, E. Gunther, R. R. Reinert, R. Luttiken, A. Marton, P. Geslin, J. Wagner, and R. Hakenbeck. 1995. Penicillin-resistant Streptococcus pneumoniae in Germany: genetic relationship to clones from other European countries. *J. Med. Microbiol.* **43:**377–385.

67. Rudolph, K. M., A. J. Parkinson, C. M. Black, and L. W. Mayer. 1993. Evaluation of polymerase chain reaction for diagnosis of pneumococcal pneumonia. *J. Clin. Microbiol.* **31:**2661–2666.

68. Saruta, K., T. Matsunaga, M. Kono, S. Hoshina, S. Kanemoto, O. Sakai, and K. Machida. 1997. Simultaneous detection of Streptococcus pneumoniae and Haemophilus influenzae by nested PCR amplification from cerebrospinal fluid samples. *FEMS Immunol. Med. Microbiol.* **19:**151–157.

69. Schlegel, L., J. M. Prostak, C. Spicq, G. Sissia, A. Fremaux, and P. Geslin. 1998. Presumptive tests and molecular hybridization for the identification of untypable strains of Streptococcus pneumoniae. *Pathol. Biol.* **46:**459–463. (In French.)

70. Scott, J. A., E. L. Marston, A. J. Hall, and K. Marsh. 2003. Diagnosis of pneumococcal pneumonia by *psaA* PCR analysis of lung aspirates from adult patients in Kenya. *J. Clin. Microbiol.* **41:**2554–2559.

71. Sibold, C., J. Henrichsen, A. Konig, C. Martin, L. Chalkley, and R. Hakenbeck. 1994. Mosaic pbpX genes of major clones of penicillin-resistant Streptococcus pneumoniae have evolved from pbpX genes of a penicillin-sensitive Streptococcus oralis. *Mol. Microbiol.* **12:**1013–1023.

72. Teng, L. J., P. R. Hsueh, J. C. Tsai, P. W. Chen, J. C. Hsu, H. C. Lai, C. N. Lee, and S. W. Ho. 2002. groESL sequence determination, phylogenetic analysis, and species differentiation for viridans group streptococci. *J. Clin Microbiol.* **40:**3172–3178.

73. Tong, H., X. Gao, and X. Dong. 2003. *Streptococcus oligofermentans* sp. nov., a novel oral isolate from caries-free humans. *Int. J. Syst. Evol. Microbiol.* **53:**1101–1104.

74. Ubukata, K., Y. Asahi, A. Yamane, and M. Konno. 1996. Combinational detection of autolysin and penicillin-binding protein 2B genes of *Streptococcus pneumoniae* by PCR. *J. Clin. Microbiol.* **34:**592–596.

75. van Haeften, R., S. Palladino, I. Kay, T. Keil, C. Heath, and G. W. Waterer. 2003. A quantitative LightCycler PCR to detect Streptococcus pneumoniae in blood and CSF. *Diagn. Microbiol. Infect. Dis.* **47:**407–414.

76. Veenhoven, R., D. Bogaert, C. Uiterwaal, C. Brouwer, H. Kiezebrink, J. Bruin, I. J. E, P. Hermans, R. de Groot, B. Zegers, W. Kuis, G. Rijkers, A. Schilder, and E. Sanders. 2003. Effect of conjugate pneumococcal vaccine followed by polysaccharide pneumococcal vaccine on recurrent acute otitis media: a randomised study. *Lancet* **361:**2189–2195.

77. Verhelst, R., T. Kaijalainen, T. De Baere, G. Verschraegen, G. Claeys, L. Van Simaey, C. De Ganck, and M. Vaneechoutte. 2003. Comparison of five genotypic techniques for identification of optochin-resistant pneumococcus-like isolates. *J. Clin. Microbiol.* **41:**3521–3525.

78. Waite, R. D., D. W. Penfold, J. K. Struthers, and C. G. Dowson. 2003. Spontaneous sequence duplications within capsule genes cap8E and tts control phase variation in Streptococcus pneumoniae serotypes 8 and 37. *Microbiology* **149:**497–504.

79. Waite, R. D., J. K. Struthers, and C. G. Dowson. 2001. Spontaneous sequence duplication within an open reading frame of the pneumococcal type 3 capsule locus causes high-frequency phase variation. *Mol. Microbiol.* **42:**1223–1232.

80. Walker, J. A., R. L. Allen, P. Falmagne, M. K. Johnson, and G. J. Boulnois. 1987. Molecular cloning, characterization, and complete nucleotide sequence of the gene for pneumolysin, the sulfhydryl-activated toxin of *Streptococcus pneumoniae. Infect. Immun.* **55:**1184–1189.

81. Wasilauskas, B. L., and K. D. Hampton. 1984. An analysis of Streptococcus pneumoniae identification using biochemical and serological procedures. *Diagn. Microbiol. Infect. Dis.* **2:**301–307.

82. **Whatmore, A. M., V. A. Barcus, and C. G. Dowson.** 1999. Genetic diversity of the streptococcal competence (*com*) gene locus. *J. Bacteriol.* **181:**3144–3154.

83. **Whatmore, A. M., and C. G. Dowson.** 1999. The autolysin-encoding gene (*lytA*) of *Streptococcus pneumoniae* displays restricted allelic variation despite localized recombination events with genes of pneumococcal bacteriophage encoding cell wall lytic enzymes. *Infect. Immun.* **67:**4551–4556.

84. **Whatmore, A. M., A. Efstratiou, A. P. Pickerill, K. Broughton, G. Woodard, D. Sturgeon, R. George, and C. G. Dowson.** 2000. Genetic relationships between clinical isolates of *Streptococcus pneumoniae*, *Streptococcus oralis*, and *Streptococcus mitis*: characterization of "atypical" pneumococci and organisms allied to *S. mitis* harboring *S. pneumoniae* virulence factor-encoding genes. *Infect. Immun.* **68:**1374–1382.

85. **Whatmore, A. M., S. J. King, N. C. Doherty, D. Sturgeon, N. Chanter, and C. G. Dowson.** 1999. Molecular characterization of equine isolates of *Streptococcus pneumoniae*: natural disruption of genes encoding the virulence factors pneumolysin and autolysin. *Infect. Immun.* **67:**2776–2782.

86. **Wheeler, J., R. Freeman, M. Steward, K. Henderson, M. J. Lee, N. H. Piggott, G. J. Eltringham, and A. Galloway.** 1999. Detection of pneumolysin in sputum. *J. Med. Microbiol.* **48:**863–866.

87. **Wuorimaa, T., and H. Kayhty.** 2002. Current state of pneumococcal vaccines. *Scand. J. Immunol.* **56:**111–129.

88. **Zhang, Y., D. J. Isaacman, R. M. Wadowsky, J. Rydquist-White, J. C. Post, and G. D. Ehrlich.** 1995. Detection of *Streptococcus pneumoniae* in whole blood by PCR. *J. Clin. Microbiol.* **33:**596–601.

COMPARATIVE GENOMICS OF *STREPTOCOCCUS PNEUMONIAE*: INTRASTRAIN DIVERSITY AND GENOME PLASTICITY

Hervé Tettelin and Susan K. Hollingshead

2

INTRODUCTION

Despite the significance of *Streptococcus pneumoniae* as a pathogen, the details of the genomic sequence of *S. pneumoniae* were slow to come. As many as eight corporate organizations generated genome sequence information from this organism for their own purposes, but these data were not available to the public. Year 2001 put an end to the long wait with the publication of three genomes in the span of only 5 months: the draft genome sequence of strain G54, a type 19F clinical isolate in early summer (7), and the complete genome sequences of strain TIGR4, a type 4 virulent isolate, in July (32) and the avirulent laboratory strain R6 in October (18). Each of the three genome sequences was generated by the whole-genome shotgun sequencing strategy.

As with many other gram-positive organisms, the genome of *S. pneumoniae* proved difficult to sequence. Protocols had to be adapted for preparation of shotgun cloning-grade genomic DNA; the most important aspect proved to be the harvesting of cells in mid-logarithmic phase to ensure that capsular polysaccharides, which hamper efficient cloning, could be separated from the DNA. Several genomic libraries were constructed, none of which were truly random: several regions of the genome remained unsequenced (170 regions were not covered by clones) and had to be amplified by PCR. Finally, assembly of the genome proved challenging as well due to the presence of a large and unprecedented fraction of repetitive sequences. Two regions of the TIGR4 strain genome in particular hampered closure. One involved a very large repetitive region that encoded a mucin-like protein (SP1772), and the other involved an invertible element (SP0505-SP0507) present in mixed forms in different clones from the shotgun libraries.

The comparison of these three pneumococcal genomes reveals a remarkable number of indels and features such as repeats that can contribute to genome plasticity. The mosaic genome structures observed in these 3 and in a survey of 13 additional genomes are a testament to the remarkably adaptive nature of this organism. The genome provides an inventory of the traits that can be used in adapting to ecological niches. While primarily colonizing the nasopharynx as a well-behaved commen-

Hervé Tettelin, Department of Microbial Genomics, The Institute for Genomic Research, 9712 Medical Center Drive, Rockville, MD 20850. *Susan K. Hollingshead*, Department of Microbiology, University of Alabama at Birmingham, 845 19th Street South, Birmingham, AL 35294.

The Pneumococcus. Volume Editor, Elaine I. Tuomanen,
© 2004 ASM Press, Washington, D.C.

sal in humans, the pneumococcus can also become a pest, by moving on to colonize the ears or upper airways, or a more formidable pathogen, by further colonizing the lungs or the blood. A comparison of strain-to-strain differences extends the inventory to provide a snapshot of the larger repertoire of traits present within the species. One of the goals of comparing genomes is to understand differences in virulence and how the repertoire of traits is maintained and selected and distributed within the species. A number of these traits, reviewed here, are not present in all isolates but can be moved by transformation. Encompassing the most recent evolutionary history of the strains, comparative genome analysis can also uncover some of the footprints of evolution and reveal some of the process of change. Combined with data from population biology studies that emphasize the importance of recombination and with data from molecular pathogenesis studies pertaining to the mechanics of transformation, the genomic footprints demonstrate a remarkable capacity for adaptation within this species.

GENOME ORGANIZATION

The pneumococcal genome contains over 2,000 genes (Table 1), of which about 60% could be assigned a predicted function. There is a strong gene orientation bias towards genes being transcribed away from the origin of replication, as has now also been observed in all other low-GC gram-positive organisms. Of the 2,236 genes in TIGR4, 1,155 are located on the right of the origin of replication and 916 (79%) of these are transcribed in the same direction as the DNA replication fork; similarly, 1,081 genes are on the left of the origin of replication and 857 (79%) of them are transcribed in the same direction.

The main repeated sequences in the genome are classified in three classes: insertion sequence (IS), BOX, and RUP elements (Table 1). The majority of IS elements have undergone insertions, deletions, and/or point mutations that result in frameshifted or otherwise nonfunctional transposase genes. RUP elements are thought to act like nonautonomous IS mobilized by the IS630-Spn1 transposase (24). RUP elements are preferentially inserted

TABLE 1 Genome features of the three sequenced strains of *S. pneumoniae*

Feature	TIGR4, virulent	R6, avirulent	G54, virulent (draft: 31 contigs)
Genome size (bp)	2,160,837	2,038,615	~2,100,000
G+C content (%)	39.7	39.7	39.5
Predicted genes	2,236	2,043	~2,046
Known function	1,440	1,289	1,134
Surface-exposed proteins			
Signal peptide	256	153	221
Lipoprotein	47	42	45
LPxTG	19	13	NA[a]
Choline binding	15	10	NA
Capsule	Type 4	7,504-bp deletion	Type 19F
rRNA genes	4	4	NA
tRNA genes	58	58	51
sRNA genes	3	3	3
IS elements	105	91	NA
BOX elements	128	115	NA
RUP elements	111	84	NA

[a] NA, not available.

in IS elements in the genome. BOX elements are composed of three subunits: *boxA, boxB* (which can be present in multiple copies), and *boxC* (20). Analysis of the R6 genome revealed that almost three times as many BOX elements and 1.5-fold more RUP elements are located between genes oriented 3′ to 3′ (where transcription termination signals are expected) than between genes oriented 5′ to 5′ (where transcriptional promoters are expected) (18). This might suggest a role for these elements in transcription termination.

However, a primary role for the numerous repeats might be their potential contribution to genomic rearrangements using the repeats as seeds for recombination events, as is discussed below.

METABOLISM, TRANSPORT, AND SURFACE-EXPOSED PROTEINS

It was first noted for *S. pneumoniae* that catabolism of carbohydrates is an important part of the lifestyle, and this was later found to be a trait shared with most of the lactic acid bacterial group (1, 3, 15, 19, 31). All members of this group lack the tricarboxylic acid cycle and depend heavily upon fermentation for their energy needs. For all of the lactic acid group, on a per-genome size basis, the relative number of genes involved in sugar catabolism compared with other organisms is high, with predicted capacity for utilization of a wide range of sugars and substituted nitrogen compounds as substrates.

TIGR4 contains pathways for the catabolism of 13 sugars (the monosaccharides glucose, mannose, fucose, fructose, galactitol, and galactose; the disaccharides sucrose, lactose, trehalose, maltose, and cellobiose; the trisaccharide raffinose; and the fructose oligosaccharide inulin), 10 amino acids (L isomers of asparagine, aspartate, cadaverine, glutamate, glutamine, isoleucine, leucine, lysine, proline, and threonine), pentitols, and N-acetylglucosamine. There are also some as-yet-uncharacterized sugar transporters. R6 lacks one of two fructose-related PTS systems but contains at least one additional sugar ABC transporter. *Haemophilus influenzae* and *Neisseria meningitidis* are also colonizers of the respiratory tract, but their panel of sugar transporters is much more limited and they are much more reliant on carboxylates and other compounds for their carbon needs. These differing nutritional needs might reduce the competition for resources between the gram-positive versus the gram-negative bacteria within the respiratory tract. Because reliance on sugar transport and metabolism appears to be a common feature of streptococci, competition with other streptococci may be likely to occur.

Extracellular enzymes identified in the genome sequence include N-acetylglucosaminidases, α- and β-galactosidases, endoglycosidases, hydrolases, hyaluronidases, and neuraminidases. Such enzymes may contribute to the degradation of the pneumococcal capsule as well as host mucins, glycolipids, and hyaluronic acid. This would provide for a synergy between host damage and availability of compounds that can be transported inside the cell. The compounds could then be utilized for nutritional purposes or potentially could be used in remodeling of surface structures.

Proteins likely to be surface exposed were predicted in TIGR4 by searching protein sequences for motifs such as choline binding, LPxTG cell wall anchor, lipoprotein, and signal peptide (32). This analysis identified several proteins (Table 1), including those characterized prior to the availability of the genome sequence, such as immunoglobulin A1 protease, hyaluronidase, sortase, choline-binding proteins, and PspA; it also identified novel sortases, a putative amylopullulanase, a putative endo-β-N-acetylglucosaminidase, and several cell wall-anchored proteins and lipoproteins. The proteins could be involved in adhesion and degradation of host polysaccharides.

We have noted an intriguing functional pattern for proteins that are surface attached by alternate means. In TIGR4, 16 proteins are predicted to be covalently attached through the action of the transpeptidase sortase A (SP1218). Predicted functions for these appear to be in the class of degradative enzymes, including proteases and hexosaminidases. While some of these may assist in adhesion by uncovering adhesive sites, or in host defense by interacting with host-derived molecules, their most immediate activities involve enzymatic cleavages. This contrasts with the functional class of enzymes attached to the surface by this means in *Staphylococcus aureus*. The term MSCRAMM (microbial surface components recognizing adhesive matrix molecules) has been applied to the latter, which consist primarily of the adhesive functional class. There are three surface proteins thought to be surface anchored by auxiliary sortases (SP466, -467, and -468) present in TIGR4, but absent in R6, that might be considered as belonging to the adhesive class, based on protein sequence similarities to other staphylococcal and streptococcal proteins. The choline-binding protein family, unique to *S. pneumoniae*, seems to contain a mixture of potential host interaction proteins, some of which are clearly involved in host defense, while others may be involved in adhesion.

COMPARISON OF THE *S. PNEUMONIAE* GENOME WITH THOSE OF OTHER MEMBERS OF THE LACTIC ACID GROUP

The genomes were aligned with PROmer (5), a suffix tree-based program that extracts protein sequences from two DNA sequences in all six frames of translation and reports regions of identical protein sequences with their coordinates in both genomes. The alignments of paired genomes from each species in the lactic acid bacterial group were arranged in order of decreasing number of matching protein sequences (Fig. 1). For most comparisons between streptococci excluding *S. pneumoniae*, characteristic signal lines near the diagonals are readily apparent. When any cross-species comparison involves *S. pneumoniae*, however, these lines are greatly blurred, indicating a loss of signal. The greater "noise" in comparisons involving *S. pneumoniae* is not a function of the genetic distance between the two sets of orthologs, as *Streptococcus mutans* is more distant

to group A or group B streptococci yet shows more evidence for synteny than is seen in the *S. pneumoniae* comparisons. A loss of signal should be expected in a genome experiencing a high frequency of recombination, especially if a certain amount of illegitimate recombination has occurred. The PROmer comparisons suggest that this is the case for the genome of *S. pneumoniae*. All of the comparisons with *Lactococcus lactis* also showed a decrease in the signal-to-noise ratio. This could be because of a greater genetic distance for this species, or it could be because the *L. lactis* genome shows a similar level of genetic diversity or genetic plasticity to the *S. pneumoniae* genome.

We also used a Blast e-value cutoff of $10e^{-15}$ to compare the genes in the pneumococcal genome to those of *Streptococcus pyogenes* (group A streptococcus [GAS]) and *Streptococcus agalactiae* (group B streptococcus [GBS]) (Color Plate 1). This analysis reveals that ~1,163 TIGR4 genes are shared with both GAS and GBS, 165 are shared with GBS only, 74 are shared with GAS only, and 696 are specific to TIGR4 (31).

COMPARISON OF COMPLETE *S. PNEUMONIAE* GENOMES

The Venn diagram in Color Plate 1B shows the Blast comparison of the TIGR4, R6, and G54 genomes. Approximately 1,802 genes, or about 80% of the total gene complement, were shared by all three genomes. TIGR4 shared 43 genes uniquely with G54 and 95 with R6, while G54 shared about 95 uniquely with R6, as well. TIGR4 had 177 unique genes relative to the other two, G54 had 178, and R6 had 86. Pairwise, each strain varied from the other two in about 10% of its total gene complement. The nonshared genes represent both single-gene insertions and deletions and also a number of nonshared regions that are discussed further below.

The PROmer alignment of the two complete genomes of TIGR4 and R6 shows a genome structure that is conserved throughout their lengths, with several insertions or deletions (indels) observed but no major inversion

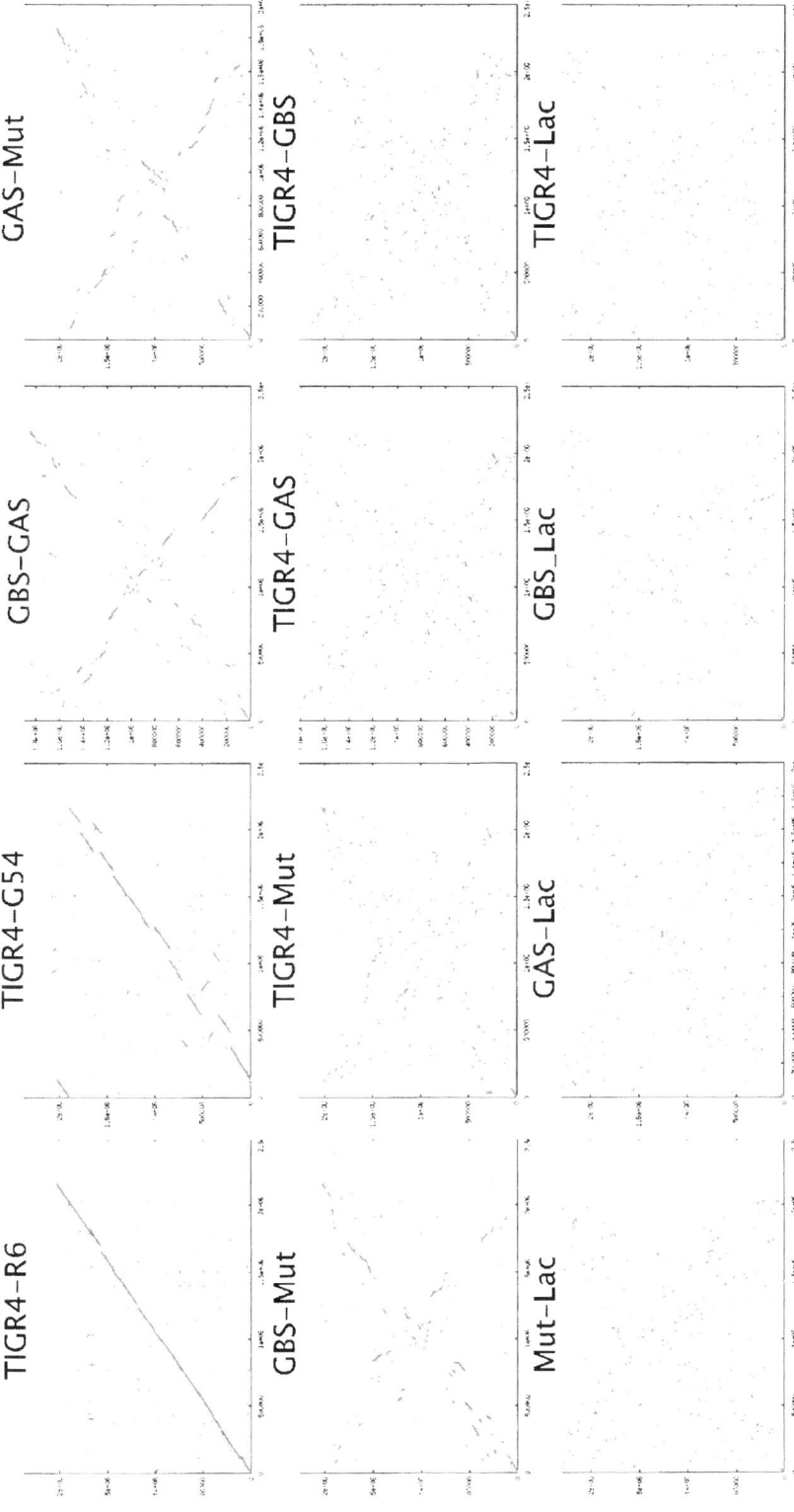

FIGURE 1 PROmer plots showing proteome-to-proteome comparisons of various species within the lactic acid group of bacteria. Genes along the diagonal represent conserved proteins, potential orthologs, that are encoded in similar positions for the cross-genome comparison. The genes encoding these proteins are still in synteny between the two genomes under comparison. Scattered dots away from the diagonal represent the detection of potential orthologs whose position has shifted since the two genomes last shared a common ancestor. The genes encoding these scattered proteins now lack synteny. While comparisons of GAS, GBS, and *S. mutans* maintain some synteny, most comparisons with *S. pneumoniae* show little synteny.

(Fig. 1). A PROmer alignment of TIGR4 was also made with proteins from the 31 contigs representing the draft genome sequence of G54 after ordering these using TIGR4 as a scaffold. In this case, the information gleaned is more tenuous because the relative order of the contigs cannot be determined with absolute certainty and there remains the possibility that some contigs could be incorrectly assembled. In this case, two possible inversions are noted (Fig. 1).

The aligned comparison of TIGR4 and R6 was further examined in some detail. In absolute terms, the genome of TIGR4 is 122,222 bp larger than that of R6, but the comparative analysis reveals that 142,844 bp of the TIGR4 genome are specific to this strain and similarly 75,080 bp of the R6 genome are specific (Table 2). This is indicative of regions inserted in the virulent isolate TIGR4 or deleted in R6, but also of regions between conserved flanks that are different between the two strains. There were six regions of indels between TIGR4 and R6 and five regions of different content between conserved flanks.

The first area is a 2,443-bp segment between conserved regions R6_C1 and R6_C2 (Table 2). It encodes two hypothetical proteins and is flanked on the right side by a BOX element. The equivalent region in TIGR4, between TIGR4_C1 and TIGR4_C2, is 9,502 bp and encodes a choline-binding protein (*cbpI*), a zinc metalloproteinase (*zmpC*), an acetyltransferase, and five hypothetical proteins. On the right flank of this region, there is a phosphorylase gene and a BOX element, both of which are also found in R6.

The second area is a 17,811-bp segment between conserved regions R6_C2 and R6_C3 (Table 2) that is absent in TIGR4. It encodes 15 hypothetical proteins and contains two truncated transporters, and it is flanked on the right by a RUP element. The equivalent area in TIGR4 is only 2,193 bp long and encodes only two hypothetical proteins. It is flanked on the left side by a truncated argininosuccinate lyase gene whose ortholog is intact in R6. On the right side is a RUP element adjacent to sequences encoding a hypothetical protein and PspA (SP0117).

The third area is 6,849 bp long in TIGR4 where a segment of only 340 bp that contains a BOX element exists in R6. This region, located between TIGR4_C3 and TIGR_C4, contains a gene for a transcriptional regulator, a flavoprotein gene, a decarboxylase gene, a macrolide efflux protein gene, a frameshifted repressor, and genes for six hypothetical proteins. It is flanked on the right by the gene for DNA mismatch repair protein HexB.

The fourth area is the capsule-encoding gene locus. In both genomes the locus is flanked on the left by α-1,6 glucosidase, IS elements, and a RUP element. Only TIGR4 displays IS elements on the right flank, but in both strains *aliA* is found downstream of the locus. In TIGR4 it is a 19,016-bp area carrying the type 4 capsule cassette (19 genes). In R6, it is an 11,413-bp area containing an incomplete set of genes for a type 2 capsule. The 7.5-kb deletion results in the loss of 7 complete genes and accounts for the strain's inability to synthesize a capsule.

The next area is a 14,392-bp segment in TIGR4 between TIGR4_C3 and TIGR4_C4 where only a RUP element is found in R6. The insertion is flanked by an IS element on both sides but no RUP element; it contains genes encoding three sortases (that are separate from the shared sortase), three cell wall surface-anchored proteins, a putative transcriptional regulator, and a hypothetical protein. These additional sortases and their substrates could contribute to differences in virulence and antigenicity.

Next is a bacteriocin-encoding locus (6) with a different structure in each genome. Both strains share the BlpT, BlpS, BlpR, and BlpH genes on the left and the BlpY and BlpZ genes on the right. The 5,675-bp R6 area has only four transporter genes, while the 11,087-bp TIGR4 region carries transporter genes and genes for an additional four bacteriocins, three immunity proteins, and a peptide pheromone. This locus is a two-component system that is closely related to quorum-sensing systems reg-

TABLE 2 Major regions of nonconserved gene order between TIGR4 and R6 identified by inspection of PROmer plots (Fig. 1)[a]

Strain R6	Indel (bp)[b]	No. of genes	Coordinates (bp)	Length (bp)	No. of genes	Strain TIGR4	Indel (bp)[b]	No. of genes	Coordinates (bp)	Length (bp)	No. of genes
R6_C1			0–72866	72,867	65	TIGR4_C1			0–71977	71,977	66
R6_NC1	2,443	2				TIGR4_NC1	9,502	8			
R6_C2			75310–109690	34,381	35	TIGR4_C2			81480–115430	33,951	39
R6_NC2	17,811	17				TIGR4_NC2	2,193	2			
R6_C3			127502–167730	40,229	39	TIGR4_C3			117624–158760	41,137	46
R6_NC3	340	0				TIGR4_NC3	6,849	11			
R6_C4			168071–313696	145,626	155	TIGR4_C4			165610–320077	154,468	174
R6_NC4	11,413	12				TIGR4_NC4	19,016	19			
R6_C5			325110–414435	89,326	90	TIGR4_C5			339094–434629	95,536	94
R6_NC5	915	1				TIGR4_NC5	14,392	10			
R6_C6			415351–470514	55,164	48	TIGR4_C6			449022–506312	57,291	57
R6_NC6	5,675	7				TIGR4_NC6	11,087	17			
R6_C7			476190–938858	462,669	486	TIGR4_C7			517400–992015	474,616	511
R6_NC7	17,516	15				TIGR4_NC7	9,258	10			
R6_C8			956375–1019624	63,250	64	TIGR4_C8			1001274–1063072	61,799	62
R6_NC8	822	0				TIGR4_NC8	11,247	19			
R6_C9			1020447–1182139	161,693	146	TIGR4_C9			1074320–1235211	160,892	160
R6_NC9	13,780	16				TIGR4_NC9	26,289	29			
R6_C10			1195920–1380025	184,106	203	TIGR4_C10			1261501–1454046	192,546	209
R6_NC10	10,372	2				TIGR4_NC10	1,005	0			
R6_C11			1390398–1444629	54,232	59	TIGR4_C11			1455052–1512324	57,273	64
R6_NC11	102	0				TIGR4_NC11	11,000	12			
R6_C12			1444732–1577354	132,623	135	TIGR4-C12			1523325–1655955	132,631	132
R6_NC12	99	0				TIGR4_NC12	37,753	19			
R6_C13			1577454–2038615	461,162	446	TIGR4_C13			1693709–2160837	467,129	466
Total	81,288	72		1,957,327	1,972	Total	159,591	156		2,001,246	2,080

[a] Because only major regions are indicated, there exist many minor rearrangements in regions reported as conserved (C). NC, nonconserved.
[b] The "Indel" column indicates the size of the area between two conserved regions; for instance, the area between regions R6_C1 and R6_C2 is 2,443 bp, while the equivalent area in TIGR4 between TIGR4_C1 and TIGR4_C2 is 9,502 bp.

ulating cell density-dependent phenotypes such as the development of genetic competence or the production of antimicrobial peptides in lactic acid bacteria.

The following area is flanked in both genomes by fragments of the streptococcal conjugative transposon Tn5252 on the left and the gene for cell division protein FtsW on the right. R6 (17,516 bp) displays genes encoding a protein similar to transcriptional regulator MutR, UDP-N-acetyl-D-mannosaminuronic acid dehydrogenase, a carboxylase involved in nikkomycin biosynthesis, a macrolide efflux permease, and hypothetical proteins. TIGR4 (9,258 bp) contains genes for an ABC transporter, a putative protein kinase, a protein similar to transcriptional regulator PlcR, and hypothetical proteins and two IS elements on the right flank.

The two IS elements at the beginning of R6_C9 and shared with TIGR4 allowed the insertion of a phage-like 11,247-bp fragment in TIGR4. It carries genes for two integrase/recombinase proteins, one of which is truncated, a transcriptional regulator, and several hypothetical proteins.

The next area exhibits two loci quite different in R6 and TIGR4 that are both flanked by IS, RUP, and BOX elements on the left and IS elements on the right. The 13,780-bp segment in R6 contains genes for an ABC transporter, two truncated N-acetylneuraminate lyase subunits, a cytidine deaminase, and N-acetylmannosamine-6-P epimerase. The 26,289-bp region in TIGR4 encodes complete V-type ATPase, an oxidoreductase, a putative neuraminidase, N-acetylneuraminate lyase, a putative N-acetylmannosamine-6-P epimerase, a putative phosphosugar-binding transcriptional regulator, and type II DNA modification methyltransferase Spn5252IP.

The 10,372-bp region between R6_C10 and R6_C11 encodes a large hypothetical protein of 2,551 amino acids and another hypothetical protein at the location where there is only a BOX element in TIGR4.

The next area is an 11,000-bp segment in TIGR4 between TIGR4_C11 and TIGR4_C12. No repeated element was found at the equivalent 102-bp region in R6, while the TIGR4 fragment is flanked on the right by an IS element and another IS element is located near its left end. The area contains genes for a phosphotransferase (PTS) transporter system, a transketolase, a ribulose-phosphate 3-epimerase family protein, and a putative transcription antiterminator.

The last area is the largest: it is a 37,753-bp fragment between TIGR4_C12 and TIGR4_C13 that is flanked on the left side by a BOX element. The right flank of the region exhibits four genes, a RUP element, another gene, and a BOX element, all of which are shared with R6. The area encodes the large cell wall-anchored 4,776-amino-acid protein SP1772 that contains 540 imperfect repeats of the motif SASTSASA and eight glycosyltransferases that could make O-linked glycosylations on the serines of SP1772. This would produce a structure similar to mucins that might coat the surface of the pneumococcus and interact with host cellular mucins. The area also encodes two preprotein translocases and a frameshifted gene for sodium/dicarboxylate symporter.

Although a comprehensive study of the G54 aligned regions was not undertaken due to uncertainty about the assembly, several of the individual regions listed above have been inspected in the G54 genome as well. For many of the regions where the conservation of gene order is disturbed between TIGR4 and R6, there is also a break in gene order between G54 and R6 as well as between TIGR4 and G54. In several cases, there are sites where each of the three genomes appears to have been interrupted by the alternate insertion of three distinct gene clusters in each of the three strains. In one case, G54 shares three of the inserted genes with R6 and four with TIGR4. While a genome arrangement such as this (sometimes referred to as a "cassette-type structure") has been understood for the capsular region for several years, the finding of multiple regions with this apparent arrangement is new.

COMPARATIVE GENOME HYBRIDIZATIONS

Nonconserved regions between TIGR4 and R6 were also detected in a comparative genome hybridization (CGH) of the two genomes using DNA microarrays (32). The analysis, which was also performed for the encapsulated serotype 2 strain D39, a parental strain to R6 (29), identified nine regions in TIGR4 that did not hybridize with the other two strains. Six of those regions displayed an atypical nucleotide composition compared to the rest of the TIGR4 genome, suggesting that these regions might have been acquired horizontally. These include the capsule locus, the V-type ATPase locus, and the SP1772 locus, which correspond to NC regions 4, 9, and 11, respectively, in Table 2.

CGH has also been used to compare the genomes of 13 additional strains (Color Plate 2). The strains were chosen primarily with an attempt to maximize their diversity within the species. Six strains (MA to MD, MF, and MG) are members of internationally recognized clones that carry multiple drug resistance traits (22, 28). These six were also chosen to span genetic diversity in pneumococci as measured by multilocus sequence typing from a study of isolates from invasive disease by Enright and Spratt (11). Isolates from carriage were shown to exhibit greater genetic diversity than invasive isolates in a study by Robinson et al. (26) which examined both types of isolates from the same community. Another seven isolates were chosen for CGH from this study (ME and MI to MK). The choice included isolates of a capsule serotype different from that of the first set to minimize the chance of same-genotype strains. Only capsular serotype 6 is represented by more than one isolate (MC, MD, and MF). Strains ME and MF are of genotypes associated with carriage, and strains MM and MI are of genotypes associated with invasive disease (26). In addition, strain MA is associated with a clone that caused meningitis in the United Kingdom.

A DNA probe from each strain was labeled with Cy3 and was hybridized in competition with a Cy5 probe prepared from TIGR4 to a microarray (Color Plate 2). Collectively, the CGH of the 13 strains showed at least 13 regions of diversity involving multiple genes, defined as three or more contiguous genes absent or divergent compared to the TIGR4 arrayed genome (Table 3). Nine of these correspond roughly to regions that were also varied in the comparison with the R6 genome as noted above. Four additional regions of diversity were held in common in the R6 and TIGR4 genomes.

A combined view of microarray CGH data and the direct genome comparisons suggests to us that these regions of diversity are indeed regions of genome plasticity. For many of these, genes are likely to be present but of varied content in different strains. G54, R6, and TIGR4 often have heterogeneous gene clusters in the regions of diversity, as was discussed above. A contrasting view might be that the regions of diversity represent only rare strain-specific insertions in TIGR4, but in this case, the variety of gene content would not be present. Moreover, there is often a subset of the genes within a given region in some strains but not in others. For example, the region encoding the SP1772 mucin-like protein was present in MB, MG, and MJ, but not all components of the cluster were always present. The secA and secY genes associated with this cluster were also present in MA, MF, and the highly divergent strain MH. Similarly, the cluster encoding sortases and cell wall proteins was present in MC, MD, and ME, although only ME contained an ortholog of SP0463. It is worthy of note that genes within the diversity regions often contribute to the virulence of an isolate containing them (16). The alternate sortase region is a case of this type (17). A link to virulence adds in the idea that the regions of diversity might alter fitness of particular clones and thus could also be considered "adaptation regions."

The general content of the variable regions is given in abbreviated form in Table 3. Most regions contain genes that would be able to provide a strain with greater adaptability to its environment. For example, RD 9 contains

TABLE 3 Regions of diversity identified in CGH of 13 strains

Region of diversity	% GC content	5' coordinate	3' coordinate	Corresponding NC region[a]	Variable SP no.	Genes included
RD 1	Typical	70942	82798	NC 1	67–74	Choline-binding protein I, immunoglobulin A1 protease, conserved hypothetical protein, acetyltransferase, CysE/LacA/LpxA/NodL family
RD 2	Typical	158322	168492	NC 3	163–171	Transcriptional regulator PlcR, flavoprotein, hypothetical protein, decarboxylase, macrolide efflux protein, ROK family protein
RD 3	Atypical	315337	339095	NC 4	346–360	Capsular polysaccharide biosynthesis genes
RD 4	Typical	432307	448763	NC 5	461–468	Regulator, cell wall anchor proteins (×3), sortase (×3)
RD 5	Typical	661715	668857	***	694–698	Conserved domain protein, HesA/MoeB/ThiF family protein, hypothetical protein (×2), ABC transporter
RD 6	Highly atypical	986871	1001275	NC 7	1047–1064	Hypothetical proteins (×5), regulator (×2), ABC transporter, phosphoesterase, protein kinase
RD 7	Atypical	1061170	1078626	NC 8	1129–1147	Phage-related island, hypothetical proteins (×10), regulator (×2)
RD 8	Atypical	1234619	1274977	NC 9	1450–1489	V-type sodium ATP synthase, oxidoreductase, neuraminidase
RD 9	Typical	1511649	1526538	NC 11	1612–1623	PTS system, transketolase, ribulose-phosphate 3-epimerase family protein, transcriptional antiterminator, cation-transporting ATPase
RD 10	Highly atypical	1655004	1694168	NC 12	1755–1773	Mucin-like protein, glycosyltransferase (×7), *secA*, *secY*, hypothetical proteins (×4)
RD 11	Atypical	1738159	1743044	***	1828–1831	UDP-glucose 4-epimerase, galactose-1-phosphate uridylyltransferase, phosphate transport system regulatory protein PhoU, hypothetical protein
RD 12	Highly atypical	1849037	1860533	***	1948–1955	Conserved domain protein, bacteriocin protein, hypothetical protein (×3), ABC transporter, serine protease
RD 13	Typical	2072290	2085649	***	2168–2498	Fucolectin-related protein, PTS system, conserved hypothetical protein, fucose operon FucU protein, L-fuculose phosphate aldolase, L-fuculose kinase

[a] Comparison to nonconserved (NC) regions in Table 1, indels between R6 and TIGR4. NC_2 and NC_6 represented insertions in R6, and NC_6 was present based on CGH but divergent between the two completed genomes. These regions, marked by triple asterisks, are conserved in the TIGR4-R6 comparison.

genes for a PTS system, a transketolase, a ribulose-phosphate 3-epimerase family protein, a transcriptional antiterminator, and a cation-transporting ATPase. RD 8 contains genes for a V-type sodium ATP synthase, an oxidoreductase, and a neuraminidase and several others. These could provide alternative routes for handling the energy needs of the cell. Other regions of diversity are potentially associated with niche preservation (bacteriocin-like molecules or mucin-like glycoprotein production). Collectively, the content of regions of diversity is associated with forms of adaptation.

Each strain had its own unique subset of adaptation genes in common with TIGR4. The capsule region between SP350 and SP360 varied from that in TIGR4 in each of the strains (all strains were of a different capsular type). Some strains also varied in *cpsB* (SP347), *cpsC* (SP348), or *cpsD* (SP349), which are more often conserved genes across multiple capsular types. The only region outside of the capsular region (capsule being diversified primarily through strain selection) that appeared to be specific to TIGR4 alone was RD 7, which contains a phage-like element. All other regions of diversity represented genes that were present in at least one additional strain from our collection of genetically diverse strains. This mosaic distribution supports the idea that these regions are associated with genome plasticity.

About half of the adaptation regions are also associated with atypical GC content, indicating that they may have initially derived from horizontal transfer from other organisms. If so, one might have expected them to be of limited presence within our chosen group of strains that spans the intraspecies diversity. However, as the regions with atypical GC showed a presence among this set of isolates almost equal to that of the regions with more typical GC content, their acquisition from another species may not have occurred recently, or they may have spread to other isolates within the species over a short time, evolutionarily. Most adaptation regions lack structural features that are associated

with pathogenicity islands. Almost all the adaptation regions had some association with repetitive elements, including insertion sequences or Box or RUP elements. Only in two cases is the same element on the flanking sides of the segment of diversity, RD 3 (IS*1167*) and RD 7 (sequence corresponding to phage-like integrase). In fact, the association of repetitive sequences with these regions is not much greater, if at all, than with all regions of the genome (see density of repeats below). Nevertheless, the occasional presence of multiple repetitive elements bordering an adaptation region suggests that the repeated sequences may assist in the incorporation of heterologous DNA in these regions through homology-assisted illegitimate recombination (25).

Outside of the clustered adaptation regions, there were numerous individual genes (total of 444) that were of variable presence in one or more strains compared to TIGR4. Roughly 8 to 10% of total genomic content differed in the comparison of each strain with TIGR4, but as discussed above, the specific 10% was unique for each strain comparison. This figure for pairwise strain comparisons agrees well with more exact strain-to-strain comparative genome analysis of the three completed genomes. Collectively, 602 genes of 1,954 genes assayed (31%) showed a variable presence in 1 or more of the 13 strains, and 1,352 were present in all strains. The 1,352 genes present in all of the 13 strains might define a core complement of genes for *S. pneumoniae*.

The noncore gene set in *S. pneumoniae* is quite large. The number (602) of different genes that varied in the total intraspecies comparisons by CGH is nearly as large as the number (total of 696) of genes that varied in interspecies Blast comparison with GAS and GBS. This noncore fraction of 31% is greater than has been seen in similar studies with *Helicobacter pylori* (6%) (27), *S. aureus* (12%) (14), and *Streptococcus agalactiae* (18%) (31) and is vastly different from the 1% observed for *Vibrio cholerae* (10). The noncore fraction of the gene content might be considered to represent adaptive, rather than core, functions of *S. pneumoniae.*

Of the 602 genes missing in at least 1 other strain, only 17 genes were found to be missing from all 13 other strains. These 17 loci included 8 genes in the capsule region (none of the strains tested were capsule type 4), two transcriptional regulator genes, SP1130 and SP1131, and one region (SP1142 to SP1145) with four genes annotated as corresponding to hypothetical proteins or conserved hypothetical proteins and three additional hypothetical proteins (SP1132, SP1139, and SP1793). Roughly half (i.e., 285) of the genes that were dispensable in at least one strain were absent in only one or two of the strains tested. The other half (i.e., 317) of the dispensable genes were missing in three or more of the strains tested. A total of 158 of these correspond to the adaptation regions as described above.

REPEATS AND THEIR POTENTIAL ROLE IN MODULATION OF GENOME EVOLUTION

One remarkable aspect of the genome of *S. pneumoniae* compared to those of other bacteria is the abundance and genome-wide density of repeats (2, 32). There are at least eight different IS elements, with seven of these in more than eight copies apiece, both intact and truncated. Altogether, the IS elements alone make up 5% of the coding capacity of the genome. The non-IS elements known as BOX elements and RUP elements are found in over 100 copies each. In addition, there were 300 paralogous genes and at least 250 repeats larger than 23 bp within genes. If repeats with expected occurrence values below 0.01 are considered, then *S. pneumoniae* has repeat families that vary in size from 23 to 1,371 bp (median size of 31 bp). The density of repeats is about 1 every 500 bp; this was the greatest density of repeats seen for 51 genomes analyzed by Aras et al. (2).

The significance of the high density of repeats is that they may play a significant role in generating the genome plasticity that is observed. We have noted that most of the regions of diversity identified by CGH and the breaks in the conserved order of genes visualized with the PROmer alignment of whole genomes are bordered on at least one side by an IS element, a RUP element, or a BOX element. Moreover, in some cases there are multiple copies of partial and intact elements directly adjacent to one another towards one end of a given region (capsule being one example). One difficulty in assigning functional significance to an association of regions with repeats is that the density of repeats is so high that it would be surprising if there were not an association of a least one repetitive element with genomic regions of this size. Nevertheless, it seems likely that repeats could be used as portable regions of homology in genome rearrangements.

The presence of repeats can facilitate both intra- and intergenomic recombination. Intragenomic recombination, which causes the loss of intervening DNA between the repeats, is known to be a high-frequency event, increasing with the proximity of the repeats and the size of the repeat unit. Intragenomic recombination might be expected to be a common event in a genome with a high density of repeats, like the pneumococcus. Intragenomic recombination causes the loss of the intervening genes between repeats, but when combined with a finite possibility for the reintroduction of the lost genes through intergenomic recombination, as can occur during transformation, it could become a powerful mechanism for sampling new genes, for modifying genes, and for testing combinations of genes.

CONCLUSIONS

The vast majority of clinical and nonclinical isolates of *S. pneumoniae* are transformable, and the frequent exchange of genetic information through transformation could permit a high degree of genetic plasticity (for a review, see reference 4). One glimpse of the adaptive nature of the pneumococcus was evident in the rapid rise in the number of isolates with reduced susceptibility to penicillin that occurred in the early 1990s; this was initially noted in a few genetic lineages but rapidly spread to other lineages (8, 9, 23, 28). A second glimpse was afforded by studies of the population structure for this organism. Population studies have revealed that the pneumococcus is much more likely to sustain gene changes during evolution that occur by recombination than it is to sustain gene changes through mutations (12, 13, 21, 30). A third remarkable glimpse of adaptability was in the molecular demonstration of the capacity to integrate nonhomologous DNA through the process of transformation (25).

Genome-to-genome comparisons, either through the in silico annotation of whole genomes or through CGH using DNA microarrays, are once again revealing the footprints of genetic plasticity in the pneumococcal genomes. One distinctive footprint is the presence of a number of regions of diversity that have different complements of genes in different isolates. The genetic content of these regions, together with the known information about the pneumococcal capacity for undertaking recombinational changes, suggests that these regions of plasticity may allow the strains to adapt to differing environmental parameters. The pneumococcal genome had the highest density of repeat regions of any genome screened in a comparison field of 51 sequenced genomes. These repeats may have facilitated genomic plasticity. On an evolutionary scale, genomic plasticity quickly blurs gene order, a footprint shown in the PROmer plots of genome-to-genome comparisons. All these findings are the hallmarks of a highly adaptable organism.

ACKNOWLEDGMENTS

We thank Scott Peterson and Alex Wolf for encouraging the inclusion of CGH work of which they were contributors prior to publication. We acknowledge the hard work and efforts of numerous contributors to the analysis and annotation of the three completed genomes at The Institute for Genomic Research, at Eli Lilly and Company, at GlaxoSmithKline, at the University of Alabama at Birmingham, and at the University of Illinois at Chicago. We are grateful to John Glass, Scott Peterson, and James Watt for helpful comments and for critical reading of the manuscript. The CGH work was supported by NIH grant UO1 AI40645. Our work is further supported by NIH grants R01 AI053749 (to S.K.H. and H.T.). The further financial support of the Merck Genome Research Institute (MGRI72) and the Incyte Genomics, Inc., Pathoseq database program in the completion of these genomes is gratefully acknowledged.

REFERENCES

1. **Ajdic, D., W. M. McShan, R. E. McLaughlin, G. Savic, J. Chang, M. B. Carson, C. Primeaux, R. Tian, S. Kenton, H. Jia, S. Lin, Y. Qian, S. Li, H. Zhu, F. Najar, H. Lai, J. White, B. A. Roe, and J. J. Ferretti.** 2002. Genome sequence of Streptococcus mutans UA159, a cariogenic dental pathogen. *Proc. Natl. Acad. Sci. USA* **99:**14434–14439.

2. **Aras, R. A., J. Kang, A. I. Tschumi, Y. Harasaki, and M. J. Blaser.** 2003. Extensive repetitive DNA facilitates prokaryotic genome plasticity. *Proc. Natl. Acad. Sci. USA* **100:**13579–13584.

3. **Bolotin, A., P. Wincker, S. Mauger, O. Jaillon, K. Malarme, J. Weissenbach, S. D. Ehrlich, and A. Sorokin.** 2001. The complete genome sequence of the lactic acid bacterium Lactococcus lactis ssp. lactis IL1403. *Genome Res.* **11:**731–753.

4. **Claverys, J. P., M. Prudhomme, I. Mortier-Barriere, and B. Martin.** 2000. Adaptation to the environment: Streptococcus pneumoniae, a paradigm for recombination-mediated genetic plasticity? *Mol. Microbiol.* **35:**251–259.

5. **Delcher, A. L., A. Phillippy, J. Carlton, and S. L. Salzberg.** 2002. Fast algorithms for large-scale genome alignment and comparison. *Nucleic Acids Res.* **30:**2478–2483.

6. **de Saizieu, A., C. Gardes, N. Flint, C. Wagner, M. Kamber, T. J. Mitchell, W. Keck, K. E. Amrein, and R. Lange.** 2000. Microarray-based identification of a novel *Streptococcus pneumoniae* regulon controlled by an autoinduced peptide. *J. Bacteriol.* **182:**4696–4703.

7. **Dopazo, J., A. Mendoza, J. Herrero, F. Caldara, Y. Humbert, L. Friedli, M. Guerrier, E. Grand-Schenk, C. Gandin, M. de Francesco, A. Polissi, G. Buell, G. Feger, E. Garcia, M. Peitsch, and J. F. Garcia-Bustos.** 2001. Annotated draft genomic sequence from a Streptococcus pneumoniae type 19F clinical isolate. *Microb. Drug Resist.* **7:**99–125.

8. **Dowson, C. G., T. J. Coffey, and B. G. Spratt.** 1994. Origin and molecular epidemiology of penicillin-binding-protein-mediated resistance to beta-lactam antibiotics. *Trends Microbiol.* **2:**361–366.

9. **Dowson, C. G., A. Hutchinson, J. A. Brannigan, R. C. George, D. Hansman, J. Liñares, A. Tomasz, J. Maynard, and B. G.**

Spratt. 1989. Horizontal transfer of penicillin-binding protein genes in penicillin-resistant clinical isolates of *Streptococcus pneumoniae*. *Proc. Natl. Acad. Sci. USA* **86:**8842–8846.

10. **Dziejman, M., E. Balon, D. Boyd, C. M. Fraser, J. F. Heidelberg, and J. J. Mekalanos.** 2002. Comparative genomic analysis of Vibrio cholerae: genes that correlate with cholera endemic and pandemic disease. *Proc. Natl. Acad. Sci. USA* **99:**1556–1561.

11. **Enright, M. C., and B. G. Spratt.** 1998. A multilocus sequence typing scheme for Streptococcus pneumoniae: identification of clones associated with serious invasive disease. *Microbiology* **144:**3049–3060.

12. **Feil, E. J., J. M. Smith, M. C. Enright, and B. G. Spratt.** 2000. Estimating recombinational parameters in Streptococcus pneumoniae from multilocus sequence typing data. *Genetics* **154:**1439–1450.

13. **Feil, E. J., and B. G. Spratt.** 2001. Recombination and the population structures of bacterial pathogens. *Annu. Rev. Microbiol.* **55:**561–590.

14. **Fitzgerald, J. R., D. E. Sturdevant, S. M. Mackie, S. R. Gill, and J. M. Musser.** 2001. Evolutionary genomics of Staphylococcus aureus: insights into the origin of methicillin-resistant strains and the toxic shock syndrome epidemic. *Proc. Natl. Acad. Sci. USA* **98:**8221–8226. (First published 10 July 2001.)

15. **Glaser, P., C. Rusniok, C. Buchrieser, F. Chevalier, L. Frangeul, T. Msadek, M. Zouine, E. Couve, L. Lalioui, C. Poyart, P. Trieu-Cuot, and F. Kunst.** 2002. Genome sequence of Streptococcus agalactiae, a pathogen causing invasive neonatal disease. *Mol. Microbiol.* **45:**1499–1513.

16. **Hava, D. L., and A. Camilli.** 2002. Large-scale identification of serotype 4 Streptococcus pneumoniae virulence factors. *Mol. Microbiol.* **45:**1389–1406.

17. **Hava, D. L., C. J. Hemsley, and A. Camilli.** 2003. Transcriptional regulation in the *Streptococcus pneumoniae rlrA* pathogenicity islet by RlrA. *J. Bacteriol.* **185:**413–421.

18. **Hoskins, J., W. E. Alborn, Jr., J. Arnold, L. C. Blaszczak, S. Burgett, B. S. DeHoff, S. T. Estrem, L. Fritz, D. J. Fu, W. Fuller, C. Geringer, R. Gilmour, J. S. Glass, H. Khoja, A. R. Kraft, R. E. Lagace, D. J. LeBlanc, L. N. Lee, E. J. Lefkowitz, J. Lu, P. Matsushima, S. M. McAhren, M. McHenney, K. McLeaster, C. W. Mundy, T. I. Nicas, F. H. Norris, M. O'Gara, R. B. Peery, G. T. Robertson, P. Rockey, P. M. Sun, M. E. Winkler, Y. Yang, M. Young-Bellido, G. Zhao, C. A. Zook, R. H. Baltz, S. R. Jasku-**

nas, P. R. Rosteck, Jr., P. L. Skatrud, and J. I. Glass.** 2001. Genome of the bacterium *Streptococcus pneumoniae* strain R6. *J. Bacteriol.* **183:**5709–5717.

19. **Kleerebezem, M., J. Boekhorst, R. van Kranenburg, D. Molenaar, O. P. Kuipers, R. Leer, R. Tarchini, S. A. Peters, H. M. Sandbrink, M. W. Fiers, W. Stiekema, R. M. Lankhorst, P. A. Bron, S. M. Hoffer, M. N. Groot, R. Kerkhoven, M. de Vries, B. Ursing, W. M. de Vos, and R. J. Siezen.** 2003. Complete genome sequence of Lactobacillus plantarum WCFS1. *Proc. Natl. Acad. Sci. USA* **100:**1990–1995.

20. **Martin, B., O. Humbert, M. Camara, E. Guenzi, J. Walker, T. Mitchell, P. Andrew, M. Prudhomme, G. Alloing, R. Hakenbeck, D. A. Morrison, G. J. Boulnois, and J. P. Claverys.** 1992. A highly conserved repeated DNA element located in the chromosome of Streptococcus pneumoniae. *Nucleic Acids Res.* **20:**3479–3483.

21. **Maynard Smith, J., N. H. Smith, M. O'Rourke, and B. G. Spratt.** 1993. How clonal are bacteria? *Proc. Natl. Acad. Sci. USA* **90:**4384–4388.

22. **McGee, L., L. McDougal, J. Zhou, B. G. Spratt, F. C. Tenover, R. George, R. Hakenbeck, W. Hryniewicz, J. C. Lefevre, A. Tomasz, and K. P. Klugman.** 2001. Nomenclature of major antimicrobial-resistant clones of *Streptococcus pneumoniae* defined by the pneumococcal molecular epidemiology network. *J. Clin. Microbiol.* **39:**2565–2571.

23. **Munoz, R., T. J. Coffey, M. Daniels, C. G. Dowson, G. Laible, J. Casal, R. Hakenbeck, M. Jacobs, J. M. Musser, and B. G. Spratt.** 1991. Intercontinental spread of a multiresistant clone of serotype 23F Streptococcus pneumoniae. *J. Infect. Dis.* **164:**302–306.

24. **Oggioni, M. R., and J. P. Claverys.** 1999. Repeated extragenic sequences in prokaryotic genomes: a proposal for the origin and dynamics of the RUP element in Streptococcus pneumoniae. *Microbiology* **145:**2647–2653.

25. **Prudhomme, M., V. Libante, and J. P. Claverys.** 2002. Homologous recombination at the border: insertion-deletions and the trapping of foreign DNA in Streptococcus pneumoniae. *Proc. Natl. Acad. Sci. USA* **99:**2100–2105.

26. **Robinson, D. A., K. M. Edwards, K. B. Waites, D. E. Briles, M. J. Crain, and S. K. Hollingshead.** 2001. Clones of *Streptococcus pneumoniae* isolated from nasopharyngeal carriage and invasive disease in young children in central tennessee. *J. Infect. Dis.* **183:**1501–1507.

27. **Salama, N., K. Guillemin, T. K. McDaniel,**

G. Sherlock, L. Tompkins, and S. Falkow. 2000. A whole-genome microarray reveals genetic diversity among Helicobacter pylori strains. *Proc. Natl. Acad. Sci. USA* **97:**14668–14673.

28. Sibold, C., J. Wang, J. Henrichsen, and R. Hakenbeck. 1992. Genetic relationships of penicillin-susceptible and -resistant *Streptococcus pneumoniae* strains isolated on different continents. *Infect. Immun.* **60:**4119–4126.

29. Smith, M. D., and W. R. Guild. 1979. A plasmid in *Streptococcus pneumoniae*. *J. Bacteriol.* **137:**735–739.

30. Spratt, B. G., W. P. Hanage, and E. J. Feil. 2001. The relative contributions of recombination and point mutation to the diversification of bacterial clones. *Curr. Opin. Microbiol.* **4:**602–606.

31. Tettelin, H., V. Masignani, M. J. Cieslewicz, J. A. Eisen, S. Peterson, M. R. Wessels, I. T. Paulsen, K. E. Nelson, I. Margarit, T. D. Read, L. C. Madoff, A. M. Wolf, M. J. Beanan, L. M. Brinkac, S. C. Daugherty, R. T. DeBoy, A. S. Durkin, J. F. Kolonay, R. Madupu, M. R. Lewis, D. Radune, N. B. Fedorova, D. Scanlan, H. Khouri, S. Mulligan, H. A. Carty, R. T. Cline, S. E. Van Aken, J. Gill, M. Scarselli, M. Mora, E. T. Iacobini, C. Brettoni, G. Galli, M. Mariani, F. Vegni, D. Maione, D. Rinaudo, R. Rappuoli, J. L. Telford, D. L. Kasper, G. Grandi, and C. M. Fraser. 2002. Complete genome sequence and comparative genomic analysis of an emerging human pathogen, serotype V Streptococcus agalactiae. *Proc. Natl. Acad. Sci. USA* **99:**12391–12396.

32. Tettelin, H., K. E. Nelson, I. T. Paulsen, J. A. Eisen, T. D. Read, S. Peterson, J. Heidelberg, R. T. DeBoy, D. H. Haft, R. J. Dodson, A. S. Durkin, M. Gwinn, J. F. Kolonay, W. C. Nelson, J. D. Peterson, L. A. Umayam, O. White, S. L. Salzberg, M. R. Lewis, D. Radune, E. Holtzapple, H. Khouri, A. M. Wolf, T. R. Utterback, C. L. Hansen, L. A. McDonald, T. V. Feldblyum, S. Angiuoli, T. Dickinson, E. K. Hickey, I. E. Holt, B. J. Loftus, F. Yang, H. O. Smith, J. C. Venter, B. A. Dougherty, D. A. Morrison, S. K. Hollingshead, and C. M. Fraser. 2001. Complete genome sequence of a virulent isolate of Streptococcus pneumoniae. *Science* **293:**498–506.

CAPSULES

Janet Yother

3

The capsular polysaccharides of *Streptococcus pneumoniae* represent a diverse group of polymers that play an essential role in the virulence of the organism. Ninety serologically distinct capsules have been recognized (55), and the structures for more than half of these have been determined (107). The polysaccharides differ with respect to both their sugar compositions and linkages. Most are complex structures containing multiple sugars, linkages, and, often, side chains (Fig. 1). A few, such as types 3 and 37, are relatively simple structures, being comprised of only one or two sugars. Some are reminiscent of the teichoic acids, containing ribitol or phosphorylcholine, as seen in the group 6 and 15 capsules. Despite this diversity, all perform the same primary function of reducing opsonophagocytosis by limiting access of phagocytic receptors to complement bound to the *S. pneumoniae* cell wall.

Recent years have seen a renewed interest in the characterization of the capsular polysaccharides of *S. pneumoniae* and many other bacteria. Central to this renaissance have been the molecular characterization and complete nucleotide sequence determinations of numerous capsule loci. These studies have not only allowed a description of the genes and proteins involved in capsule synthesis but have also led to a new understanding of the mechanisms and regulation of polysaccharide production, genetic exchange of capsule loci, and virulence properties associated with the capsule.

Many of the recent advances described in this chapter have their roots in classic studies of *S. pneumoniae* that have historical significance both for the pneumococcal field and, often, for biology as a whole. Some of those studies are briefly described below, and further details can be found in previous reviews (7, 74, 125). As will be evident, the observations regarding virulence, linkage, and exchange of capsule genes, as well as a role for unlinked genes in capsule synthesis, have been borne out and extended in the molecular genetic analyses.

S. PNEUMONIAE CAPSULES—A HISTORICAL PERSPECTIVE

Since Pasteur's description in 1881 of light microscopic observations of a substance, or "aureole," surrounding *S. pneumoniae* (90), the capsule has played a central role in many historic observations regarding genetics, immunology, and pathogenesis. By 1916, the essential nature of the capsule had become apparent through the characterization of spon-

Janet Yother, Department of Microbiology, University of Alabama at Birmingham, Birmingham, AL 35294.

The Pneumococcus. Volume Editor, Elaine I. Tuomanen,
© 2004 ASM Press, Washington, D.C.

Type   Repeating Unit

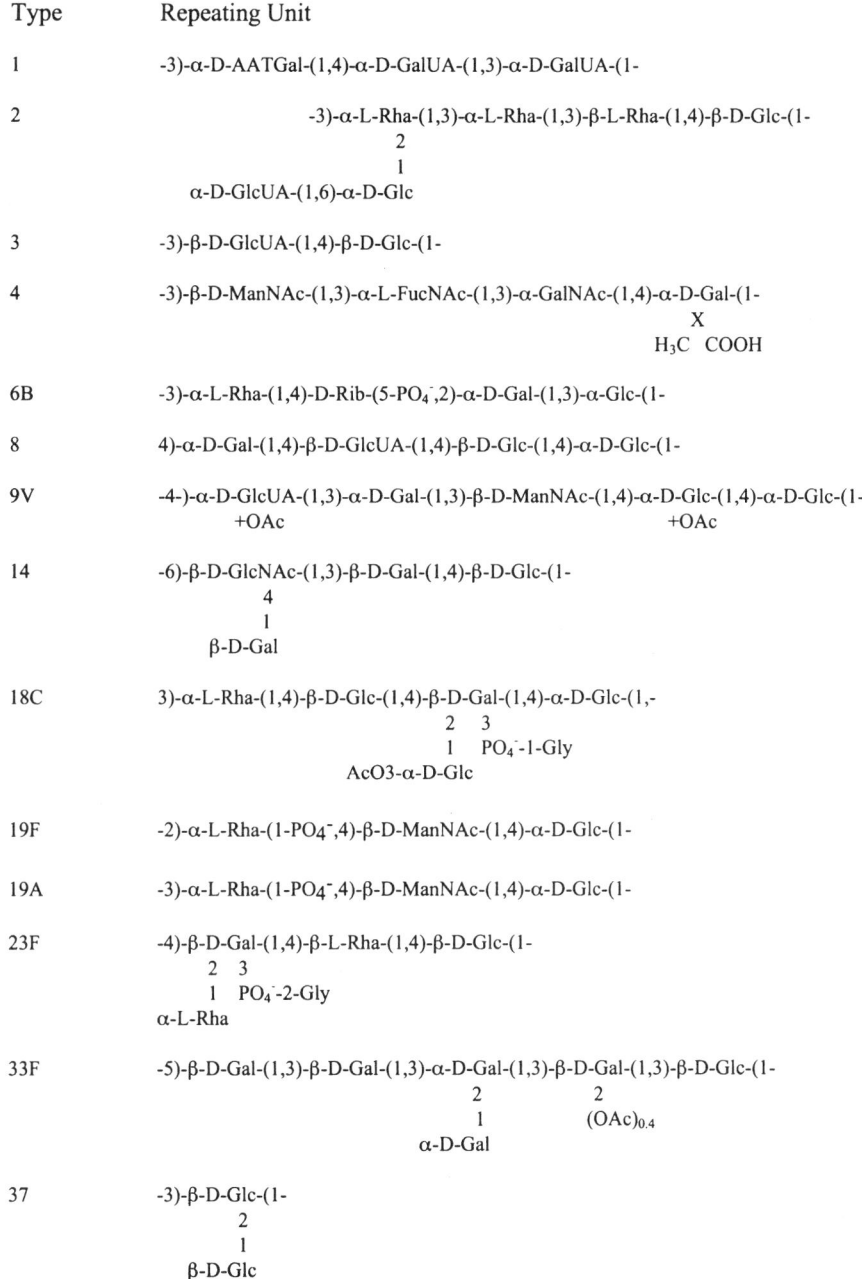

1   -3)-α-D-AATGal-(1,4)-α-D-GalUA-(1,3)-α-D-GalUA-(1-

2   -3)-α-L-Rha-(1,3)-α-L-Rha-(1,3)-β-L-Rha-(1,4)-β-D-Glc-(1-
                                            2
                                            1
                        α-D-GlcUA-(1,6)-α-D-Glc

3   -3)-β-D-GlcUA-(1,4)-β-D-Glc-(1-

4   -3)-β-D-ManNAc-(1,3)-α-L-FucNAc-(1,3)-α-GalNAc-(1,4)-α-D-Gal-(1-
                                                                    X
                                                                H₃C COOH

6B   -3)-α-L-Rha-(1,4)-D-Rib-(5-PO₄⁻,2)-α-D-Gal-(1,3)-α-Glc-(1-

8   4)-α-D-Gal-(1,4)-β-D-GlcUA-(1,4)-β-D-Glc-(1,4)-α-D-Glc-(1-

9V   -4-)-α-D-GlcUA-(1,3)-α-D-Gal-(1,3)-β-D-ManNAc-(1,4)-α-D-Glc-(1,4)-α-D-Glc-(1-
                        +OAc   +OAc

14   -6)-β-D-GlcNAc-(1,3)-β-D-Gal-(1,4)-β-D-Glc-(1-
                            4
                            1
                        β-D-Gal

18C   3)-α-L-Rha-(1,4)-β-D-Glc-(1,4)-β-D-Gal-(1,4)-α-D-Glc-(1,-
                                    2   3
                                    1   PO₄⁻-1-Gly
                            AcO3-α-D-Glc

19F   -2)-α-L-Rha-(1-PO₄⁻,4)-β-D-ManNAc-(1,4)-α-D-Glc-(1-

19A   -3)-α-L-Rha-(1-PO₄⁻,4)-β-D-ManNAc-(1,4)-α-D-Glc-(1-

23F   -4)-β-D-Gal-(1,4)-β-L-Rha-(1,4)-β-D-Glc-(1-
                        2   3
                        1   PO₄⁻-2-Gly
                α-L-Rha

33F   -5)-β-D-Gal-(1,3)-β-D-Gal-(1,3)-α-D-Gal-(1,3)-β-D-Gal-(1,3)-β-D-Glc-(1-
                                    2   2
                                    1   (OAc)₀.₄
                            α-D-Gal

37   -3)-β-D-Glc-(1-
                    2
                    1
                β-D-Glc

FIGURE 1 Repeating units of capsule structures. Capsule types for which the complete nucleotide sequence of the capsule locus has been published are shown. Where possible, the expected biological repeating unit is shown, based on initiation with glucose. The alternate glucose may be used in capsules containing more than one glucose. AATGal, 2-acetamido-4-amino-2,4,6-trideoxy-D-galactose; FucNAc, *N*-acetylfucosamine; Gal, galactose; Glc, glucose; GlcNAc, *N*-acetylglucosamine; GlcUA, glucuronic acid; Gly, glycerol; ManNAc, *N*-acetylmannosamine; OAc, acetate; Rha, rhamnose; Rib, ribitol. References for chemical repeating units are as in reference 107.

taneous mutants that had correspondingly lost capsule, resistance to phagocytosis, and virulence in animal models (43, 67, 102). Serologically distinct capsules were recognized in the early 1900s, and the polysaccharide nature of the capsule was reported in 1925 (12). The classic experiments of Griffith, reported in 1928, were the first to demonstrate genetic exchange in bacteria and emphasized the importance of the capsule in virulence (49). There, nonencapsulated (or rough), avirulent derivatives were transformed to a virulent, encapsulated (or smooth) phenotype following coinfection in mice. A direct role for capsule in virulence was demonstrated by the enzymatic removal of the type 3 capsule and subsequent avirulence in mouse infections (10).

Griffith's work laid the foundation for others to study the exchange of capsule types, a phenomenon that would later be achieved in vitro using either heat-killed cultures or cell extracts as donors and involving numerous serotypes (2, 36, 68). The ability to exchange capsule types led to the discovery by Avery, MacLeod, and McCarty that Griffith's "transforming principle" was DNA and hence the genetic material (11). The concept that capsule genes for different types might exist as exchangeable units and that homologous DNA outside a type-specific region provided the recombination sites necessary for exchange was postulated by Ephrussi-Taylor and later expanded on by Austrian et al. (9, 41, 42). Biochemical evidence supporting a replacement mechanism came from experiments in which strains containing uronic acids in their capsules were transformed with DNA from strains lacking these sugars (9). Only the former possess a UDP-Glc dehydrogenase activity, which was lost upon transformation to the donor capsule type. Confirmation of linkage of the genes involved in capsule synthesis and the first maps of capsule loci were obtained through recombination experiments (93). Rarely, binary strains expressing more than one capsule type were obtained during transformations. In stable binaries, the donor capsule genes inserted at a site unlinked to the recipient capsule locus,

whereas unstable binaries resulted from integration of the donor capsule genes near the recipient capsule locus (8, 9, 16, 17). The involvement of unlinked genes in capsule synthesis was suggested when certain type 1 recipients transformed to type 3 encapsulation were found to produce small type 3 capsules (72). Transformation of the type 3 genes from the latter strains to other recipients resulted in normal levels of capsule, indicating preservation of the type 3 genes and influence on capsule production by an unlinked factor. Many of the enzymes involved in capsule synthesis, including the UDP-Glc dehydrogenases of types 1, 2, and 3, the UDP-GalUA epimerase of type 1, and the type 3 synthase, were characterized in biochemical studies in the 1950s and 1960s (9, 18, 75, 97–99).

GENETIC ORGANIZATION OF CAPSULE LOCI

At present, the complete nucleotide sequences for the types 1, 2, 3, 4, 6B, 8, 9V, 14, 18C, 19A, 19F, 23F, 33F, and 37 loci have been published (Fig. 2), and more than 65 others have been determined (http://www.sanger.ac.uk/Projects/S_pneumoniae/CPS/). In addition, the genetic bases for serotype distinctions within the group 15 and 19 capsules have been described (81, 110). Despite the diversity of the *S. pneumoniae* capsular polysaccharides, the genetic loci for all exhibit a similar organization in which genes required for synthesis of a specific capsule type are flanked by genes common to all types (Fig. 2). The sequences defining the type-specific and common regions were originally recognized through hybridization studies and subsequently identified through sequence analyses (3, 38, 39, 46, 50, 64). Genes contained within the type-specific regions are unique to a given serotype or serogroup. The proteins encoded by these regions consist of glycosyltransferases, polymerases, transporters, and enzymes necessary for the synthesis of nucleotide sugars unique to a given capsule. There are some occurrences of homologous sequences within the type-specific regions of more than one type. In addition to the serogroup-specific genes that

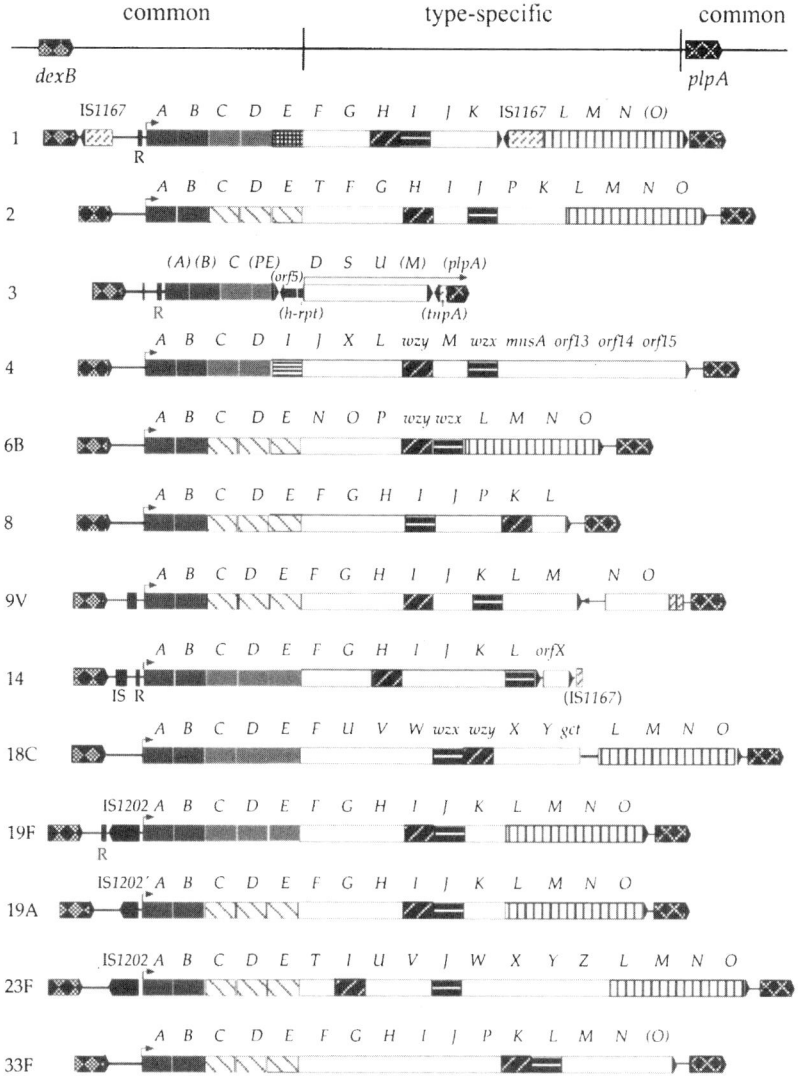

FIGURE 2 Capsule genetic loci. The loci are shown for capsule types for which the complete nucleotide sequences have been published. For most loci, genes were named alphabetically in the order in which they occurred in the locus. Therefore, similarly named genes may not be homologous. Open boxes indicate type-specific genes, which encode glycosyltransferases, nucleotide sugar biosynthetic enzymes, and epimerases. Similarly shaded boxes are homologous. Sequences in parentheses are mutated compared to their respective homologs. The flanking gene in the downstream common region was originally identified as *plpA* (38) and later designated *aliA* (80) due to an apparent allelic difference between the two genes. However, *plpA* and *aliA* are the same; it is only in the type 3 locus that the gene is altered. *cpsD*, *cpsS*, and *cpsU* in the type 3 locus are also referred to as *capA*, *capB*, and *capC*, respectively (3). *cpsP* in the type 3 locus is homologous to *cpsD* in the other loci. The type 4, 6B, and 18C genes were named according to the bacterial polysaccharide gene nomenclature system (59) and are not referred to by the *cps* designations. For clarity, *cps* designations are shown here for the upstream common sequences and the TDP-Rha biosynthetic genes in these loci. The remainder of the letters are from the original designations. Differences in the *cps19fI*- and *cps19aI*-encoded polymerases are responsible for the different linkages in the polysaccharides (81). Symbols: ▦, conserved common sequences; ▤, class I common sequences; ◨, class II common sequences; ▦ and ▥, initiating glycosyltransferases not homologous to class I or II; ▨, Wzy homolog, putative polymerase; ▭, Wzx homolog, putative flippase; ▥, TDP-Rha biosynthesis. Classes of common sequences are based on reference 79. Arrows indicate putative or known transcription start sites and, where demonstrated, length of transcript. Some of the many IS elements and repetitive sequences (R) are indicated. Maps are derived from references 3, 23, 38, 39, 58, 59, 66, 70, 78, 80, 81, 84, 85, and 109.

occur among all types within a group, these homologous sequences may encode proteins involved in shared pathways, such as those necessary for TDP-rhamnose synthesis in capsules containing rhamnose. TDP-rhamnose-related genes occurring at the ends of the type-specific regions have been identified in types or groups 2, 6, 18, 19, and 23 (Fig. 1 and 2). Functional homologs, such as those necessary for polymer transport (Wzx homologs) and polymerization (Wzy homologs), are also present in most type-specific regions (Fig. 2).

The capsule-specific regions are flanked upstream by *dexB* and downstream by *plpA* (alternately referred to as *aliA*) (3, 38, 50, 80). Neither of these genes appears to have a role in capsule synthesis. The region between *dexB* and the first type-specific gene contains the upstream common genes *cpsA*, *cpsB*, *cpsC*, *cpsD*, and, in most types, *cpsE*. The *cpsA* and *cpsB* homologs are highly conserved among all types, whereas those of *cpsC*, *cpsD*, and *cpsE* can be divided into two distinct classes (Fig. 2) (79). *cpsA*, *cpsB*, *cpsC*, and *cpsD* encode proteins that are involved in the modulation of capsule production, whereas *cpsE* encodes the glycosyltransferase required to initiate capsule synthesis (see below). The capsule locus is predicted to be transcribed as a single operon, with initiation occurring at a promoter located upstream of *cpsA* (66, 80, 85). This organization also occurs in the capsule loci of many other bacteria, including *Streptococcus agalactiae* (group B streptococcus), *Staphylococcus aureus*, and *Lactococcus lactis* (28, 95, 108, 125). In addition, the organization is similar to that found in group 1 capsules of *Escherichia coli* and other gram-negative bacteria, where homologs of the common sequences *cpsC* and *cpsD*, as well as functional equivalents of other enzymes, occur (117). As described below, this organization is reflective of a similar mechanism of capsule synthesis.

While exceptions to the organization described above are few, they are significant in terms of the capsule types involved and the resulting differences in biosynthetic mechanism. In particular, most genes in the upstream common region of the type 3 capsule locus are truncated or otherwise mutated, and this region is neither transcribed nor required for capsule synthesis. Here, transcription begins upstream of the first type-specific gene, *cps3D* (not homologous to the *cpsD* gene found in the common regions of other serotypes), and continues through *plpA* (3, 23). Within this operon are additional truncated genes that are not required for capsule synthesis: *cps3M*, which encodes a C-terminally truncated phosphoglucomutase (PGM); *tmpA*, a partial transposon sequence with homology to IS*1167*; and *plpA*, in which 40% of the 5′ end is deleted (23, 126). Of the remaining genes, only two (*cps3D*, encoding a UDP-Glc dehydrogenase, and *cps3S*, encoding the type 3 synthase) are required for synthesis of the GlcUA-Glc capsule; the third, *cps3U*, is a homolog of *galU* (38, 39). Both *cps3U* and *galU* encode Glc-1-P uridylyltransferases, but the former is neither required nor sufficient for capsule production (38, 76). A similar organization occurs in the hyaluronan (HA) capsule locus of *Streptococcus pyogenes* (group A streptococcus), where homologs of *cps3D*, *cps3S*, and *cps3U* (*hasB*, *hasA*, and *hasC*, respectively) are transcribed as a single operon (*hasABC*), and *hasC* is not required for capsule synthesis (6, 34). As described below, the type 3 polysaccharide and the HA capsule are synthesized by a similar mechanism that is distinct from that of the majority of *S. pneumoniae* capsules and other streptococcal and staphylococcal capsules. The type 37 capsule may also be synthesized by a similar mechanism. Here, a single gene, *tts*, encodes the synthase necessary to synthesize the homopolymeric Glc capsule. *tts* lies outside the usual capsule locus, which is occupied by a cryptic type 33F locus that contains multiple mutations and is not required for type 37 synthesis (69, 71).

The presence of insertion sequence (IS) elements and nonfunctional genes (or gene fragments) is a common finding in the capsule loci, as already noted for the type 3 locus and the cryptic type 33F locus in type 37 strains. Other examples include the type 1 locus, which is flanked by IS*1167* elements oriented in the same direction. These elements occur in multiple locations in the *S. pneumoniae* chromosome (128). In the type 1 locus, a cryptic set of genes normally involved in TDP-rhamnose synthesis is located between the downstream IS*1167* and *plpA*, even though rhamnose does not occur in either the type 1 capsule or type 1 strains. In addition, a 115-bp element is located upstream of *cpsA* and occurs in many capsule loci as well as several other regions of the *S. pneumoniae* chromosome (85). Other loci with IS-like elements or their remnants include types 9V (IS*1167* and others), 14 (IS*1167*), 19F (IS*1202*), and 23F (IS*1202*) (50, 66, 92, 109). The type 3 locus also contains a partial H-rpt element upstream of the first type-specific gene (126). Homologs of H-rpt elements are frequently found in polysaccharide loci of gram-negative bacteria, and it has been proposed that they may be involved in rearrangements of adjacent sequences and the formation of complex loci. The frequent occurrences of these many different IS-like elements and truncated sequences suggest that transposition has played a role in the evolution of the capsule loci and possibly the generation of binary strains observed in earlier studies (23, 38, 85). Additionally, the remnants of ancestral capsule types may be present in many capsule loci.

GENETIC EXCHANGE OF CAPSULE LOCI

Flanking regions common among serotypes provide the sites for the genetic exchange of capsule loci that was originally observed in Griffith's experiments. Molecular evidence for the replacement of recipient type-specific genes with those of the donor was obtained using linkage to an antibiotic resistance marker placed in the donor *plpA* (39). Hybridization with recipient and donor type-specific genes demonstrated the expected loss and gain, respectively, that followed selection for the antibiotic marker. The many truncations of sequences in the type 3 locus also made it possible to demonstrate exchange in the common regions.

Exchange of capsule loci among clinical isolates is evident from population analyses using pulsed-field gel electrophoresis, restriction fragment length polymorphism, multilocus sequence typing, and repetitive-element primer PCR analyses. In these studies, isolates expressing different capsular serotypes were found to represent the same genetic background (13, 32, 33, 86). From sequence analyses, points of recombinational exchange were identified both within the immediate common flanking sequences and at more distant sites. Crossovers between the type 19F and 23F loci were observed upstream of *dexB* and either downstream of *plpA* or within the conserved TDP-Rha synthesis genes (33). Crossovers between the types 9V and 14 loci occurred within *cpsA* and *pbp1A*, located 5.8 kb downstream of *plpA* (31). These types of crossovers, which involve DNA replacements from 21 to more than 25 kb, challenge the limits of the *S. pneumoniae* transformation system, where the median size of DNA fragments taken into the cell is approximately 7 kb (91). Such crossovers may therefore occur relatively rarely, and selection for the event using an antibiotic resistance marker located in *plpA* yields a low frequency of capsule-switched transformants, except in the case of type 3 donors, where the size of the locus is considerably smaller (39).

MECHANISMS OF CAPSULE SYNTHESIS

Block-Type (Wzy-Dependent) Pathway

The similar organization seen in most *S. pneumoniae* capsule loci is reflected in a similar mechanism of synthesis. For these capsule types, synthesis involves initiation on a lipid acceptor followed by subunit assembly on the cytoplasmic face of the membrane, transfer to the extracellular face of the membrane, subunit polymerization, and finally linkage to the peptidoglycan (Fig. 3). Growth is proposed to occur at the reducing end of the polymer. This block-type mechanism of synthesis is similar to that described for some lipopolysaccharide (LPS) O antigens (116) and is also referred to as Wzy dependent because of the Wzy polymerase that is utilized in O-antigen synthesis.

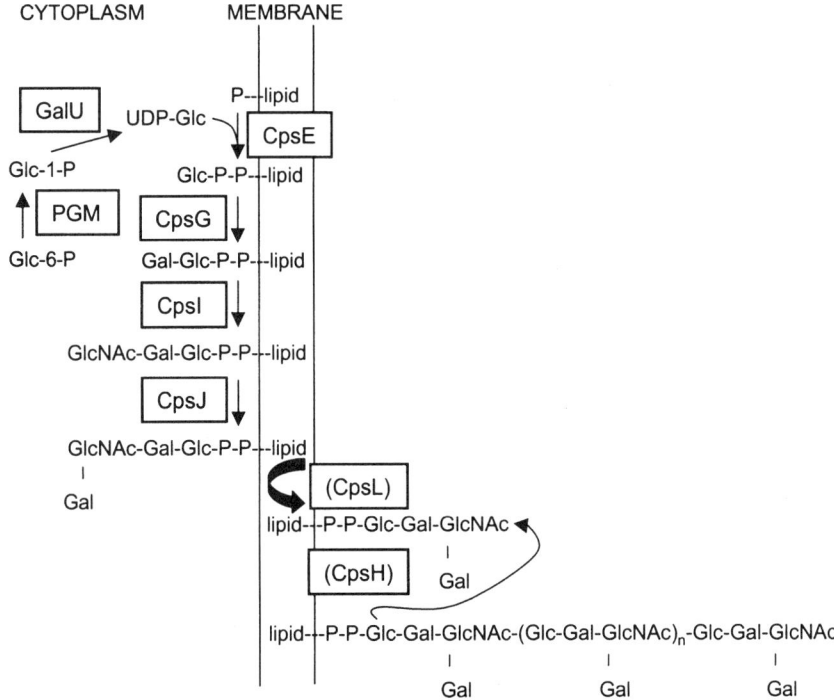

FIGURE 3 Block-type (Wzy-dependent) synthesis. The repeat unit is for type 14. Functions for the glycosyltransferases CpsE, CpsG, CpsI, and CpsJ have been experimentally determined in the type 14 system (65, 66). CpsL and CpsH are the type 14 Wzx and Wzy homologs, respectively. PGM and GalU are the cellular enzymes and are not capsule specific. Functions of proteins in parentheses have not been experimentally demonstrated in the *S. pneumoniae* system. PGM, GalU, CpsG, CpsI, and CpsJ are located in the cytoplasm, whereas CpsE, CpsL, and CpsH are membrane associated. Following transfer of the chain, the lipid-P-P is hydrolyzed to lipid-P and recycled to the cytoplasmic face of the membrane.

It also occurs with the *E. coli* group 1 capsules and others noted above.

Initiation of synthesis usually involves transfer of Glc-1-P from UDP-Glc onto a lipid acceptor and is catalyzed by a glycosyl-1-P transferase. For capsules that contain Gal but not Glc, synthesis initiates by transfer of Gal-1-P from UDP-Gal to the lipid carrier (64). In multiple serotypes containing Glc in their capsules, initiation of the repeat unit is dependent on CpsE (64, 109), and for types 2 and 14, the activity of the recombinant enzyme has been demonstrated (65; R. T. Cartee and J. Yother, unpublished data). As noted above, homologs of CpsE are encoded by most capsule loci in this group, and the *cpsE* genes can be divided into two distinct groups based on DNA iden-

tity. In the type 1 and 4 loci, the genes encoding the putative initiating glycosyltransferases are not highly homologous to either *cpsE* class (Fig. 2). In contrast to the other capsules shown, the type 1 and 4 capsules do not contain glucose, and the initiating glycosyltransferase must recognize a different UDP-sugar. In eubacteria, the lipid carrier utilized in most polysaccharide syntheses, including that of capsules, teichoic acid, peptidoglycan, and LPS, is the C_{55} undecaprenyl-P. This lipid is likely the carrier for *S. pneumoniae* capsule synthesis, which for type 2 has been found to initiate on a polyprenyl-P (Cartee and Yother, unpublished).

Following initiation, additional glycosyltransferases encoded within the type-specific regions catalyze transfer of the remaining sugars from their respective nucleotide-sugar precursors to the nonreducing end of the growing subunit. These steps have been characterized in type 14, where the glycosyltransferases Cps14G, Cps14I, and Cps14J sequentially transfer Gal, GlcNAc, and a side chain Gal, respectively, from their UDP-sugar precursors to the membrane-associated lipid-P-P-Glc (Fig. 3) (65, 66). In capsules containing rhamnose, such as type 19F, a conserved pathway consisting of Cps19fL, Cps19fN, Cps19fM, and Cps19fO and their homologs in other types is responsible for the conversion of Glc-1-P to TDP-Rha (Fig. 4A) (80). The syntheses of other sugars unique to the capsule and not found in other cellular structures, such as GlcUA, are also catalyzed by enzymes encoded by type-specific genes. For example, genes confirmed or predicted to encode UDP-Glc dehydrogenases necessary for the conversion of UDP-Glc to UDP-GlcUA occur in the type 1, 2, 8, and 9 loci (reference 85 and references cited in the legend for Fig. 2). The type 1 capsule does not contain GlcUA, but UDP-GlcUA is converted to the necessary UDP-GalUA precursor by a UDP-GlcUA epimerase encoded by *cap1J* (83). The syntheses of precursors for constituents that occur in other cellular structures, such as Glc, the GlcNAc of peptidoglycan, and the AATGal, PC, and ribitol of the teichoic acids, are catalyzed by enzymes involved in ba-

sic cellular metabolic functions and are encoded by genes outside the capsule locus. These components are then utilized directly in capsule synthesis or are epimerized to the necessary capsule constituents by enzymes encoded in the capsule loci. The syntheses of probably all *S. pneumoniae* capsules require the activity of the cellular PGM and Glc-1-P uridylyltransferase (GalU) for the conversions of Glc-6-P to Glc-1-P and Glc-1-P to UDP-Glc, respectively (Fig. 3). The genes encoding PGM and GalU, and their roles in capsule synthesis, have been characterized in strains expressing the type 3 capsule (see below).

Although many of the enzymes involved in the remaining biosynthetic steps can be predicted based on homology to proteins involved in LPS O-antigen synthesis (117), none of these functions have been demonstrated for *S. pneumoniae* capsules. Transport of lipid-linked subunits across the cytoplasmic membrane is proposed to be mediated by "flippases," which contain approximately 12 membrane-spanning domains. Polymerization occurs by transfer of the reducing end of the growing polysaccharide chain onto the nonreducing end of the new repeat unit (Fig. 3). By analogy to LPS O-antigen synthesis, these steps involve homologs of the Wzx repeat unit transporter and the Wzy polymerase, respectively. These proteins have been identified in all characterized loci in this group (Fig. 2).

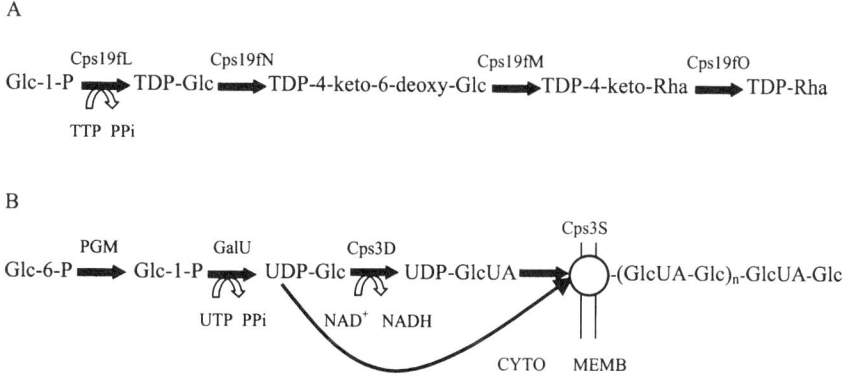

FIGURE 4 Biosynthetic pathways for (A) dTDP-Rha synthesis in type 19F (80) and (B) type 3 synthesis (38). The type 3 synthase (Cps3S) is located in the membrane.

The final step in capsule synthesis involves covalent linkage to the cell wall, which has been demonstrated for types 2, 4, 6A, 6B, 7F, 8, 14, 19F, and 23F (100) and probably occurs for all capsules in this group. Neither the enzyme(s) involved nor the precise point when transfer occurs has been determined, and at least a small percentage of the total polymer can be detected in culture supernatants. Examinations of capsular polysaccharides contained in isolated membranes and cell walls by immunoblot analyses have shown that transfer is independent of polymer size (14). For the type III capsule of *S. agalactiae*, whose synthesis occurs by a similar mechanism, the polysaccharide is linked to GlcNAc residues of the peptidoglycan through a phosphodiester bond and an oligosaccharide linker (37).

Synthase-Dependent Pathway

Synthesis of the type 3 capsule requires only a single glycosyltransferase to catalyze formation of the glycosidic linkages necessary to form the polymer from UDP-Glc and UDP-GlcUA (Fig. 4B) (5, 38). The synthase, encoded by *cps3S*, is a processive β-glycosyltransferase that belongs to the glycosyltransferase family 2, which also includes cellulose synthases, chitin synthases, the Nod factor synthase NodC from *Rhizobium*, and the HA synthases of eukaryotes and prokaryotes (including *S. pyogenes*) (24, 38, 60). These enzymes are proposed to have binding sites for the nucleotide sugars and the growing polysaccharide and to transport the polymer across the cytoplasmic membrane as it is being synthesized (60). Based on the crystal structure of SpsA from *Bacillus subtilis*, it has been proposed that these enzymes contain only a single nucleotide binding site. For enzymes like the type 3 and HA synthases, which catalyze the addition of two different sugars, specificity of the binding site is thought to be changed by addition of the previous sugar to the polymer (29). Synthesis of these polymers does not involve undecaprenol or another polyprenol. For the type 3 polysaccharide, initiation occurs on a glycerophosphate lipid acceptor (25). Polymer formation proceeds by alternate addition of Glc and GlcUA to the nonreducing end of the growing chain, followed by translocation through the polysaccharide binding site (26). In the presence of adequate levels of UDP-Glc and UDP-GlcUA, the polymer remains bound to the synthase, allowing for a highly processive reaction and formation of high-molecular-weight polymer (26). The size of the polymer decreases dramatically as the concentration of UDP-sugars is decreased. During in vitro reactions using isolated membranes containing the synthase, addition of UDP-Glc when the concentration of UDP-GlcUA is insufficient to bind to its respective binding site results in termination of chain extension and rapid ejection of the polymer from the enzyme (45). The ejection mechanism is postulated to result from failure of the polymer to be recognized by the polysaccharide binding site due to premature translocation in the absence of addition of both sugars. Regulation of UDP-GlcUA concentration may thus be critical in controlling the processivity of the type 3 synthase and thereby modulating polymer formation and size.

Synthesis of the type 3 precursor nucleotide sugars requires the activities of a PGM and a Glc-1-P uridylyltransferase that are encoded outside the capsule locus by *pgm* and *galU*, respectively, to generate UDP-Glc (52, 76). The UDP-Glc dehydrogenase activity necessary for generation of UDP-GlcUA from UDP-Glc is encoded by *cps3D*, located in the type 3 locus (4, 38, 39). No additional enzymes required for synthesis or transport of the type 3 polysaccharide have been identified, and synthase-mediated transport apparently occurs during chain extension. The synthase and its homologs lack the large number of transmembrane domains that are found in transporters such as Wzx. In the streptococcal HA synthases, interactions with multiple cardiolipin molecules may be involved in the formation of a pore through which extrusion occurs (105).

Unlike *S. pneumoniae* capsules formed by the Wzy-dependent mechanism, the type 3 polysaccharide is not covalently linked to the cell wall (100), and approximately half the total polymer can be detected in culture supernatants (52). Although the mechanism for cell association is not clear, interactions with the lipid primer and the synthase could play a role.

REGULATION AND MODULATION OF CAPSULE SYNTHESIS

The ability to regulate capsule expression is probably critical to the survival of *S. pneumoniae* in different host niches. Reductions in capsule expression in the nasopharyngeal cavity may be important for exposing adhesins necessary for colonization, whereas high-level capsule expression is essential for avoiding complement-mediated opsonophagocytosis during systemic infections. In cell culture, nonencapsulated isolates adhere more firmly than their isogenic encapsulated counterparts (104), and reductions in capsule expression in the in vivo environment are therefore expected. Weiser et al. observed a reduced amount of capsule and enhanced colonization in transparent-phase variants, as compared to opaque-phase variants, which produce elevated levels of capsule and are enhanced in systemic virulence (62, 114). High-frequency phase variation required the presence of a BOX element that may alter expression of a downstream regulatory gene (94). As described below, a requirement for capsule production during sustained colonization has been demonstrated (73), and the ability of antibodies against the capsule to reduce carriage indicates that capsule is expressed in this environment (35). In animal models of systemic infection, increases in capsule gene expression over that occurring during laboratory culture have been observed (88).

Phase variation due to spontaneous sequence duplications within the capsule genes has been observed in strains of serotypes 3, 8, and 37. For type 3 strains grown under biofilm conditions, approximately one-third of the recovered isolates were nonencapsulated, and approximately 20% of these were able to revert to the encapsulated phenotype upon subculture on blood agar. Sequence analyses revealed tandem duplications ranging in size from 11 to 239 bp that disrupted *cps3D* at different sites in the different isolates. Restoration of the encapsulated phenotype resulted from precise excision of the duplication (112). Similar results were obtained with the type 8 and 37 isolates, but here the duplications were in *cpsE* (which encodes the initiating glycosyltransferase) and *tts* (which encodes the type 37 synthase), respectively (111). Thus, multiple mechanisms for phase variation may be involved in controlling capsule synthesis.

The common genes found in most *S. pneumoniae* capsule loci encode proteins involved in modulation of capsule production. CpsA is a membrane protein with homology to LytR, a transcriptional regulator in *B. subtilis* (50). A precise role for CpsA in *S. pneumoniae* capsule synthesis has not been determined, but mutations in *cpsA* cause reductions in capsule amounts (82). In *S. agalactiae*, deletion of the *cpsA* homolog results in reduced levels of capsule transcript (30). CpsB, CpsC, and CpsD are part of a phosphoregulatory pathway that modulates capsule production and chain length. CpsC and CpsD are the transmembrane and ATP-binding domains, respectively, of an autophosphorylating tyrosine kinase (15, 82). Phosphorylation occurs on tyrosine residues in CpsD (82), and the initial phosphorylation requires CpsC (15). Phosphorylated CpsD is capable of transferring its phosphate to either a dephosphorylated CpsD or other exogenous substrates, indicating that it may be capable of modulating additional proteins (15). CpsC and CpsD are homologous to the N and C termini, respectively, of autophosphorylating tyrosine kinases involved in modulation of polysaccharide chain length in *Sinorhizobium meliloti* (ExoP) and *E. coli* (Wzc) (87, 124). In *S. pneumoniae*, deletion of either *cpsC* or *cpsD* results in the production of only low levels of capsule (82) in the form of short-chain polymers that can still be transferred to the cell wall (14).

CpsB is a phosphotyrosine phosphatase and a kinase inhibitor that can interact with CpsD and prevent its initial phosphorylation (15). Although tyrosine phosphorylation was originally described as a mechanism for negatively regulating capsule expression (82), more recent information indicates a positive correlation between CpsD phosphorylation and capsule production (14). Deletion of *cpsB* results in increases in both phosphorylation and capsule production (14), and mutation of the tyrosine phosphorylation residues in CpsD results in decreases in capsule levels (77). Thus, interaction between CpsB, CpsC, and CpsD may result in reduced phosphorylation of CpsD and reduced levels of capsule production. An additional factor may be involved, as deletion of *cpsB* in Rx1 derivatives that have been transformed with either the type 2 locus or the type 19F locus results in high levels of phosphorylation but low levels of capsule, in contrast to the high levels of capsule that are observed in *cpsB* deletion mutants in the type 2 D39 background (14, 82). Rx1, a derivative of D39 containing multiple uncharacterized mutations, apparently lacks, or has an alteration in, a factor important in the control of capsule synthesis (14).

A positive correlation between capsule levels and CpsD phosphorylation was also observed with clinical isolates (115). Here, paired patient isolates exhibited the transparent phenotype and no phosphorylation of CpsD when cultured from the nasopharynx, but opaque isolates with phosphorylated CpsD were recovered from the blood. Both CpsD phosphorylation and capsule production were increased under anaerobic, compared to aerobic, conditions, indicating that oxygen may play an important role in capsule regulation.

VIRULENCE

The essential nature of the capsule in virulence was established in early studies of *S. pneumoniae*, as described above. Those studies relied mainly on spontaneous mutants and generally involved transformation of nonencapsulated intermediates. More recent studies have taken advantage of the molecular characterizations of the capsule loci to construct defined mutations and, with the ready ability to transform encapsulated isolates (54, 127), to directly manipulate encapsulated, virulent strains. The resulting studies have confirmed the requirement for

capsule using strains altered only in type 3 production due to mutations in *cps3D* or *cps3S* (53, 73) or in type 2 production due to mutations in *cpsC* or *cpsD* (15). In addition, the need to properly regulate capsule production during infection is suggested by the effect of *cpsB* mutations in the type 2 strain D39 (14). As noted above, these mutants exhibit increased CpsD phosphorylation and capsule levels. However, they are reduced in both the ability to colonize the nasopharynx and the ability to cause lethal infection following intravenous (i.v.) inoculation. In contrast, virulence following intraperitoneal (i.p.) inoculation is unaltered by the loss of CpsB. These results may reflect the effect of the phosphotyrosine regulatory system not only on capsule production but also on the integration of capsule expression with that of other factors important in virulence.

As observed with the *cpsB* mutant, the effects of alterations in capsule are not always easily predictable. A derivative of the type 3 strain A66 that contains a defined point mutation in *cps3D* exhibits a five-fold reduction in type 3 capsule levels. Although only modestly reduced in virulence following i.p. inoculation, the mutant was essentially avirulent when inoculated by the i.v. route (73). In contrast, a similar reduction in capsule level in the type 3 strain WU2 due to a defined point mutation in *pgm* resulted in avirulence when inoculated by both the i.p. and i.v. routes, suggesting that alterations in additional pathways affected by this mutation may contribute to the reduction in virulence (53). Both mutants, regardless of route of infection, exhibited high levels of virulence in immunodeficient XID mice (53, 73), which have reduced innate antibody to phosphocholine (21, 118). Thus, the effects of alterations in capsule amount are dependent on multiple factors, including the pathway leading to the capsule alteration, the route of inoculation, and the immune status of the host.

Colonization and Adherence

A role for capsule in colonization was demonstrated using the type 2 and type 3 strains described above that contain defined mutations in the capsule loci and in *pgm*. Derivatives producing either no capsule or amounts of capsule less than 6% of the parental level could not be recovered from nasal washes of mice 7 days after intranasal inoculation. In contrast, both the *cps3D* mutant of A66 and the *pgm* mutant of WU2, which produce 20% of the parental levels of capsule, were recovered in numbers equivalent to those of their parent strains (73). Longer-term colonization, however, was affected by the reductions in capsule (A. D. Magee and J. Yother, unpublished data). Thus, reduced levels of capsule are sufficient for colonization, and strains of low virulence, such as the *pgm* mutant, may retain the ability to colonize the nasopharynx. As described below, the type of capsule expressed may also influence colonization.

Role of Capsule Type

Although 90 capsular serotypes have been recognized (55), the majority of invasive *S. pneumoniae* infections are caused by strains representing only a small number of types. The prevalent serotypes vary among age groups and geographic locations (96). In adults, types 1, 3, 4, 5, 6A, 6B, 7F, 8, 9N, 9V, 11A, 12F, 14, 18C, 19A, 19F, 22F, and 23F predominate, and these serotypes are included in the 23-valent vaccine. Among these, types 3, 7F, 9V, 14, and 23F occur frequently, and types 3, 6B, and 19F are especially common in fatal cases (56). In children, types 1, 4, 5, 6A, 6B, 9V, 14, 18C, 19A, 19F, and 23F are among the predominant types causing invasive disease (27). Type 3 strains occur frequently in otitis media but rarely cause invasive disease in children. Formulations of the pneumococcal capsular polysaccharide conjugate vaccine are based on the prevalence of specific serotypes in the target population.

In animal models of pneumococcal disease, a similar restriction of types is observed, with virulence being closely correlated with the capsule type expressed (20, 113). In general, serotypes causing disease in adults are virulent in mice, whereas the pediatric serotypes can colonize but do not cause invasive disease in these models (20, 123). In vitro assays, including complement activation and phagocytosis, have also shown a clear association between capsule type and the outcomes of these measures of virulence (19, 44, 47, 57).

The predominance of particular serotypes suggests a role for the capsular polysaccharide structure in determining virulence. Genetic exchange of capsule loci has been used to construct isogenic strains for the purpose of examining this role. When the type 3 locus from a highly virulent strain was used to replace the capsule loci of a similarly virulent type 2 strain, a highly virulent type 5 strain, or a relatively avirulent type 6B strain, the resulting capsule-switched isolates produced type 3 capsule at levels comparable to the donor strain. Virulence following i.p. inoculation varied among the derivatives, however, and was unchanged in the type 2 recipient, decreased in the type 5 recipient, and increased in the type 6B recipient (61). In clinical isolates exhibiting the same genetic background but different capsule types, isolates expressing the type 3 capsule, but not the type 23F capsule, were virulent in mice (86). The capsule type may also affect the ability to colonize, as capsule-switched isolates expressing the type 6B, 7F, 14, and 19F capsules were recovered more frequently from mice inoculated intranasally with these strains than with their isogenic type 4 parent (106). Thus, the effect of capsule type is significant, but it can be influenced by the genetic background in which it is expressed. This observation may explain, in part, why invasive strains expressing some capsule types, including 3, 7F, 9V, 12F and 14, exhibit a high degree of genetic relatedness, whereas others, such as types 6A, 6B, 19F, and 23F, are more diverse (40, 51, 89, 103).

Complement, Phagocytosis, and Protection of the Cell Surface

The major role of the *S. pneumoniae* capsule is to provide protection against complement-mediated opsonophagocytosis by blocking access to cell wall-localized C3b. Earlier studies showed that activation of the alternative complement pathway by the cell wall leads to deposition of similar amounts of C3b fragments on the walls of encapsulated and nonencapsulated derivatives (22, 119, 120, 121). When isogenic strains in which type 3 capsule production was eliminated by mutation of *cps3D* were used, however, the deposition of approximately 50% more C3 was observed on the mutant than on the encapsulated parent, but binding of C3-specific antibody was six-fold higher for the nonencapsulated strain (1). Thus, while the capsule acts primarily to limit access to C3, it can also affect the amount of complement deposited.

Among clinical isolates of different capsular serotypes, levels of complement activation and deposition, as well as phagocytosis, vary in in vitro assays (19, 44, 47, 48, 57, 63, 101, 119, 120, 122). Comparison of complement deposition on the type 2 strain D39 and the type 3 strain WU2 revealed two-fold more total C3, and eight-fold greater binding of C3-specific antibody, on the former. Genetic replacement of the D39 type 2 locus with the type 3 locus of WU2 resulted in levels of C3 deposition and binding of C3-specific antibody that were intermediate between those of the two parents. In contrast, antibodies to noncapsular surface antigens exhibited high levels of binding to the type 2 D39 parent and reduced but equivalent levels of binding to the type 3 WU2 and the type 3 derivative of D39 (1). Thus, the capsule structure is a major determinant of accessibility to the surface, but the deposition of and access to bound complement are influenced both by capsule type and by other factors in the genetic background.

CONCLUSIONS

Studies over the past 10 years have revealed the underlying genetic diversity that is responsible for the wide variety of capsule structures originally recognized through serologic reactions. They have also provided molecular explanations of many earlier studies and have made possible the further characterization of biosynthetic mechanisms, roles of the capsule in virulence, and the evolution of serotypes. Many features, such as the mechanisms regulating capsule expression in different environments,

the role of different capsule structures in virulence, and the mechanisms for generating diversity, are just beginning to be explored. The recognition of new roles in virulence, novel regulatory mechanisms, and two distinct mechanisms of synthesis provide important avenues of research for the coming years. As has long been the case, the *S. pneumoniae* capsules will continue to serve as paradigms for studies of virulence factors, immune responses, and polysaccharide biochemistry.

ACKNOWLEDGMENTS

Work in my laboratory is supported by Public Health Service grants GM53017 and AI28457 from the National Institutes of Health.

REFERENCES

1. **Abeyta, M., G. G. Hardy, and J. Yother.** 2003. Genetic alteration of capsule type but not PspA type affects accessibility to surface-bound complement and surface antigens in *Streptococcus pneumoniae. Infect. Immun.* **71:**218–225.
2. **Alloway, J. L.** 1932. The transformation *in vitro* of R pneumococci into S forms of different specific types by the use of filtered pneumococcal extracts. *J. Exp. Med.* **55:**91–99.
3. **Arrecubieta, C., E. Garcia, and R. Lopez.** 1995. Sequence and transcriptional analysis of a DNA region involved in the production of capsular polysaccharide in *Streptococcus pneumoniae* type 3. *Gene* **167:**1–7.
4. **Arrecubieta, C., R. Lopez, and E. Garcia.** 1994. Molecular characterization of *cap3A*, a gene from the operon required for the synthesis of the capsule of *Streptococcus pneumoniae* type 3: sequencing of mutations responsible for the unencapsulated phenotype and localization of the capsular cluster on the pneumococcal chromosome. *J. Bacteriol.* **176:**6375–6383.
5. **Arrecubieta, C., R. Lopez, and E. Garcia.** 1996. Type 3-specific synthase of *Streptococcus pneumoniae* (Cap3B) directs type 3 polysaccharide biosynthesis in *Escherichia coli* and in pneumococcal strains of different serotypes. *J. Exp. Med.* **184:**449–455.
6. **Ashbaugh, C. D., S. Albertí, and M. R. Wessels.** 1998. Molecular analysis of the capsule gene region of group A streptococcus: the *hasAB* genes are sufficient for capsule expression. *J. Bacteriol.* **180:**4955–4959.
7. **Austrian, R.** 1981. Pneumococcus: the first one hundred years. *Rev. Infect. Dis.* **3:**183–189.

8. **Austrian, R., and H. P. Bernheimer.** 1959. Simultaneous production of two capsular polysaccharides by pneumococcus. I. Properties of a pneumococcus manifesting binary capsulation. *J. Exp. Med.* **110:**571–584.

9. **Austrian, R., H. P. Bernheimer, E. E. B. Smith, and G. T. Mills.** 1959. Simultaneous production of two capsular polysaccharides by pneumococcus. II. The genetic and biochemical basis of binary capsulation. *J. Exp. Med.* **110:**585–602.

10. **Avery, O. T., and R. Dubos.** 1931. The protective action of a specific enzyme against type III pneumococcus infection in mice. *J. Exp. Med.* **54:**73–89.

11. **Avery, O. T., C. M. MacLeod, and M. McCarty.** 1944. Studies on the chemical nature of the substance inducing transformation of pneumococcal types. Induction of transformation by a desoxyribonucleic acid fraction isolated from pneumococcus Type III. *J. Exp. Med.* **79:**137–158.

12. **Avery, O. T., and H. J. Morgan.** 1925. Immunological reactions of the isolated carbohydrate and protein of pneumococcus. *J. Exp. Med.* **42:**347–353.

13. **Barnes, D. M., S. Whittier, P. H. Gilligan, S. Soares, A. Tomasz, and F. W. Henderson.** 1995. Transmission of multidrug-resistant serotype 23F *Streptococcus pneumoniae* in group day care: evidence suggesting capsular transformation of the resistant strain in vivo. *J. Infect. Dis.* **171:**890–896.

14. **Bender, M. H., R. T. Cartee, and J. Yother.** 2003. Positive correlation between tyrosine phosphorylation of CpsD and capsular polysaccharide production in *Streptococcus pneumoniae. J. Bacteriol.* **185:**6057–6066.

15. **Bender, M. H., and J. Yother.** 2001. CpsB is a modulator of capsule-associated tyrosine kinase activity in *Streptococcus pneumoniae. J. Biol. Chem.* **276:**47966–47974.

16. **Bernheimer, H. P., and I. E. Wermundsen.** 1972. Homology in capsular transformations in *Pneumococcus. Mol. Gen. Genet.* **116:**68–83.

17. **Bernheimer, H. P., and I. E. Wermundsen.** 1969. Unstable binary capsulated transformants in pneumococcus. *J. Bacteriol.* **98:**1073–1079.

18. **Bernheimer, H. P., I. E. Wermundsen, and R. Austrian.** 1968. Mutation in pneumococcus type 3 affecting multiple cistrons concerned with the synthesis of capsular polysaccharide. *J. Bacteriol.* **96:**1099–1102.

19. **Branconier, J. H., and H. Odeberg.** 1982. Granulocyte phagocytosis and killing of virulent and avirulent serotypes of *Streptococcus pneumoniae. J. Lab. Clin. Med.* **100:**279–287.

20. **Briles, D. E., M. J. Crain, B. M. Gray, C. Forman, and J. Yother.** 1992. Strong association between capsular type and virulence for mice among human isolates of *Streptococcus pneumoniae. Infect. Immun.* **60:**111–116.

21. **Briles, D. E., M. Nahm, K. Schoroer, J. Davie, P. Baker, J. Kearney, and R. Barletta.** 1981. Antiphosphocholine antibodies found in normal mouse serum are protective against intravenous infection with type 3 *Streptococcus pneumoniae. J. Exp. Med.* **153:**694–705.

22. **Brown, E. J., K. A. Joiner, R. M. Cole, and M. Berger.** 1983. Localization of complement component 3 on *Streptococcus pneumoniae*: anticapsular antibody causes complement deposition on the pneumococcal capsule. *Infect. Immun.* **39:**403–409.

23. **Caimano, M. J., G. G. Hardy, and J. Yother.** 1998. Capsule genetics in *Streptococcus pneumoniae* and a possible role for transposition in the generation of the type 3 locus. *Microb. Drug Resist.* **4:**11–23.

24. **Campbell, J. A., G. J. Davies, V. Bulone, and B. Henrissat.** 1997. A classification of nucleotide-diphospho-sugar glycosyltransferases based on amino acid sequence similarities. *Biochem. J.* **326:**929–942.

25. **Cartee, R. T., W. T. Forsee, J. W. Jensen, and J. Yother.** 2001. Expression of the *Streptococcus pneumoniae* type 3 synthase in *Escherichia coli*: assembly of type 3 polysaccharide on a lipid primer. *J. Biol. Chem.* **276:**48831–48839.

26. **Cartee, R. T., W. T. Forsee, J. S. Schutzbach, and J. Yother.** 2000. Mechanism of type 3 capsular polysaccharide synthesis in *Streptococcus pneumoniae. J. Biol. Chem.* **275:**3907–3914.

27. **Centers for Disease Control and Prevention.** 1997. Prevention of pneumococcal disease: recommendations of the Advisory Committee on Immunization Practices (ACIP). *Morb. Mortal. Wkly. Rep.* **46:**1–24.

28. **Chaffin, D., S. Beres, H. Yim, and C. Rubens.** 2000. The serotype of type Ia and III group B streptococci is determined by the polymerase gene within the polycistronic capsule operon. *J. Bacteriol.* **182:**4466–4477.

29. **Charnock, S. J., B. Henrissat, and G. J. Davies.** 2001. Three-dimensional structures of UDP-sugar glycosyltransferases illuminate the biosynthesis of plant polysaccharides. *Plant Physiol.* **125:**527–531.

30. **Cieslewicz, M. J., D. L. Kasper, Y. Wang, and M. R. Wessels.** 2001. Functional analysis in the Ia group B *Streptococcus* of a cluster of genes

involved in extracellular polysaccharide production by diverse species of streptococci. *J. Biol. Chem.* **276:**139–146.

31. **Coffey, T. J., M. Daniels, M. C. Enright, and B. G. Spratt.** 1999. Serotype 14 variants of the Spanish penicillin-resistant serotype 9V clone of *Streptococcus pneumoniae* arose by large recombinational replacements of the *cpsA-pbp1a* region. *Microbiology* **145:**2023–2031.

32. **Coffey, T. J., C. G. Dowson, M. Daniels, J. Zhou, C. Martin, B. G. Spratt, and J. M. Musser.** 1991. Horizontal transfer of multiple penicillin-binding protein genes, and capsular biosynthetic genes, in natural populations of *Streptococcus pneumoniae. Mol. Microbiol.* **5:**2255–2260.

33. **Coffey, T. J., M. C. Enright, M. Daniels, J. K. Morona, R. Morona, W. Hryniewicz, J. C. Paton, and B. G. Spratt.** 1998. Recombinational exchanges at the capsular polysaccharide biosynthetic locus lead to frequent serotype changes among natural isolates of *Streptococcus pneumoniae. Mol. Microbiol.* **27:**73–83.

34. **Crater, D. L., and I. van de Rijn.** 1995. Hyaluronic acid synthesis operon (*has*) expression in group A streptococci. *J. Biol. Chem.* **270:**18452–18458.

35. **Dagan, R., M. Muallem, R. Melamed, O. Leroy, and P. Yagupsky.** 1997. Reduction of pneumococcal nasopharyngeal carriage in early infancy after immunization with tetravalent pneumococcal vaccines conjugated to either tetanus toxoid or diphtheria toxoid. *Pediatr. Infect. Dis. J.* **16:**1060–1064.

36. **Dawson, M. H., and R. H. P. Sia.** 1931. *In vitro* transformation of pneumococcal types. I. A technique for inducing transformation of pneumococcal types *in vitro. J. Exp. Med.* **54:**681–699.

37. **Deng, L., D. L. Kasper, T. P. Krick, and M. R. Wessels.** 2000. Characterization of the linkage between the type III capsular polysaccharide and the bacterial cell wall of group B *Streptococcus. J. Biol. Chem.* **275:**7497–7504.

38. **Dillard, J. P., M. W. Vandersea, and J. Yother.** 1995. Characterization of the cassette containing genes for type 3 capsular polysaccharide biosynthesis in *Streptococcus pneumoniae. J. Exp. Med.* **181:**973–983.

39. **Dillard, J. P., and J. Yother.** 1994. Genetic and molecular characterization of capsular polysaccharide biosynthesis in *Streptococcus pneumoniae* type 3. *Mol. Microbiol.* **12:**959–972.

40. **Enright, M. C., and B. G. Spratt.** 1998. A multilocus sequence typing scheme for *Streptococcus pneumoniae*: identification of clones associated with serious invasive disease. *Microbiology* **144:**3049–3060.

41. **Ephrussi-Taylor, H.** 1949. Additive effects of certain transforming agents from some variants of pneumococcus. *J. Exp. Med.* **89:**399–424.

42. **Ephrussi-Taylor, H.** 1951. Genetic aspects of transformations of pneumococci. *Cold Spring Harbor Symp. Quant. Biol.* **16:**445–456.

43. **Eyre, J. W., and J. W. Washbourn.** 1897. Resistant forms of the pneumococcus. *J. Pathol. Bacteriol.* **4:**394–400.

44. **Fine, D. P.** 1975. Pneumococcal type-associated variability in alternate complement pathway activation. *Infect. Immun.* **12:**772–778.

45. **Forsee, W. T., R. T. Cartee, and J. Yother.** 2000. Biosynthesis of type 3 capsular polysaccharide in *Streptococcus pneumoniae*: enzymatic chain release by an abortive translocation process. *J. Biol. Chem.* **275:**25972–25978.

46. **Garcia, E., P. Garcia, and R. Lopez.** 1993. Cloning and sequencing of a gene involved in the synthesis of the capsular polysaccharide of *Streptococcus pneumoniae* type 3. *Mol. Gen. Genet.* **239:**188–195.

47. **Giebink, G. S., J. Verhoef, P. K. Peterson, and P. G. Quie.** 1977. Opsonic requirements for phagocytosis of *Streptococcus pneumoniae* types VI, XVIII, XXIII, and XXV. *Infect. Immun.* **18:**291–297.

48. **Gordon, D. L., G. M. Johnson, and M. K. Hostetter.** 1986. Ligand-receptor interactions in the phagocytosis of virulent *Streptococcus pneumoniae* by polymorphonuclear leukocytes. *J. Infect. Dis.* **154:**619–626.

49. **Griffith, F.** 1928. The significance of pneumococcal types. *J. Hyg.* **27:**113–159.

50. **Guidolin, A., J. K. Morona, R. Morona, D. Hansman, and J. C. Paton.** 1994. Nucleotide sequence analysis of genes essential for capsular polysaccharide biosynthesis in *Streptococcus pneumoniae* type 19F. *Infect. Immun.* **62:**5384–5396.

51. **Hall, L. M., R. A. Whiley, B. Duke, R. C. George, and A. Efstratiou.** 1996. Genetic relatedness within and between serotypes of *Streptococcus pneumoniae* from the United Kingdom: analysis of multilocus enzyme electrophoresis, pulsed-field gel electrophoresis, and antimicrobial resistance patterns. *J. Clin. Microbiol.* **34:**853–859.

52. **Hardy, G. G., M. J. Caimano, and J. Yother.** 2000. Capsule biosynthesis and basic metabolism in *Streptococcus pneumoniae* are linked through the cellular phosphoglucomutase. *J. Bacteriol.* **182:**1854–1863.

53. **Hardy, G. G., A. D. Magee, C. L. Ventura, M. J. Caimano, and J. Yother.** 2001. Essential role for cellular phosphoglucomutase in virulence of type 3 *Streptococcus pneumoniae. Infect. Immun.* **69:**2309–2317.

54. **Havarstein, L. S., G. Coomaraswamy, and D. A. Morrison.** 1995. An unmodified heptadecapeptide pheromone induces competence for genetic transformation in *Streptococcus pneumoniae. Proc. Natl. Acad. Sci. USA* **92:**11140–11144.

55. **Henrichsen, J.** 1995. Six newly recognized types of *Streptococcus pneumoniae. J. Clin. Microbiol.* **33:**2759–2762.

56. **Henriques, B., M. Kalin, A. Ortqvist, B. Olsson Liljequist, M. Almela, T. J. Marrie, M. A. Mufson, A. Torres, M. A. Woodhead, S. B. Svenson, and G. Kallenius.** 2000. Molecular epidemiology of *Streptococcus pneumoniae* causing invasive disease in 5 countries. *J. Infect. Dis.* **182:**833–839.

57. **Hostetter, M. K.** 1986. Serotypic variations among virulent pneumococci in deposition and degradation of covalently bound C3b: implications for phagocytosis and antibody production. *J. Infect. Dis.* **153:**682–693.

58. **Iannelli, F., B. J. Pearce, and G. Pozzi.** 1999. The type 2 capsule locus of *Streptococcus pneumoniae. J. Bacteriol.* **181:**2652–2654.

59. **Jiang, S.-M., L. Wang, and P. R. Reeves.** 2001. Molecular characterization of *Streptococcus pneumoniae* type 4, 6B, 8, and 18C capsular polysaccharide gene clusters. *Infect. Immun.* **69:**1244–1255.

60. **Keenleyside, W. J., and C. Whitfield.** 1996. A novel pathway of O-polysaccharide biosynthesis in *Salmonella enterica* serovar Borreze. *J. Biol. Chem.* **271:**28581–28592.

61. **Kelly, T., J. P. Dillard, and J. Yother.** 1994. Effect of genetic switching of capsular type on virulence of *Streptococcus pneumoniae. Infect. Immun.* **62:**1813–1819.

62. **Kim, J. O., and J. N. Weiser.** 1998. Association of intrastrain phase variation in quantity of capsular polysaccharide and teichoic acid with the virulence of *Streptococcus pneumoniae. J. Infect. Dis.* **177:**368–377.

63. **Knecht, J. C., G. Schiffman, and R. Austrian.** 1970. Some biological properties of Pneumococcus type 37 and the chemistry of its capsular polysaccharide. *J. Exp. Med.* **132:**475–487.

64. **Kolkman, M. A., B. A. van der Zeijst, and P. J. Nuijten.** 1998. Diversity of capsular polysaccharide synthesis gene clusters in *Streptococcus pneumoniae. J. Biochem. (Tokyo)* **123:**937–945.

65. **Kolkman, M. A., B. A. van der Zeijst, and P. J. Nuijten.** 1997. Functional analysis of glycosyltransferases encoded by the capsular polysaccharide biosynthesis locus of *Streptococcus pneumoniae* serotype 14. *J. Biol. Chem.* **272:**19502–19508.

66. **Kolkman, M. A., W. Wakarchuk, P. J. Nuijten, and B. A. van der Zeijst.** 1997. Capsular polysaccharide synthesis in *Streptococcus pneumoniae* serotype 14: molecular analysis of the complete *cps* locus and identification of genes encoding glycosyltransferases required for the biosynthesis of the tetrasaccharide subunit. *Mol. Microbiol.* **26:**197–208.

67. **Kruse, W., and S. Pansini.** 1891. Untersuchungen uber den *Diplococcus pneumoniae* und verwandte Streptokokken. *Z. Hyg. Infektionskr.* **11:**279–280.

68. **Langvad-Nielson, A.** 1944. Change of capsule in the pneumococcus. *Acta Pathol. Microbiol. Scand.* **21:**362–369.

69. **Llull, D., E. García, and R. López.** 2001. Tts, a processive β-glucosyltransferase of *Streptococcus pneumoniae*, directs the synthesis of the branched type 37 capsular polysaccharide in pneumococcus and other gram-positive species. *J. Biol. Chem.* **276:**21053–21061.

70. **Llull, D., R. Lopez, E. Garcia, and R. Munoz.** 1998. Molecular structure of the gene cluster responsible for the synthesis of the polysaccharide capsule of *Streptococcus pneumoniae* type 33F. *Biochim. Biophys. Acta* **1443:**217–224.

71. **Llull, D., R. Munoz, R. Lopez, and E. Garcia.** 1999. A single gene (*tts*) located outside the *cap* locus directs the formation of *Streptococcus pneumoniae* type 37 capsular polysaccharide. Type 37 pneumococci are natural, genetically binary strains. *J. Exp. Med.* **190:**241–251.

72. **MacLeod, C. M., and M. R. Krauss.** 1953. Control by factors distinct from the S transforming principle of the amount of capsular polysaccharide produced by type III pneumococci. *J. Exp. Med.* **97:**767–771.

73. **Magee, A. D., and J. Yother.** 2001. Requirement for capsule in colonization by *Streptococcus pneumoniae. Infect. Immun.* **69:**3755–3761.

74. **Mäkelä, P., and B. A. D. Stocker.** 1969. Genetics of polysaccharide biosynthesis. *Annu. Rev. Genet.* **3:**291–322.

75. **Mills, G. T., and E. B. Smith.** 1962. Biosynthetic aspects of capsule formation in the pneumococcus. *Br. Med. Bull.* **18:**27–30.

76. **Mollerach, M., R. Lopez, and E. Garcia.** 1998. Characterization of the *galU* gene of *Streptococcus pneumoniae* encoding a uridine diphosphoglucose pyrophosphorylase: a gene essential for capsular polysaccharide biosynthesis. *J. Exp. Med.* **188:**2047–2056.

77. **Morona, J., R. Morona, D. C. Miller, and J. C. Paton.** 2003. Mutational analysis of the carboxy-terminal $(YGX)_4$ repeat domain of CpsD, an autophosphorylating tyrosine kinase

required for capsule biosynthesis in *Streptococcus pneumoniae. J. Bacteriol.* **185:**3009–3019.

78. **Morona, J. K., D. C. Miller, T. J. Coffey, C. J. Vindurampulle, B. G. Spratt, R. Morona, and J. C. Paton.** 1999. Molecular and genetic characterization of the capsule biosynthetic locus of *Streptococcus pneumoniae* type 23F. *Microbiology* **145:**781–789.

79. **Morona, J. K., R. Morona, and J. C. Paton.** 1999. Analysis of the 5′ portion of the type 19A capsule locus identifies two classes of *cpsC, cpsD,* and *cpsE* genes in *Streptococcus pneumoniae. J. Bacteriol.* **181:**3599–3605.

80. **Morona, J. K., R. Morona, and J. C. Paton.** 1997. Characterization of the locus encoding the *Streptococcus pneumoniae* type 19F capsular polysaccharide biosynthetic pathway. *Mol. Microbiol.* **23:**751–763.

81. **Morona, J. K., R. Morona, and J. C. Paton.** 1999. Comparative analysis of capsular biosynthesis in *Streptococcus pneumoniae* types belonging to serogroup 19. *J. Bacteriol.* **181:**5355–5364.

82. **Morona, J. K., J. C. Paton, D. C. Miller, and R. Morona.** 2000. Tyrosine phosphorylation of CpsD negatively regulates capsular polysaccharide biosynthesis in *Streptococcus pneumoniae. Mol. Microbiol.* **35:**1431–1442.

83. **Munoz, R., R. Lopez, M. de Frutos, and E. Garcia.** 1999. First molecular characterization of a uridine diphosphate galacturonate 4-epimerase: an enzyme required for capsular biosynthesis in Streptococcus pneumoniae type 1. *Mol. Microbiol.* **31:**703–713.

84. **Muñoz, R., M. Mollerach, R. Lopez, and E. Garcia.** 1999. Characterization of the type 8 capsular gene cluster of *Streptococcus pneumoniae. J. Bacteriol.* **181:**6214–6219.

85. **Muñoz, R., M. Mollerach, R. Lopez, and E. Garcia.** 1997. Molecular organization of the genes required for the synthesis of type 1 polysaccharide of *Streptococcus pneumoniae:* formation of binary encapsulated pneumococci and identification of cryptic dTDP-rhamnose biosynthesis genes. *Mol. Microbiol.* **25:**79–92.

86. **Nesin, M., M. Ramirez, and A. Tomasz.** 1998. Capsular transformation of a multidrug-resistant *Streptococcus pneumoniae* in vivo. *J. Infect. Dis.* **177:**707–713.

87. **Niemeyer, D., and A. Becker.** 2001. The molecular weight distribution of succinoglycan produced by *Sinorhizobium meliloti* is influenced by specific tyrosine phosphorylation and ATPase activity of the cytoplasmic domain of the ExoP protein. *J. Bacteriol.* **183:**5163–5170.

88. **Ogunniyi, A. D., P. Giammarinaro, and J. C. Paton.** 2002. The genes encoding virulence-associated proteins and the capsule of *Streptococcus pneumoniae* are upregulated and differentially expressed in vivo. *Microbiology* **148:**2045–2053.

89. **Overweg, K., D. Bogaert, M. Sluijter, J. Yother, J. Dankert, R. de Groot, and P. W. Hermans.** 2000. Genetic relatedness within serotypes of penicillin-susceptible *Streptococcus pneumoniae* isolates. *J. Clin. Microbiol.* **38:**4548–4553.

90. **Pasteur, L.** 1881. Note sur la maladie nouvelle provoquee par la salive d'un enfant mort de la rage. *Bull. Acad. Med. (Paris)* **10:**94–103.

91. **Puyet, A., B. Greenberg, and S. A. Lacks.** 1990. Genetic and structural characterization of *endA.* A membrane-bound nuclease required for transformation of *Streptococcus pneumoniae. J. Mol. Biol.* **213:**727–738.

92. **Ramirez, M., and A. Tomasz.** 1998. Molecular characterization of the complete 23F capsular polysaccharide locus of *Streptococcus pneumoniae. J. Bacteriol.* **180:**5273–5278.

93. **Ravin, A. W.** 1960. Linked mutations borne by deoxyribonucleic acid controlling the synthesis of capsular polysaccharide in pneumococcus. *Genetics* **45:**1387–1403.

94. **Saluja, S. K., and J. N. Weiser.** 1995. The genetic basis of colony opacity in *Streptococcus pneumoniae:* evidence for the effect of BOX elements on the frequency of phenotypic variation. *Mol. Microbiol.* **16:**215–227.

95. **Sau, S., N. Bhasin, E. R. Wann, J. C. Lee, T. J. Foster, and C. Y. Lee.** 1997. The *Staphylococcus aureus* allelic genetic loci for serotype 5 and 8 capsule expression contain the type-specific genes flanked by common genes. *Microbiology* **143:**2395–2405.

96. **Scott, J., A. Hall, R. Dagan, J. Dixon, S. Eykyn, A. Fenoll, M. Hortal, L. Jette, J. Jorgensen, F. Lamothe, C. Latorre, J. MacFarlane, D. Shales, L. Smart, and A. Taunay.** 1996. Serogroup-specific epidemiology of *Streptococcus pneumoniae:* association with age, sex and geography in 7,000 episodes of invasive disease. *Clin. Infect. Dis.* **22:**973–981.

97. **Smith, E. B., G. T. Mills, and H. P. Bernheimer.** 1961. Biosynthesis of pneumococcal polysaccharides. I. Properties of the system synthesizing type III capsular polysaccharide. *J. Biol. Chem.* **236:**2179–2182.

98. **Smith, E. E. B., and G. T. Mills.** 1960. Uridine pyrophosphoglucose dehydrogenase in capsulated and non-capsulated strains of pneumococcus type I. *Microbiology* **22:**265–271.

99. **Smith, E. E. B., G. T. Mills, H. P. Bernheimer, and R. Austrian.** 1958. The formation of uridine pyrophosphoglucuronic acid

from uridine pyrophosphoglucose by extracts of a noncapsulated strain of pneumococcus. *Biochim. Biophys. Acta* **28**:211–212.

100. **Sorensen, U. B., J. Henrichsen, H. C. Chen, and S. C. Szu.** 1990. Covalent linkage between the capsular polysaccharide and the cell wall peptidoglycan of *Streptococcus pneumoniae* revealed by immunochemical methods. *Microb. Pathog.* **8**:325–334.

101. **Stephens, C. G., R. C. Williams, Jr., and W. P. Reed.** 1977. Classical and alternative complement pathway activation by pneumococci. *Infect. Immun.* **17**:296–302.

102. **Stryker, L. M.** 1916. Variations in the pneumococcus induced by growth in immune serum. *J. Exp. Med.* **24**:49–68.

103. **Takala, A. K., J. Vuopio-Varkila, E. Tarkka, M. Leinonen, and J. M. Musser.** 1996. Subtyping of common pediatric pneumococcal serotypes from invasive disease and pharyngeal carriage in Finland. *J. Infect. Dis.* **173**:128–135.

104. **Talbot, U. M., A. W. Paton, and J. C. Paton.** 1996. Uptake of *Streptococcus pneumoniae* by respiratory epithelial cells. *Infect. Immun.* **64**:3772–3777.

105. **Tlapak-Simmons, V. L., E. S. Kempner, B. A. Baggenstoss, and P. H. Weigel.** 1998. The active streptococcal hyaluronan synthases (HASs) contain a single HAS monomer and multiple cardiolipin molecules. *J. Biol. Chem.* **273**:26100–26109.

106. **Trzcinski, K., C. M. Thompson, and M. Lipsitch.** 2003. Construction of otherwise isogenic serotype 6B, 7F, 14, and 19F capsular variants of *Streptococcus pneumoniae* strain TIGR4. *Appl. Environ. Microbiol.* **69**:7364–7370.

107. **van Dam, J. E., A. Fleer, and H. Snippe.** 1990. Immunogenicity and immunochemistry of *Streptococcus pneumoniae* capsular polysaccharides. *Antonie Leeuwenhoek* **58**:1–47.

108. **van Kranenburg, R., J. D. Marugg, I. I. van Swam, N. J. Willem, and W. M. de Vos.** 1997. Molecular characterization of the plasmid-encoded *eps* gene cluster essential for exopolysaccharide biosynthesis in *Lactococcus lactis*. *Mol. Microbiol.* **24**:387–397.

109. **van Selm, S., M. A. B. Kolkman, B. A. M. van der Zeijst, K. A. Zwaagstra, W. Gaastra, and J. P. M. van Putten.** 2002. Organization and characterization of the capsule biosynthesis locus of *Streptococcus pneumoniae* serotype 9V. *Microbiology* **148**:1747–1755.

110. **van Selm, S., L. M. van Cann, M. A. Kolkman, B. A. van der Zeijst, and J. P. van Putten.** 2003. Genetic basis for the structural difference between *Streptococcus pneumoniae* serotype 15B and 15C capsular polysaccharides. *Infect. Immun.* **71**:6192–6198.

111. **Waite, R. D., D. W. Penfold, J. K. Struthers, and C. G. Dowson.** 2003. Spontaneous sequence duplications within capsule genes cap8E and tts control phase variation in *Streptococcus pneumoniae* serotypes 8 and 37. *Microbiology* **149**:497–504.

112. **Waite, R. D., J. K. Struthers, and C. G. Dowson.** 2001. Spontaneous sequence duplication within an open reading frame of the pneumococcal type 3 capsule locus causes high-frequency phase variation. *Mol. Microbiol.* **42**:1223–1232.

113. **Walter, A. W., V. H. Guerin, M. W. Beattie, H. Y. Cotler, and H. B. Bucca.** 1941. Extension of the separation of types among pneumococci: description of 17 types in addition to types 1 to 32 (Cooper). *J. Immunol.* **41**:279.

114. **Weiser, J. N., R. Austrian, P. K. Sreenivasan, and H. R. Masure.** 1994. Phase variation in pneumococcal opacity: relationship between colonial morphology and nasopharyngeal colonization. *Infect. Immun.* **62**:2582–2589.

115. **Weiser, J. N., D. Bae, H. Epino, S. B. Gordon, M. Kapoor, L. A. Zenewicz, and M. Shchepetov.** 2001. Changes in availability of oxygen accentuate differences in capsular polysaccharide expression by phenotypic variants and clinical isolates of *Streptococcus pneumoniae*. *Infect. Immun.* **69**:5430–5439.

116. **Whitfield, C.** 1995. Biosynthesis of lipopolysaccharide O antigens. *Trends Microbiol.* **3**:178–185.

117. **Whitfield, C., and A. Paiment.** 2003. Biosynthesis and assembly of Group 1 capsular polysaccharides in *Escherichia coli* and related extracellular polysaccharides in other bacteria. *Carbohydr. Res.* **338**:2491–2502.

118. **Wicker, L. S., and I. Scher.** 1986. X-linked immune deficiency (*xid*) of CBA/N mice. *Curr. Top. Microbiol. Immunol.* **124**:87–101.

119. **Winkelstein, J. A., A. S. Abramovitz, and A. Tomasz.** 1980. Activation of C3 via the alternative complement pathway results in fixation of C3b to the pneumococcal cell wall. *J. Immunol.* **124**:2502–2506.

120. **Winkelstein, J. A., J. J. A. Bocchini, Jr., and G. Schiffman.** 1976. The role of the capsular polysaccharide in the activation of the alternative pathway by the pneumococcus. *J. Immunol.* **116**:367–370.

121. **Winkelstein, J. A., and A. Tomasz.** 1977. Activation of the alternative pathway by pneumococcal cell walls. *J. Immunol.* **118**:451–454.

122. **Wood, W. B., Jr., and M. R. Smith.** 1949. The inhibition of surface phagocytosis by the

capsular "slime layer" of pneumococcus type III. *J. Exp. Med.* **90:**85–96.

123. **Wu, H. Y., A. Virolainen, B. Mathews, J. King, M. W. Russell, and D. E. Briles.** 1997. Establishment of a *Streptococcus pneumoniae* nasopharyngeal colonization model in adult mice. *Microb. Pathog.* **23:**127–137.

124. **Wugeditsch, T., A. Paiment, J. Hocking, J. Drummelsmith, C. Forrester, and C. Whitfield.** 2001. Phosphorylation of Wzc, a tyrosine autokinase, is essential for assembly of group 1 capsular polysaccharides in *Escherichia coli. J. Biol. Chem.* **276:**2361–2371.

125. **Yother, J.** 1999. Common themes in the genetics of streptococcal capsular polysaccharides, p. 161–184. *In* J. B. Goldberg (ed.), *Genetics of Bacterial Polysaccharides.* CRC Press, Boca Raton, Fla.

126. **Yother, J., K. D. Ambrose, and M. J. Caimano.** 1997. Association of a partial H-rpt element with the type 3 capsule locus of *Streptococcus pneumoniae. Mol. Microbiol.* **25:**201–204.

127. **Yother, J., L. S. McDaniel, and D. E. Briles.** 1986. Transformation of encapsulated *Streptococcus pneumoniae. J. Bacteriol.* **168:**1463–1465.

128. **Zhou, L., F. M. Hui, and D. A. Morrison.** 1995. Characterization of IS*1167,* a new insertion sequence in *Streptococcus pneumoniae. Plasmid* **33:**127–138.

CHOLINE-BINDING PROTEINS

Edwin Swiatlo, Larry S. McDaniel, and David E. Briles

4

STRUCTURAL FEATURES OF CHOLINE-BINDING PROTEINS

A special characteristic of *Streptococcus pneumoniae* is its requirement for choline, or closely related analogs, for growth. Pneumococci covalently link phosphorylcholine to teichoic and lipoteichoic acids found in the peptidoglycan and cytoplasmic membrane, respectively. Although other human-pathogenic bacteria contain choline in their cell walls, only pneumococci express surface proteins which specifically bind choline as a mechanism of attachment to the cell surface. The functions of most of these choline-binding proteins (CBPs) are unknown, but a few have been studied in some depth and it is apparent that they have a role in pathogenesis and can be protective immunogens. In addition to CBPs encoded by chromosomal genes, there are at least five pneumococcal phages which express lysozyme and amidases that are activated by binding choline residues in lipoteichoic acid. This review is restricted to those CBPs that are encoded by genes on the pneumococcal chromosome.

The first protein to be implicated as a CBP was autolysin, which was found to interact with choline in teichoic acids during hydrolysis of peptidoglycan (20). Although autolysin must bind choline for full enzymatic activity, it does not appear to be bound to the cell surface choline residues. The first surface-exposed protein found to bind choline was pneumococcal surface protein A (PspA). In the unencapsulated strain Rx1 (derived from the capsular type 2 strain D39) this protein was noted to have 10 direct repeats, each consisting of 20 highly conserved amino acids, at the carboxyl terminus (48) (Fig. 1). Subsequently, PspA from strain EF5668 was found to have nine direct repeats in the choline-binding domain (25) and PspA from TIGR4 was also found to have nine repeats (43). The repeat regions of PspA are homologous to those of the major pneumococcal autolysin (LytA) and numerous pneumococcal phage lysins. These amidases were known to be activated by binding choline in cell wall and membrane teichoic acids by way of these C-terminal repeats. Experimental proof of the interaction of PspA with choline followed shortly after the nucleotide sequence and structural characteristics of PspA were described (50). The binding of PspA to choline has been ex-

Edwin Swiatlo, Departments of Medicine and Microbiology, University of Mississippi Medical Center, VA Medical Center Research Service, Jackson, MS 39216. *Larry S. McDaniel*, Departments of Microbiology, Surgery, and Medicine, University of Mississippi Medical Center, Jackson, MS 39216. *David E. Briles*, Departments of Microbiology and Pediatrics, University of Alabama at Birmingham, Birmingham, AL 35294.

The Pneumococcus. Volume Editor, Elaine I. Tuomanen,
© 2004 ASM Press, Washington, D.C.

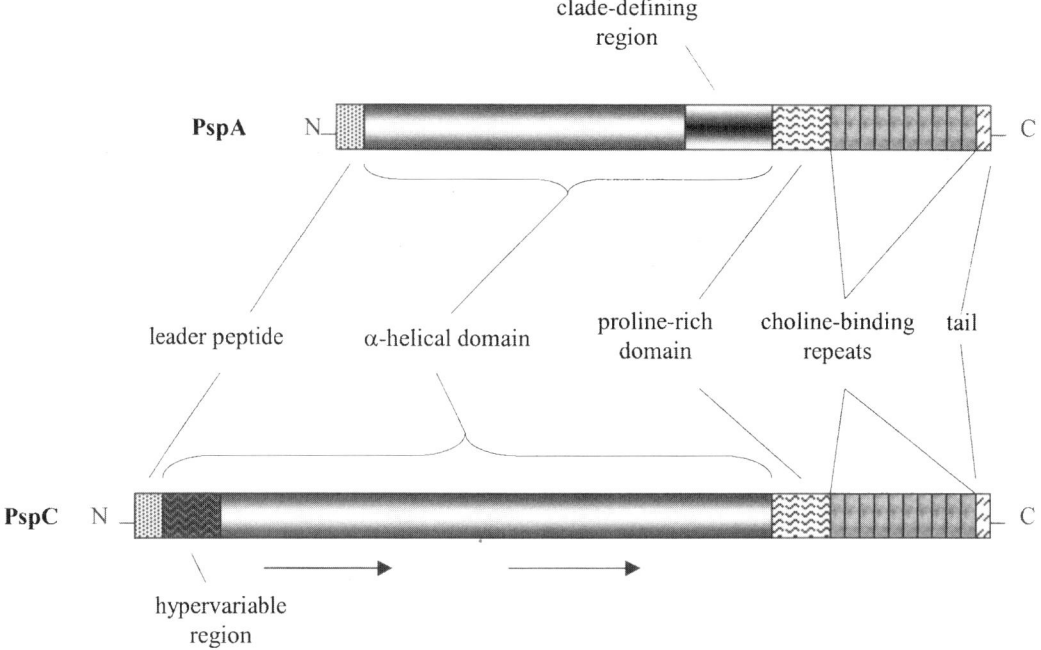

FIGURE 1 General organization of PspA and PspC from the nonencapsulated strain Rx1. PspA contains 619 amino acid residues, and PspC contains 923. The strains contain an identical 31-amino-acid leader peptide at the amino terminus, followed by an α-helical domain that varies in length. In PspA a sequence of about 100 amino acids directly adjacent to the proline-rich domain defines specific clades. Arrows below PspC mark the locations of direct repeats in the α-helical domain. In both molecules a region rich in prolines follows the α-helix and is thought to interact with the cell wall. At the carboxyl terminus is a choline-binding domain containing 10 repeats of 20 amino acids each and a short "tail" of uncharged amino acids. The proline-rich and choline-binding domains of PspA and PspC have identical amino acid sequences as determined by all alleles sequenced to this point.

ploited to purify native PspA by affinity chromatography with resins containing either choline or ethanolamine (9, 50).

Studies using Southern blotting and probes containing sequences from the choline-binding domain of PspA revealed the existence of several other chromosomal loci which potentially encoded CBPs (41). One of these sequences, designated the *pspA*-like sequence, was found to be an open reading frame and was submitted to GenBank as PspC (also called CbpA [Fig. 1]). Since the initial description of PspA, two pneumococcal genome sequences have been completed and putative choline-binding motifs have been found in additional proteins. Pneumococcal strain TIGR4 is a capsular type 4 strain isolated from blood during the course of an invasive infection. The genome of this strain

has been completely sequenced, and 15 proteins with potential choline-binding domains have been identified (43). Two of these proteins appear to have either a frameshift mutation or a significantly degenerated sequence, neither of which is a sequencing artifact. The structural characteristics of potential CBPs encoded by the 13 open reading frames other than *pspA* and *pspC* in TIGR4 are summarized in Fig. 2. The complete genome sequence is also known for laboratory strain R6, an unencapsulated derivative of D39. This strain has only 10 potential CBPs and lacks CbpC, CbpE, CbpI, CbpJ, and PcpA. Additionally, R6 contains a phosphocholine esterase, designated LytD, which is not present in TIGR4.

The CBPs of TIGR4 contain a 20-amino-acid choline-binding domain repeated three to nine times. These repeats occur at the carboxyl terminus of the protein, except for LytB, in which the choline-binding domains occur at the amino terminus. The number of these repeated motifs in any one protein varies among strains, and inter- and intragenic rearrangement of these genetic modules has contributed to significant heterogeneity, especially in PspA and PspC (see chapter 2). PcpA from strain R6 contains 11 choline-binding domains and is the largest CBP described for pneumococci (38). Three of the CBPs do not have an identifiable signal peptide at the amino terminus, although one of these proteins contains a domain with homology to amidase enzymes, suggesting that its physiological function is related to cell wall structure and that the protein is, therefore, secreted. The presence of choline-binding domains implies that any protein expressing these domains is secreted, since choline is a constituent of teichoic and lipoteichoic acids, which are cell surface polymers. However, it is possible that some proteins containing choline-binding domains remain intracellular and are regulated by intracellular choline levels, or are involved in the intracellular synthesis of teichoic acids. It is also possible that not all proteins containing choline-binding domains are actually able to bind choline.

Other than the choline-binding domains and signal peptides, there is little structural similarity among CBPs. PspA and PspC have α-helical motifs which vary in their primary amino acid sequence (10, 19) (Fig. 1). This variability among strains is thought to arise from selective pressure of host immune responses to these surface proteins. Four of the CBPs in TIGR4 have domains that are homologous to endopeptidases and probably are important in maintaining the structural integrity of the peptidoglycan cell wall. PcpA has been postulated to be an adhesin based on its content of leucine-rich repeats (38). Four CBPs from TIGR4 have only choline-binding domains and no other homology to known protein families. The choline-binding domains of these proteins occupy most of the length of the peptide, leaving only very short stretches of amino acids available for other functions. The genes for these proteins may serve primarily as sources for modular rearrangement of domains in other CBPs. If they are functional proteins they certainly have novel domains and will undoubtedly provide important insights into pneumococcal physiology or pathogenesis.

ATTACHMENT OF CBPs

Most pneumococcal proteins with choline-binding domains contain a leader sequence typical of gram-positive bacteria and are probably translocated to the cell surface by signal peptidase I (33). Once on the cell surface, CBPs associate with choline covalently attached to teichoic or lipoteichoic acids by phosphodiester bonds. The mechanism of this interaction is incompletely understood, and only recently has a three-dimensional structure of a CBP-choline interaction become available. The enzyme acetylcholinesterase has been crystallized, and its choline-binding domain is well characterized (40). This enzyme contains a pocket rich in aromatic amino acids which is thought to bind the quaternary amino group of choline. The choline-binding domain in pneumococcal proteins contains a conserved Trp-Tyr-Tyr sequence which occurs in the middle of the 20-amino-acid sequence. Although direct experimental evidence is lacking, this short stretch of aromatic acids may be critical for choline binding.

A potential three-dimensional structure of the choline-binding domain of LytA has been derived from crystallized, recombinant protein overexpressed in *Escherichia coli* (13). This structure has been described as a series of β-hairpin turns which stack to form a left-handed superhelix. This structure is stabilized by choline molecules present at the hydrophobic interface of consecutive hairpin loops. For LytA, the enzymatically active configuration is postulated to be a homodimer which relies on choline to hold together the final β-hairpins formed by the 11 C-terminal amino acid residues (12). The critical role of choline in maintaining the monomer and dimer structure

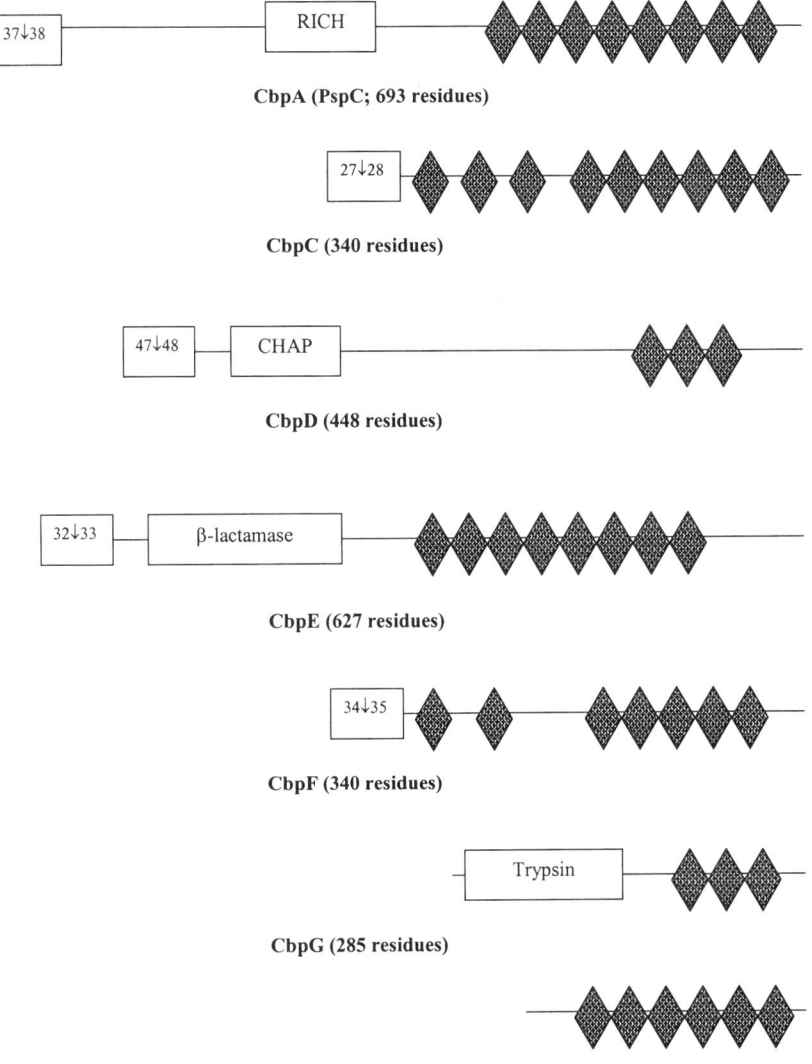

FIGURE 2 Schematic representation of CBPs in TIGR4, with the amino terminus at the far left. Diamonds represent choline-binding domains with homology to the consensus sequence, GWVKDNGTWYYLNSSGAMAT. Boxes within the structures represent domains with a high degree of homology to protein families in the Pfam database. RICH, rich in charged residues; CHAP, cysteine, histidine-dependent, amidohydrolase/peptidase. Boxes at the amino termini of some proteins designate potential cleavage sites of putative leader sequences.

of LytA may also function to anchor nonenzymatic CBPs to the pneumococcal cell surface.

The number of choline-binding repeats in pneumococcal CBPs appears to be important in the avidity of CBPs for teichoic acids. PspA in strain Rx1 contains 10 choline-binding repeats but will no longer strongly attach to the cell surface if the five repeats at the carboxyl terminus are deleted (50). This truncated PspA molecule will, however, still bind to a DEAE-

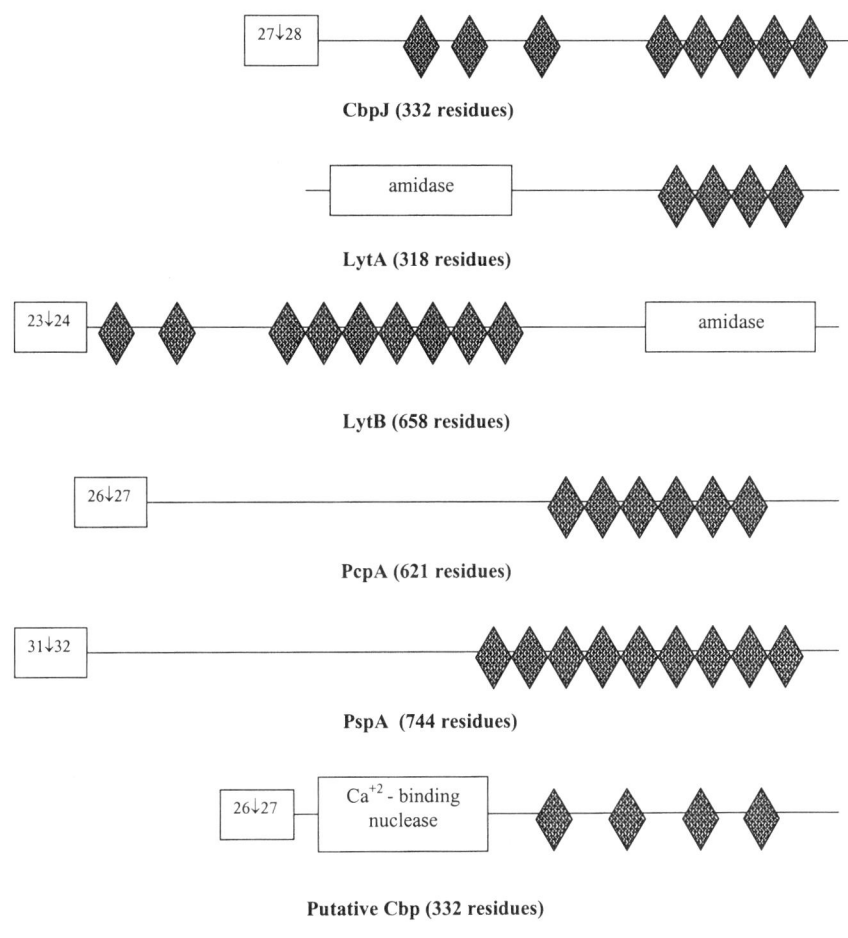

CbpJ (332 residues)

LytA (318 residues)

LytB (658 residues)

PcpA (621 residues)

PspA (744 residues)

Putative Cbp (332 residues)

FIGURE 2 (*Continued*)

cellulose column, suggesting that its choline-binding properties are not completely eliminated. A separate PspA mutant was constructed that lacks six internal repeats but still has the 10th repeat and the 17-amino-acid C-terminal tail. This molecule remains attached to the pneumococcal surface even though it had only four repeats. This finding suggested that the 17-amino-acid C-terminal tail, and possibly 10th repeat, is important for surface attachment of PspA (35). The major pneumococcal autolysin LytA contains four to six choline-binding domains but is not found in eluates of pneumococcal strain Rx1 incubated with choline-containing buffers. Although autolysin requires choline for enzymatic activity,

it appears that its surface attachment is not primarily choline dependent. One hypothesis is that CBPs which bind choline for enzymatic activity do not require as great a number of choline-binding domains as those structural proteins which are firmly anchored to the cell surface. Alternatively, CBPs which cannot be eluted from intact cells with excess choline may assume unique tertiary conformations which render them unable to bind to free choline in solution. Clearly, at this point, the only conclusion which can be supported with experimental data is that the presence of a conserved domain in multiple copies correlates with an ability to bind choline residues on pneumococcal teichoic acids.

REGULATION OF CBP EXPRESSION

Like all bacteria, the pneumococcus has the ability to regulate gene expression in response to environmental signals. While there is a lack of information about specific signals that affect expression of CBPs, there is evidence that the pneumococcus regulates expression of the various CBPs at different stages of growth and in different environments. Phase variation (transparent and opaque colony types) has been described for the pneumococcus. Pneumococci in the nasopharynx have been shown to express predominantly the transparent phenotype (45), which appears to have increased adherence and less capsular polysaccharide and which expresses more PspC and less PspA. Transparent colony cells also express more LytA on their surface (46). In contrast, the opaque phase predominates when pneumococci are in the blood. The opaque phenotype is associated with increased capsular polysaccharide and PspA and less PspC (3, 31).

Several studies have demonstrated that both PspA and PspC are expressed in vivo in a mouse model. Analysis of PspA mRNA levels by northern blotting demonstrates similar levels of expression of PspA in pneumococci grown in vitro and in pneumococci growing in blood in a mouse bacteremia model. Additionally, while antibody against PspA has been established as protective in murine models of infection, recent studies demonstrate that antibodies to PspA are protective even after infection has been ongoing for up to 12 h (42). When PspC mRNA was studied, it was observed that PspC is expressed in both the nasopharynx and lungs of infected mice (3). While there may be less PspC expressed by pneumococci growing in the blood than on mucosal surfaces, the PspC that is expressed in blood is still functional since factor H, which binds PspC (see below), can be detected on the surface of pneumococci growing in blood (L. Quinn, S. Dave, S. Carmicle, and L. S. McDaniel, unpublished results).

The effect of environmental stress on pneumococcal genes has been examined. It was demonstrated that gene expression for PspC but not PspA was increased in response to heat shock at 42°C. It has also been suggested that heat shock proteins, specifically ClpP, may be negative regulators of PspC (22).

BIOLOGICAL FUNCTION AND RELEVANCE TO DISEASE

Pneumococcal CBPs have been shown to play a role in pathogenesis in various murine models of disease (3, 14, 27). It is likely that there is some redundancy in the function of CBPs, and this, along with the large effect of the pneumococcal capsule, may explain why more is not known about the function of CBPs in pneumococcal disease.

Mutants that lack CbpD, CbpE, or CbpG show a reduced capacity for colonization in an infant rat model (14). Other proteins, including LytB and LytC, that interact with choline on the pneumococcal surface may contribute to colonization of the nasopharynx by undetermined mechanisms. Mutants that lack CbpG were also shown to have reduced virulence in a sepsis model. Based on sequence analysis, CbpG has homology to the large family of serine proteases. While the function of CbpG has not been established biochemically, it appears to be involved in both pneumococcal colonization and survival in the blood. PcpA has been found to be important for virulence in a lung infection model (18). PcpA has a motif suggesting adhesin function, although this has not been proven functionally (38).

PspA and PspC are the two most well-characterized CBPs in terms of their biological functions and roles in disease. While these two proteins can be considered paralogs based on sequence homology, they make distinct contributions to pneumococcal virulence. Mutants that lack PspA on their surface are more rapidly cleared from the blood of infected mice (27). This suggests that PspA functions to inhibit the uptake of pneumococci by phagocytic cells. It is likely that the role of PspA in bacteremia and sepsis is the result, in part, of its inhibition of complement deposition and activation (35, 44). PspA has also been shown to bind lactoferrin (16), and this interaction accounts for virtually all surface-bound lactoferrin on pneumococci (15). However, PspA binds lactoferrin, which is produced as apo-lactoferrin (lacking bound iron) in high concentrations in secretions and at foci of infection. Apolactoferrin is bactericidal for many bacteria, and recent data indicate that PspA significantly inhibits the killing of pneumococci by apolactoferrin (M. Shaper, S. K. Hollingshead, W. H. Benjamin, Jr., and D. E. Briles, submitted for publication).

Additional evidence of the functional importance of PspA was provided in studies that examined surface attachment requirements for PspA function. Clearance of PspA mutants from blood cannot be restored to rates seen for wild-type strains by addition of exogenous PspA. Moreover, strains secreting truncated PspA lacking the choline-binding domain are as attenuated as strains expressing no PspA (35, 49). Therefore, PspA must contain the choline-binding domain and be attached to the pneumococcal surface to contribute to pneumococcal virulence in a bacteremia and sepsis model.

PspC has a choline-binding domain that is virtually identical to that of PspA. PspC appears to be a multifunctional pneumococcal surface protein, as three different human proteins have been shown to bind within its α-helical domain. PspC binds the secretory component of immunoglobulin A (sIgA) (17), complement component C3 (39), and complement factor H (11). An allelic variant of *pspC* designated Hic contains an LPXTG motif and is bound through the activity of sortase rather than a choline-binding domain. Hic also binds factor H and has been shown to inhibit complement deposition on pneumococci (21).

Mutants that do not express PspC show reduced virulence in mouse models of bacteremia, pneumonia, and nasal colonization (3, 5), and there is evidence that PspC acts as an adhesin (37). While the functional mechanisms by which PspC contributes to virulence are not completely understood, some clues are suggested by its interactions with host proteins.

For example, the PspC-sIgA interaction may facilitate the adherence/invasion phase of infection via the polymeric Ig receptor on mucosal epithelial cells (24, 51). The binding of factor H and the inhibition of complement deposition are suspected to play a role in invasive infections such as pneumonia and bacteremia. PspC has been shown to bind factor H during pneumococcal bacteremia in mice (Quinn et al., unpublished), indicating that this interaction is of biological importance. PspC has also been shown to bind factor H and sIgA simultaneously in vitro (S. Dave, unpublished data) (Fig. 3).

It is clear that CBP functions are important in the natural history of pneumococcal infection and colonization. The variety of amino-terminal domains that have been observed in this family of surface proteins suggests that they provide a diverse complement of phenotypic traits to pneumococci. A clear understanding of the functional significance of CBPs in host-bacterium interactions will come from further detailed studies.

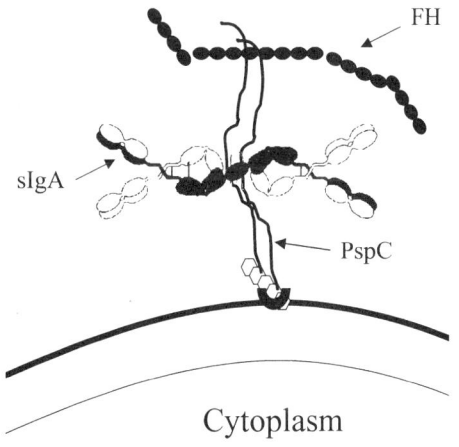

FIGURE 3 Diagram of binding of sIgA via the secretory component and factor H (FH) to PspC. The host proteins bind to separate sites on the α-helical portion of PspC. Both are relatively large host proteins but do not compete for binding in in vitro assays, indicating that they bind separate sites. Simultaneous binding in vivo has not been demonstrated.

IMMUNOLOGY OF CBPs

Because CBPs are predicted to be exposed on the bacterial cell surface, they have attracted attention as potential immunogens and vaccine candidates. Despite the attractiveness of these surface proteins as vaccine components, only three, PspA, PspC, and autolysin, have been studied for their ability to induce protective immune responses in animals.

Noncapsular antigens of pneumococci have been known since at least 1928, and the first of these to be characterized were teichoic (C-polysaccharide) and lipoteichoic (F-antigen) acids. Both of these polymers contain phosphorylcholine as the immunodominant epitope. In 1949 a serologically variable protein antigen on pneumococcal cells was described (2). The first description of an immunogenic protein expressed by pneumococci that could induce protective immune responses involved monoclonal antibodies generated against protease-sensitive antigens of the unencapsulated laboratory strain Rx1 (26). The protein was subsequently named pneumococcal surface protein A (PspA) and shown to contain a choline-binding motif similar to that of LytA and to bind choline residues in teichoic acids (48, 50). Concurrently, pneumolysin was shown to be a protective antigen in a mouse model (32) and, together with PspA, offered the first evidence that protein antigens could protect against pneumococcal infection.

PspA can induce antibodies in mice, rabbits, monkeys, and humans which can passively protect against otherwise fatal pneumococcal infections in mice (9a). Active immunization with PspA can protect mice from sepsis, pneumonia, and carriage (8, 29, 47). Combined immunization with PspA and pneumolysin was more effective than either antigen alone in protecting against sepsis and pneumonia (7, 29).

PspA is serologically variable, and there is evidence that significant recombination has led to mosaic genes distributed among strains independent of capsule type (19). This antigenic diversity is strong evidence that host immune responses have applied selective pressure for variability of epitopes. Despite the serologic variability of PspA, certain conserved epitopes can be identified across clades and families (7, 28). Ongoing work is aimed at identifying the minimum number of protective epitopes which will induce the most comprehensive array of protective antibodies (36).

Immunity to fatal bacteremia in a mouse model is readily achieved with subcutaneous injection of PspA. Immunity to nasopharyngeal carriage requires mucosal delivery of PspA with an adjuvant such as the B subunit of cholera toxin or interleukin 12 (1, 47). Mucosal immunization also provides protection against systemic disease. Genetic immunization with a eukaryotic expression plasmid containing *pspA* can induce specific antibodies which are protective against pneumococcal sepsis (6).

A largely overlooked aspect of pneumococcal pathogenesis is that the primary manifestation of pneumococcal disease in adults is focal pneumonia without bacteremia, a syndrome which has received little attention by vaccine developers. PspA immunization, either alone or in combination with a detoxified derivative of pneumolysin, can significantly attenuate the progression of pneumococcal pneumonia in a murine model (8). This model is significant in that it more closely reflects the natural history of pneumonia in humans by not progressing to bacteremia in most cases. Small phase I trials in human volunteers have demonstrated that PspA is immunogenic and human immune sera can passively protect mice against an otherwise fatal pneumococcal challenge (7). Naturally occurring antibodies to PspA can be detected in children after either infection or asymptomatic carriage (34) and are evidence that PspA is expressed in vivo in humans. Pneumococci collected from bacteremic mice transcribe *pspA*, and antibodies to PspA can protect bacteremic mice after infection has been established for up to 12 h (42). These studies were crucial to confirm that PspA is expressed in vivo, since all studies up to that time had used pneumococci grown in vitro as the infectious challenge. Together, these studies have moved PspA to the forefront as a leading candidate for a protein-based pneumococcal vaccine in humans.

After *pspA* was sequenced and the protein structure was deduced, it became apparent that there were other pneumococcal genes which contained a high degree of homology to the sequence of *pspA* encoding the choline-binding domain. One of these genes was sequenced and submitted to GenBank as the PspC gene from strain EF6796 in 1996. The PspC protein was subsequently discovered and sequenced independently and designated CbpA, SpsA, and PspC (10, 17, 37). CBPs from a mutant strain that does not produce PspA were affinity purified with a choline-containing affinity matrix and used to immunize mice. Hyperimmune serum was shown to passively protect mice from pneumococcal bacteremia with a capsule type 3 strain. A protein designated choline-binding protein A (CbpA) was found to be the predominant protein in this mixture, and it was postulated that antibodies to CbpA in the immune sera were primarily responsible for its protective efficacy. Additionally, this protein was shown to be involved in pneumococcal binding to activated type II pneumocytes (37). A second report described a protein named SpsA which was shown to bind the secretory component of human polymeric IgA (17). A third report described the close structural relationship between PspA and CbpA/SpsA and designated the protein PspC (10). This study focused on the immunological aspects of PspC and noted extensive serologic cross-reactivity between PspA and some variants of PspC. Immunization of animals with purified PspC was able to protect against lethal infection with a pneumococcal mutant which does not express PspC.

Because of its presumed surface location and the structural similarity between PspC and PspA an obvious hypothesis to test was that PspC is an effective immunogen. PspC was compared with pneumolysin toxoid for its ability to elicit protective antibodies in a murine sepsis model. Immunization with PspC was shown to confer greater protection against sepsis than the pneumolysin toxoid, and the combination of antigens afforded no greater protection than PspC alone (30). Because PspC may be an important adhesin, it has been used as a mucosal immunogen with the cholera toxin B subunit. Intranasal immunization with PspC protects animals against intranasal challenge with a virulent pneumococcal strain that typically causes pneumonia and sepsis when given by this route (3).

The major autolysin of pneumococci, designated LytA, is known to be important in remodeling the cell wall of dividing pneumococci and is the common final point for many processes which lead to cell lysis. Mutations in *lytA* greatly reduce virulence, and immunization of mice with LytA provides partial protection against sepsis (5, 23). It was postulated that antibodies to LytA were protective because they inhibit pneumococcal cell lysis and reduce the release of intracellular pneumolysin and proinflammatory cell wall fragments. This hypothesis was supported by the observation that immunization with autolysin provides protection comparable to that seen with pneumolysin immunization, but the two proteins together are no better than either one alone (23). It is possible that LytA and pneumolysin have independent contributions to virulence, and indeed, some pneumococcal strains can release pneumolysin without a functional autolysin (4). Autolysis of pneumococci does not occur at neutral pH, but at sites of focal infection where the microenvironment is highly acidic, autolysin may be an important mechanism for release of pneumolysin.

The exact number of CBPs probably varies for each strain, but a significant subset of CBPs is probably expressed by all pneumococcal strains which are associated with human disease. Encouraging results in animal models of pneumococcal infection continue to recommend PspA and PspC as potential vaccine candidates. Further investigation of additional CBPs may provide insights into even more potent vaccine components, either as single proteins or as part of a multivalent peptide mixture.

REFERENCES

1. **Arulanandam, B. P., J. M. Lynch, D. E. Briles, S. Hollingshead, and D. W. Metzger.** 2001. Intranasal vaccination with pneumococcal surface protein A and interleukin-12 augments antibody-mediated opsonization and protective immunity against *Streptococcus pneumoniae* infection. *Infect. Immun.* **69:**6718–6724.

2. **Austrian, R., and O. M. Macleod.** 1949. A type specific protein from pneumococcus. *J. Exp. Med.* **89:**439–450.

3. **Balachandran, P., A. Brooks-Walter, A. Virolainen-Julkunen, S. Hollingshead, and D. E. Briles.** 2002. Role of pneumococcal surface protein C in nasopharyngeal carriage and pneumonia and its ability to elicit protection against carriage of *Streptococcus pneumoniae*. *Infect. Immun.* **70:**2526–2534.

4. **Balachandran, P., S. K. Hollingshead, J. C. Paton, and D. E. Briles.** 2001. The autolytic enzyme LytA of *Streptococcus pneumoniae* is not responsible for releasing pneumolysin. *J. Bacteriol.* **183:**3108–3116.

5. **Berry, A. M., and J. C. Paton.** 2000. Additive attenuation of virulence of *Streptococcus pneumoniae* by mutation of the genes encoding pneumolysin and other putative pneumococcal virulence proteins. *Infect. Immun.* **68:**133–140.

6. **Bosarge, J. R., J. M. Watt, D. O. McDaniel, E. Swiatlo, and L. S. McDaniel.** 2001. Genetic immunization with the region encoding the α-helical domain of PspA elicits protective immunity against *Streptococcus pneumoniae*. *Infect. Immun.* **69:**5456–5463.

7. **Briles, D. E., S. K. Hollingshead, J. King, A. Swift, P. A. Braun, M. K. Park, L. M. Ferguson, M. H. Nahm, and G. S. Nabors.** 2000. Immunization of humans with recombinant pneumococcal surface protein A (rPspA) elicits antibodies that passively protect mice from fatal infection with *Streptococcus pneumoniae* bearing heterologous PspA. *J. Infect. Dis.* **182:**1694–1701.

8. **Briles, D. E., S. K. Hollingshead, J. C. Paton, E. W. Ades, L. Novak, F. W. van Ginkel, and W. H. Benjamin.** 2003. Immunizations with pneumococcal surface protein A and pneumolysin are protective against pneumonia in a murine model of pulmonary infection with *Streptococcus pneumoniae*. *J. Infect. Dis.* **188:**339–348.

9. **Briles, D. E., J. D. King, M. A. Gray, L. S. McDaniel, E. Swiatlo, and K. A. Benton.** 1996. PspA, a protection-eliciting pneumococcal protein: immunogenicity of isolated native PspA in mice. *Vaccine* **14:**858–867.

9a. **Briles, D. E., J. C. Paton, and S. K. Hollingshead.** 1997. Pneumococcal, common proteins and other strategies. *In* M. M. Levine, J. B. Kaper, R. Rappuoli, M. Liu, and M. Good (ed.), *New Generation Vaccines*. Marcel Dekker, Inc., New York, N.Y.

10. **Brooks-Walter, A., D. E. Briles, and S. K. Hollingshead.** 1999. The *pspC* gene of *Streptococcus pneumoniae* encodes a polymorphic protein, PspC, which elicits cross-reactive antibodies to PspA and provides immunity to pneumococcal bacteremia. *Infect. Immun.* **67:**6533–6542.

11. **Dave, S., A. Brooks-Walter, M. K. Pangburn, and L. S. McDaniel.** 2001. PspC, a pneumococcal surface protein, binds human factor H. *Infect. Immun.* **69:**3435–3437.

12. **Fernandez-Tornero, C., E. Garcia, R. Lopez, G. Gimenez-Gallego, and A. Romero.** 2002. Two new crystal forms of the choline-binding domain of the major pneumococcal autolysin: insights into the dynamics of the active homodimer. *J. Mol. Biol.* **321:**163–173.

13. **Fernandez-Tornero, C., R. Lopez, E. Garcia, G. Gimenez-Gallego, and A. Romero.** 2001. A novel solenoid fold in the cell wall anchoring domain of the pneumococcal virulence factor LytA. *Nat. Struct. Biol.* **8:**1020–1024.

14. **Gosink, K. K., E. R. Mann, C. Guglielmo, E. I. Tuomanen, and H. R. Masure.** 2000. Role of novel choline binding proteins in virulence of *Streptococcus pneumoniae*. *Infect. Immun.* **68:**5690–5695.

15. **Hakansson, A., H. Roche, S. Mirza, L. S. McDaniel, A. Brooks-Walter, and D. E. Briles.** 2001. Characterization of binding of human lactoferrin to pneumococcal surface protein A. *Infect. Immun.* **69:**3372–3381.

16. **Hammerschmidt, S., G. Bethe, P. H. Remane, and G. S. Chhatwal.** 1999. Identification of pneumococcal surface protein A as a lactoferrin-binding protein of *Streptococcus pneumoniae*. *Infect. Immun.* **67:**1683–1687.

17. **Hammerschmidt, S., S. R. Talay, P. Brandtzaeg, and G. S. Chhatwal.** 1997. SpsA, a novel pneumococcal surface protein with specific binding to secretory immunoglobulin A and secretory component. *Mol. Microbiol.* **25:**1113–1124.

18. **Hava, D. L., C. J. Hemsley, and A. Camilli.** 2003. Transcriptional regulation in the *Streptococcus pneumoniae rlrA* pathogenicity islet by RlrA. *J. Bacteriol.* **185:**413–421.

19. **Hollingshead, S. K., R. Becker, and D. E. Briles.** 2000. Diversity of PspA: mosaic genes and evidence for past recombination in *Streptococcus pneumoniae*. *Infect. Immun.* **68:**5889–5900.

20. **Holtje, J.-V., and A. Tomasz.** 1975. Specific recognition of choline residues in the cell wall tei-

choic acid by *N*-acetylmuramyl-L-alanine amidase of pneumococcus. *J. Biol. Chem.* **250:**6072–6075.

21. **Jarva, H., R. Janulczyk, J. Hellwage, P. F. Zipfel, L. Bjorck, and S. Meri.** 2002. *Streptococcus pneumoniae* evades complement attack and opsonophagocytosis by expressing the *pspC* locus-encoded Hic protein that binds to short consensus repeats 8–11 of factor H. *J. Immunol.* **168:**1886–1894.

22. **Kwon, H.-Y., S.-W. Kim, M.-H. Choi, D. A. Ogunniyi, J. C. Paton, S.-H. Park, S.-N. Pyo, and D.-K. Rhee**. 2003. Effect of heat shock and mutations in ClpL and ClpP on virulence gene expression in *Streptococcus pneumoniae*. *Infect. Immun.* **71:**3757–3765.

23. **Lock, R. A., D. Hansman, and J. C. Paton.** 1992. Comparative efficacy of autolysin and pneumolysin as immunogens protecting mice against infection by *Streptococcus pneumoniae*. *Microb. Pathog.* **12:**137–143.

24. **Lu, L., M. E. Lamm, H. Li, B. Corthesy, and J.-R. Zhang.** 2003. The human polymeric immunoglobulin receptor binds to *Streptococcus pneumoniae* via domains 3 and 4. *J. Biol. Chem.* **278:**48178–48187.

25. **McDaniel, L. S., D. O. McDaniel, S. K. Hollinghead, and D. E. Briles.** 1998. Comparison of the PspA sequence from *Streptococcus pneumoniae* EF5668 to the previously identified PspA sequence from strain Rx1 and ability of PspA from EF5668 to elicit protection against pneumococci of different capsular types. *Infect. Immun.* **66:**4748–4754.

26. **McDaniel, L. S., G. Scott, J. F. Kearney, and D. E. Briles.** 1984. Monoclonal antibodies against protease-sensitive pneumococcal antigens can protect mice from fatal infection with *Streptococcus pneumoniae*. *J. Exp. Med.* **160:**386–397.

27. **McDaniel, L. S., J. Yother, M. Vijayakumar, L. McGarry, W. R. Guild, and D. E. Briles.** 1987. Use of insertional inactivation to facilitate studies of biological properties of pneumococcal surface protein A (PspA). *J. Exp. Med.* **165:**381–394.

28. **Nabors, G. S., P. A. Braun, D. J. Herrmann, M. L. Heise, D. J. Pyle, S. Gravenstein, M. Schilling, L. M. Ferguson, S. K. Hollingshead, D. E. Briles, and R. S. Becker.** 2000. Immunization of healthy adults with a single recombinant pneumococcal surface protein A (PspA) variant stimulates broadly cross-reactive antibodies to heterologous PspA molecules. *Vaccine* **18:**1743–1754.

29. **Ogunniyi, A. D., R. L. Folland, D. E. Briles, S. K. Hollingshead, and J. C. Paton.** 2000. Immunization of mice with combinations of pneumococcal virulence proteins elicits enhanced protection against challenge with *Streptococcus pneumoniae*. *Infect. Immun.* **68:**3028–3033.

30. **Ogunniyi, A. D., M. C. Woodrow, J. T. Poolman, and J. C. Paton.** 2001. Protection against *Streptococcus pneumoniae* elicited by immunization with pneumolysin and CbpA. *Infect. Immun.* **69:**5997–6003.

31. **Overweg, K., C. D. Pericone, G. G. C. Verhoef, J. N. Weiser, H. D. Meiring, A. P. J. M. de Jong, and R. de Groot.** 2000. Differential protein expression in phenotypic variants of *Streptococcus pneumoniae*. *Infect. Immun.* **68:**4604–4610.

32. **Paton, J. C., R. A. Lock, and D. J. Hansman.** 1983. Effect of immunization with pneumolysin on survival time of mice challenged with *Streptococcus pneumoniae*. *Infect. Immun.* **40:**548–552.

33. **Peng, S.-B., L. Wang, J. Moomaw, R. B. Peery, P.-M. Sun, R. B. Johnson, and J. Lu.** 2001. Biochemical characterization of signal peptidase I from Gram-positive *Streptococcus pneumoniae*. *J. Bacteriol.* **183:**621–627.

34. **Rapola, S., V. Jantti, R. Haikala, R. Syrjanen, G. M. Carlone, J. S. Sampson, and D. E. Briles.** 2000. Natural development of antibodies to pneumococcal surface protein A, pneumococcal surface adhesin A, and pneumolysin in relation to pneumococcal carriage and acute otitis media. *J. Infect. Dis.* **182:**1146–1152.

35. **Ren, B., A. J. Szalai, O. Thomas, S. K. Hollingshead, and D. E. Briles.** 2003. Both family 1 and family 2 PspA proteins can inhibit complement deposition and confer virulence to a capsular serotype 3 strain of *Streptococcus pneumoniae*. *Infect. Immun.* **71:**75–85.

36. **Roche, H., B. Ren, L. S. McDaniel, A. Hakansson, and D. E. Briles.** 2003. Relative roles of genetic background and variation in PspA in the ability of antibodies to PspA to protect against capsular type 3 and 4 strains of *Streptococcus pneumoniae*. *Infect. Immun.* **71:**4498–4505.

37. **Rosenow, C., P. Ryan, J. N. Weiser, S. Johnson, P. Fontan, A. Ortqvist, and H. R. Masure.** 1997. Contribution of novel choline-binding proteins to adherence, colonization and immunogenicity of *Streptococcus pneumoniae*. *Mol. Microbiol.* **25:**819–829.

38. **Sanchez-Beato, A. R., R. Lopez, and J. L. Garcia.** 1998. Molecular characterization of PcpA: a novel choline-binding protein of *Streptococcus pneumoniae*. *FEMS Microbiol. Lett.* **164:**207–214.

39. **Smith, B. L., and M. K. Hostetter.** 2000. C3 as a substrate for adhesion of *Streptococcus pneumoniae*. *J. Infect. Dis.* **182:**497–508.

40. **Sussman, J. L., and I. Silman.** 1992. Acetylcholinesterase: structure and use as a model for specific cation-protein interactions. *Curr. Opin. Struct. Biol.* **2:**721–729.

41. **Swiatlo, E., A. Brooks-Walter, D. E. Briles, and L. S. McDaniel.** 1997. Oligonucleotides identify conserved and variable regions of *psp*A and *psp*A-like sequences of *Streptococcus pneumoniae*. *Gene* **188:**279–284.

42. **Swiatlo, E., J. King, G. S. Nabors, B. Mathews, and D. E. Briles.** 2003. Pneumococcal surface protein A is expressed in vivo, and antibodies to PspA are effective for therapy in a murine model of pneumococcal sepsis. *Infect. Immun.* **71:**7149–7153.

43. **Tettelin, H., K. E. Nelson, I. T. Paulsen, J. A. Eisen, T. D. Read, S. Peterson, J. Heidelberg, R. T. DeBoy, D. H. Haft, R. J. Dodson, A. S. Durkin, M. Gwinn, J. T. Kolonay, W. C. Nelson, J. D. Peterson, L. A. Umayam, O. White, S. L. Salzberg, M. R. Lewis, D. Radune, E. Holtzapple, H. Khouri, A. M. Wolf, T. R. Utterback, C. L. Hansen, L. A. McDonald, T. V. Feldblyum, S. Angiuoli, T. Dickinson, E. K. Hickey, I. E. Holt, B. J. Loftus, F. Yang, H. O. Smith, J. C. Venter, B. A. Dougherty, D. A. Morrison, S. K. Hollingshead, and C. M. Fraser.** 2001. Complete genome sequence of a virulent isolate of *Streptococcus pneumoniae*. *Science* **293:**498–506.

44. **Tu, A. H., R. L. Fulgham, M. A. McCrory, D. E. Briles, and A. J. Szalai.** 1999. Pneumococcal surface protein A inhibits complement activation by *Streptococcus pneumoniae*. *Infect. Immun.* **67:**4720–4724.

45. **Weiser, J. N., R. Austrian, P. K. Sreenivasan, and H. R. Masure.** 1994. Phase variation in pneumococcal opacity: relationship between colonial morphology and nasopharyngeal colonization. *Infect. Immun.* **62:**2582–2589.

46. **Weiser, J. N., Z. Markiewicz, E. I. Tuomanen, and J. H. Wani.** 1996. Relationship between phase variation in colony morphology, intrastrain variation in cell wall physiology, and nasopharyngeal colonization by *Streptococcus pneumoniae*. *Infect. Immun.* **64:**2240–2245.

47. **Wu, H.-Y., M. H. Nahm, Y. Guo, M. W. Russell, and D. E. Briles.** 1997. Intranasal immunization of mice with PspA (pneumococcal surface protein A) can prevent intranasal carriage, pulmonary infection, and sepsis with *Streptococcus pneumoniae*. *J. Infect. Dis.* **175:**839–846.

48. **Yother, J., and D. E. Briles.** 1992. Structural properties and evolutionary relationships of PspA, a surface protein of *Streptococcus pneumoniae*, as revealed by sequence analysis. *J. Bacteriol.* **174:**601–609.

49. **Yother, J., G. L. Handsome, and D. E. Briles.** 1992. Truncated forms of PspA that are secreted from *Streptococcus pneumoniae* and their use in functional studies and cloning of the *psp*A gene. *J. Bacteriol.* **174:**610–618.

50. **Yother, J., and J. M. White.** 1994. Novel surface attachment mechanism of the *Streptococcus pneumoniae* protein PspA. *J. Bacteriol.* **176:**2976–2985.

51. **Zhang, J. R., K. E. Mostov, M. E. Lamm, M. Nanno, S. Shimida, M. Ohwaki, and E. I. Tuomanen.** 2000. The polymeric immunoglobulin receptor translocates pneumococci across human nasopharyngeal epithelial cells. *Cell* **128:**827–837.

PNEUMOLYSIN AND OTHER VIRULENCE PROTEINS

Tim J. Mitchell

5

INTRODUCTION

Streptococcus pneumoniae (the pneumococcus) makes a range of molecules that can be considered virulence factors. The polysaccharide capsule is essential for virulence (87), and the cell wall plays a crucial role in the inflammation induced by the organism. The organism also makes a range of protein virulence factors, including surface proteins (choline-binding proteins, LPXTG-anchored proteins, and lipoproteins) as well as a range of enzymes (superoxide dismutase, NADH oxidase, zinc metalloproteinases) and the pore-forming toxin pneumolysin (PLY). The capsule, cell wall, and choline-binding proteins are considered elsewhere in this volume (chapters 3 to 5). This chapter focuses on PLY and some of the other protein virulence factors of the pneumococcus.

STRUCTURE-FUNCTION RELATIONSHIPS OF PLY

PLY is a 53-kDa protein made by all clinical isolates of the pneumococcus. It belongs to the family of pore-forming toxins produced by more than 20 species of gram-positive bacteria.

These toxins were originally described as hemolysins but can lyse any eukaryotic cell with cholesterol in its membrane. An exception seems to be intermedilysin produced by *Streptococcus intermedius*, which is specific for human cells and has a protein receptor rather than cholesterol. PLY is expressed during the late log phase of growth of the pneumococcus (10). Analysis of the primary amino acid sequence of PLY does not reveal a typical secretion signal sequence. It was previously suggested that PLY is released only when cells undergo autolysis, and this explains, at least in part, the role of autolysin in virulence (14). However, it has now been demonstrated that PLY can be released independently of the major autolysin (8). The lytic activity of PLY involves the formation of large pores (up to 30 nm in diameter) in cell membranes caused by the oligomerization of up to 50 toxin monomers (66). The initial interaction of the toxin with the membrane probably occurs through binding to membrane cholesterol. As well as being lytic, PLY can have a range of effects on cells and tissues at sublytic concentrations (discussed below) and can activate the classical complement pathway in the absence of specific antibody (64).

The regions of the toxin molecule responsible for the biological activities have been defined at least in part. Analysis of the primary

Tim J. Mitchell, Division of Infection and Immunity, Institute of Biomedical and Life Sciences, Joseph Black Building, University of Glasgow, Glasgow G12 8QQ, United Kingdom.

The Pneumococcus. Volume Editor, Elaine I. Tuomanen,
© 2004 ASM Press, Washington, D.C.

amino acid sequence gives no real clues to the mechanism of pore formation by PLY, as there are no large hydrophobic regions that could be considered as membrane-spanning domains. The primary sequence contains only a single cysteine residue. This observation was initially surprising, as PLY was termed a thiol-activated toxin that was believed to represent reduction of an intramolecular disulfide bond. The single cysteine residue is contained within the unde-capeptide region that is conserved in many members of this family (Color Plate 3). The undecapeptide region in PLY consists of amino acid residues 427 to 437. The cysteine was thought to be essential for activity of the toxin, but site-directed mutagenesis studies showed that this residue could be substituted with alanine without affecting the hemolytic activity of the toxin (80). The undecapeptide region is essential in the activity of the toxin, and mutations in this region can have dramatic effects. For example, substitution of the tryp-tophan at position 433 with phenylalanine re-duces the hemolytic activity of the toxin by 99%.

PLY is able to activate the complement sys-tem in the absence of specific antitoxin anti-bodies (64, 71). The ability of the toxin to ac-tivate complement is mediated through binding of the Fc fragment of immunoglobu-lin. The region responsible for this activity has been identified (Color Plate 3) and mutated to examine the basis of this activity (64). The as-partic acid at position 385 of PLY is important in complement activation, and mutants altered at this amino acid have been used to determine the importance of complement activation in the pathology induced by the toxin.

Studies of the structure and function of PLY have been greatly enhanced by the availability of the crystal structure for the closely related toxin perfringolysin produced by *Clostridium perfringens* (73). The availability of this struc-ture allowed a model of PLY to be built (74) (Color Plate 3). The protein consists of four domains, although only one of these (domain 4) is contiguous within the primary amino acid sequence. Domain 4 is the C-terminal domain of the protein and contains the conserved un-decapeptide sequence. Domain 4 has been cloned and expressed separately from the rest of the molecule and retains its ability to bind to cells and to cholesterol (6). Residues important in cholesterol binding also map to domain 4. The binding of the intact toxin to membrane cholesterol induces a conformational change in the toxin monomer driving pore formation. Domain 4 has been suggested to be the major membrane insertion domain, and binding of cholesterol to domain 4 displaces the loop composed primarily of the undecapeptide to form a "hydrophobic dagger" that initiates membrane insertion (73).

The conserved undecapeptide region is in-timately involved in membrane insertion (73). Mutations of the tryptophan residues of this region of PLY cause significant reductions in hemolytic activity (21, 58). The greatest re-duction is caused by mutation of the first tryp-tophan (W433) of PLY, which causes greater than 99% reduction in hemolytic activity. The crucial role of this residue may be related to its position at the tip of the hydrophobic dagger (73). Using planar lipid bilayers revealed that the substitution of this tryptophan residue al-tered the properties of the pore such that it had a higher conductance and less sensitivity to di-valent cations (58).

INFLAMMATORY PROPERTIES OF PLY

PLY lyses all cells with cholesterol in their membranes. The toxin can have other effects on cells and has been shown to affect the pro-duction of proinflammatory mediators such as tumor necrosis factor alpha (TNF-α), inter-leukin 1β (IL-1β), and IL-6. The toxin can stimulate nitric oxide production from macrophages as well as increased transcription of the genes for COX-2 (22). COX-2 is the induced, rate-limiting enzyme involved in prostaglandin synthesis. The pathway by which PLY activates mouse macrophages is dependent on gamma interferon (IFN-γ), as macrophages with a genetically inactivated IFN-γ signaling pathway are unable to produce NO in response to the toxin (22). Braun et al. propose that PLY has the ability to act as a general activator of macrophages (22), possibly through activation of NF-κB.

PLY can activate phospholipase A in pulmonary epithelial cells (77). The activated phospholipase has a broad substrate specificity for cellular membrane phospholipids. Activation of phospholipase during an infection could contribute directly to lung damage by the release of free fatty acids and lysophosphatides. Release of arachidonic acid by the activated phospholipase could promote chemotaxis and respiratory burst of neutrophils (7, 33). Arachidonic acid could also be metabolized through the eicosanoid pathway, leading to the production of leukotrienes and platelet-activating precursor (48). Products of the eicosanoid pathway are also major chemotaxins for neutrophils (24).

The toxin interacts with neutrophils to cause increased production of superoxide, increased production of elastase, increased expression of β_2 integrins, and increased production of prostaglandin E_2 and leukotriene B_4 (29, 30). The toxin is therefore extremely proinflammatory. Cell recruitment may also be affected by PLY, as it has been shown to stimulate the production of IL-8 from human neutrophils (28). Thus, PLY has the potential to cause increased recruitment of inflammatory cells and increased production of proinflammatory mediators from these cells. Activation of inflammatory pathways explains the ability of the toxin to induce pneumonitis similar to that induced by the whole organism (40). It seems somewhat of a paradox that the organism contributes to its own demise by producing this toxin and inducing inflammation. However, it may be that the inflammation induced is not effective in actually clearing the organism, as the toxin has also been shown to reduce killing of pneumococci by neutrophils in vitro (70).

The effects of PLY on mouse macrophages involve an interaction with Toll-like receptor 4 (TLR4) (61). Malley and coworkers (61) showed that the stimulation of macrophages to produce TNF-α and IL-6 is dependent on the cytoplasmic TLR-adaptor molecule, myeloid differentiation factor 88 (MyD88), and macrophages from mice with the targeted deletion of MyD88 did not respond to PLY. Purified PLY has a synergistic interaction with pneumococcal cell walls which mediates inflammatory action via TLR2 (91). Furthermore, macrophages from mice carrying a spontaneous mutation in TLR4 were less responsive to both PLY alone and PLY in conjunction with pneumococcal cell wall than were wild-type macrophages. Interestingly, TLR4 mutant mice were more susceptible to lethal infection after intranasal colonization with pneumococci than were control mice. These findings show that interaction of PLY with the innate immune system via TLR4 protects colonized animals from disease and suggest that inappropriate interaction with TLR4 could be responsible for some of the pathology seen in pneumococcal diseases such as pneumonia and meningitis.

The effects of PLY on gene expression have been examined using microarrays to compare the effect of wild-type pneumococci with PLY-negative (PLY⁻) organisms on THP-1 mononuclear cells. Genes found to be up-regulated in response to wild-type but not PLY⁻ organisms include mannose binding lectin 1, lysozyme, α-1 catenin, cadherin 17, caspases 4 and 6, macrophage inflammatory protein 1β (MIP-1β), interleukin 8 monocyte chemotactic protein 3, IL-2 receptor β, IL-15 receptor α, IFN receptor 2, and prostaglandin E synthase. Down-regulated genes include those for complement receptor 2/CD21, platelet-activating factor acetylhydrolase, and oxidized low-density lipoprotein receptor 1. These genes include representatives of functional genes involved in the antipathogen response, cell adhesion, cell cycle control, apoptosis, cell-cell signaling, lipid metabolism, protein degradation and folding, protein translocation, and signal transduction. PLY therefore has dramatic effects on mononuclear cell function, some of which overlap with known effects of the toxin (stimulation of IL-8 production, for example). The significance of most of these changes to the disease process remains to be determined.

THE ROLE OF PLY IN APOPTOSIS

S. pneumoniae has been reported to induce apoptosis in a number of cell types, including neutrophils, macrophages, and neuronal cells (23, 35, 83, 93). The contribution of PLY to the apoptotic process has been investigated. Zysk et al. (93) report that PLY is responsible for mainly necrosis in human neutrophils and that apoptosis is induced by heat-killed bacteria. Dockrell et al. (35) used a PLY$^-$ mutant of the pneumococcus to show that apoptosis of macrophages infected with pneumococci was reduced in the absence of PLY.

As *S. pneumoniae* is a major cause of meningitis, a crucial interaction is between the organism and the blood-brain barrier. Using endothelial cells of cerebral origin as a model of the blood-brain barrier, it has been shown that the majority of the damage inflicted on these cells is caused by PLY (94). The damage induced by the toxin was dependent on protein synthesis, tyrosine phosphorylation, and caspase activity (94).

The most detailed studies of the role of PLY in apoptosis have been done using neuronal tissues. Braun et al. (23) report that PLY can induce an apoptosis-inducing factor (AIF)-dependent form of apoptosis. Using a combination of bacterial mutants it was shown that both PLY and hydrogen peroxide are responsible for the apoptosis of human microglial cells induced by *S. pneumoniae*. Ultrastructural analysis of human microglial cells showed that blocking the activity of PLY and hydrogen peroxide reduced the mitochondrial damage caused by the pneumococcus. The effects of the whole pneumococcus on microglial cells could be reproduced by addition of purified PLY, and the effect was dependent on the pore-forming activity of the toxin. PLY induces the increase of intracellular calcium levels and triggers the release of AIF from mitochondria in primary rat neurons. Pretreatment of microglial cells with the intra- and

extracellular calcium chelator BAPTA-AM [1,2-bis (*o*-aminophenoxy) ethane *N,N,N′,N′*-tetraacetic acid tetra(acetomethoxyl)ester] blocked the release of AIF induced by PLY, suggesting that this event is mediated by increases in intracellular calcium levels. Finally, it was shown that PLY contributes to neuronal damage in a rabbit model of pneumococcal meningitis.

An effect of PLY on calcium levels has also been demonstrated in neuroblastoma cells, in which purified PLY was shown to induce apoptosis in a calcium-dependent manner (83). The changes in calcium levels were due to pore formation by the toxin rather than opening of voltage-gated Ca^{2+} channels. This study also showed that mitogen-activated protein kinase p38 was important in PLY-induced cell death.

INTERACTION OF PLY WITH ANIMAL TISSUES

The effect of PLY on a number of tissues has been evaluated. The toxin has been instilled directly into the rat lung, where it induces pathology very similar to that seen during infection with the pneumococcus (40). Use of altered versions of the toxin in this system showed that both the cell lytic and complement-activating activities contribute to the pathology in the lung. Organ culture models have also been used to determine the effect of PLY on the lung. PLY increases the alveolar permeability of perfused rat lungs and is toxic to type II rat alveolar epithelial cells (76). Perturbation of this barrier may play an important role in the development of pneumococcal disease. Changes in alveolar permeability have also been observed in whole-animal models of pneumococcal pneumonia (56).

The toxin may also serve to reduce nonspecific host defense mechanisms in the lung, as it causes slowing of the cilial beat of human nasal epithelium and can disrupt the integrity of the tissue at higher concentration (38, 39). PLY also has dramatic effects on the ciliated epithelial cells from the brains of rats. Brain cells are more sensitive to the toxin than respiratory cilia (65). Ciliated cells line the ventricular surface of the brain and cerebral aqueducts and form a barrier between the cerebrospinal fluid (CSF) and the brain tissue (3) which may protect the tissue from damage when the CSF is infected. Perturbation of these cilia by PLY may therefore play a role in pathogenesis of meningitis.

A common complication of pneumococcal meningitis is hearing loss (42), and PLY has been shown to play a role in this process. When the toxin is perfused through the scala tympani of the ear, it causes widespread histological and electrophysiological damage (31). Scanning electron microscopy of toxin-treated tissue showed widespread damage to both the inner and outer hair cells and supporting cells of the tissue. Damage to hair cells includes disruption and splaying of the sterocilia and complete dissolution of hair bundles. Use of various metabolic inhibitors to block the effects of PLY in the ear suggests that its activity is mediated by the excessive stimulation of *N*-methyl-D-aspartate receptors within the cochlea, which leads to the overproduction of nitric oxide (4). PLY may also perturb the membrane of the round window during pneumococcal otitis media and allow access of the toxin from the middle ear to the cochlea (37). The importance of PLY in hearing loss has also been confirmed in an infection model of meningitis (89).

PLY plays a role in ocular infections with *S. pneumoniae*. When purified PLY is instilled into the rabbit eye an inflammatory response is generated (51). The nature of this response is similar to that induced by infection with the pneumococcus. Rabbits made neutropenic prior to challenge showed a much reduced inflammatory response to the toxin, indicating that neutrophils may be a source of tissue-damaging enzymes (44). The role of PLY in ocular infection has been confirmed using a rat model of endophthalmitis (67). Injection of PLY intravitreally induced many of the features of endophthalmitis.

THE ROLE OF PLY IN PATHOGENESIS

The role of PLY in the pathogenesis of infection has been studied using mutants of the pneumococcus in which the gene for the toxin has been interrupted or deleted. Early experiments with PLY⁻ strains showed that lack of the toxin reduces the virulence of the organism in both intranasal and systemic routes of infection (18). When instilled into the lung, PLY⁻ pneumococci induce much less inflammation (26, 54). Cell recruitment in the lung in response to respiratory tract infections with PLY⁻ mutants is delayed and reduced in comparison to infection with wild-type bacteria (54). Neutrophil responses were the most affected, and the redistribution of T and B lymphocytes in and around inflamed bronchioles was also delayed and reduced. Whether the reduction in inflammation was due to a poorer inflammatory stimulus by the reduced bacterial loads in the lungs infected with PLY⁻ or due to direct damage by wild-type bacteria is unclear. The PLY⁻ mutant also causes bacteremia later than the wild-type organism following intranasal infection. After intranasal infection, lack of PLY also causes reduced numbers of bacteria in the nasopharynx, trachea, and lungs (55), although Rubins and coworkers show that PLY has no role in colonization (78).

The role of PLY in bacteremia appears to be to protect the pneumococcus from the infection-induced host response. In a bacteremia model, wild-type bacteria showed exponential growth until bacterial numbers as high as 10^{10} CFU/ml of blood were reached, when the animals died (9). When PLY⁻ bacteria were used, the initial growth rate of bacteria was the same until numbers of about 10^7 CFU/ml were reached, when the increase in CFU per milliliter ceased and the numbers of organisms in the blood remained constant for several days. Interestingly, if PLY⁻ bacteria were coinjected with wild-type organisms they demonstrated wild-type characteristics, suggesting the toxin mediates its effects at a distance.

In mice infected with PLY⁻ bacteria there is evidence of an inflammatory response, and a study was undertaken to determine the role of proinflammatory cytokines in the resistance to pneumococcal infection (12). This study shows that the resistance to PLY⁻ strains is dependent on TNF-α but independent of IL-1β or IL-6. If wild-type organisms are given subsequent to challenge with PLY⁻ bacteria they do not cause disease. Therefore, an inability to produce PLY is associated with chronic bacteremia and development of resistance to wild-type bacteria. This resistance is mediated by the production of TNF-α.

The virulence of PLY⁻ mutants has also been studied in models of eye infection. PLY⁻ bacteria used in model of ocular infection show greatly reduced virulence (52). A nonhemolytic strain of the pneumococcus produced by chemical mutagenesis shows the same virulence as the parent strain, suggesting that an activity other than cell lysis is important in the pathogenesis of this infection (see below).

The role of PLY in other infections is more controversial. PLY appears not to play a role in the inflammation associated with otitis media in the chinchilla model (79). Intracisternal injection of PLY into rabbit brain causes a rapid inflammatory response, but in this study there was no evidence of a contribution of the toxin to the inflammation caused by the whole organism as determined by a comparison of wild-type and PLY⁻ pneumococci (43). The lack of effect of PLY on inflammation of the brain has been confirmed in a mouse model (88). In this study, although the toxin did not contribute to inflammation of the meninges, animals infected by the intracerebral route with PLY⁻ organisms did show lower bacterial titers in the blood and increased survival times (88). Similar results with regard to inflammation were seen in a guinea pig model of infection (89). Use of wild-type and PLY⁻ organisms in this system showed there to be no difference in the levels of inflammation induced. The PLY⁻ organism does induce less protein influx into the CSF, less ultrastructural damage to the cochlea

in infected animals, and less associated hearing loss (89).

PLY has been shown to play a key role in brain damage induced during pneumococcal meningitis in the rabbit model of infection (23). Also in the rabbit model of pneumococcal meningitis the levels of PLY released have been measured and shown to be around 20 ng ml of CSF⁻¹ (82). This is similar to the level of toxin found in the CSF of patients suffering from pneumococcal meningitis, which ranges from about 1 to 180 ng ml of CSF⁻¹. The level of toxin released into the CSF during antibiotic treatment is dependent on the type of antibiotic used. Much more PLY is found in the CSF if ceftriaxone is used to treat the infection in rabbits compared to the level with a non-bacteriolytic antibody such as rifampin (82). This emphasizes the importance of bacterial lysis in the release of PLY and has important consequences when considering treatment of pneumococcal meningitis.

The two main activities of PLY are its lytic activity and its ability to activate complement, and regions of the protein responsible for these activities have been identified (21). The ability of the toxin to cause stimulation of cytokine production has been linked to its lytic activity. However, there is also biological activity associated with the toxin that is not contained within these regions. For example, neutralization of the lytic action of PLY with cholesterol does not prevent the stimulation of IFN-γ and nitric oxide production from spleen cells by the toxin, and a truncated form of PLY that lacks pore-forming activity is also still active in this assay. In order to understand how the various activities of PLY are involved in the virulence of the whole organism, a series of isogenic mutants has been constructed expressing versions of the toxin that carry various amino acid substitutions which affect the activity of the toxin (13, 15). The contribution of lytic activity and complement activation by PLY has been evaluated using strains producing PLY with reduced ability to activate complement (H⁺/C⁻), reduced lytic activity (H⁻/C⁺), or reduced activities in both

(H^-/C^-). An isogenic mutant of the pneumococcus in which both lytic activity and the ability of the toxin to activate complement were removed (H^-/C^-) is still more virulent than a toxin-negative mutant, showing that PLY has an additional property that is not abolished by point mutations which reduce cytotoxic activity and complement activation to virtually undetectable levels (15).

In intraperitoneal infection of mice there is no contribution of complement activation to virulence, as the H^+/C^- strain was no different from the wild type in this model. However, strains expressing toxin with reduced lytic activity (H^-/C^+) had reduced virulence. The contribution of PLY to virulence in this model is therefore dependent on cytolytic properties of the toxin.

Examination of the role of the toxin in intravenous challenge (11) shows that elimination of either activity of PLY alone does not affect the virulence of the strain and only when the mutations in the two regions are combined is there reduced virulence, suggesting that PLY has another activity other than lytic or complement activity that can contribute to virulence in this model. Alcantara and coworkers studied the effects of pneumococcal bacteremia on complement levels in normal and cirrhotic rats (which have lower levels of complement) (1). The complement-depleting activity of PLY had only minimal effect on host defense in animals with normal complement levels but was detrimental in cirrhotic animals, being associated with impaired phagocytosis and bloodstream clearance. This may explain why cirrhotic patients are at increased risk of pneumococcal bacteremia.

In bronchopneumonia following intranasal infection with pneumococci both lytic and complement activity are important, as isogenic mutants with reductions in these activities have reduced virulence (2). Both activities of PLY contribute to the pathology in the lung, as well as the timing of the onset of bacteremia. Histological changes in the lungs are delayed after infection with either H^+/C^- or H^-/C^+ strains compared to the wild type. The effect of both activities presumably reflects the fact that complement activation is linked to the accumulation of T cells in the lung, whereas lytic activity is involved in neutrophil recruitment (53). These findings are similar to those seen in lobar pneumonia (75), where the absence of either complement activation or lytic activity of PLY renders isogenic mutants less virulent. However, in contrast to the case for bronchopneumonia, these two mutations were not additive in their effect. The lack of an additive effect in lobar pneumonia is proposed to show that the lytic activity of the toxin is involved in several steps during pathogenesis, whereas complement activation has a more limited role in reducing bacterial clearance in the lung. The lytic activity of the toxin is associated with acute lung injury and bacterial growth for up to 6 h after inoculation, whereas the complement-activating activity of the toxin is needed for bacterial growth and bacteremia at 24 h postinoculation. Complement activation by PLY in lobar pneumonia plays a single role in pulmonary infection by promoting survival of pneumococci in the lung tissues and facilitating invasion of the blood (75).

In summary, PLY plays several roles in infection. The toxin appears to have no role in the inflammation associated with meningitis (43, 88, 89) but has a role in deafness associated with meningitis (88) and in bacteremia (9) and pneumonia (18). Where PLY does have a role, the contribution of complement activation and the lytic activities of the toxin to the disease process differ according to the type of infection. A comparison of strains with point mutations that abolish the complement activation and lytic activities of PLY with a toxin-negative mutant suggests that PLY has other activities that contribute to the disease process. These activities await further characterization.

OTHER VIRULENCE FACTORS

LPXTG-Anchored Surface Proteins

Analysis of the pneumococcal genome sequence indicates the presence of some 17 putative LPXTG-anchored proteins, although this may vary among strains (85). The LPXTG motif is normally near the C terminus of the protein but in some of the pneumococcal proteins is near the N terminus. The significance of this is unclear. The C-terminal consensus is a recognition sequence for sortase enzymes, which recognize LPXTG and anchor the protein to the cell surface by covalent linkage of the threonine of the motif to the pentaglycine linkage of the peptidoglycan of the cell wall. The role of several of these LPXTG proteins in virulence has been investigated. These include hyaluronidase, neuraminidase, and serine protease PrtA.

Hyaluronidase breaks down the hyaluronic acid component of mammalian connective tissue and extracellular matrix and is secreted by 99% of clinical isolates (49). The degradation of hyaluronic acid may aid bacterial spread and colonization, as demonstrated for other microorganisms (41). In addition, hyaluronidase may potentiate pulmonary inflammation during pneumococcal pneumonia by complex interactions with proinflammatory cytokines and chemokines. TNF-α and IL-1β are able to induce the production of hyaluronic acid by fibroblasts (50). Hyaluronic acid can promote further cytokine secretion by binding to CD44 on host cells. The system is further complicated by the ability of IL-1 to release host hyaluronidase.

Breakdown products of hyaluronic acid stimulate chemokine expression by macrophages (63), which in vivo would induce cell recruitment and potentiate inflammation. Thus the expression of hyaluronidase by *S. pneumoniae* during pneumococcal pneumonia might disrupt normal hyaluronic acid metabolism, leading to direct tissue disruption, up-regulation of proinflammatory cytokines, and cell recruitment. Hyaluronidase has also been shown to promote the ability of pneumococcus to cause meningitis when instilled along with bacteria into the nasopharynx in mouse models (59, 92). However, a hyaluronidase-negative mutant does not show reduced virulence in a mouse model unless the gene for PLY is also deleted (16).

Neuraminidase cleaves *N*-acetylneuraminic acid from glycolipids, lipoproteins, and oligosaccharides on cell surfaces and in body fluids (25). This may cause direct damage to the host, or it may unmask potential binding sites for the organism. Loss of sialic acid due to neuraminidase activity accompanies the advance of pneumococci up the eustachian tube to the middle ear (60). Pneumococci have the genes to make several enzymes with neuraminidase activity, and the role of neuraminidase A in pathogenesis has been investigated. It plays a role in colonization and development of otitis media in the chinchilla model (86) but does not play a role in meningitis-associated deafness (89).

PrtA is a serine protease that is highly conserved in clinical isolates. Genetic knockouts of this gene cause decreased virulence of the pneumococcus in a mouse intraperitoneal infection model (19) by a mechanism that is as yet unclear.

LPXTG proteins are anchored to the cell surface by the activity of a sortase enzyme. The role of sortase in virulence has been investigated. Deletion of the SrtA sortase gene caused neuraminidase to be secreted in the growth medium rather than anchored to the cell surface, confirming the role of this enzyme in anchoring LPXTG proteins to the cell (57). This mutant was not less virulent when injected into mice but showed less adherence to tissue culture cells. Another sortase enzyme (SrtD) was identified as a virulence factor in the pneumonia model during a signature-tagged mutagenesis screen for virulence factors of *S. pneumoniae* (45).

Lipoproteins

In a screen for pneumococcal adhesins Cundell and coworkers (32) identified mutations in genes identified as peptide permeases that reduce adherence to resting lung or endothelial cells. Although the role of these lipoproteins in infection has not yet been evaluated, a homolog of these permeases, pneumococcal surface antigen A (PsaA), has been assigned a role in pneumococcal virulence (84). PsaA is part of an ABC transporter in which *psaA* is the substrate-binding lipoprotein, *psaB* is the ATP-binding protein, and *psaC* is the permease (34). The complex transports manganese ions. Mutations in *psaA* cause decreased adhesion to cells, decreased virulence, and increased sensitivity to oxidative stress (17, 62, 69). It is assumed that PsaA itself is not an adhesin but regulates the expression or assembly of other molecules needed for adhesion.

Hydrogen Peroxide

Hydrogen peroxide production plays an important role in the virulence of *S. pneumoniae*. The amounts of hydrogen peroxide produced are similar to those produced by polymorphonuclear cells during respiratory burst (72). Toxicity has been demonstrated in rat epithelial cells (36). Peroxide is produced by the action of pyruvate oxidase, which is important in the virulence of the pneumococcus (81). Hydrogen peroxide can injure respiratory epithelium (38) and rat brain (46, 47) and plays a crucial role in the induction of brain injury during meningitis (23).

SOD

The pneumococcus contains two types of superoxide dismutase (SOD), MnSOD and FeSOD (90). Inactivation of *sodA*, the gene for MnSOD, shows that MnSOD plays a role in the virulence of the pneumococcus in animal models (90). The pattern of inflammation in the lungs infected with the knockout mutant is different from that seen with wild-type organisms. After infection with the mutant, neutrophils are packed around the bronchioles, in contrast to the wild-type infection, in which neutrophils are more diffusely localized.

NADH Oxidase

A gene termed *nox* has been identified that encodes a soluble flavoprotein that reoxidizes NADH and reduces molecular oxygen to water (5). Disruption of this gene shows that Nox is involved in genetic competence and virulence. The virulence and persistence in mice of a blood isolate of *S. pneumoniae* are attenuated by a *nox* insertion mutation. It has been proposed that the enzymatic activity of the NADH oxidase is probably involved in transducing a signal related to oxygen availability into the cell and so behaves as an oxygen sensor. Oxygen may therefore play a role in the regulation of pneumococcal transformability and virulence by a mechanism involving Nox (5).

Zinc Metalloproteases

The pneumococcal genome contains three putative zinc metalloprotease genes, encoding the ZmpB, ZmpC, and immunoglobulin A (IgA) proteases (68). The three proteins contribute differently to virulence in mice (27), with ZmpB and IgA playing a major role and ZmpC playing a minor role (20, 27, 45). A substrate for ZmpC is human matrix metalloproteinase 9 (MMP-9), which is known to be involved in a variety of physiological and pathological matrix-degrading processes, including tissue invasion of metastases and opening of the blood-brain barrier. ZmpC may therefore activate the MMP-9 proprotease to give active MMP-9 which is involved in the disease process. ZmpB null mutations of the pneumococcus are not attenuated in the ability to colonize the lung during pneumonia but are attenuated in the ability to cause invasive disease (20). ZmpB is associated with the ability of the pneumococcus to induce production of TNF-α in the lungs (20).

CONCLUDING REMARKS

The role of PLY in the virulence of *S. pneumoniae* is now well established, and this molecule is being considered as a potential vaccine component. The availability of genome sequences of the several pneumococcal strains and genome-wide mutagenesis using signature-tagged mutagenesis has identified many other potential virulence factors. The detailed role of many of these factors in the disease process is still to be worked out, but it is clear that the roles they play and their contributions differ among strains. As virulence is a multifactorial process, we need to understand not only the mechanism of these factors but also how they are combined to form the overall virulence "type" of the organism. The overall virulence gene content of the pneumococcus will presumably dictate its ability to cause disease. Whether there is a set of genes that allow the pneumococcus to cause the various types of invasive infection remains open to question and highlights the need to study complete genotypes rather than individual virulence factors in order to understand the disease-causing potential of a particular strain of pneumococcus.

REFERENCES

1. **Alcantara, R. B., L. C. Preheim, and M. J. Gentry-Nielsen.** 2001. Pneumolysin-induced complement depletion during experimental pneumococcal bacteremia. *Infect. Immun.* **69:** 3569–3575.

2. **Alexander, J. E., A. M. Berry, J. C. Paton, J. B. Rubins, P. W. Andrew, and T. J. Mitchell.** 1998. The course of pneumococcal pneumonia is altered by amino acid changes affecting the activity of pneumolysin. *Microb. Pathog.* **24:**167–174.

3. **Alfzelius, B. A.** 1979. The immotile-cilia syndrome and other ciliary diseases. *Int. Rev. Exp. Pathol.* **19:**1–43.

4. **Amee, F. Z., S. D. Comis, and M. P. Osbourne.** 1995. N^{G}-methyl-L-arginine protects the guinea pig cochlea from the cytotoxic effects of pneumolysin. *Acta Oto-Laryngol.* **115:**386–391.

5. **Auzat, I., S. Chapuy-Regaud, G. Le Bras, D. Dos Santos, A. D. Ogunniyi, I. Le Thomas, J.-R. Garel, J. C. Paton, and M.-C. Trombe.** 1999. The NADH oxidase of Streptococcus pneumoniae: its involvement in competence and virulence. *Mol. Microbiol.* **34:**1018–1028.

6. **Baba, H., I. Kawamura, C. Kohda, T. Nomura, Y. Ito, T. Kimoto, I. Watanabe, S. Ichiyama, and M. Mitsuyama.** 2001. Essential role of domain 4 of pneumolysin from Streptococcus pneumoniae in cytolytic activity as determined by truncated proteins. *Biochem. Biophys. Res. Commun.* **281:**37–44.

7. **Badwey, J. A., J. T. Curnutte, J. M. Robinson, C. B. Berde, M. J. Karnovsky, and M. L. Karnovsky.** 1984. Effects of free fatty acids on release of superoxide and on change of shape by human neutrophils. Reversibility by albumin. *J. Biol. Chem.* **259:**7870–7877.

8. **Balachandran, P., S. K. Hollingshead, J. C. Paton, and D. E. Briles.** 2001. The autolytic enzyme LytA of Streptococcus pneumoniae is not responsible for releasing pneumolysin. *J. Bacteriol.* **183:**3108–3116.

9. **Benton, K., M. Everson, and D. Briles.** 1995. A pneumolysin-negative mutant of *Streptococcus pneumoniae* causes chronic bacteremia rather than acute sepsis in mice. *Infect. Immun.* **63:**448–455.

10. **Benton, K., J. Paton, and D. Briles.** 1997. Differences in virulence for mice among *Streptococcus pneumoniae* strains of capsular types 2, 3, 4, 5, and 6 are not attributable to differences in pneumolysin production. *Infect. Immun.* **65:**1237–1244.

11. **Benton, K. A., J. C. Paton, and D. E. Briles.** 1997. The hemolytic and complement-activating properties of pneumolysin do not contribute individually to virulence in a pneumococcal bacteremia model. *Microb. Pathog.* **23:**201–209.

12. **Benton, K. A., J. L. VanCott, and D. E. Briles.** 1998. Role of tumor necrosis factor alpha in the host response of mice to bacteremia caused by pneumolysin-deficient *Streptococcus pneumoniae*. *Infect. Immun.* **66:**839–842.

13. **Berry, A., J. Alexander, T. Mitchell, P. Andrew, D. Hansman, and J. Paton.** 1995. Effect of defined point mutations in the pneumolysin gene on the virulence of *Streptococcus pneumoniae*. *Infect. Immun.* **63:**1969–1974.

14. **Berry, A. M., R. A. Lock, D. Hansman, and J. C. Paton.** 1989. Contribution of autolysin to virulence of *Streptococcus pneumoniae*. *Infect. Immun.* **57:**2324–2330.

15. **Berry, A. M., A. D. Ogunniyi, D. C. Miller, and J. C. Paton.** 1999. Comparative virulence of *Streptococcus pneumoniae* strains with insertion-duplication, point, and deletion mutations in the pneumolysin gene. *Infect. Immun.* **67:**981–985.

16. **Berry, A. M., and J. C. Paton.** 2000. Additive attenuation of virulence of *Streptococcus pneumoniae* by mutation of the genes encoding pneumolysin and other putative pneumococcal virulence proteins. *Infect. Immun.* **68:**133–140.

17. **Berry, A. M., and J. C. Paton.** 1996. Sequence heterogeneity of PsaA, a 37-kilodalton putative adhesin essential for virulence of *Streptococcus pneumoniae*. *Infect. Immun.* **64:**5255–5262.

18. **Berry, A. M., J. Yother, D. E. Briles, D. Hansman, and J. C. Paton.** 1989. Reduced virulence of a defined pneumolysin-negative mutant of *Streptococcus pneumoniae*. *Infect. Immun.* **57:**2037–2042.

19. **Bethe, G., R. Nau, A. Wellmer, R. Hakenbeck, R. R. Reinert, H.-P. Heinz, and G. Zysk.** 2001. The cell wall-associated serine protease PrtA: a highly conserved virulence factor of

Streptococcus pneumoniae. *FEMS Microbiol. Lett.* **205:**99–103.

20. **Blue, C. E., G. K. Paterson, A. R. Kerr, M. Berge, J. P. Claverys, and T. J. Mitchell.** 2003. ZmpB, a novel virulence factor of *Streptococcus pneumoniae* that induces tumor necrosis factor alpha production in the respiratory tract. *Infect. Immun.* **71:**4925–4935.

21. **Boulnois, G. J., J. C. Paton, T. J. Mitchell, and P. W. Andrew.** 1991. Structure and function of pneumolysin, the multifunctional, thiol-activated toxin of *Streptococcus pneumoniae*. *Mol. Microbiol.* **5:**2611–2616.

22. **Braun, J. S., R. Novak, G. Gao, P. J. Murray, and J. L. Shenep.** 1999. Pneumolysin, a protein toxin of *Streptococcus pneumoniae*, induces nitric oxide production from macrophages. *Infect. Immun.* **67:**3750–3756.

23. **Braun, J. S., J. E. Sublett, D. Freyer, T. J. Mitchell, J. L. Cleveland, E. I. Tuomanen, and J. R. Weber.** 2002. Pneumococcal pneumolysin and H_2O_2 mediate brain cell apoptosis during meningitis. *J. Clin. Investig.* **109:**19–27.

24. **Cabellos, C., D. E. MacIntyre, M. Forrest, M. Burroughs, S. Prasad, and E. Tuomanen.** 1992. Differing roles for platelet-activating factor during inflammation of the lung and sub-arachnoid space: the special case of *Streptococcus pneumoniae*. *J. Clin. Investig.* **90:**612–618.

25. **Camara, M., G. J. Boulnois, P. W. Andrew, and T. J. Mitchell.** 1994. A neuraminidase from *Streptococcus pneumoniae* has the features of a surface protein. *Infect. Immun.* **62:**3688–3695.

26. **Canvin, J. R., A. P. Marvin, M. Sivakumaran, J. C. Paton, G. J. Boulnois, P. W. Andrew, and T. J. Mitchell.** 1995. The role of pneumolysin and autolysin in the pathology of pneumonia and septicemia in mice infected with a type-2 pneumococcus. *J. Infect. Dis.* **172:**119–123.

27. **Chiavolini, D., G. Memmi, T. Maggi, F. Iannelli, G. Pozzi, and M. Oggioni.** 2003. The three extra-cellular zinc metalloproteinases of Streptococcus pneumoniae have a different impact on virulence in mice. *BMC Microbiol.* **3:**14.

28. **Cockeran, R., C. Durandt, C. Feldman, T. J. Mitchell, and R. Anderson.** 2002. Pneumolysin activates the synthesis and release of interleukin-8 by human neutrophils in vitro. *J. Infect. Dis.* **186:**562–565.

29. **Cockeran, R., H. C. Steel, T. J. Mitchell, C. Feldman, and R. Anderson.** 2001. Pneumolysin potentiates production of prostaglandin E_2 and leukotriene B_4 by human neutrophils. *Infect. Immun.* **69:**3494–3496.

30. **Cockeran, R., A. J. Theron, H. C. Steel, N. M. Matlola, T. J. Mitchell, C. Feldman, and R. Anderson.** 2001. Proinflammatory interactions of pneumolysin with human neutrophils. *J. Infect. Dis.* **183:**604–611.

31. **Comis, S. D., M. P. Osborne, J. Stephen, M. J. Tarlow, T. L. Hayward, T. J. Mitchell, P. A. Andrew, and G. J. Boulnois.** 1993. Cytotoxic effect on hair cells of the guinea pig cochlea produced by pneumolysin, the thiol activated toxin of *Streptococcus pneumoniae*. *Acta Oto-Laryngol.* **113:**152–159.

32. **Cundell, D. R., B. J. Pearce, J. Sandros, A. M. Naughton, and H. R. Masure.** 1995. Peptide permeases from *Streptococcus pneumoniae* affect adherence to eucaryotic cells. *Infect. Immun.* **63:**2493–2498.

33. **Curnutte, J. T., J. M. Badwey, J. M. Robinson, M. J. Karnovsky, and M. L. Karnovsky.** 1984. Studies on the mechanism of superoxide release from human neutrophils stimulated with arachidonate. *J. Biol. Chem.* **259:**11851–11857.

34. **Dintilhac, A., G. Alloing, C. Granadel, and J. P. Claverys.** 1997. Competence and virulence of Streptococcus pneumoniae: Adc and PsaA mutants exhibit a requirement for Zn and Mn resulting from inactivation of putative ABC metal permeases. *Mol. Microbiol.* **25:**727–739.

35. **Dockrell, D. H., M. Lee, D. H. Lynch, and R. C. Read.** 2001. Immune-mediated phagocytosis and killing of Streptococcus pneumoniae are associated with direct and bystander macrophage apoptosis. *J. Infect. Dis.* **184:**713–722.

36. **Duane, P. G., J. B. Rubins, H. R. Weisel, and E. N. Janoff.** 1993. Identification of hydrogen peroxide as a *Streptococcus pneumoniae* toxin for rat alveolar epithelial cells. *Infect. Immun.* **61:**4392–4397.

37. **Engel, F., R. Blatz, J. Kellner, M. Palmer, U. Weller, and S. Bhakdi.** 1995. Breakdown of the round window permeability barrier evoked by streptolysin O: possible etiologic role in development of sensorineural hearing loss in acute otitis media. *Infect. Immun.* **63:**1305–1310.

38. **Feldman, C., R. Anderson, R. Cockeran, T. Mitchell, P. Cole, and R. Wilson.** 2002. The effects of pneumolysin and hydrogen peroxide, alone and in combination, on human ciliated epithelium in vitro. *Respir. Med.* **96:**580–585.

39. **Feldman, C., T. J. Mitchell, P. W. Andrew, G. J. Boulnois, R. C. Read, H. C. Todd, P. J. Cole, and R. Wilson.** 1990. The effect of Streptococcus pneumoniae pneumolysin on human respiratory epithelium in vitro. *Microb. Pathog.* **9:**275–284.

40. **Feldman, C., N. C. Munro, P. K. Jeffery, T. J. Mitchell, P. W. Andrew, G. J. Boulnois, D. Guerreiro, J. A. L. Rohde, H. C. Todd, P. J. Cole, and R. Wilson.** 1991. Pneumolysin induces the salient histologic features of pneumo-

coccal infection in the rat lung *in vivo. Am. J. Respir. Cell Mol. Biol.* **5:**416–423.

41. Fitzgerald, T. J., and L. A. Repesh. 1987. The hyaluronidase associated with *Treponema pallidum* facilitates treponemal dissemination. *Infect. Immun.* **55:**1023–1028.

42. Fortnum, H. M. 1992. Hearing impairment after bacterial meningitis: a review. *Arch. Dis. Child.* **67:**1128–1133.

43. Friedland, I. R., M. M. Paris, S. Hickey, S. Shelton, K. Olsen, J. C. Paton, and G. H. McCracken. 1995. The limited role of pneumolysin in the pathogenesis of pneumococcal meningitis. *J. Infect. Dis.* **172:**805–809.

44. Harrison, J. C., Z. A. Karcioglu, and M. K. Johnson. 1982–1983. Response of leukopenic rabbits to pneumococcal toxin. *Curr. Eye Res.* **2:**705–710.

45. Hava, D., and A. Camilli. 2002. Large-scale identification of serotype 4 Streptococcus pneumoniae virulence factors. *Mol. Microbiol.* **45:**1389–1406.

46. Hirst, R. A., A. Rutman, K. Sikand, P. W. Andrew, T. J. Mitchell, and C. Ocallaghan. 2000. Effect of pneumolysin on rat brain ciliary function: comparison of brain slices with cultured ependymal cells. *Pediatr. Res.* **47:**381–384.

47. Hirst, R. A., K. S. Sikand, A. Rutman, T. J. Mitchell, P. W. Andrew, and C. Ocallaghan. 2000. Relative roles of pneumolysin and hydrogen peroxide from *Streptococcus pneumoniae* in inhibition of ependymal ciliary beat frequency. *Infect. Immun.* **68:**1557–1562.

48. Holtzman, M. J. 1991. Arachidonic acid metabolism. Implications of biological chemistry for lung function and disease. *Am. Rev. Respir. Dis.* **143:**188–203.

49. Humphrey, J. H. 1948. Hyaluronidase production by pneumococci. *J. Pathol. Bacteriol.* **55:**273–275.

50. Irwin, C. R., S. L. Schor, and M. W. Ferguson. 1994. Effects of cytokines on gingival fibroblasts in vitro are modulated by the extracellular matrix. *J. Periodontal Res.* **29:**309–317.

51. Johnson, M. K., and J. H. Allen. 1975. The role of cytolysin in pneumococcal ocular infection. *Am. J. Ophthalmol.* **80:**518–520.

52. Johnson, M. K., J. A. Hobden, M. Hagenah, R. J. OCallaghan, J. M. Hill, and S. Chen. 1990. The role of pneumolysin in ocular infections with Streptococcus pneumoniae. *Curr. Eye Res.* **9:**1107–1114.

53. Jounblat, R., A. Kadioglu, T. J. Mitchell, and P. W. Andrew. 2003. Pneumococcal behavior and host responses during bronchopneumonia are affected differently by the cytolytic and complement-activating activities of pneumolysin.

Infect. Immun. **71:**1813–1819.

54. Kadioglu, A., N. A. Gingles, K. Grattan, A. Kerr, T. J. Mitchell, and P. W. Andrew. 2000. Host cellular immune response to pneumococcal lung infection in mice. *Infect. Immun.* **68:**492–501.

55. Kadioglu, A., S. Taylor, F. Iannelli, G. Pozzi, T. J. Mitchell, and P. W. Andrew. 2002. Upper and lower respiratory tract infection by *Streptococcus pneumoniae* is affected by pneumolysin deficiency and differences in capsule type. *Infect. Immun.* **70:**2886–2890.

56. Kerr, A. R., J. J. Irvine, J. J. Search, N. A. Gingles, A. Kadioglu, P. W. Andrew, W. L. McPheat, C. G. Booth, and T. J. Mitchell. 2002. Role of inflammatory mediators in resistance and susceptibility to pneumococcal infection. *Infect. Immun.* **70:**1547–1557.

57. Kharat, A. S., and A. Tomasz. 2003. Inactivation of the *srtA* gene affects localization of surface proteins and decreases adhesion of *Streptococcus pneumoniae* to human pharyngeal cells in vitro. *Infect. Immun.* **71:**2758–2765.

58. Korchev, Y. E., C. L. Bashford, C. Pederzolli, C. A. Pasternak, P. J. Morgan, P. W. Andrew, and T. J. Mitchell. 1998. A conserved tryptophan in pneumolysin is a determinant of the characteristics of channels formed by pneumolysin in cells and planar lipid bilayers. *Biochem. J.* **329:**571–577.

59. Kostyukova, N. N., M. O. Volkova, V. V. Ivanova, and A. S. Kvetnaya. 1995. A study of pathogenic factors of Streptococcus pneumoniae strains causing meningitis. *FEMS Immunol. Med. Microbiol.* **10:**133–137.

60. Linder, T. E., R. L. Daniels, D. J. Lim, and T. F. DeMaria. 1994. Effect of intranasal inoculation of *Streptococcus pneumoniae* on the structure of the surface carbohydrates of the chinchilla eustachian tube and middle ear mucosa. *Microb. Pathog.* **16:**435–441.

61. Malley, R., P. Henneke, S. C. Morse, M. J. Cieslewicz, M. Lipsitch, C. M. Thompson, E. Kurt-Jones, J. C. Paton, M. R. Wessels, and D. T. Golenbock. 2003. Recognition of pneumolysin by Toll-like receptor 4 confers resistance to pneumococcal infection. *Proc. Natl. Acad. Sci. USA* **100:**1966–1971.

62. Marra, A., S. Lawson, J. S. Asundi, D. Brigham, and A. E. Hromockyj. 2002. In vivo characterization of the psa genes from Streptococcus pneumoniae in multiple models of infection. *Microbiology* **148:**1483–1491.

63. McKee, C. M., M. B. Penno, M. Cowman, M. D. Burdick, R. M. Strieter, C. Bao, and P. W. Noble. 1996. Hyaluronan (HA) fragments induce chemokine gene-expression in alveolar

macrophages—the role of HA size and CD44. *J. Clin. Investig.* **98:**2403–2413.

64. **Mitchell, T. J., P. W. Andrew, F. K. Saunders, A. N. Smith, and G. J. Boulnois.** 1991. Complement activation and antibody binding by pneumolysin via a region of the toxin homologous to a human acute-phase protein. *Mol. Microbiol.* **5:**1883–1888.

65. **Mohammed, B. J., T. J. Mitchell, P. W. Andrew, R. A. Hirst, and C. Ocallaghan.** 1999. The effect of the pneumococcal toxin, pneumolysin on brain ependymal cilia. *Microb. Pathog.* **27:**303–309.

66. **Morgan, P. J., S. C. Hyman, A. J. Rowe, T. J. Mitchell, P. W. Andrew, and H. R. Saibil.** 1995. Subunit organisation and symmetry of pore-forming oligomeric pneumolysin. *FEBS Lett.* **371:**77–80.

67. **Ng, E. W. M., N. Samiy, J. B. Rubins, F. V. Cousins, K. L. Ruoff, A. S. Baker, and D. J. D'Amico.** 1997. Implication of pneumolysin as a virulence factor in *Streptococcus pneumoniae* endophthalmitis. *Retina* **17:**521–529.

68. **Oggioni, M. R., G. Memmi, T. Maggi, D. Chiavolini, F. Iannelli, and G. Pozzi.** 2003. Pneumococcal zinc metalloproteinase ZmpC cleaves human matrix metalloproteinase 9 and is a virulence factor in experimental pneumonia. *Mol. Microbiol.* **49:**795–805.

69. **Ogunniyi, A. D., R. L. Folland, D. E. Briles, S. K. Hollingshead, and J. C. Paton.** 2000. Immunization of mice with combinations of pneumococcal virulence proteins elicits enhanced protection against challenge with *Streptococcus pneumoniae.* *Infect. Immun.* **68:**3028–3033.

70. **Paton, J. C., and A. Ferrante.** 1983. Inhibition of human polymorphonuclear leukocyte respiratory burst, bactericidal activity, and migration by pneumolysin. *Infect. Immun.* **41:**1212–1216.

71. **Paton, J. C., B. Rowan-Kelly, and A. Ferrante.** 1984. Activation of human complement by the pneumococcal toxin pneumolysin. *Infect. Immun.* **43:**1085–1087.

72. **Pericone, C. D., K. Overweg, P. W. M. Hermans, and J. N. Weiser.** 2000. Inhibitory and bactericidal effects of hydrogen peroxide production by *Streptococcus pneumoniae* on other inhabitants of the upper respiratory tract. *Infect. Immun.* **68:**3990–3997.

73. **Rossjohn, J., S. C. Feil, W. J. McKinstry, R. K. Tweten, and M. W. Parker.** 1997. Structure of a cholesterol-binding, thiol-activated cytolysin and a model of its membrane form. *Cell* **89:**685–692.

74. **Rossjohn, J., R. J. C. Gilbert, D. Crane, P. J. Morgan, T. J. Mitchell, A. J. Rowe, P. W. Andrew, J. C. Paton, R. K. Tweten, and M.** W. Parker. 1998. The molecular mechanism of pneumolysin, a virulence factor from *Streptococcus pneumoniae.* *J. Mol. Biol.* **284:**449–461.

75. **Rubins, J. B., D. Charboneau, C. Fasching, A. M. Berry, J. C. Paton, J. E. Alexander, P. W. Andrew, T. J. Mitchell, and E. N. Janoff.** 1996. Distinct role for pneumolysin's cytotoxic and complement activities in the pathogenesis of pneumococcal pneumonia. *Am. J. Respir. Crit. Care Med.* **153:**1339–1346.

76. **Rubins, J. B., P. G. Duane, D. Clawson, D. Charboneau, J. Young, and D. E. Niewoehner.** 1993. Toxicity of pneumolysin to pulmonary alveolar epithelial cells. *Infect. Immun.* **61:**1352–1358.

77. **Rubins, J. B., T. J. Mitchell, P. W. Andrew, and D. E. Niewoehner.** 1994. Pneumolysin activates phospholipase A in pulmonary artery endothelial cells. *Infect. Immun.* **62:**3829–3836.

78. **Rubins, J. B., A. H. Paddock, D. Charboneau, A. M. Berry, J. C. Paton, and E. N. Janoff.** 1998. Pneumolysin in pneumococcal adherence and colonization. *Microb. Pathog.* **25:** 337–342.

79. **Sato, K., M. Quartey, C. Liebeler, C. Le, and G. Giebink.** 1996. Roles of autolysin and pneumolysin in middle ear inflammation caused by a type 3 *Streptococcus pneumoniae* strain in the chinchilla otitis media model. *Infect. Immun.* **64:**1140–1145.

80. **Saunders, F. K., T. J. Mitchell, J. A. Walker, P. W. Andrew, and G. J. Boulnois.** 1989. Pneumolysin, the thiol-activated toxin of *Streptococcus pneumoniae,* does not require a thiol group for in vitro activity. *Infect. Immun.* **57:**2547–2552.

81. **Spellerberg, B., D. R. Cundell, J. Sandros, B. J. Pearce, I. Idanpaan-Heikkila, C. Rosenow, and H. R. Masure.** 1996. Pyruvate oxidase, as a determinant of virulence in *Streptococcus pneumoniae.* *Mol. Microbiol.* **19:**803–813.

82. **Spreer, A., H. Kerstan, T. Bottcher, J. Gerber, A. Siemer, G. Zysk, T. J. Mitchell, H. Eiffert, and R. Nau.** 2003. Reduced release of pneumolysin by *Streptococcus pneumoniae* in vitro and in vivo after treatment with nonbacteriolytic antibiotics in comparison to ceftriaxone. *Antimicrob. Agents Chemother.* **47:**2649–2654.

83. **Stringaris, A. K., J. Geisenhainer, F. Bergmann, C. Balshusemann, U. Lee, G. Zysk, T. J. Mitchell, B. U. Keller, U. Kuhnt, and J. Gerber.** 2002. Neurotoxicity of pneumolysin, a major pneumococcal virulence factor, involves calcium influx and depends on activation of p38 mitogen-activated protein kinase. *Neurobiol. Dis.* **11:**355–368.

84. **Talkington, D. F., B. G. Brown, J. A. Tharpe, A. Koenig, and H. Russell.** 1996. Pro-

tection of mice against fatal pneumococcal challenge by immunization with pneumococcal surface adhesin A (PsaA). *Microb. Pathog.* 21:17–22.

85. Tettelin, H., K. E. Nelson, I. T. Paulsen, J. A. Eisen, T. D. Read, S. Peterson, J. Heidelberg, R. T. DeBoy, D. H. Haft, R. J. Dodson, A. S. Durkin, M. Gwinn, J. F. Kolonay, W. C. Nelson, J. D. Peterson, L. A. Umayam, O. White, S. L. Salzberg, M. R. Lewis, D. Radune, E. Holtzapple, H. Khouri, A. M. Wolf, T. R. Utterback, C. L. Hansen, L. A. McDonald, T. V. Feldblyum, S. Angiuoli, T. Dickinson, E. K. Hickey, I. E. Holt, B. J. Loftus, F. Yang, H. O. Smith, J. C. Venter, B. A. Dougherty, D. A. Morrison, S. K. Hollingshead, and C. M. Fraser. 2001. Complete genome sequence of a virulent isolate of Streptococcus pneumoniae. *Science* 293:498–506.

86. Tong, H. H., L. E. Blue, M. A. James, and T. F. DeMaria. 2000. Evaluation of the virulence of a *Streptococcus pneumoniae* neuraminidase-deficient mutant in nasopharyngeal colonization and development of otitis media in the chinchilla model. *Infect. Immun.* 68:921–924.

87. Watson, D. A., and D. M. Musher. 1990. Interuption of capsule production in *Streptococcus pneumoniae* type serotype 3 by insertion of Tn916. *Infect. Immun.* 58:3135–3138.

88. Wellmer, A., G. Zysk, J. Gerber, T. Kunst, M. von Mering, S. Bunkowski, H. Eiffert, and R. Nau. 2002. Decreased virulence of a pneumolysin-deficient strain of *Streptococcus pneumoniae* in murine meningitis. *Infect. Immun.* 70:6504–6508.

89. Winter, A. J., S. D. Comis, M. P. Osborne, M. J. Tarlow, J. Stephen, P. W. Andrew, J. Hill, and T. J. Mitchell. 1997. A role for pneumolysin but not neuraminidase in the hearing loss and cochlear damage induced by experimental pneumococcal meningitis in guinea pigs. *Infect. Immun.* 65:4411–4418.

90. Yesilkaya, H., A. Kadioglu, N. Gingles, J. E. Alexander, T. J. Mitchell, and P. W. Andrew. 2000. Role of manganese-containing superoxide dismutase in oxidative stress and virulence of *Streptococcus pneumoniae. Infect. Immun.* 68:2819–2826.

91. Yoshimura, A., E. Lien, R. R. Ingalls, E. Tuomanen, R. Dziarski, and D. Golenbock. 1999. Cutting edge: recognition of gram-positive bacterial cell wall components by the innate immune system occurs via toll-like receptor 2. *J. Immunol.* 163:1–5.

92. Zwijnenburg, P. J. G., T. van der Poll, S. Florquin, S. Akira, K. Takeda, J. J. Roord, and A. M. van Furth. 2003. Interleukin-18 gene-deficient mice show enhanced defense and reduced inflammation during pneumococcal meningitis. *J. Neuroimmunol.* 138:31–37.

93. Zysk, G., L. Bejo, B. K. Schneider-Wald, R. Nau, and H.-P. Heinz. 2000. Induction of necrosis and apoptosis of neutrophil granulocytes by Streptococcus pneumoniae. *Clin. Exp. Immunol.* 122:61–66.

94. Zysk, G., B. K. Schneider-Wald, J. H. Hwang, L. Bejo, K. S. Kim, T. J. Mitchell, R. Hakenbeck, and H.-P. Heinz. 2001. Pneumolysin is the main inducer of cytotoxicity to brain microvascular endothelial cells caused by *Streptococcus pneumoniae. Infect. Immun.* 69:845–852.

CELL WALL HYDROLASES

Rubens López, Ernesto García, Pedro García, and José Luis García

6

INTRODUCTION

Bacterial murein hydrolases (MHs) are enzymes that specifically cleave covalent bonds of the cell wall peptidoglycan. Some MHs can, eventually, cause cell lysis and are also designated autolysins. The wide distribution of MHs in bacteria, and particularly of autolysins, has led to the generalized idea that these enzymes participate in a variety of fundamental biological functions, such as the synthesis of the cell wall, separation of the daughter cells at the end of cell division, and genetic transformation (63). Moreover, lytic enzymes are responsible for the irreversible effects caused by β-lactam antibiotics (62, 67). The involvement of the lytic enzymes in those essential functions requires a strict and efficient regulation, otherwise some MHs can act as a Trojan horse to destroy the cell. It is generally believed that MHs are fundamental pacemaker enzymes for growth. Many bacterial species possess more than one enzyme that hydrolyzes the same bond, a fact that complicates the determination of their biological role(s). This redundancy also contributes to the common thought of assigning a basic role to lytic enzymes in the biology of bacteria. For *Escherichia coli*, 18 different MHs have been described, and 13 of these enzymes can act as autolysins. Construction of mutants lacking multiple MHs has failed to achieve final proof of the essential role of these enzymes in general growth (under laboratory conditions at least), although some MHs are involved in cell separation after cell division (27).

The activity of some pneumococcal MHs appears to be constrained by the membrane lipoteichoic acid (LTA) at the posttranslational level (29). Pneumococcal teichoic acid (TA) and LTA contain in their structure choline (Ch), an aminoalcohol that plays a fundamental biological role in the physiology of pneumococcus: Ch is considered to be the surface signature of *Streptococcus pneumoniae* (67). Ch residues of the cell wall act as binding ligands of a family of pneumococcal proteins known as choline-binding proteins (CBPs). Up to 15 CBPs have been found to be encoded in the genome of the TIGR4 strain, which recently has been sequenced (65). The three MHs identified so far in *S. pneumoniae* are CBPs (65).

For historical, structural, and functional reasons (see below), we update here not only our current knowledge on the "true" pneumococcal autolysins (the LytA amidase and LytC

Rubens López, Ernesto García, Pedro García, and José Luis García, Departamento de Microbiología Molecular, Centro de Investigaciones Biológicas, CSIC, E-28040 Madrid, Spain.

The Pneumococcus. Volume Editor, Elaine I. Tuomanen.
© 2004 ASM Press, Washington, D.C.

lysozyme) but also our knowledge on the glucosaminidase LytB, a nonautolytic MH that participates in cell separation, and on Pce, another CBP that releases phosphorylcholine residues from TA. TAs and the peptidoglycan constitute the two major components of the cell wall of *S. pneumoniae*. The four above-mentioned proteins are referred to as cell wall hydrolases (CWHs) hereafter.

THE CELL WALL HYDROLASES OF *S. PNEUMONIAE*

Gene Structure, Regulation, and Secretion

CWHs show both substrate and bond specificities. The former characteristic is related to their interaction with the insoluble substrate (the cell wall), whereas the latter determines the site of action. The bond specificity allows the classification of pneumococcal CWHs as follows: (i) Glycosidases, which include the LytC β-*N*-acetylmuramidase (lysozyme) and LytB β-*N*-acetylglucosaminidase, which hydrolyze the β-1-4 glycosidic bond between MurNAc and GlcNAc or vice versa, respectively (transglycosylases, which cleave the same bond as lysozymes but transfer the glycosyl moiety not onto water but onto the C-6 hydroxyl group of muramic acid, have not been reported yet for *S. pneumoniae*); (ii) the LytA *N*-acetylmuramoyl-L-alanine amidase (amidase), which cleaves the amide bond between the lactyl group of MurNAc and the α-amino group of L-alanine (the first amino acid of the stem peptide of the cell wall); and (iii) the Pce phosphorylcholine esterase acting at the TA side chains (Fig. 1). Endopeptidases cut within the peptide moiety of the peptidoglycan, but since no pneumococcal CBP endopeptidases have been described so far, this type of CWH is not considered in this chapter.

LytA

The *lytA* gene encodes the major *S. pneumoniae* autolysin (amidase) and represents the first example of a bacterial autolytic gene that was cloned and expressed (19). The *lytA* gene codes for a 318-amino-acid protein with a pre-

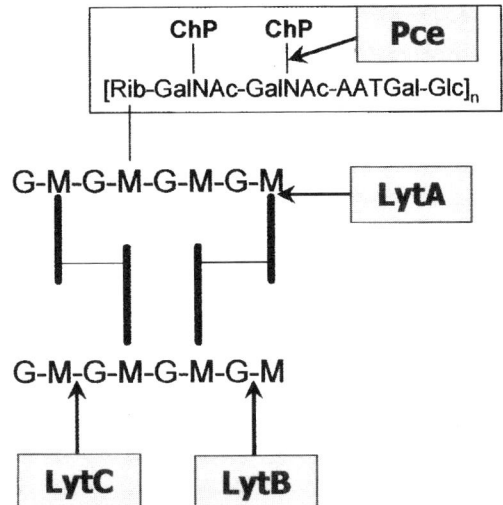

FIGURE 1 Diagrammatic sketch of the cell wall of pneumococci, indicating the chemical bonds cleaved by different CWHs. The peptidoglycan chains and a repeat unit of TA are shown. Abbreviations: G and M, *N*-acetylglucosamine and *N*-acetylmuramic acid residues, respectively; Rib, ribitol-5-phosphate; GalNAc, *N*-acetyl-D-galactosamine; AATGal, 2-acetamido-4-amino,2,4,6-trideoxy-D-galactose; Glc, D-glucose; ChP, phosphorylcholine.

dicted M_r of 36,532. The translation product of *lytA* is the low active form (E-form) of the amidase (19) that is converted, in vitro and in vivo, to the fully active C-form in a process named conversion (68). The promoter of the *lytA* gene has been described previously (11). The *lytA* gene has a rather long (240-bp) leader sequence with a high A+T content (70%). Although two open reading frames (ORFs) have been found in the leader region, it seems unlikely that these sequences can be translated due to the absence of appropriate ribosome-binding sites. A transcriptional terminator consisting of a hairpin structure (−20.8 kcal/mol) typical of rho-independent prokaryotic terminators was also localized.

A peculiar *lytA* gene has been identified in some bile (deoxycholate)-insoluble streptococcal clinical isolates and formerly classified as atypical pneumococci (13, 44). These atypical strains, which autolyze at the end of the stationary phase of growth, contain highly divergent *lytA* alleles (pairwise evolutionary distances higher than 20%). Furthermore, the *lytA* alleles from atypical isolates exhibited a remarkable polymorphism and high divergence values (pairwise evolutionary distance ≤ 7%) compared with the limited genetic variation (0.11 to 3.2%) that exists among otherwise unrelated *S. pneumoniae* isolates (73). The most noticeable feature of the atypical alleles is the presence of a 6-bp deletion, located between nucleotide positions 868 and 873, coding for Thr_{290}-Gly_{291} in the P6 motif of the wild-type amidase. The atypical amidases have a reduced specific activity. In addition, and in sharp contrast with that characteristic of the typical LytA amidase, the atypical LytA enzymes were inhibited by 1% deoxycholate. However, the atypical LytA enzymes can still be activated by 1% Triton X-100, a detergent that could be used as an alternative diagnostic test for this kind of strain (44).

Seto and Tomasz first reported an increase in autolytic activity during competence development (60). Although our preliminary analyses had suggested that the gene was constitutively expressed during the exponential phase of growth (11), more recent data have revealed that the 5.7-kb competence-specific *cinA-recA* transcript also includes the *lytA* gene and that transcription of *lytA* is strongly induced during competence from a promoter located upstream of *cinA* (9, 42) (Fig. 2A). Moreover, in noncompetent cells, *lytA* is expressed from both the *recA* and the *lytA* promoters. More recently, the kinetics of global changes in transcription patterns during competence development in *S. pneumoniae* were analyzed by using macro- and microarrays (47, 50). Competence was induced by the addition of competence-stimulating peptide to *S. pneumoniae* cultures grown to the early exponential phase. These analyses have fully confirmed that the *lytA* gene is overexpressed during competence. The reason(s) for this finding is still unclear, since it is well known that the LytA amidase is not required for transformation, as every *lytA* mutant tested is fully transformable (18, 57). Charpentier and coworkers recently claimed that a mu-

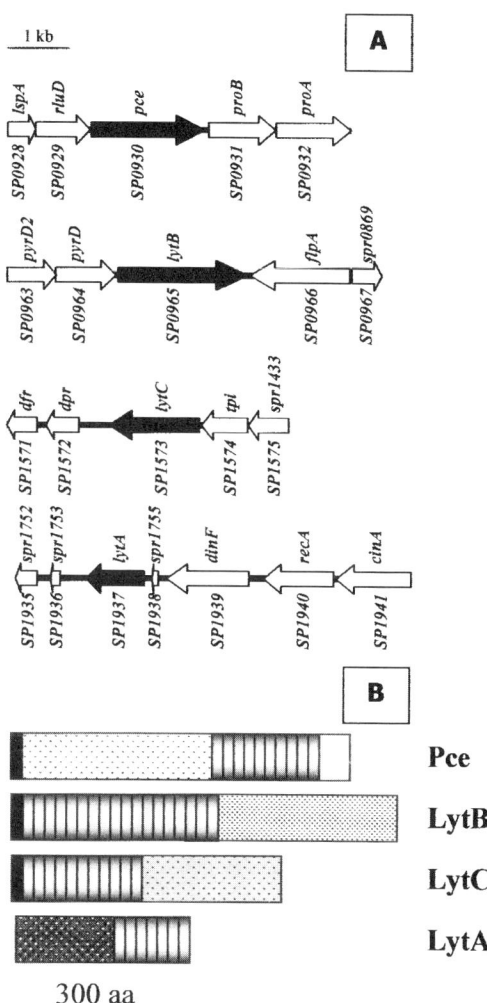

FIGURE 2 Schematic representation of the structures of *lytA*, *lytB*, *lytC*, and *pce*. (A) The genes are labeled as in references 32 and 65 above and below the arrows, respectively. Arrows indicate the direction of transcription of the ORFs. (B) The CWHs are drawn from the NH_2 end to the COOH end. Black and dotted bars indicate the parts of the proteins corresponding to the signal peptide and the active center, respectively. Shaded rectangles correspond to the CBRs. aa, amino acids.

tant deficient in the ClpC ATPase, a subfamily of HSP100/Clp molecular chaperone–regulators of proteolysis, formed long chains and failed to undergo lysis after treatment with penicillin or vancomycin (4). Moreover, the *clpC* mutant apparently failed to express LytA and other CBPs, suggesting that the heat shock protein ClpC might play an essential complex pleiotropic role in pneumococcal physiology. Nevertheless, more recent results using a ΔclpC mutant constructed by gene deletion/replacement have established that ClpC does not play a major role in autolysis (5). Besides, it has also been reported that transcription of the lytA gene is induced in an S. pneumoniae clpP mutant (ClpP, ATP-dependent protease) (51), although no significant variations in autolysis and/or LytA levels were found in this and other clp mutants affected in the heat shock response (5, 34, 52). Most likely, the induction of lytA in the clpP mutant can be ascribed to the induction of the cinA-recA-dinF-lytA operon (see above).

Available data suggest that LytA associates with the inner face of the membrane and then is translocated through it, where its interaction with LTA should contribute to its stabilization. Immunocytochemical analyses demonstrated that the amidase was mainly localized in S. pneumoniae at the equatorial division zone where the septum starts to be produced (8, 10). In experiments carried out with recombinant E. coli cells expressing lytA, the amidase was found loosely bound to the outer face of the cytoplasmic (inner) membrane (10). A key point of the regulation of LytA activity will be, most likely, linked to the elucidation of the timing of transport, as well as the triggering event of the general disorganization of the LytA-LTA complex leading to the uncontrolled action of LytA and subsequent lysis. This mechanism appears to control the activity of the enzyme even in two extreme situations such as the hyperproduction of LytA (from inside) or the addition of the enzyme to the culture medium (from outside). In these cases, culture lysis only takes place shortly after cells enter the stationary phase, in contrast with the normal lysis-prone behavior observed at the late stationary phase (54). The translocation step is carried out by a still-unknown mechanism, since LytA lacks a standard cleavable signal peptide. Many phage lytic enzymes also lack signal peptides and gain access to the peptidoglycan with the cooperation of small proteins (holins) that produce membrane openings that enable the passage of large molecules (75). An analogous mechanism has been recently reported for Staphylococcus aureus, where a 14.7-kDa protein (CidA) acts as a holin by increasing both extracellular MH activity and sensitivity to penicillin-induced killing (49). Although cidA homologs appear to exist in many gram-positive and gram-negative bacteria, a pneumococcal protein with a significant sequence similarity to CidA has not been found. Recently, it has been proposed that a zinc metalloprotease, ZmpB, was a key factor in controlling translocation to the cell surface of several CBPs, including LytA (43). The authors claimed that in a zmpB background, LytA formed a sodium dodecyl sulfate-resistant 80-kDa complex with CinA, a protein induced when cells become competent for genetic transformation (see above). Nevertheless, this assumption proved inconsistent when the zmpB "mutant" from Novak and coworkers was demonstrated to exhibit characteristic traits of the viridans group streptococci, including lack of capsule and resistance to optochin (3).

LytB

The lytB gene encodes a 76.4-kDa putative glucosaminidase (658 amino acid residues). The LytB enzyme contains a 23-amino-acid, cleavable signal peptide (predicted M_r of the processed protein, 73,800) (22). The lytB gene is preceded by the pyrD and pyrD2 genes, encoding, respectively, a dihydroorotate dehydrogenase and the dihydroorotate dehydrogenase electron transfer subunit (Fig. 2A). These genes are transcribed from the same strand as lytB, and due to the short distance between the start codon of LytB and the stop codon of pyrD (98 nucleotides), it is possible that they could form part of the same transcriptional unit. Moreover, a stable hairpin structure that could form a rho-independent terminator was not detected between the two genes. Nevertheless, the putative existence of a promoter between the pyrD and lytB genes that might control the independent expression of *lytB* remains to be investigated. The *lytB* gene is flanked at its 3′ end by a putative rho-independent terminator. Downstream of the *lytB* gene, but in the complementary strand, the *flpA* gene (also named *pavA*), which encodes a fibronectin-binding protein, is located. Inactivation of *pvaA* does not appear to affect LytB biosynthesis or function, since *pavA* pneumococcal mutants are not affected in cell morphology (28). Reverse transcription-PCR experiments have shown that transcription of the *lytB* gene is high during the early exponential phase of growth and decays as the culture enters the stationary phase (6).

LytB is most probably a glucosaminidase capable of degrading Ch-containing cell walls. Nevertheless, the in vitro degradation of this substrate was rather low and has precluded until now the precise determination of the enzymatic specificity. The construction of *lytB* mutants by insertion-duplication mutagenesis resulted in the formation of long chains (more than 100 cells), directly demonstrating the fundamental role that this enzyme plays in cell separation. These long chains could be dispersed, in a dose-dependent way, by the addition of purified LytB (8). The preparation of a translational fusion between the *gfp* gene, coding for the green fluorescent protein, and *lytB* allowed the purification of a fusion protein, GFP-LytB. This protein was used to demonstrate the specific localization of LytB at the poles of the cell (Fig. 3). An interesting observation from these experiments was that LytB presented a wide range of substrate recognition, in contrast to the absolute requirement of LytA for Ch to bind and hydrolyze the pneumococcal cell walls. It has been proposed that poles provide a mechanism by which proteins, once inserted, are maintained at this location. In this case, GFP-LytB would bind only to the specific target (the old wall zones or poles), to which it remained bound for a long time. The presence of Ch is essential for the successful cell septation by LytB, but initial binding to ethanolamine (EA)-containing cell walls could also be achieved (8).

LytC

The *lytC* gene is 1,506 bp long and encodes a lysozyme (LytC) of 501 amino acid residues with a predicted M_r of 58,682 (21a). LytC has a cleavable signal peptide of 33 amino acid residues, as demonstrated when the mature protein (about 55 kDa) was purified from *S. pneumoniae*. The *lytC* gene is preceded by an ORF whose product showed strong similarity (66% identity, 78% similarity) to the Tpi (triosephosphate isomerase) gene of *Lactobacillus delbrueckii* (Fig. 2A). The gene immediately preceding *tpi* (*spr1433/SP0967*) appears to code for DnaD, a putative primosome compo-

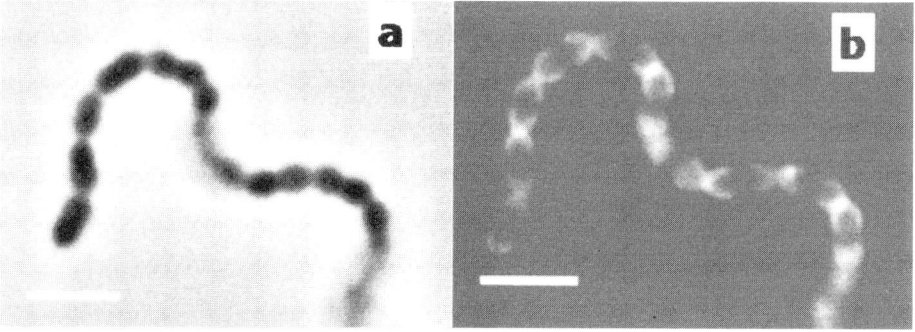

FIGURE 3 Localization of GFP-LytB in the surface of *S. pneumoniae*. A culture of a *lytB* mutant of the R6 strain received fusion protein (9.5 μg/ml) and was incubated for 1 min at 37°C. Pictures were taken with a Nikon Eclipse inverted microscope in phase-contrast (a) and using fluorescence (b). Bars, 4 μm. (Reprinted from the *Journal of Bacteriology* [8] with permission of the publisher.)

nent. Further upstream we found the *metA* gene, encoding a homoserine O-succinyltransferase (not shown). All these genes are so close that they appear to form part of the same transcriptional unit, but the existence and the real physiological role of this putative transcription unit containing the *lytC* gene remains to be elucidated.

In sharp contrast with what occurs at 37°C, cells of the Δ*lytA* strain M31 lysed when incubated overnight at 30°C (56). The insertional inactivation of *lytC* in the Δ*lytA* strain M31 (M31C) makes the cells resistant to autolysis upon incubation at 30°C. Addition of purified LytC lysozyme to this deficient strain showed that this enzyme was kept under regulatory control of the "cured" cells until the culture reached the stationary phase of growth, and then the culture started to lyse. Besides, the *lytC* mutants exhibited a normal growth rate and an average chain length similar to that of the parental strains. Combined cell fractionation and Western blot analysis showed that the unprocessed LytC protein is located in the cytoplasm, whereas the processed, active form of LytC is tightly bound to the cell envelope (21a). This binding strength is probably related to the number of repeating motifs, and this type of binding could be a mechanism for the regulation and control of the potentially suicidal activity of this type of MH.

Pce

The *pce* gene is 1,884 bp long and codes for a protein of 627 amino acid residues with a predicted M_r of 72,104. According to the amino-terminal amino acid analysis, Pce contains a typical signal peptide of 25 amino acids that renders, after its cleavage, a mature protein of 69,426 Da (602 amino acids). The start codon of the *pce* gene is located 2 nucleotides downstream of the stop codon of the *rluD* gene, which putatively encodes a 23S rRNA pseudouridine synthase (Fig. 2A), suggesting that *pce* is probably transcribed with other genes located upstream. In sharp contrast with that reported for *lytA* and *lytB*, transcription of the *pce* gene reaches a maximum in the stationary phase of growth, as revealed by using reverse transcription-PCR (6). Besides, a palindromic sequence forming a hairpin loop with a Δ*G* of −7.7 kcal/mol that might act as a putative rho-independent terminator was found just downstream of the stop codon of the *pce* gene. Genes responsible for proline biosynthesis are located further downstream of *pce*. The complete *pce* gene has been cloned and expressed in *E. coli* (7, 71). The purified Pce protein was expressed in the processed mature form, indicating that the signal peptide was functional in *E. coli*. In vitro experiments carried out with purified Pce confirmed that this enzyme is a TA phosphorylcholine esterase that removes only a maximum of 20% of phosphorylcholine residues from the cell wall TA, in agreement with early results (30). As it is also bound to the envelope, Pce should play its role only after its secretion through the membrane, although it is currently difficult to assign a defined function to the protein. It has been suggested that the roughly 20% of residues removable by the enzyme either might exist in an anatomically unique position in the cell wall or might represent terminal residues in the TA chains (7, 71). Some data on the Ch-independent strains displaying noticeable physiological and morphological changes have suggested that TA chains may actually block the access of the enzyme molecules to the peptidoglycan substrate (61, 74). Also, this esterase activity might regulate the availability of Ch residues required for activity and/or attachment of this and other CBPs.

Pneumococcal CWHs Are CBPs

All the pneumococcal CWHs described above have been shown to possess an absolute requirement for the presence of Ch for activity. The first direct experimental evidence came from the early work carried out with the LytA amidase at the laboratory of Höltje and Tomasz (31). However, it was not until 1988 that this requirement could be explained on structural bases once the gene encoding a Ch-dependent lysozyme (Cpl-1) from the pneumococcal bacteriophage Cp-1 was characterized (17). Sequence comparisons revealed that the C-terminal parts of Cpl-1 and LytA shared extensive sequence similarities, whereas the respective N-terminal moieties were completely different. As the two enzymes shared only the Ch requirement, it was assumed that the carboxyl-terminal domains of both proteins were involved in Ch binding, whereas the active center should reside at the N-terminal domain. Further experimental support was obtained with additional research on several new phage lytic enzymes and with construction of functional chimeric proteins (37). The C-terminal domain of CBPs is constituted by a tandem of about 20 residue repeats, known as Ch-binding repeats (CBRs) (Fig. 2B). Similar repeats (or motifs) have now been found in a large family of cell wall-binding proteins (Pfam ID code PF01473; CW_binding_1 repeat) that currently has more than 1,400 members (2), although only a few of them appear to be related to Ch binding.

Before the simultaneous description of LytB and LytC, all previously described CBPs (including those from phage origin) possessed the Ch-binding domain at their C-terminal part. LytB and LytC, however, have changed this general building plan, as their Ch-binding domain is located N terminally (Fig. 2B). Pce appears to represent an intermediate situation, since the CBRs are followed by a tail of 85 amino acid residues (Fig. 2B). Also, the numbers of CBRs vary among the pneumococcal CWHs: 7, 18, 11, and 10 in LytA, LytB, LytC, and Pce, respectively. A minimum of four CBRs appears to be required for efficient binding to the cell wall (20), and it is conceivable that the binding strength increases with the number of repeats. In the case of LytB, about half of the CBRs (those located C terminally at the Ch-binding domain) appear to be better conserved than the other half. This difference suggests that the "variant" motifs might play a different or, at least, complementary role in cell substrate recognition. This is in agreement with the observation that LytB is able to bind to EA-grown cells, in sharp contrast with what has been reported for all other pneumococcal CWHs (8).

STRUCTURE-FUNCTION RELATIONSHIPS OF THE LytA AMIDASE

Cloning and expression of the DNA regions encoding the N- or C-terminal moieties of LytA have been used to demonstrate that LytA has a modular structure composed of the catalytic N-terminal domain and a C-terminal domain (C-LytA) responsible for Ch binding (58, 59). Analytical ultracentrifugation studies revealed that LytA exhibits a monomer-dimer association equilibrium, through the C-terminal part of the molecule, and that dimerization is enhanced by Ch binding in such a way that at 140 mM choline only dimers are present (69). On the basis of sequence comparison it was originally proposed that C-LytA was constructed by six tandem repeats of about 20 amino acids (P1 to P6) and an 11-amino-acid C-terminal tail (17). Nevertheless, this tail, although divergent, actually is part of a seventh repeat, as demonstrated by crystallographic studies (see below). It has already been mentioned that LytA must contain at least four CBRs to bind efficiently to the Ch residues of the cell wall. Nevertheless, the catalytic efficiency decreases by 90% upon deletion of the final tail (55). The structural implications and binding modifications of deleting step-by-step CBR5 and CBR6 (originally designated p5 and p6, respectively) have been examined by comparing the amidases purified from four different truncated mutants (70). Removal of this region has minor effects on secondary-structure content but significantly affects the stability of native conformations. The last 11 amino acids and CBR5 serve to stabilize C-LytA; each increases the domain transition temperature by about 6°C. Moreover, CBR5 participates, in a Ch-dependent way, in the stabilization of the N-terminal domain, indicating a cooperative pathway providing molecular communication between the Ch-binding domain and the N-terminal region. Although CBR5 and CBR6 favor the Ch-amidase interaction, the tail is an essential factor in the monomer-dimer self-association equilibrium of LytA and its regulation by Ch. CBR6 appears to scarcely affect the amidase stability but could provide the proper three-dimensional orientation of the C-terminal amino acids.

Physicochemical techniques such as analytical ultracentrifugation or differential scanning calorimetry have been used to study the structure of LytA (40, 69). On the basis of sedimentation velocity analysis the shape of the LytA dimer appears to be a prolate ellipsoid of 13 by 190 Å, which can be visualized as a stalk, made of the C-terminal domains, bearing at each end a catalytic site (Color Plate 4A). The secondary structure of this autolysin consists of 47% β-sheets, 19% α-helices, 23% turns, and 11% irregular structures. The secondary structure of the whole LytA molecule is very similar to that of C-LytA, with a characteristic high β-sheet content (see below). The composition of the N-terminal part, therefore, was concluded to also have a secondary structure similar to that of the whole LytA molecule or its C-terminal domain. The binding of LytA to Ch residues does not cause any significant change in the secondary-structure composition, which is indicative of the lack of major structural changes for the molecule during this event. However, two modes of Ch binding to C-LytA have been detected; one is low affinity and another is high affinity. Saturation of Ch binding to the high-affinity sites induces dimerization and subsequent increase in affinity for the substrate. The dimerization involves primarily the C-terminal part of the molecule, with preferential binding of two Ch molecules to the dimer (40). On the other hand, saturation of the low-affinity sites requires a Ch concentration similar to that necessary for inhibition of this amidase in in vivo assays. The presence of these two Ch-binding sites, together with inducible dimerization, might play an essential role in LytA cellular targeting by causing a preferential location of the enzyme at the sites of its action on the cell wall. Furthermore, deconvolution of calorimetric curves has revealed a folding of the polypeptide chain in several independent or quasi-independent cooperative domains. Elementary transitions in the isolated C-LytA are similar, but not identical, to those assigned to the C-terminal

domain in the complete amidase, particularly in the absence of Ch. These results indicate that the N-terminal region of the protein is important for attaining the native tertiary fold of the C terminus. These results taken together provided evidence that LytA consists of at least four domains per subunit (two per module), designated N1, N2, C1, and C2 (69) (Color Plate 4A).

THE CRYSTAL STRUCTURE OF THE CHOLINE-BINDING DOMAIN (C-LytA)

The recently reported crystal structure of a recombinant, shortened form of C-LytA (C-LytA*) in complex with Ch revealed a protein fold, the β-3-solenoid spiral staircase, which consists exclusively of β-hairpins that stack to form a left-handed superhelix maintained by Ch molecules at hydrophobic cavities on the protein surface (15). The overall shape of the C-LytA* monomer is approximately cylindrical, with a diameter of 25 Å and a height of ~60 Å (Color Plate 4B). The secondary structure is comprised of six independent β-hairpins, each consisting of two antiparallel β-strands connected by a short internal loop region. All the β-strands have the same length and character (five residues and predominantly hydrophobic). Consecutive hairpins are connected by loops of 8 to 10 residues that contain a type I+G1 β-bulge turn, plus 4 to 6 residues mostly in an extended conformation. The hairpins extend perpendicularly from the axis towards the surface of the cylinder. Ch molecules at the hydrophobic interface of consecutive hairpins maintain this unique structure. The C-terminal hairpin in the solenoid is responsible for the formation of the catalytically active homodimer, which has a boomerang-like shape with arm lengths of 50 Å and an angle between the arms of ~85° (Color Plate 4B and C). Inspection of the buried surface area in the crystal showed that aromatic and hydrophobic residues of repeats 5 and 6 make up the most extensive interactions in both monomers. In this way, the dimer is formed by a tail-to-tail association of two monomers where hairpins 5 and 6 extend perpendicularly in an antiparallel fashion to one another. As commented above, LytA forms dimers in solution. Dimerization has been related to regulation of the catalytic activity of LytA because when it lacks the C-terminal hairpin, it becomes monomeric and undergoes a significant decrease (90%) in catalytic efficiency. CBPs with longer Ch-binding domains seem to lack the last hairpin, probably because their affinity for the cell wall might be enough, as they may contain a higher number of Ch-binding sites. In a similar way, the Ch-binding domains of the other pneumococcal CWHs (LytB, LytC, and Pce), whose function must also be tightly regulated, contain different amino acid changes that are likely to be related to the regulation of their biological activities (see above).

More recently, two new crystal forms of C-LytA* were obtained, and their structures were solved and refined to 2.4 and 2.8-Å resolution (14). Although the general fold in the structures derived from both crystal forms is essentially the same, two different conformations of the basic homodimer were observed. Although the monomers are highly similar overall, significant variations arise from a different orientation of hairpin 6 with respect to the rest of the structure. The distortion is caused by the different conformation of the two-residue insertion (Arg$_{304}$–Pro$_{305}$) between hairpin 6 and its preceding loop, which is not present in the other CBRs found in LytA. As a consequence of the different arrangement of hairpin 6 in the tetragonal crystals, the resulting boomerang is much more open than in the other case, with an internal angle of ~110° (14). As the boomerang is thought to carry the N-terminal catalytic domain of LytA at the end of its arms, the degree of flexibility described above may affect the relative position of these domains, which in turn is likely to be crucial in the regulation of the catalytic activity of LytA in vivo.

Four Ch-binding sites per monomer of C-LytA* were found in all the crystal forms, although the existence of an additional Ch-binding site has been inferred from preliminary X-ray diffraction studies on the complete C-LytA protein (16). Each site is formed by the interface between consecutive hairpin pairs. The forces contributing to the interaction between C-LytA* and Ch molecules are hydrophobic and electrostatic. The hydrophobic component arises from the direct contact of the three Ch methyl groups with the four hydrophobic residues that constitute the site: three aromatic residues from the hairpins surrounding the site (Trp, Phe, or Tyr) and a hydrophobic residue (Met or Leu) from an 8- to 10-residue connecting loop. A cation–π interaction between the electron-rich systems of the aromatic rings and the positive charge of Ch enhances the binding.

BIOLOGICAL ROLES OF PNEUMOCOCCAL CWHs

As already mentioned, CWHs are thought to be implicated in a series of important biological functions. Among them, the relationship between autolysins and tolerance against β-lactam antibiotics is discussed in another chapter of this book and is not reviewed here. A role for CWHs in genetic transformation has been repeatedly proposed. Genetic transformation is an important mechanism for gene exchange between streptococci in nature that implies the ability to take up free DNA from the medium and the incorporation of this DNA into the receptor cell genome by genetic recombination. Under natural conditions, the pneumococcus might participate in gene exchange when located in biofilms in places such as the nasopharynx in the common carrier state of this microorganism (24, 72). As early as in 1960, release of transforming DNA to the medium was reported to take place in pneumococci (45, 46). However, we do not know yet whether the source of donor DNA is the autolyzed dead cell or implies the existence of an active process. Since the pores in the cell wall in gram-positive organisms are too small to allow DNA to pass through, it has been assumed that the liberation of these molecules from the cytoplasm to the medium involves degradation of the peptidoglycan. Also, a role for several pneumococcal CWHs, mainly the LytA amidase, in DNA release from a subfraction (5 to 20%) of the actively growing, bacterial population, probably by cell lysis, has been proposed. The other fraction will act as recipient of the DNA in the medium (64). It is noteworthy that release of transforming DNA is greatly reduced in antibiotic-treated cultures compared to that in untreated controls (23). Most interestingly, competence development and cell lysis appear to be regulated by the same quorum-sensing mechanism (ComCDE) (see above), and DNA release and competence development appear to be causally related (M. Moscoso and J. P. Claverys, Abstr. ASM Conf. Streptococcal Genet., abstr. 43, 2002). Moreover, release is affected by the inactivation of the lytic amidase, since *lytA* mutants exhibited a two- to fourfold decrease in the amount of DNA available in the medium (48). Besides, CWHs do not appear to be required for transformation, since *S. pneumoniae* mutants lacking all the four CWHs are still transformable at, or near, normal levels (6).

The involvement of CWHs in cell separation has also been repeatedly claimed. The formation of long chains of cells when pneumococci are grown in EA medium was attributed to LytA, the only autolysin known at that time (66). However, it is now clear that *lytA* mutants grow forming only small chains (8 to 10 cells) (57). The LytB enzyme is the CWH directly responsible for chain formation and requires the presence of Ch in the TA for activity, although, in remarkable contrast to other pneumococcal CBPs, it is also able to bind to EA-grown cells (see above). Chain formation was not observed either in *lytC* or *pce* mutants.

The implication of the LytA autolysin in the pathogenesis of pneumococcal infection and its potential as a carrier protein in conjugate pneumococcal vaccines have been recently reviewed (33, 35). Several studies have shown that *lytA* mutants have attenuated virulence in animal models compared to that of isogenic wild-type pneumococci. Bacteriolysis caused by CWHs (such as LytA and LytC) plays a very important role in the pathophysiology of inflammation (25). It has been suggested that the cell wall degradation products released as a re-

sult of the action of LytA mediate severe inflammation reactions. Peptidoglycan stem tripeptides released by LytA digestion of isolated *S. pneumoniae* cell wall showed a high capacity to trigger cytokine release from peripheral blood monocytes (38), whereas although TA is an important part of the cell wall in gram-positive organisms, it did not participate in the cytokine-releasing activity (39). In addition, diverse experiments in animal models concluded that LytA contributes to pathogenesis by catalyzing the release of pneumococcal cytoplasmic contents, including virulence factors such as pneumolysin (41). Although this is the case in a majority of *S. pneumoniae* strains, it has been recently reported that extracellular release of pneumolysin may be independent of pneumococcal lysis in some strains (1).

Insertion-duplication mutagenesis was employed to construct *lytB*, *lytC*, or *pce* (*cbpE*) mutants of a type 4 pneumococcal strain. All of them showed a significant reduction in the colonization of the nasopharynx using an infant rat model, but no changes in virulence were observed when tested in a model of pneumococcus-induced sepsis (26). Moreover, loss of function of LytC or Pce reduced adherence to Detroit cells at 30°C to about 70%. On the other hand, Vollmer and Tomasz have reported that inactivation of Pce in pneumococci producing type 3 capsule caused a significant increase in virulence when tested in an intraperitoneal mouse model (71). It may be that the increase in the number of Ch residues (due to the inactivation of the *pce* gene) could facilitate interactions with the platelet-activating factor receptor during infection.

CONCLUDING REMARKS

The cloning of *lytA* has facilitated the isolation of the genes encoding the cell wall lytic enzymes from pneumococcal bacteriophages based on sequence homologies (17). We have analyzed in detail the lytic enzymes encoded by five different bacteriophages (Cp-1, HB-3, EJ-1, Dp-1, and MM1) infecting *S. pneumoniae* (P. García, J. L. García, R. López, and E. García, submitted for publication). Sequence comparison revealed that the host and phage-coded enzymes exhibit a modular organization represented by two domains. There is a high degree of similarity between the domain found in bacterial and phage lytic enzymes that is responsible for binding to the Ch residues present in the TA of the cell wall substrate, whereas the other domain localized the active enzymatic center. We proposed that Ch served as an element of selective pressure to preserve the substrate recognition domain of the pneumococcal lytic enzymes characterized so far (36).

This global analysis led us to propose that pneumococcal cell wall lytic enzymes could be the result of the fusion of two independent functional domains. The enzyme coded by Cp-1 has been identified as a lysozyme (Cpl-1), whereas those coded by Dp-1, HB-3 and EJ-1 (Pal, Hbl, and Ejl, respectively) are amidases. Interestingly, a lytic enzyme isolated from phage Cp-7 and identified as lysozyme can degrade cell walls containing Ch analogs, and consequently, Cpl-7 lysozyme is not a CBP sensu stricto (21).

The construction of active chimeric proteins between lysins of phage and bacteria led to new enzymes exhibiting novel properties that were, as expected, a combination of those showed by the parental enzymes (12, 37). This experimental approach has provided important clues to the organization of the lytic enzymes as well as to the coevolution of some genes through the relationships between host and phage. On the other hand, the recent elucidation of the first crystal structure of a Ch-binding domain has fully confirmed the prediction on the fundamental role of Ch in the mechanism of attachment of CBPs to the bacterial surface (15).

It has been well documented that the repeated switch between the lytic and lysogenic states of temperate phages results, in some cases, in the incorporation of new ORFs. In the pneumococcal model, the genes coding for lytic enzymes illustrate on these prophage genome excisions. However, most of these genes yield interrupted ORFs designated as remnants (53), and it has been proposed that this biological scenario might favor gene rearrangement, through recombination, that can be selected during the course of evolution under appropriate environmental conditions.

REFERENCES

1. **Balachandran, P., S. K. Hollingshead, J. C. Paton, and D. E. Briles.** 2001. The autolytic enzyme LytA of *Streptococcus pneumoniae* is not responsible for releasing pneumolysin. *J. Bacteriol.* **183:**3108–3116.

2. **Bateman, A., E. Birney, L. Cerruti, R. Durbin, L. Etwiller, S. R. Eddy, S. Griffiths-Jones, K. L. Howe, M. Marshall, and E. L. L. Sonnhammer.** 2002. The Pfam protein families database. *Nucleic Acids Res.* **30:**276–280.

3. **Bergé, M., P. García, F. Iannelli, M. F. Prere, C. Granadel, A. Polissi, and J. P. Claverys.** 2001. The puzzle of *zmpB* and extensive chain formation, autolysis defect and nontranslocation of choline-binding proteins in *Streptococcus pneumoniae*. *Mol. Microbiol.* **39:**1651–1660.

4. **Charpentier, E., R. Novak, and E. Tuomanen.** 2000. Regulation of growth inhibition at high temperature, autolysis, transformation and adherence in *Streptococcus pneumoniae* by *clpC*. *Mol. Microbiol.* **37:**717–726.

5. **Chastanet, A., M. Prudhomme, J. P. Claverys, and T. Msadek.** 2001. Regulation of *Streptococcus pneumoniae clp* genes and their role in competence development and stress survival. *J. Bacteriol.* **183:**7295–7307.

6. **de las Rivas, B.** 2002. Isolation and characterization of new choline-binding proteins from *Streptococcus pneumoniae*. Ph.D. thesis. Universidad Complutense de Madrid, Madrid, Spain.

7. **de las Rivas, B., J. L. García, R. López, and P. García.** 2001. Molecular characterization of the pneumococcal teichoic acid phosphorylcholine esterase. *Microb. Drug Resist.* **7:**213–222.

8. **de las Rivas, B., J. L. García, R. López, and P. García.** 2002. Purification and polar localization of pneumococcal LytB, a putative endo-β-N-acetylglucosaminidase: the chain-dispersing murein hydrolase. *J. Bacteriol.* **184:**4988–5000.

9. **de Saizieu, A., U. Certa, J. Warrington, C. Gray, W. Keck, and J. Mous.** 1998. Bacterial transcripts imaging by hybridization of total RNA by oligonucleotide arrays. *Nat. Biotechnol.* **16:**45–48.

10. **Díaz, E., E. García, C. Ascaso, E. Méndez, R. López, and J. L. García.** 1989. Subcellular localization of the major pneumococcal autolysin: a peculiar mechanism of secretion in *Escherichia coli*. *J. Biol. Chem.* **264:**1238–1244.

11. **Díaz, E., and J. L. García.** 1990. Characterization of the transcription unit encoding the major pneumococcal autolysin. *Gene* **90:**157–162.

12. **Díaz, E., R. López, and J. L. García.** 1990. Chimeric phage-bacterial enzymes: a clue to the modular evolution of genes. *Proc. Natl. Acad. Sci. USA* **87:**8125–8129.

13. **Díaz, E., R. López, and J. L. García.** 1992. Role of the major pneumococcal autolysin in the atypical response of a clinical isolate of *Streptococcus pneumoniae. J. Bacteriol.* **174:**5508–5515.

14. **Fernández-Tornero, C., E. García, R. López, G. Giménez-Gallego, and A. Romero.** 2002. Two new crystal forms of the choline-binding domain of the major pneumococcal autolysin: insights into the dynamics of the active homodimer. *J. Mol. Biol.* **321:**163–173.

15. **Fernández-Tornero, C., R. López, E. García, G. Giménez-Gallego, and A. Romero.** 2001. A novel solenoid fold in the cell wall anchoring domain of the pneumococcal virulence factor LytA. *Nat. Struct. Biol.* **8:**1020–1024.

16. **Fernández-Tornero, C., A. Ramón, C. Fernández-Cabrera, G. Giménez-Gallego, and A. Romero.** 2002. Expression, crystallization and preliminary X-ray diffraction studies on the complete choline-binding domain of the major pneumococcal autolysin. *Acta Crystallogr. D Biol. Crystallogr.* **58:**556–558.

17. **García, E., J. L. García, P. García, A. Arrarás, J. M. Sánchez-Puelles, and R. López.** 1988. Molecular evolution of lytic enzymes of *Streptococcus pneumoniae* and its bacteriophages. *Proc. Natl. Acad. Sci. USA* **85:**914–918.

18. **García, E., J. L. García, P. García, C. Ronda, J. M. Sánchez-Puelles, and R. López.** 1987. Molecular genetics of the pneumococcal amidase: characterization of *lytA* mutants, p. 189–192. *In* J. J. Ferretti and R. Curtiss III (ed.), *Streptococcal Genetics*. ASM Press, Washington, D.C.

19. **García, E., J. L. García, C. Ronda, P. García, and R. López.** 1985. Cloning and expression of the pneumococcal autolysin gene in *Escherichia coli. Mol. Gen. Genet.* **201:**225–230.

20. **García, J. L., E. Díaz, A. Romero, and P. García.** 1994. Carboxy-terminal deletion analysis of the major pneumococcal autolysin. *J. Bacteriol.* **176:**4066–4072.

21. **García, P., J. L. García, E. García, J. M. Sánchez-Puelles, and R. López.** 1990. Modular organization of the lytic enzymes of *Streptococcus pneumoniae* and its bacteriophages. *Gene* **86:**81–88.

21a.**García, P., M. P. González, E. García, J. L. García, and R. López.** 1999. The molecular characterization of the first autolytic lysozyme of *Streptococcus pneumoniae* reveals evolutionary mobile domains. *Mol. Microbiol.* **33:**128–138.

22. **García, P., M. P. González, E. García, R. López, and J. L. García.** 1999. LytB, a novel pneumococcal murein hydrolase essential for cell separation. *Mol. Microbiol.* **31:**1275–1277.

23. **Gerber, J., H. Eiffert, H. Fleischer, A. Wellmer, U. Munzel, and R. Nau.** 2001. Reduced release of DNA from *Streptococcus pneumoniae* after treatment with rifampin in comparison to spontaneous growth and ceftriaxone treatment. *Eur. J. Clin. Microbiol. Infect. Dis.* **20:**490–493.

24. **Ghigo, J. M.** 2003. Are there biofilm-specific physiological pathways beyond a reasonable doubt? *Res. Microbiol.* **154:**1–8.

25. **Ginsburg, I.** 2002. The role of bacteriolysis in the pathophysiology of inflammation, infection and post-infectious sequelae. *APMIS* **110:**753–770.

26. **Gosink, K. K., E. R. Mann, C. Guglielmo, E. I. Tuomanen, and H. R. Masure.** 2000. Role of novel choline binding proteins in virulence of *Streptococcus pneumoniae. Infect. Immun.* **68:**5690–5695.

27. **Heidrich, C., A. Ursinus, J. Berger, H. Schwarz, and J.-V. Höltje.** 2002. Effects of multiple deletions of murein hydrolases on viability, septum cleavage, and sensitivity to large toxic molecules in *Escherichia coli. J. Bacteriol.* **184:**6093–6099.

28. **Holmes, A. R., R. McNab, K. W. Millsap, M. Rohde, S. Hammerschmidt, J. L. Mawdsley, and H. F. Jenkinson.** 2001. The *pavA* gene of *Streptococcus pneumoniae* encodes a fibronectin-binding protein that is essential for virulence. *Mol. Microbiol.* **41:**1395–1408.

29. **Höltje, J.-V., and A. Tomasz.** 1975. Lipoteichoic acid: a specific inhibitor of autolysin activity in pneumococcus. *Proc. Natl. Acad. Sci. USA* **72:**1690–1694.

30. **Höltje, J.-V., and A. Tomasz.** 1974. Teichoic acid phosphorylcholine esterase. A novel enzyme activity in pneumococcus. *J. Biol. Chem.* **249:**7032–7034.

31. **Höltje, J.-V., and A. Tomasz.** 1975. Specific recognition of choline residues in the cell wall teichoic acid by *N*-acetylmuramic acid L-alanine amidase of pneumococcus. *J. Biol. Chem.* **250:**6072–6076.

32. **Hoskins, J., W. E. Alborn, J. Arnold, L. C. Blaszczak, S. Burgett, B. S. DeHoff, S. T. Estrem, L. Fritz, D.-J. Fu, W. Fuller, C. Geringer, R. Gilmour, J. S. Glass, H. Khoje, A. R. Kraft, R. E. Lagace, D. J. LeBlanc, L. N. Lee, E. J. Lefkowitz, J. Lu, P. Matsushima, S. M. McAhren, M. McHenney, K. McLeaster, C. W. Mundy, T. I. Nicas, F. H. Norris, M. O'Gara, R. B. Peery, G. T. Robertson, P. Rockey, P.-M. Sun, M. E. Winkler, Y. Yang, M. Young-Bellido, G. Zhao, C. A. Zook, R. H. Baltz, R. Jaskunas, P. R. J. Rosteck, P. L. Skatrud, and J. I.**

Glass. 2001. Genome of the bacterium *Streptococcus pneumoniae* strain R6. *J. Bacteriol.* **183:**5709–5717.

33. **Jedrzejas, M. J.** 2001. Pneumococcal virulence factors: structure and function. *Microbiol. Mol. Biol. Rev.* **65:**187–207.

34. **Kwon, H.-Y., S.-W. Kim, M.-H. Choi, A. D. Ogunniyi, J. C. Paton, S.-H. Park, S.-N. Pyo, and D.-K. Rhee.** 2003. Effect of heat shock and mutations in ClpL and ClpP on virulence gene expression in *Streptococcus pneumoniae. Infect. Immun.* **71:**3757–3765.

35. **Lee, C. J., T. R. Wang, and C. E. Frasch.** 2001. Immunogenicity in mice of pneumococcal glycoconjugate vaccines using pneumococcal protein carriers. *Vaccine* **19:**3216–3225.

36. **López, R., E. García, P. García, and J. L. García.** 1997. The pneumococcal cell wall degrading enzymes: a modular design to create new lysins? *Microb. Drug Resist.* **3:**199–211.

37. **López, R., E. García, P. García, and J. L. García.** 2000. The pneumococcal cell wall degrading enzymes: a modular design to create new lysins? p. 197–209. *In* A. Tomasz (ed.), *Streptococcus pneumoniae—Molecular Biology & Mechanisms of Disease.* Mary Ann Liebert, Inc., Larchmont, N.Y.

38. **Majcherczyk, P. A., H. Langen, D. Heumann, M. Fountoulakis, M. P. Glauser, and P. Moreillon.** 1999. Digestion of *Streptococcus pneumoniae* cell walls with its major peptidoglycan hydrolase releases branched stem peptides carrying proinflammatory activity. *J. Biol. Chem.* **274:**12537–12543.

39. **Majcherczyk, P. A., E. Rubli, D. Heumann, M. P. Glauser, and P. Moreillon.** 2003. Teichoic acids are not required for *Streptococcus pneumoniae* and *Staphylococcus aureus* cell walls to trigger the release of tumor necrosis factor by peripheral blood monocytes. *Infect. Immun.* **71:**3707–3713.

40. **Medrano, F. J., M. Gasset, C. López-Zúmel, P. Usobiaga, J. L. García, and M. Menéndez.** 1996. Structural characterization of the unligated and choline-bound forms of the major pneumococcal autolysin LytA amidase. Conformational transitions induced by temperature. *J. Biol. Chem.* **271:**29152–29161.

41. **Mitchell, T. J., J. E. Alexander, P. J. Morgan, and P. W. Andrew.** 1997. Molecular analysis of virulence factors of *Streptococcus pneumoniae. Soc. Appl. Bacteriol. Symp. Ser.* **26:** 62S–71S.

42. **Mortier-Barrière, I., A. de Saizieu, J. P. Claverys, and B. Martin.** 1998. Competence-specific induction of *recA* is required for full recombination proficiency during transformation in *Streptococcus pneumoniae. Mol. Microbiol.* **27:** 159–170.

43. **Novak, R., E. Charpentier, J. S. Braun, E. Park, S. Murti, E. Tuomanen, and R. Masure.** 2000. Extracellular targeting of choline-binding proteins in *Streptococcus pneumoniae* by a zinc metalloprotease. *Mol. Microbiol.* **36:**366–376.

44. **Obregón, V., P. García, E. García, A. Fenoll, R. López, and J. L. García.** 2002. Molecular peculiarities of the *lytA* gene isolated from clinical pneumococcal strains that are bile insoluble. *J. Clin. Microbiol.* **40:**2545–2554.

45. **Ottolenghi, E., and R. D. Hotchkiss.** 1960. Appearance of genetic transforming activity in pneumococcal cultures. *Science* **132:**1257–1258.

46. **Ottolenghi, E., and R. D. Hotchkiss.** 1962. Release of genetic transforming agent from pneumococcal cultures during growth and disintegration. *J. Exp. Med.* **116:**491–519.

47. **Peterson, S., R. T. Cline, H. Tettelin, V. Sharov, and D. A. Morrison.** 2000. Gene expression analysis of the *Streptococcus pneumoniae* competence regulons by use of DNA microarrays. *J. Bacteriol.* **182:**6192–6202.

48. **Ramirez, M.** 1998. DNA exchange in natural populations of *Streptococcus pneumoniae.* Ph.D. thesis. Universidade Nova de Lisboa, Lisbon, Portugal.

49. **Rice, K. C., B. A. Firek, J. B. Nelson, S. J. Yang, T. G. Patton, and K. W. Bayles.** 2003. The *Staphylococcus aureus cidAB* operon: evaluation of its role in regulation of murein hydrolase activity and penicillin tolerance. *J. Bacteriol.* **185:**2635–2643.

50. **Rimini, R., B. Jansson, G. Feger, T. C. Roberts, M. de Francesco, A. Gozzi, F. Faggioni, E. Domenici, D. M. Wallace, N. Frandsen, and A. Polissi.** 2000. Global analysis of transcription kinetics during competence development in *Streptococcus pneumoniae* using high density DNA arrays. *Mol. Microbiol.* **36:**1279–1292.

51. **Robertson, G. T., W. L. Ng, J. Foley, R. Gilmour, and M. E. Winkler.** 2002. Global transcriptional analysis of *clpP* mutations of type 2 *Streptococcus pneumoniae* and their effects on physiology and virulence. *J. Bacteriol.* **184:**3508–3520.

52. **Robertson, G. T., W.-L. Ng, R. Gilmour, and M. E. Winckler.** 2003. Essentiality of *clpX,* but not *clpP, clpL, clpC,* or *clpE,* in *Streptococcus pneumoniae* R6. *J. Bacteriol.* **185:**2961–2966.

53. **Romero, A., R. López, and P. García.** 1992. The insertion site of the temperate phage HB-746 is located near the phage remnant in the pneumococcal host chromosome. *J. Virol.* **66:**2860–2864.

54. **Ronda, C., J. L. García, E. García, J. M. Sánchez-Puelles, and R. López.** 1987. Biological role of the pneumococcal amidase. Cloning of the *lytA* gene in *Streptococcus pneumoniae. Eur. J. Biochem.* **164:**621–624.

55. Sánchez-Puelles, J. M., J. L. García, R. López, and E. García. 1987. 3′-end modifications of the *Streptococcus pneumoniae lytA* gene: role of the carboxy terminus of the pneumococcal autolysin in the presence of enzymatic activation (conversion). *Gene* **61**:13–19.

56. Sánchez-Puelles, J. M., C. Ronda, E. García, E. Méndez, J. L. García, and R. López. 1986. A new peptidoglycan hydrolase in *Streptococcus pneumoniae*. *FEMS Microbiol. Lett.* **35**:163–166.

57. Sánchez-Puelles, J. M., C. Ronda, J. L. García, P. García, R. López, and E. García. 1986. Searching for autolysin functions. Characterization of a pneumococcal mutant deleted in the *lytA* gene. *Eur. J. Biochem.* **158**:289–293.

58. Sánchez-Puelles, J. M., J. M. Sanz, J. L. García, and E. García. 1990. Cloning and expression of gene fragments encoding the choline-binding domain of pneumococcal murein hydrolases. *Gene* **89**:69–75.

59. Sanz, J. M., E. Díaz, and J. L. García. 1992. Studies on the structure and function of the N-terminal domain of the pneumococcal murein hydrolases. *Mol. Microbiol.* **6**:921–931.

60. Seto, H., and A. Tomasz. 1975. Protoplast formation and leakage of intramembrane cell components: induction by the competence activator substance of pneumococci. *J. Bacteriol.* **121**:344–353.

61. Severin, A., D. Horne, and A. Tomasz. 1997. Autolysis and cell wall degradation in a choline-independent strain of *Streptococcus pneumoniae*. *Microb. Drug Resist.* **3**:391–400.

62. Severin, A., and A. Tomasz. 2000. The peptidoglycan of *Streptococcus pneumoniae*, p. 179–195. *In* A. Tomasz (ed.), *Streptococcus pneumoniae— Molecular Biology & Mechanisms of Disease*. Mary Ann Liebert, Inc., Larchmont, N.Y.

63. Shockman, G. D., and J.-V. Höltje. 1994. Microbial peptidoglycan (murein) hydrolases, p. 131–166. *In* J.-M. Ghuysen and R. Hakenbeck (ed.), *Bacterial Cell Wall*. Elsevier, Amsterdam, The Netherlands.

64. Steinmoen, H., E. Knutsen, and L. S. Håvarstein. 2002. Induction of natural competence in *Streptococcus pneumoniae* triggers lysis and DNA release from a subfraction of the cell population. *Proc. Natl. Acad. Sci. USA* **99**:7681–7686.

65. Tettelin, H., K. E. Nelson, I. T. Paulsen, J. A. Eisen, T. D. Read, S. Peterson, J. Heidelberg, R. T. DeBoy, D. H. Haft, R. J. Dodson, A. S. Durkin, M. Gwinn, J. F. Kolonay, W. C. Nelson, J. D. Peterson, L. A. Umayam, O. White, S. L. Salzberg, M. R. Lewis, D. Radune, E. Holtzapple, H. Khouri, A. M. Wolf, T. R. Utterback, C. L. Hansen, L. A. McDonald, T. V. Feldblyum, S. Angiuoli, T. Dickinson, E. K. Hickey, I. E. Holt, B. J. Loftus, F. Yang, H. O. Smith, J. C. Venter, B. A. Dougherty, D. A. Morrison, S. K. Hollingshead, and C. M. Fraser. 2001. Complete genome sequence of a virulent isolate of *Streptococcus pneumoniae*. *Science* **293**: 498–506.

66. Tomasz, A. 1968. Biological consequences of the replacement of choline by ethanolamine in the cell wall of Pneumococcus: chain formation, loss of transformability, and loss of autolysis. *Proc. Natl. Acad. Sci. USA* **59**:86–93.

67. Tomasz, A., and W. Fischer. 2000. The cell wall of *Streptococcus pneumoniae*, p. 191–200. *In* V. A. Fischetti, R. P. Novick, J. J. Ferretti, D. A. Portnoy, and J. I. Rood (ed.), *Gram-Positive Pathogens*. ASM Press, Washington, D.C.

68. Tomasz, A., and M. Westphal. 1971. Abnormal autolytic enzyme in a pneumococcus with altered teichoic acid composition. *Proc. Natl. Acad. Sci. USA* **68**:2627–2630.

69. Usobiaga, P., F. J. Medrano, M. Gasset, J. L. Garcia, J. L. Saiz, G. Rivas, J. Laynez, and M. Menendez. 1996. Structural organization of the major autolysin from *Streptococcus pneumoniae*. *J. Biol. Chem.* **271**:6832–6838.

70. Varea, J., J. Saiz, C. López-Zumel, B. Monterroso, F. J. Medrano, J. L. Arrondo, I. Iloro, J. Laynez, J. L. García, and M. Menéndez. 2000. Do sequence repeats play an equivalent role in the choline-binding module of pneumococcal LytA amidase? *J. Biol. Chem.* **275**: 26842–26855.

71. Vollmer, W., and A. Tomasz. 2001. Identification of the teichoic acid phosphocholine esterase in *Streptococcus pneumoniae*. *Mol. Microbiol.* **39**:1610–1622.

72. Waite, R. D., D. W. Penfold, J. K. Struthers, and C. G. Dowson. 2003. Spontaneous sequence duplications within capsule genes *cap8E* and *tts* control phase variation in *Streptococcus pneumoniae* serotypes 8 and 37. *Microbiology* **149**:497–504.

73. Whatmore, A. M., and C. G. Dowson. 1999. The autolysin-encoding gene (*lytA*) of *Streptococcus pneumoniae* displays restricted allelic variation despite localized recombination events with genes of pneumococcal bacteriophage encoding cell wall lytic enzymes. *Infect. Immun.* **67**:4551–4556.

74. Yother, J., K. Leopold, J. White, and W. Fischer. 1998. Generation and properties of a *Streptococcus pneumoniae* mutant which does not require choline or analogs for growth. *J. Bacteriol.* **180**:2093–2101.

75. Young, R. 2002. Bacteriophage holins: deadly diversity. *J. Mol. Microbiol. Biotechnol.* **4**:21–36.

TRANSFORMATION

Sanford A. Lacks

7

INTRODUCTION

Transformation, which alters the genetic makeup of an individual, is a concept that intrigues the human imagination. In *Streptococcus pneumoniae* such transformation was first demonstrated. Perhaps our fascination with genetics derived from our ancestors observing their own progeny, with its retention and assortment of parental traits, but such interest must have been accelerated after the dawn of agriculture. It was in pea plants that Gregor Mendel in the late 1800s examined inherited traits and found them to be determined by physical elements, or genes, passed from parents to progeny. In our day, the material basis of these genetic determinants was revealed to be DNA by the lowly bacteria, in particular, the pneumococcus. For this species, transformation by free DNA is a sexual process that enables cells to sport new combinations of genes and traits.

Genetic transformation of the type found in *S. pneumoniae* occurs naturally in many species of bacteria (70), but initially only a few other transformable species were found, namely, *Haemophilus influenzae*, *Neisseria*

meningitidis, *Neisseria gonorrhoeae*, and *Bacillus subtilis* (95). Natural transformation, which requires a set of genes evolved for the purpose, contrasts with artificial transformation, which is accomplished by shocking cells either electrically, as in electroporation, or by ionic and temperature shifts. Although such artificial treatments can introduce very small amounts of DNA into virtually any type of cell, the amounts introduced by natural transformation are a millionfold greater, and *S. pneumoniae* can take up as much as 10% of its cellular DNA content (40).

HISTORY

Despite its small size and relative simplicity, the pneumococcus, as the major causative agent of pneumonia, has been a scourge of humanity, particularly before the advent of antibiotics. In the early 1900s, pneumococci were the object of study in many laboratories worldwide, including those of Fred Griffith at the Public Health Ministry in London, England, and of Oswald Avery at the Rockefeller Institute in New York, N.Y. At the time, numerous serologic types had been identified in *S. pneumoniae*, and the variation in capsule responsible for the serologic differences was the focus of study in Griffith's laboratory.

Sanford Lacks, Biology Department, Brookhaven National Laboratory, Upton, NY 11973.

The Pneumococcus. Volume Editor, Elaine I. Tuomanen,
© 2004 ASM Press, Washington, D.C.

Discovery of Transformation

In 1928, Griffith reported that heat-killed encapsulated pneumococci could transfer the ability to make a capsule and, hence, to infect mice when injected together with live, unencapsulated (hence nonpathogenic) pneumococci (31). Griffith termed the phenomenon "transformation." He conjectured that a seed of capsular polysaccharide was perhaps transferred from the heat-killed bacteria, but he also wondered whether an enzymatically active protein might be the agent transferred. No one at the time supposed that bacteria contained genes, let alone that DNA was the genetic material.

DNA Is the Transforming Principle

Soon after Griffith's discovery, Oswald Avery et al. at the Rockefeller Institute took up the problem. Subsequent developments, such as a procedure to transform bacterial cells in vitro as opposed to in mice (17), and the extraction in soluble form of the active principle from heat-killed cells, enabled its further resolution (2). However, proof that the transforming principle was DNA awaited the landmark paper of 1944 (3). That work had enormous impact in demonstrating that the genetic material consisted of DNA.

Quantitation of Transformation

Although capsular transformation was effective in proving that bacteria have genes and that genes are composed of DNA, it depended on screening for smooth (encapsulated) and rough (unencapsulated) colonies as opposed to selective growth of transformed cells. For precise quantitation in bacterial genetics, selectable markers are preferable because they allow counting of small numbers of transformants in a largely untransformed population. Rollin Hotchkiss, who continued the work of Avery's group at Rockefeller, obtained drug resistance markers and devised methods for the quantitative measurement of transformation frequencies. The value of quantitative measurements for understanding molecular mechanisms of transformation, regulation of recipient cell competence, processes of DNA degradation, and genetic mapping by transformation is apparent from the results of Hotchkiss and those influenced by his laboratory. In 1957, using a streptomycin resistance marker, he showed that transformants increased linearly with added marker DNA, until a saturation level was obtained, thereby demonstrating a discrete number of DNA uptake sites on the recipient cells (41). Competence for DNA uptake, measured as transformation frequency, was found to vary systematically during the culture growth cycle (40). Later it was shown that competence depended on accumulation of an extracellular polypeptide (34, 108).

DNA Structure

The intense interest in DNA that followed the discovery of Avery et al. led ultimately to the structural model for DNA proposed by Watson and Crick (113), which was based on the crystallographic data of Franklin and Gosling (27) and the chemical data of Chargaff (12). In this model, two complementary strands of DNA are helically wound around each other and attached by hydrogen bonds between complementary DNA bases. This structure was soon supported by additional studies of pneumococcal transformation. Using quantitative measurements of transforming activity, Marmur and Lane demonstrated strand separation on thermal denaturation of the native, double-stranded DNA and the renaturation of its transforming activity by annealing the separated strands at submelting temperatures (73). Thus, there was back-and-forth interplay between results from the transformation of *S. pneumoniae* and our knowledge of the role and structure of DNA.

Our current understanding of the mechanism of transformation and the genetics of *S. pneumoniae* has depended on a variety of experimental approaches: tracing of the fate of isotopically labeled DNA, analysis of genetic recombination frequencies, isolation and characterization of transformation-defective and other mutants, DNA cloning and sequencing, and identification and use of the competence-inducing peptide to characterize the regulatory aspects of transformation.

MOLECULAR FATE OF DNA IN TRANSFORMATION

Eclipse and the Entry Nuclease

By labeling DNA with radioactive P atoms, it was shown that transforming DNA is physically incorporated into the cells of S. pneumoniae (24, 67). The donor DNA must be in a native, double-stranded form; although single-stranded DNA can be taken up to a slight extent, its ability to transform is <0.1% of that of native DNA (5, 78). A curious and important finding, however, was the eclipse of donor markers immediately after DNA uptake; that is, DNA extracted from newly transformed cells was itself devoid of transforming activity when tested on other cells (25). Examination of the molecular state of the newly introduced DNA showed it to be present as single strands (44). An amount of donor DNA equivalent to the entered strands was released as oligonucleotides outside the cell (48). The eclipse in transforming activity is thus ex-

plained by the conversion of DNA upon entry to single strands, which are unable to efficiently enter tester cells. A membrane-located nuclease, EndA, was implicated in entry of the DNA strands (51–53), as shown in Fig. 1. endA mutants are reduced in transformation to ~0.1% of the wild type (52, 94). Terminal labeling of donor DNA has shown that the 3′ end of the incoming strand enters first (77).

Bound DNA

Instead of taking DNA into the cells, mutants lacking EndA bind large amounts of DNA to the outside of the cell (51). This DNA is irreversibly bound; it can be removed by treatment with DNase but not by washing. Such externally bound DNA is always found to have undergone at least one single-strand break (47, 49). Although the DNA might be initially bound reversibly, it quickly undergoes strand breaks, perhaps at the sites of DNA entry, which may bind the DNA to a surface protein and render it irreversibly bound. These breaks limit the weighted of DNA strand segments entering the cell to a weighted average of ~5 kb. When EndA is present, it initially acts

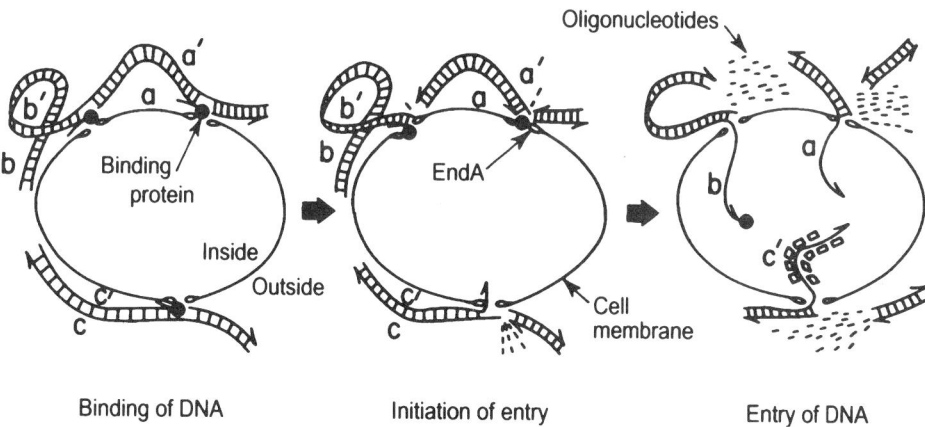

FIGURE 1 Model for DNA uptake in transformation of S. pneumoniae. Double-stranded DNA is irreversibly bound to the cell surface and undergoes single-strand cleavage at random sites, possibly by action of a binding protein. A membrane-located nuclease, EndA, initiates entry of the bound strand by endonucleolytic cleavage of the complementary strand to give a double-strand break. Processive action of EndA 5′ to 3′ degrades the complementary strand to oligonucleotides, which remain outside the cell, while donor strands enter from their 3′ end (half-arrowhead). It is not known whether the strand enters without (a) or with (b) a pilot protein. The entering DNA is covered with a single-strand binding protein (c′).

on the strand opposite the break made on DNA binding to give a double-strand break (49), as shown in Fig. 1. If the internal strand segments are homologous to the chromosome, they are rapidly integrated, largely intact, into the recipient chromosome (26, 44). DNA lacking homology is taken up equally well into the cell, but it fails to be integrated and is eventually degraded (50).

MODES OF GENETIC TRANSFORMATION

Chromosomal Transformation

The pneumococcal chromosome corresponds to a circular genome containing 3.2 million base pairs (39, 105). Spontaneous and chemically induced mutations in many genes have been obtained; they correspond to various single-site base changes and deletions and insertions of all sizes (45). Some chemical agents preferentially produce specific base changes (45, 60). UV light is not mutagenic in *S. pneumoniae*, unlike in other bacteria (28). With current technology, designated mutations can be tailor-made in vitro, using synthetic oligonucleotides and PCR, and introduced into the bacteria by transformation. Any selectable mutation or its wild-type allele constitutes a marker that can be ana-lyzed by transformation. Classical transformation was concerned with changes in the recipient cell chromosome. For this to occur, single-stranded DNA must synapse with the double-stranded chromosomal DNA (45), as shown in Fig. 2A. As in most recombination systems, the protein RecA is required (74, 82, 87).

For chromosomal transformation, frequencies increase linearly with increasing DNA concentration, which indicates that entry of a single DNA fragment suffices (41). Markers located nearby on the chromosome will exhibit linkage, that is, show a cotransformation frequency greater than expected for two separate entry events. Recombination frequencies increase linearly with the genomic distance between the markers. Fine-structure analysis of recombination at two genetic loci—*mal* and *ami*—revealed a frequency of recombination of 0.02% per nucleotide (22, 45). This analysis allowed a linear mapping of these two loci, which was confirmed by analysis of overlapping deletions and, subsequently, by DNA sequencing (15, 60). The physical basis of recombination derives from double-strand breaks during DNA extraction, single-strand breaks on binding of DNA, and possible breaks or exchanges during strand integration (54).

FIGURE 2 (A) Chromosomal transformation. Heavy line, donor DNA strand segment. Thin line, chromosomal DNA. M and m, marker difference between donor and recipient. For plasmid transformation, substitute resident plasmid for chromosomal DNA. (1) Linear synapsis; (2) integration intermediate; (3) covalent joining. (B) Plasmid establishment. (1) Annealing of complementary strand fragments that entered separately; (2) repair synthesis; (3) completed replicon. (C) Chromosomal facilitation of plasmid establishment. (1) Circular synapsis followed by repair synthesis and ligation to close the plasmid strand; (2) synthesis of the complementary strand from the plasmid origin of replication; (3) release of established plasmid. (D) Ectopic integration of the *mal* marker in the vicinity of the *sul* locus. (1) Donor DNA consists of separately cloned *mal* and *sul* genes ligated together; (2) circular synapsis of the donor strand fragment at the *sul* chromosomal locus (a gap is filled by repair synthesis); (3) a single-strand crossover integrates the donor strand into the chromosome; (4) replication of the chromosome converts the integrated single-strand segment to a duplex form, giving a *mal* segment inserted between duplicated *sul* segments. (E) Mutagenesis of the *ami* gene by additive insertion of a nonreplicating plasmid. (1) Donor DNA consists of the *ami* gene joined to an *E. coli* plasmid containing an *erm* gene expressible in *S. pneumoniae*; (2) circular synapsis of the donor strand at the *ami* chromosomal locus and repair synthesis; (3) a single-strand crossover integrates the donor strand into the chromosome; (4) replication of the chromosome converts the integrated single-strand segment to a duplex form so that the *E. coli* plasmid segment is inserted between duplicated *ami* segments, thereby producing an aminopterin resistance mutation. Letters a to d and a′ to d′ in panels D and E designate parts of the *sul* and *ami* loci, respectively.

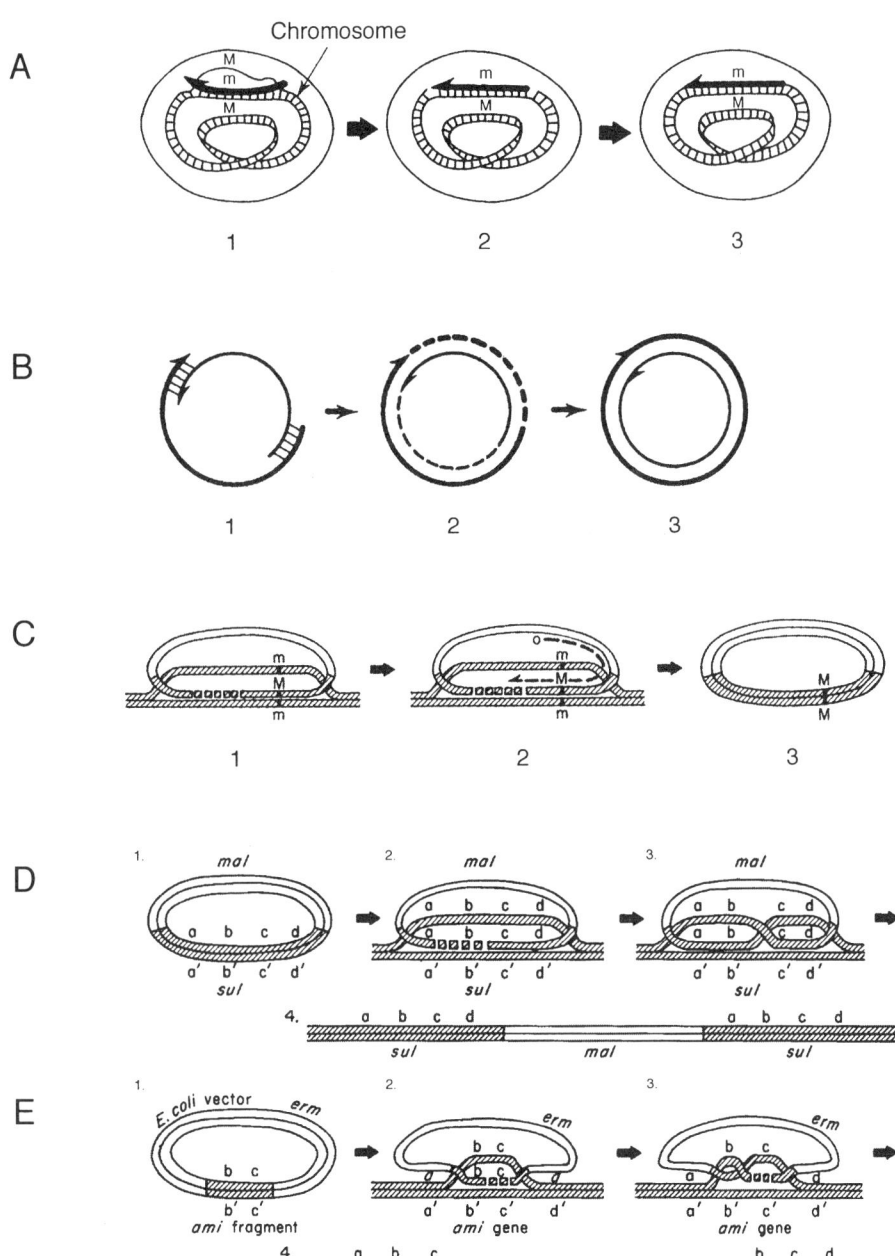

Plasmid Transfer and Transformation

Chromosomal transformation is only one of several types of genetic transforming events; additional types relate to plasmids and other circular donor DNA, as shown in Fig. 2B. Although plasmids are found infrequently in natural strains of *S. pneumoniae*, they can be introduced from other streptococcal species. Plasmid transformation occurs in a cell containing a plasmid when it is treated with plasmid DNA containing a genetic marker. This is a process akin to chromosomal transformation, and it also has a linear dependence on DNA concentration. It is distinct from plasmid transfer, which follows introduction of monomeric plasmid DNA into a cell lacking the plasmid. Because a single entry event can provide only a linear strand, two entry events are required in this case to give a complementing pair of strands that can circularize into a plasmid capable of replication, and the frequency of establishment thus depends on the square of the DNA concentration (98).

Cloning in *S. pneumoniae* and Chromosomal Facilitation of Plasmid Transfer

Recombinant plasmids containing chromosomal DNA can be made and propagated in *S. pneumoniae*; the vector pMV158 and its derivatives have been particularly useful in this regard (102). Such cloning of pneumococcal DNA in *S. pneumoniae* itself has been helpful because many genes, such as *malM*, *hexA*, and *endA*, could not be cloned in systems using *Escherichia coli*, in which they exhibited toxic effects (101). Interestingly, transfer of a recombinant plasmid requires only a single entry event. Establishment of the recombinant plasmid apparently results from a circular synapsis of the linear plasmid strand with the chromosome, which allows replication of an intact complementary strand (Fig. 2C). This interaction, which elevates the frequency of plasmid establishment, is called chromosomal facilitation of plasmid transfer (69). During circular synapsis, markers may be introduced from the chromosome into the cloned segment in the plasmid. This may

or may not be desirable but can be easily monitored in the product.

Circular Integration

Circular synapsis is also the basis for circular integration into the chromosome of nonreplicating circular DNA, which could be artifically constructed de novo, or a recombinant plasmid incapable of replication in *S. pneumoniae* (Fig. 2D and E). Such integration, also called additive transformation, has been very useful for introducing marked mutations into genes to analyze their function and facilitate their cloning (56). Although the recombination event generally depends on homology, illegitimate recombination, at points lacking extensive homology, may occur more frequently than in linear chromosomal transformation and may produce chromosomal deletions adjacent to the point of insertion (109). Circular integration also allows the ectopic insertion of a pneumococcal or foreign gene at a place in the genome where it is not normally found (72). This can be accomplished by ligating a chromosomal DNA segment at the desired location to DNA containing either a pneumococcal or foreign gene and circularizing the product prior to its use as donor DNA (Fig. 2D).

Conjugative Transposons

Although drug resistance genes are not usually found on plasmids in *S. pneumoniae*, they are frequently located on conjugative transposons, which are large chromosomal elements ranging from 15 to 60 kb that contain mobilization factors for their self-transmission to other cells (110). As inserts in the chromosomal DNA, these elements can also be transferred by the transformation mechanism.

COMPETENCE AND ITS REGULATION

Initially, investigators found it difficult to reproduceably transform *S. pneumoniae*, because variations in strains used, growth media, and timing of DNA treatment affected results. Thus, it became evident that cells of *S. pneumoniae* are not always competent to take up DNA. Early observation of temporal variation of competence during a culture growth cycle (40) and of competence-stimulating activity of cell extracts (85, 108) suggested that competence for transformation was under regulatory control. An important finding was that a specific set of proteins is induced during the development of competence (79). Extension of these studies coupled with analysis of transformation-negative mutants has identified a two-tiered regulatory mechanism for controlling competence.

Quorum Sensing

A quorum-sensing mechanism constitutes the first tier of regulation, as illustrated in Fig. 3. It involves the products of five genes contained in two separate operons. ComC, the product of the first gene in one operon, is a polypeptide containing 41 amino acid residues and a GG motif past which cleavage occurs to give a carboxyl-terminal 17-mer oligopeptide (Fig. 4A) that is excreted from the cell (34). Similar leader sequences with GG motifs are commonly found in propeptide bacteriocins destined for excretion from the bacterial cell (34). The products of the other operon, ComA and ComB, which are, respectively, a transmembrane protein and an ATP-binding cytosolic protein of the ABC (ATP-Binding Cassette) transporter family, are responsible for processing and excreting the 17-mer (34, 42). At sufficiently high concentrations in the medium (which are nevertheless low in molar terms, that is, ~10 nM) this competence-stimulating polypeptide (CSP) can induce high levels of competence in an incompetent culture (34).

Roughly half of all pneumococcal strains isolated from patients encode the CSP1 sequence shown in Fig. 4A; nearly all the rest encode a distinct but similar sequence, CSP2, that differs in eight amino acid residues (92). Corresponding to these two alternative forms of CSP, the sequences of the corresponding *comD* products differ in 13 of the first hundred

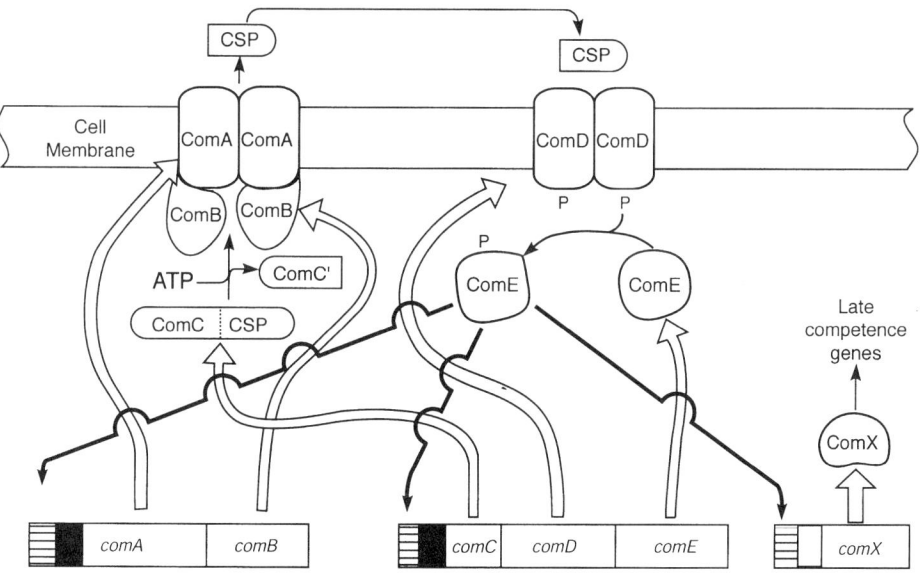

FIGURE 3 Model of quorum sensing in the regulation of competence for transformation. Accumulated extracellular CSP signals ComD to phosphorylate ComE, which then enhances synthesis of CSP and ComX. ComX is needed to transcribe genes required for transformation. Relevant genes are shown at the bottom. Open arrows point to gene products. Solid arrows show effects on promoters. Operon control elements: black, SigA promoter; white, weak SigA promoter; horizontal hatch, binding site for ComE enhancer. Other designations: P, protein phosphate; ComC', residual *comC* product after removal of CSP.

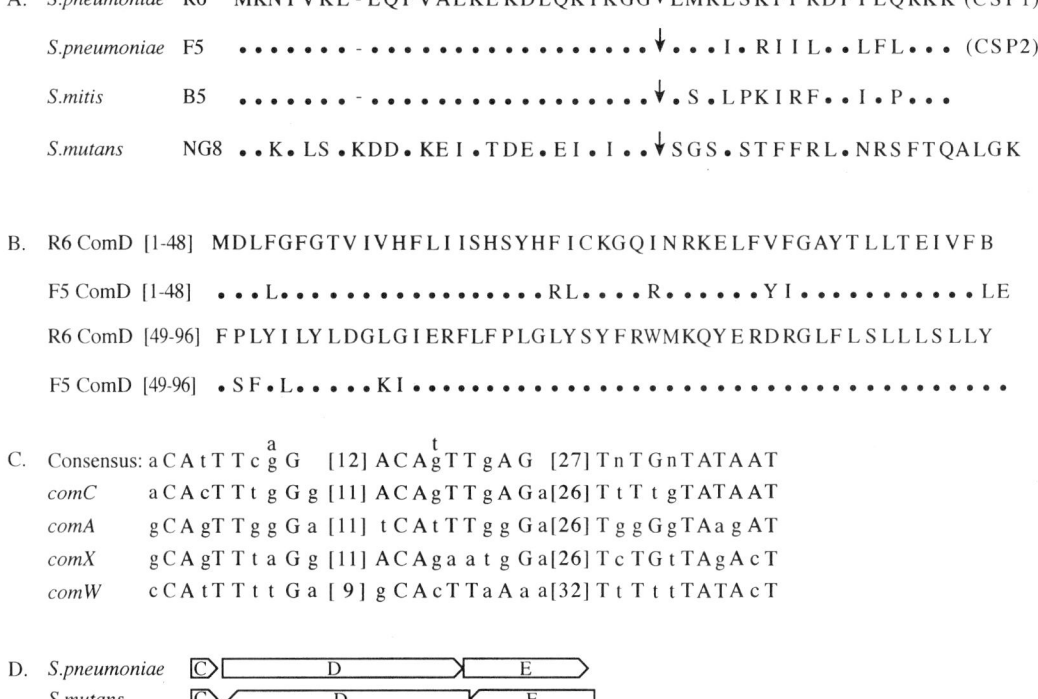

FIGURE 4 Variation in competence regulatory components of streptococci. (A) CSPs. Species and strain are indicated. Arrows indicate point of cleavage from the precursor. Dots indicate identity to the peptide sequence above. (B) CSP receptor regions of ComD in *S. pneumoniae* strains. The first 96 amino-terminal residues are compared. (C) Binding sites for ComE in *S. pneumoniae*. The consensus sequence is from reference 114. Uppercase letters indicate correspondence to relatively invariant bases of the consensus. Numbers in brackets give distance between ComE-binding sequence repeats and between the second repeat and the extended −10 promoter site (97). (D) Arrangement of the *comCDE* genes. Boxes depicting genes point in the direction of transcription.

amino-terminal residues in the two classes (115), as shown in Fig. 4B. This supports the idea that residues in the amino-terminal half of ComD specifically recognize CSP (35). From its sequence, ComD is an integral membrane protein with six transmembrane helices in its amino-terminal half. Its carboxyl-terminal half corresponds to a histidine kinase. Together ComD and ComE function as a two-component signal transduction system frequently found in bacteria (13, 88). In the present case, CSP acting externally on the ComD receptor presumably triggers the transfer of a phosphate residue from histidine in ComD to ComE, the response regulator, which would alter ComE and enable it to ac-

tivate the genes under its control. CSP acts to increase transcription from the *comAB* and *comCDE* operons themselves, thereby resulting in its autocatalytic production and a surge of competence.

The consensus binding site in DNA for ComE action, which was deduced by comparing the promoter regions of the *comAB* and *comCDE* operons (114), is shown in Fig. 4C. Both operons contain an extended −10 promoter motif typical for *S. pneumoniae* (97), which alone can afford a low basal level of transcription (59). This ensures the production and release of CSP at a low rate so that it will accumulate in the medium. When external CSP reaches a sufficient concentration, it acts back on ComD, which in turn activates ComE. The latter acts as an enhancer to greatly increase transcription from both operons, which lack a typical −35 promoter motif. Instead, binding of activated ComE to the directly repeated sequences in the DNA elicits a high level of transcription. When the population of bacteria reaches a suitable density to allow interaction between released DNA and cells, that is, when a quorum is present, this system provides a surge of competence to facilitate transformation. Enabling genetic exchange during late exponential growth allows the selection of new genotypes at a moment late enough to recognize increasing stress on the population, yet early enough to synthesize a more adaptive response.

Alternative Sigma Factor

A most important third operon affected by ComE (Fig. 4C) contains the single gene *comX*. In *S. pneumoniae* there are two copies of this operon, each of which contains a single *comX* gene that encodes an alternative sigma factor (66). During competence, this sigma factor, ComX, replaces SigA in RNA polymerase. ComX, therefore, serves as the link between the two tiers of competence regulation. It does not recognize the usual promoter of *S. pneumoniae*, but rather a different −10 sequence upstream from the mRNA start site, TACGAATA. This sequence, called the "combox," is found upstream of operons containing nearly all other genes required for competence for DNA uptake and other functions associated with transformation (10), as shown in Fig. 5. The sequential expression of the two tiers causes the early competence gene transcripts of the first tier, which depend on the response regulator ComE, to reach a peak ∼7.5 min after addition of CSP to a noncompetent culture, whereas the late competence gene transcripts of the second tier, which depend on ComX, reach a peak ∼12.5 min after such addition (1, 90a, 96).

In addition to early and late classes, in which mRNA levels are generally increased ∼50-fold and decline after reaching their peak values, induction of competence affects two additional classes of genes (90a, 96). Genes in the "delayed" response class are expressed more gradually, and mRNA increases only 2- to 10-fold. None of these genes appear to be involved in transformation. Rather, they encode chaperonins, heat shock proteins, and proteases implicated in protein folding or elimination of misfolded proteins. Thus, they may represent a response to stressful conditions. Genes in the "repressed" class, although numerous (over 60 have been reported), are transiently reduced only 2- to 4-fold in expression; they encode ribosomal proteins and various enzymes and transport systems (90a, 96). Regulation of the delayed and repressed genes is not affected by elimination of ComX (90a), and the mechanisms of their control are unknown.

Approximately a dozen additional operons that show transcriptional behavior similar to early competence genes have been reported (6, 90a, 96). Only half of them are preceded by a repeat sequence corresponding to the ComE binding site. With the exception of one gene, *comW* (originally called CPIP912 [6]), preceded by the binding site (Fig. 4C), none of these early genes are essential for transformation (90a). The function of ComW is unknown, but it could be the factor additional to ComX that enhances synthesis of late competence gene products (71, 90a). Why the other operons are part of the first-tier competence regulon and whether they have any function related to transformation has not been ascertained.

Combox Regulon

Sixteen late competence genes that are required for DNA uptake or subsequent functions in transformation have been identified (for a review, see reference 58). These genes, to be discussed in detail in "Molecular mechanism of DNA uptake," below, are found in eight operons, each of which is preceded by a combox (Fig. 5). A search for additional combox sequences in *S. pneumoniae* revealed eight

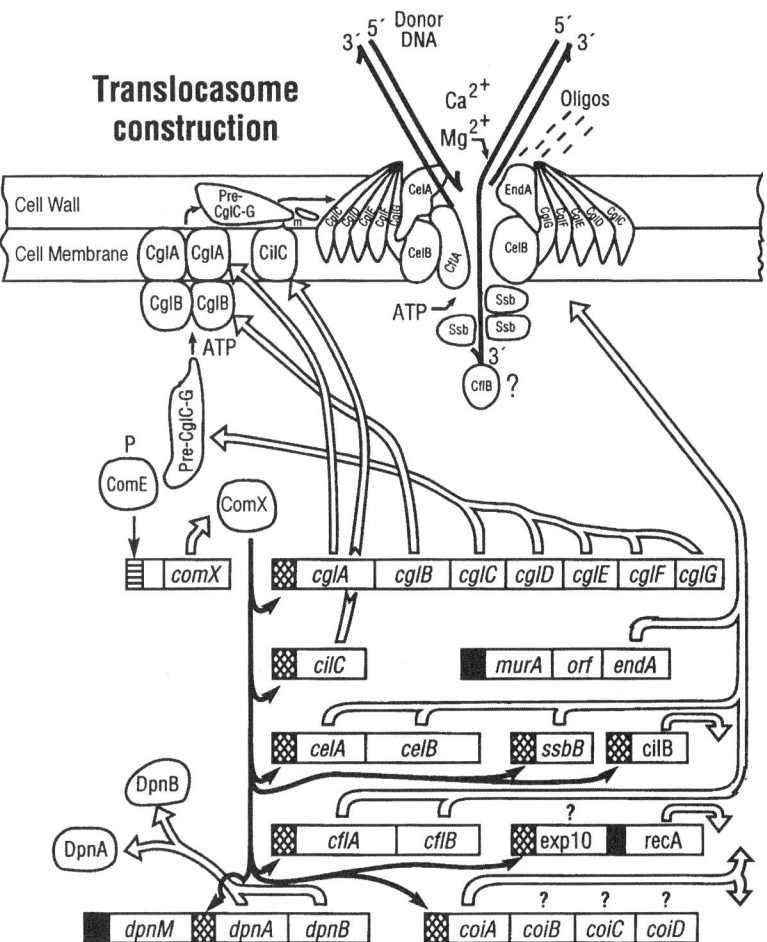

FIGURE 5 Late competence genes and construction and function of the DNA uptake apparatus in *S. pneumoniae*. DNA is depicted by heavy lines, with half-arrow-
heads indicating the 3′ direction. The "translocasome" is a hypothetical structure ex-
truding through the cell wall and formed by CglC-G proteins, which are exported
by the CglA-CglB complex and processed by CilC. Other components of the
translocasome are CelA, which binds DNA; EndA, which degrades one strand; CelB,
which forms a membrane pore for entry of the other strand; and CflA, which may
unwind donor DNA. CoiA and CflB also may function in DNA uptake, possibly by
nicking and attaching to DNA prior to entry. Calcium and magnesium ions are required for DNA uptake, with the latter needed by EndA. Upon entry, single strands
are coated with Ssb. CilB, DpnA, and RecA act subsequent to DNA uptake. Relevant genes are shown at the bottom. Open arrows point to gene products. Solid arrows show effects on promoters. Operon control elements: black, SigA promoter;
horizontal hatch, ComE enhancer; crosshatch, ComX promoter. Other designations:
P, protein phosphate; m, methyl group on CglC, CglD, and CglF after processing by
CilC. Question marks indicate an uncertain role in transformation.

new operons expressed as late competence genes (90). However, they encode proteins that show no obvious relevance to transformation. Similarly, searches of transcripts elicited by CSP revealed eight more operons with kinetics similar to those of late competence genes, but again deletion of the genes or the associated combox did not affect transformation (6, 90a, 96). Only four of the additional late operons induced by CSP lacked a combox. Further investigation will be needed to see whether these putative members of the second-tier competence regulon do function in transformation or whether they have roles in another cellular process, perhaps one requiring similar quorum sensing.

Strain Variation and Constitutive Competence

In the laboratory strain Rx, in which competence regulation has been most intensively studied, competence appears as a sharp peak of ~30 min in duration. The upsurge of competence is clearly due to the autocatalytic nature of the first-tier regulation and its positive induction of competence genes. However, it is not known why synthesis abruptly ceases. It has been hypothesized that one of the late competence genes may block competence development, but no evidence for such a function has yet been adduced. Another laboratory strain, R6, exhibits a broad peak of competence lasting over several generations of bacterial growth. It is possible that the latter strain is deficient in the blocking function, but it is not known which expression pattern is typical of wild strains. It is conceivable that such strain variations occur naturally, with a short-lived, high spike of competence being desirable under some circumstances and a more moderate but longer-lived level of competence under others. In the former case, rapid synthesis of competence proteins may interfere with normal protein synthesis, and this was in fact observed (79). However, it has long been known that a derivative of R6, the *trt* mutant, is constitutively competent (48).

The *trt* mutation is located in *comD*; it corresponds to a change in ComD of Asp_{299} to Asn, which apparently alters the histidine kinase so that it can phosphorylate ComE even without activation of ComD by CSP (61). The *trt* mutant was selected on the basis of its transformability in the presence of trypsin, a proteolytic enzyme that degrades CSP. Additional mutations have recently been found in both *comD* (75) and *comE* (20) which similarly render their products constitutively active.

MOLECULAR MECHANISM OF DNA UPTAKE

In addition to the constitutive *endA* gene discussed above, four inducible operons containing late competence genes, *celAB*, *cflAB*, *cglABCDEFG*, and *cilC*, are uniquely involved in DNA uptake. The genes in these operons are homologous to those genes found responsible for DNA uptake in another transformable gram-positive bacterium, *B. subtilis*, in which their protein products have been characterized with respect to primary sequence, proteolytic processing, membrane location, and other properties (reviewed in reference 19). Mutations in the pneumococcal genes affect transformation similarly to those in *B. subtilis* (7, 10, 89), so the role of the encoded proteins can be assumed to correspond in the two species (Table 1).

The *cgl* operon of *S. pneumoniae* contains seven genes, which are adjacent to each other or slightly overlapping, as in *B. subtilis*. Together with *cilC*, the *cglABCDEFG* operon encodes a set of proteins similar to those responsible for the extrusion of type IV pilins and other proteins through the outer membrane and cell wall of gram-negative bacteria (38). Mutations in any of these genes prevent binding and entry of DNA and reduce transformability to <0.01% of normal. Based on their homology, the *cglA* and *cglB* products are, respectively, an energy-transducing protein and a membrane-spanning protein responsible for transporting through the cell membrane the products of the five down-

TABLE 1 Genes of *S. pneumoniae* implicated in transformation

Gene[a]	Regulation[b]	Homolog[c]	Product	Function	R6 genome	TIGR4 genome
endA	C		Membrane nuclease	Degrade donor strand	spr1779	SP1964
recA	C, L	Eco *recA*	Rec protein	DNA recombination	spr1757	SP1940
comA	E		ABC transporter	Process/export CSP	spr0043	SP0042
comB	E		ABC transporter	Process/export CSP	spr0044	SP0043
comC	E		CSP precursor	Quorum sensing	spr2043	SP2237
comD	E		Histidine kinase	Detect CSP signal	spr2042	SP2236
comE	E		Response regulator	Early gene transcripts	spr2041	SP2235
comW	E		81-aa[d] protein	Unknown	spr0020	SP0018
comX1	E		Sigma factor	Late gene transcripts	spr0013	SP0014
comX2	E		Sigma factor	Late gene transcripts	spr1819	SP2006
celA/cilE	L	Bsu *comEA*	DNA-binding protein	Bind donor DNA	spr0856	SP0954
celB	L	Bsu *comEB*	Transmembrane protein	DNA entry pore?	spr0857	SP0955
cflA	L	Bsu *comFA*	Putative helicase	Assist DNA entry?	spr2013	SP2208
cflB	L	Bsu *comFB*	221-aa protein	DNA pilot protein?	spr2012	SP2207
cglA/cilD	L	Bsu *comGA*	ABC transporter	Export CglC–CglG	spr1864	SP2053
cglB	L	Bsu *comGB*	ABC transporter	Export CglC–CglG	spr1863	SP2052
cglC	L	Bsu *comGC*	Membrane protein	Cell wall channel?	spr1862	SP2051
cglD	L	Bsu *comGD*	Membrane protein	Cell wall channel?	spr1861	SP2050
cglE	L	Bsu *comGE*	Membrane protein	Cell wall channel?	spr1860	SP2049
cglF	L	Bsu *comGF*	Membrane protein	Cell wall channel?	spr1859	SP2048
cglG	L	Bsu *comGG*	Membrane protein	Cell wall channel?	spr1858	SP2047
cilB/smf	L	Hin *dprA*	287-aa protein	Postentry processing	spr1144	SP1266
cilC/pilD	L	Bsu *comC*	Peptidase-Mtase	Process CglC, CglD, CglF	spr1628	SP1808
coiA	L		318-aa protein	Unknown	spr0881	SP0978
dpnA	C, L		Methyltransferase	Protect donor DNA	—[e]	—[e]
exp10/cinA	L		Membrane protein	Unknown	spr1758	SP1941
lytA	C, L		Cell wall lysin	Release donor DNA?	spr1754	SP1937
ssbB/cilA	L	Eco *ssb*	Strand-binding protein	Protect donor strands	spr1724	SP1908

[a] Alternative name indicated after slash.
[b] C, constitutive expression; E, early inducible expression; L, late inducible expression.
[c] Species designations: Bsu, *B. subtilis*; Eco, *E. coli*; Hin, *H. influenzae*.
[d] aa, amino acid.
[e] —, absent in strains R6 and TIGR4, which are DpnI strains.

stream *cgl* genes, which are all smaller polypeptides with hydrophobic segments at their N termini. In most bacterial species, including *B. subtilis*, the amino acid residues KGFT precede the hydrophobic segment, and homologs of CilC cleave the leader peptide between the G and F residues (19) during its transport. However, in *S. pneumoniae*, a similar sequence, KAFT, precedes the segment in CglC, CglD and CglF (58), and cleavage presumably occurs between A and F. The product of *cilC* and its homologs is a peptidase-methyltransferase that processes extruded proteins by cleaving the polypeptide at the F residue and methylating the new N terminus (104).

How the extruded *cgl* products act in DNA uptake is conjectural, for their homologs in *B. subtilis* have not shown binding or other action on DNA, but possible functions can be envisioned (14). By analogy to their counterparts in gram-negative bacteria, they may form an appendage outside the cell membrane and passing through the cell wall. This structure may act both as a pore in the peptidoglycan layer of the cell wall through which external DNA can pass and as a scaffold on which other proteins that bind and process DNA for entry are arranged. It is also possible that the Cgl protein *complex* can bind external DNA, perhaps reversibly.

The *celA* and *celB* genes are essential for DNA uptake, and mutations in them reduce transformability to <0.01% (10, 89). CelA is a membrane protein of ~20 kDa with a long stretch of hydrophobic residues at its N terminus. Mutations of its *B. subtilis* homolog prevent binding of DNA to the cell, and the protein in vitro binds tightly to DNA (93). Binding is to the C-terminal portion of the protein, which is external to the cell. In *S. pneumoniae*, *celA* mutations greatly reduce DNA binding and completely eliminate strand degradation by EndA (7). The CelA protein thus may be responsible for nicking DNA, thereby binding it irreversibly, or possibly binding it to another protein, such as CflB. CelA may also recruit EndA to degrade the complementary strand.

CelB is an 80-kDa protein with multiple hydrophobic stretches corresponding to transmembrane segments. The protein is not required for donor DNA binding or degradation, but it is necessary for DNA entry (7, 19). It is likely that this protein forms a channel in the membrane for passage of single-stranded DNA.

The two genes in the *cfl* operon of *S. pneumoniae* correspond to the two genes of the *comF* operon of *B. subtilis* that are essential for transformation. CflA (and ComFA) mutants are reduced in transformability to 0.1% of normal (19, 65). They fail to take up DNA, but they do show binding and degradation (7). CflA and ComFA, both 50-kDa proteins, show sequence similarity to ATP-dependent DNA helicases (19). Although they lack hydrophobic regions, ComFA was associated with the cytoplasmic face of the cell membrane (19). With respect to function, it has been suggested that the protein acts as a helicase to help separate the DNA strands and propel one strand into the cell. However, it may have other, unknown functions. CflB is homologous to ComFC of *B. subtilis*. Mutations in the gene encoding ComFC reduce transformability only to 10% of normal, and the role of the gene product in transformation has not been elucidated.

A provisional model of how EndA and the various components encoded by late competence genes may act to bring DNA into the cell and process it for chromosomal integration or plasmid establishment is shown in Fig. 5.

Release of Donor DNA

For transformation to occur under natural conditions, DNA must be released from donor cells as well as taken up by recipient cells. Early studies indicated that transforming DNA was released from cells in culture (84). Furthermore, an ability to form spheroplasts, which is indicative of cell wall fragility, was correlated with competence (53, 100). Treatment of cells with crude preparations of CSP increased fragility (100); treatment with trypsin decreased fragility, except in the constitutively competent *trt* mutant (53). This fragility depended on the *lytA* gene, since it was blocked by the *cwl-1* mutation in that gene (53). Recently, it was shown that treatment of non-competent cells with pure CSP can release up to 20% of a normally internal enzyme and also significant amounts of chromosomal DNA (103). It still is not known whether this external material leaks out of all competent cells or whether it represents the complete lysis of a small proportion of them. Also, it has not been definitively shown that the *lytA* product, and not other cell wall lytic enzymes encoded by *lytB* or *lytC*, is uniquely required. In this connection, it is of interest that a late competence operon contains the *lytA* gene. This operon contains *exp10* (or *cinA*), a gene that encodes a membrane protein of unknown function, *recA*, *dinF*, a gene homologous to a DNA damage-inducible gene of *B. subtilis*, and *lytA* (74, 87). Although *lytA* and *recA* have their own promoters and their products are constitutively produced, they are nevertheless upregulated ~5-fold during competence. Thus, competence regulation based on quorum sensing may enable both release and uptake of DNA by pneumococcal cells under conditions of cell density when they are most likely to achieve genetic exchange and to benefit from it.

FATE OF DNA WITHIN THE CELL AND RECOMBINATION

DNA Bound to Protein

What happens when donor strands enter the cell? First, they are covered by a single-stranded DNA-binding protein, which may protect them from degradation by nucleases (81). *S. pneumoniae* harbors two *ssb* loci that encode proteins homologous to the single-stranded DNA-binding protein of *E. coli*. One, *ssbA*, is expressed constitutively and presumably functions in DNA replication, and the other, *ssbB*, is a late competence gene. The additional production is important because mutants defective in *ssbB* (=*cilA*) are reduced to 3% of normal transformability (10). Whether such binding also assists the uptake process is unknown.

Recombination

Binding of SsbB may facilitate recombination, as such proteins do in other systems. Essential to recombination, however, is the *recA* gene. As mentioned above, *recA* expression is increased during competence. The single-stranded donor DNA may synapse with chromosomal DNA to initially form a three-stranded structure (45), as has been proposed for RecA-mediated recombination in general (9). Subsequently, the recipient DNA

segment corresponding to the donor is replaced and eliminated to give hybrid chromosomal DNA containing donor segments of varied size, with a weighted average of ~5 kb (26). Details of the mechanism of this recombination are not known. Finally, the donor segment, which initially has free ends (55), is ligated into the chromosome. Another competence-induced gene that may play a role in chromosomal recombination is *cilB*, which is homologous to the *H. influenzae* gene *dprA*, in which mutations reduce chromosomal transformation to 0.01%, with no effect on plasmid transfer (43). This behavior is reminiscent of mutations in *S. pneumoniae* called *recP*, in which chromosomal transformation, but not plasmid transfer, was blocked (80). The genes corresponding to the *recP* and *recQ* mutations were not fully characterized. It is possible that *recP* corresponds to *cilB*, and *recQ*, in which mutations blocked both chromosomal transformation and plasmid establishment, may correspond to *recA*. Action of the *cilB* gene product may be confined to chromosomal integration, inasmuch as plasmid establishment is a simpler process, not requiring displacement of existing chromosomal DNA.

Whether other competence-induced genes affecting transformation act in the recombination process is not known, but several late competence genes have not been characterized with respect to either DNA uptake or recombination. They include the aforementioned *cflB* gene and *coiA* (89). Mutations in these genes reduce transformability only 10- to 100-fold. As deduced from the genomic sequence of *S. pneumoniae*, the *coiA* gene is the first in an operon containing four genes. The products of *coiA* and *coiB* both appear to be peptidases, *coiC* encodes a putative methyltransferase, and *coiD* encodes a putative cell wall serine proteinase. It is conceivable that these enzymes play a role in cell wall remodeling during the development of competence, but singly mutating the three downstream genes does not affect transformation (90a).

Gene Redundancy and Functional Analysis

Many important cellular systems contain redundant genes or alternative pathways to ensure retention of the system function despite the occurrence of spontaneous mutations. For example, there are duplicate *comX* genes in *S. pneumoniae* (66). Genes such as *cflB* and *exp10*, in which mutations reduce transformability minimally or not at all, may represent redundant or alternative paths. A fortiori, the numerous newly reported genes that are induced during the development of competence (90a, 96) might correspond to this class of redundant or alternative genes. These genes were not identified in screening for transformation-defective mutations, presumably because single mutations have little impact. To determine whether such genes are required for transformation, strains doubly mutant in pairs of the genes must be tested.

Restriction and Transformation

The case of *dpnA* is unusual in that it is regulated as a late competence gene (59), but it is not normally required for either chromosomal or plasmid transformation. It encodes a DNA methyltransferase that protects unmodified incoming plasmid DNA by methylating it while it is in a single-stranded form, thereby allowing plasmid establishment by unmodified donor DNA in a cell containing the DpnII restriction system (11), as shown in Fig. 6B. *S. pneumoniae* contains one or the other of two complementary restriction systems, DpnI and DpnII, which appear to be designed to prevent infection by bacterial viruses that inject double-stranded DNA. Chromosomal transformation, because it is mediated by DNA converted to single strands on entry, is not affected by restriction, as the nucleases act only on double-stranded DNA (111). In the case of plasmid establishment, two donor strands anneal to form the replicon. The facts that the pneumococcal restriction systems are designed to avoid degradation of transforming DNA and that provision is made (i.e., the *dpnA* gene) even to

counter restriction in plasmid transfer indicate that genetic transformation plays an important role in survival of *S. pneumoniae* as a species.

Mismatch Repair

Subsequent to uptake and synapsis with chromosomal DNA, but prior to its ligation to recipient DNA, the donor DNA is subject to mismatch repair. DNA mismatch repair is a mechanism, universally present in cells, that was first discovered in *S. pneumoniae* by analysis of transformation frequencies given by different mutational markers at the *malA* (45) and *amiA* (22) loci. DNA mismatches correspond to noncomplementary base pairings or deletions (or additions) in one strand to produce an aberration in the double-helical DNA structure. In chromosomal transformation, a single-stranded segment of donor DNA replaces the homologous segment of host DNA, thereby producing a mismatch wherever a genetic marker is present. It was found that different types of mismatch result in characteristic transformation efficiencies ranging from 0.05 to 1.0 (22, 45). A repair system that differentially recognizes the various types of mismatch and eliminates the donor contribution to the mismatched heteroduplex accounts for the differences in integration efficiency (21, 45, 46). Products of the unlinked *hexA* and *hexB* genes are necessary for this repair; mutations in either gene give high integration efficiencies for all markers (4, 16, 46). Because the Hex system recognizes and corrects a variety of DNA mismatches, it is called a *generalized* mismatch repair system (55). *S. pneumoniae* also contains a specialized repair system that recognizes only the mismatched sequence 5'-ATT<u>A</u>AT/TAA<u>G</u>TA-5' (mismatched bases are underlined) and converts the A to C (86). For a more detailed description of mismatch repair in *S. pneumoniae* with references, see reference 57.

Mutations in the *hex* genes also have a mutator effect (107). Thus, the system normally acts also to correct errors in newly replicated DNA. This is probably the main purpose of the repair system, and it explains why the *hex*

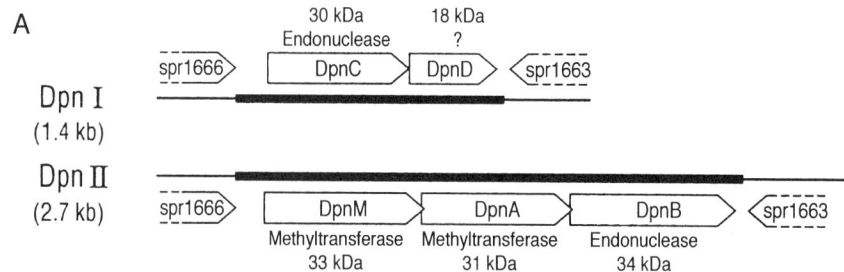

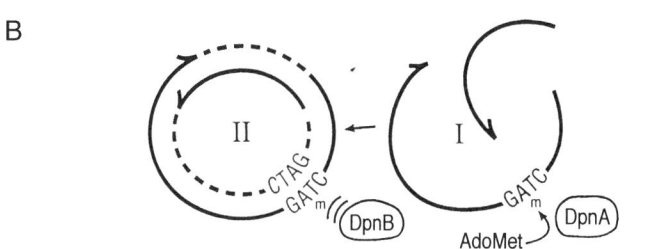

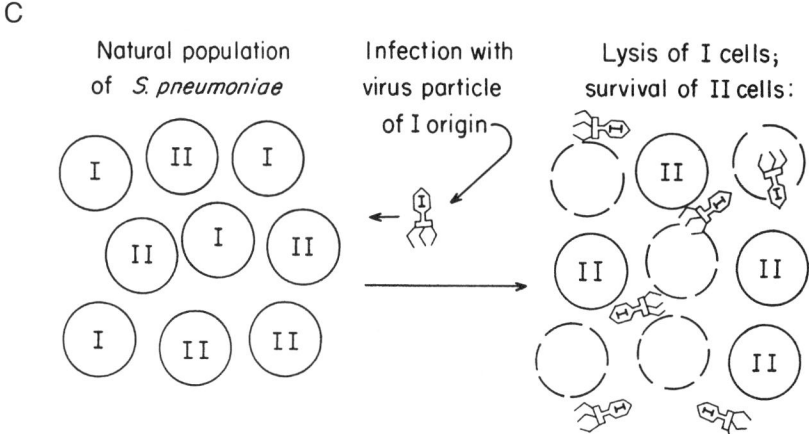

FIGURE 6 Restriction enzyme systems of *S. pneumoniae*. (A) Restriction gene cassettes of *S. pneumoniae* and their products. Symbols: thin bar, *S. pneumoniae* chromosome; thick bar, Dpn cassette; open boxes, genes in the cassettes or in the adjacent chromosome, showing direction of transcription. (B) Role of DpnA methylase in enabling unmethylated plasmid transfer into cells containing the DpnII restriction system. The degradative processing of DNA entering the cell by the transformation pathway requires the reconstitution of a plasmid from complementary strands that separately enter the cell. In a host lacking the DpnA methyltransferase, unmethylated plasmid DNA, upon reconstitution to a double-stranded form, would be cleaved by the DpnII endonuclease. In a host containing DpnA, single strands are methylated upon entry, so that the reconstituted plasmid is protected from the DpnII endonuclease. (C) Possible survival value of complementary restriction systems. I and II, cells making DpnI and DpnII, respectively. Infection of a mixed population by a single viral particle would destroy only part of the population.

genes are constitutively expressed. Their action after transformation may simply be incidental to the process and have no survival value. Sets of genes homologous to the *hex* genes of *S. pneumoniae* are found in nearly all living species. The generalized mismatch repair system that they encode reduces spontaneous mutation rates a thousandfold. As a consequence, the homologous system in human cells prevents various types of cancer (23).

COMPARISON WITH OTHER GRAM-POSITIVE AND GRAM-NEGATIVE SPECIES

Homologs to the genes responsible for competence and DNA uptake in *S. pneumoniae* are found in many other species of bacteria. These species can be considered in four groups depending on their degree of divergence from *S. pneumoniae*. In group I are closely related streptococcal species, such as *Streptococcus mitis* and *Streptococcus gordonii*, among others, that contain homologs of both early and late genes, as well as *endA*, and they are all transformable. In this group, the arrangement of the *comCDE* genes is identical to that in *S. pneumoniae*, but the CSPs and their cognate receptor portions in the histidine kinase are different (36), as indicated in Fig. 4A for *S. mitis*. Group II comprises more distantly related streptococcal species, such as *Streptococcus mutans* and *Streptococcus pyogenes*, which also contain early and late genes, with the late genes under combox control. *S. mutans* is transformable (68), but there are no reports of transformation of *S. pyogenes*. Homologs of *comX* are found in groups I and II. *S. mutans* contains close homologs of *endA* and *comCDE*, except the arrangement of the latter genes differs from that in *S. pneumoniae* and group I (68), as shown in Fig. 4D. All the transformable species in groups I and II are similarly regulated by quorum sensing and depend on similar gene products for DNA uptake and processing. However, it is conceivable that some of these products and others induced by the same regulatory system may have additional physiological roles, including response to stress.

Bacterial species outside of the genus *Streptococcus*, such as *Lactococcus lactis* and *Enterococcus faecalis*, which are closely related to *Streptococcus*, comprise group III. In these species, homologs of early genes cannot be clearly discerned, but the late genes required for DNA uptake and processing are present; however, they are not under combox control. For example, in *L. lactis*, the *cglABCDEFG* operon is present (58), but it is apparently transcribed from a typical streptococcal SigA promoter (97), TTGAat..n = 11..TnatnTATAtT. Although no transformation of *L. lactis* has been reported, if it does occur, it must be regulated by a system very different from that in *S. pneumoniae*. Alternatively, these late gene homologs may have an entirely different physiological function. One possibility is that they serve to respond to physiological stresses, such as a dearth of nutrients, that may result from crowded conditions or other causes. Some genes induced during the development of competence have no role in transformation but may act to ameliorate the effects of stresses such as heat shock (90a, 96). It is possible either that the genes required for transformation acquired stress-related functions during evolution or that transformation co-opted structures previously evolved for such functions.

Distantly related bacteria that are naturally competent for transformation constitute group IV. There are many transformable species (70), but gram-positive *B. subtilis* and gram-negative *H. influenzae* have been the most thoroughly studied. Interestingly, they both have regulatory mechanisms for induction of competence under specified conditions, but these mechanisms are different from the quorum sensing system of *S. pneumoniae* and from each other. However, for the most part, the late competence gene products are homologous throughout the bacterial kingdom (reviewed in reference 58). In fact, the pioneering work with *B. subtilis* (reviewed in reference 19) facilitated the characterization of these products in other species. Table 1 lists the genes in *B. subtilis* and *H. influenzae* that correspond to the genes in *S.*

pneumoniae believed to be part of the transformation mechanism.

RELATIONSHIP OF TRANSFORMATION TO OTHER REGULONS AND TO VIRULENCE

Mutational Analysis of Function

On a practical level, transformation has provided the means for determining the functions of many genes in *S. pneumoniae*. Stable disruption of a gene can be accomplished by chromosomal transformation with DNA from that gene into which a drug resistance marker is inserted. Circular integration of a nonreplicating plasmid containing a fragment of the gene and a selectable marker can similarly produce sufficiently stable insertion mutations, although the original plasmid is released at a low frequency. Properties of cells bearing these mutations can then be studied. However, the insertion may have polar effects on downstream genes in an operon. Introduction into the construct of a reporter gene can allow measurement of the target gene expression under various conditions. The possibilities of altering recipient cell genes are vastly increased by the use of oligonucleotide synthesis and PCR and with the availability of the genomic sequence of *S. pneumoniae* (references 39 and 105; access at http://www.ncbi.nlm.nih.gov/genomes/MICROBES/Complete.html).

Two-Component Signal Transduction Systems

The quorum-sensing system for eliciting competence is an example of a two-component signaling (TCS) system, in which a sensor histidine kinase (HK) phosphorylates a response regulator protein (RR) that carries out the effect. In most TCS systems, the histidine kinase protein contains an N-terminal transmembrane domain that allows it to respond to external signals, and the response regulator interacts with DNA to stimulate or inhibit transcription of one or more operons, which constitute a regulon (30). TCS systems are found in many bacteria; the histidine kinase domains and response regulator proteins are sufficiently conserved to

enable their recognition from sequence alone. Also, on the basis of sequence, they can be divided into subgroups, even across species. *S. pneumoniae* contains 13 TCS systems and a single isolated response regulator (63, 106), which fall into four subgroups, as indicated in Table 2. All of its sensor proteins appear to have transmembrane domains (63).

With respect to function, only ComDE has been well characterized. In the 13 TCS systems, only one gene, the response regulator of TCS02/492, is essential for viability (63, 106). Induction of this regulator affects expression of fatty acid biosynthetic genes, and the system may be required for regulation of membrane fluidity (M. L. Mohedano, K. Overweg, A. de la Fuente, M. Reuter, F. Mulholland, P. Lopez, and J. Wells, submitted for publication). Another system, TCS13/486, on the basis of its sequence similarity to ComDE and the presence of an adjacent operon encoding both a putative ABC transporter system and a polypeptide with a GG cleavage site, was surmised to be a quorum-sensing system similar to that of ComDE (63, 106). On the basis of homology to PhoPR of *B. subtilis*, TCS04/481 may be involved in phosphate regulation (83). CiaRH is an interesting TCS system that can affect both competence and virulence. Its signal for kinase action and the DNA binding sequence for its regulator are unknown, but from the experimentally determined binding sites for the regulator that affect transcription, it appears that at least 10 operons are in the regulon (76). Activation of the regulator increases transcription in seven operons; elimination of the regulator increases transcription in the other three. Competence falls into the latter group, being inhibited by the regulator, which apparently acts at a binding site upstream from *comCDE*. The *C306* mutation (Thr$_{230}$ to Pro) in CiaH, which (like *trt* in ComD) may render the kinase constitutively active, inhibits transformability (76). This mutation also increases beta-lactam resistance slightly (76). It was suggested that the CiaRH regulon mediates the transition from expo-

TABLE 2 Two-component signal transduction systems in *S. pneumoniae*

Group[a]	R6 genome	TIGR4 genome	Reference 62 name	Reference 106 name	Component	Virulence[b]	Regulatory function
AgrA/Agr	spr2042	SP2236	ComDE	498	HK	ND	Quorum sensing (88)
	spr2041	SP2235			RR	ND	
	spr0464	SP527	TCS13	486	HK	ND	
	spr0463	SP526			RR	−4	
LuxR/Nar	spr1815	SP2001	TCS11	479	HK	ND	
	spr1814	SP2000			RR	WT	
	spr0343	SP386	TCS03	474	HK	WT[c]	
	spr0344	SP387			RR	—[c]	
AraC/Lyt	spr0153	SP155	TCS07	539	HK	−4[c]	
	spr0154	SP156			RR	—[c]	
	spr0579	SP662	TCS09	488	HK	−1[c]	
	spr0578	SP661			RR	—[c]	
OmpR/Pho	spr1473	SP1632	TCS01	480	HK	−5[c]	
	spr1474	SP1633			RR	—[c]	
	spr1106	SP1226	TCS02	492	HK	WT	
	spr1107	SP1227			RR	E	
	spr1894	SP2083	TCS04	481	HK	ND	Phosphate metabolism (83)
	spr1893	SP2082			RR	−6	
	spr1997	SP2192	TCS06	478	HK	−3[c]	
	spr1998	SP2193			RR	—[c]	
	spr0077	SP0084	TCS08	484	HK	ND	
	spr0076	SP0083			RR	−3	
	spr0529	SP0604	TCS10	491	HK	ND	
	spr0528	SP0603			RR	−1	
	spr0708	SP0799	CiaRH	494	HK	−5[c]	Stationary phase? (76)
	spr0707	SP0798			RR	—[c]	
	spr0336	SP0376		489	RR	−4	

[a] Group designations in references 62 and 106, respectively.

[b] Virulence measured as $\log_{10}$ reduction in bacterial count in lung tissue 48 h after intranasal instillation in mice (106), with mutant compared to wild-type strain 0100993. ND, not determined; E, essential for growth; WT, no reduction from the wild-type strain.

[c] —, Mutant with deletions in both the HK and RR genes.

nential to stationary phase (76). Interestingly, the *ciaRH* operon shows delayed hyperexpression after CSP treatment (90a).

TCS Systems and Virulence

The effect on bacterial virulence of mutations in components of the various TCS systems was examined with two different infection models. A bacteremia model, in which pneumococci were injected into the peritoneum of mice, revealed no attenuation in any of the mutants tested (63). However, a pneumonia model, in which pneumococci were inoculated intranasally in mice and the bacterial load in lung tissue was analyzed after 48 h, showed considerable attenuation for some mutants (106), as indicated in Table 2. Apparently, the pneumonia model, which requires nasal carriage and adhesion and colonization in the lung as well as growth and resistance to host defenses, is a more stringent and superior method for revealing virulence factors. For 9 of the 13 two-component systems and the isolated response regulator, deletion of the response regulator reduced the yield of bacteria in the lung by from 1 to 6 log units. The most attenuated strain was mutated in TCS04/481, which is presumably involved in phosphate metabolism. Inasmuch as the response regulator of TCS02/492 was essential for growth, it could not be deleted, but overexpression of the kinase decreased virulence in a bacteremia model (112). Deletion of *ciaRH* gave a 5-log unit attenuation. A gene, *htrA*, which appears to be normally up-regulated by the CiaRH TCS, has been implicated in nasopharyngeal carriage of *S. pneumoniae* (99). This gene may encode a serine protease.

Virulence Genes

Two large-scale investigations to identify virulence genes in *S. pneumoniae* were carried out using insertional mutagenesis of pathogenic strains. Mutants were produced by circular integration of a nonreplicating plasmid containing a drug resistance marker. Both the pioneering study (91) and the later one (64) relied on signature tagging of individual mutants with a short, variable DNA sequence (37). This method allowed identification of bacteria that could not survive in a mixed infection with as many as 100 different mutant strains pooled into a single inoculum. Screening of large numbers of mutants was thereby greatly facilitated. Both studies examined survival after intranasal inoculation by analyzing output from the lung, and one study also used a bacteremia model and analyzed output from the spleen (64). In the latter study, screening of 1,786 mutants yielded 186 that were attenuated, divided approximately equally among those defective only in the pneumonia model, those defective in the bacteremia model, and those defective in both. Among 56 mutants selected for further study, 46 different genes were implicated, and they were characterized with respect to genomic location and putative

function. Similarly, in the earlier study (91), screening of 1,250 mutants yielded 200 attenuated strains mutated in 126 different genes, which were further characterized.

Results of the two independent studies of virulence are in surprisingly good agreement, even in some unexpected ways. Perhaps it is not surprising that as many as 10% of nonessential genes in *S. pneumoniae* are required for virulence. However, quite a few of these genes are implicated in rather mundane functions of transport, biosynthesis, and other types of metabolism. For example, of 46 genes implicated in one study, 11 were ABC-type transporters. Important metabolic systems in *S. pneumoniae* appear to be redundant. As an example, the bacterium has seven different transport systems for uptake of glutamine (39), its likely source of nitrogen. Both studies identified the glutamine transport gene *glnQ* as a virulence factor. Under particular conditions, as in infection, only one of the redundant functional forms may be operative and hence essential for growth only under those conditions. Such regulatory effects might explain the contribution to virulence of so many genes not required for growth in culture. Both studies also identified several genes previously known to affect virulence, among them those encoding proteases (*iga*, *clpC*) and adhesins (*pavA*).

In both of the above-described studies, it was concluded that pathogenicity islands, that is, clusters of virulence genes, do not occur in *S. pneumoniae*. However, this conclusion may be questioned, because neither study obtained mutations in the capsular genes, which are known to be required for virulence (8). And the capsular gene cluster, consisting of ~15,000 bp, constitutes a hefty target. In fact, a later study using in vitro transposon mutagenesis, followed by insertion into the chromosome via transformation, revealed a possible small pathogenicity island of seven genes bracketed by insertion elements (33). The cluster encoded three surface proteins, three sortases for attaching the proteins to the cell wall, and a regulatory protein. Mutations in the regulatory gene or one sortase caused attenuation of virulence, apparently by interfering with nasal carriage (33). Otherwise, the results of the transposon study did not differ appreciably from those of the earlier studies. More complete characterization of the numerous virulence genes and their effects in vivo will be necessary to understand their role.

Quorums versus Crowds

Several observations suggest a connection between competence for transformation and virulence, but the evidence is weak. Although the CiaRH TCS can affect both processes, virulence depends on action of the response regulator (106), whereas competence depends on inhibition of its action (76). Mutations in *comB* (64) and *comD* (6) were reported to reduce virulence but only gave 10-fold attenuation. It is possible that many genes in the competence regulons function in stress response rather than transformation. The quorum-sensing mechanism that facilitates transformation can also serve as a "crowd-sensing" mechanism to facilitate growth in adverse conditions, which may include infected tissues. For this reason, genes affecting transformation, stress response, and virulence may overlap in function.

CASSETTE MECHANISMS

Population Genetics

Transformation plays an important role in the population genetics of *S. pneumoniae*. First and foremost, it allows the spread of mutant forms of genes among populations. It can produce mosaic genes, as shown for drug resistance (18, 32). Second, it can introduce novel genes from genetically distant sources either in plasmids or in chromosomal fragments. A surprising number of pneumococcal genes appear to have originated in gram-negative bacteria (39). For example, genes SP1467 and SP1468 exhibit 76 and 88% identity with genes from *H. influenzae* (105), but in this case transformation could have been in either direction. Third, transformation facilitates the maintenance of diverse populations of *S. pneumoniae* with respect to certain traits. Several systems of biological importance exist in two or more states in populations of pneumococci. This population diver-

sity must have survival value for the species. In the following three cases, a similar mechanism of allelic substitution by a multigenic cassette is responsible for changes of state.

Restriction Systems

Although the capsular genes, which constitute such a cassette, have been investigated since the time of Griffith (31), insight into the mechanism by which transformation transfers these cassettes came from analysis of the Dpn restriction systems of *S. pneumoniae* (62). Cells of *S. pneumoniae* contain either the DpnI or DpnII system (Fig. 6A). The DpnI endonuclease recognizes and cleaves the methylated DNA sequence 5′-GmATC; cells that produce it contain unmodified DNA. The DpnII system is complementary to the DpnI system in that it recognizes the unmethylated sequence 5′-GATC. Unlike other restriction systems, it encodes *two* methyltransferases, DpnM and DpnA, which methylate double- and single-stranded DNA, respectively. The DpnII endonuclease cleaves unmethylated, double-stranded DNA. Both endonucleases cleave only double-stranded DNA, so incoming single strands of susceptible transforming DNA are not degraded. Furthermore, DpnA can methylate such strands, thereby protecting reconstituted plasmids (Fig. 6B), as described above. Thus, these systems are designed to block bacterial virus infection but not to interfere with genetic transformation between cells with different restriction systems. The dual systems may prevent viral epidemics from wiping out an entire pneumococcal population, inasmuch as an initial infecting agent would be either methylated or not (Fig. 6C). The genes encoding the two restriction systems are clustered at the same locus in the chromosome. Since the genes bordering them are identical, one system can replace the other by chromosomal transformation (62).

Capsular Genes

The polysaccharide capsule that surrounds the pneumococcal cell is essential for its virulence. At least 90 different capsule types exist (8). The capsule protects the pneumococcus from destruction by phagocytes. Immunity to pneumococcal infection is directed mainly to the capsule, so the multiplicity of capsule types is clearly beneficial to the pathogen. Genetic investigation of several capsular types revealed that the genes for their biosynthesis are present at the same genetic locus, between *plpA* and *dexB* (105). Thus, capsular gene clusters of different specificities can be transferred between cells by chromosomal transformation just like the restriction gene cassettes. As many as 2% of pneumococcal genes appear to be truncated (39); these defective genes may be remnants of past transformations. In the case of the capsular locus, many such fragments at the capsular cassette borders correspond to genes of other capsular types (8). Such surrounding capsular gene remnants could facilitate transformation to those specific capsule types.

Competence Systems

With respect to the regulation of competence, *S. pneumoniae* has two distinct, but closely related, systems for quorum sensing. As described above, the competence-stimulating peptides of the two systems, CSP1 and CSP2, differ by several amino acids, and the cognate receptor portions of the transmembrane histidine kinase, ComD, also differ. Approximately half of wild strains have the CSP1 system, and the other half have the CSP2 system. With mixed populations in nature, the result would be to remove part of the pneumococcal population from quorum sensing. Such inhibition of genetic exchange, however, should be deleterious rather than beneficial for the species. A possible resolution of this quandary is deducible from the prior examples of cassette mechanisms. In both of the above cases, diversity protected the species from noxious agents, bacteriophages and immune antibodies, respectively. It is conceivable that CSP1 or CSP2 can prompt an antibody response from host organisms that would inactivate it and block the development of competence with respect to only cells of that type in the population. The persistence of two states of CSP in the species argues for the importance of competence induction for pathogenesis. Although other stress response functions could be helpful, the ability of genetic exchange may be the most important stress-related response.

Methionyl–tRNA Synthetases

All pneumococcal strains contain the *metS1* gene, which encodes a methionyl-tRNA synthetase, an essential enzyme. Approximately half of strains isolated from patients contain a second such enzyme encoded by *metS2* (29). The second enzyme is resistant to certain synthetic inhibitors of the first enzyme. It was suggested that similar inhibitors occur as natural antibiotics (29). If so, this may be another case of population diversity conferring a selective advantage on the species.

Despite the considerable information presented in this chapter and additional information that space limitation precluded from consideration, many questions remain concerning the regulation of competence, DNA uptake, and genetic recombination, and the ramifications of these processes, especially for virulence of the pneumococcus.

ACKNOWLEDGMENTS

Donald Morrison and Paloma Lopez generously provided information prior to publication. This chapter was written at Brookhaven National Laboratory during a Guest Appointment in the Biology Department, which operates under the auspices of the U.S. Department of Energy Office of Biological and Environmental Research.

REFERENCES

1. **Alloing, G., B. Martin, C. Granadel, and J. P. Claverys**. 1998. Development of competence in *Streptococcus pneumoniae*: pheromone autoinduction and control of quorum sensing by the oligopeptide permease. *Mol. Microbiol.* **29**:75–83.
2. **Alloway, J. L.** 1931. The transformation in vitro of R pneumococci into S forms of different specific types by the use of filtered pneumococcus extracts. *J. Exp. Med.* **55**:91–99.
3. **Avery, O. T., C. M. MacLeod, and M. McCarty.** 1944. Studies on the chemical nature of the substance inducing transformation of pneumococcal types. Induction of transformation by a desoxyribonucleic acid fraction isolated from pneumococcus type III. *J. Exp. Med.* **89**:137–158.

4. **Balganesh, T. S., and S. A. Lacks.** 1985. Heteroduplex DNA mismatch repair system of *Streptococcus pneumoniae*: cloning and expression of the *hexA* gene. *J. Bacteriol.* **162**:979–984.

5. **Barany, F., and J. D. Boeke.** 1983. Genetic transformation of *Streptococcus pneumoniae* by DNA cloned into the single-stranded bacteriophage f1. *J. Bacteriol.* **153**:200–210.

6. **Bartilson, M., A. Marra, J. Christine, J. S. Asundi, W. P. Schneider, and A. E. Hromockyj.** 2001. Differential fluorescence induction reveals *Streptococcus pneumoniae* loci regulated by competence stimulatory peptide. *Mol. Microbiol.* **39**:126–135.

7. **Berge, M., M. Moscoso, M. Prudhomme, B. Martin, and J. P. Claverys.** 2002. Uptake of transforming DNA in Gram-positive bacteria: a view from *Streptococcus pneumoniae*. *Mol. Microbiol.* **45**:411–421.

8. **Caimano, M. J., G. G. Hardy, and J. Yother.** 1998. Capsule genetics in *Streptococcus pneumoniae* and a possible role for transposition in the generation of the type 3 locus. *Microb. Drug Resist.* **4**:11–23.

9. **Camerini-Otero, R., and P. Hsieh.** 1995. Homologous recombination proteins in prokaryotes and eukaryotes. *Annu. Rev. Genet.* **29**:509–552.

10. **Campbell, E. A., S. Y. Choi, and H. R. Masure.** 1998. A competence regulon in *Streptococcus pneumoniae* revealed by genome analysis. *Mol. Microbiol.* **27**:929–939.

11. **Cerritelli, S., S. S. Springhorn, and S. A. Lacks.** 1989. DpnA, a methylase for single-strand DNA in the *Dpn*II restriction system, and its biological function. *Proc. Natl. Acad. Sci. USA* **86**:9223–9227.

12. **Chargaff, E.** 1951. Structure and function of nucleic acids as cell constituents. *Fed. Proc.* **10**:654–659.

13. **Cheng, Q., E. A. Campbell, A. M. Naughton, S. Johnson, and H. R. Masure.** 1997. The *com* locus controls genetic transformation in *Streptococcus pneumoniae*. *Mol. Microbiol.* **23**:683–692.

14. **Chung, Y. S., F. Breidt, and D. Dubnau.** 1998. Cell surface localization and processing of the ComG proteins required for DNA binding during transformation of *Bacillus subtilis*. *Mol. Microbiol.* **29**:905–913.

15. **Claverys, J. P., V. Mejean, A. M. Gasc, and A. M. Sicard.** 1983. Mismatch repair in *Streptococcus pneumoniae*: relationship between base mismatches and transformation efficiencies. *Proc. Natl. Acad. Sci. USA* **80**:5956–5960.

16. **Claverys, J. P., H. Prats, H. Vasseghi, and M. Gherardi.** 1984. Identification of *Streptococcus pneumoniae* mismatch repair genes by an additive transformation approach. *Mol. Gen. Genet.* **196**:91–96.

17. **Dawson, M. H., and R. H. P. Sia.** 1931. In vitro transformation of pneumococcal types. I. A technique for inducing transformation of pneumococcal types in vitro. *J. Exp. Med.* **54**:681–699.

18. **Dowson, C. G., T. J. Coffey, C. Kell, and R. A. Whiley.** 1993. Evolution of penicillin resistance in *Streptococcus pneumoniae*; the role of *Streptococcus mitis* in the formation of a low affinity PBP2B in *S. pneumoniae*. *Mol. Microbiol.* **9**:635–643.

19. **Dubnau, D.** 1997. Binding and transport of transforming DNA by *Bacillus subtilis*: the role of type-IV pilin-like proteins—a review. *Gene* **192**:191–198.

20. **Echenique, J. R., S. Chapuy-Regaud, and M.-C. Trombe.** 2000. Competence regulation by oxygen in *Streptococcus pneumoniae*: involvement of *ciaRH* and *comCDE*. *Mol. Microbiol.* **36**:688–696.

21. **Ephrussi-Taylor, H., and T. C. Gray.** 1966. Genetic studies of recombining DNA in pneumococcal transformation. *J. Gen. Physiol.* **49**(Part 2):211–231.

22. **Ephrussi-Taylor, H., A. M. Sicard, and R. Kamen.** 1965. Genetic recombination in DNA-induced transformation of pneumococcus. I. The problem of relative efficiency of transforming factors. *Genetics* **51**:455–475.

23. **Fishel, R., M. K. Lescoe, M. R. S. Rao, N. G. Copeland, N. A. Jenkins, J. Garber, M. Kane, and R. Kolodner.** 1993. The human mutator gene homolog *MSH2* and its association with hereditary nonpolyposis colon cancer. *Cell* **75**:1027–1038.

24. **Fox, M. S.** 1957. Deoxyribonucleic acid incorporation by transformed bacteria. *Biochim. Biophys. Acta* **26**:83–85.

25. **Fox, M. S.** 1960. Fate of transforming deoxyribonucleate following fixation by transforming bacteria. II. *Nature* **187**:1004–1006.

26. **Fox, M. S., and M. K. Allen.** 1964. On the mechanism of deoxyribonucleate integration in pneumococcal transformation. *Proc. Natl. Acad. Sci. USA* **52**:412–419.

27. **Franklin, R., and R. Gosling.** 1953. Molecular configuration in sodium thymonucleate. *Nature* **171**:740–741.

28. **Gasc, A. M., N. Sicard, J. P. Claverys, and A. M. Sicard.** 1980. Lack of SOS repair in *Streptococcus pneumoniae*. *Mutat. Res.* **70**:157–165.

29. **Gentry, D. R., K. A. Ingraham, M. J. Stanhope, S. Rittenhouse, R. L. Jarvest, P. J.**

O'Hanlon, J. R. Brown, and D. J. Holmes. 2003. Variable sensitivity to bacterial methionyl-tRNA synthetase inhibitors reveals subpopulations of *Streptococcus pneumoniae* with two distinct methionyl-tRNA synthetase genes. *Antimicrob. Agents Chemother.* **47**:1784–1789.

30. Grebe, T. W., and J. B. Stock. 1999. The histidine protein kinase superfamily. *Adv. Microb. Physiol.* **41**:139–227.

31. Griffith, F. 1928. The significance of pneumococcal types. *J. Hyg.* **27**:113–159.

32. Hakenbeck, R., N. Balmelle, B. Weber, C. Gardes, W. Keck, and A. de Saizieu. 2001. Mosaic genes and mosaic chromosomes: intra- and interspecies genomic variation of *Streptococcus pneumoniae*. *Infect. Immun.* **69**:2477–2486.

33. Hava, D. L., and A. Camilli. 2002. Large-scale identification of serotype 4 *Streptococcus pneumoniae* virulence factors. *Mol. Microbiol.* **45**:1389–1406.

34. Håvarstein, L. S., G. Coomaraswamy, and D. A Morrison. 1995. An unmodified heptadecapeptide induces competence for genetic transformation in *Streptococcus pneumoniae*. *Proc. Natl. Acad. Sci. USA* **92**:11140–11144.

35. Håvarstein, L. S., P. Gaustad, I. F. Nes, and D. A Morrison. 1996. Identification of the streptococcal competence-pheromone receptor. *Mol. Microbiol.* **21**:863–869.

36. Håvarstein, L. S., R. Hakenbeck, and P. Gaustad. 1997. Natural competence in the genus *Streptococcus*: evidence that streptococci can change pherotype by interspecies recombinational exchanges. *J. Bacteriol.* **179**:6589–6594.

37. Hensel, M., J. E. Shea, C. Gleeson, M. D. Jones, E. Dalton, and D. W. Holden. 1995. Simultaneous identification of bacterial virulence genes by negative selection. *Science* **269**:400–403.

38. Hobbs, M., and J. S. Mattick. 1993. Common components in the assembly of type 4 fimbriae, DNA transfer systems, filamentous phage and protein-secretion apparatus: a general system for the formation of surface-associated protein complexes. *Mol. Microbiol.* **10**:233–243.

39. Hoskins, J., W. E. Alborn, Jr., J. Arnold, L. C. Blaszczak, S. Burgett, B. S. DeHoff, S. T. Estrem, L. Fritz, D. J. Fu, W. Fuller, C. Geringer, R. Gilmour, J. S. Glass, H. Khoja, A. R. Kraft, R. E. Lagace, D. J. LeBlanc, L. N. Lee, E. J. Lefkowitz, J. Lu, P. Matsushima, S. M. McAhren, M. McHenney, K. McLeaster, C. W. Mundy, T. I. Nicas, F. H. Norris, M. O'Gara, R. B. Peery, G. T. Robertson, P. Rockey, P. M. Sun, M. E. Winkler, Y. Yang, M. Young-Bellido, G. Zhao, C. A. Zook, R. H. Baltz, S. R. Jaskunas, P. R. Rosteck, Jr., P. L. Skatrud, and J.

I. Glass. 2001. Genome of the bacterium *Streptococcus pneumoniae* strain R6. *J. Bacteriol.* **183**:5709–5717.

40. Hotchkiss, R. D. 1954. Cyclical behavior in pneumococcal growth and transformability occasioned by environmental changes. *Proc. Natl. Acad. Sci. USA* **40**:49–55.

41. Hotchkiss, R. D. 1957. Criteria for the quantitative genetic transformation of bacteria, p. 321–335. *In* W. D. McElroy and B. Glass (ed.), *The Chemical Basis of Heredity*. Johns Hopkins Press, Baltimore, Md.

42. Hui, F. M., and D. A. Morrison. 1991. Genetic transformation in *Streptococcus pneumoniae*: nucleotide sequence analysis shows *comA*, a gene required for competence induction, to be a member of the bacterial ATP-dependent transport protein family. *J. Bacteriol.* **173**:372–381.

43. Karudapuram, S., X. Zhao, and G. J. Barcak. 1995. DNA sequence and characterization of *Haemophilus influenzae dprA+*, a gene required for chromosomal but not plasmid DNA transformation. *J. Bacteriol.* **177**:3235–3240.

44. Lacks, S. 1962. Molecular fate of DNA in genetic transformation of pneumococcus. *J. Mol. Biol.* **5**:119–131.

45. Lacks, S. 1966. Integration efficiency and genetic recombination in pneumococcal transformation. *Genetics* **53**:207–235.

46. Lacks, S. 1970. Mutants of *Diplococcus pneumoniae* that lack deoxyribonucleases and other activities possibly pertinent to genetic transformation. *J. Bacteriol.* **101**:373–383.

47. Lacks, S. 1979. Uptake of circular deoxyribonucleic acid and mechanism of deoxyribonucleic acid transport in genetic transformation of *Streptococcus pneumoniae*. *J. Bacteriol.* **138**:404–409.

48. Lacks, S., and B. Greenberg. 1973. Competence for deoxyribonucleic acid uptake and deoxyribonuclease action external to cells in the genetic transformation of *Diplococcus pneumoniae*. *J. Bacteriol.* **114**:152–163.

49. Lacks, S., and B. Greenberg. 1976. Single-strand breakage on binding of DNA to cells in the genetic transformation of *Diplococcus pneumoniae*. *J. Mol. Biol.* **101**:255–275.

50. Lacks, S., B. Greenberg, and K. Carlson. 1967. Fate of donor DNA in pneumococcal transformation. *J. Mol. Biol.* **29**:327–347.

51. Lacks, S., B. Greenberg, and M. Neuberger. 1974. Role of a deoxyribonuclease in the genetic transformation of *Diplococcus pneumoniae*. *Proc. Natl. Acad. Sci. USA* **71**:2305–2309.

52. Lacks, S., B. Greenberg, and M. Neuberger. 1975. Identification of a deoxyribonuclease impli-

cated in genetic transformation of *Diplococcus pneumoniae. J. Bacteriol.* **123**:222–232.

53. **Lacks, S., and M. Neuberger.** 1975. Membrane location of a deoxyribonuclease implicated in the genetic transformation of *Diplococcus pneumoniae. J. Bacteriol.* **124**:1321–1329.

54. **Lacks, S. A.** 1977. Binding and entry of DNA in bacterial transformation, p. 179–232. *In* J. L. Reissig (ed.), *Microbial Interactions, Receptors and Recognition.* Chapman and Hall, London, England.

55. **Lacks, S. A.** 1989. Generalized DNA mismatch repair—its molecular basis in *Streptococcus pneumoniae* and other organisms, p. 325–339. *In* L. O. Butler, C. Harwood, and B. E. B. Moseley (ed.), *Genetic Transformation and Expression.* Intercept, Andover, England.

56. **Lacks, S. A.** 1997. Cloning and expression of pneumococcal genes in *Streptococcus pneumoniae. Microb. Drug Resist.* **3**:327–337.

57. **Lacks, S. A.** 1998. DNA repair and mutagenesis in *Streptococcus pneumoniae*, p. 263–286. *In* J. A. Nickoloff and M. F. Hoekstra (ed.), *DNA Damage and Repair*, vol. 1. *DNA Repair in Prokaryotes and Lower Eukaryotes.* Humana Press, Totowa, N.J.

58. **Lacks, S. A.** 1999. DNA uptake by transformable bacteria, p. 138–168. *In* J. K. Broome-Smith, S. Baumberg, C. J. Stirling, and F. B. Ward (ed.), *Transport of Molecules across Microbial Membranes.* Cambridge University Press, Cambridge, England.

59. **Lacks, S. A., S. Ayalew, A. G. de la Campa, and B. Greenberg.** 2000. Regulation of competence for genetic transformation in *Streptococcus pneumoniae*: expression of *dpnA*, a late competence gene encoding a DNA methyltransferase of the *Dpn*II restriction system. *Mol. Microbiol.* **35**:1089–1098.

60. **Lacks, S. A., J. J. Dunn, and B. Greenberg.** 1982. Identification of base mismatches recognized by the heteroduplex-DNA-repair system of *Streptococcus pneumoniae. Cell* **31**:327–336.

61. **Lacks, S. A., and B. Greenberg.** 2001. Constitutive competence for genetic transformation in *Streptococcus pneumoniae* caused by mutation of a transmembrane histidine kinase. *Mol. Microbiol.* **42**:1035–1045.

62. **Lacks, S. A., B. M. Mannarelli, S. S. Springhorn, and B. Greenberg.** 1986. Genetic basis of the complementary DpnI and DpnII restriction systems of S. pneumoniae: an intercellular cassette mechanism. *Cell* **46**:993–1000.

63. **Lange, R., C. Wagner, A. de Saizieu, N. Flint, J. Molnos, M. Stieger, P. Caspers, M. Kamber, W. Keck, and K. E. Amrein.** 1999. Domain organization and molecular characterization of 13 two-component systems identified by genome sequencing of *Streptococcus pneumoniae. Gene* **237**:223–234.

64. **Lau, G. W., S. Haataja, M. Lonetto, S. E. Kensit, A. Marra, A. P. Bryant, D. McDevitt, D. A. Morrison, and D. W. Holden.** 2001. A functional genomic analysis of type 3 *Streptococcus pneumoniae* virulence. *Mol. Microbiol.* **40**:555–571.

65. **Lee, M. S., B. A. Dougherty, A. C. Madeo, and D. A. Morrison.** 1999. Construction and analysis of a library for random insertional mutagenesis in *Streptococcus pneumoniae*: use for recovery of mutants defective in genetic transformation and for identification of essential genes. *Appl. Environ. Microbiol.* **65**:1883–1890.

66. **Lee, M. S., and D. A. Morrison.** 1999. Identification of a new regulator in *Streptococcus pneumoniae* linking quorum sensing to competence for genetic transformation. *J. Bacteriol.* **181**:5004–5016.

67. **Lerman, R. S., and L. J. Tolmach.** 1957. Genetic transformation. I. Cellular incorporation of DNA accompanying transformation in pneumococcus. *Biochim. Biophys. Acta* **28**:68–82.

68. **Li, Y. H., P. C. Lau, J. H. Lee, R. P. Ellen, and D. G. Cvitkovitch.** 2001. Natural genetic transformation of *Streptococcus mutans* growing in biofilms. *J. Bacteriol.* **183**:897–908.

69. **Lopez, P., M. Espinosa, D. L. Stassi, and S. A. Lacks.** 1982. Facilitation of plasmid transfer in *Streptococcus pneumoniae* by chromosomal homology. *J. Bacteriol.* **150**:692–701.

70. **Lorenz, M. G., and W. Wackernagel.** 1994. Bacterial gene transfer by natural genetic transformation in the environment. *Microbiol. Rev.* **58**:563–602.

71. **Luo, P., H. Li, and D. A. Morrison.** 2003. ComX is a unique link between multiple quorum sensing outputs and competence in *Streptococcus pneumoniae. Mol. Microbiol.* **50**:623–633.

72. **Mannarelli, B. M., and S. A. Lacks.** 1984. Ectopic integration of chromosomal genes in *Streptococcus pneumoniae. J. Bacteriol.* **160**:867–873.

73. **Marmur, J., and D. Lane.** 1960. Strand separation and specific recombination in deoxyribonucleic acids: biological studies. *Proc. Natl. Acad. Sci. USA* **46**:453–461.

74. **Martin, B., P. Garcia, M. P. Castanie, and J. P. Claverys.** 1995. The *recA* gene of *Streptococcus pneumoniae* is part of a competence-induced operon and controls lysogenic induction. *Mol. Microbiol.* **15**:367–379.

75. **Martin, B., M. Prudhomme, G. Alloing, C. Granadel, and J. P. Claverys.** 2000. Cross-regulation of competence pheromone production and export in the early control of transformation

in *Streptococcus pneumoniae. Mol. Microbiol.* **38**:867–878.

76. **Mascher, T., D. Zahner, M. Merai, N. Balmelle, A. B. de Saizieu, and R. Hakenbeck.** 2003. The *Streptococcus pneumoniae cia* regulon: CiaR target sites and transcription profile analysis. *J. Bacteriol.* **185**:60–70.

77. **Mejean, V., and J. P. Claverys.** 1988. Polarity of DNA entry in transformation of *Streptococcus pneumoniae. Mol. Gen. Genet.* **213**:444–448.

78. **Miao, R., and W. R. Guild.** 1970. Competent *Diplococcus pneumoniae* accept both single- and double-stranded deoxyribonucleic acid. *J. Bacteriol.* **101**:361–364.

79. **Morrison, D. A., and M. F. Baker.** 1979. Competence for genetic transformation in pneumococcus depends on synthesis of a small set of proteins. *Nature* **282**:215–217.

80. **Morrison, D. A., S. A. Lacks, W. G. Guild, and J. M. Hageman.** 1983. Isolation and characterization of three new classes of transformation-deficient mutants of *Streptococcus pneumoniae* that are defective in DNA transport and genetic recombination. *J. Bacteriol.* **156**:281–290.

81. **Morrison, D. A., and B. Mannarelli.** 1979. Transformation in pneumococcus: nuclease resistance of deoxyribonucleic acid in the eclipse complex. *J. Bacteriol.* **140**:655–665.

82. **Mortier-Barriere, I., A. de Saizieu, J. P. Claverys, and B. Martin.** 1998. Competence-specific induction of *recA* is required for full recombination proficiency during transformation in *Streptococcus pneumoniae. Mol. Microbiol.* **27**:159–170.

83. **Novak, R., A. Cauwels, E. Charpentier, and E. Tuomanen.** 1999. Identification of a *Streptococcus pneumoniae* gene locus encoding proteins of an ABC phosphate transporter and a two-component regulatory system. *J. Bacteriol.* **181**:1126–1133.

84. **Ottolenghi, E., and R. D. Hotchkiss.** 1962. Release of genetic transforming agent from pneumococcal cultures during growth and disintegration. *J. Exp. Med.* **116**:491–519.

85. **Pakula, R., and W. Walczak.** 1963. On the nature of competence of transformable streptococci. *J. Gen. Microbiol.* **31**:125–133.

86. **Pasta, F., and M. A. Sicard.** 1994. Hyperrecombination in pneumococcus: A/G to C.G repair and requirement for DNA polymerase I. *Mutat. Res.* **315**:113–122.

87. **Pearce, B. J., A. M. Naughton, E. A. Campbell, and H. R. Masure.** 1995. The *rec* locus, a competence-induced operon in *Streptococcus pneumoniae. J. Bacteriol.* **177**:86–93.

88. **Pestova, E. V., L. S. Håvarstein, and D. A. Morrison.** 1996. Regulation of competence for genetic transformation in *Streptococcus pneumoniae* by an auto-induced peptide pheromone and a two-component regulatory system. *Mol. Microbiol.* **21**:853–862.

89. **Pestova, E. V., and D. A. Morrison.** 1998. Isolation and characterization of three *Streptococcus pneumoniae* transformation-specific loci by use of a *lacZ* reporter insertion vector. *J. Bacteriol.* **180**:2701–2710.

90. **Peterson, S., R. T. Cline, H. Tettelin, V. Sharov, and D. A. Morrison.** 2000. Gene expression analysis of the *Streptococcus pneumoniae* competence regulons by use of DNA microarrays. *J. Bacteriol.* **182**:6192–6202.

90a. **S. N. Peterson, C. K. Sung, R. Cline. B. V. Desai, E. Snesrud, P. Luo, J. Walling, H. Li, M. Mintz, G. Tsegaye, P. Burr, Y. Do, S. Ahn, J. Gilbert, R. Fleischmann, and D. A. Morrison.** 2004. Identification of competence pheromone responsive genes in *Streptococcus pneumoniae* by use of DNA microarrays. *Mol. Microbiol.* **51**:1051–1070.

91. **Polissi, A., A. Pontiggia, G. Feger, M. Altieri, H. Mottl, L. Ferrari, and D. Simon.** 1998. Large-scale identification of virulence genes from *Streptococcus pneumoniae. Infect. Immun.* **66**:5620–5629.

92. **Pozzi, G., L. Masala, F. Iannelli, R. Manganelli, L. S. Håvarstein, L. Piccoli, D. Simon, and D. A. Morrison.** 1996. Competence for genetic transformation in encapsulated strains of *Streptococcus pneumoniae*: two allelic variants of the peptide pheromone. *J. Bacteriol.* **178**:6087–6090.

93. **Provvedi, R., and D. Dubnau.** 1999. ComEA is a DNA receptor for transformation of competent *Bacillus subtilis. Mol. Microbiol.* **31**:271–280.

94. **Puyet, A., B. Greenberg, and S. A. Lacks.** 1990. Genetic and structural characterization of EndA, a membrane-bound nuclease required for transformation of *Streptococcus pneumoniae. J. Mol. Biol.* **213**:727–738.

95. **Ravin, A. W.** 1960. The genetics of transformation. *Adv. Genet.* **10**:61–163.

96. **Rimini, R., B. Jansson, G. Feger, T. C. Roberts, M. de Francesco, A. Gozzi, F. Faggioni, E. Domenici, D. M. Wallace, N. Frandsen, and A. Polissi.** 2000. Global analysis of transcription kinetics during competence development in *Streptococcus pneumoniae* using high density DNA arrays. *Mol. Microbiol.* **36**:1279–1292.

97. **Sabelnikov, A. G., B. Greenberg, and S. A. Lacks.** 1995. An extended −10 promoter alone directs transcription of the *DpnII* operon of *Streptococcus pneumoniae. J. Mol. Biol.* **250**:144–155.

98. **Saunders, C. W., and W. R. Guild.** 1980. Monomer plasmid DNA transforms *Streptococcus pneumoniae*. *Mol. Gen. Genet.* **181**:57–62.

99. **Sebert, M. E., L. M. Palmer, M. Rosenberg, and J. N. Weiser.** 2002. Microarray-based identification of *htrA*, a *Streptococcus pneumoniae* gene that is regulated by the CiaRH two-component system and contributes to nasopharyngeal colonization. *Infect. Immun.* **70**:4059–4067.

100. **Seto, H., and A. Tomasz.** 1975. Protoplast formation and leakage of intramembrane cell components: induction by the competence activator substance of pneumococci. *J. Bacteriol.* **121**:344–353.

101. **Stassi, D. L., and S. A. Lacks.** 1982. Effect of strong promoters on the cloning in *Escherichia coli* of DNA fragments from *Streptococcus pneumoniae*. *Gene* **18**:319–328.

102. **Stassi, D. L., P. Lopez, M. Espinosa, and S. A. Lacks.** 1981. Cloning of chromosomal genes in *Streptococcus pneumoniae*. *Proc. Natl. Acad. Sci. USA* **78**:7028–7032.

103. **Steinmoen, H., E. Knutsen, and L. S. Håvarstein.** 2002. Induction of natural competence in *Streptococcus pneumoniae* triggers lysis and DNA release from a subfraction of the cell population. *Proc. Natl. Acad. Sci. USA* **99**:7681–7686.

104. **Strom, M. S., D. N. Nunn, and S. Lory.** 1993. A single bifunctional enzyme, PilD, catalyzes cleavage and N-methylation of proteins belonging to the type IV pilin family. *Proc. Natl. Acad. Sci. USA* **90**:2404–2408.

105. **Tettelin, H., K. E. Nelson, I. T. Paulsen, J. A. Eisen, T. D. Read, S. Peterson, J. Heidelberg, R. T. DeBoy, D. H. Haft, R. J. Dodson, A. S. Durkin, M. Gwinn, J. F. Kolonay, W. C. Nelson, J. D. Peterson, L. A. Umayam, O. White, S. L. Salzberg, M. R. Lewis, D. Radune, E. Holtzapple, H. Khouri, A. M. Wolf, T. R. Utterback, C. L. Hansen, L. A. McDonald, T. V. Feldblyum, S. Angiuoli, T. Dickinson, E. K. Hickey, I. E. Holt, B. J. Loftus, F. Yang, H. O. Smith, J. C. Venter, B. A. Dougherty, D. A. Morrison, S. K. Hollingshead, and C. M. Fraser.** 2001. Complete genome sequence of a virulent isolate of *Streptococcus pneumoniae*. *Science* **293**: 498–506.

106. **Throup, J. P., K. K. Koretke, A. P. Bryant, K. A. Ingraham, A. F. Chalker, Y. Ge, A. Marra, N. G. Wallis, J. R. Brown, D. J. Holmes, M. Rosenberg, and M. K. Burnham.** 2000. A genomic analysis of two-component signal transduction in *Streptococcus pneumoniae*. *Mol. Microbiol.* **35**: 566–576.

107. **Tiraby, G., and M. S. Fox.** 1973. Marker discrimination in transformation and mutation of pneumococcus. *Proc. Natl. Acad. Sci. USA* **70**:3541–3545.

108. **Tomasz, A., and R. D. Hotchkiss.** 1964. Regulation of the transformability of pneumococcal cultures by macromolecular cell products. *Proc. Natl. Acad. Sci. USA* **51**:480–487.

109. **Vasseghi, H., J. P. Claverys, and A. M. Sicard.** 1981. Mechanism of integrating foreign DNA during transformation in *Streptococcus pneumoniae*, p. 137–154. *In* M. Polsinelli and G. Mazza (ed.), *Transformation 1980*. Cotswold Press, Oxford, England.

110. **Vijayakumar, M. N., S. D. Priebe, and W. R. Guild.** 1986. Structure of a conjugative element in *Streptococcus pneumoniae*. *J. Bacteriol.* **166**:978–984.

111. **Vovis, G. F., and S. Lacks.** 1977. Complementary action of restriction enzymes Endo R.*Dpn*I and Endo R.*Dpn*II on bacteriophage fl DNA. *J. Mol. Biol.* **115**:525–538.

112. **Wagner, C., A. de Saizieu, H. J. Schonfeld, M. Kamber, R. Lange, C. J. Thompson, and M. G. Page.** 2002. Genetic analysis and functional characterization of the *Streptococcus pneumoniae vic* operon. *Infect. Immun.* **70**:6121–6128.

113. **Watson, J. D., and F. H. C. Crick.** 1953. Molecular structure of nucleic acids. *Nature* **171**:737–738.

114. **Ween, O., P. Gaustad, and L. S. Håvarstein.** 1999. Identification of DNA binding sites for ComE, a key regulator of natural competence in *Streptococcus pneumoniae*. *Mol. Microbiol.* **33**:817–827.

115. **Whatmore, A. M., V. A. Barcus, and C. G. Dowson.** 1999. Genetic diversity of the streptococcal competence (*com*) gene locus. *J. Bacteriol.* **181**:3144–3154.

HOST-MICROBE INTERACTIONS

EVOLUTIONARY AND POPULATION BIOLOGY OF *STREPTOCOCCUS PNEUMONIAE*

Brian G. Spratt, William P. Hanage,
and Angela B. Brueggemann

8

INTRODUCTION

Bacteria exist within populations, whether it is the population of pneumococci in the nasopharynx of an individual child or the isolates circulating within a local community, within a country, or globally. Studies of bacterial populations attempt to describe the genetic variation that is present within the population and how it has arisen, in terms of the known evolutionary processes of mutation, recombination, selection, drift, ecological isolation, etc. For a bacterial pathogen there is additional interest in the isolates that cause disease. How do they relate to the total population and what selective forces led to the emergence of the ability to cause disease, and what are the likely consequences of perturbations of the population by intervention measures, be they antibiotics or vaccines? In this chapter we cover some of the basic population and evolutionary biology of the pneumococcus; several aspects of the population biology of the pneumococcus are covered in other chapters, including those that discuss carriage (chapter 9) and the impact of the

conjugate vaccines (chapter 18). Another central issue in population biology—the concept of species—is particularly murky in the case of the bacteria, and our ability to cleanly delineate that population of streptococcal isolates which should be given the name *Streptococcus pneumoniae* is addressed in chapter 1.

CHARACTERIZING PNEUMOCOCCI

Accurate characterization of pneumococcal isolates is crucial for molecular epidemiology and also for addressing aspects of their population and evolutionary biology. Pneumococci have traditionally been characterized by serology, which divides the population into only 90 immunologically distinct types. Serotyping is important since conjugate vaccines are targeted against the capsular polysaccharides of those serotypes most commonly associated with invasive disease in children (22), and the serotype is considered to be a major determinant in the varying potential of pneumococci to cause invasive disease (2, 26, 27, 36). However, for studies of pneumococcal populations, a much more discriminatory method is required. Pulsed-field gel electrophoresis has been widely used, but the differences in DNA fragment sizes are of little utility for addressing the population and evolutionary biology of the pneumococcus, which requires a procedure

Brian G. Spratt and William P. Hanage, Department of Infectious Disease Epidemiology, Imperial College London, St. Mary's Hospital, London W2 1PG, United Kingdom. *Angela B. Brueggemann,* Academic Department of Microbiology and Infectious Disease, John Radcliffe Hospital, University of Oxford, Oxford OX3 9DU, United Kingdom.

that indexes genetic variation whose origins are well understood, which is considered to be selectively neutral, and which accumulates at a relatively low rate.

Multilocus enzyme electrophoresis (MLEE) is such a technique, and it has provided key insights into the population biology of many bacterial pathogens (43, 44), although it has not been used extensively for characterizing pneumococcal populations. MLEE has now been superseded by multilocus sequence typing (MLST), which uses the same principles as MLEE but assigns alleles at multiple housekeeping genes directly by sequencing internal fragments of those genes, rather than indirectly from the electrophoretic mobilities of their gene products on starch gels (31). The pneumococcal MLST scheme (13) uses sequence variation within internal fragments (about 500 bp) of seven housekeeping genes, and for each gene fragment, sequences that differ (even at a single nucleotide) are assigned different allele numbers. The allele numbers at the seven loci provide the allelic profile of an isolate, and each unique allelic profile is assigned as a sequence type (ST). The great advantage of MLST is that the data are precise, allowing different laboratories to recognize unambiguously if they have the same strain, or different strains, by interrogating the pneumococcal MLST website and database (over 2,000 strains) at Imperial College London (http://spneumoniae.mlst.net).

The sequences of ~500-bp fragments from the seven housekeeping genes also can be used to address aspects of the population and evolutionary biology of pneumococci and this chapter focuses on recent work that uses this precise tool for characterizing pneumococcal isolates.

IDENTIFYING CLONES AND CLONAL COMPLEXES WITHIN THE PNEUMOCOCCAL POPULATION

MLST defines each isolate by its allelic profile, and the clustering among isolates may be displayed as a dendrogram (tree), using the matrix of pairwise differences between the allelic profiles of the isolates (31, 47). A tree is useful for recognizing those isolates that have identical or very closely related allelic profiles (clones and clonal complexes), but this approach has serious limitations. One practical problem is that the pneumococcal MLST database is now so large that it is impossible to display the relatedness of all isolates on a tree that can be read when printed on a single page. Another problem is that as isolates diverge, they will relatively rapidly (on evolutionary time scales) accumulate one or more nucleotide differences at each of the seven MLST loci. The relationships among major pneumococcal lineages cannot therefore be probed, since isolates of the different lineages often will differ at all loci. Furthermore, even within a lineage, a tree is unsatisfactory, as it provides no information about how clusters of similar isolates may have emerged.

The relationships among major lineages could be explored using a tree constructed from the concatenated sequences of all seven MLST loci. However, this approach will only reconstruct the "true" evolutionary relationships among the major lineages if the genetic variation that accumulates as the lineages diverge is predominantly the result of point mutation. Genetic variation arises within pneumococcal housekeeping genes much more frequently by recombination than point mutation (reference 16; see also below), and the relationships among major lineages implied by a tree constructed using the concatenated sequences are likely to be highly suspect. This view is supported by an analysis of a set of distantly related pneumococcal isolates which showed that the similarities among the seven trees obtained for these isolates using the sequences of the individual MLST loci were no better than those among random trees (17).

Fortunately, epidemiological studies are typically concerned with outbreaks of disease, or the spread between countries of antibiotic-resistant or virulent strains. Over these very short evolutionary time scales, of weeks to a few hundred years, recombination is unlikely to prevent the recognition of clones and clonal complexes within most bacterial populations. Thus, although the phylogenetic complexities introduced by frequent homologous recombination at the MLST loci may be problematic over long periods of evolutionary time, given an appropriate model of bacterial evolution, it should be possible to accurately reconstruct evolutionary events that occur over short time scales.

Recent evolutionary events are poorly represented by a tree but a useful procedure (based upon related sequence types [BURST]) has been developed for this purpose (http://eburst.mlst.net). BURST divides a bacterial population into a number of nonoverlapping clonal complexes and predicts the likely founding genotype (ST) of each clonal complex and the most likely patterns of evolutionary descent of all STs from the founder. Relationships among clonal complexes are ignored, as the long-term impact of recombination renders these uncertain (17a).

BURST is based on a simple evolutionary scenario (Fig. 1A) in which a founding genotype increases in frequency in the population, under selection or random genetic drift, and gradually diversifies to produce a number of minor variants of the founder (a clonal complex). In terms of MLST, the descendants of the founding genotype are initially identical in their allelic profile (same ST), and the initial products of clonal diversification are variants of the founder that differ at only one of the seven loci examined by MLST (single-locus

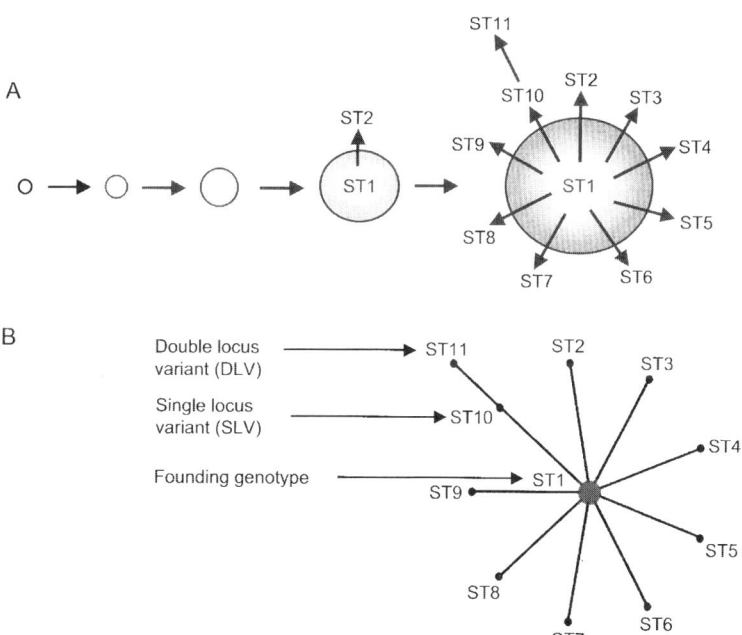

FIGURE 1 Emergence of a clonal complex. (A) A pneumococcal isolate increases in frequency in the population (shown by the increasing size of the circle) and starts to diversify. Initially the isolates appear uniform by MLST (ST1), but diversification results in variants (ST2 to ST10) that have a different allele at one of the seven MLST loci (SLVs); one of these SLVs (ST10) has diversified further to form a DLV (ST11). (B) The isolates resulting from the diversification of ST1 are shown as an eBURST diagram. The founder of the clonal complex identified by eBURST is placed centrally with radial links to its derived SLVs. The SLV that diversified to form a DLV is also linked.

variants [SLVs]). Subsequently, SLVs diversify to produce double-locus variants (DLVs) and triple-locus variants, and so on. This simple mode of diversification allows the predicted founder to be assigned as the ST within a clonal complex with the largest number of SLVs.

The current implementation of the BURST algorithm (eBURST [17a]) displays the predicted patterns of descent within each clonal complex, placing the predicted founding ST at the center, with radial links to all of its SLVs and on to DLVs, etc. (Fig. 1B). The eBURST diagram is far more informative than

a tree. Figure 2 shows a tree constructed from the matrix of pairwise differences in the allelic profiles of isolates that share alleles at four or more of the MLST loci with the multiply antibiotic-resistant Spain[23F]-1 clone (ST81; [34]). The tree identifies the large number of indistinguishable isolates in the MLST database with the typical allelic profile of the Spain[23F]-1 clone and a number of very similar isolates, which are all multiply antibiotic resistant. Isolates that are linked to ST81 at a genetic distance of >0.4 are not multiply antibiotic resistant and are relatively distantly related to the Spain[23F]-1 clone. However, there is no indica-

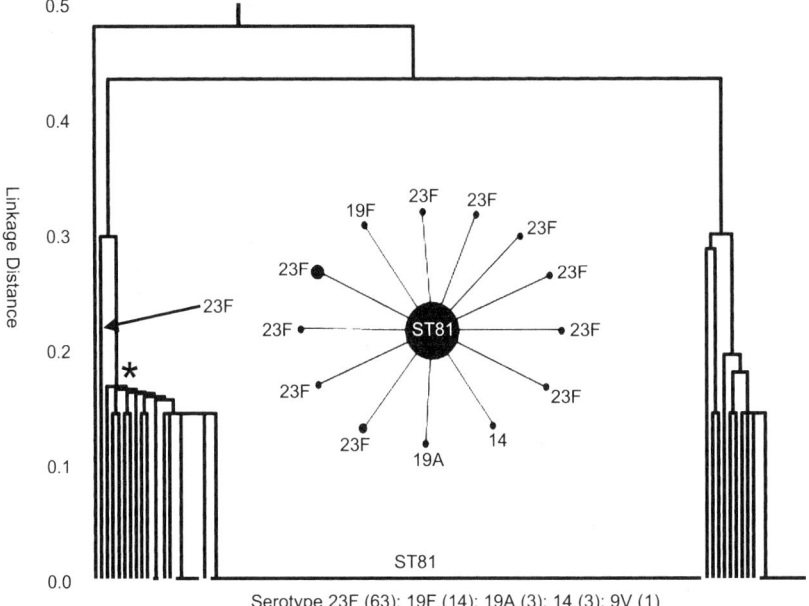

FIGURE 2 Relatedness among isolates of the Spain[23F]-1 (ST81) clonal complex. The tree was constructed from the matrix of pairwise differences between the allelic profiles of isolates in the pneumococcal MLST database that shared alleles at four or more loci with ST81. The pneumococcal MLST database was analyzed by eBURST, and the clonal complex including ST81 was displayed as an eBURST diagram. The areas of the circles representing the individual STs (except for ST81, ST numbers are not shown) indicate the prevalence of each ST in the current database. The isolate on the tree indicated by an arrow is a DLV of ST81. eBURST is conservative, and DLVs of the founding genotype for which the linking SLV is not represented in the input data are not shown (17a). For this reason the DLV of ST81 is not shown in the eBURST diagram. The serotypes of the isolates of ST81 in the MLST database are shown on the tree (the number of isolates of each serotype is shown in parentheses), as are the serotypes of its SLVs on the eBURST diagram. Except for ST81, ST numbers are not shown on the tree or eBURST diagram.

tion that ST81 is the founding genotype, or that (with the exception of a single DLV) all of the cluster of closely related isolates are SLVs of ST81. The eBURST diagram strongly predicts that ST81 is the founding genotype (bootstrap confidence level of 100%) and shows that all of the closely related resistant isolates are SLVs of ST81 which are radially linked to this founder (Fig. 2).

The extent of clustering among all genotypes within the current pneumococcal MLST database can also be displayed as a single eBURST diagram (a "population snapshot" [17a]). Color Plate 5 shows such a snapshot, illustrating the major and minor clonal complexes, and the large number of unlinked STs, which differ from all other STs in the population at two or more of the seven MLST loci. The area of the circle that represents each ST indicates the number of isolates of that ST in the input data. The population snapshot therefore provides information about both the clusters of closely related STs and the prevalence of each ST. This visualization of the population will obviously reflect the nature of the database that is being analyzed (which in the case of the pneumococcal MLST database is biased, as it depends on the isolates that are submitted) and would be most useful for MLST data from isolates that have been sampled in a systematic way—for example, national population-based surveillance of invasive disease isolates.

The ability to display the likely patterns of diversification of isolates within a clonal complex from their predicted founder allows changes of serotype or of putative virulence genes, or the acquisition of antibiotic resistance, to be explored. It also provides the data required to explore the mode of diversification of pneumococcal clones.

MECHANISM AND RATE OF DIVERSIFICATION OF PNEUMOCOCCAL CLONES

Pneumococcal clones diversify relatively rapidly, due mainly to the substantial impact of homologous recombination, presumably mediated by genetic transformation, which results in small segments of the chromosome of a recipient pneumococcus being replaced with the corresponding region from a distinct strain (16). MLST data have been used to quantify the extent of recombination in bacterial species. eBURST first identifies the founders of each clonal complex and the derived SLVs. A comparison of the sequence of the allele in the founder with those of the altered alleles in each of the derived SLVs can then be used to distinguish those allelic changes that have occurred by recombination from those that have arisen by point mutation. In pneumococci the ratio is about 10:1 in favor of recombination (16), and although this relatively high rate of recombination does not prevent the emergence of clones, it makes pneumococcal clones relatively unstable compared to those in many other bacterial species.

The relatively rapid diversification of pneumococcal genotypes can be observed for antibiotic-resistant clones for which the maximum age of the clone is known. Multiply antibiotic-resistant isolates of S. pneumoniae cannot have existed prior to the introduction of antibiotics into medicine and are probably less than 30 years old, yet considerable variation in the allelic profiles of the major resistant clones is observed (50). As shown in Fig. 2, the Spain^{23F}-1 clone (ST81) has already diversified to form a cluster of SLVs and one DLV. Since these SLVs, and the DLV of ST81, are multiply antibiotic resistant, we can be confident that they have arisen by the diversification of this pneumococcal clone into a clonal complex over a period of about 30 years.

INTERSPECIES RECOMBINATION

Pneumococci colonizing the nasopharynx live alongside many other bacterial species. Among these are the viridans group streptococci and bacteria of uncertain taxonomic status that are clearly very closely related to, but slightly distinct from, pneumococci (49; see also chapter 1). The housekeeping genes of viridans group streptococci, and of the latter isolates, are distinguishable from those of authentic pneumococci, and there appears to be relatively little genetic exchange between these groups within typical housekeeping loci. Thus, the sequences of most of the pneumococcal alleles at the MLST loci (except *ddl*; see below) are very similar, differing on average at about 1% of sites (Fig. 3). However, the genetic barrier between pneumococci and closely related

"species" is to some extent porous; in a small number of serotypeable (i.e., unambiguously authentic) pneumococci we can identify divergent (typically about 5%) alleles at MLST loci that have almost certainly been introduced from a closely related species (Fig. 3). Such interspecies recombinational events in pneumococcal housekeeping genes are unlikely to provide any selective advantage. However, this

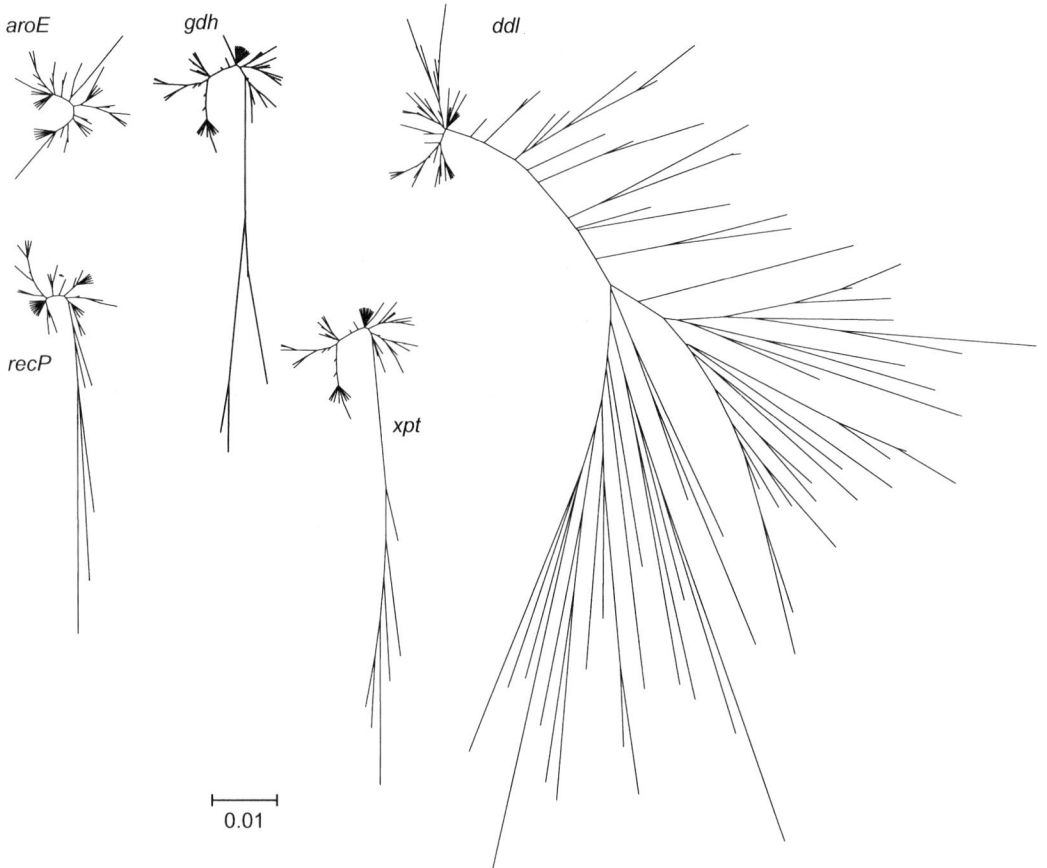

FIGURE 3 Relatedness of the alleles at MLST loci. A neighbor-joining tree was constructed from the sequences of the alleles at five of the MLST loci (http://spneumoniae.mlst.net). Only those alleles present in serotypeable pneumococci were included, as these almost certainly are authentic pneumococci. The alleles at one of these housekeeping loci (*aroE*) are all very similar in sequence, as are the great majority of alleles at *gdh*, *recP*, and *xpt* (and *gki* and *spi* [data not shown]), but a few alleles of the latter loci are about 5% diverged from the typical pneumococcal alleles. The rare divergent alleles have presumably been introduced by interspecies recombination from very closely related streptococcal isolates. The large number of more highly divergent *ddl* alleles (typically 10% divergent), which occur in penicillin-resistant isolates, are also the result of interspecies recombination, which in this case is driven by its close genetic linkage to the penicillin-binding protein 2b gene (see text). The scale bar corresponds to a genetic distance of 0.01 (1%).

phenomenon has also been observed in situations where pneumococci gain a strong selective advantage from the replacement, notably among penicillin-resistant pneumococci, in which altered penicillin-binding proteins with reduced affinity for penicillin have been acquired by interspecies recombination (6).

The *ddl* gene, which is used in the pneumococcal MLST scheme, lies next to the penicillin-binding protein 2b gene. The *ddl* alleles fall into two very different clusters: those that are highly divergent (typically about 10%), and which are found in many penicillin-resistant pneumococci, and those characteristic of penicillin-susceptible isolates, which are relatively uniform in sequence (Fig. 3). The highly divergent alleles arose as the DNA replacements that introduced altered penicillin-binding protein 2b genes from related streptococci have in many cases extended into or through *ddl* (15).

Any understanding of pneumococcal population genetics should therefore take into account the existence of an available gene pool that extends beyond the boundaries of the "species." The ability to acquire divergent alleles from closely related streptococci has certainly been important for the emergence of target-mediated resistance to antibiotics (1, 6, 20); it may also be relevant for potential pneumococcal vaccines based on conserved protein antigens, where selection for variation in the target antigen, imposed by the vaccine, may select for divergent forms of the antigen (escape variants) that arise by interspecies recombination.

SEROTYPES AND CLONES

Until relatively recently, a pneumococcal serotype was implicitly assumed to represent a group of closely related isolates with similar biological properties. The introduction of molecular typing showed that a single serotype typically includes a number of divergent genotypes (Fig. 4). The diversity of genotypes within a single serotype could in some cases simply be due to extensive divergence in the genotypes of the descendants of the ancient ancestor of the serotype. This possibility is diffi-

cult to address since a history of frequent recombination prevents the construction of a reliable phylogeny. There is good evidence that changes of serotype occur relatively frequently, by recombination at the capsular locus (7, 8, 40), and the major reason for the diversity of genotypes within a serotype is likely to be the horizontal spread of the capsular genes among divergent pneumococcal lineages. Evidence for the horizontal movement of capsular genes is clear where isolates have indistinguishable genotypes but different serotypes (7, 8). Many of the reported examples of serotype changes have been among isolates of the major antibiotic-resistant clones, but this is almost certainly because these have been characterized more extensively than susceptible clones.

The Spain[23F]-1 (ST81) clonal complex (Fig. 1) clearly shows a history of serotype switching; isolates of ST81 are predominantly serotype 23F, but variants expressing several other capsular polysaccharides have arisen (types 19F, 19A, 14, and 9V). Similarly, SLVs of ST81 vary in serotype. Most of the larger clonal complexes include isolates of more than one serotype. Figure 5 shows an example of a clonal complex consisting of penicillin-susceptible isolates, where the predicted founder (serotype 11A; ST62) appears to have diversified to produce a number of SLVs, one of which has become serotype 8 and has generated a number of its own SLVs.

Some capsular genes appear to be more commonly transferred by recombination among lineages than others. There are several reasons why this may occur, including differences in the abilities of isolates to be transformed or in the genetic organization within the capsular regions, which makes recombination among the capsular loci of some serotypes more likely than others, and differences in the frequencies with which isolates of different serotypes coexist in the nasopharynx. Almost all well-documented changes of serotype have been between vaccine serotypes or vaccine-related serotypes. This is not surprising given the recombinational mechanism of serotype changes, since isolates have to meet each other, and these are the serotypes most commonly found in the nasopharynx in children. However, these are also the serotypes that have been most commonly characterized. Rarely carried serotypes (notably serotype 1) should be less likely to change serotype or to donate their capsular genes to other lineages.

If opportunities for contact between isolates of different serotypes are a major reason for the observed patterns of serotype changes, these may change following the widespread use of the conjugate pneumococcal vaccines, since nonvaccine serotypes may become more commonly carried (chapter 18) and thus more

likely to be donors of capsular genes to other pneumococci (48). Similarly, a different pattern of serotype changes would be predicted in developing countries where carriage of nonvaccine types may be more common.

Concurrent carriage of multiple serotypes or isolates is a prerequisite for changes of serotypes or the horizontal spread of antibiotic resistance genes. We know that multiple carriage exists, both through direct observation (2; see also chapter 9), and indirectly through the documentation of the consequences of such events (serotype changes and the predominant role of recombination in clonal diversification). It has

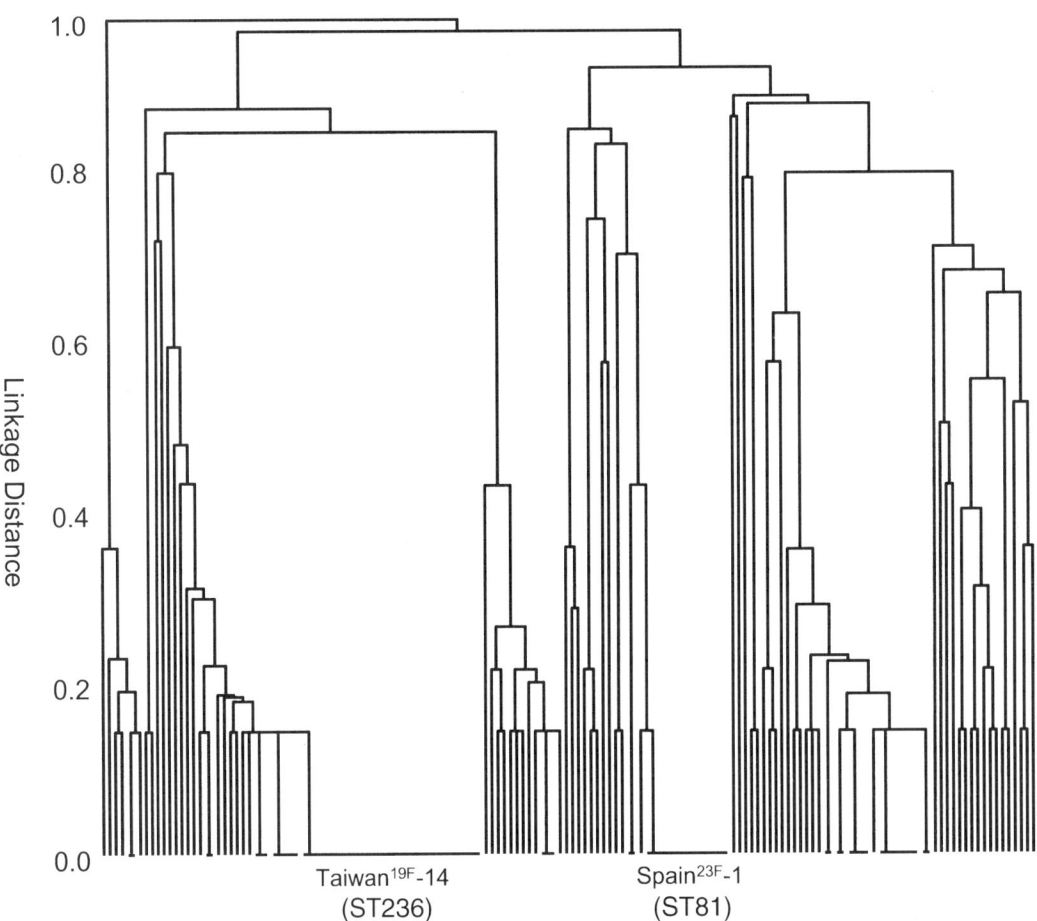

FIGURE 4 Genetic diversity of isolates of serotype 19F. The allelic profiles of all isolates of serotype 19F were extracted from the MLST database, and a dendrogram was constructed based on the pairwise differences in the profiles. The two most prevalent serotype 19F clones in the database are labeled; one of these corresponds to ST81, the Spain[23F]-1 clone, as serotype 19F variants of this clone are now prevalent.

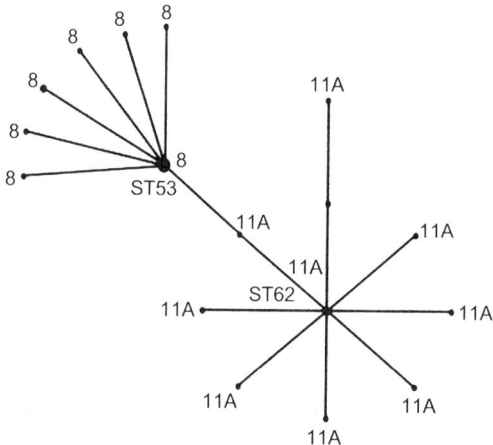

FIGURE 5 Serotype changes within a penicillin-susceptible clonal complex. The clonal complex was identified within the MLST database and was displayed using eBURST; the serotypes of isolates of each ST were obtained from the pneumococcal MLST database.

been hard to study multiple carriage directly, however, because of the difficulty of detecting even moderately rare types by conventional methods of serotyping (24). The development of molecular serotyping techniques (3, 28), which directly assay serotype-specific capsular genes, may go a long way towards solving this problem. Although the incidence of multiple carriage is uncertain (most studies suggest about 10%), it is likely to vary depending on the extent of carriage. For example, studies from the developing world show much higher carriage rates (19) and high levels of multiple carriage (38), which may impact on the extent of serotype change, the stability of pneumococcal clones, or the rate of horizontal spread of antibiotic resistance.

CARRIAGE VERSUS DISEASE

An analysis of a bacterial population should be based on an appropriate sample of the natural population. Until recently, very little attention has been applied to sampling, and many collections of pneumococci that have been analyzed are highly biased towards antibiotic-resistant isolates or invasive isolates or are from a variety of different countries, decades, or age groups. For pneumococci the natural population corresponds closely to that residing in the nasopharynx, since asymptomatic carriage is vastly more common than invasive disease and is also far more common than more prevalent pneumococcal diseases such as acute otitis media and nonbacteremic pneumonia. At any time the great majority of the pneumococcal population is within the nasopharynges of healthy carriers, mostly children.

The carriage state is central to all aspects of pneumococcal biology (2, 38). Transmission of pneumococci to new hosts occurs from the nasopharynx, and natural selection therefore operates on the carried population. Isolates with increased transmission to new hosts, or which have an increased ability to be transmitted within a new epidemiological niche (e.g., day care facilities [10]), or which have acquired antibiotic resistance are selected from among the nasopharyngeal flora. Additionally, of course, it is from the nasopharynx that pneumococci occasionally gain access to other body sites where they may initiate an episode of mucosal or invasive pneumococcal disease.

The ability of pneumococci to cause invasive disease probably has little or no impact on the population biology of the species. Isolates from disease are of great interest to clinicians and to those who develop vaccines, but natural selection that increases the ability of isolates to cause invasive disease is of no consequence unless disease is linked to transmission, which seems very unlikely in the cases of septicemia, meningitis, and otitis media; it is hard to think of any way in which access to the blood, cerebrospinal fluid, or middle ear could aid transmission to new hosts. The selective forces that determine why isolates vary in the ability to cause invasive disease are still very unclear for pathogens such as the pneumococcus for which, except perhaps in the case of pneumonia, causing disease is unlikely to increase transmission (29). A possible contributor to the emergence of more virulent strains is competition among isolates for carriage in the nasopharynx, which may select for aggressive interactions with the nasopharyngeal mucosa and may coincidentally increase the chance of some isolates gaining access to the blood.

The carried population is thus of primary interest to the population biologist, and disease isolates need to be considered in the context of carriage. There is an understandable tendency to focus on isolate collections from serious disease which identify those clones (isolates with identical allelic profiles) most associated with disease within a country or internationally. However, it is only by comparing the prevalence with which a clone or serotype causes a particular disease in children, in a particular geographic region, and over a defined time period, with its prevalence in the nasopharynges of children from the same region during the same time period, that we can observe how disease relates to carriage. That is, we need to relate the amount of disease caused by a clone or serotype to the extent of exposure of the same population to that clone or serotype.

The amount of disease caused by a strain may greatly over- or underestimate its "invasiveness." For example, in countries that have a high prevalence of antibiotic resistance, multiply antibiotic-resistant clones are commonly recovered from cases of invasive disease (14), but this does not imply that they are particularly invasive, as exposure to these strains is likely to be unusually high. Conversely, serotype 1 pneumococci may cause very few cases of invasive disease in a country, but there is little doubt that this is a highly invasive serotype. Serotype 1 pneumococci are very rarely found in the nasopharynx, even in areas where the serotype is prevalent among invasive isolates (2, 25, 39), implying that serotype 1 pneumococci have a high attack rate, and they would be expected to cause high levels of disease if they became commonly carried in a community. Indeed this frequently happened in the past: outbreaks of serotype 1 disease occurred in hospitals, prisons, and shelters, where carriage and transmission of serotype 1 became common during the outbreak (11, 45). Thus, highly invasive serotypes may be low in the rank order of prevalence of serotypes in disease if they are rarely acquired (or carried, as acqui-

sition is very rarely measured), and serotypes that are weakly invasive may be high in the rank order if they are very commonly acquired. Careful sampling is therefore required to understand whether a serotype (or clone) is causing much disease in a community because it is highly invasive or because it is frequently acquired.

It is not clear how strains of serotype 1 are maintained in the community if they are very rarely carried, but increases in transmission of serotypes with high attack rates can have a major impact, as seen by the recent increase in serotype 1 disease in Sweden (23).

The relationship between disease and carriage is also important in understanding what we mean by an invasive serotype. For example, consider a serotype that is commonly recovered from pediatric invasive disease. If pneumococci of this serotype are sampled exclusively from invasive disease and these isolates belong to only two clones, we lack some important information. Are there many other clones of this serotype that are commonly found in carriage, but rarely in disease, implying that the invasiveness of the serotype is due to the impact of these two hypervirulent clones and that most clones of this serotype rarely cause disease? Alternatively, pneumococci of this serotype from carriage may belong to the same two clones as those from disease, and there may be no clones that are commonly carried and yet rarely cause disease. The answer to this question, which has yet to be adequately addressed, is relevant to the relative importance of the serotype, compared to the overall genotype, in determining the ability of pneumococci to cause disease. If highly divergent clones of the same serotype have the same invasiveness, it would suggest that the serotype is the key determinant, whereas if they vary greatly in invasiveness, it suggests that the overall genotype plays a major role.

There have been several attempts to investigate the pneumococcal serotypes or isolates from childhood invasive disease in the context of carriage (4, 35, 41, 46), although some of these used inadequate sampling procedures. A recent study has attempted to measure the prevalence of the major serotypes and individual clones in invasive disease compared to carriage in a group of children in the Oxford (United Kingdom) region (4). Several serotypes (14, 18C, 4, and 1), and some individual clones, were found to be significantly overrepresented in invasive disease compared to carriage, and others were found to be significantly underrepresented (e.g., serotype 23F). There was also no evidence of any difference in invasiveness for two highly divergent invasive clones of serotype 14 (ST9 and ST124), providing some preliminary support for the primacy of serotype over overall genotype.

A similar study of the prevalence of serotypes and clones causing acute otitis media in children, in relation to carriage, has been carried out in a region of Finland, and in contrast to the study of invasive disease, it showed little variation in the ability of serotypes or clones to cause otitis media (21). In the latter study the pneumococci causing otitis media were an approximate reflection of those that were carried by the children, whereas in the study of invasive disease, the pneumococci from disease were significantly less diverse than those that were carried, and some serotypes and clones were much more likely to cause invasive disease than others. The extent of the differences among serotypes in the ability to cause uncomplicated pneumococcal pneumonia is unknown due to difficulties in obtaining sufficient numbers of isolates by aspiration from the infected lung (33).

Differences between serotypes in the ability to cause the different types of pneumococcal disease may have important consequences for the long-term efficacy of the conjugate vaccines, as widespread vaccination of infants is expected to increase the carriage and transmission of nonvaccine serotypes (i.e., serotype replacement [9, 30, 32, 37]). If the disease potential of nonvaccine serotypes is similar to that of vaccine serotypes, which may be the case for otitis media, there may be little overall reduction in the disease burden. Fortunately, for invasive disease, it is likely that most nonvaccine

types (excluding serotypes 1 and 5, which will be in the higher-valency vaccines) are considerably less invasive than the most invasive types covered by the conjugate vaccines, but some nonvaccine types are occasionally recovered from invasive disease and could still cause a substantial amount of disease if they become very commonly carried.

MAJOR CLONES CAUSING INVASIVE DISEASE

Many studies have examined the serotypes associated with invasive disease in different countries and age groups, but there have been relatively few that examine which clones within these serotypes are associated with invasive disease, and in the vast majority of these studies it is impossible to compare the results among studies. Those few studies that have used MLST allow the major clones (STs) associated with invasive disease in multiple countries to be identified (4, 5, 12, 13, 18, 23, 42; see also http://spneumoniae.mlst.net).

Table 1 shows the major STs within the serotypes most commonly associated with invasive disease. For many of these serotypes, one or two STs are recovered from invasive disease in different countries or continents. For example, two major unrelated clones expressing a serotype 14 capsule (ST9 and ST124) were found in several countries (13), and this observation was confirmed by subsequent studies that identified the same two clones causing invasive disease in the Oxford region (4) and across the United States (18). In other cases (e.g., serotypes 6A and 19A) the population appears to be more heterogeneous, and no single ST has been identified that is the predominant cause of invasive disease in several countries. When considering these data, we should note that several of the major invasive clones (for example, ST81) are antibiotic resistant and therefore may be overrepresented in the MLST database. Nonetheless, in some countries these resistant clones do appear to be important causes of invasive disease (14). This does not necessarily mean that they are particularly invasive, as local antibiotic selective pressures may

TABLE 1 Major clones of the serotypes most commonly associated with invasive disease

Serotype	Major invasive clone[a]	Country(ies)[b]	Allelic profile[c]	Comment
1	ST217	Den, Fra, Ger, Isr, Ken, SA, USA	10-18-4-1-7-19-9	
1	ST227	Can, Den, Neth, UK, USA	12-5-13-5-17-4-20	
1	ST306	Can, Den, Fra, Ger, Neth, Nor, Pol, Spa, USA	12-8-13-5-16-4-20	
3	ST180	Can, Den, Neth, Spa, Tw, UK, USA,	7-15-2-10-6-1-22	
4	ST205	Aus, Den, Swe, UK, USA,	10-5-4-5-13-10-18	
4	ST247	Den, Neth, Spa	16-13-4-5-6-10-14	
4	ST695	USA	16-13-4-4-6-113-18	
5	ST289	Col, UK	16-12-9-1-41-33-33	Columbia[5]-19
6A	Heterogeneous			
6B	ST90	Aus, Ice, Por, Spa, Tw, UK, USA	5-6-1-2-6-3-4	Spain[6B]-2
6B	ST138	Den, Swe, UK, USA	7-5-8-5-10-6-14	
6B	ST315	Ita, Pol	20-28-1-1-15-14-14	Poland[6B]-20
7F	ST191	Bra, Den, Fin, Neth, Nor, UK, Uru, USA	8-9-2-1-6-1-17	
8	ST53	Bra, Neth, Spa, UK	2-5-1-11-16-3-14	
9V	ST156[d]	Can, Cze, Fra, Ita, Pol, Por, Spa, UK, USA	7-11-10-1-6-8-1	Spain[9V]-3 clone
9V	ST162[d]	Bra, Can, Ita, UK	7-11-10-1-6-8-14	
12F	ST218	Can, Den, Spa, UK, Uru, USA	10-20-14-1-6-1-29	
14	ST9	Arg, Aus, Ger, Ita, Por, UK, USA	1-5-4-5-5-1-8	England[14]-9 clone
14	ST15	Bra, Ger, Ita, Neth, Tw, UK	1-5-4-5-5-3-8	
14	ST124	Aus, Can, Den, Fin, Ger, Neth, Nor, Swe, UK, USA	7-5-1-8-14-11-14	
15B[e]	Heterogeneous (ST199)	(Neth, USA)	(8-13-14-4-17-4-14)	
18C	ST113	Neth, Spa, UK, USA	7-2-1-1-10-1-21	
19A[e]	Heterogeneous (ST199)	(Ice, UK, USA)	(8-13-14-4-17-4-14)	
19F	ST81	Aus, Mal, Por, Spa, Tw, UK, USA	4-4-2-4-4-1-1	Spain[23F]-1 clone
19F	ST236	Aus, Kor, Mal, Tw, UK, USA, Vn	15-16-19-15-6-20-26	Taiwan[19F]-14
23F	ST37	(Den)[f], USA	1-8-6-2-6-4-6	Tennessee[23F]-4 clone
23F	ST81	Aus, Ger, Ice, Pol, Sing, Spa, Tw, UK, USA, Vn	4-4-2-4-4-1-1	Spain[23F]-1 clone
23F	ST242	Bra, Ita, Tw, USA	15-29-4-21-30-1-14	Taiwan[23F]-14

[a] Some of the major invasive clones are commonly encountered with more than one serotype. Data are taken from MLST studies of isolates from invasive disease (4, 5, 12, 13, 18, 23, 42; see also http://spneumoniae.mlst.net); isolates from studies using other typing procedures are not included.

[b] Arg, Argentina; Aus, Australia; Bra, Brazil; Can, Canada; Cze, Czech Republic; Col, Colombia; Den, Denmark; Fin, Finland, Fra, France; Ger, Germany; Ice, Iceland; Isr, Israel; Ita, Italy, Ken, Kenya; Kor, S. Korea; Mal, Malaysia; Neth, The Netherlands; Nor, Norway; Pol, Poland; Por, Portugal; SA, South Africa; Sing, Singapore; Sp, Spain; Swe, Sweden; Tw, Taiwan; UK, United Kingdom; Uru, Uruguay; USA, United States of America; Vn, Vietnam.

[c] The allelic profile shows the allele numbers at the seven MLST loci in the order *aroE-gdh-gki-recP-spi-xpt-ddl*.

[d] ST156 is an SLV of ST162 (the Spain[9V]-3 clone) and is considered to be its penicillin-susceptible parent, an alteration at the *ddl* locus occurring at the time penicillin resistance emerged as a result of the very close genetic linkage of *ddl* to the penicillin-binding protein 2b gene.

[e] The most prevalent clone of this serotype is shown in parentheses.

[f] The isolates from invasive disease in Denmark are penicillin susceptible, in contrast to those from the United States, which are highly cephalosporin resistant.

have led to a high level of antibiotic resistance, thus increasing the prevalence of these clones in the community and thus the opportunities for them to cause serious infections.

Antibiotic resistance might be expected to have arisen in those clones that were commonly carried in the nasopharynges of children. This appears to be the case for some of the major resistant clones. Thus, penicillin-susceptible isolates of the highly cephalosporin-resistant Tennessee[23F]-4 clone (ST37 [34]) are commonly found to be carried in Finnish children (21) and have also been associated with invasive disease in Denmark (13). In several other cases, resistance appears to have emerged in a rare genotype that has subsequently become common due to the selective advantage of resistance. For example, penicillin-susceptible isolates of the Spain[23F]-1 clone (34, 50), or similar genotypes, have never been reported among samples from carriage or disease.

Further studies using MLST will undoubtedly provide a much better view of the major clones of the pneumococcal population associated with invasive disease and of any associations of particular clones with different forms of invasive disease (e.g., meningitis) or with poor outcomes, and will identify any major geographic differences that exist. They may also detect any significant differences in the pneumococci of the same serotype causing disease in young children compared to adults or the elderly. However, the need for careful sampling, and the diversity of the pneumococcal population, makes it difficult to obtain samples of sufficient size to obtain statistically robust answers to many of these questions.

International invasive clones are defined as those that have been recovered from multiple countries. In reality, of course, they are the clones encountered in those countries that have frequent international links. As carriage is common among children, the movements of people (recent and historical) provide frequent opportunities for the introduction of new isolates into a community. The rapid global spread of some of the antibiotic-resistant clones testifies to the potential for strains that emerge in one country to become rapidly established in others. The extent to which the same clones were found in different countries before the era of frequent international travel is unknown, as large collections of old strains have not been examined. Geographic differences might be expected to be most pronounced for pneumococci of those serotypes that are rarely carried, and there are indications of such differences among the major clones of serotype 1 (5).

We have very little information regarding the clones that are prevalent in developing countries, especially in remote regions. It is an open question whether these isolates will be the same as those that are found elsewhere. We might expect that they will exhibit allelic profiles different from those of isolates in developed countries, as even if the alleles and serotypes are the same, recombination will have shuffled them into new combinations. The high carriage rates among children in these regions (38) afford ample opportunities for multiple carriage and for frequent recombination at both the MLST and capsular loci.

Serotypes that are rarely encountered in developed countries appear to cause a substantial amount of disease in some developing countries (22). An understanding of the prevalence of these serotypes (and clones) in carriage, compared to disease, could distinguish any serotypes or clones that have high attack rates from those that cause substantial invasive disease because they are very commonly carried.

CONCLUDING REMARKS

The ability of MLST to compare the genotypes of isolates that are characterized in different laboratories should have a major impact on our knowledge of the pneumococcal population. However, to address many of the key questions, more attention has to be applied to appropriate sampling and to relating isolates from disease to those from carriage. One of the most interesting (and important) questions relates to the long-term effects of changes in the pneumococcal population imposed by the widespread introduction into developed countries of conjugate vaccines, which are imperfect in the sense that they prevent disease caused by most, but not all, of the pneumococcal isolates that cause disease. Shifts in the serotypes being carried, or causing disease, can readily be monitored, but understanding the basis of any shifts requires precise isolate characterization by MLST.

It is clear that recombination has a predominant role in clonal diversification and that the pneumococcus has access to extensive genetic variation within the diverse population of streptococci that are closely related to the pneumococcus, which has implications for the routes by which resistance to antibiotics or escape from vaccines may occur.

Carriage is central to pneumococcal population biology, and several features of the carrier state, reviewed elsewhere in this volume (chapter 9), are important for a better understanding of pneumococcal population biology, for instance, the duration of carriage, heterogeneity of duration among serotypes or clones, and the incidence of carriage of multiple genotypes allowing opportunities for recombination, to mention but a few. An understanding of these features of carriage is a required foundation for any attempt to combine epidemiology and population biology. Equally important, carried serotypes and carriage rates vary in different parts of the world and with age, and the impact of this population subdivision has yet to be studied. For some age groups and populations there are almost no data.

The pneumococcal population contains many clonal complexes and many divergent lineages, and isolates with the same capsular serotype may be only distantly related in their overall genotype. Whether divergent clones of the same serotype differ widely in the ability to cause the different types of pneumococcal disease is still unclear, as is the invasiveness of those nonvaccine serotypes that are excluded from current and forthcoming conjugate vaccines but which are high in the rank order of prevalence in serious disease in some countries of the developing world. Finally, strain choice for genome projects, for studying virulence, or for microarray studies often appears somewhat arbitrary. A better understanding of the pneumococcal population, and of the differences among isolates in the ability to cause disease or to be carried, will allow the molecular determinants of virulence to be addressed through a more informed choice of comparator strains. A reference set of strains for this purpose has recently been assembled in the laboratory of B. G. S. In this way the important genetic differences that provide some pneumococcal isolates with increased or decreased ability to be carried or to cause serious disease may be understood, although for pathogens such as the pneumococcus, with which carriage rarely leads to disease, this is unlikely to be easy.

ACKNOWLEDGMENTS

We are grateful to the Wellcome Trust for support and to all those within the pneumococcal community who submit data to the MLST public database. B.G.S. is a Wellcome Trust Principal Research Fellow.

REFERENCES

1. **Adrian, P. V, and K. P. Klugman.** 1997. Mutations in the dihydrofolate reductase gene of trimethoprim-resistant isolates of *Streptococcus pneumoniae. Antimicrob. Agents Chemother.* **41:** 2406–2413.
2. **Austrian, R.** 1986. Some aspects of the pneumococcal carrier state. *J. Antimicrob. Chemother.* **18**(Suppl. A):35–45.
3. **Brito, D. A., M. Ramirez, and H. de Lencastre.** 2003. Serotyping *Streptococcus pneumoniae* by multiplex PCR. *J. Clin. Microbiol.* **41:**2378–2384.
4. **Brueggemann, A. B., D. T. Griffiths, E. Meats, T. Peto, D. W. Crook, and B. G. Spratt.** 2003. Clonal relationships between invasive and carriage *Streptococcus pneumoniae,* and serotype- and clone-specific differences in invasive disease potential. *J. Infect. Dis.* **187:**1424–1432.
5. **Brueggemann, A. B., and B. G. Spratt.** 2003. Geographic distribution and clonal diversity of *Streptococcus pneumoniae* serotype 1 isolates. *J. Clin. Microbiol.,* **41:**386–392.
6. **Coffey, T. J., C. G. Dowson, M. Daniels, and B. G. Spratt.** 1995. Genetics and molecular biology of β-lactam-resistant pneumococci. *Microb. Drug Resist.* **1:**25–30.
7. **Coffey, T. J., M. C. Enright, M. Daniels, J. K. Morona, R. Morona, W. Hryniewicz, J.**

C. Paton, and B. G. Spratt. 1998. Recombinational exchanges at the capsular polysaccharide biosynthetic locus lead to frequent serotype changes among natural isolates of *Streptococcus pneumoniae*. *Mol. Microbiol.* **27**:73–83.

8. Coffey, T. J., M. Daniels, M. C. Enright, and B. G. Spratt. 1999. Serotype 14 variants of the Spanish penicillin-resistant serotype 9V clone of *Streptococcus pneumoniae* arose by large recombinational replacements of the *cpsA-pbp1a* region. *Microbiology* **145**:2023–2031.

9. Dagan, R., N. Givon-Lavi, O. Zamir, M. Sikuler-Cohen, L. Guy, J. Janco, P. Yagupsky, and D. Fraser. 2002. Reduction of nasopharyngeal carriage of *Streptococcus pneumoniae* after administration of a 9-valent pneumococcal conjugate vaccine to toddlers attending day care centers. *J. Infect. Dis.* **185**:927–936.

10. de Lencastre, H., and A. Tomasz. 2002. From ecological reservoir to disease: the nasopharynx, day-care centres and drug-resistant clones of *Streptococcus pneumoniae*. *J. Antimicrob. Chemother.* **50**(Suppl. S2):75–81.

11. DeMaria, A., K. Browne, S. L. Berk, E. J. Sherwood, and W. McCabe. 1980. An outbreak of type 1 pneumococcal pneumonia in a men's shelter. *JAMA* **244**:1446–1449.

12. Dicuonzo, G., G. Gherardi, R. E. Gertz, F. D'Ambrosio, A. Goglio, G. Lorino, S. Recchia, A. Pantosti, and B. Beall. 2002. Genotypes of invasive pneumococcal isolates recently recovered from Italian patients. *J. Clin. Microbiol.* **40**:3660–3665.

13. Enright, M. C., and B. G. Spratt. 1998. A multilocus sequence typing scheme for *Streptococcus pneumoniae*: identification of clones associated with serious invasive disease. *Microbiology* **144**:3049–3060.

14. Enright, M. C., A. Fenoll, D. Griffiths, and B. G. Spratt. 1999. The three major Spanish clones of penicillin-resistant *Streptococcus pneumoniae* are the most common clones recovered from recent cases of meningitis in Spain. *J. Clin. Microbiol.* **37**:3210–3216.

15. Enright, M. C., and B. G. Spratt. 1999. Extensive variation in the *ddl* gene of penicillin-resistant *Streptococcus pneumoniae* results from a hitchhiking effect driven by the penicillin-binding protein 2b gene. *Mol. Biol. Evol.* **16**:1687–1695.

16. Feil, E. J., J. Maynard Smith, M. C. Enright, and B. G. Spratt. 2000. Estimating recombinational parameters in *Streptococcus pneumoniae* from multilocus sequence typing data. *Genetics* **154**:1439–1450.

17. Feil, E. J., E. C. Holmes, D. E. Bessen, M.-S. Chan, N. P. J. Day, M. C. Enright, R. Goldstein, D. W. Hood, A. Kalia, C. E. Moore, J. Zhou, and B. G. Spratt. 2001. Recombination within natural populations of pathogenic bacteria: short-term empirical estimates and long-term phylogenetic consequences. *Proc. Natl. Acad. Sci. USA* **98**:182–187.

17a. Feil, E. J., B. C. Li, D. M. Aanensen, W. P. Hanage, and B. G. Spratt. 2004. eBURST: inferring patterns of evolutionary descent among clusters of related bacterial genotypes from multilocus sequence typing data. *J. Bacteriol.* **186**:1518–1530.

18. Gertz, R. E., M. C. McEllistrem, D. J. Boxrud, Z. Li., V. Sakota, T. A. Thompson, R. R. Facklam, J. M. Besser, L. H. Harrison, C. G. Whitney, and B. Beall. 2003. Clonal distribution of invasive pneumococcal isolates from children and selected adults in the United States prior to 7-valent conjugate vaccine introduction. *J. Clin. Microbiol.* **41**:4194–4216.

19. Greenwood, B. 1999. The epidemiology of pneumococcal infection in children in the developing world. *Philos. Trans. R. Soc. Lond. B* **354**:777–785.

20. Hakenbeck, R., N. Balmelle, B. Weber, C. Gardès, W. Keck, and A. de Saizieu. 2001. Mosaic genes and mosaic chromosomes: intra- and interspecies genomic variation of *Streptococcus pneumoniae*. *Infect. Immun.* **69**:2477–2486.

21. Hanage, W. P., K. Auranen, R. Syrjänen, E. Herva, P. H. Mäkelä, T. Kilpi, and B. G. Spratt. 2004. Ability of pneumococcal serotypes and clones to cause acute otitis media: implications for the prevention of otitis media by conjugate vaccines. *Infect. Immun.* **72**:76–81.

22. Hausdorff, W. P., J. Bryant, P. R. Paradiso, and G. R. Siber. 2000. Which pneumococcal serogroups cause the most invasive disease: implications for conjugate vaccine formulation and use, part I. *Clin. Infect. Dis.* **30**:100–121.

23. Henriques Normark, B., M. Kalin, Å. Örtqvist, T. Åkerlund, B. Olsson Liljequist, J. Hedlund, S. B. Svenson, J. Zhou, B. G. Spratt, S. Normark, and G. Källenius. 2001. Dynamics of penicillin-susceptible clones in invasive pneumococcal disease. *J. Infect. Dis.* **184**:861– 869.

24. Huebner, R. E., R. Dagan, N. Porath, A. D. Wasas, and K. P. Klugman. 2000. Lack of utility of serotyping multiple colonies for detection of simultaneous nasopharyngeal carriage of different pneumococcal serotypes. *Pediatr. Infect. Dis. J.* **19**:1017–1020.

25. Jebaraj, R., T. Cherian, P. Raghupathy, K. N. Brahmadathan, M. K. Lalitha, K. Thomas, and M. C. Steinhoff. 1999. Nasopharyngeal colonization of infants in southern

India with *Streptococcus pneumoniae*. *Epidemiol. Infect.* **123:**383–388.

26. **Kadioglu, A., S. Taylor, F. Iannelli, G. Pozzi, T. J. Mitchell, and P. W. Andrew.** 2002. Upper and lower respiratory tract infection by *Streptococcus pneumoniae* is affected by pneumolysin deficiency and differences in capsule type. *Infect. Immun.* **70:**2886–2890.

27. **Kelly, T., J. P. Dillard, and J. Yother.** 1994. Effect of genetic switching of capsular type on virulence of *Streptococcus pneumoniae*. *Infect. Immun.* **62:**1813–1819.

28. **Lawrence, E. R., D. B. Griffiths, S. A. Martin, R. C. George, and L. M. C. Hall.** 2003. Evaluation of semiautomated multiplex PCR assay for determination of *Streptococcus pneumoniae* serotypes and serogroups. *J. Clin. Microbiol.* **41:**601–607.

29. **Lipsitch, M., and E. R. Moxon.** 1997. Virulence and transmissibility of pathogens: what is the relationship? *Trends Microbiol.* **5:**1–37.

30. **Lipsitch, M.** 1999. Bacterial vaccines and serotype replacement: lessons from *Haemophilus influenzae* and prospects for *Streptococcus pneumoniae*. *Emerg. Infect. Dis.* **5:**336–345.

31. **Maiden, M. C. J., J. A. Bygraves, E. Feil, G. Morelli, J. E. Russell, R. Urwin, Q. Zhang, J. Zhou, K. Zurth, D. A. Caugant, I. M. Feavers, M. Achtman, and B. G. Spratt.** 1998. Multilocus sequence typing: a portable approach to the identification of clones within populations of pathogenic microorganisms. *Proc. Natl. Acad. Sci. USA* **95:**3140–3145.

32. **Mbelle, N., R. E. Huebner, A. D. Wasas, A. Kimura, I. Chang, and K. P. Klugman.** 1999. Immunogenicity and impact on nasopharyngeal carriage of a nonavalent pneumococcal conjugate vaccine. *J. Infect. Dis.* **180:**1171–1176.

33. **McCracken, G. H.** 2000. Etiology and treatment of pneumonia. *Pediatr. Infect. Dis. J.* **19:**373–377.

34. **McGee, L., L. McDougal, J. Zhou, B. G. Spratt, F. C. Tenover, R. George, R. Hakenbeck, W. Hryniewicz, J.-C. Lefévre, A. Tomasz, and K. P. Klugman.** 2001. Nomenclature of major antimicrobial-resistant clones of *Streptococcus pneumoniae* defined by the Pneumococcal Molecular Epidemiology Network (PMEN). *J. Clin. Microbiol.* **39:**2565–2571.

35. **Muller-Graf, C. D., A. M. Whatmore, S. J. King, K. Trzcinski, A. P. Pickerill, N. Doherty, J. Paul, D. Griffiths, D. Crook, and C. G. Dowson.** 1999. Population biology of *Streptococcus pneumoniae* isolated from oropharyngeal carriage and invasive disease. *Microbiology* **145:**3283–3293.

36. **Nesin, M., M. Ramirez, and A. Tomasz.** 1998. Capsular transformation of a multidrug-resistant *Streptococcus pneumoniae* in vivo. *J. Infect. Dis.* **177:**707–713.

37. **Obaro, S. K., R. Adegbola, W. A. Banya, and B. M. Greenwood.** 1996. Carriage of pneumococci after pneumococcal vaccination. *Lancet* **348:**271–272.

38. **Obaro, S., and R. J. Adegbola.** 2002. The pneumococcus: carriage, disease and conjugate vaccines. *Med. Microbiol.* **51:**98–104.

39. **Porat, N., R. Trefler, and R. Dagan.** 2001. Persistence of two invasive *Streptococcus pneumoniae* clones of serotypes 1 and 5 in comparison to that of multiple clones of serotypes 6B and 23F among children in southern Israel. *J. Clin. Microbiol.* **39:**1827–1832.

40. **Ramirez, M., and A. Tomasz.** 1999. Acquisition of new capsular genes among clinical isolates of antibiotic-resistant *Streptococcus pneumoniae*. *Microb. Drug Resist.* **5:**241–246.

41. **Robinson, D. A, K. M. Edwards, K. B. Waites, D. E. Briles, M. J. Crain, and S. K. Hollingshead.** 2001. Clones of *Streptococcus pneumoniae* isolated from nasopharyngeal carriage and invasive disease in young children in central Tennessee. *J. Infect. Dis.* **183:**1501–1507.

42. **Sa-Leao, R., S. E. Vilhelmsson, H. de Lencastre, K. G. Kristinsson, and A. Tomasz.** 2002. Diversity of penicillin-nonsusceptible *Streptococcus pneumoniae* circulating in Iceland after the introduction of the penicillin-resistant clone Spain[6B]-2. *J. Infect. Dis.* **186:**966–975.

43. **Selander, R. K., D. A. Caugant, H. Ochman, J. M. Musser, M. N. Gilmour, and T. S. Whittam.** 1986. Methods of multilocus enzyme electrophoresis for bacterial population genetics and systematics. *Appl. Environ. Microbiol.* **51:**873–884.

44. **Selander, R. K., J. M. Musser, D. A. Caugant, M. N. Gilmour, and T. S. Whittam.** 1987. Population genetics of pathogenic bacteria. *Microb. Pathog.* **3:**1–7.

45. **Smillie, W. G., G. H. Warnock, and H. J. White.** 1938. A study of a type I pneumococcus epidemic at the State Hospital at Worcester, Mass. *Am. J. Public Health* **28:**293–302.

46. **Smith, T., D. Lehman, J. Montgomery, M. Grattan, I. D. Riley, and M. P. Alpers.** 1993. Acquisition and invasiveness of different serotypes of *Streptococcus pneumoniae* in young children. *Epidemiol. Infect.* **111:**27–39.

47. **Spratt, B. G.** 1999. Multilocus sequence typing: molecular typing of bacterial pathogens in an era of rapid DNA sequencing and the internet. *Curr. Opin. Microbiol.* **2:**312–316.

48. **Spratt, B. G., and B. M. Greenwood**. 2000. Prevention of pneumococcal disease by vaccination: does serotype replacement matter? *Lancet* **356:**1210–1211.

49. **Whatmore, A. M., A. Efstratiou, A. P. Pickerill, K. Broughton, G. Woodard, D. Sturgeon, R. George, and C. G. Dowson.** 2000. Genetic relationships between clinical isolates of *Streptococcus pneumoniae, Streptococcus oralis,* and *Streptococcus mitis:* characterization of "atypical" pneumococci and organisms allied to *S. mitis* harboring *S. pneumoniae* virulence factor-encoding genes. *Infect. Immun.* **68:**1374–1382.

50. **Zhou, J., M. C. Enright, and B. G. Spratt.** 2000. Identification of the major Spanish clones of penicillin-resistant pneumococci via the Internet using multilocus sequence typing. *J. Clin. Microbiol.* **38:**977–986.

PNEUMOCOCCAL CARRIAGE

Derrick W. Crook, Angela B. Brueggemann, Karen L. Sleeman, and Timothy E. A. Peto

9

INTRODUCTION

The human nasopharynx is the principal ecological niche for the heterogeneous population of *Streptococcus pneumoniae* (the pneumococcus), which exists as 90 different capsular types or serotypes (27). The horse is a possible alternative host for only a capsular type 3 variant clone of a human lineage of the pneumococcus (56). Survival of the pneumococcus, therefore, depends on successful transmission among humans, and a failure in this would likely result in disappearance of the organism from the human population. The focus of this chapter is to understand the behavior of *S. pneumoniae* and the factors that affect it in its normal habitat, the human nasopharynx.

DEFINITION AND MEASUREMENT OF CARRIAGE

The process of gaining access to and setting up residence in the human nasopharynx involves three discrete phases, acquisition, carriage, and termination of carriage, which are often collectively referred to as "carriage or colonization"; however, recognizing that these are discrete phases is important in understanding pneumococcal biology. Acquisition is the point when *S. pneumoniae* is introduced to and establishes itself within the host. This leads to a period, called carriage or colonization, when the pneumococcus is residing in the host. The length of time the organism persists in the host is varied and is dependent upon factors related to both the host and the organism. Termination of carriage is the point when the organism is lost from the human host. Measuring each of these discrete phases—acquisition, carriage, and termination of carriage—is understandably imprecise and is one of the major factors that has restricted a thorough understanding of the host-pneumococcus relationship in nature. Furthermore, the relationship between the onset of pneumococcus-related disease and the timing of these discrete phases is unclear. For example, does acquisition represent the moment when occurrence of disease is most likely? Does prolonged carriage protect against disease caused by the colonizing strain of *S. pneumoniae*?

Clearly the first step to understanding the dynamics of acquisition and carriage is to have the ability to detect the presence of

Derrick W. Crook, Angela B. Brueggemann, and Timothy E. A. Peto, Interdepartmental Academic Unit of Microbiology and Infectious Disease, John Radcliffe Hospital, University of Oxford, Headington, Oxford OX3 9DU, United Kingdom. *Karen L. Sleeman*, Oxford Vaccine Group, University of Oxford, Centre for Clinical Vaccinology & Tropical Medicine, Churchill Hospital, Old Rd., Headington, Oxford OX3 7LJ, United Kingdom.

The Pneumococcus. Volume Editor, Elaine I. Tuomanen,
© 2004 ASM Press, Washington, D.C.

pneumococci in the nasopharynx; however, the optimal detection method is unresolved. Historically, the method for isolating the organism was to inoculate a mouse with human nasopharyngeal secretions and then culture samples taken from the mouse for evidence of pneumococcal infection (2). However, this is a slow, laborious, and costly procedure that is unsuitable for large-scale studies and has largely been abandoned. Consequently, without an accepted "gold standard" method, published studies have used many different methods to investigate pneumococcal acquisition and carriage (2, 42), which has unfortunately resulted in serious limitations in study comparability.

A 1975 report by Hendley and colleagues showed there was no statistical difference between the isolation of pneumococci by culturing nasopharyngeal secretions onto blood agar that contained gentamicin and isolation using the mouse inoculation method (26). This made the use of high-throughput culture-based methods attractive when conducting large-scale pneumococcal carriage studies. However, a method for sampling and isolating pneumococci was only recently agreed upon and published by the World Health Organization (42). This consensus report offers detailed, systematic methods for sampling, culturing, and characterizing pneumococcal isolates recovered from the upper respiratory tract, methods that ideally would be used by all pneumococcus investigators worldwide to ensure future comparability between studies. A number of sampling and laboratory operating procedures are specified, three of which are particularly important: sampling the nasopharynx using a calcium alginate or Dacron polyester-tipped nasopharyngeal swab, transporting the sample in skim milk-tryptone-glucose-glycerin, and culturing the sample on blood (horse, sheep or goat) agar containing 2.5 or 5.0 μg of gentamicin, colistin-nalidixic acid, or colistin-oxolinic acid per ml.

However, although this is an approach that works very well for most pneumococcal studies, it is important to remember that it still does not overcome several inherent limitations of swabbing and culture technology. First, culture and recognition of pneumococci in a microbiological sample which may also contain colonies that have many of the same morphological characteristics as pneumococci (e.g., viridans group streptococci) confound efforts to recognize and isolate pneumococci. Secondly, a swab of the nasopharynx is limited because it may only sample a fraction of the pneumococci potentially occupying this niche, that is, it is likely that this technique will detect pneumococci only when they are present at a density sufficient to be sampled and cultured from a swab. As a result, pneumococci present in the nasopharynx below this threshold density will not be detected. This also raises issues about whether a negative culture result is truly negative or whether the organism was simply below the limit of detection. Therefore, determining the prevalence of pneumococci in cross-sectional studies is likely to be an underestimate of the true prevalence, and when undertaking longitudinal carriage studies, variation in the density of pneumococci above and below the threshold of detection may lead to overestimates of acquisition and loss, respectively.

Another major limitation to consider when investigating pneumococcal colonization is the dependence on interval sampling and the possibility that the chosen sampling interval is outside the bounds of important colonization events. For example, a serotype or serogroup that inhabits the nasopharynx for only a few days or weeks (e.g., serotypes 20 and 31 [48]) may seldom be detected, whereas serotypes that reside in the nasopharynx for long periods (e.g., serogroups 6 and 19 [48]) are likely to be detected when using sampling intervals of weeks or months.

A further issue to consider when conducting carriage studies is the existence of 90 known pneumococcal serotypes and the possibility that more than 1 serotype may be colonizing an individual at the same time. The sampling and culture techniques are generally inadequate for detecting carriage of multiple serotypes in a single individual, at least when sampling large numbers of people (29, 42). This is a major limitation in longitudinal studies of pneumococcal acquisition and carriage or when investigating serotype replacement after widespread pneumococcal vaccination. Furthermore, because there are so many different capsular types, very large studies are required in order to achieve the statistical power necessary to determine biologically important serotype-specific differences related to transmission, disease potential, or genetic diversity.

HISTORICAL BACKGROUND

The initial detection and first accounts of carriage of *S. pneumoniae* were independently made in the early 1880s by Pasteur, who inoculated a rabbit with saliva from a rabies patient (43), and Sternberg, who inoculated a rabbit with his own saliva (51). In 1939, Heffron provided a comprehensive summary of the earliest large carriage studies (25). An important limitation of these early carriage studies was that serotyping as a laboratory technique was in development; therefore, investigators were only able to differentiate serotypes 1, 2, and 3 (previously referred to as serotypes I, II, and III), and all other serotypes that we now recognize were collectively referred to as group IV. The differentiation of the heterogeneous group of type IV organisms into unique serotypes only started in the 1930s.

The main findings of these early studies can be summarized in several points. (i) Pneumococci were asymptomatically carried by 25 to 70% of the population. (ii) The proportion of carriers to adult pneumonia cases varied for serotypes 1, 2, and 3. While serotype 1 caused most cases of pneumonia (33%), it was rarely carried in the general population (0.4%). The reverse was true for serotype 3, which had a higher prevalence among carriers (11%) than among patients with pneumonia (11%). Serotype 2, a serotype rarely seen today, had a carriage rate of 0.8% and caused 20% of pneumonia cases. These data suggested that there were major

serotype-specific differences in the ability to cause disease and/or in carriage dynamics. (iii) Pneumococcal transmission occurred between pneumococcal carriers or patients with pneumococcal disease and noncolonized people, and transmission was particularly likely to occur in families and through close contact. (iv) Duration of carriage did not clearly differ among individual serotypes. (v) There was a lower carriage rate in infants and children, 35%, than in adults, 65%. (vi) A higher proportion of both children and adults with respiratory infections such as bronchitis were carriers than were healthy children and adults. (vii) Cold months of the year (i.e., winter season) were associated with higher pneumococcal carriage rates than were the warmer months.

RISK FACTORS FOR CARRIAGE

It is important to clarify that studies which have explored risk factors for pneumococcal carriage have not made a distinction between acquisition and carriage; therefore, data from these studies reflect risk factors related to the prevalence of pneumococci in carriage. No studies that assess risk factors for acquisition of pneumococci have been published. Since acquisition in part measures transmission, factors affecting acquisition would be better determinations of which factors affect transmission of pneumococci, so caution needs to be exercised when interpreting carriage data as a proxy for pneumococcal transmission.

Age

In most longitudinal studies, all of which have been conducted with young children, there is a significant age-specific increase in carriage over the first year of life (1, 10, 18, 31–33, 40, 53). Carriage prevalence varies markedly among studies; however, there appears to be a progressive rise during the first year from a prevalence rate of <10% in the weeks after birth to nearly 100% at 1 year of life in some studies. The data on age-specific changes in older age groups is less clear. Recent reports suggest that carriage decreases in children over

4 years of age (24, 31–33) and decreases further among adults (24, 32). Some pediatric studies have not found a significant change in age-specific carriage prevalence with increasing age (12, 34, 39, 49), and the explanation for this is unclear. It may be a feature of differing transmission characteristics and acquisition frequencies between populations, or it may be related to the sampling and culture techniques used by different investigators. Interestingly, the results of recent pediatric studies are the reverse of those made by investigators in the early 1900s, who found that approximately 65% of adults carried pneumococci, compared to 35% of children (25), which may indicate a temporal change in pneumococcal colonization over the past century.

A markedly greater proportion of individuals in low-income countries such as India (4), The Gambia (32), and Papua New Guinea (39) are colonized with pneumococci than are individuals in high-income countries; however, no robust comparative investigation has been undertaken to conclusively demonstrate that economic development in a country is a risk factor when adjusted for other risk factors. If national economic development is an independent risk factor for carriage, it might actually be a reflection of genetic and/or social factors related to differences in the host-pneumococcus relationship among persons from different countries.

Siblings

One of the strongest risk factors for pneumococcal carriage in an infant is the presence of a colonized sibling, and this is observed in both high-income (10, 18, 31, 55) and low-income (4) countries. The infant has been shown to acquire pneumococci from an older sibling (17, 31), suggesting that children at their peak age of pneumococcal carriage (i.e., 2 to 5 years) are a main source of pneumococcal transmission.

Respiratory Illness

The extent to which pneumococci may associate with mild nonspecific illness is uncertain, as studies have differed in their findings. Gray and colleagues reported that pneumococcal acquisition was commonly followed by disease, mainly acute otitis media (18). This publication did not report on the association of carriage (as distinct from acquisition) with nonspecific disease; however, two other studies have shown an association between mild, nonspecific upper respiratory disease and carriage (3, 15). Other studies have predictably shown associations with pneumococcal carriage and acute otitis media (59) or sinusitis (34). These observations suggest a role for pneumococci in a wider spectrum of mild respiratory diseases besides specific conditions such as acute otitis media or sinusitis.

Other Possible Risk Factors

Differences in carriage during certain months of the year, particularly the winter months, were recognized in the studies from the early 1900s and in more recent studies (12, 22), suggesting that season is a potential risk factor for carriage. Some studies have identified day care attendance as a prominent risk factor for carriage (3, 10, 14), although other studies have not (30, 31); therefore, exposure to day care is not universally identified as a risk factor. Finally, identifiable associations with factors such as breast-feeding (3, 4, 49), smoking (3, 30, 49), and crowding (12) have not consistently been found. One study found that breast-feeding reduced carriage (13), while another study demonstrated that parental smoking was a risk factor for carriage (52).

RISK FACTORS FOR ACQUISITION

Recently, data collected from a large longitudinal study of a birth cohort of 214 children in Oxfordshire, United Kingdom, who were swabbed twice weekly for 3 months and then monthly up to 6 months of age were analyzed (K. Sleeman, I. Daniels, D. Griffiths, J. Deeks, T. Peto, D. Crook, and R. Moxon, 3rd Int. Symp. Pneumococci Pneumococcal Dis., 2002, Session R-03, abstr. no. 8). Acquisition was defined as isolation of an *S. pneumoniae* serotype that had not been detected at a previous visit. All other pneumococcal isolations were defined as carriage. The data were analyzed for associations between pneumococcal acquisition and the following factors: breast-feeding, number of siblings, day care attendance, exposure to cigarette smoke, consultations to general practitioners for infections, antibiotic treatment, and season. Using a multivariate model, only number of siblings (hazard ratio of 1.5; 95% confidence interval [CI], 1.3 to 1.8; $P < 0.001$) and visits to a general practitioner for mild upper respiratory disease (hazard ratio of 1.8; 95% CI, 1.1 to 2.9; $P = 0.02$) were independently associated with acquisition of pneumococci. By contrast, there was no association between carriage and visits to a general practitioner for mild upper respiratory disease, indicating that the association with general practitioner visits was specific to acquisition. The other factors—breast-feeding, day care attendance, exposure to cigarette smoke, antibiotic use, and season—were not independently associated with acquisition of pneumococci in a multivariate analysis. This analysis, using incidence of acquisition, eliminates the effect of duration of carriage as a factor that may bias the analysis when using carriage (i.e., prevalence) data.

CARRIAGE OF ANTIBIOTIC-RESISTANT PNEUMOCOCCI

The evolution of antibiotic-resistant strains of pneumococci probably occurs de novo only rarely; however, once in existence and under the selective pressure of antibiotic use, these strains are able to spread widely (38). A number of factors that affect the carriage of antibiotic-resistant pneumococci have been identified, the single most important of which is exposure to antibiotics (3, 10, 45, 54, 57). The most compelling data come from a study of independent risk factors of carriage in Israeli children (45). Among these children (both carriers and noncarriers of pneumococci), the following factors were independently associated with carriage of penicillin-resistant pneumococci: antibiotic course taken during the previous 3 months, odds ratio (OR) of 1.52 (95%

CI, 1.25 to 1.85), $P < 0.0001$; age of <24 months, OR of 2.24 (95% CI, 1.2 to 4.2), $P = 0.01$; and day care attendance, OR of 3.8 (95% CI, 1.9 to 7.47), $P < 0.0001$. Among carriers of pneumococci, only prior exposure to antibiotics was an independent risk factor for carriage of penicillin-resistant isolates, with an OR of 2.24 (95% CI, 1.64 to 3.05), $P < 0.0001$. In another Israeli study, female gender, young age, and cold season were independently associated with carriage of penicillin-resistant strains (57).

Another feature of penicillin resistance is that the resistant clones are principally found among the capsular types included in the pneumococcal conjugate vaccines (10, 54) (see Table 1); thus, vaccine effectiveness has had a direct effect on the reduction of carriage of antibiotic-resistant clones (5, 9, 11).

The selection of antibiotic-resistant strains in children who received antibiotic therapy was directly investigated by Dagan et al. (8) in a study which showed that antibiotic administration reduced the carriage of susceptible, but not resistant, isolates. There was also a significant dose-dependent relationship between the MIC of the antibiotic required to inhibit growth of an isolate and isolate survival; that is, more strains for which the MICs were higher survived antibiotic treatment than those strains for which the MICs were lower. This finding is consistent with the in situ selection of preexisting resistant clones over susceptible clones.

DISTRIBUTION OF CARRIED SEROTYPES

The distribution of carried serotypes varies among different studies. Five studies from Alabama (19), The Gambia (32), Papua New Guinea (48), Greece (54), and Oxford, United Kingdom (Sleeman et al., 3rd ISPPD), have been chosen to illustrate this point (Table 1). The percentages of serotypes contained in the 7-valent conjugate vaccine are 56, 58, 82, 66, and 67%, respectively. The heterogeneity between these studies is significant: χ^2 (df = 4), 43; $P < 0.0001$. There are many possible explanations for this heterogeneity, including

TABLE 1 Rank order of serogroups of *S. pneumoniae* from five carriage studies[a]

Rank order	Alabama, 1975–1978 (n = 235)		The Gambia, 1989–1991 (n = 98)		Papua New Guinea, 1985–1987 (n = 1,965)		Greece, 1997–1999 (n = 522)		Greece (antibiotic resistant), 1997–1999 (n = 239)		Oxford, 1999–2002 (n = 518)	
	Serogroup	% of total	Serogroup	% of total	Serogroup	% of total	Serogroup	% of total	Serogroup	% of total	Serogroup	% of total
1	**6**	**14.5**	**19**	**28.6**	**6**	**19.6**	**6**	**23.8**	**23**	**27.2**	**6**	**26.4**
2	3	13.1	**6**	**23.5**	**19**	**17.2**	**19**	**15.9**	**6**	**22.2**	**19**	**16.0**
3	**19**	**12.3**	**23**	**12.2**	**23**	**10.4**	**23**	**15.1**	**19**	**17.2**	**23**	**11.0**
4	**23**	**11.5**	9	7.1	NT	8.1	15	6.3	NT	10.0	**14**	**6.9**
5	**18**	**8.9**	**4**	**6.1**	**14**	**6.0**	**14**	**5.7**	**14**	**8.8**	10	5.4
6	11	6.4	**4**	**5.1**	15	4.1	NT	4.2	11	3.8	**18**	**4.4**
7	37	5.5	**14**	**5.1**	33	3.6	11	4.0	15	3.3	15	4.2
8	15	5.1	3	3.1	35	2.9	**18**	**3.4**	**9**	**2.5**	11	4.2
9	**14**	**3.4**	13	2.0	**9**	**2.3**	9	2.5	**18**	**0.8**	3	4.1
10	NT	2.6	5	1.0	**9**	**2.3**	3	2.5	10	0.8	**9**	**3.7**
11	13	2.1	20	1.0	13	2.3	7	2.5	1	0.8	22	3.5
12	8	1.7	33	1.0	21	2.3	34	2.5	Others[c]	2.5	35	2.1
13	**9**	**1.7**	21	1.0	34	2.2	10	1.5			21	1.7
14	Others[b]	7.7	11	1.0	22	1.8	33	1.0			33	1.5
15			22	1.0	**4**	**1.4**	24	1.0			16	1.4

[a] Data in boldface type indicate serotypes included in the heptavalent conjugate vaccine.
[b] Other serotypes or serogroups (≤1.3% of total) include 1, 7, 16, 17, 20, 33, 35, and 38.
[c] Other serotypes or serogroups (0.4% each of total) include 8, 20, 21, 22, 24, and 33.

different countries, ages of study subjects, study designs, and sampling and culture methodologies. This limits the extent to which comparisons can be made among carriage studies. The extent of antimicrobial resistance in a country may also have an impact on vaccine coverage. In Table 1, Greek data (54) are presented illustrating the difference in the percentage of vaccine serotypes among antibiotic-resistant isolates (78%) compared to carried pneumococci found in the general pediatric population who have not recently been treated with antibiotics (66%).

RELATIONSHIP AMONG ACQUISITION, DURATION OF CARRIAGE, AND INVASIVE DISEASE

Few studies have directly measured serotype-specific carriage duration. Gray and colleagues measured the duration of pneumococcal carriage by nasopharyngeal sampling of 79 children monthly from birth to 2 years (18). Carriage duration was approximately 4 months for serotypes or serogroups 6, 14, 19, and 23, compared to 2.7 months for all other serotypes combined. However, these estimates of duration were biased because the analysis did not adjust for different observation periods following acquisition. In Papua New Guinea, Smith and colleagues estimated serotype-specific duration of carriage by analyzing paired nasopharyngeal samples at different time intervals and assumed a constant rate of loss from the nasopharynx (48). Although the sample was small, it appeared that some serotypes or serogroups (e.g., 6 and 19) were carried for about 60 days, while others (e.g., 12, 20, 31, and 33) persisted for only about 12 days. Mean duration of carriage for each serotype or serogroup was associated with the frequency of isolation of that serotype causing invasive disease. However, caution is needed when attempting to infer a biological characteristic of a serotype or serogroup from data which are dependent on whether that serotype is a rare or common serotype. No association was found after the data were adjusted for the prevalence of each serotype.

Recently the duration of carriage of different serotypes was measured in two cohorts of newborn children monitored for 6 months and 2 years in Oxfordshire (K. Sleeman, L. Daniels, S. Gupta, M. Maiden, E. Miller, R. George, K. Knox, D. Griffiths, T. Peto, R. Moxon, and D. Crook, 3rd Int. Symp. Pneumococci, Pneumococcal Dis., 2002, Session R-10, abstr. no. 14). Children were swabbed weekly, monthly, or on alternate months. Termination of carriage was defined as two negative swabs taken in succession from a study subject. The median time to clearance for each serotype was estimated from survival statistics. The median duration of carriage for each serotype varied significantly, ranging between 6 and 22 weeks. The data from these cohorts of children were also used to calculate serotype-specific acquisition rates. An attack rate was calculated for each serotype by dividing the incidence of invasive pneumococcal disease (as reported to the Health Protection Agency, London, United Kingdom) with the incidence of pneumococcal acquisition. Clear heterogeneity in serotype-specific attack rates was noted, ranging from less than 5 to more than 50 cases per 100,000 acquisitions. The serotype-specific relationship between duration of carriage and attack rates following acquisition is displayed in Fig. 1, and these data suggest that attack rates are higher with those serotypes which are carried for shorter periods of time.

DEVELOPMENT OF ANTIBODIES IN RESPONSE TO CARRIAGE

A homotypic anticapsular serum antibody response to colonizing pneumococcal serotypes was described in three studies (20, 21, 23), but it was detectable only in a proportion of the carriers. These observations were explored further in large systematic studies in Finland (44, 46, 47, 50), where it was shown that serum antibodies to pneumococcal capsular polysaccharides appeared in response to carriage but varied by individual serotype. For example, carriage of serotypes 11A and 14 produced an antibody response as early as 6 months of age, while a significant antibody re-

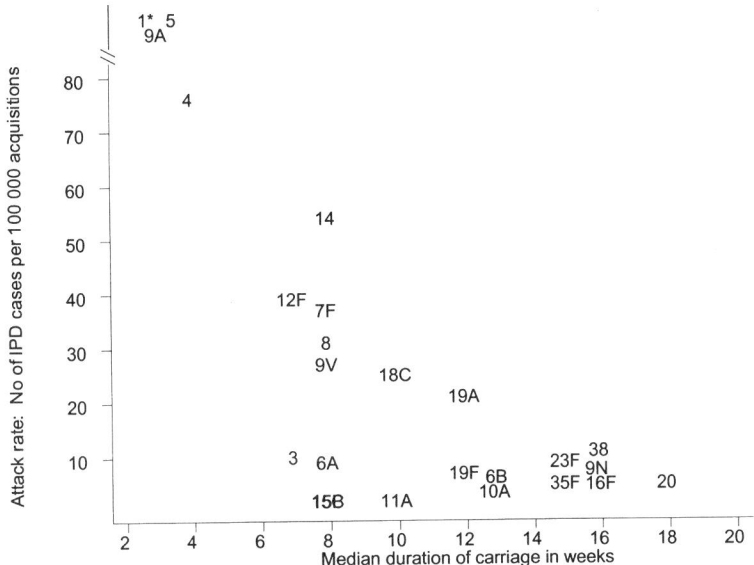

FIGURE 1   The attack rate is the incidence of invasive pneumococcal disease (IPD) per incidence of pneumococcal acquisition. The median duration for each serotype was estimated from survival analysis of interval swabbing of the nasopharynx. The invasive disease incidence data were derived from national United Kingdom data from children under 2 years of age. Acquisition data and duration data were derived from a carriage study of two Oxford birth cohorts of children studied for 6 months (214 children) and 2 years (100 children). 1* is a crude estimate for serotype 1. As serotype 1 strains were only isolated from cases of invasive disease and not from the nasopharynx, the duration of carriage was assumed to be <6 weeks and the attack rate was estimated to be >60/100,000.

sponse to carriage of serotypes 6B, 19F, and 23F was not detectable (50).

Serum antibody responses to pneumococcal surface protein A (PspA), pneumococcal surface adhesion protein A (PsaA), and pneumolysin (Ply) elicited by colonizing pneumococci have also been measured by the same investigators (44), and more recently it was shown that salivary immunoglobulin G (IgG) and secretory IgA to PspA, PsaA, Ply (47), and pneumococcal capsular polysaccharide antigens were produced in response to carried pneumococci (46). An adult human challenge model showed that PspA and choline-binding protein A (CbpA) produced antibody rises on successful challenge, while no antibody response to lipoteichoic acid, IgA1 protease, Ply, proteinase maturation protein A, or PsaA was detected (37). These observations are consistent with carriage playing a

major role in the development of adaptive immunity to the pneumococcus, which may in turn play a major role in immune selection of pneumococci, determine the distribution of serotypes in the human population, and influence pneumococcal transmission.

EFFECT OF NATURAL ANTIBODY OR VACCINE-INDUCED ANTIBODY ON CARRIAGE

There are little direct epidemiological data which show that acquired immunity plays a role in modulating carriage of pneumococci. The increase and subsequent decrease in pneumococcal carriage between 0 and 4 years of age is consistent with acquired immunity playing a role in reducing carriage, but the only direct evidence for natural immunity playing a role in preventing acquisition and thereby reducing carriage comes from a recent human challenge study (36). In this study the protective effect was associated with antibodies directed at the hypervariable N terminus of PspA; however, no protective effect was detectable for preexisting antibody to the homotypic capsular polysaccharide of the challenge strain.

Several studies have shown a dramatic effect of various pneumococcal conjugate vaccine formulations on acquisition or carriage of vaccine-associated and non-vaccine-associated pneumococci (5, 6, 11, 16, 35, 41). A range of protein conjugate (outer membrane protein complex of *Neisseria meningitidis* serogroup B, diphtheria toxin CRM197, and tetanus toxoid) and different valency (from 4 to 9 polysaccharide antigens) vaccines have been used and have demonstrated similarly significant effects. These studies showed that a significant reduction in vaccine serotypes is discernible not only when immunizing infants (11, 35) but also when immunizing children around 2 years of age (5, 6, 9, 41). A significant reduction in carriage of vaccine-related serotype 6A, a serotype not contained in the multivalent conjugate vaccines (which contain the type 6B capsular antigen), was shown in one study (6) but not in another (35), and an antibody response to vaccine-related serotype 19A (serotype 19F is included in the conjugate vaccines) was not detected in either of these two studies (6, 35). This suggests that the immunological effect of the vaccine on carriage of vaccine-related serotypes is varied.

The conjugate vaccines have a marked and reciprocal effect on the acquisition and carriage of nonvaccine serotypes. A significant increase in nonvaccine serotypes has been observed in several studies (5, 6, 35); however, two other studies have not shown a similar vaccine effect. One study was probably underpowered to detect the effect (58), and the other aimed to determine the effect 2 to 5 years after immunization (30). The second study raises the question as to whether any increase in acquisition and carriage of nonvaccine serotypes after widespread vaccination will be detectable in older children or at a long interval postvaccination. The possibility that a less pronounced vaccine effect occurs in older children is supported by an observation by Dagan and colleagues that children over 36 months of age exhibit a less pronounced reduction in carriage of vaccine-associated serotypes (6). The validity of these study observations will become apparent with time following the widespread implementation of conjugate pneumococcal vaccines by many developed countries.

Two additional beneficial effects have been observed in the human population following administration of the conjugate vaccine. The first is a significant reduction in carriage of antibiotic-resistant pneumococci (5, 35), which is perhaps not surprising given that the major antibiotic-resistant clones are mainly of vaccine serotypes; the second is a reduction in the acquisition and carriage of vaccine-associated pneumococci in unimmunized younger siblings of vaccine recipients (16).

The effect of polyvalent pneumococcal polysaccharide vaccines on nasopharyngeal carriage is markedly different from that of the conjugate vaccines. No reduction in the carriage of pneumococci was observed in a number of studies conducted with older children who were known to have a responsive immune system (7, 9, 28). The reason for this is unclear but is most likely related to the concentration or avidity of antibody generated by the two different vaccine formulations, as the antibody elicited by pure polysaccharide is generally lower in concentration and avidity than the antibodies elicited by conjugate vaccines.

Widespread implementation of the conjugate vaccine may have similar effects on the overall pneumococcal population (both carried and disease-causing pneumococci) as have been observed in the vaccine clinical trials referred to above, that is, a reduction in the prevalence of vaccine serotypes with a corresponding increase in the prevalence of nonvaccine serotypes. Issues related to serotype replacement and restructuring of the pneumococcal population are explored further in chapter 18.

REFERENCES

1. **Aniansson, G., B. Alm, B. Andersson, P. Larsson, O. Nylen, H. Peterson, P. Rigner, M. Svanborg, and C. Svanborg.** 1992. Nasopharyngeal colonization during the first year of life. *J. Infect. Dis.* **165**(Suppl 1):S38–S42.

2. **Austrian, R.** 1986. Some aspects of the pneumococcal carrier state. *J. Antimicrob. Chemother.* **18**(Suppl. A):35–45.

3. **Ciftci, E., U. Dogru, D. Aysev, E. Ince, H. Guriz, and U. D. Aysev.** 2001. Investigation of risk factors for penicillin-resistant *Streptococcus pneumoniae* carriage in Turkish children. *Pediatr. Int.* **43**:385–390.

4. **Coles, C. L., L. Rahmathullah, R. Kanungo, R. D. Thulasiraj, J. Katz, M. Santosham, and J. M. Tielsch.** 2002. Nasopharyngeal carriage of resistant pneumococci in young South Indian infants. *Epidemiol. Infect.* **129**:491–497.

5. **Dagan, R., N. Givon-Lavi, O. Zamir, and D. Fraser.** 2003. Effect of a nonavalent conjugate vaccine on carriage of antibiotic-resistant *Streptococcus pneumoniae* in day-care centers. *Pediatr. Infect. Dis. J.* **22**:532–540.

6. **Dagan, R., N. Givon-Lavi, O. Zamir, M. Sikuler-Cohen, L. Guy, J. Janco, P. Yagupsky, and D. Fraser.** 2002. Reduction of nasopharyngeal carriage of *Streptococcus pneumoniae* after administration of a 9-valent pneumococcal conjugate vaccine to toddlers attending day care centers. *J. Infect. Dis.* **185**:927–936.

7. **Dagan, R., S. Gradstein, I. Belmaker, N. Porat, Y. Siton, G. Weber, J. Janco, and P. Yagupsky.** 2000. An outbreak of *Streptococcus pneumoniae* serotype 1 in a closed community in southern Israel. *Clin. Infect. Dis.* **30**:319–321.

8. **Dagan, R., E. Leibovitz, D. Greenberg, P. Yagupsky, D. M. Fliss, and A. Leiberman.** 1998. Dynamics of pneumococcal nasopharyngeal colonization during the first days of antibiotic treatment in pediatric patients. *Pediatr. Infect. Dis J.* **17**:880–885.

9. **Dagan, R., R. Melamed, M. Muallem, L. Piglansky, D. Greenberg, O. Abramson, P. M. Mendelman, N. Bohidar, and P. Yagupsky.** 1996. Reduction of nasopharyngeal carriage of pneumococci during the second year of life by a heptavalent conjugate pneumococcal vaccine. *J. Infec. Dis.* **174**:1271–1278.

10. **Dagan, R., R. Melamed, M. Muallem, L. Piglansky, and P. Yagupsky.** 1996. Nasopharyngeal colonization in southern Israel with antibiotic-resistant pneumococci during the first 2 years of life: relation to serotypes likely to be included in pneumococcal conjugate vaccines. *J. Infect. Dis.* **174**:1352–1355.

11. **Dagan, R., M. Muallem, R. Melamed, O. Leroy, and P. Yagupsky.** 1997. Reduction of pneumococcal nasopharyngeal carriage in early infancy after immunization with tetravalent pneumococcal vaccines conjugated to either tetanus toxoid or diphtheria toxoid. *Pediatr. Infect. Dis. J.* **16**:1060–1064.

12. **Dowling, J. N., P. R. Sheehe, and H. A. Feldman.** 1971. Pharyngeal pneumococcal acquisitions in "normal" families: a longitudinal study. *J. Infect. Dis.* **124**:9–17.

13. **Duffy, L. C., H. Faden, R. Wasielewski, J. Wolf, and D. Krystofik.** 1997. Exclusive breastfeeding protects against bacterial colonization and day care exposure to otitis media. *Pediatrics* **100**:E7.

14. **Dunais, B., C. Pradier, H. Carsenti, M. Sabah, G. Mancini, E. Fontas, and P. Dellamonica.** 2003. Influence of child care on nasopharyngeal carriage of *Streptococcus pneumoniae* and *Haemophilus influenzae*. *Pediatr. Infect. Dis. J.* **22**:589–592.

15. **Faden, H., M. J. Waz, J. M. Bernstein, L. Brodsky, J. Stanievich, and P. L. Ogra.** 1991. Nasopharyngeal flora in the first three years of life in normal and otitis-prone children. *Ann. Otol. Rhinol. Laryngol.* **100**:612–615.

16. **Givon-Lavi, N., D. Fraser, and R. Dagan.** 2003. Vaccination of day-care center attendees reduces carriage of *Streptococcus pneumoniae* among their younger siblings. *Pediatr. Infect. Dis. J.* **22**:524–532.

17. **Givon-Lavi, N., D. Fraser, N. Porat, and R. Dagan.** 2002. Spread of *Streptococcus pneumoniae* and antibiotic-resistant *S. pneumoniae* from day-care center attendees to their younger siblings. *J. Infect. Dis.* **186**:1608–1614.

18. **Gray, B. M., G. M. Converse III, and H. C. Dillon, Jr.** 1980. Epidemiologic studies of *Streptococcus pneumoniae* in infants: acquisition, carriage, and infection during the first 24 months of life. *J. Infect. Dis.* **142**:923–933.

19. **Gray, B. M., G. M. Converse III, and H. C. Dillon, Jr.** 1979. Serotypes of *Streptococcus pneumoniae* causing disease. *J. Infect. Dis.* **140**:979–983.

20. **Gray, B. M., G. M. Converse III, N. Huhta, R. B. Johnston, Jr., M. E. Pichichero, G. Schiffman, and H. C. Dillon, Jr.** 1981. Epidemiologic studies of *Streptococcus pneumoniae* in infants: antibody response to nasopharyngeal carriage of types 3, 19, and 23. *J. Infect. Dis.* **144**:312–318.

21. **Gray, B. M., and H. C. Dillon, Jr.** 1988. Epidemiological studies of *Streptococcus pneumoniae* in infants: antibody to types 3, 6, 14, and 23 in the first two years of life. *J. Infect. Dis.* **158**:948–955.

22. **Gray, B. M., M. E. Turner, and H. C. Dillon, Jr.** 1982. Epidemiologic studies of *Streptococcus pneumoniae* in infants. The effects of season and

age on pneumococcal acquisition and carriage in the first 24 months of life. *Am. J. Epidemiol.* **116**: 692–703.

23. **Gwaltney, J. M., Jr., M. A. Sande, R. Austrian, and J. O. Hendley.** 1975. Spread of *Streptococcus pneumoniae* in families. II. Relation of transfer of *S. pneumoniae* to incidence of colds and serum antibody. *J. Infect. Dis.* **132**:62–68.

24. **Hansman, D., S. Morris, M. Gregory, and B. McDonald.** 1985. Pneumococcal carriage amongst Australian aborigines in Alice Springs, Northern Territory. *J. Hyg.* **95**:677–684.

25. **Heffron, R.** 1939. Pneumonia. A Commonwealth Fund Book. Copyright 1939, The Commonwealth Fund. Reprinted 1979, by Harvard University Press, Cambridge, Mass.

26. **Hendley, J. O., M. A. Sande, P. M. Stewart, and J. M. Gwaltney, Jr.** 1975. Spread of *Streptococcus pneumoniae* in families. I. Carriage rates and distribution of types. *J. Infect. Dis.* **132**:55–61.

27. **Henrichsen, J.** 1995. Six newly recognized types of *Streptococcus pneumoniae. J. Clin. Microbiol.* **33**: 2759–2762.

28. **Herva, E., J. Luotonen, M. Timonen, M. Sibakov, P. Karma, and P. H. Makela.** 1980. The effect of polyvalent pneumococcal polysaccharide vaccine on nasopharyngeal and nasal carriage of *Streptococcus pneumoniae. Scand. J. Infect. Dis.* **12**:97–100.

29. **Huebner, R. E., R. Dagan, N. Porath, A. D. Wasas, and K. P. Klugman.** 2000. Lack of utility of serotyping multiple colonies for detection of simultaneous nasopharyngeal carriage of different pneumococcal serotypes. *Pediatr. Infect. Dis. J.* **19**: 1017–1020.

30. **Lakshman, R., C. Murdoch, G. Race, R. Burkinshaw, L. Shaw, and A. Finn.** 2003. Pneumococcal nasopharyngeal carriage in children following heptavalent pneumococcal conjugate vaccination in infancy. *Arch. Dis. Child.* **88**: 211–214.

31. **Leino, T., K. Auranen, J. Jokinen, M. Leinonen, P. Tervonen, and A. K. Takala.** 2001. Pneumococcal carriage in children during their first two years: important role of family exposure. *Pediatr. Infect. Dis. J.* **20**:1022–1027.

32. **Lloyd-Evans, N., T. J. O'Dempsey, I. Baldeh, O. Secka, E. Demba, J. E. Todd, T. F. McArdle, W. S. Banya, and B. M. Greenwood.** 1996. Nasopharyngeal carriage of pneumococci in Gambian children and in their families. *Pediatr. Infect. Dis. J.* **15**:866–871.

33. **Loda, F. A., A. M. Collier, W. P. Glezen, K. Strangert, W. A. Clyde, Jr., and F. W. Denny.** 1975. Occurrence of *Diplococcus pneumoniae* in the upper respiratory tract of children. *J. Pediatr.* **87**:1087–1093.

34. **Marchisio, P., S. Esposito, G. C. Schito, A. Marchese, R. Cavagna, and N. Principi.** 2002. Nasopharyngeal carriage of *Streptococcus pneumoniae* in healthy children: implications for the use of heptavalent pneumococcal conjugate vaccine. *Emerg. Infect. Dis.* **8**:479–484.

35. **Mbelle, N., R. E. Huebner, A. D. Wasas, A. Kimura, I. Chang, and K. P. Klugman.** 1999. Immunogenicity and impact on nasopharyngeal carriage of a nonavalent pneumococcal conjugate vaccine. *J. Infect. Dis.* **180**:1171–1176.

36. **McCool, T. L., T. R. Cate, G. Moy, and J. N. Weiser.** 2002. The immune response to pneumococcal proteins during experimental human carriage. *J. Exp. Med.* **195**:359–365.

37. **McCool, T. L., T. R. Cate, E. I. Tuomanen, P. Adrian, T. J. Mitchell, and J. N. Weiser.** 2003. Serum immunoglobulin G response to candidate vaccine antigens during experimental human pneumococcal colonization. *Infect. Immun.* **71**:5724–5732.

38. **McGee, L., L. McDougal, J. Zhou, B. G. Spratt, F. C. Tenover, R. George, R. Hakenbeck, W. Hryniewicz, J. C. Lefevre, A. Tomasz, and K. P. Klugman.** 2001. Nomenclature of major antimicrobial-resistant clones of *Streptococcus pneumoniae* defined by the pneumococcal molecular epidemiology network. *J. Clin. Microbiol.* **39**:2565–2571.

39. **Montgomery, J. M., D. Lehmann, T. Smith, A. Michael, B. Joseph, T. Lupiwa, C. Coakley, V. Spooner, B. Best, and I. D. Riley.** 1990. Bacterial colonization of the upper respiratory tract and its association with acute lower respiratory tract infections in Highland children of Papua New Guinea. *Rev. Infect. Dis.* **12**(Suppl. 8):S1006–S1016.

40. **Norris, C. F., S. R. Mahannah, K. Smith-Whitley, K. Ohene-Frempong, and K. L. McGowan.** 1996. Pneumococcal colonization in children with sickle cell disease. *J. Pediatr.* **129**:821–827.

41. **Obaro, S. K., R. A. Adegbola, W. A. Banya, and B. M. Greenwood.** 1996. Carriage of pneumococci after pneumococcal vaccination. *Lancet* **348**:271–272.

42. **O'Brien, K. L., and H. Nohynek.** 2003. Report from a WHO Working Group: standard method for detecting upper respiratory carriage of *Streptococcus pneumoniae. Pediatr. Infect. Dis. J.* **22**:e1–e11.

43. **Pasteur, L.** 1881. Sur une maladie nouvelle, provoqué par la salive d'un enfant mort de la rage. *C. R. Acad. Sci.* **92**:159.

44. **Rapola, S., V. Jantti, R. Haikala, R. Syrjanen, G. M. Carlone, J. S. Sampson, D. E. Briles, J. C. Paton, A. K. Takala, T. M. Kilpi, and H. Kayhty.** 2000. Natural develop-

ment of antibodies to pneumococcal surface protein A, pneumococcal surface adhesin A, and pneumolysin in relation to pneumococcal carriage and acute otitis media. *J. Infect. Dis.* **182:**1146–1152.

45. **Regev-Yochay, G., M. Raz, B. Shainberg, R. Dagan, M. Varon, M. Dushenat, and E. Rubinstein.** 2003. Independent risk factors for carriage of penicillin-non-susceptible *Streptococcus pneumoniae. Scand. J. Infect. Dis.* **35:**219–222.

46. **Simell, B., T. M. Kilpi, and H. Kayhty.** 2002. Pneumococcal carriage and otitis media induce salivary antibodies to pneumococcal capsular polysaccharides in children. *J. Infect. Dis.* **186:** 1106–1114.

47. **Simell, B., M. Korkeila, H. Pursiainen, T. M. Kilpi, and H. Kayhty.** 2001. Pneumococcal carriage and otitis media induce salivary antibodies to pneumococcal surface adhesin a, pneumolysin, and pneumococcal surface protein a in children. *J. Infect. Dis.* **183:**887–896.

48. **Smith, T., D. Lehmann, J. Montgomery, M. Gratten, I. D. Riley, and M. P. Alpers.** 1993. Acquisition and invasiveness of different serotypes of *Streptococcus pneumoniae* in young children. *Epidemiol. Infect.* **111:**27–39.

49. **Soewignjo, S., B. D. Gessner, A. Sutanto, M. Steinhoff, M. Prijanto, C. Nelson, A. Widjaya, and S. Arjoso.** 2001. *Streptococcus pneumoniae* nasopharyngeal carriage prevalence, serotype distribution, and resistance patterns among children on Lombok Island, Indonesia. *Clin. Infect. Dis.* **32:**1039–1043.

50. **Soininen, A., H. Pursiainen, T. Kilpi, and H. Kayhty.** 2001. Natural development of antibodies to pneumococcal capsular polysaccharides depends on the serotype: association with pneumococcal carriage and acute otitis media in young children. *J. Infect. Dis.* **184:**569–576.

51. **Sternberg, G.** 1881. A fatal form of septicaemia in the rabbit, produced by subcutaneous injection of human saliva; an experimental research. National Board of Health annual report, 1882. Government Printing Office, Washington, D.C.

52. **Sung, R. Y., J. M. Ling, S. M. Fung, S. J. Oppenheimer, D. W. Crook, J. T. Lau, and A. F. Cheng.** 1995. Carriage of *Haemophilus influenzae* and *Streptococcus pneumoniae* in healthy Chinese and Vietnamese children in Hong Kong. *Acta Paediatr.* **84:**1262–1267.

53. **Syrjanen, R. K., T. M. Kilpi, T. H. Kaijalainen, E. E. Herva, and A. K. Takala.** 2001. Nasopharyngeal carriage of *Streptococcus pneumoniae* in Finnish children younger than 2 years old. *J. Infect. Dis.* **184:**451–459.

54. **Syrogiannopoulos, G. A., G. D. Katopodis, I. N. Grivea, and N. G. Beratis.** 2002. Antimicrobial use and serotype distribution of nasopharyngeal *Streptococcus pneumoniae* isolates recovered from Greek children younger than 2 years old. *Clin. Infect. Dis.* **35:**1174–1182.

55. **Vives, M., M. E. Garcia, P. Saenz, M. A. Mora, L. Mata, H. Sabharwal, and C. Svanborg.** 1997. Nasopharyngeal colonization in Costa Rican children during the first year of life. *Pediatr. Infect. Dis. J.* **16:**852–858.

56. **Whatmore, A. M., S. J. King, N. C. Doherty, D. Sturgeon, N. Chanter, and C. G. Dowson.** 1999. Molecular characterization of equine isolates of *Streptococcus pneumoniae:* natural disruption of genes encoding the virulence factors pneumolysin and autolysin. *Infect. Immun.* **67:**2776–2782.

57. **Yagupsky, P., N. Porat, D. Fraser, F. Prajgrod, M. Merires, L. McGee, K. P. Klugman, and R. Dagan.** 1998. Acquisition, carriage, and transmission of pneumococci with decreased antibiotic susceptibility in young children attending a day care facility in southern Israel. *J. Infect. Dis.* **177:**1003–1012.

58. **Yeh, S. H., K. M. Zangwill, H. Lee, S. J. Chang, V. I. Wong, D. P. Greenberg, and J. I. Ward.** 2003. Heptavalent pneumococcal vaccine conjugated to outer membrane protein of *Neisseria meningitidis* serogroup b and nasopharyngeal carriage of *Streptococcus pneumoniae* in infants. *Vaccine* **21:**2627–2631.

59. **Zenni, M. K., S. H. Cheatham, J. M. Thompson, G. W. Reed, A. B. Batson, P. S. Palmer, K. L. Holland, and K. M. Edwards.** 1995. *Streptococcus pneumoniae* colonization in the young child: association with otitis media and resistance to penicillin. *J. Pediatr.* **127:**533–537.

EPIDEMIOLOGY OF PNEUMOCOCCAL DISEASE

Jay C. Butler

10

INTRODUCTION

Transmission of *Streptococcus pneumoniae* infection occurs via respiratory droplets from persons with pneumococcal disease or, more commonly, persons without disease who carry the organism in the nasopharynx (145). Following exposure, the organism may establish itself in the nasopharynx of its new host. Most commonly, this results in asymptomatic colonization and the organism is carried for a period of weeks to months (60, 133). However, in some instances, the newly acquired organism evades host defensive mechanisms and causes illness. Data from infants suggest that the risk for progression from asymptomatic colonization to disease appears to be greatest soon after exposure and acquisition of the organism in the nasopharynx (89), but disease may also develop after months of colonization (9).

For many persons, colonization is an immunizing event resulting in production of antibodies directed against the capsular polysaccharide of the colonizing strain (91, 147). Thus, in the vast majority of instances, colonization can be characterized as a peaceful, albeit transient, coexistence between host and

potential pathogen. However, even in the absence of disease, this situation is not entirely benign. Pneumococci carried in the nasopharynx can be transmitted to others and can cause disease in susceptible persons, and if the colonized person receives antimicrobial agents, the carried strains may develop drug resistance. Although disease occurs in only a small proportion of persons who become colonized, the ubiquity of the organism in human populations results in a large burden of disease. A number of factors characteristic of specific pneumococcal strains and the status of host defenses are associated with progression from asymptomatic colonization to disease. This chapter focuses on the epidemiology of pneumococcal disease in terms of pathogen characteristics and host risk factors. The mechanism and epidemiology of carriage, the epidemiology of drug-resistant pneumococci, and the pathogenesis of invasive disease are described in detail elsewhere in this book.

The clinical manifestations of *S. pneumoniae* infection are protean but can be classified into two major categories: *invasive infections*, where the organism is isolated from a normally sterile body site, such as the bloodstream or central nervous system, and *mucosal infections*, most often involving the upper respiratory tract. *S. pneumoniae* is the most commonly identified

Jay C. Butler, Arctic Investigations Program, Centers for Disease Control and Prevention, Anchorage, AK 99508.

The Pneumococcus. Volume Editor, Elaine I. Tuomanen,
© 2004 ASM Press, Washington, D.C.

cause of community-acquired pneumonia (67, 135, 136), bacterial meningitis (174), acute otitis media (16, 53), and acute bacterial sinusitis (129, 200, 201). Less common manifestations of pneumococcal infection include pyogenic arthritis (170), osteomyelitis (1, 4, 124), pyomyositis (22, 44), necrotizing fasciitis (11), endocarditis (5, 45), pericarditis (119, 172), mycotic aneurysm (150), central nervous system abscess (31, 93), bacterial peritonitis (34, 58), urinary and genital tract infections (180, 203), parotitis (81), epiglottitis (51, 120), buccal or periorbital cellulitis (82), mastoiditis (118), endophthalmitis (43, 144), and conjunctivitis (190). Pneumococcal infections have rarely been associated with parainfectious conditions that are classically associated with other infectious agents, such as hemolytic-uremic syndrome (20, 33, 88), thrombotic thrombocytopenic purpura (149), Waterhouse-Friderichsen syndrome (159, 179), and rhabdomyolysis (186).

In temperate regions, cases of pneumococcal disease are most common during the colder weather months, when transmission rates are the highest (57, 89, 121). Despite the annual winter "epidemics" of pneumococcal disease, most infections appear to occur sporadically, i.e., are not part of a recognized disease outbreak or cluster. However, outbreaks of pneumonia, meningitis, and conjunctivitis caused by a single strain of pneumococcus are occasionally reported. Most of these outbreaks occur in institutional settings, such as nursing homes and residential care facilities (36, 83, 152), hospital wards (14, 188), military units (92, 110, 157), a jail (111), homeless shelters (55, 141), day care centers for children (41, 48, 163), and schools and colleges (38, 137). Outbreaks have also been reported from small, closed, or remote communities (50, 162). Crowding, which facilitates transmission of the organism to large numbers of susceptible persons, is a common thread in all of these outbreaks.

EPIDEMIOLOGICAL METHODS

Prospective surveillance that is population based, i.e., counting cases (numerator) from a defined population (denominator), is the best method of determining incidence of disease and is the most scientifically rigorous method for determining factors associated with increased risk of pneumococcal disease. However, surveillance that is not population based, such as hospital-based case series, can also provide insight into factors associated with the risk of disease, particularly if appropriate comparison groups and controls are selected. Additionally, data collected during investigation of pneumococcal disease outbreaks contribute further to the understanding of pneumococcal disease epidemiology.

For all epidemiological methods, accuracy of the data is dependent on the specificity of the case definition of pneumococcal disease, the sensitivity and specificity of diagnostic methods, and clinical practices in utilizing diagnostic tests. Data based on rates of pneumonia and influenza hospitalizations and deaths can be quickly obtained through administrative databases but are not specific for pneumococcal disease because of the large number of other causes of pneumonia and inaccuracy of coding (94). The specificity of the case definition can be improved by requiring laboratory evidence of pneumococcal infection; however, the lack of a sensitive and specific diagnostic "gold standard" for laboratory diagnosis of pneumococcal infection is a challenge. The most specific case definition is based on isolation of S. pneumoniae from a normally sterile body site (invasive infection). However, such a case definition is insensitive, and surveillance based on isolation of the pneumococcus from a normally sterile body site will underestimate the overall disease burden. For example, bacteremia is documented in only 15 to 30% of cases of pneumococcal pneumonia (143, 146). Additionally, changes in laboratory and clinical practices will influence observed rates of disease. In North Jutland County, Denmark, the number of cases of pneumococcal bacteremia reported each year increased more than fourfold between 1981 and 1996 (173). The greater number of cases reported during later years was associated with a greater number of blood cultures submitted to the laboratory and a larger required volume of blood because of new blood culture laboratory methods. However, not all such changes can be attributed to methodological artifact. A >4-fold increase in rates of pneumococcal bacteremia in Sweden from 1987 to 1997 could be attributed in part to introduction and spread of strains with increased virulence (105). The majority of pneumococcal infections, particularly mucosal infections, occur without collection of specimens for microbiological assessment. Therefore, data on pneumococcal otitis media and sinusitis are gathered primarily during the course of clinical research studies rather than public health surveillance.

Our understanding of pneumococcal disease epidemiology is further refined through studies of specific strains of *S. pneumoniae*. Differences in the polysaccharide structure of the pneumococcal capsule have permitted identification of at least 90 serotypes of *S. pneumoniae* (103, 117). Serotyping has been the primary method to understand the epidemiology of specific strains of *S. pneumoniae* for decades (104). More recently, additional subtyping techniques and methods for genetic characterization have provided powerful tools for understanding the epidemiology of pneumococcal disease outbreaks and the global spread of specific pathogenic clones. Genetic typing methods utilized for strain characterization in epidemiological studies of pneumococcal disease include pulsed-field gel electrophoresis (PFGE), ribotyping, penicillin-binding protein fingerprinting, multilocus sequence typing (MLST), BOX-PCR, and restriction fragment end labeling (48, 62, 106, 107, 123, 126, 184, 198).

EPIDEMIOLOGY OF PNEUMOCOCCAL SEROTYPES

The vast majority of pneumococcal disease is caused by a relatively small number of serotypes. The number of pneumococcal serotypes found colonizing persons in a given community is generally greater than the number causing invasive infection in the same community. It has long been observed that certain serotypes, particularly 1 and 2, appear to be more "invasive" than other serotypes because they are rarely isolated from asymptomatically colonized persons (187). In analysis of serotypes from children with invasive pneumococcal disease and nasopharyngeal colonization in Oxford, England, serotypes 1, 4, 14, and 18C were significantly more common among the invasive disease isolates and serotype 23F was more common among the colonizing strains (25).

Because pneumococcal polysaccharide and conjugate vaccines provide protection by inducing antibodies directed against the polysaccharide capsule, understanding the epidemiology of pneumococcal serotypes is crucial to optimal vaccine formulation and for targeting those serotypes most likely to cause disease. The distribution of serotypes among disease-causing pneumococci is influenced by the age and immune status of the host, the type of disease, and geographic region. A smaller number of serotypes cause a larger proportion of invasive infections among young children than among adults (168). As a result, the proportion of isolates covered by the heptavalent conjugate vaccine is much greater for young children than for adults (168). After the first decade of life, there is a dramatic decline in the relative importance of some serogroups (6, 14, and 19, and to a lesser extent 23), while there appears to be a more gradual decline through adulthood for serotype 1 (176). Among persons with invasive infection, serotypes 3 and 8 appear to have a relative preference for adults (176).

Compared with healthy adults, those with immunocompromising medical conditions may be more likely to develop disease caused by serotypes more commonly seen among children (202). Among adults with invasive pneumococcal disease in South Africa, those who were infected with human immunodeficiency virus (HIV) were more likely to be infected with serogroups 6, 14, 19, and 23 (49, 70, 116). In surveillance data for invasive infection for adults aged 18 to 64 years in the United States, persons with HIV infection were more likely to be infected with serotypes 6A, 6B, 9N, 9V, 18C, 19A, 19F, and 23F than were adults with no underlying disease and were less likely to be infected with serotypes 1, 7F, and 12F (76). Persons with advanced HIV infection in the United States often receive trimethoprim-sulfamethoxazole to prevent opportunistic infection with *Pneumocystis jiroveci*, which could select for drug-resistant serotypes, but these differences in serotype distribution were noted regardless of trimethoprim-sulfamethoxazole susceptibility of the pathogen. Moreover, the serotype distribution among persons with HIV infection was similar to that for persons with other immunocompromising medical conditions, suggesting that host immunity may be the most important factor contributing to the risk of infection with these "immune-dependent serotypes" (76).

Pneumococcal polysaccharide and polysaccharide conjugate vaccines protect against disease by inducing serotype-specific antibodies (7, 148). Thus, the relative proportion of disease caused by each serotype is influenced by successful immunization. This effect has been exploited for estimating vaccine effectiveness using the indirect cohort method, in which the relative proportions of disease caused by serotypes included and not included in the polysaccharide vaccines are compared between vaccinated and unvaccinated persons (23, 29). The validity of the indirect cohort method rests on the assumption that there are no indirect effects of polysaccharide vaccine on the distribution of serotypes causing disease in unvaccinated persons. Future studies using the indirect cohort method for estimating polysaccharide vaccine effectiveness may require adjustment for the indirect effect of conjugate vaccine immunization of infants and young children on the distribution of serotypes causing disease in adults who have not received conjugate vaccine (205).

The distribution of serotypes causing mucosal infections differs from that of invasive pneumococcal disease, with serogroups 3, 19, and 23 more often found in middle ear fluid than in blood or cerebrospinal fluid (CSF)

(97). In national, hospital laboratory-based surveillance in the United States, the seven most common serotypes causing invasive disease in children aged <6 years (4, 6B, 9V, 14, 18C, 19F, and 23F), plus serologically related types (6A, 9A, 9L, 18B, and 18F), accounted for 86% of 3,169 isolates from blood, 83% of 401 isolates from CSF, but only 65% of isolates from middle ear fluid (30). Serotype 3 accounted for 9% of middle ear fluid isolates but ≤1% of blood or CSF isolates. The importance of serotype 3 in mucosal infections is further supported by several studies of middle ear fluid isolates. Serotype 3 accounted for 11% of 396 middle ear fluid isolates in Birmingham, Ala., from 1975 to 1978, 11% of 777 middle ear fluid isolates from children with acute otitis media at a pediatric clinic in rural Kentucky, and 10% of 500 middle ear fluid isolates collected during the course of acute otitis media and surveillance studies in the United States from 1996 to 1999 (17, 90, 115). In an analysis of data sets representing 3,232 children with acute otitis media in Europe, Israel, the United States, and Argentina and *S. pneumoniae* isolated from middle ear fluid, the most common serotypes were 19F, 23F, 14, and 6B, in descending order of frequency (100). Although 3 was the eighth most commonly encountered serotype and accounted for only 4% of all isolates in this review, it accounted for 10% of isolates from children aged ≥24 months and 22% of isolates from those aged 60 months and older. Vaccination with pneumococcal conjugate vaccine causes a shift in serotypes causing otitis media. In a large trial conducted in Finland, vaccination reduced the overall number of episodes of pneumococcal otitis media by preventing infection caused by serotypes included in the vaccine and serologically cross-reactive types, but vaccinated children had more episodes caused by serotypes not included in the vaccine (64).

Fewer data are available for serotypes causing other mucosal infections. In a series of 34 children with mastoiditis at eight children's hospitals from 1993 through 1998, more than half of the cases were caused by serogroup 19 (118).

Compared with isolates from children with invasive infection, serotype 3 and serogroup 19 isolates were disproportionately represented among the children with mastoiditis.

It has been suggested that serotypes 1 through 3 and 5 may be more prone to cause pneumococcal disease epidemics because they are rarely isolated from the nasopharynges of asymptomatically colonized persons and because most outbreaks during the first half of the 20th century involved these serotypes (68, 109, 110, 142, 157, 182, 183, 191). While these serotypes may more often result in disease after acquisition, the earlier outbreaks occurred at a time when most sporadic disease was caused by the same serotypes (6, 24, 26, 72, 192). Of 26 reported pneumococcal disease outbreaks occurring in the 1990s, only 3 were caused by serotype 1 (50, 83, 86). The majority of recent outbreaks in the United States were caused by serotypes 23F, 14, and 4 (36, 37, 48, 73, 83, 152, 177).

Serotypes that most commonly cause invasive pneumococcal infection appear to differ somewhat in various regions of the world. Serotype 1 is common in children and adults in Asia, Africa, Latin America, and parts of Europe, but it is uncommon in the United States, Canada, Scandinavia, and Oceania (19, 65, 98). Serotype 5 accounts for >9% of invasive disease in Latin America and Asia but a smaller proportion in Europe and Africa. Serotype 5 infections are quite rare in the United States, Canada, and Oceania (98). Currently, pneumococcal conjugated vaccines under evaluation provide protection against a limited number of serotypes (generally ≤11), and these differences pose a challenge to designing one vaccine that will provide optimal protection anywhere in the world. Nonetheless, serogroups 6, 14, and 19 are among the leading causes of invasive infection among young children in all regions of the world (98, 176, 185), and these serotypes are invariably included in all vaccines considered for infant immunization. Moreover, the validity of the observed geographic variability in serotype distribution has been questioned. Based on the observation that pneumococcal serotypes differ in the severity of disease they produce (7), geographic differences in serotype distribution may be explained in part by differences in blood culture collection practices in various parts of the world (99). Prospective studies controlling for patient age, disease severity, and blood culture collection practices will be required to test this hypothesis.

Differences in serotype distribution among racial or ethnic groups within the same region have been reported, particularly for serotype 1. In Israel, serotype 1 accounted for 14% of invasive infections among Jewish children, compared with 32% among Bedouins (75). In multivariable analysis, serotype 1 infections were threefold more common among Bedouins than among Jews. Serotype 1 caused 12% of invasive infections among Alaska Natives from 1991 through 1998, compared with 3% among non-Native residents (171). Serotype 1 has also been noted to be a leading serotype causing invasive disease in Navajo Indians and among Aboriginal people in central Australia (13, 87). An explanation for differences in serotype distribution among ethnic groups remains enigmatic.

In general, the distribution of serotypes causing pneumococcal disease appears to change little over periods of a least four decades, but substantial changes may occur over longer intervals (10, 68). For example, serotypes 1, 2, and 3 accounted for roughly one-half of isolates at Boston City Hospital prior to 1950, but from 1979 to 1982 they accounted for <5%, while serogroups 4, 6, 14, 18, and 19 became increasingly more common (10, 12, 71, 193). Of 3,644 isolates from blood of patients at 10 U.S. hospitals serotyped by Austrian et al. between 1967 and 1975, serotypes 1, 2, and 3 accounted for 8.5, 0.3, and 7.1%, respectively (8). Of over 7,000 blood isolates submitted from clinical laboratories to the Centers for Disease Control and Prevention (CDC) from 1978 through 1995, only 1 was serotype 2 (CDC, unpublished data). In 1998, the proportions of serotype 1 and 3 isolates from blood submitted to the

CDC through population-based surveillance were down to 2.4 and 3.3%, respectively (168). Possible explanations for changes in serotype distribution over time include the impact of increasing antimicrobial drug use during the latter half of the 20th century, improved living conditions, the growing number of immunocompromised persons, and changes in blood culture collection and processing practices (68).

EPIDEMIOLOGY OF INVASIVE MOLECULAR CLONES

Genetic characterization of disease-causing pneumococcal isolates has provided further insight into the epidemiology of *S. pneumoniae* beyond that provided by serotyping alone. Multiple genetic subtypes may be found within a single serotype. Additionally, *S. pneumoniae* is able to undergo spontaneous change in serotype such that isolates of different serotypes may be indistinguishable by genotyping techniques. Reported serotype switches include 23F to 19F, 23F to 14, 23F to 19B, 19A to 11, 14 to 19A, 14 to 9V, 14 to 9A, and 14 to 24F (54, 156). Surveillance for pneumococcal genotypes will be needed to determine whether serotype switching of pathogenic strains will occur under selective immunological pressure from conjugate vaccines.

A major challenge to delineating the epidemiology of disease-causing pneumococcal clones is determining clonality. To this end, the Pneumococcal Molecular Epidemiology Network (PMEN) was established in 1997 to standardize laboratory methods and epidemiological definitions for identifying clones of antimicrobial drug-resistant *S. pneumoniae* using PFGE, BOX-PCR, and MLST (140). The system of nomenclature utilizes the country or region where the clone was first identified and, in superscript, the serotype, followed by a dash and a numeral reflecting the sequential recognition of the clone by PMEN. For example, Tennessee^{23F}-4 was first identified in Memphis, Tenn., was serotype 23F when first identified, and was the fourth clone recognized by PMEN. If serotype switching is identified

within a clone, a dash and the new serotype(s) are added to differentiate the variant. For example, Spain^{23F}-1 expressing serotypes 14 and 19 would be designated Spain^{23F}-1-14,19. As of September 2003, 26 multidrug-resistant clones had been recognized by PMEN.

Genotyping of antimicrobial drug-resistant strains has documented the global spread of a limited number of clones that account for most penicillin-resistant pneumococcal isolates (139). PFGE analysis demonstrated that 12 clones accounted for 74 to 80% of isolates with high-level penicillin resistance (MIC $\geq$ 2.0 μg/ml) collected during three surveillance studies in the United States from 1994 to 2000 (166). In a study of another national surveillance in the United States, 51% of 328 penicillin-resistant isolates from persons with respiratory disease in 39 states belonged to one of two clones: Spain^{23F}-1-14,19 (39%) and Spain^{9V}-3-14,19 (12%) (46). Of 127 Spain^{23F}-1-14,19 isolates, 39 (30%) expressed one of the variant serotypes (14, 19, or nontypeable). Thus, genotyping provides insight into global spread of pneumococcal strains that could not have been obtained on the basis of antimicrobial susceptibility testing and serotyping alone.

Although genes coding for a number of virulence factors may influence the invasiveness of *S. pneumoniae* (125, 161), the polysaccharide capsule appears to be the most important factor. A study of 150 invasive isolates and 351 nasopharyngeal carriage isolates from children using MLST identified two distinct clones of serotype 14, the most common serotype, but each clone was similarly represented among disease-causing and colonizing isolates (25). No difference in invasiveness was identified for any genotype. However, the study may have been underpowered to exclude the possibility that certain clones within a serotype may have greater potential to cause invasive disease. A study of 212 invasive and colonizing serogroup 6 isolates from around the world found that certain genotypes identified by MLST were more frequently associated with invasive disease and that the putative ancestral genotypes were more frequently found among colonizing strains (167).

HOST FACTORS ASSOCIATED WITH PNEUMOCOCCAL DISEASE

Age

All published reports from population based surveillance of invasive pneumococcal disease show that the incidence is greatest among the very young and the elderly (Fig. 1). The highest rates of disease occur among children aged 6 to 11 months (Fig. 2). Neonatal infections (occurring in the first 30 days of life) are unusual in industrialized countries (84, 203). The increased risk of disease among children younger than 2 years of age is likely related to immature immunological response to the polysaccharide capsule and high prevalence of colonization (Fig. 3).

Although prevalence of colonization declines with increasing age, incidence of invasive infection progressively increases after age 50 to 65 years (Fig. 1 and 3). The reported annual incidence of invasive pneumococcal disease among persons aged 65 years and older in the general population in North America and Europe ranges from 25 to 90 cases/100,000 (32, 181). While rates of invasive pneumococcal disease among the elderly are similar to those among young children, disease manifestations differ: the elderly are more likely to have pneumonia and less likely to have primary bacteremia (bacteremia without identified focus). Additionally, the elderly are much more likely to die from invasive infection (69). Case fatality rates are generally less than 2% among young children in industrialized countries but exceed 20% among the elderly despite antimicrobial therapy and intensive care support (160, 168). Persons aged 65 years and older account for less than 30% of cases in the United States but over half of deaths from invasive pneumococcal disease (32). Factors contributing to the higher rates of pneumococcal infection and death among the elderly include higher rates of underlying medical conditions and age-related immune dysfunction (35).

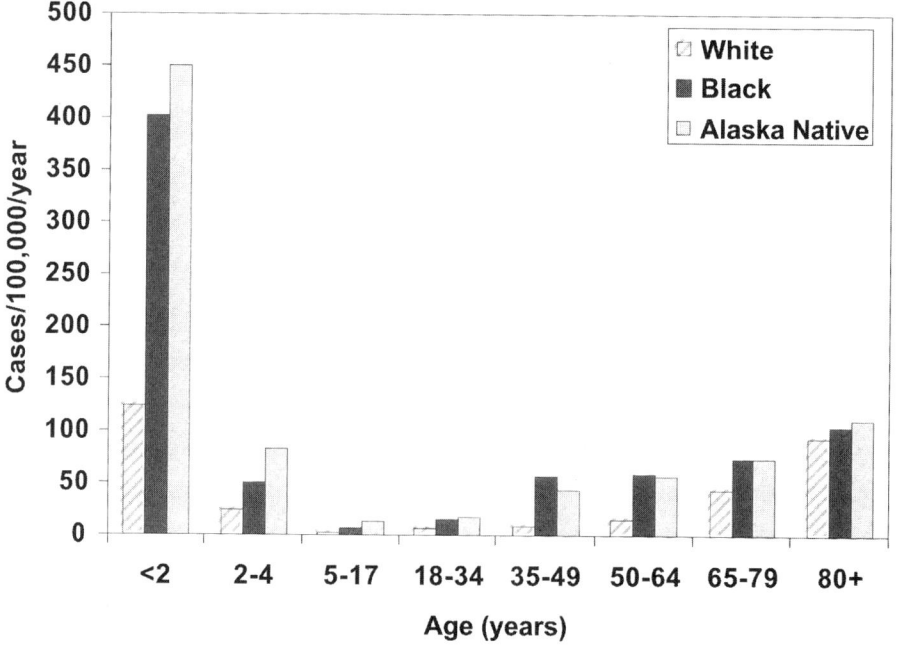

FIGURE 1 Incidence of invasive pneumococcal disease by age and race, United States. Data for whites and blacks are from CDC surveillance conducted in the San Francisco Bay area, Calif.; Connecticut; metropolitan Atlanta, Ga.; Baltimore, Md.; Minneapolis-St. Paul, Minn.; Rochester, N.Y.; Portland, Oreg.; five urban counties in Tennessee; and San Antonio, Tex., 1998 (168). Data for Alaska Natives are from CDC surveillance in Alaska, 1996 to 2000 (CDC, unpublished).

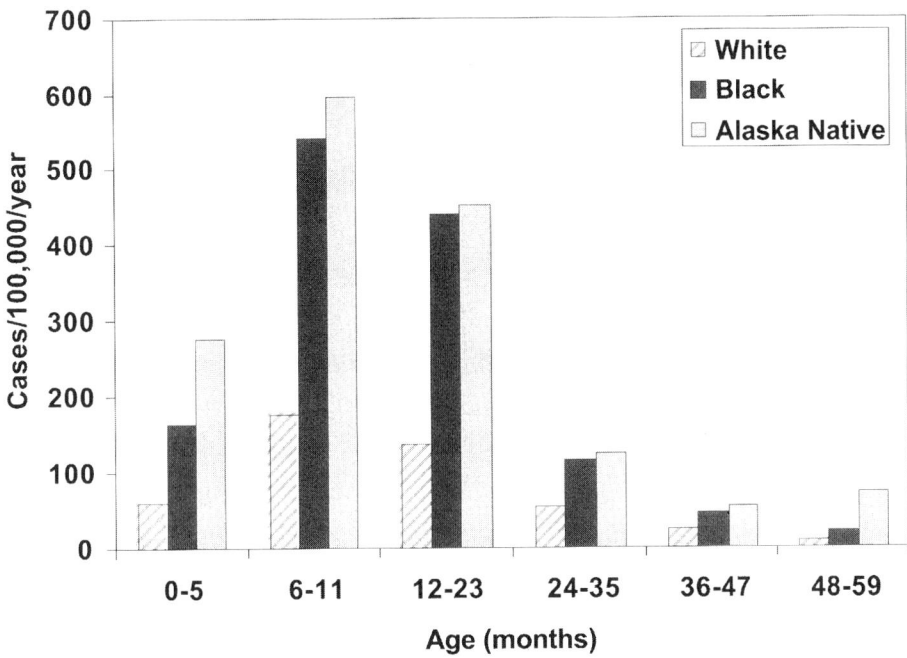

FIGURE 2 Incidence of invasive pneumococcal disease by age group and race among children aged <5 years, United States. Data are from CDC surveillance (39).

Sex, Race, and Ethnicity

Males are overrepresented among infants, children, and adults with pneumococcal disease (122, 168). Among immunocompetent adults aged 18 to 64 with invasive disease identified through population-based laboratory surveillance in metropolitan Atlanta (Ga.), Baltimore (Md.), and Toronto (Canada), male sex was an independent risk factor for invasive infection in case control analysis after controlling for

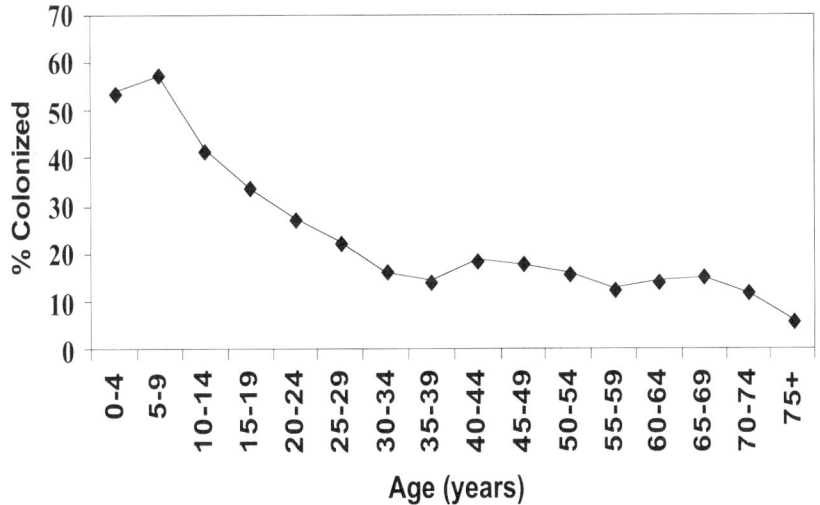

FIGURE 3 Prevalence of nasopharyngeal colonization by age among residents of rural communities in Alaska, 2000 to 2001 (CDC, unpublished).

race, underlying illnesses, smoking, education level, and exposure to children (153). Certain racial and ethnic groups appear to be at higher risk for pneumococcal infection. In the United States, rates of invasive pneumococcal disease are higher for blacks and Native Americans than for whites and persons of other racial backgrounds (Fig. 1) (47, 158, 168, 171, 207). One aspect of the differing epidemiology among racial groups with implications for prevention of infection through vaccination programs is the observation that the age-related increase in rates occurs earlier among blacks and Alaska Natives than among whites (52, 96). Similarly, rates of invasive pneumococcal disease among Aborigines in central Australia are some 20 to 30 times those reported for non-Aborigines living in Australia (195, 196). Factors contributing to these disparities in disease rates are poorly defined and probably multifactorial but may include differences in living conditions and higher prevalence of immunocompromising conditions. High rates of HIV infection among blacks in urban areas have been proposed as a major contributing factor, but even among HIV-infected persons, the risk of infection is higher for blacks than for whites (21, 59, 78, 154).

Socioeconomic Status

Pneumococcal infection disproportionately affects the poor. Rates of invasive pneumococcal disease in Dallas County, Tex., were higher among persons living in census tracts with median annual household incomes of <$20,000 than among those living in areas with median incomes of ≥$40,000 for both whites and blacks (158). Among persons living in the lowest-income areas in metropolitan Atlanta, rates of invasive disease were up to 10-fold higher in blacks than in whites, but the differences between races become less pronounced in areas with greater median incomes, and rates are essentially the same at the highest income levels (40). These findings, based on ecological measures of socioeconomic status, suggest that factors contributing to the higher rates of disease

among blacks are not significant among persons with the presumably better living conditions attained with higher income. A case control study found lower individual household income among cases, but this association was not found after controlling for confounding factors, and lower level of education was the only independent measure of socioeconomic status associated with disease in multivariable analysis (153).

Living Conditions

As noted earlier in this chapter, crowding of susceptible persons is a common factor in outbreaks of pneumococcal infection in institutional settings. In the absence of recognized outbreaks, day care attendance has been strongly associated with increased risk of invasive pneumococcal infection in North America and Scandinavia (80, 128, 189). Within households, colonization frequently leads to spread of the organism to other family members, and clusters of illnesses in families caused by the same serotypes of pneumococcus have been reported (27, 95, 112, 127, 132, 178, 194). In recent case control studies, household crowding was not more common among children with invasive pneumococcal disease compared with controls; household crowding was more common among infected adults but was not a risk factor after controlling for possible confounding variables (128, 153).

Exposure to children appears to increase the risk of pneumococcal disease in adults. Children are more commonly colonized with pneumococci than are adults, and adults with preschool-aged children in the household have been found in some studies to be more likely to be colonized than those who do not live with children but not in others (18, 102). Multivariable analysis of data from a case control study of immunocompetent adults with invasive infection indicated greater risk of disease among persons living with young children who attend day care than among persons without children in the home (153). A case control study of HIV-infected adults (84% of cases and controls were male) found contact with a child aged <10 years to be independently associated with increased risk of pneumococcal infection (21).

Chronic Medical Conditions

The risk and severity of pneumococcal infection is increased for persons with certain chronic medical conditions, including functional or anatomic asplenia, HIV infection, chronic obstructive pulmonary disease, asthma, cirrhosis, diabetes mellitus, chronic renal failure, cancer (particularly hematological malignancies), organ or bone marrow transplantation, nephritic syndrome, and hypogammaglobulinemia and for persons taking immunosuppressive medications, such as corticosteroids (2, 3, 28, 42, 56, 63, 85, 113, 130, 131, 151, 153, 154, 169, 197, 206). The strength of the evidence documenting the risk associated with each of these conditions varies. Persons with sickle cell disease leading to functional asplenia and persons with HIV infection are clearly at particularly high risk. The annual attack rate of pneumococcal bacteremia for adults with HIV infection approaches 1%, and it exceeds 5% for HIV-infected children (134, 154, 164). Among adults aged 18 to 64 years in San Francisco, Calif., with invasive pneumococcal disease, 43% had documented HIV infection (154). Rates of recurrent invasive pneumococcal infection are more than six-fold greater for persons with HIV infection than for those without HIV infection (74, 138). Although the risk of pneumococcal infection may be greater among persons with more advanced HIV disease (CD4 lymphocyte count of <200 cells/mm^3) (78, 114), it is common for pneumococcal disease to be the first opportunistic infection in the course of HIV infection (77). Therefore, HIV testing has been recommended for otherwise healthy young and middle-aged adults presenting with pneumococcal infection (77).

Persons with congenital or traumatic CSF fistulae and children who have received cochlear implants are at greatly increased risk of bacterial meningitis, and *S. pneumoniae* is the pathogen most commonly isolated in these cases (61, 165, 204). In a retrospective analysis of 160 patients with traumatic CSF fistulae, 49 (30.6%) developed meningitis prior to surgical repair of the dural defect (61). In 2002, the U.S. Food and Drug Administration received reports of bacterial meningitis among children with cochlear implants. Among 4,264 children less than 6 years of age who received implants from 1997 to 2002, 29 cases of bacterial meningitis involving 26 children were identified. Over 60% of cases were associated with pneumococci, representing an incidence of 138.2 cases/100,000 person-years—a rate more than 30-fold greater than that observed among children in the general U.S. population (165).

Alcohol and Tobacco Use

Heavy alcohol use and cigarette smoking have been noted to be common among otherwise healthy persons with pneumococcal infection (28, 158). A community-based case control study of immunocompetent adults aged 18 to 64 years found that just over half of all cases of invasive pneumococcal infection were attributable to cigarette smoking (153). Compared with controls, persons with invasive infection were over four times more likely to be current cigarette smokers after controlling for sex, race, chronic illnesses, level of education, and exposure to children. A dose-response relationship between disease and the number of cigarettes smoked per day and pack-years of cigarette smoking provides further support for this association. Increased risk of infection for former smokers persisted for at least 5 years after they quit smoking, but risk returned to baseline by 10 years. Among HIV-infected adults, cigarette smoking was nearly twice as common among those with invasive pneumococcal infection than among controls after controlling for race, CD4 lymphocyte count, vaccination history, and contact with children (21). Secondhand smoke exposure also has been associated with increased risk of invasive infection (153). Among young children, living with at least one cigarette smoker was associated with invasive infection in rural Alaska but not after controlling for day care center attendance and the protective effect of breast-feeding (80).

Preceding or Coincident Infection
In temperate areas, pneumococcal infections are more common during the winter months, when respiratory viral infections are most common (57). Acute otitis media in children often develops within days after onset of upper respiratory tract infection (101). Case control studies of persons with invasive pneumococcal disease show that recent otitis media is reported more frequently among preschool children, and upper respiratory tract infection during the preceding month is reported more frequently among adults, compared with age-matched controls (128, 153). One outbreak of pneumococcal pneumonia in a long-term care facility may have been associated with antecedent parainfluenza virus type 1 infection (73). Adults experimentally infected with influenza A virus appear to have increased susceptibility to nasopharyngeal colonization with *S. pneumoniae* (66, 199), and a number of lines of epidemiological evidence suggest an association between influenza A and pneumococcal disease. In a study at four tertiary care hospitals in Houston, Tex., rates of invasive pneumococcal infection increased within a few weeks of increased circulation of respiratory viruses, particularly influenza virus and respiratory syncytial viruses, in the community during the fall and winter (121). Among persons developing pulmonary complications during the influenza A (H3N2) pandemic during the late 1960s, *S. pneumoniae* was the most commonly identified bacterial pathogen (15, 175). Compared with age-matched controls, preschool- and school-aged children hospitalized with severe pneumococcal pneumonia and their family members were more likely to report influenza-like illness during the 28 days before admission and were more likely to have serological evidence of recent influenza A (H1N1) infection during the 1995–1996 season (155).

TRENDS IN EPIDEMIOLOGY OF PNEUMOCOCCAL DISEASE
Future trends in pneumococcal disease epidemiology likely will be influenced by selective factors in the environment, such as use of antimicrobial drugs, and by changes in the immune status of the organism's human host. The aging of the global population will have a major impact on the epidemiology of pneumococcal disease in the 21st century. It is estimated that from 1998 to 2025, the world's elderly population (aged ≥65 years) will more than double while the population younger than 15 years will increase by only 6% (108), resulting in a larger proportion of the population at increased risk of infection because of immune senescence (35). Just as the global epidemic of AIDS has been the major determinant in changes in pneumococcal disease epidemiology during the final decade of the 20th century, the success of efforts to prevent the spread of HIV infection and to prevent opportunistic infections in persons already infected will be reflected in rates of pneumococcal disease, particularly in young and middle-aged adults. Finally, the future trends in the epidemiology will be influenced by technology. The number of persons at greater risk of pneumococcal infection because of immunosuppressive medications will likely continue to increase. Vaccination promises to have the most profound impact on preventing illness and death from pneumococcal disease since the advent of antimicrobial drug therapy (205). Continued surveillance and application of molecular typing techniques, such as MLST, will be required to understand the public health impact of current and future pneumococcal vaccines and to ensure that the health benefits from vaccination are long-lasting (79).

REFERENCES
1. **Abbasi, S., S. L. Orlicek, I. Almohsen, G. Luedtke, and B. K. English.** 1996. Septic arthritis and osteomyelitis caused by penicillin and cephalosporin-resistant *Streptococcus pneumoniae* in a children's hospital. *Pediatr. Infect. Dis. J.* **15:**78–83.

2. **Abrahamsen, A. F., E. A. Hoiby, E. Hannisdal, O. G. Jorgensen, H. Holte, V. Hasseltvedt, and H. Host.** 1997. Systemic pneumococcal disease after staging splenectomy for Hodgkins disease 1969–1980 without pneumococcal vaccine protection: a follow-up study. *Eur. J. Haematol.* **58:**73–77.

3. **Amber, I. J., E. M. Gilbert, G. Schiffman, and J. A. Jacobson.** 1990. Increased risk of pneumococcal infections in cardiac transplant recipients. *Transplantation* **49:**122–125.

4. **Arlievsky, N., K. I. Li, and J. L. Munoz.** 1998. Septic arthritis with osteomyelitis due to *Streptococcus pneumoniae* in human immunodeficiency virus-infected children. *Clin. Infect. Dis.* **27:**898–899.

5. **Aronin, S. I., S. K. Mukherjee, J. C. West, and E. L. Cooney.** 1998. Review of pneumococcal endocarditis in adults in the penicillin era. *Clin. Infect. Dis.* **26:**165–171.

6. **Austrian, R.** 1977. Prevention of pneumococcal infection by immunization with capsular polysaccharides of *Streptococcus pneumoniae:* current status of polyvalent vaccines. *J. Infect. Dis.* **136:**S38–S42.

7. **Austrian, R.** 1981. Some observations on the pneumococcus and on the current status of pneumococcal disease and its prevention. *Rev. Infect. Dis.* **3:**S1–S17.

8. **Austrian, R., R. M. Douglas, G. Schiffman, A. M. Coetzee, H. J. Koornhof, S. Hayden-Smith, and R. D. Reid.** 1976. Prevention of pneumococcal pneumonia by vaccination. *Trans. Assoc. Am. Phys.* **89:**184–194.

9. **Austrian, R., V. M. Howie, and J. H. Ploussard.** 1977. The bacteriology of pneumococcal otitis media. *Johns Hopkins Med. J.* **141:**104–111.

10. **Babl, F. E., S. I. Pelton, S. Theodore, and J. O. Klein.** 2001. Constancy of distribution of serogroups of invasive pneumococcal isolates among children: experience during 4 decades. *Clin. Infect. Dis.* **32:**1155–1161.

11. **Ballon-Landa, G. R., G. Gherardi, B. Beall, S. Krosner, and V. Nizet.** 2001. Necrotizing fasciitis due to penicillin-resistant *Streptococcus pneumoniae:* case report and review of the literature. *J. Infect.* **42:**272–277.

12. **Barry, M. A., D. E. Craven, and M. Finland.** 1984. Serotypes of *Streptococcus pneumoniae* isolated from blood cultures at Boston City Hospital between 1979 and 1982. *J. Infect. Dis.* **149:**449–452.

13. **Benin, A. L., K. L. O'Brien, J. P. Watt, R. Reid, E. R. Zell, S. Katz, C. Donaldson, A. Parkinson, A. Schuchat, M. Santosham, and C. G. Whitney.** 2003. Effectiveness of the 23-valent polysaccharide vaccine against invasive pneumococcal disease in Navajo adults. *J. Infect. Dis.* **188:**81–89.

14. **Berk, S. L., K. A. Gage, S. A. Holtsclaw-Berk, and J. K. Smith.** 1985. Type 8 pneumococcal pneumonia: an outbreak on an oncology ward. *South. Med. J.* **78:**159–161.

15. **Bisno, A. L., J. P. Griffin, K. A. Van Epps, H. B. Niell, and M. W. Rytel.** 1971. Pneumonia and Hong Kong influenza: a prospective study of the 1968–1969 epidemic. *Am. J. Med. Sci.* **261:**251–263.

16. **Block, S. L.** 1997. Causative pathogens, antibiotic resistance and therapeutic considerations in acute otitis media. *Pediatr. Infect. Dis. J.* **16:**449–456.

17. **Block, S. L., J. Hedrick, C. J. Harrison, R. Tyler, A. Smith, R. Findlay, and E. Keegan.** 2002. Pneumococcal serotypes from acute otitis media in rural Kentucky. *Pediatr. Infect. Dis. J.* **21:**859–865.

18. **Borer, A., H. Meirson, N. Peled, N. Porat, R. Dagan, D. Fraser, J. Gilad, N. Zehavi, and P. Yagupsky.** 2001. Antibiotic-resistant pneumococci carried by young children do not appear to disseminate to adult members of a closed community. *Clin. Infect. Dis.* **33:**436–444.

19. **Brandileone, M. C., A. L. de Andrade, J. L. Di Fabio, M. L. Guerra, and R. Austrian.** 2003. Appropriateness of a pneumococcal conjugate vaccine in Brazil: potential impact of age and clinical diagnosis, with emphasis on meningitis. *J. Infect. Dis.* **187:**1206–1212.

20. **Brandt, J., C. Wong, S. Mihm, J. Roberts, J. Smith, E. Brewer, R. Thiagarajan, and B. Warady.** 2002. Invasive pneumococcal disease and hemolytic uremic syndrome. *Pediatrics* **110:**371–376.

21. **Breiman, R. F., D. W. Keller, M. A. Phelan, D. H. Sniadack, D. S. Stephens, D. Rimland, M. M. Farley, A. Schuchat, and A. L. Reingold.** 2000. Evaluation of effectiveness of the 23-valent pneumococcal capsular polysaccharide vaccine for HIV-infected patients. *Arch. Intern. Med.* **160:**2633–2638.

22. **Breton, J. R., G. Pi, L. Lacruz, I. Calvo, I. Rodriguez, A. Sanchez, J. J. Camarena, and R. Hernandez.** 2001. Pneumococcal pyomyositis. *Pediatr. Infect. Dis. J.* **20:**85–87.

23. **Broome, C. V., R. R. Facklam, and D. W. Fraser.** 1980. Pneumococcal disease after pneumococcal vaccination: an alternative method to estimate the efficacy of pneumococcal vaccine. *N. Engl. J. Med.* **303:**549–552.

24. **Brown, J. W., and J. C. Lockhart.** 1949. Pneumococcic infections in adults at the San Francisco hospital from 1932 to 1946. *Am. Pract.* (Philadelphia) **4:**214–217.

25. **Brueggemann, A. B., D. T. Griffiths, E. Meats, T. Peto, D. W. Crook, and B. G. Spratt.** 2003. Clonal relationships between inva-

sive and carriage Streptococcus pneumoniae and serotype- and clone-specific differences in invasive disease potential. *J. Infect. Dis.* **187:**1424–1432.

26. **Bullowa, J. G. M., and C. Wilcox.** 1935. Incidence of bacteremia in the pneumonias and its relation to mortality. *Arch. Intern. Med.* **55:** 558–573.

27. **Bunim, J. J., and J. D. Trask.** 1935. Familial studies on lobar pneumonia: epidemiologic studies on families of children with pneumococcic lobar pneumonia. *Am. J. Dis. Child.* **50:**626–635.

28. **Burman, L. A., R. Norrby, and B. Trollfors.** 1985. Invasive pneumococcal infections: incidence, predisposing factors, and prognosis. *Rev. Infect. Dis.* **7:**133–142.

29. **Butler, J. C., R. F. Breiman, J. F. Campbell, H. B. Lipman, C. V. Broome, and R. R. Facklam.** 1993. Pneumococcal polysaccharide vaccine efficacy: an evaluation of current recommendations. *JAMA* **270:**1826–1831.

30. **Butler, J. C., R. F. Breiman, H. B. Lipman, J. Hofmann, and R. R. Facklam.** 1995. Serotype distribution of *Streptococcus pneumoniae* infections among preschool children in the United States, 1978–1994: implications for development of a conjugate vaccine. *J. Infect. Dis.* **171:**885–889.

31. **Butler, J. C., J. L. Lennox, L. K. McDougal, J. A. Sutcliffe, A. Tait-Kamradt, and F. C. Tenover.** 2003. Macrolide-resistant pneumococcal endocarditis and epidural abscess that develop during erythromycin therapy. *Clin. Infect. Dis.* **36:**e19–e25.

32. **Butler, J. C., and A. Schuchat.** 1999. Epidemiology of pneumococcal infections in the elderly. *Drugs Aging* **15:**11–19.

33. **Cabrera, G. R., J. D. Fortenberry, B. L. Warshaw, C. R. Chambliss, J. C. Butler, and B. G. Cooperstone.** 1998. Hemolytic uremic syndrome associated with invasive *Streptococcus pneumoniae* infection. *Pediatrics* **101:**699–703.

34. **Capdevila, O., R. Pallares, I. Grau, F. Tubau, J. Linares, J. Ariza, and F. Gudiol.** 2001. Pneumococcal peritonitis in adult patients: report of 64 cases with special reference to emergence of antibiotic resistance. *Arch. Intern. Med.* **161:**1742–1748.

35. **Castle, S. C.** 2000. Clinical relevance of age-related immune dysfunction. *Clin. Infect. Dis.* **31:**578–585.

36. **Centers for Disease Control and Prevention.** 2001. Outbreak of pneumococcal pneumonia among unvaccinated residents of a nursing home—New Jersey, April 2001. *Morb. Mortal. Wkly. Rep.* **50:**707–710.

37. **Centers for Disease Control and Prevention.** 1997. Outbreaks of pneumococcal pneumonia among unvaccinated residents of chronic care facilities—Massachusetts, October 1995, Oklahoma, February 1996, and Maryland, May–June 1996. *Morb. Mortal. Wkly. Rep.* **46:**60–62.

38. **Centers for Disease Control and Prevention.** 2003. Pneumococcal conjunctivitis at an elementary school—Maine, September 20–December 6, 2002. *Morb. Mortal. Wkly. Rep.* **52:**64–66.

39. **Centers for Disease Control and Prevention.** 2000. Preventing pneumococcal disease among infants and young children: recommendations of the Advisory Committee on Immunization Practices (ACIP). *Morb. Mortal. Wkly. Rep.* **49**(RR-9):1–35.

40. **Chen, F. M., R. F. Breiman, M. Farley, B. Plikaytis, K. Deaver, and M. S. Cetron.** 1998. Geocoding and linking data from population-based surveillance and the US Census to evaluate the impact of median household income on the epidemiology of invasive *Streptococcus pneumoniae* infections. *Am. J. Epidemiol.* **148:** 1212–1218.

41. **Cherian, T., M. C. Steinhoff, L. H. Harrison, D. Rohn, L. K. McDougal, and J. Dick.** 1994. A cluster of invasive pneumococcal disease in young children in child care. *JAMA* **271:** 695–697.

42. **Chilcote, R. R., R. L. Baehner, and D. Hammond.** 1976. Septicemia and meningitis in children splenectomized for Hodgkin's disease. *N. Engl. J. Med.* **295:**798–800.

43. **Cid, M., and N. R. Sabates.** 1997. Penicillin-resistant *Streptococcus pneumoniae* endophthalmitis. *Am. J. Ophthalmol.* **123:**133–135.

44. **Collazos, J., A. Fernandez, E. Martinez, J. Mayo, and J. M. de la Viuda.** 1996. Pneumococcal pyomyositis: case report, review of the literature, and comparison with classic pyomyositis caused by other bacteria. *Arch. Intern. Med.* **156:**1470–1474.

45. **Collazos, J., M. Garcia-Cuevas, E. Martinez, J. Mayo, and I. Lekuona.** 1996. Prosthetic valve endocarditis due to *Streptococcus pneumoniae.* *Arch. Intern. Med.* **156:**2141, 2146, 2148.

46. **Corso, A., E. P. Severina, V. F. Petruk, Y. R. Mauriz, and A. Tomasz.** 1998. Molecular characterization of penicillin-resistant *Streptococcus pneumoniae* isolates causing respiratory disease in the United States. *Microb. Drug Resist.* **4:**325–337.

47. **Cortese, M. M., M. Wolff, J. Almeido-Hill, R. Reid, J. Ketcham, and M. Santosham.** 1992. High incidence rates of invasive pneumococcal disease in the White Mountain Apache population. *Arch. Intern. Med.* **152:**2277–2282.

48. Craig, A. S., P. C. Erwin, W. Schaffner, J. A. Elliott, W. L. Moore, X. T. Ussery, L. Patterson, A. D. Dake, S. G. Hannah, and J. C. Butler. 1999. Carriage of multidrug-resistant *Streptococcus pneumoniae* and impact of chemoprophylaxis during an outbreak of meningitis at a day care center. *Clin. Infect. Dis.* **29:**1257–1264.

49. Crewe-Brown, H. H., A. S. Karstaedt, G. L. Saunders, M. Khoosal, N. Jones, A. Wasas, and K. P. Klugman. 1997. *Streptococcus pneumoniae* blood culture isolates from patients with and without human immunodeficiency virus infection: alterations in penicillin susceptibilities and in serogroups or serotypes. *Clin. Infect. Dis.* **25:**1165–1172.

50. Dagan, R., S. Gradstein, I. Belmaker, N. Porat, Y. Siton, G. Weber, J. Janco, and P. Yagupsky. 2000. An outbreak of *Streptococcus pneumoniae* serotype 1 in a closed community in southern Israel. *Clin. Infect. Dis.* **30:**319–321.

51. Daum, R. S., J. P. Nachman, C. D. Leitch, and F. C. Tenover. 1994. Nosocomial epiglottitis associated with penicillin- and cephalosporin-resistant *Streptococcus pneumoniae* bacteremia. *J. Clin. Microbiol.* **32:**246–248.

52. Davidson, M., A. J. Parkinson, L. R. Bulkow, M. A. Fitzgerald, H. V. Peters, and D. J. Parks. 1994. The epidemiology of invasive pneumococcal disease in Alaska, 1986–1990—ethnic differences and opportunities for prevention. *J. Infect. Dis.* **170:**368–376.

53. Del Beccaro, M. A., P. M. Mendelman, A. F. Inglis, M. A. Richardson, N. O. Duncan, C. R. Clausen, and T. L. Stull. 1992. Bacteriology of acute otitis media: a new perspective. *J. Pediatr.* **120:**81–84.

54. De Lencastre, H., and A. Tomasz. 2002. From ecological reservoir to disease: the nasopharynx, day-care centres and drug-resistant clones of *Streptococcus pneumoniae*. *J. Antimicrob. Chemother.* **50:**75–81.

55. DeMaria, A., Jr., K. Browne, S. L. Berk, E. J. Sherwood, and W. R. McCabe. 1980. An outbreak of type 1 pneumococcal pneumonia in a men's shelter. *JAMA* **244:**1446–1449.

56. Donaldson, S. S., E. Glatstein, and K. L. Vosti. 1978. Bacterial infections in pediatric Hodgkin's disease: relationship to radiotherapy, chemotherapy and splenectomy. *Cancer* **41:**1949–1958.

57. Dowell, S. F., C. G. Whitney, C. Wright, C. E. Rose, Jr., and A. Schuchat. 2003. Seasonal patterns of invasive pneumococcal disease. *Emerg. Infect. Dis.* **9:**573–579.

58. Dugi, D. D., III, D. M. Musher, J. E. Clarridge III, and R. Kimbrough. 2001. Intraabdominal infection due to *Streptococcus pneumoniae*. *Medicine* **80:**236–244.

59. Dworkin, M. S., J. W. Ward, D. L. Hanson, J. L. Jones, and J. E. Kaplan. 2001. Pneumococcal disease among human immunodeficiency virus-infected persons: incidence, risk factors, and impact of vaccination. *Clin. Infect. Dis.* **32:**794–800.

60. Ekdahl, K., I. Ahlinder, H. B. Hansson, E. Melander, S. Molstad, M. Soderstrom, and K. Persson. 1997. Duration of nasopharyngeal carriage of penicillin-resistant *Streptococcus pneumoniae*: experiences from the South Swedish Pneumococcal Intervention Project. *Clin. Infect. Dis.* **25:**1113–1117.

61. Eljamel, M. S., and P. M. Foy. 1990. Acute traumatic CSF fistulae: the risk of intracranial infection. *Br. J. Neurosurg.* **4:**381–385.

62. Enright, M. C., and B. G. Spratt. 1998. A multilocus sequence typing scheme for *Streptococcus pneumoniae*: identification of clones associated with serious invasive disease. *Microbiology* **144:**3049–3060.

63. Eraklis, A. J., S. V. Kevy, L. K. Diamond, and R. E. Gross. 1967. Hazard of overwhelming infection after splenectomy in childhood. *N. Engl. J. Med.* **276:**1225–1229.

64. Eskola, J., T. Kilpi, A. Palmu, J. Jokinen, J. Haapakoski, E. Herva, A. Takala, H. Kayhty, P. Karma, R. Kohberger, G. Siber, and P. H. Makela. 2001. Efficacy of a pneumococcal conjugate vaccine against acute otitis media. *N. Engl. J. Med.* **344:**403–409.

65. Eskola, J., A. K. Takala, E. Kela, E. Pekkanen, R. Kalliokoski, and M. Leinonen. 1992. Epidemiology of invasive pneumococcal infections in children in Finland. *JAMA* **268:**3323–3327.

66. Fainstein, V., D. M. Musher, and T. R. Cate. 1980. Bacterial adherence to pharyngeal cells during viral infection. *J. Infect. Dis.* **141:**172–176.

67. Fang, G. D., M. Fine, J. Orloff, D. Arisumi, V. L. Yu, W. Kapoor, J. T. Grayston, S. P. Wang, R. Kohler, R. R. Muder, Y. C. Yee, J. D. Rihs, and R. M. Vickers. 1990. New and emerging etiologies for community-acquired pneumonia with implications for therapy. A prospective multicenter study of 359 cases. *Medicine* **69:**307–316.

68. Feikin, D. R., and K. P. Klugman. 2002. Historical changes in pneumococcal serogroup distribution: implications for the era of pneumococcal conjugate vaccines. *Clin. Infect. Dis.* **35:**547–555.

69. Feikin, D. R., A. Schuchat, M. Kolczak, N. L. Barrett, L. H. Harrison, L. Lefkowitz, A.

McGeer, M. M. Farley, D. J. Vugia, C. Lexau, K. R. Stefonek, J. E. Patterson, and J. H. Jorgensen. 2000. Mortality from invasive pneumococcal pneumonia in the era of antibiotic resistance, 1995–1997. *Am. J. Public Health* **90:**223–229.

70. Feldman, C., M. Glatthaar, R. Morar, A. G. Mahomed, S. Kaka, M. Cassel, and K. P. Klugman. 1999. Bacteremic pneumococcal pneumonia in HIV-seropositive and HIV-seronegative adults. *Chest* **116:**107–114.

71. Finland, M., and M. W. Barnes. 1977. Acute bacterial meningitis at Boston City Hospital during 12 selected years, 1935–1972. *J. Infect. Dis.* **136:**400–415.

72. Finland, M., and M. W. Barnes. 1977. Changes in occurrence of capsular serotypes of Streptococcus pneumoniae at Boston City Hospital during selected years between 1935 and 1974. *J. Clin. Microbiol.* **5:**154–166.

73. Fiore, A. E., C. Iverson, T. Messmer, D. Erdman, S. M. Lett, D. F. Talkington, L. J. Anderson, B. Fields, G. M. Carlone, R. F. Breiman, and M. S. Cetron. 1998. Outbreak of pneumonia in a long-term care facility: antecedent human parainfluenza virus 1 infection may predispose to bacterial pneumonia. *J. Am. Geriatr. Soc.* **46:**1112–1117.

74. Frankel, R. E., M. Virata, C. Hardalo, F. L. Altice, and G. Friedland. 1996. Invasive pneumococcal disease: clinical features, serotypes, and antimicrobial resistance patterns in cases involving patients with and without human immunodeficiency virus infection. *Clin. Infect. Dis.* **23:**577–584.

75. Fraser, D., N. Givon-Lavi, N. Bilenko, and R. Dagan. 2001. A decade (1989–1998) of pediatric invasive pneumococcal disease in 2 populations residing in 1 geographic location: implications for vaccine choice. *Clin. Infect. Dis.* **33:**421–427.

76. Fry, A. M., R. R. Facklam, C. G. Whitney, B. D. Plikaytis, and A. Schuchat. 2003. Multistate evaluation of invasive pneumococcal diseases in adults with human immunodeficiency virus infection: serotype and antimicrobial resistance paterns in the United States. *J. Infect. Dis.* **188:**643–652.

77. Garcia-Leoni, M. E., S. Moreno, P. Rodeno, E. Cercenado, T. Vicente, and E. Bouza. 1992. Pneumococcal pneumonia in adult hospitalized patients infected with the human immunodeficiency virus. *Arch. Intern. Med.* **152:**1808–1812.

78. Gebo, K. A., R. D. Moore, J. C. Keruly, and R. E. Chaisson. 1996. Risk factors for pneumococcal disease in human immunodeficiency virus-infected patients. *J. Infect. Dis.* **173:**857–862.

79. Gertz, R. E., Jr., M. C. McEllistrem, D. J. Boxrud, Z. Li, V. Sakota, T. A. Thompson, R. R. Facklam, J. M. Besser, L. H. Harrison, C. G. Whitney, and B. Beall. 2003. Clonal distribution of invasive pneumococcal isolates from children and selected adults in the United States prior to 7-valent conjugate vaccine introduction. *J. Clin. Microbiol.* **41:**4194–4216.

80. Gessner, B. D., X. T. Ussery, A. J. Parkinson, and R. F. Breiman. 1995. Risk factors for invasive disease caused by Streptococcus pneumoniae among Alaska Native children younger than two years of age. *Pediatr. Infect. Dis. J.* **14:**123–128.

81. Giglio, M. S., M. Landaeta, and M. E. Pinto. 1997. Microbiology of recurrent parotitis. *Pediatr. Infect. Dis. J.* **16:**386–390.

82. Givner, L. B., E. O. Mason, Jr., W. J. Barson, T. Q. Tan, E. R. Wald, G. E. Schutze, K. S. Kim, J. S. Bradley, R. Yogev, and S. L. Kaplan. 2000. Pneumococcal facial cellulitis in children. *Pediatrics* **106:**e61.

83. Gleich, S., Y. Morad, R. Echague, J. R. Miller, J. Kornblum, J. S. Sampson, and J. C. Butler. 2000. *Streptococcus pneumoniae* serotype 4 outbreak in a home for the aged: report and review of recent outbreaks. *Infect. Control Hosp. Epidemiol.* **21:**711–717.

84. Gomez, M., S. Alter, M. L. Kumar, S. Murphy, and M. H. Rathore. 1999. Neonatal *Streptococcus pneumoniae* infection: case reports and review of the literature. *Pediatr. Infect. Dis. J.* **18:**1014–1018.

85. Gopal, V., and A. L. Bisno. 1977. Fulminant pneumococcal infections in 'normal' asplenic hosts. *Arch. Intern. Med.* **137:**1526–1530.

86. Gratten, M., F. Morey, J. Dixon, K. Manning, P. Torzillo, R. Matters, J. Erlich, J. Hanna, V. Asche, and I. Riley. 1993. An outbreak of serotype 1 *Streptococcus pneumoniae* infection in central Australia. *Med. J. Aust.* **158:**340–342.

87. Gratten, M., P. Torzillo, F. Morey, J. Dixon, J. Erlich, J. Hagger, and J. Henrichsen. 1996. Distribution of capsular types and antibiotic susceptibility of invasive *Streptococcus pneumoniae* isolated from Aborigines in central Australia. *J. Clin. Microbiol.* **34:**338–341.

88. Gray, B. M. 2001. Is pneumococcal hemolytic-uremic syndrome a new disease? *Infect. Med.* **18:**251–258.

89. Gray, B. M., G. M. Converse III, and H. C. Dillon, Jr. 1980. Epidemiologic studies of *Streptococcus pneumoniae* in infants: acquisition, carriage, and infection during the first 24 months of life. *J. Infect. Dis.* **142:**923–933.

90. **Gray, B. M., G. M. Converse III, and H. C. Dillon, Jr.** 1979. Serotypes of *Streptococcus pneumoniae* causing disease. *J. Infect. Dis.* **140:** 979–983.

91. **Gray, B. M., and H. C. Dillon, Jr.** 1988. Epidemiological studies of *Streptococcus pneumoniae* in infants: antibody to types 3, 6, 14, and 23 in the first two years of life. *J. Infect. Dis.* **158:** 948–955.

92. **Gray, G. C., J. D. Callahan, A. W. Hawksworth, C. A. Fisher, and J. C. Gaydos.** 1999. Respiratory diseases among U.S. military personnel: countering emerging threats. *Emerg. Infect. Dis.* **5:**379–385.

93. **Grigoriadis, E., and W. L. Gold.** 1997. Pyogenic brain abscess caused by *Streptococcus pneumoniae*: case report and review. *Clin. Infect. Dis.* **25:**1108–1112.

94. **Guevara, R. E., J. C. Butler, B. J. Marston, J. F. Plouffe, T. M. File, Jr., and R. F. Breiman.** 1999. Accuracy of ICD-9-CM codes in detecting community-acquired pneumococcal pneumonia for incidence and vaccine efficacy studies. *Am. J. Epidemiol.* **149:**282–289.

95. **Gwaltney, J. M., Jr., M. A. Sande, R. Austrian, and J. O. Hendley.** 1975. Spread of *Streptococcus pneumoniae* in families. II. Relation of transfer of *S. pneumoniae* to incidence of colds and serum antibody. *J. Infect. Dis.* **132:**62–68.

96. **Harrison, L. H., D. M. Dwyer, L. Billmann, M. S. Kolczak, and A. Schuchat.** 2000. Invasive pneumococcal infection in Baltimore, MD: implications for immunization policy. *Arch. Intern. Med.* **160:**89–94.

97. **Hausdorff, W. P., J. Bryant, C. Kloek, P. R. Paradiso, and G. R. Siber.** 2000. The contribution of specific pneumococcal serogroups to different disease manifestations: implications for conjugate vaccine formulation and use, part II. *Clin. Infect. Dis.* **30:**122–140.

98. **Hausdorff, W. P., J. Bryant, P. R. Paradiso, and G. R. Siber.** 2000. Which pneumococcal serogroups cause the most invasive disease: implications for conjugate vaccine formulation and use, part I. *Clin. Infect. Dis.* **30:**100–121.

99. **Hausdorff, W. P., G. Siber, and P. R. Paradiso.** 2001. Geographical differences in invasive pneumococcal disease rates and serotype frequency in young children. *Lancet* **357:**950–952.

100. **Hausdorff, W. P., G. Yothers, R. Dagan, T. Kilpi, S. I. Pelton, R. Cohen, M. R. Jacobs, S. L. Kaplan, C. Levy, E. L. Lopez, E. O. Mason, Jr., V. Syriopoulou, B. Wynne, and J. Bryant.** 2002. Multinational study of pneumococcal serotypes causing acute otitis media in children. *Pediatr. Infect. Dis. J.* **21:**1008–1016.

101. **Heikkinen, T., and T. Chonmaitree.** 2003. Importance of respiratory viruses in acute otitis media. *Clin. Microbiol. Rev.* **16:**230–241.

102. **Hendley, J. O., M. A. Sande, P. M. Stewart, and J. M. Gwaltney, Jr.** 1975. Spread of *Streptococcus pneumoniae* in families. I. Carriage rates and distribution of types. *J. Infect. Dis.* **132:**55–61.

103. **Henrichsen, J.** 1995. Six newly recognized types of *Streptococcus pneumoniae*. *J. Clin. Microbiol.* **33:**2759–2762.

104. **Henrichsen, J.** 1999. Typing of *Streptococcus pneumoniae*: past, present, and future. *Am. J. Med.* **107:**50S–54S.

105. **Henriques Normark, B., M. Kalin, A. Ortqvist, T. Akerlund, B. O. Liljequist, J. Hedlund, S. B. Svenson, J. Zhou, B. G. Spratt, S. Normark, and G. Kallenius.** 2001. Dynamics of penicillin-susceptible clones in invasive pneumococcal disease. *J. Infect. Dis.* **184:**861–869.

106. **Hermans, P. W., M. Sluijter, S. Dejsirilert, N. Lemmens, K. Elzenaar, A. van Veen, W. H. Goessens, and R. de Groot.** 1997. Molecular epidemiology of drug-resistant pneumococci: toward an international approach. *Microb. Drug Resist.* **3:**243–251.

107. **Hermans, P. W., M. Sluijter, T. Hoogenboezem, H. Heersma, A. van Belkum, and R. de Groot.** 1995. Comparative study of five different DNA fingerprint techniques for molecular typing of *Streptococcus pneumoniae* strains. *J. Clin. Microbiol.* **33:**1606–1612.

108. **High, K. P.** 2002. Infection in an ageing world. *Lancet Infect. Dis.* **2:**655.

109. **Hirsch, E. F., and M. McKinney.** 1919. An epidemic of pneumococcus bronchopneumonia. *J. Infect. Dis.* **24:**594–617.

110. **Hodges, R., and C. MacLeod.** 1946. Epidemic pneumococcal pneumonia. V. Final considerations of the factors underlying the epidemic. *Am. J. Hyg.* **44:**237–243.

111. **Hoge, C. W., M. R. Reichler, E. A. Dominguez, J. C. Bremer, T. D. Mastro, K. A. Hendricks, D. M. Musher, J. A. Elliott, R. R. Facklam, and R. F. Breiman.** 1994. An epidemic of pneumococcal disease in an overcrowded, inadequately ventilated jail. *N. Engl. J. Med.* **331:**643–648.

112. **Holle, H. A., and J. G. M. Bullowa.** 1940. Pneumococcal cross-infection in the home and hospital. *N. Engl. J. Med.* **223:**887–890.

113. **Janoff, E. N., R. F. Breiman, C. L. Daley, and P. C. Hopewell.** 1992. Pneumococcal disease during HIV infection. Epidemiologic,

clinical, and immunologic perspectives. *Ann. Intern. Med.* **117:**314–324.

114. **Janoff, E. N., J. O'Brien, P. Thompson, J. Ehret, G. Meiklejohn, G. Duvall, and J. M. Douglas, Jr.** 1993. *Streptococcus pneumoniae* colonization, bacteremia, and immune response among persons with human immunodeficiency virus infection. *J. Infect. Dis.* **167:**49–56.

115. **Joloba, M. L., A. Windau, S. Bajaksouzian, P. C. Appelbaum, W. P. Hausdorff, and M. R. Jacobs.** 2001. Pneumococcal conjugate vaccine serotypes of *Streptococcus pneumoniae* isolates and the antimicrobial susceptibility of such isolates in children with otitis media. *Clin. Infect. Dis.* **33:**1489–1494.

116. **Jones, N., R. Huebner, M. Khoosal, H. Crewe-Brown, and K. Klugman.** 1998. The impact of HIV on *Streptococcus pneumoniae* bacteraemia in a South African population. *AIDS* **12:**2177–2184.

117. **Kamerling, J. P.** 2000. Pneumococcal polysaccharides: a chemical view, p. 81–114. *In* A. Tomasz (ed.), Streptococcus pneumoniae: *Molecular Biology and Mechanisms of Disease.* Mary Anne Liebert, Inc., Larchmont, N.Y.

118. **Kaplan, S. L., E. O. Mason, Jr., E. R. Wald, K. S. Kim, L. B. Givner, J. S. Bradley, W. J. Barson, T. Q. Tan, G. E. Schutze, and R. Yogev.** 2000. Pneumococcal mastoiditis in children. *Pediatrics* **106:** 695–699.

119. **Kauffman, C. A., C. Watanakunakorn, and J. P. Phair.** 1973. Purulent pneumococcal pericarditis. A continuing problem in the antibiotic era. *Am. J. Med.* **54:**743–750.

120. **Kessler, H. A., R. Schade, G. M. Trenholme, J. E. Jupa, and S. Levin.** 1980. Acute pneumococcal epiglottitis in immunocompromised adults. *Scand. J. Infect. Dis.* **12:**207–210.

121. **Kim, P. E., D. M. Musher, W. P. Glezen, M. C. Rodriguez-Barradas, W. K. Nahm, and C. E. Wright.** 1996. Association of invasive pneumococcal disease with season, atmospheric conditions, air pollution, and the isolation of respiratory viruses. *Clin. Infect. Dis.* **22:**100–106.

122. **Klein, J. O.** 1981. The epidemiology of pneumococcal disease in infants and children. *Rev. Infect. Dis.* **3:**246–253.

123. **Koeuth, T., J. Versalovic, and J. R. Lupski.** 1995. Differential subsequence conservation of interspersed repetitive *Streptococcus pneumoniae* BOX elements in diverse bacteria. *Genome Res.* **5:**408–418.

124. **Kutas, L. M., J. M. Duggan, and C. A. Kauffman.** 1995. Pneumococcal vertebral osteomyelitis. *Clin. Infect. Dis.* **20:**286–290.

125. **Lau, G. W., S. Haataja, M. Lonetto, S. E. Kensit, A. Marra, A. P. Bryant, D. McDevitt, D. A. Morrison, and D. W. Holden.** 2001. A functional genomic analysis of type 3 *Streptococcus pneumoniae* virulence. *Mol. Microbiol.* **40:**555–571.

126. **Lefevre, J. C., G. Faucon, A. M. Sicard, and A. M. Gasc.** 1993. DNA fingerprinting of *Streptococcus pneumoniae* strains by pulsed-field gel electrophoresis. *J. Clin. Microbiol.* **31:** 2724–2728.

127. **Leino, T., K. Auranen, J. Jokinen, M. Leinonen, P. Tervonen, and A. K. Takala.** 2001. Pneumococcal carriage in children during their first two years: important role of family exposure. *Pediatr. Infect. Dis. J.* **20:**1022–1027.

128. **Levine, O. S., M. Farley, L. H. Harrison, L. Lefkowitz, A. McGeer, and B. Schwartz.** 1999. Risk factors for invasive pneumococcal disease in children: a population-based case-control study in North America. *Pediatrics* **103:**e28.

129. **Lew, D., H. Hausler, F. Southwick, and A. Baker.** 1984. Sphenoid sinusitis. *Biomed. Pharmacother.* **38:**214–216.

130. **Linnemann, C. C., Jr., and M. R. First.** 1979. Risk of pneumococcal infections in renal transplant patients. *JAMA* **241:**2619–2621.

131. **Lipsky, B. A., E. J. Boyko, T. S. Inui, and T. D. Koepsell.** 1986. Risk factors for acquiring pneumococcal infections. *Arch. Intern. Med.* **146:**2179–2185.

132. **Lloyd-Evans, N., T. J. O'Dempsey, I. Baldeh, O. Secka, E. Demba, J. E. Todd, T. F. McArdle, W. S. Banya, and B. M. Greenwood.** 1996. Nasopharyngeal carriage of pneumococci in Gambian children and in their families. *Pediatr. Infect. Dis. J.* **15:**866–871.

133. **Loda, F. A., A. M. Collier, W. P. Glezen, K. Strangert, W. A. Clyde, and F. W. Denny.** 1975. Occurrence of *Diplococcus pneumoniae* in the upper respiratory tract of children. *J. Pediatr.* **87:**1087–1093.

134. **Mao, C., M. Harper, K. McIntosh, C. Reddington, J. Cohen, R. Bachur, B. Caldwell, and H. W. Hsu.** 1996. Invasive pneumococcal infections in human immunodeficiency virus-infected children. *J. Infect. Dis.* **173:**870–876.

135. **Marrie, T. J.** 2001. Etiology of community-acquired pneumonia, p. 131–141. *In* T. J. Marrie (ed.), *Community-Acquired Pneumonia.* Plenum Publishers, New York, N.Y.

136. **Marston, B. J., J. F. Plouffe, T. M. File, Jr., B. A. Hackman, S. J. Salstrom, H. B. Lipman, M. S. Kolczak, and R. F. Breiman.** 1997. Incidence of community-acquired pneumonia requiring hospitalization.

Results of a population-based active surveillance study in Ohio. *Arch. Intern. Med.* **157:**1709–1718.

137. **Martin, M., J. H. Turco, M. E. Zegans, R. R. Facklam, S. Sodha, J. A. Elliott, J. H. Pryor, B. Beall, D. D. Erdman, Y. Y. Baumgartner, P. A. Sanchez, J. D. Schwartzman, J. Montero, A. Schuchat, and C. G. Whitney.** 2003. An outbreak of conjunctivitis due to atypical *Streptococcus pneumoniae*. *N. Engl. J. Med.* **348:**1112–1121.

138. **McEllistrem, M. C., A. B. Mendelsohn, M. A. Pass, J. A. Elliott, C. G. Whitney, J. A. Kolano, and L. H. Harrison.** 2002. Recurrent invasive pneumococcal disease in individuals with human immunodeficiency virus infection. *J. Infect. Dis.* **185:**1364–1368.

139. **McGee, L., K. P. Klugman, and A. Tomasz.** 2000. Serotypes and clones of antibiotic-resistant pneumococci. *In* A. Tomasz (ed.), *Streptococcus pneumoniae: Molecular Biology and Mechanism of Disease.* Mary Ann Liebert, Inc., Larchmont, N.Y.

140. **McGee, L., L. McDougal, J. Zhou, B. G. Spratt, F. C. Tenover, R. George, R. Hakenbeck, W. Hryniewicz, J. C. Lefevre, A. Tomasz, and K. P. Klugman.** 2001. Nomenclature of major antimicrobial-resistant clones of *Streptococcus pneumoniae* defined by the pneumococcal molecular epidemiology network. *J. Clin. Microbiol.* **39:**2565–2571.

141. **Mercat, A., J. Nguyen, and B. Dautzenberg.** 1991. An outbreak of pneumococcal pneumonia in two men's shelters. *Chest* **99:**147–151.

142. **Miller, J. L., and F. B. Lusk.** 1918. Epidemic of streptococcal pneumonia and empyema at Camp Dodge, Iowa. *JAMA* **71:**702–704.

143. **Mufson, M. A.** 1981. Pneumococcal infections. *JAMA* **246:**1942–1948.

144. **Mulhern, M. G., P. I. Condon, and M. O'Keefe.** 1997. Endophthalmitis after astigmatic myopic laser in situ keratomileusis. *J. Cataract Refract. Surg.* **23:**948–950.

145. **Musher, D. M.** 2003. How contagious are common respiratory tract infections? *N. Engl. J. Med.* **348:**1256–1266.

146. **Musher, D. M.** 1992. Infections cause by *Streptococcus pneumoniae*: clinical spectrum, pathogenesis, immunity, and treatment. *Clin. Infect. Dis.* **14:**801–809.

147. **Musher, D. M., J. E. Groover, M. R. Reichler, F. X. Riedo, B. Schwartz, D. A. Watson, R. E. Baughn, and R. F. Breiman.** 1997. Emergence of antibody to capsular polysaccharides of *Streptococcus pneumoniae* during outbreaks of pneumonia: association with na-

sopharyngeal colonization. *Clin. Infect. Dis.* **24:**441–446.

148. **Musher, D. M., H. M. Phan, D. A. Watson, and R. E. Baughn.** 2000. Antibody to capsular polysaccharide of *Streptococcus pneumoniae* at the time of hospital admission for pneumococcal pneumonia. *J. Infect. Dis.* **182:**158–167.

149. **Myers, K. A., and T. J. Marrie.** 1993. Thrombotic microangiopathy associated with *Streptococcus pneumoniae* bacteremia: case report and review. *Clin. Infect. Dis.* **17:**1037–1040.

150. **Naktin, J., and J. DeSimone.** 1999. Lumbar vertebral osteomyelitis with mycotic abdominal aortic aneurysm caused by highly penicillin-resistant *Streptococcus pneumoniae*. *J. Clin. Microbiol.* **37:**4198–4200.

151. **Notter, D. T., P. L. Grossman, S. A. Rosenberg, and J. S. Remington.** 1980. Infections in patients with Hodgkin's disease: a clinical study of 300 consecutive adult patients. *Rev. Infect. Dis.* **2:**761–800.

152. **Nuorti, J. P., J. C. Butler, J. M. Crutcher, R. Guevara, D. Welch, P. Holder, and J. A. Elliott.** 1998. An outbreak of multidrug-resistant pneumococcal pneumonia and bacteremia among unvaccinated nursing home residents. *N. Engl. J. Med.* **338:**1861–1868.

153. **Nuorti, J. P., J. C. Butler, M. M. Farley, L. H. Harrison, A. McGeer, M. S. Kolczak, and R. F. Breiman.** 2000. Cigarette smoking and invasive pneumococcal disease. *N. Engl. J. Med.* **342:**681–689.

154. **Nuorti, J. P., J. C. Butler, L. Gelling, J. L. Kool, A. L. Reingold, and D. J. Vugia.** 2000. Epidemiologic relation between HIV and invasive pneumococcal disease in San Francisco County, California. *Ann. Intern. Med.* **132:**182–190.

155. **O'Brien, K. L., M. I. Walters, J. Sellman, P. Quinlisk, H. Regnery, B. Schwartz, and S. F. Dowell.** 2000. Severe pneumococcal pneumonia in previously healthy children: the role of preceding influenza infection. *Clin. Infect. Dis.* **30:**784–789.

156. **Pantosti, A., G. Gherardi, M. Conte, F. Faella, G. Dicuonzo, and B. Beall.** 2002. A novel, multiple drug-resistant, serotype 24F strain of *Streptococcus pneumoniae* that caused meningitis in patients in Naples, Italy. *Clin. Infect. Dis.* **35:**205–208.

157. **Park, J. H., Jr., and H. T. Chickering.** 1919. Type I pneumococcal lobal pneumonia among Puerto Rican laborers of Camp Jackson, South Carolina. *JAMA* **73:**183–186.

158. **Pastor, P., F. Medley, and T. V. Murphy.** 1998. Invasive pneumococcal disease in Dallas

County, Texas: results from population-based surveillance in 1995. *Clin. Infect. Dis.* **26:**590–595.

159. **Piccioli, A., G. Chini, M. Mannelli, and M. Serio.** 1994. Bilateral massive adrenal hemorrhage due to sepsis: report of two cases. *J. Endocrinol. Investig.* **17:**821–824.

160. **Plouffe, J. F., R. F. Breiman, R. R. Facklam, and the Franklin County Pneumonia Study Group.** 1996. Bacteremia with *Streptococcus pneumoniae.* Implications for therapy and prevention. *JAMA* **275:**194–198.

161. **Polissi, A., A. Pontiggia, G. Feger, M. Altieri, H. Mottl, L. Ferrari, and D. Simon.** 1998. Large-scale identification of virulence genes from *Streptococcus pneumoniae. Infect. Immun.* **66:**5620–5629.

162. **Proulx, J. F., S. Dery, L. P. Jette, J. Ismael, M. Libman, and P. De Wals.** 2002. Pneumonia epidemic caused by a virulent strain of *Streptococcus pneumoniae* serotype 1 in Nunavik, Quebec. *Canada Communicable Dis. Rep.* **28:** 129–131.

163. **Rauch, A. M., M. O'Ryan, R. Van, and L. K. Pickering.** 1990. Invasive disease due to multiply resistant *Streptococcus pneumoniae* in a Houston, Tex., day-care center. *Am. J. Dis. Child.* **144:**923–927.

164. **Redd, S. C., G. W. Rutherford III, M. A. Sande, A. R. Lifson, W. K. Hadley, R. R. Facklam, and J. S. Spika.** 1990. The role of human immunodeficiency virus infection in pneumococcal bacteremia in San Francisco residents. *J. Infect. Dis.* **162:**1012–1017.

165. **Reefhuis, J., M. A. Honein, C. G. Whitney, S. Chamany, E. A. Mann, K. R. Biernath, K. Broder, S. Manning, S. Avashia, M. Victor, P. Costa, O. Devine, A. Graham, and C. Boyle.** 2003. Risk of bacterial meningitis in children with cochlear implants. *N. Engl. J. Med.* **349:**435–445.

166. **Richter, S. S., K. P. Heilmann, S. L. Coffman, H. K. Huynh, A. B. Brueggemann, M. A. Pfaller, and G. V. Doern.** 2002. The molecular epidemiology of penicillin-resistant *Streptococcus pneumoniae* in the United States, 1994–2000. *Clin. Infect. Dis.* **34:**330–339.

167. **Robinson, D. A., D. E. Briles, M. J. Crain, and S. K. Hollingshead.** 2002. Evolution and virulence of serogroup 6 pneumococci on a global scale. *J. Bacteriol.* **184:**6367–6375.

168. **Robinson, K. A., W. Baughman, G. Rothrock, N. L. Barrett, M. Pass, C. Lexau, B. Damaske, K. Stefonek, B. Barnes, J. Patterson, E. R. Zell, A. Schuchat, and C. G. Whitney.** 2001. Epi-demiology of invasive *Streptococcus pneumoniae* infections in the United States, 1995–1998: opportunities for prevention in the conjugate vaccine era. *JAMA* **285:**1729–1735.

169. **Robinson, M. G., and R. J. Watson.** 1966. Pneumococcal meningitis in sickle-cell anemia. *N. Engl. J. Med.* **274:**1006–1008.

170. **Ross, J. J., C. L. Saltzman, P. Carling, and D. S. Shapiro.** 2003. Pneumococcal septic arthritis: review of 190 cases. *Clin. Infect. Dis.* **36:**319–327.

171. **Rudolph, K. M., A. J. Parkinson, A. L. Reasonover, L. R. Bulkow, D. J. Parks, and J. C. Butler.** 2000. Serotype distribution and antimicrobial resistance patterns of invasive isolates of *Streptococcus pneumoniae*: Alaska, 1991–1998. *J. Infect. Dis.* **182:**490–496.

172. **Saenz, R. E., C. V. Sanders, K. E. Aldridge, and M. M. Patel.** 1998. Purulent pericarditis with associated cardiac tamponade caused by a *Streptococcus pneumoniae* strain highly resistant to penicillin, cefotaxime, and ceftriaxone. *Clin. Infect. Dis.* **26:**762–763.

173. **Schonheyder, H. C., H. T. Sorensen, B. Kristensen, and B. Korsager.** 1997. Reasons for increase in pneumococcal bacteraemia. *Lancet* **349:**1554.

174. **Schuchat, A., K. Robinson, J. D. Wenger, L. H. Harrison, M. Farley, A. L. Reingold, L. Lefkowitz, and B. A. Perkins.** 1997. Bacterial meningitis in the United States in 1995. *N. Engl. J. Med.* **337:**970–976.

175. **Schwarzmann, S. W., J. L. Adler, R. J. Sullivan, Jr., and W. M. Marine.** 1971. Bacterial pneumonia during the Hong Kong influenza epidemic of 1968–1969. *Arch. Intern. Med.* **127:**1037–1041.

176. **Scott, J. A., A. J. Hall, R. Dagan, J. M. Dixon, S. J. Eykyn, A. Fenoll, M. Hortal, L. P. Jette, J. H. Jorgensen, F. Lamothe, C. Latorre, J. T. Macfarlane, D. M. Shlaes, L. E. Smart, and A. Taunay.** 1996. Serogroup-specific epidemiology of *Streptococcus pneumoniae*: associations with age, sex, and geography in 7,000 episodes of invasive disease. *Clin. Infect. Dis.* **22:**973–981.

177. **Sheppard, D. C., K. A. Bartlett, and H. W. Lampiris.** 1998. *Streptococcus pneumoniae* transmission in chronic-care facilities: description of an outbreak and review of management strategies. *Infect. Control Hosp. Epidemiol.* **19:** 851–853.

178. **Shimada, J., N. Yamanaka, M. Hotomi, M. Suzumoto, A. Sakai, K. Ubukata, T. Mitsuda, S. Yokota, and H. Faden.** 2002. Household transmission of *Streptococcus pneumo-*

niae among siblings with acute otitis media. *J. Clin. Microbiol.* **40**:1851–1853.

179. **Simolte, S., E. Smith, and D. M. Musher.** 1996. Waterhouse-Friderichsen syndrome due to *Streptococcus pneumoniae* in an otherwise health adult. *Infect. Dis. Clin. Pract.* **5**:398–400.

180. **Sirotnak, A. P., S. C. Eppes, and J. D. Klein.** 1996. Tuboovarian abscess and peritonitis caused by *Streptococcus pneumoniae* serotype 1 in young girls. *Clin. Infect. Dis.* **22**:993–996.

181. **Sleeman, K., K. Knox, R. George, E. Miller, P. Waight, D. Griffiths, A. Efstratiou, K. Broughton, R. T. Mayon-White, E. R. Moxon, and D. W. Crook.** 2001. Invasive pneumococcal disease in England and Wales: vaccination implications. *J. Infect. Dis.* **183**:239–246.

182. **Smillie, W. G.** 1936. A study of an outbreak of type II pneumococcus pneumonia in the Veterans' Administration hospital at Bedford, Massachusetts. *Am. J. Hyg.* **24**:522–535.

183. **Smillie, W. G., G. Warnock, and H. White.** 1938. Study of a type I pneumococcus epidemic at the State Hospital at Worcester, Massachusetts. *Am. J. Public Health* **28**:293–302.

184. **Smith, A. M., K. P. Klugman, T. J. Coffey, and B. G. Spratt.** 1993. Genetic diversity of penicillin-binding protein 2B and 2X genes from *Streptococcus pneumoniae* in South Africa. *Antimicrob. Agents Chemother.* **37**:1938–1944.

185. **Sniadack, D. H., B. Schwartz, H. Lipman, J. Bogaerts, J. C. Butler, R. Dagan, G. Echaniz-Aviles, N. Lloyd-Evans, A. Fenoll, N. I. Girgis, J. Henrichsen, K. Klugman, D. Lehmann, A. K. Takala, J. Vandepitte, S. Gove, and R. F. Breiman.** 1995. Potential interventions for the prevention of childhood pneumonia: geographic and temporal differences in serotype and serogroup distribution of sterile site pneumococcal isolates from children—implications for vaccine strategies. *Pediatr. Infect. Dis. J.* **14**:503–510.

186. **Spataro, V., and C. Marone.** 1993. Rhabdomyolysis associated with bacteremia due to *Streptococcus pneumoniae*: case report and review. *Clin. Infect. Dis.* **17**:1063–1064.

187. **Stillman, E. G.** 1917. Further studies on the epidemiology of lobar pneumonia. *J. Exp. Med.* **26**:513–535.

188. **Subramanian, D., J. A. Sandoe, V. Keer, and M. H. Wilcox.** 2003. Rapid spread of penicillin-resistant *Streptococcus pneumoniae* among high-risk hospital inpatients and the role of molecular typing in outbreak confirmation. *J. Hosp. Infect.* **54**:99–103.

189. **Takala, A. K., J. Jero, E. Kela, P. R. Ronnberg, E. Koskenniemi, and J. Eskola.** 1995. Risk factors for primary invasive pneumococcal disease among children in Finland. *JAMA* **273**:859–864.

190. **Taylor, S. N., and C. V. Sanders.** 1999. Unusual manifestations of invasive pneumococcal infection. *Am. J. Med.* **107**:12S–27S.

191. **Tenny, C. F., and W. T. Riverburgh.** 1919. A group of 68 cases of type I pneumonia occurring in 30 days at Camp Upton. *Arch. Intern. Med.* **24**:545–552.

192. **Thompson, R. T., J. M. Ruegsegger, M. A. Blankenhorn, and M. Hamburger.** 1951. Primary pneumococcic pneumonia at the Cincinnati General Hospital, 1936–1950. *J. Lab. Clin. Med.* **37**:73–87.

193. **Tilghman, F. C., and M. Finland.** 1937. Clinical significance of bacteremia in pneumococcal pneumonia. *Arch. Intern. Med.* **59**:602–619.

194. **Tilghman, F. C., and M. Finland.** 1936. Pneumococcic infections in families. *J. Clin. Investig.* **15**:493–499.

195. **Torzillo, P. J., J. N. Hanna, F. Morey, M. Gratten, J. Dixon, and J. Erlich.** 1995. Invasive pneumococcal disease in central Australia. *Med. J. Aust.* **162**:182–186.

196. **Trotman, J., B. Hughes, and L. Mollison.** 1995. Invasive pneumococcal disease in central Australia. *Clin. Infect. Dis.* **20**:1553–1556.

197. **Twomey, J. J.** 1973. Infections complicating multiple myeloma and chronic lymphocytic leukemia. *Arch. Intern. Med.* **132**:562–565.

198. **van Belkum, A., M. Sluijuter, R. de Groot, H. Verbrugh, and P. W. Hermans.** 1996. Novel BOX repeat PCR assay for high-resolution typing of *Streptococcus pneumoniae* strains. *J. Clin. Microbiol.* **34**:1176–1179.

199. **Wadowsky, R. M., S. M. Mietzner, D. P. Skoner, W. J. Doyle, and P. Fireman.** 1995. Effect of experimental influenza A virus infection on isolation of *Streptococcus pneumoniae* and other aerobic bacteria from the oropharynges of allergic and nonallergic adult subjects. *Infect. Immun.* **63**:1153–1157.

200. **Wald, E. R.** 1992. Sinusitis in children. *N. Engl. J. Med.* **326**:319–323.

201. **Wald, E. R., G. J. Milmoe, A. Bowen, J. Ledesma-Medina, N. Salamon, and C. D. Bluestone.** 1981. Acute maxillary sinusitis in children. *N. Engl. J. Med.* **304**:749–754.

202. **Weisholtz, S. J., B. J. Hartman, and R. B. Roberts.** 1983. Effect of underlying disease and age on pneumococcal serotype distribution. *Am. J. Med.* **75**:199–205.

203. **Westh, H., L. Skibsted, and B. Korner.** 1990. *Streptococcus pneumoniae* infections of the female genital tract and in the newborn child. *Rev. Infect. Dis.* **12:**416–422.

204. **Whitecar, J. P., Jr., J. L. Reddin, and W. W. Spink.** 1966. Recurrent pneumococcal meningitis: a review of the literature and studies on a patient who recovered from eleven attacks caused by five serotypes of *Diplococcus pneumoniae. N. Engl. J. Med.* **274:**1285–1289.

205. **Whitney, C. G., M. M. Farley, J. Hadler, L. H. Harrison, N. M. Bennett, R. Lynfield, A. Reingold, P. R. Cieslak, T. Pilishvili, D. Jackson, R. R. Facklam, J. H. Jorgensen, A. Schuchat, and Active Bacterial Core Surveillance of the Emerging Infec-tions Program Network.** 2003. Decline in invasive pneumococcal disease after the intro-duction of protein-polysaccharide conjugate vaccine. *N. Engl. J. Med.* **348:**1737–1746.

206. **Winston, D. J., G. Schiffman, D. C. Wang, S. A. Feig, C. H. Lin, E. L. Marso, W. G. Ho, L. S. Young, and R. P. Gale.** 1979. Pneumo-coccal infections after human bone-marrow trans-plantation. *Ann. Intern. Med.* **91:**835–841.

207. **Zangwill, K. M., C. M. Vadheim, A. M. Vannier, L. S. Hemenway, D. P. Green-berg, and J. I. Ward.** 1996. Epidemiology of invasive pneumococcal disease in southern Cali-fornia: implications for the design and conduct of a pneumococcal conjugate vaccine efficacy trial. *J. Infect. Dis.* **174:**752–759.

MECHANISMS OF CARRIAGE

Jeffrey N. Weiser

11

GENERAL CONSIDERATIONS

Streptococcus pneumoniae is one of many closely related oral streptococci of the mitis phylogenetic group that colonize the human oro- and nasopharynx (88). The trait that distinguishes the pneumococcus from these other species is, after colonization is established and depending on certain host and bacterial factors, the ability to cause disease in a small proportion of its hosts. Colonization, therefore, is the initial step in the pathogenesis of all pneumococcal disease.

Host Specificity

One of the major unresolved questions about pneumococcal colonization is what molecular factors account for its narrow host specificity. Few pneumococcal enzymes or adhesins have been shown to have specificity for human targets. Examples include the immunoglobulin A1 (IgA1) protease, which acts to circumvent the protective effects of this major immunoglobulin subclass of the mucosal surface of the nasopharynx, and choline-binding protein A (CbpA), which binds to secretory component on the polymeric immunoglobulin receptor (35, 59, 80). Whether one of these specific interactions is sufficient to explain why humans are the only significant natural reservoir for this species is currently not known. Hence, the highly specific adaptation of the pneumococcus to humans presents many challenges and limitations to fully defining the mechanisms of carriage. Several animal models, including adult mice, infant rats, rabbits, and chinchillas, have been used to study pneumococcal carriage (68, 74, 85, 91). In many circumstances there may be transmission and sustained colonization of the oro- and nasopharynx persisting for a period of days to weeks following an oral or intranasal inoculation. These models, which resemble natural carriage in many respects, have been useful in determining the role in colonization of numerous host and bacterial factors. However, in few instances is the contribution to carriage of any of these factors fully defined. Successful colonization and infection of small animals with this pathogen also point to our lack of understanding of the requirements for natural carriage, which is restricted to humans (and rare equine outbreaks for limited pneumococcal types) (89). As a result, many fundamental questions about natural pneumococcal carriage, such as its precise anatomical location and the density of organisms on the pharyngeal

Jeffrey N. Weiser, Department of Microbiology and Pediatrics, University of Pennsylvania, 402A Johnson Pavilion, Philadelphia, PA 19104-6076.

The Pneumococcus. Volume Editor, Elaine I. Tuomanen,

mucosa, remain unanswered. Recently, experimental carriage studies performed in healthy adults have offered the prospect of utilizing the natural host to further investigate this fundamental aspect of pneumococcal biology (49).

Requirements of Colonization versus Infection

The ability of the pneumococcus to colonize the mucosal surface of the nasopharynx without apparent inflammation and to cause disease at other sites correlates with the varied expression and selection of certain cell surface properties under these different conditions. This implies that there are certain characteristics that promote carriage but are not permissive for survival during the inflammatory response that accompanies infection. It has been suggested that the characteristics that enable the organism's "invasive" behavior may somehow enhance persistence during carriage with disease, a relatively rare consequence which, in the case of the pneumococcus, does not enhance transmission. A further consideration is that the characterization of variably expressed pneumococcal surface features may provide insight into the specific molecular requirements of colonization. For the majority of isolates, two states are easily differentiated by their opaque and transparent colony morphologies when viewed on transparent medium with oblique, transmitted illumination (85). The transparent colony form has been shown to be better adapted for colonization when tested in several animal models (76, 84). Moreover, when type-matched pneumococci are isolated from both the nasopharynx and blood from the same individual and directly compared, the predominant phenotypes are transparent and opaque, respectively (81). A number of bacterial characteristics vary in association with colony opacity through a switching mechanism that remains unknown. For transparent pneumococci, there is a decreased density of capsular polysaccharide and increased amount of teichoic acid in comparison to those for the opaque variant of the same strain (43). In addition, colony variants differ in relative expression of surface proteins such as the autolysin, LytA. The contributions to colonization of many of these variably expressed cell surface characteristics are reviewed in this chapter.

Regulation of Gene Expression

In contrast to the stochastic changes associated with opacity variation, other studies demonstrate that successful colonization requires regulatory systems to control expression of certain genes. For instance, the *ciaRH* two-component signal transduction system is required for efficient nasopharyngeal colonization in an animal model, suggesting a role in carriage for sensing environmental signals and increased expression of genes such as *htrA*, a putative serine protease gene, regulated by this locus (63). Other genes regulated by the *ciaRH* system are also involved in quorum sensing associated with the induction of competence for genetic transformation. However, the precise nature of the contribution of quorum sensing to colonization has not yet been determined.

Transmission

Finally, the ultimate success of the organism in colonization of a population will also depend on the efficiency of human-to-human transmission. Specific characteristics of the pneumococcus that facilitate its transmission, which is thought to occur through intimate contact between individuals or exposure to large-droplet secretions, have not been explored (54).

ADHERENCE TO HOST STRUCTURES

One of the critical steps in colonization is bacterial attachment to host cells or tissues. Most studies of relevance to colonization have focused on the adherence of the pneumococcus to epithelial cells, since there is little evidence to suggest that the organism interacts with mucous layer components such as mucin or other host cell types during carriage (19). Analysis of adherence to cells in culture has led to the identification of several putative cellular adhesins. However, no single bacterial structure appears to account completely for adherence, suggesting that multiple receptor-ligand interactions contribute to pneumococcal attachment in these assays.

Pneumococcal Adhesins

Surface molecules that have been shown to function as adhesins to human epithelial cells include phosphorylcholine (ChoP) and CbpA (15, 61). ChoP, an otherwise unusual prokaryotic structural component, is common to several other genera residing primarily in the upper respiratory tract, such as *Haemophilus*, *Actinobacillus*, *Mycoplasma*, and *Neisseria* (83). Since many ChoP-expressing species occupy a similar biological niche, this observation suggests that ChoP may be a particularly important factor in their interaction with their host during carriage. In *Haemophilus influenzae*, ChoP expression appears to promote persistence in the airway (86). For the pneumococcus as well as other species using choline for structural synthesis, the incorporation of choline from environmental sources and its synthesis as ChoP linked to the cell surface requires the *licA* through *licD* genes (64, 87, 93). ChoP, a feature common to both the cell wall-associated and lipoteichoic acids, has been shown to be necessary for pneumococcal binding to the receptor for platelet-activating factor (rPAF) (15, 26). Since the natural ligand for this receptor also contains ChoP, this suggests that the pneumococcus mimics PAF to utilize its receptor. Other ChoP-expressing species are now known to bind to this same receptor, which is widely distributed on host tissues, including the epithelial surface of the human nasopharynx (29). ChoP is present in larger amounts on transparent pneumococci, which could account for the increased adherence and more efficient colonization of this variant (43). The specific in vivo contribution of the ChoP-rPAF interaction to colonization, however, has been difficult to establish because pneumococci are relatively growth deficient in the absence of choline and colonization of rPAF knockout mice has not yet been assessed.

CbpA (also referred to as PspC or SpsA), another putative adhesin, is one of a family of surface proteins referred to as choline-binding proteins that are noncovalently anchored to ChoP (10, 34, 61). Therefore, factors such as opacity associated with alterations in the expression of ChoP are likely to have secondary effects on CbpA as well as other choline-binding proteins. As noted above, CbpA binds to secretory component, which is found on the polymeric immunoglobulin receptor and secretory forms of IgA and IgM on mucosal surfaces. Carriage of a CbpA mutant is reduced by 100-fold compared to that of its parent in a mouse model (61). Since the binding of CbpA appears to be highly specific for the human but not murine form of secretory component, this result suggests that other functions of the protein may be important in vivo (92). In this regard, the binding of CbpA to immobilized sialic acid, lacto-*N*-neotetraose, and, in a separate study, complement component C3 on the surface of epithelial cells has been proposed as an additional function for this pneumococcal protein (61, 66).

Glycoconjugate Receptors

Other studies provide evidence for adhesive interactions with a number of glycoconjugates on host cells or tissues. Pneumococcal adherence to epithelial cells obtained directly from the human pharynx is inhibited by *N*-acetylglucosamine-β-1-3-galatose (GlcNacβ1→3Galβ) (1). A portion of the adhesive interaction with human lung epithelial cells in culture (type II pneumocytes) is also inhibited by micromolar amounts of either GalNAcβ1→4Gal or GalNAcβ1→3Gal in a nonadditive manner (16). Further development of antiadhesive strategies revealed that binding of clinical isolates to a conjunctival cell line and primary human bronchial epithelial cells was inhibited preferentially by sialylated oligosaccharides that terminate with the disaccharide NeuAc α2–3(or 6)Galβ1 and weakly by oligosaccharides that terminate with lactosamine (Galβ1-4GlcNAcβ1) (5). The pneumococcus is also one of many pathogens reported to bind to *N*-acetylglucosamine-β-1-

4-galactose (GlcNacβ1→4Galβ) found on some human glycosphingolipids (44). Although such glycosphingolipids are present within the respiratory tract, the in vivo relevance of this later study has not been further explored.

Adherence to ECM Components

There is, moreover, evidence that binding to extracellular matrix (ECM) components may be important once the organism gains access to basement membranes. The expression of a surface-attached hyaluronidase (a hyaluronate lyase), Hyl, which could facilitate spread through a matrix of hyaluronan, a major polysaccharide component of host connective tissues, suggests that such a strategy may contribute to pneumococcal pathogenesis (41). PavA and Eno, specific surface adhesins that bind ECM components fibronectin and plasminogen, respectively, have been described (6, 36, 78). Although pavA contributes to virulence in a murine model of sepsis, there is no information concerning a role in colonization of any pneumococcal proteins that interact with ECM components.

Inhibitory Effects of Capsule

Adherence to host structures may be particularly problematic for an encapsulated organism like the pneumococcus. Some degree of encapsulation appears to be essential for colonization, although even small amounts of capsular polysaccharide effectively block attachment to host cells (46, 60). Many of the adhesive interactions described above have been documented only with nonencapsulated mutants. In the nasopharynx, the inhibitory effect on adherence of the negatively charged capsule may be mitigated by the binding of cationic host molecules (27, 82). For example, type-specific IgA1 markedly enhances rather than inhibits bacterial attachment to host epithelial cells in culture. This event is dependent on cleavage by IgA1 protease, resulting in the release of cationic Fab_α which retains antigen binding properties (82). Coating of pneumococci with anti-capsular polysaccharide Fab_α appears to unmask the bacterial ChoP ligand,

allowing for increased adherence mediated by binding to rPAF.

METABOLIC AND PHYSIOLOGICAL CONSIDERATIONS

The pneumococcus is an aerotolerant, fermentative organism with complex growth requirements. Although the sources of its energy in vivo are not known, it appears to rely extensively on sugar metabolism for its nutrition. Of its genes that are similar to genes for known transporters, over 30% are predicted to encode sugar transporters, among the most of any prokaryote whose entire genome has been sequenced (73). Its ability to utilize a large number of different sugars has been proposed as a mechanism allowing it to exist in its niche in the nasopharynx. Despite its generally fermentative metabolism, in an oxygen-rich environment pyruvate generated through the Embden-Meyerhof-Parnas pathway will be oxidized to acetate (rather than converted to lactate as occurs under anaerobic conditions), allowing for additional ATP generation. The relative importance of this pathway is demonstrated by the requirement for pyruvate oxidase (SpxB) during nasopharyneal colonization in an animal model even though this enzyme does not provide a significant advantage for growth in vitro (58, 68).

Oxidative Stress

The oxygen-rich environment of the airway presents a challenge to the survival of the organism. As a by-product of the aerobic metabolism of pyruvate by SpxB, the pneumococcus, like other oral streptococci, generates levels of H_2O_2 as high as 2 mM in its culture supernatant under routine in vitro growth conditions (58, 68). This level of H_2O_2, which is equivalent to that generated by neutrophils, has been shown to be highly toxic to other species because it is converted to the highly reactive hydroxyl radical in the presence of Fe^{2+} (via the Fenton reaction) (39). Endogenously generated H_2O_2 may contribute to a high rate of spontaneous mutation in S. pneumoniae due to oxidative damage (57). The ability of the pneumococcus to take up DNA fragments and to incorporate homologous sequences is observed only during aerobic growth, when it can provide a means of correcting mutations that may result, at least in part, from its oxidative lifestyle (21). The pneumococcus expresses an Mn^{2+} superoxide dismutase for avoiding the effects of superoxide, but it lacks catalase as well as many of the key mechanisms known in other bacterial species to permit survival under oxidative stress. Optimal growth in vitro requires an exogenous source of catalase that is usually provided by supplementation of the medium with erythrocytes. It is unknown how the pneumococcus avoids the lethal effects of oxidative stress or whether the host provides a source of H_2O_2-neutralizing catalase on the mucosal surface. Another potentially important consequence of growth under atmospheric levels of oxygen is diminished expression of the capsular polysaccharide (81). This decrease in amounts of capsular polysaccharide triggered by increasing levels of environmental oxygen correlates with inhibition of tyrosine phosphorylation of the capsular polysaccharide regulatory gene, *cpsD*. The down-regulation of capsule expression in the presence of oxygen, which appears to be common to many pneumococcal types, may be more permissive for attachment of the organism to host structures during colonization (see "Adherence to host structures" above).

Acquisition of Iron

Another well-recognized challenge to mucosal organisms is the acquisition of iron, an essential micronutrient that is sequestered by the host and available in its free form at levels generally insufficient for bacterial growth. The pneumococcus does not secrete a siderophore, although several iron uptake ABC transporters (Pit, Piu, and Pia) have been identified and shown to be necessary for survival in vivo and in growth conditions of reduced iron availability (11, 12). It has been suggested that PiuA is a lipoprotein that binds hemin and hemoglobin and that these substrates can re-store pneumococcal growth under iron-limiting conditions (72). In addition, the iron-containing glycoprotein lactoferrin, one of the chief sources of iron on mucosal surfaces, is bound by pneumococcal surface protein A (PspA) (33). There is no evidence, however, that the pneumococcus is able to utilize the abundance of lactoferrin (or transferrin) as a source of iron. Rather, it is proposed that the interaction with PspA circumvents the antimicrobial activities of lactoferrin (32a).

Requirement for Choline

As previously discussed, one distinguishing feature of normal pneumococcal growth is the need for choline for use in teichoic acid biosynthesis. Choline stores in the host are found predominantly in membranes in the form of phosphatidylcholine. *H. influenzae*, which is also able to incorporate choline using a similar pathway (described above), has evolved a mechanism for removal of choline from the degradation products of host membrane lipids, suggesting that amounts of free choline on the mucosal surface may be limiting (24). *S. pneumoniae* contains several putative choline permeases, including LicB, but the source(s) of this unusual bacterial nutrient in vivo has not been established (25).

INTERACTIONS WITH INNATE HOST DEFENSES

The success of the pneumococcus in colonization indicates that it has effective mechanisms for mitigating innate host defense mechanisms that protect the upper respiratory tract. For example, one of the major factors contributing to the antibacterial activity of airway surface fluid is the abundance of the murein hydrolase lysozyme. The pneumococcus, however, expresses a peptidoglycan N-acetylglucosamine deacetylase (PgdA) which leaves many glucosamine residues of the pneumococcal peptidoglycan unacetylated and thus resistant to lysozyme (79). Other well-recognized components of host defense of the airway such as surfactant, which includes proteins with antimicrobial properties, and alveolar macrophages are found in the lower but not upper respiratory tract.

Evasion of Complement

The complement system is a major component of innate immunity. Available evidence points to roles for both the alternative and classical pathways of complement activation in innate immunity against *S. pneumoniae*. Pneumococcal cell wall and lipoteichoic acids have been noted to activate directly the alternative complement pathway, although these structures may be effectively obscured by the capsular polysaccharide (38, 90). It has recently been suggested that although the alternative pathway is typically associated with innate immunity, the classical pathway may be the more dominant pathway through interaction with naturally acquired IgM and C-reactive protein (CRP) (see below) (13). Complement components have been detected on the surface of the airway. Complement activity on mucosal surfaces, however, is poorly characterized (2). Nonetheless, a number of circumstantial observations suggest that evasion of the antimicrobial effects of complement is a requirement for successful carriage. Capsule, a well-established virulence factor necessary to inhibit opsonization and phagocytosis due to the deposition of complement (and antibody) during invasive infection, also appears to be important during colonization. In a murine model, expression of at least some capsular polysaccharide was found to be essential for colonization (46). Other major surface constituents such as PspA have also been shown to diminish the effects of complement, although PspA may be dispensable during colonization in part because of overlapping function with other choline-binding proteins. These include CbpA and in some strains a homologous factor H-binding protein, Hic (4, 40, 77). Other pneumococcal surface components may function to enzymatically inactivate complement components (3, 94). Although these characteristics have enabled the organism to survive the more potent complement system found in serum, it seems unlikely that the pneumococcus developed a highly evolved strategy to evade complement if this did not contribute to its survival on the mucosal surface.

CRP

Of particular relevance to the pneumococcus is the host molecule CRP. CRP is an acute-phase reactant that interacts specifically with ChoP and, when bound, activates complement through C1q and the classical pathway (70). Mice, which express constitutively low levels of CRP, express increased levels of this protein and are more resistant to invasive pneumococcal infection when carrying the human transgene for CRP (71). CRP had been considered exclusively as a serum component, but it has recently been shown to be expressed by the epithelium of the human nasopharynx and to be present in airway surface fluid at significant concentrations, even in the absence of an inflammatory stimulus (28). In addition to its role in promoting complement deposition, CRP at the concentrations found in the human airway has been shown to block adherence that is mediated by ChoP binding to rPAF on epithelial cells in culture (29). If similar events occur in vivo, CRP may be a component of innate immunity that targets ChoP-expressing bacteria such as the pneumococcus and limits their colonization.

Antimicrobial Peptides

Another innate defense mechanism of the airway involves small peptides with broad antimicrobial activity. At least three defensins as well as the cathelicidin LL-37 have been identified in the human respiratory tract. Although the antipneumococcal activity of these molecules has not been extensively studied, it would seem likely that the organism has evolved the ability to evade their protective effects. In this regard, other defensins have been noted to stimulate adherence of several mucosal pathogens, including *S. pneumoniae*, to human epithelial cells, such as the NCI-H292 cell line, in culture (27).

Inflammatory Signaling

Pneumococcal components such as peptido-glycan/teichoic acid and pneumolysin have been reported to activate immune cells via toll-like receptors 2 and 4 (and MD2), respectively (47, 62). Moreover, TLR4 mutant mice were significantly more susceptible to lethal infection than controls following intranasal colonization with pneumolysin-positive pneu-mococci (47). This might indicate a role for toll-like receptor-mediated innate immune responses in resistance to the spread of bacteria during colonization. On the other hand, observations that the organism does not appear to stimulate inflammation during natural colonization suggest that such responses may be limited on the epithelial surface of the human nasopharynx.

EVASION OF ADAPTIVE IMMUNITY

Another challenge faced by the pneumococcus during carriage is the protective effect of host immunoglobulins secreted onto the mucosal surface. One of the capsule's key functions in enhancing colonization may be that although it is itself immunogenic, it obscures underlying protein and cell wall antigens. Although colonization is thought to cause an increase in type-specific serum antibody, it remains unclear whether carriage is an immunizing event for a given pneumococcal type (31, 55, 67). Several studies have suggested that vaccination with capsular polysaccharide results in a reduction in nasopharyngeal carriage in children (17, 18, 48). How, then, might the pneumococcus evade the protective effect of naturally acquired type-specific antibody? Diminished carriage following immunization has been associated primarily with the induction of serum IgG, whereas natural carriage results in a prominently mucosal IgA response. As noted above, the pneumococcus is one of many respiratory tract pathogens that expresses an IgA1 protease that cleaves human IgA1. This leaves its surface antigens blocked by Fab_α fragments without triggering inflammation through host recognition of bound Fc_α. Since >90% of the IgA in the human nasopharynx is of this sub-class, this characteristic of the organism may allow it to evade the protective effective of this aspect of the mucosal immune response. Clearance may occur only when sufficient amounts of other classes and subclasses of type-specific antibody are generated. The ability of the pneumococcus to inactivate IgA1 in normal hosts could account for the generally unaltered incidence of infection observed in most IgA-deficient individuals.

Although there has been a focus on capsular polysaccharide as the immunodominant surface structure, there is now considerable evidence that other surface components are sufficiently antigenic to trigger an immune response leading to diminished carriage. Immunization with several proteins, including PsaA, CbpA, and PspA, has been shown to elicit a protective immune response against nasopharyngeal carriage in mice (4, 9, 42). Experimental human carriage studies showed that following colonization there was a significant rise in levels of serum IgG to PspA and CbpA but not PsaA or other surface structures tested, including type-specific capsular polysaccharide (50). In this study, the susceptibility of the subjects to become colonized was associated with lower preinoculation levels of antibody to PspA, providing further evidence of the key role of immunity to this protein during carriage (49). The potentially protective immune response, however, appeared to be directed to highly variable domains in the antigenically variable PspA (and CbpA) (50). This suggests that, like other successful pathogens, the pneumococcus evades adaptive immunity through strain-to-strain heterogeneity in surface structures, in particular those that are most immunogenic during carriage. These antigenically variable components are now known to include capsular polysaccharide, PspA, and CbpA. The ability of the population of pneumococci to vary these structures at a sufficient frequency in response to immune pressure may depend on a number of factors. These factors may include efficient horizontal gene transfer between simultaneously carried strains mediated by natural competence and generally high rates of spontaneous mutation. Evidence for the former is found in the evolution of genes and genomic regions as occurs in the switching of capsular type, while the latter has been observed in *pspA* during experimental human carriage (49, 53).

COMPETITION WITH OTHER MEMBERS OF THE MICROFLORA

In contrast to the lower respiratory tract, which is generally considered to be sterile in the absence of disease, the upper respiratory tract of humans is heavily colonized. Recent evidence based on analysis of DNA obtained from airway surface fluids suggests that there may be >500 different species inhabiting the human pharynx (56). Only a relatively small proportion of these organisms have been cultured and characterized. Although it is not known how many of these species compete for the same niche as the pneumococcus, the density and diversity of the microflora would suggest that success at colonization of this mucosal surface would require adaptive mechanisms for dealing with coexisting microorganisms.

Most of the studies addressing the question of how the pneumococcus interacts with other members of the upper respiratory tract flora have focused on other opportunistic pathogens that occupy a similar niche rather than members of the commensal flora. One reason for understanding these interactions is the possibility that manipulation of the microflora through the use of antibiotics or vaccination might exacerbate problems caused by species that are in competition with the pneumococcus. For instance, a trial of the pneumococcal conjugate vaccine in children showed that immunization was associated with a decrease in the incidence of otitis media for serotypes included in the vaccine. The recipients of the vaccine, however, were more likely to experience otitis media attributed to *H. influenzae*, raising the possibility that diminished carriage of the pneumococcus following vaccination allowed for an increased presence of another opportunistic pathogen (22).

Common Pathways for Colonization

In fact, it is not surprising that the pneumococcus may compete with species such as *H. influenzae*, also a common resident of the human nasopharynx and a source of a similar set of diseases involving the respiratory tract. It is now appreciated that *S. pneumoniae* and *H. influenzae* may share a similar niche because of common evolutionary strategies affecting interaction with the human host. Although distantly related (one is a gram-positive species and the other a gram-negative species), both organisms express ChoP on their cell surface using a highly homologous set of *lic* genes that may have been exchanged at some time in the past between these naturally competent species (87, 93). In fact, whole-genome comparisons reveal that other than genes widely shared among bacteria, the *lic* genes are some of the few shared by these otherwise dissimilar pathogens that occupy the same microenvironment (83). For both pathogens, expression of cell surface ChoP enables adhesive interactions with the same receptor (rPAF) (15, 69). In addition, a number of other genera residing primarily in the respiratory tract express this cell surface feature, which is an otherwise unusual structural component in bacteria (83). These observations suggest that common mechanisms, such as the ability to bind particular host receptors, could be a source of competition between species.

Bacterium-Bacterium Interactions

In addition to competitive interactions that involve their host, it also appears that the pneumococcus has the capacity to act directly on species sharing its environment. When grown in coculture, there is a bactericidal effect of the pneumococcus on *H. influenzae* observed with as few as 10^6 pneumococci/ml (58). This effect, caused by the pneumococcal production of hydrogen peroxide, is sufficient to overwhelm the protective activity of catalase expressed by potential competitors, such as members of the genera *Haemophilus*, *Neisseria*, and *Moraxella*, resulting in inhibition of growth or death in vitro. It is unknown whether there is a similar effect in vivo, since the *spxB* mutant of *S. pneumoniae*, which produces only minimal amounts of H_2O_2, is also rapidly cleared from the nasopharynx in animal models of carriage. This observation could be explained by either an inability to handle competitors or decreased metabolic fitness associated with the loss of the pyruvate oxidase (68). A second way in which the pneumococcus directly impacts its potential competitors is the expression of neuraminidase on its surface. NanA, one of at least two pneumococcal neuraminidases, is able to remove terminal sialic acid residues from host cell substrates and has been demonstrated to promote colonization, possibly by unmasking receptors or facilitating the utilization of sialic acid as a nutrient (7, 74, 75). In addition, sialic acid obtained from the host is used to decorate terminal lipopolysaccharide structures on *H. influenzae* and *Neisseria meningitidis*, where it is thought to contribute to survival through molecular mimicry (8, 23, 37). Coculture of either of these species with *S. pneumoniae* expressing NanA results in loss of sialic acid from their surfaces (65). Finally, whole-genome analysis has revealed that the pneumococcus contains a gene cluster encoding putative double glycine-type bacteriocins and associated immunity proteins which generally function in intraspecies competition (20). While these putative bacteriocins have not yet been analyzed for their antimicrobial activity or specificity, there is evidence in animal models of carriage of competition among strains of pneumococci (45). In humans, however, simultaneous colonization with multiple strains of different pneumococcal types is not uncommon during childhood (30).

The characteristics of the pneumococcus described above may promote the ability of the organism to effectively compete on the heavily colonized surface of the human pharynx. It is also possible that there may be cooperative interactions that are permissive for pneumococcal survival in this environment as well. For example, many of the species colonizing the human nasopharynx secrete IgA1 proteases that cleave IgA1 within its hinge region. The IgA1 protease of *H. influenzae*, for example,

has been shown to modify IgA1 so as to increase adherence of an *S. pneumoniae* strain deficient in IgA1 protease activity (82).

Effects of Viral Infection

A final consideration is that pneumococcal infection frequently occurs in the setting of a recent or concurrent upper respiratory infection from common viruses. Although the precise mechanisms for this clinically important phenomenon are not yet understood, experimental evidence suggests that an antecedent syndrome such as occurs with influenza facilitates pneumococcal colonization, possibly by increasing adherence or through the selection of variants more likely to cause infection (32, 76). Viral infection could augment pneumococcal attachment by a direct effect on host tissues or by triggering cytokine release and subsequent up-regulation of receptors (14, 51, 52). An example of the former is the neuraminidase of influenza virus, which may serve to expose pneumococcal receptors that are otherwise inaccessible because of sialylation (52). Alternatively, higher rates of pneumococcal infection during a viral syndrome could result from inhibition of the normal clearance mechanisms of the airway or more efficient transmission between hosts.

REFERENCES

1. **Andersson, B., J. Dahmen, T. Frejd, H. Leffler, G. Magnusson, G. Noori, and C. S. Eden.** 1983. Identification of an active dissaccharide unit of a glycoconjugate receptor for pneumococci attaching to human pharyngeal epithelial cells. *J. Exp. Med.* **158:**559–570.
2. **Andoh, A., Y. Fujiyama, T. Kimura, H. Uchihara, H. Sakumoto, H. Okabe, and T. Bamba.** 1997. Molecular characterization of complement components (C3, C4, and factor B) in human saliva. *J. Clin. Immunol.* **17:**404–407.
3. **Angel, C., M. Ruzek, and M. Hostetter.** 1994. Degradation of C3 by *Streptococcus pneumoniae. J. Infect. Dis.* **170:**600–608.
4. **Balachandran, P., A. Brooks-Walter, A. Virolainen-Julkunen, S. Hollingshead, and D. Briles.** 2002. Role of pneumococcal surface protein C in nasopharyngeal carriage and pneumonia and its ability to elicit protection against

carriage of *Streptococcus pneumoniae. Infect. Immun.* **70:**2526–2534.

5. **Barthelson, R., A. Mobasseri, D. Zopf, and P. Simon.** 1998. Adherence of *Streptococcus pneumoniae* to respiratory epithelial cells is inhibited by sialylated oligosaccharides. *Infect. Immun.* **66:** 1439–1444.

6. **Bergmann, S., M. Rohde, G. Chhatwal, and S. Hammerschmidt.** 2001. α-Enolase of *Streptococcus pneumoniae* is a plasmin(ogen)-binding protein displayed on the bacterial cell surface. *Mol. Microbiol.* **40:**1273–1287.

7. **Berry, A., R. Lock, and J. Paton.** 1996. Cloning and characterization of *nanB*, a second *Streptococcus pneumoniae* neuraminidase gene, and purification of the NanB enzyme from recombinant *Escherichia coli. J. Bacteriol.* **178:**4854–4860.

8. **Bouchet, V., D. Hood, J. Li, J. Brisson, G. Randle, A. Martin, Z. Li, R. Goldstein, E. Schweda, S. Pelton, J. Richards, and E. Moxon.** 2003. Host-derived sialic acid is incorporated into *Haemophilus influenzae* lipopolysaccharide and is a major virulence factor in experimental otitis media. *Proc. Natl. Acad. Sci. USA* **100:**8898–8903.

9. **Briles, D. E., E. Ades, J. C. Paton, J. S. Sampson, G. M. Carlone, R. C. Huebner, A. Virolainen, E. Swiatlo, and S. K. Hollingshead.** 2000. Intranasal immunization of mice with a mixture of the pneumococcal proteins PsaA and PspA is highly protective against nasopharyngeal carriage of *Streptococcus pneumoniae. Infect. Immun.* **68:**796–800.

10. **Brooks-Walter, A., D. E. Briles, and S. K. Hollingshead.** 1999. The *pspC* gene of *Streptococcus pneumoniae* encodes a polymorphic protein, PspC, which elicits cross-reactive antibodies to PspA and provides immunity to pneumococcal bacteremia. *Infect. Immun.* **67:**6533–6542.

11. **Brown, J., S. Gilliland, and D. Holden.** 2001. A *Streptococcus pneumoniae* pathogenicity island encoding an ABC transporter involved in iron uptake and virulence. *Mol. Microbiol.* **40:**572–585.

12. **Brown, J. S., S. M. Gilliland, J. Ruiz-Albert, and D. W. Holden.** 2002. Characterization of Pit, a *Streptococcus pneumoniae* iron uptake ABC transporter. *Infect. Immun.* **70:**4389–4398.

13. **Brown, J. S., T. Hussell, S. M. Gilliland, D. W. Holden, J. C. Paton, M. R. Ehrenstein, M. J. Walport, and M. Botto.** 2002. The classical pathway is the dominant complement pathway required for innate immunity to *Streptococcus pneumoniae* infection in mice. *Proc. Natl. Acad. Sci. USA* **99:**16969–16974.

14. **Cundell, D., C. Gerard, I. Idanpaan-Heikkila, E. Tuomanen, and N. Gerard.** 1996. PAF receptor anchors *Streptococcus pneumoniae* to activated human endothelial cells. *Adv. Exp. Med. Biol.* **416:**89–94.

15. **Cundell, D. R., N. P. Gerard, C. Gerard, I. Idanpaan-Heikkila, and E. I. Tuomanen.** 1995. *Streptococcus pneumoniae* anchor to activated human cells by the receptor for platelet-activating factor. *Nature* **377:**435–438.

16. **Cundell, D. R., and E. I. Tuomanen.** 1994. Receptor specificity of adherence of *Streptococcus pneumoniae* to human type II pneumocytes and vascular endothelial cells in vitro. *Microb. Pathog.* **17:**361–374.

17. **Dagan, R., N. Givon-Lavi, O. Zamir, M. Sikuler-Cohen, L. Guy, J. Janco, P. Yagupsky, and D. Fraser.** 2002. Reduction of nasopharyngeal carriage of *Streptococcus pneumoniae* after administration of a 9-valent pneumococcal conjugate vaccine to toddlers attending day care centers. *J. Infect. Dis.* **185:**927–936.

18. **Dagan, R., M. Muallem, R. Melamed, O. Leroy, and P. Yagupsky.** 1997. Reduction of pneumococcal nasopharyngeal carriage in early infancy after immunization with tetravalent pneumococcal vaccines conjugated to either tetanus toxoid or diphtheria toxoid. *Pediatr. Infect. Dis. J.* **16:**1060–1064.

19. **Davies, J., I. Carlstedt, A.-K. Nilsson, A. Hakansson, H. Sabharwal, L. van Alphen, M. van Ham, and C. Svanborg.** 1995. Binding of *Haemophilus influenzae* to purified mucins from the human respiratory tract. *Infect. Immun.* **63:**2485–2492.

20. **de Saizieu, A., C. Gardes, N. Flint, C. Wagner, M. Kamber, T. Mitchell, W. Keck, K. Amrein, and R. Lange.** 2000. Microarray-based identification of a novel *Streptococcus pneumoniae* regulon controlled by an autoinduced peptide. *J. Bacteriol.* **182:**4696–4703.

21. **Echenique, J., and M. Trombe.** 2001. Competence repression under oxygen limitation through the two-component MicAB signal-transducing system in *Streptococcus pneumoniae* and involvement of the PAS domain of MicB. *J. Bacteriol.* **183:**4599–4608.

22. **Eskola, J., T. Kilpi, A. Palmu, J. Jokinen, J. Haapakoski, E. Herva, A. Takala, H. Kayhty, P. Karma, A. Kohberger, G. Siber, P.H. Makela, and the Finnish Otitis Media Study Group.** 2001. Efficacy of a pneumococcal conjugate vaccine against acute otitis media. *N. Engl. J. Med.* **344:**403–409.

23. **Estabrook, M. M., J. M. Griffiss, and G. A. Jarvis.** 1997. Sialylation of *Neisseria meningitidis* lipooligosaccharide inhibits serum bactericidal activity masking lacto-*N*-neotraose. *Infect. Immun.* **65:**4436–4444.

24. Fan, X., H. Goldfine, E. Lysenko, and J. Weiser. 2001. The transfer of choline from the host to the bacterial cell surface requires glpQ in *Haemophilus influenzae*. *Mol. Microbiol.* **41:**1029–1036.

25. Fan, X., C. D. Pericone, E. Lysenko, H. Goldfine, and J. N. Weiser. 2003. Multiple mechanisms for choline transport and utilization in *Haemophilus influenzae*. *Mol. Microbiol.* **50:** 537–548.

26. Fischer, W., T. Behr, R. Hartmann, K. C. J. Peter, and H. Egge. 1993. Teichoic acid and lipoteichoic acid of *Streptococcus pneumoniae* possess identical chain structures. A reinvestigation of teichoid acid (C polysaccharide). *Eur. J. Biochem.* **215:**851–857.

27. Gorter, A., P. Hiemstra, S. de Bentzmann, S. van Wetering, J. Dankert, and L. van Alphen. 2000. Stimulation of bacterial adherence by neutrophil defensins varies among bacterial species but not among host cell types. *FEMS Immunol. Med. Microbiol.* **28:**105–111.

28. Gould, J., and J. Weiser. 2001. Expression of C-reactive protein in the human respiratory tract. *Infect. Immun.* **69:**1747–1754.

29. Gould, J., and J. Weiser. 2002. The inhibitory effect of C-reactive protein on bacterial phosphorylcholine-platelet activating factor receptor mediated adherence is blocked by surfactant *J. Infect. Dis.* **186:**361–371.

30. Gundel, M. 1933. Bakteriologische und Epidemiologische Untersuchungen über die Besiedlung der oberen Atmungswege Gesunder mit Pneumokokken. *Z. Hyg. Infektionskr.* **114:**659–704.

31. Gwaltney, J. J., M. Sande, R. Austrian, and J. Hendley. 1975. Spread of *Streptococcus pneumoniae* in families. II. Relation of transfer of *S. pneumoniae* to incidence of colds and serum antibody. *J. Infect. Dis.* **132:**62–68.

32. Hakansson, A., A. Kidd, G. Wadell, H. Sabharwal, and C. Svanborg. 1994. Adenovirus infection enhances in vitro adherence of *Streptococcus pneumoniae*. *Infect. Immun.* **62:**2707 2714.

32a. Hakansson, A., H. Roche, S. Mirza, L. S. McDaniel, A. Brooks-Walter, and D. E. Briles. 2001. Characterization of binding of human lactoferrin to pneumococcal surface protein A. *Infect. Immun.* **69:**3372–3381.

33. Hammerschmidt, S., G. Bethe, P. H. Remane, and G. S. Chhatwal. 1999. Identification of pneumococcal surface protein A as a lactoferrin-binding protein of *Streptococcus pneumoniae*. *Infect. Immun.* **67:**1683–1687.

34. Hammerschmidt, S., S. R. Talay, P. Brandtzaeg, and G. S. Chhatwal. 1997. SpsA, a novel pneumococcal surface protein with specific binding to secretory immunoglobulin A and secretory component. *Mol. Microbiol.* **25:**1113–1124.

35. Hammerschmidt, S., M. Tillig, S. Wolff, J. Vaerman, and G. Chhatwal. 2000. Species-specific binding of human secretory component to SpsA protein of *Streptococcus pneumoniae* via a hexapeptide motif. *Mol. Microbiol.* **36:**726–736.

36. Holmes, A., R. McNab, K. Millsap, M. Rohde, S. Hammerschmidt, J. Mawdsley, and H. Jenkinson. 2001. The pavA gene of *Streptococcus pneumoniae* encodes a fibronectin-binding protein that is essential for virulence. *Mol. Microbiol.* **41:**1395–1408.

37. Hood, D., K. Makepeace, M. Deadman, R. Rest, P. Thibault, A. Martin, J. Richards, and E. Moxon. 1999. Sialic acid in the lipopolysaccharide of Haemophilus influenzae: strain distribution, influence on serum resistance and structural characterization. *Mol. Microbiol.* **33:**679–692.

38. Hummell, D. S., A. J. Swift, A. Tomasz, and J. A. Winkelstein. 1985. Activation of the alternative complement pathway by pneumococcal lipoteichoic acid. *Infect. Immun.* **47:** 384–387.

39. Imlay, J., and S. Linn. 1988. DNA damage and oxygen radical toxicity. *Science* **240:**1302–1309.

40. Janulczyk, R., F. Iannelli, A. Sjöholm, G. Pozzi, and L. Björck. 2000. Hic, a novel surface protein of *Streptococcus pneumoniae* that interferes with complement function. *J. Biol. Chem.* **275:**37257–37263.

41. Jedrzejas, M., L. Mello, B. de Groot, and S. Li. 2002. Mechanism of hyaluronan degradation by *Streptococcus pneumoniae* hyaluronate lyase. Structures of complexes with the substrate. *J. Biol. Chem.* **277:**28287–28297.

42. Johnson, S. E., J. K. Dykes, D. L. Jue, J. S. Sampson, G. M. Carlone, and E. W. Ades. 2002. Inhibition of pneumococcal carriage in mice by subcutaneous immunization with peptides from the common surface protein pneumococcal surface adhesin a. *J. Infect. Dis.* **185:**489–496.

43. Kim, J., and J. Weiser. 1998. Association of intrastrain phase variation in quantity of capsular polysaccharide and teichoic acid with the virulence of *Streptococcus pneumoniae*. *J. Infect. Dis.* **177:**368–377.

44. Krivan, H. C., D. D. Roberts, and V. Ginsberg. 1988. Many pulmonary pathogenic bacteria bind specifically to the carbohydrate sequence GalNAcβ1–4Gal found in some glycolipids. *Proc. Natl. Acad. Sci. USA* **85:**6157 6161.

45. Lipsitch, M., J. Dykes, S. Johnson, E. Ades, J. King, D. Briles, and G. Carlone. 2000. Competition among *Streptococcus pneumoniae* for

intranasal colonization in a mouse model. *Vaccine* **18:**2895–2901.

46. **Magee, A., and J. Yother.** 2001. Requirement for capsule in colonization by *Streptococcus pneumoniae. Infect. Immun.* **69:**3755–3761.

47. **Malley, R., P. Henneke, S. Morse, M. Cieslewicz, M. Lipsitch, C. Thompson, E. Kurt-Jones, J. Paton, M. Wessels, and D. Golenbock.** 2003. Recognition of pneumolysin by Toll-like receptor 4 confers resistance to pneumococcal infection. *Proc. Natl. Acad. Sci. USA* **100:**1966–1971.

48. **Mbelle, N., R. Huebner, A. Wasas, A. Kimura, I. Chang, and K. Klugman.** 1999. Immunogenicity and impact on nasopharyngeal carriage of a nonavalent pneumococcal conjugate vaccine. *J. Infect. Dis.* **180:**1171–1176.

49. **McCool, T. L., T. R. Cate, G. Moy, and J. N. Weiser.** 2002. The immune response to pneumococcal proteins during experimental human carriage. *J. Exp. Med.* **195:**359–365.

50. **McCool, T. L., T. R. Cate, E. I. Tuomanen, P. Adrian, T. J. Mitchell, and J. N. Weiser.** 2003. Serum immunoglobulin G response to candidate vaccine antigens during experimental human pneumococcal colonization. *Infect. Immun.* **71:**5724–5732.

51. **McCullers, J., and K. Bartmess.** 2003. Role of neuraminidase in lethal synergism between influenza virus and *Streptococcus pneumoniae. J. Infect. Dis.* **187:**1000–1009.

52. **McCullers, J., and J. Rehg.** 2002. Lethal synergism between influenza virus and *Streptococcus pneumoniae*: characterization of a mouse model and the role of platelet-activating factor receptor. *J. Infect. Dis.* **186:**341–350.

53. **Muller-Graf, C., A. Whatmore, S. King, K. Trzcinski, A. Pickerill, N. Doherty, J. Paul, D. Griffiths, D. Crook, and C. Dowson.** 1999. Population biology of *Streptococcus pneumoniae* isolated from oropharyngeal carriage and invasive disease. *Microbiology* **145:**3283–3293.

54. **Musher, D.** 2003. How contagious are common respiratory tract infections? *N. Engl. J. Med.* **348:**1256–1266.

55. **Musher, D., J. Groover, M. Reichler, F. Riedo, B. Schwartz, D. Watson, R. Baughn, and R. Breiman.** 1997. Emergence of antibody to capsular polysaccharides of *Streptococcus pneumoniae* during outbreaks of pneumonia: association with nasopharyngeal colonization. *Clin. Infect. Dis.* **24:**441–446.

56. **Paster, B., S. Boches, J. Galvin, R. Ericson, C. Lau, V. Levanos, A. Sahasrabudhe, and F. Dewhirst.** 2001. Bacterial diversity in human subgingival plaque. *J. Bacteriol.* **183:**3770–3783.

57. **Pericone, C., D. Bae, M. Shchepetov, T. McCool, and J. Weiser.** 2002. Short-sequence tandem and nontandem DNA repeats and endogenous hydrogen peroxide production contribute to genetic instability of *Streptococcus pneumoniae. J. Bacteriol.* **184:**4392–4399.

58. **Pericone, C. D., K. Overweg, P. W. M. Hermans, and J. N. Weiser.** 2000. Inhibitory and bactericidal effects of hydrogen peroxide production by *Streptococcus pneumoniae* on other inhabitants of the upper respiratory tract. *Infect. Immun.* **68:**3990–3997.

59. **Qiu, J., G. P. Brackee, and A. G. Plaut.** 1996. Analysis of the specificity of bacterial immunoglobulin A (IgA) proteases by comparative study of ape serum IgAs as substrate. *Infect. Immun.* **64:**933–937.

60. **Ring, A., J. N. Weiser, and E. I. Tuomanen.** 1998. Pneumococcal penetration of the blood-brain barrier: molecular analysis of a novel re-entry path. *J. Clin. Investig.* **102:**347–360.

61. **Rosenow, C., P. Ryan, J. N. Weiser, S. Johnson, P. Fontan, A. Ortqvist, and H. R. Masure.** 1997. Contribution of novel choline-binding proteins to adherence, colonization and immunogenicity of *Streptococcus pneumoniae. Mol. Microbiol.* **25:**819–829.

62. **Schroder, N., S. Morath, C. Alexander, L. Hamann, T. Hartung, U. Zahringer, U. Gobel, J. Weber, and R. Schumann.** 2003. Lipoteichoic acid (LTA) of *Streptococcus pneumoniae* and *Staphylococcus aureus* activates immune cells via Toll-like receptor (TLR)-2, lipopolysaccharide-binding protein (LBP), and CD14, whereas TLR-4 and MD-2 are not involved. *J. Biol. Chem.* **278:**15587–15594.

63. **Sebert, M. E., L. M. Palmer, M. Rosenberg, and J. N. Weiser.** 2002. Microarray-based identification of *htrA*, a *Streptococcus pneumoniae* gene that is regulated by the CiaRH two-component system and contributes to nasopharyngeal colonization. *Infect. Immun.* **70:**4059–4067.

64. **Serino, L., and M. Virji.** 2000. Phosphorylcholine decoration of lipopolysaccharide differentiates commensal Neisseriae from pathogenic strains: identification of licA-type genes in commensal Neisseriae. *Mol. Microbiol.* **35:**1550–1559.

65. **Shakhnovich, E., S. King, and J. Weiser.** 2002. Neuraminidase expressed by *Streptococcus pneumoniae* desialylates the lipopolysaccharide of *Neisseria meningitidis* and *Haemophilus influenzae*: a paradigm for interbacterial competition among pathogens of the human respiratory tract. *Infect. Immun.* **70:**7161–7164.

66. **Smith, B., and M. Hostetter.** 2000. C3 as substrate for adhesion of *Streptococcus pneumoniae. J. Infect. Dis.* **182:**497–508.

67. **Soininen, A., H. Pursiainen, T. Kilpi, and H. Kayhty.** 2001. Natural development of antibodies to pneumococcal capsular polysaccharides depends on the serotype: association with pneumococcal carriage and acute otitis media in young children. *J. Infect. Dis.* **184:**569–576.

68. **Spellerberg, B., D. R. Cundell, J. Sandros, B. J. Pearce, I. Idanpaan-Heikkila, C. Rosenow, and H. R. Masure.** 1996. Pyruvate oxidase, as a determinant of virulence in *Streptococcus pneumoniae. Mol. Microbiol.* **19:**803–813.

69. **Swords, W. E., B. A. Buscher, K. Ver Steeg Ii, A. Preston, W. A. Nichols, J. N. Weiser, B. W. Gibson, and M. A. Apicella.** 2000. Non-typeable *Haemophilus influenzae* adhere to and invade human bronchial epithelial cells via an interaction of lipooligosaccharide with the PAF receptor. *Mol. Microbiol.* **37:**13–27.

70. **Szalai, A. J., A. Agrawal, T. J. Greenhough, and J. E. Volanakis.** 1997. C-reactive protein. *Immunol. Res.* **16:**127–136.

71. **Szalai, A. J., D. E. Briles, and J. E. Volanakis.** 1995. Human C-reactive protein is protective against fatal *Streptococcus pneumoniae* infection in transgenic mice. *J. Immunol.* **155:**2557–2563.

72. **Tai, S. S., C. Yu, and J. K. Lee.** 2003. A solute binding protein of *Streptococcus pneumoniae* iron transport. *FEMS Microbiol. Lett.* **220:**303–308.

73. **Tettelin, H., K. E. Nelson, I. T. Paulsen, J. A. Eisen, T. R. Read, S. Peterson, J. Heidelberg, R. T. DeBoy, D. H. Haft, R. J. Dodson, A. S. Durkin, M. Gwinn, J. F. Kolonay, W. C. Nelson, J. D. Peterson, L. A. Umayam, O. White, S. L. Salzberg, M. R. Lewis, D. Radune, E. Holtzapple, H. Khouri, A. M. Wolf, T. R. Utterback, C. L. Hansen, L. A. McDonald, T. V. Feldblyum, S. Angiuoli, T. Dickinson, E. K. Hickey, I. E. Holt, B. J. Loftus, F. Yang, H. O. Smith, J. C. Venter, B. A. Dougherty, D. A. Morrison, S. K. Hollingshead, and C. M. Fraser.** 2001. Complete genome sequence of a virulent isolate of *Streptococcus pneumoniae. Science* **293:**498–506.

74. **Tong, H., L. Blue, M. James, and T. DeMaria.** 2000. Evaluation of the virulence of a *Streptococcus pneumoniae* neuraminidase-deficient mutant in nasopharyngeal colonization and development of otitis media in the chinchilla model. *Infect. Immun.* **68:**921–924.

75. **Tong, H., M. James, I. Grants, X. Liu, G. Shi, and T. DeMaria.** 2001. Comparison of structural changes of cell surface carbohydrates in the eustachian tube epithelium of chinchillas infected with a *Streptococcus pneumoniae* neu-raminidase-deficient mutant or its isogenic parent strain. *Microb. Pathog.* **31:**309–317.

76. **Tong, H., J. Weiser, M. James, and T. DeMaria.** 2001. Effect of influenza A virus infection on nasopharyngeal colonization and otitis media induced by transparent or opaque phenotypic variants of *Streptococcus pneumoniae* in the chinchilla model. *Infect. Immun.* **69:**602–606.

77. **Tu, A., R. Fulgham, M. McCrory, D. Briles, and A. Szalai.** 1999. Pneumococcal surface protein A inhibits complement activation by *Streptococcus pneumoniae. Infect. Immun.* **67:**4720–4724.

78. **van der Flier, M., N. Chhun, T. M. Wizemann, J. Min, J. B. McCarthy, and E. I. Tuomanen.** 1995. Adherence of *Streptococcus pneumoniae* to immobilized fibronectin. *Infect. Immun.* **63:**4317–4322.

79. **Vollmer, W., and A. Tomasz.** 2002. Peptidoglycan *N*-acetylglucosamine deacetylase, a putative virulence factor in *Streptococcus pneumoniae. Infect. Immun.* **70:**7176–7178.

80. **Wani, J., J. Gilbert, A. Plaut, and J. Weiser.** 1996. Identification, cloning, and sequencing of the immunoglobulin A1 protease gene of *Streptococcus pneumoniae. Infect. Immun.* **64:**3967–3974.

81. **Weiser, J., D. Bae, H. Epino, S. Gordon, M. Kapoor, L. Zenewicz, and M. Shchepetov.** 2001. Changes in availability of oxygen accentuate differences in capsular polysaccharide expression by phenotypic variants and clinical isolates of *Streptococcus pneumoniae. Infect. Immun.* **69:**5430–5439.

82. **Weiser, J., D. Bae, C. Fasching, R. Scamurra, A. Ratner, and E. Janoff.** 2003. Antibody-enhanced pneumococcal adherence requires IgA1 protease. *Proc. Natl. Acad. Sci. USA* **100:**415–420.

83. **Weiser, J., and E. Tuomanen.** 2002. A disease-oriented approach to the discovery of novel vaccine, p. 139–148. *In* B. Bloom and P.-H. Lambert (ed.), *The Vaccine Book.* Academic Press, New York, N.Y.

84. **Weiser, J. N.** 1999. Phase variation of *Streptococcus pneumoniae,* p. 225–231. *In* V. Fischetti (ed.), *Gram-Positive Pathogens.* ASM Press, Washington, D.C.

85. **Weiser, J. N., R. Austrian, P. K. Sreenivasan, and H. R. Masure.** 1994. Phase variation in pneumococcal opacity: relationship between colonial morphology and nasopharyngeal colonization. *Infect. Immun.* **62:**2582–2589.

86. **Weiser, J. N., N. Pan, K. L. McGowan, D. Musher, A. Martin, and J. C. Richards.** 1998. Phosphorylcholine on the lipopolysaccharide of *Haemophilus influenzae* contributes to persistence in the respiratory tract and sensitivity to serum

killing mediated by C-reactive protein. *J. Exp. Med.* **187:**631–640.

87. **Weiser, J. N., M. Shchepetov, and S. T. Chong.** 1997. Decoration of lipopolysaccharide with phosphorylcholine: a phase-variable characteristic of *Haemophilus influenzae. Infect. Immun.* **65:**943–950.

88. **Whatmore, A., A. Efstratiou, A. Pickerill, K. Broughton, G. Woodard, D. Sturgeon, R. George, and C. Dowson.** 2000. Genetic relationships between clinical isolates of *Streptococcus pneumoniae, Streptococcus oralis*, and *Streptococcus mitis*: characterization of "atypical" pneumococci and organisms allied to *S. mitis* harboring *S. pneumoniae* virulence factor-encoding genes. *Infect. Immun.* **68:**1374–1382.

89. **Whatmore, A., S. King, N. Doherty, D. Sturgeon, N. Chanter, and C. Dowson.** 1999. Molecular characterization of equine isolates of *Streptococcus pneumoniae*: natural disruption of genes encoding the virulence factors pneumolysin and autolysin. *Infect. Immun.* **67:**2776–2782.

90. **Winkelstein, J. A., and A. Tomasz.** 1978. Activation of the alternative complement pathway by pneumococcal cell wall teichoic acid. *J. Immunol.* **120:**174–178.

91. **Wu, H., A. Virolainen, B. Mathews, J. King, M. Russell, and D. Briles.** 1997. Establishment of a *Streptococcus pneumoniae* nasopharyngeal colonization model in adult mice. *Microb. Pathog.* **23:**127–137.

92. **Zhang, J., K. Mostov, M. Lamm, M. Nanno, S. Shimida, M. Ohwaki, and E. Tuomanen.** 2000. The polymeric immunoglobulin receptor translocates pneumococci across human nasopharyngeal epithelial cells. *Cell* **102:**827–837.

93. **Zhang, J.-R., I. Idanpaan-Heikkila, W. Fischer, and E. Tuomanen.** 1999. Pneumococcal licD2 gene is involved in phosphorylcholine metabolism. *Mol. Microbiol.* **31:**1477–1488.

94. **Zhang, Y., A. Masi, V. Barniak, K. Mountzouros, M. Hostetter, and B. Green.** 2001. Recombinant PhpA protein, a unique histidine motif-containing protein from *Streptococcus pneumoniae*, protects mice against intranasal pneumococcal challenge. *Infect. Immun.* **69:**3827–3836.

INFLAMMATION AND HOST DEFENSE

Paul Anthony Majcherczyk and Philippe Moreillon

12

Bacterial products triggering inflammation are often considered virulence factors. This assumes that pathogens are willingly equipped with molecular tools aimed at harming the host. However, the relationship between prokaryotes and eukaryotes is not so Manichean. It is rather a constant effort to survive at the frontier of two living kingdoms that share close to a billion years of coevolution. The goal of the microbe is to pass unnoticed by the host while living and feeding on it. In symmetry, the goal of the host immune system is to block exuberant intruders that might threaten its physiology. Disease results from inadvertently breaking this delicate equilibrium.

To detect potential harmful microorganisms, higher eukaryotes have evolved two types of systems, i.e., innate immunity and adaptive immunity. Innate immunity is present in newborns and does not need preexposure to foreign microbial materials to react. It has an intrinsic ability to recognize molecules that are highly conserved in the microbial world, also referred to as pathogen-associated molecular patterns (PAMPs), and thus can directly react against a large number of viruses, bacteria, fungi, and parasites (for reviews, see references 4, 23, 36, and 74).

However, evolution has selected for pathogens that can circumvent the system in a number of ways. These include hiding their surface from host recognition, e.g., by surrounding themselves with polysaccharidic capsules or peculiar surface proteins (50, 51, 85), or invading eukaryotic cells to pass unnoticed by the host (12, 69). It is probably such an escalade that drove the development of adaptive immunity. After having digested a pathogen, professional phagocytes present small fragments of it to cognate T lymphocytes, from adaptive immunity, in the context of their surface major histocompatibility complex class II receptor. The cognate T lymphocytes then expand and can promote the production of either cytotoxic cells (Th1 pathway) or antibody production (Th2 pathway) to get rid of the invader.

This sophisticated defense network allows both responding to unknown intruders and preventing reinfection with a previously known one. Nevertheless, innate immunity is always involved in the primary detection of invading microbes and in the inflammatory response that aims at calling in other players of

Paul Anthony Majcherczyk and Philippe Moreillon, Department of Fundamental Microbiology, University of Lausanne, Biology Building, CH-1015 Lausanne, Switzerland.

The Pneumococcus. Volume Editor, Elaine I. Tuomanen,
© 2004 ASM Press, Washington, D.C.

the host defense system. In the following, we attempt to clarify the implication of certain surface components of gram-positive organisms in triggering a response by the human innate immune system.

THE PLAYERS OF INNATE IMMUNITY

The molecular and cellular components in innate immunity have been extensively reviewed (for instance, see references 3 and 4). Here we briefly recall some salient features of them to help explain their role in bacterial recognition and destruction. Cellular factors primarily include macrophages and dendritic cells, which are abundant in tissues, and polymorphonuclear neutrophils and monocytes (the precursors of macrophages), which are abundant in the blood. Soluble factors comprise all the cellular mediators (cytokines and chemokines) that amplify the inflammatory response to infection, the complement cascade, and alternative soluble components such as the C-reactive protein, the mannose binding lectin, platelet-activating factor (PAF), and lipopolysaccharide (LPS)-binding protein (LBP).

Microbial Recognition by Cellular Factors

Professional phagocytes are equipped with numerous molecules that are able to recognize conserved microbial products and trigger a host response (Table 1). Activated phagocytes produce both cytokines, which activate inflammatory partner cells at a distance, and chemokines, which attract them to the infected site. An important surface molecule is CD14, which is linked to the surface of the membrane via a glycerophosphoinositol (GPI) anchor. CD14 recognizes both LPS of gram-negative organisms and peptidoglycan and lipoteichoic acids of gram-positive organisms (67, 89). However, since CD14 does not traverse the plasma membrane, it cannot transmit a signal to the intracytoplasmic space. The gap is fulfilled by members of a recently described family of proteins known as Toll-like receptors (TLRs) (58, 59). TLRs are transmem-

brane species that can signal to the cytosolic milieu. Microbial molecules cross-link CD14 and TLRs, which then activate transcriptional factors such as NF-κB and trigger a cellular response. Moreover, they also recognize PAMPs. At least 10 TLRs have been described; their abundance ensures a broad spectrum of microbial molecule recognition (Table 1) (49, 56, 63, 95). In certain cases TLRs may be directly cross-linked between themselves—e.g., between TLR2 and TLR1 and between TLR2 and TLR6—and transmit a transmembrane signal (64).

Monocytes and macrophage express both CD14 and TLRs on their surfaces. Polymorphonuclear cells were shown to express CD14 and at least TLR2 (49). Endothelial and epithelial cells may also do so (44, 45, 74). Thus, PAMP-recognizing molecules may respond in different ways to both trigger an inflammatory response and kill invading microbes.

The Cytosolic Nod Receptors

Recently a series of reports identified the Nod molecules—a family of intracellular proteins including Nod1/CARD4 and Nod2/CARD15—as being able to recognize peptidoglycan motifs within the cytosolic compartment (8, 30–32, 41). The Nod proteins have remarkable homologies with plant cytosolic disease-resistant proteins containing leucine-rich repeats. Leucine-rich repeats recognize a variety of microbial components and are responsible for metabolic changes or apoptosis in response to infection (17). Cultured epithelial cells expressing Nod proteins were tested for their response to several bacterial products. Nod proteins were originally thought to detect LPS of gram-negative organisms (41). However, it is now established that they detect small fragments of peptidoglycan. The interaction between Nod and peptidoglycan activates NF-κB and elicits the production of cytokines such as interleukin-8 (IL-8) (8, 30–32). Nod proteins are very specific. Nod1 detects only N-acetylglucosamine-N-acetylmuramate-tripeptide (GM-tripeptide) of gram-negative organisms, which contains mesodiaminopimelic acid in position 3 of the

TABLE 1 Major host molecules recognizing PAMPs[a]

Host factor[b]	Cell location	Bacterial PAMP	Microorganisms	Activity
CD14	Plasma membrane	LPS, LTA, peptidoglycan	Gram+/gram−	Associates with TLRs to activate NF-κB and cytokine release
TLR2 and TLR2/ TLR1	Plasma membrane	Peptidoglycan, LTA, lipoproteins, HSP, glycolipids, mannans, fimbriae, porins	Gram+/gram−	Associate with CD14 or other TLRs to activate NF-κB and cytokine release
		Zymosan	Yeast	
		GPI anchors	*Trypanosoma*	
TLR3		dsRNA	Viruses	
TLR4		LPS, LTA, HSP, bacterial proteins	Gram+/gram−	
TLR5			Retroviruses	
TLR2/ TLR6		Flagellin	Gram+/gram−	
		Lipoproteins, heat-labile soluble factor, phenol-soluble modulin	*Mycoplasma,* group B *Streptococcus, Staphylococcus*	
TLR9		Unmethylated CpG DNA	Bacteria, viruses, insects	
Nod1	Cytosol	GM-tripeptide with meso-DAP	Gram−	Activate NF-κB and cytokine release
Nod2		MDP	Gram+/gram−	
PGRPs	Plasma membrane, cytosol, plasma (?)	Peptidoglycan	Gram+/gram−	Antibacterial activity (triggers production of antibacterial molecules)
LBP	LPS, LTA, peptidoglycan	Plasma	Gram+/gram−	Presents bacterial molecules to CD14 and triggers cytokine release

[a] Nod proteins are intracellular PAMP recognition proteins. LTA, lipoteichoic acid; GPI, glycerol-phosphoinositol; dsRNA, double-stranded RNA; HSP, heat shock proteins; CpG, cytosine-phosphate-guanine; meso-DAP, mesodiaminopimelic acid; MDP, muramyl-dipeptide. Adapted from references 67, 88, and 89.
[b] The slashes between TLRs (e.g., TLR2/TLR1) indicate functional heterodimers.

stem peptide (Table 1). It does not recognize closely related structures such as GM-tripeptide containing a lysine instead, as in streptococci and staphylococci (Fig. 1 and 2), or GM-dipeptide or GM-tetrapeptide (30). In contrast, Nod2 recognizes muramyl-dipeptide, the smallest common structure of all bacterial peptidoglycans, and thus reacts against all bacteria (31). The Nod system is likely to signal pathogens that could successfully invade the host cells and escape extracellular defenses.

PGRPs

Peptidoglycan-recognizing proteins (PGRPs) are a new family of PAMP-recognizing molecules that are conserved from insects to mammals (92). There are ca. 36 known PGRPs, among which 17 are in *Drosophila*

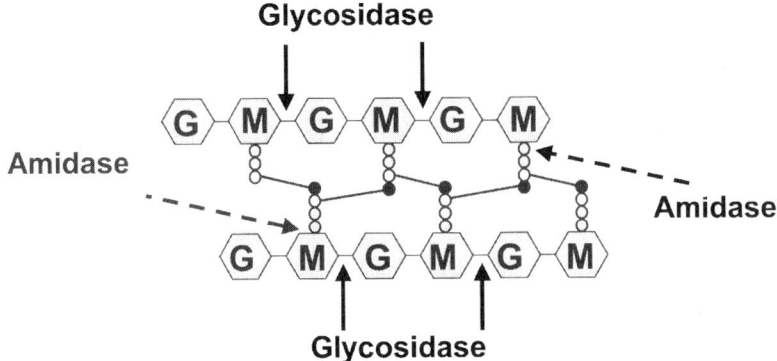

FIGURE 1 Diagram presenting the pneumococcal peptidoglycan and the sites of hydrolysis of naturally occurring N-acetylmuramic-L-alanine amidase (amidase) and muramidase (glycosidase), respectively. Mature pneumococci usually do not contain the terminal L-alanine. It is noteworthy that numerous bacteria, including both gram-positive and gram-negative genera, contain mesodiaminopimelic acid or even ornithine instead of lysine in position 3 of the stem peptide. G, N-acetylglucosamine; M, N-acetylmuramic acid. The two first (gray) circles hooked to M represent L-alanine and D-isoglutamine. The third (white) circle represents L-lysine. The fourth (black) circle represents the penultimate D-alanine. (Reproduced with permission from reference 63.)

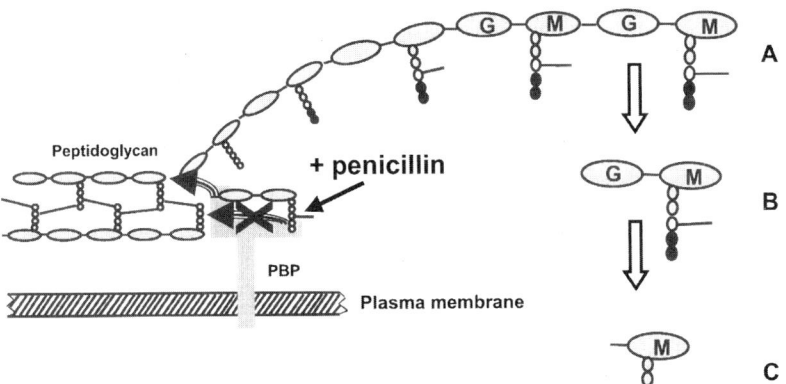

FIGURE 2 Assembly of the bacterial cell wall and production of penicillin-induced soluble peptidoglycan. Cell wall precursor disaccharide-pentapeptides are translocated through the plasma membrane and processed by membrane-anchored penicillin-binding proteins (PBP). High-molecular-weight PBPs ensure both a transglycosidase function (upper curved arrow) that elongates the glycan chain and a transpeptidase activity (lower curved arrow) that transfers the peptide bond of the penultimate D-alanine (penultimate closed circle) to a diamino acid acceptor at position 3 of the stem peptide (open circle; lysine or lysine-bound glycine side chains in the case of *S. aureus*). Subinhibitory concentrations of penicillin block transpeptidation (see lower curved arrow) but not transglycosylation. As a result, large uncross-linked glycan chains, i.e., soluble peptidoglycan (A), are released in the supernatant. Polymeric soluble peptidoglycan can be hydrolyzed by muramidase (see Fig. 1) to disaccharide-pentapeptide subunits (B). Muramyl-dipeptide (C) containing only N-acetylmuramic acid and the two first amino acids L-alanine and D-isoglutamate (or isoglutamine) is the simplest common structure of all bacterial peptidoglycans. Details are as in Fig. 1. The bars linking L-lysines and L-alanines represent the pentaglycine side chain typical of staphylococcal peptidoglycan. (Reproduced with permission from reference 63.)

melanogaster, 9 are in *Anopheles gambia*, and 4 are in mammals (10). PGRPs have several attributed functions, including direct antimicrobial activity or triggering the production of antimicrobial molecules (75), signaling via the TLR system (61), and peptidoglycan degradation (83). Among the four mammalian PGRPs (namely, PGRP-S, PGRP-L, PGRP-1α, and PGRP-1β), only two (PGRP-S and PGRP-L) have a known function. PGRP-S is an antimicrobial peptide present in polymorphonuclear neutrophils, and PGRP-S-deficient mice were shown to have an increased susceptibility to infection by nonpathogenic gram-positive bacteria (21). PGRP-L is an *N*-acetylmuramyl-L-alanine-amidase (referred to as amidase) which cleaves the amide bond between *N*-acetyl-muramic acid and L-alanine in the peptidoglycan backbone (Fig. 1). It was shown to hydrolyze staphylococcal peptidoglycan (83). It carries a membrane-anchoring domain and was detected both in cytosolic vesicles and on the surfaces of cells. RNA splicing might also yield a plasma-soluble version of it.

While the amidase PGRP-L degrades peptidoglycan in vitro, it does not lyse or kill whole bacteria. Thus, it is not responsible for direct bacterial elimination. On the other hand, it could act as a "presenting" molecule that dismantles peptidoglycan to produce appropriate peptide fragments to the CD4-TLR system (see "Structural constraints," below). Alternatively, it could function as a scavenger enzyme that degrades bacterial leftovers. Whether intracellular PGRP-L interacts positively or negatively with Nod1 and Nod2 remains to be determined.

Soluble Components

Cell-secreted cytokines and chemokines have no direct role in pathogen detection. In contrast, complement and acute-phase proteins can detect bacterial structures. The first component in the complement cascade is C1q. It is a calcium-dependent sugar-binding protein belonging to the collectin family, because it contains both collagen-like and lectin-like domains. It has six globular heads (lectins) linked together by collagen-like tails. C1q binds more or less nonspecifically to a number of microbial polysaccharides, including teichoic acids, and triggers the complement cascade toward opsonization, via C4b, and toward membrane perforation by activating the C5-C9 membrane attack complex (MAC). MAC kills gram-negative bacteria by perforating their outer membrane lipid bilayer. It does not kill gram-positive bacteria because they do not have an outer membrane, and MAC cannot assemble through their thick peptidoglycan to attack the plasma membrane.

Acute-phase proteins comprise at least C-reactive protein, mannan-binding lectin, and LBP. C-reactive protein is a member of the "pentraxin" family of proteins, so called because they are formed by five identical subunits. It binds to phosphorylcholine, which is present on the surfaces of a few important pathogens, such as the teichoic acids of pneumococci, and on the surface of *Haemophilus influenzae* (40). C-reactive protein bound to the bacterial surface opsonizes the microbe and triggers the classical cascade by binding C1q.

Mannan-binding lectin resembles C1q both structurally (it is a collectin) and functionally. Like C1q it is associated with proteases (MASP-1 and MASP-2) which, upon binding to mannan polysaccharides, cleave C4 and C2 to activate the complement cascade. Mannan-binding lectin is likely to be involved in the defense against fungi. Two other members of the collectin family are represented by pulmonary surfactants A and D. They are important in recognizing pulmonary pathogens such as *Pneumocystis carinii* and *Mycoplasma pneumoniae* (9).

LBP is a soluble polypeptide that recognizes several components of the bacterial envelope and presents them to CD14 (3, 4, 58, 59). It was initially believed that LBP specifically mediated recognition of LPS of gram-negative organisms but not walls of gram-positive organisms (54). However, recent work indicates that LBP also binds to lipoteichoic acid of gram-positive organisms and increases their inflammatory activity (24, 70, 93). In the case of lipoteichoic acid, recognition is directed toward the lipid anchor of lipoteichoic acid, thus probably mimicking recognition of LPS. In the case of peptidoglycan, binding might occur via the glycan backbone (93). It is suggested that LBP stabilizes multimeric peptidoglycan fragments to improve receptor recognition and activation.

Acute-phase proteins are produced by the liver in response to cytokines such as tumor necrosis factor α (TNF-α) and IL-1. Thus, they are not the very first response to infection and rather represent one of the many defense pathways triggered by initial recognition via complement and/or phagocytes.

PAF and Its Receptor

PAF is a potent lipid chemokine decorated with a choline residue (glycerophosphorylcholine) (35, 65). It is secreted by several kinds of leukocytes during inflammation and is responsible for a number of actions, including platelet activation, leukocyte and macrophage activation, increased vascular permeability, hypotension, stimulation of uterine contraction, stimulation of glycogenolysis, and other symptoms resembling severe allergy. PAF activates target cells via a specific receptor, which is expressed in response to cytokine (e.g., TNF-α) stimulation (39, 48). PAF receptor is a G protein that triggers its internalization into the cell upon stimulation.

Because the receptor recognizes the phosphorylcholine residue of PAF, it also recognizes phosphorylcholine residues present on pneumococcal teichoic acids. In this way it promotes the adherence of pneumococci to the surfaces of activated endothelia (14). Adherent pneumococci trigger the G-protein activity of PAF receptor and eventually become internalized into the cells. Cundell et al. (14) showed that only ≤0.1% of pneumococci overlaid on nonactivated endothelial cultures were internalized. In contrast, pretreatment of the cells with TNF-α and/or IL-1 induced both the expression of PAF receptor and a 20- to 40-fold increase in pneumococcal internalization. Addition of a specific inhibitor of PAF receptor blocked the uptake of bacteria (14). Thus,

pneumococci can trick the PAF and PAF receptor inflammatory components in order to invade the intracellular milieu.

USE OF *S. PNEUMONIAE* TO STUDY INFLAMMATION INDUCED BY GRAM-POSITIVE ORGANISMS

S. pneumoniae is the first cause of bacterial community-acquired respiratory tract infection, including sinusitis, otitis media, and pneumonia (6). It is a frequent cause of bacteremia and has become the primary cause of bacterial meningitis since the anti-*Haemophilus influenzae* type B vaccine has virtually eradicated *Haemophilus* as a causative agent (71). Neurologic sequelae of meningitis result from the overwhelming inflammatory response to bacterial products rather than from the bacteria themselves. Indeed, early treatment with anti-inflammatory agents decreases sequelae both in animal experiments and in humans (15, 76). Thus, studying pneumococcus-induced inflammation in meningitis is relevant.

Treatment of pneumococci and other bacteria with cell wall inhibitors triggers cell lysis and wall solubilization, releasing great numbers of fragments into the cerebrospinal fluid. This is accompanied by an increased inflammatory response in meningitis (77, 79). The disease can be modeled in animals. Molecules can be injected into the cerebrospinal fluid followed by sampling for leukocyte influx and cytokine and chemokine responses (60). Moreover, biochemical dissection of the pneumococcal envelope has been achieved (28, 29, 52, 80). Biochemical libraries of the pneumococcal walls were injected into rabbits with experimental meningitis. The results indicated that not all components were equally active in triggering an inflammatory response (Table 2) (80, 81, 87). Whole insoluble cell walls were active but required ca. 100 times more material than LPS of gram-negative organisms (in a weight/volume ratio) to produce a similar inflammation. Autolysin-solubilized walls (by amidase [Fig. 1]) were poorly active in vivo but contained highly inflammatory subcomponents when tested for cytokine release by

TABLE 2 Threshold bioactivities of cell wall components from different organisms tested in different models[a]

Wall component(s)	Result for:		
	Staphylococcus[b]	Pneumococcus[b]	Pneumococcus[c]
Intact bacteria (CFU/ml)	$<10^8$ (73, 93)	10^5 (27, 93)	10^5 (27, 81)
LTA (ng/ml)	100 (34)	10,000 (34)	<500 (80)
TA (ng/ml)	Not done	>100 (53)	
Insoluble PG + TA (ng/ml)	100 (53)	20–100 (52, 53, 72, 86)	20 (27, 81)
Insoluble PG (ng/ml)	100 (53)	100 (53)	
Soluble PG (ng/ml)	100 (73, 93)	Not done	<500 (80)

[a] Cell wall components were tested for the ability to induce meningeal inflammation upon injection into the subarachnoid space or for the ability to induce cytokine gene expression in cells from blood. <, lowest concentration tested that is still active. LTA, lipoteichoic acid; PG, peptidoglycan; TA, teichoic acid. Adapted with permission from reference 87.
[b] Cytokine release.
[c] Meningitis.

monocytes in vitro (52). Similarly, the simplest conserved peptidoglycan molecule muramyl-dipeptide was poorly active in triggering cytokine release in vitro but was highly active in rabbits with experimental meningitis (13). Similar kinds of observations were made with wall prepared from *Staphylococcus aureus* (Table 2). Moreover, muramyl-dipeptide is also detected by the intracellular Nod2 protein (31). Therefore, both the bacterial components and the host milieu are important. They should ideally be analyzed in an integrated fashion.

COMPARISON OF INFLAMMATORY BACTERIAL STRUCTURES
Table 3 presents the principal structures of gram-negative and gram-positive bacteria in-

volved in an inflammatory response. Secreted molecules and outermost polysaccharides and proteins are strain or species specific and are usually not recognized by the innate immune system. Typically, the polysaccharidic capsule of *S. pneumoniae* and the M protein of *Streptococcus pyogenes* are inert to professional phagocytes (50, 51, 85). Mounting first a specific antibody response against these structures is necessary to further activate complement and/or phagocytosis (42).

Below the surface are LPS in gram-negative bacteria and teichoic acids, peptidoglycan, and lipoteichoic acids in gram-positive organisms. LPS contains both a conserved lipid A moiety and variable polysaccharidic side chains (66). Conserved lipid A is highly reactive toward

TABLE 3 Anatomy of gram-negative and gram-positive bacteria and major conserved and strain-specific proinflammatory determinants[a]

Cell anatomy	Structures in:		Strain specificity
	Gram-negative bacteria	Gram-positive bacteria	
Extracellular	Toxins	Toxins	Specific
	Superantigens	Superantigens	Specific
Cell envelope	Capsule	Capsule	Specific
	Proteins	Proteins	Specific
	LPS (lipid A)	TA and LTA	Conserved
	Peptidoglycan	Peptidoglycan	Conserved
Intracellular	N-formyl-methionine	N-formyl-methionine	Conserved
	Unmethylated CpG DNA	Unmethylated CpG DNA	Conserved

[a] Abbreviations are as in Table 1. Adapted with permission from reference 63.

innate immunity, whereas variable polysaccharides are not. LPS triggers the release of cytokines by peripheral blood monocytes (PBMCs) and other responsive cells (1, 66, 67). This involves a cascade of events starting with LPS binding to the serum acute-phase protein LBP. The LBP-LPS complex then attaches to CD14, which eventually become activated via the transmembrane coreceptor TLR4 (3, 4, 58, 59).

In gram-positive bacteria the outermost teichoic acids are covalently attached to the N-acetylmuramic acid of peptidoglycan. They are constituted of either polyglycerol phosphate or polyribitol phosphate units decorated or not with D-alanine side chains (e.g., S. aureus) (16, 43) or choline (e.g., S. pneumoniae) (26). A homolog of teichoic acids are the lipoteichoic acids, which are attached to the plasma membrane via a diacyl-glycerophosphate anchor and traverse the whole peptidoglycan. Teichoic acids and lipoteichoic acids are involved in cell wall metabolism and may also serve as anchors for certain noncovalently linked proteins at the bacterial surface (e.g., choline-bound surface proteins in S. pneumoniae) (11, 26). Although vital for the bacterium, purified teichoic acids are not inflammatory (53) and purified lipoteichoic acids are moderately inflammatory compared to the LPS of gram-negative organisms (70).

The major and most conserved constituent of the envelope of gram-positive organisms is peptidoglycan. It is constituted of glycan chains made of N-acetylglucosamine and N-acetylmuramic acid disaccharide subunits, in which the N-acetylmuramate moiety is linked to highly conserved pentapeptide or tetrapeptide stems (L-alanine–D-isoglutamine–L-lysine or diaminopimelic acid–D-alanine–[D-alanine]) (see Fig. 1 and 2). The chains of disaccharide-peptide are cross-linked via peptide bridges between the penultimate D-alanine and a diamino acid (L-lysine or mesodiaminopimelic acid) located in position 3 of a neighboring stem peptide. Both gram-negative and gram-positive bacteria harbor a peptidoglycan of grossly similar features. But the structure in gram-negative bacteria is thin (1 or 2 layers), whereas it is thick (≥10 layers) in gram-positive organisms (18).

Peptidoglycan and lipoteichoic acid of gram-positive organisms may trigger cytokine release from innate-immunity cells (5, 37, 46, 47, 55, 70, 78, 82). There are some differences between cytokine release mediated by LPS of gram-negative organisms and peptidoglycan of gram-positive organisms. Like LPS, peptidoglycan prepared from penicillin-treated S. aureus can bind to the CD14 receptor of target cells (67, 88, 89). Even its minimal immunomodulatory subunit muramyl-dipeptide can do so (89). Yet, CD14 contains discrete binding sites that are different for LPS and peptidoglycan (22, 33). Moreover, although soluble CD14 (sCD14) binds both of these bacterial molecules, activation of the CD14-lacking endothelial and epithelial cells depends on sCD14 for LPS but not for peptidoglycan (45). Finally, unlike LPS, peptidoglycan does not activate intracellular signal transcription via TLR4. It uses its TLR2 homolog instead (Table 1) (70, 73). It also generates a unique pattern of gene expression when human monocytes are tested by the technique of cDNA arrays (84). Thus, LPS and peptidoglycan do not share entirely similar host recognition mechanisms.

Eventually, intracytoplasmic bacterial molecules such as unprocessed bacterial proteins containing a formyl-methionine amino-terminal (25), unmethylated DNA motifs and double-stranded RNA (3, 57) can trigger a variety of innate immune responses. However, the intracellular nature of these components makes them unlikely to be at the front line of the host-bacterium relationship.

INFLAMMATORY SUBCOMPONENTS OF WALLS OF GRAM-POSITIVE ORGANISMS

All the structures described in Table 3 can be biochemically purified and tested for the ability to trigger the release of inflammatory mediators from innate-immunity responsive cells. However, the specific activities of these various components may be very different. LPS elicits a cytokine response at concentrations as low as 1 ng/ml, whereas 100 to 1,000 times more peptidoglycan or lipoteichoic acid is required to produce the same kind of effect (Table 4). This suggests that either the wall components of gram-positive organisms are not important for inflammation or that only part of them, i.e., specific motifs, are important, whereas the rest acts as mere ballast in the stimulation assay.

One possible explanation is the dissimilar natures of the bacterial components used. LPS is made of noncovalently linked glycolipid subunits that are emulsified by plasma lipoproteins and LBP, which presents them to the cell receptor CD14 (54, 66). In contrast, purified peptidoglycan of gram-positive organisms is a highly insoluble megamolecule that may hide its most reactive motifs inside its complex

structure. In order to circumvent this problem, peptidoglycan was solubilized by at least two ways. One technique consists of growing S. aureus in subinhibitory concentrations of penicillin (Fig. 2) (88). Penicillin blocks the transpeptidation reaction responsible for hydrolyzing the last D-alanine bond of the disaccharide-pentapeptide precursor and transferring it to an acceptor (L-lysine or diaminopimelic acid) of a neighboring peptidoglycan chain. Low concentrations of penicillin block cross-linkage between adjacent glycan chains and release the nascent polymers into the supernatant.

This material has been used in a number of studies and helped sort out CD14- and TLR2-responsive receptors (Table 4) (33, 73, 88, 89). However, soluble peptidoglycan is still of high molecular mass (up to 125 kDa) and has the same low specific activity as insoluble peptido-

TABLE 4 Some landmark studies in understanding the structure-activity relationship of inflammation induced by walls of gram-positive organisms[a]

Wall material	Read-out system	Amt of reagent	Result	Reference(s)
sPGN	IL-1, IL-6 human monocytes	sPGN ≥ 1 μg/ml LPS = 1 ng/ml	sPGN binds to CD14	88
PGN and sPGN	sIGM and NF-κB in CD14⁻/⁺ cells	sPGN ≥ 1 μg/ml LPS = 1 ng/ml	sPGN binds to CD14	33
sPGN and MDP	IL-1, IL-6 human monocytes	sPGN ≥ 1 μg/ml MDP ≥ 1 μg/ml	Binds to CD14	89
Disaccharide-dipeptide and LTA	NO in J774-2 cell line	Disaccharide-peptide ≥ 1 μg/ml LTA = 100 ng/ml (S. aureus)	Synergism between disaccharide-peptide	47
GM-tetra (Escherichia coli)	G-CSF human endothelial cells	GM-tetra ≥ 1 μg/ml (no control LPS)	Stimulatory activity	19, 20
PGN, sPGN, and LTA	NF-κB in TLR2⁻/⁺ HEK293 cells	PGN ≥ 1 μg/ml LTA = 0.1–1 μg/ml	TLR2 mediates activation	73
S. aureus and S. pneumoniae	NF-κB in TLR2⁻/⁺ CHO fibroblasts	Whole bacteria	TLR2 mediates activation	94
S. pneumoniae stem peptides	TNF in PBMCs or rabbit blood	Stem peptides = 10 ng/ml LPS = 1 ng/ml	Activity depends on polymers	52

[a] sPGN, soluble peptidoglycan; MDP, muramyl-dipeptide; LTA, lipoteichoic acid; GM-tetra, N-acetylglucosamine-N-muramic acid-tetrapeptide; sIGM, cell surface-bound immunoglobulin M; NO, nitric oxide; G-CSF, granulocyte colony-stimulating factor. Adapted with permission from reference 63.

glycan. Moreover, further hydrolysis of soluble peptidoglycan to monomers results in a complete loss of activity, indicating that some degree of polymerization is necessary for inflammation (Fig. 2) (88, 89, 94). Even the smallest conserved motif of peptidoglycan, i.e., muramyl-dipeptide, is inactive in many systems, except for experimental meningitis (13). Thus, soluble peptidoglycan did not reveal a unique minimal structural constraint responsible for inflammation induced by gram-positive organisms. Moreover, neither penicillin-induced soluble peptidoglycan nor muramyl-dipeptide is a natural product spontaneously released by bacteria.

In an alternative approach, the pneumococcal peptidoglycan was dissected using wall-hydrolyzing enzymes that are physiologically used by the bacteria to reshape their wall during growth and division, i.e., N-acetylmu-ramic-L-alanine amidase and muramidase (Fig. 1) (52, 53). After solubilization, the wall fragments were separated by high-pressure liquid chromatography. The technique revealed the existence of an array of subcomponents (28, 29), some of which were totally inert in terms of inflammation and others of which were almost as active as LPS of gram-negative organisms. Amino acid analysis and mass spectrometry identified single and branched stem peptides, without contaminating amino-sugars or teichoic acids (52, 53). When tested for the ability to trigger the release of TNF from PBMCs, simple monomers and dimers were inactive, whereas trimers and more complex structures were highly active (Fig. 3) (52). Thus, the peptidoglycan of gram-positive organisms does indeed contain highly inflammatory substructures. These motifs are likely to be hidden in the inert and insoluble peptido-

Fragment Molecular weight M_r (M + H)$^+$	Amino acid composition & ratio		Predicted structure	Minimal stimulatory concentration (µg/ml)
744.39	Ala iGln Lys	1.43 1 0.92	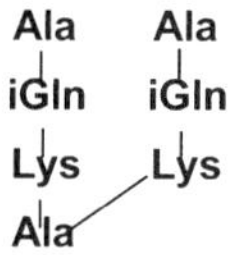 Ala Ala iGln iGln Lys Lys Ala	> 0.1
1142.6	Ala iGln Lys	1.8 1 0.9	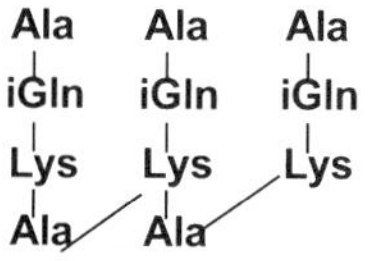 Ala Ala Ala iGln iGln iGln Lys Lys Lys Ala Ala	0.01
1285.01	Ala iGln Lys	1.87 1 0.83	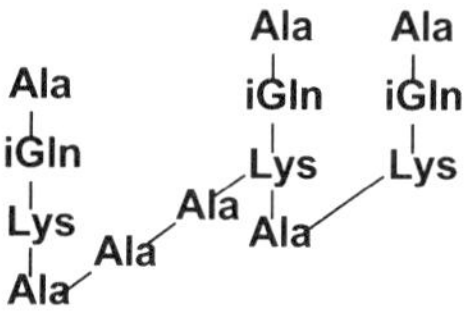 Ala Ala iGln iGln Ala Lys Lys iGln Lys Ala Ala Ala	0.001

FIGURE 3 Molecular structures of one inactive and two active internal fragments of the pneumococcal peptidoglycan. Pneumococcal peptidoglycan was solubilized with amidase, subjected to high-pressure liquid chromatography fractionation, and analyzed by a combination of mass spectrometry, amino acid determination, and TNF-releasing activity on human PBMCs. The TNF-triggering activity is expressed as the minimal concentration of wall materials required to increase the release of TNF by ≥10 times over background. The stem-peptide dimer was poorly active (>0.1 µg/ml), whereas the two stem-peptide trimers were almost as active (0.01 to 0.001 mg/ml) as LPS. (Reproduced with permission from reference 63.)

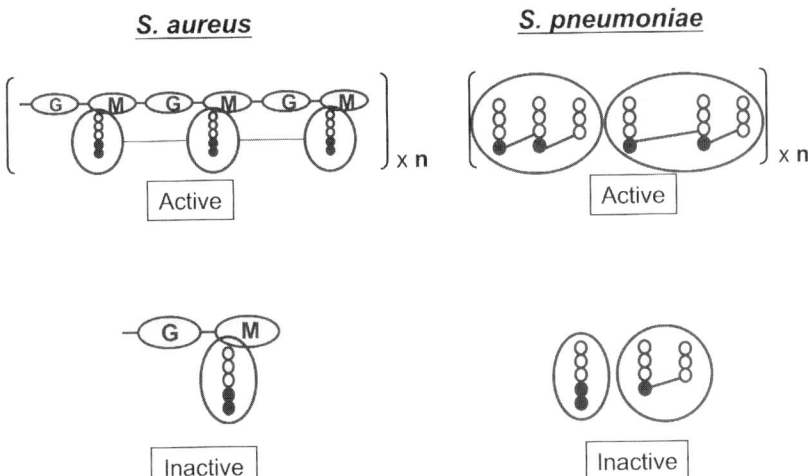

FIGURE 4 Possible structural constraints ensuring high TNF-releasing activity of peptidoglycan of gram-positive organisms. On the left side, *S. aureus* soluble peptidoglycan is a large multimer (triplets times **n**) that is not more inflammatory than whole insoluble peptidoglycan. Hydrolyzing soluble peptidoglycan to disaccharide-pentapeptide results in complete loss of TNF release. On the right side, trimers of pneumococcal stem peptides are the minimal structures conferring high TNF-triggering activity. Shorter structures are inactive, and larger polymers are less active in a weight-to-weight ratio. Thus, trimeric stem peptides might be the most active complexes, whether they are cross-linked via glycan bonds, as in *S. aureus* soluble peptidoglycan, or peptide bonds, as in pneumococcal stem peptides. Circles highlight the peptide structures. Details are as in Fig. 1. (Reproduced with permission from reference 63.)

glycan preparation in vitro. In contrast, they might be revealed in vivo during endogenous wall remodeling or during hydrolysis by exogenous host lysozyme or PGRP-L (83), and/or antibiotic therapy.

STRUCTURAL CONSTRAINTS

Monomers and Multimers
The work on pneumococcal walls indicates that some minimal structure, such as cross-linked tripeptides, is necessary to induce inflammation (for a review, see reference 63). Hence, multimers of stem peptides might be one of the minimal conserved structures recognized by innate immunity. Figure 4 exemplifies such a hypothesis. In pneumococci, three or more stem peptides cross-linked via peptide bonds are inflammatory (52). In staphylococcal soluble peptidoglycan, several stem peptides connected via glycan bonds are inflammatory as well. Since cell activation requires cross-linking between CD14 and TLRs, it is tempting to

postulate that tripeptides and larger structures allow receptor bridging and elicit a cell response, whereas monomeric and dimeric peptides are too small to do so. This possibility is reminiscent of triplet binding by mannose binding protein (38).

Secondary Structure
The inflammatory activity might also depend on other constraints, including the secondary structure of the components and/or the stereochemistry of their amino acid constituents. Some hints regarding these issues are becoming available. The secondary structure is likely to be critical. Let's consider a single stem peptide and concentrate on its conserved diamino acid L-lysine in position 3. Figure 5 reveals that its epsilon-carbon can take up an infinite number of spatial conformations. Attaching a second stem peptide to this epsilon-carbon will impose some physical constraint, but each single amino acid in the structure will keep a cer-

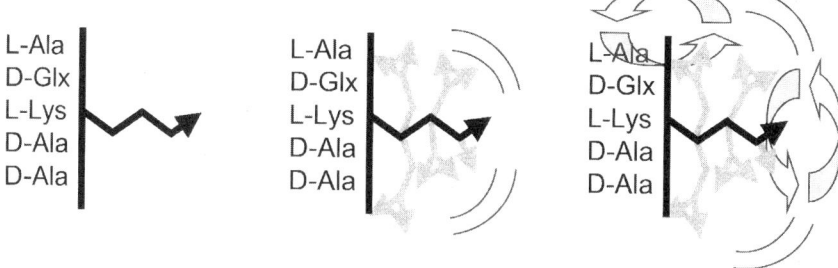

FIGURE 5 Stem peptides are highly unstable structures. The diagram depicts a simple stem peptide with the four-carbon side chain of the diamino acid lysine implicated in peptide cross-links (right) and the multiple possible planar and axial rotations of this very side chain, as well as the amino acids in the stem peptide. Cross-linkage to a second or a third stem peptide may limit free movement to a certain extent, but not enough to stabilize the structure in a fixed position. (Reproduced with permission from reference 63.)

tain flexibility that is hardly compatible with a ligand-receptor interaction. Even a third stem peptide would not stabilize the molecule completely, thus making two-dimensional nuclear magnetic resonance analysis or even bioinformatic conformation modeling elusive. Therefore, one must assume that some kind of "induced-fit" mechanism is operative for innate immunity receptors to recognize these structures.

Amino Acid Chirality

The importance of amino acid conformation was recently studied in several systems (13, 16, 31). Rabbits with experimental meningitis were inoculated with muramyl-dipeptide preparations containing various D- or L-amino acid isomers. Replacing the first L-alanine with D-alanine completely abrogated the capability of muramyl-dipeptide to induce both TNF and leukocyte influx in the meningeal space. In contrast, modifying the chirality of the second residue (L-isoglutamine to D-isoglutamine) did not affect reactivity. The chirality also appeared important for peptidoglycan detection by cytosolic Nod proteins. The native muramyl-dipeptide-L-alanine-D-glutamine (MDP-LD) triggered an NF-κBa response, whereas the MDP-LL enanthiomer did not (31). Finally, the structure-activity relationship of S. aureus lipoteichoic acids was recently studied using synthetic molecules (16). Certain

discrete chemical modifications could decrease or even completely block its inflammatory activity. This included replacing the decorating D-alanine with an L-alanine (10-fold decrease in activity) and removing the membrane lipid anchor from the molecule (100-fold decrease). In contrast, other modifications, such as replacing the sugar "gentiobiose" that links the teichoic acid moiety to its lipid-membrane anchor or modifying an N-acetylglucosamine decorating sugar, had no effect.

The fact that discrete differences in bacterial structures can profoundly affect the innate immune response has at least two implications. First, although often considered a broad-spectrum pattern-recognizing system, innate immunity is acutely specialized and capable of detecting very subtle differences between genuine bacterial molecules and modified analogs. Second, this high specificity underlines the importance of using well-characterized microbial material in investigating the system.

PROINFLAMMATORY ACTIVITIES OF WALLS FROM PATHOGENIC AND NONPATHOGENIC BACTERIA

Most studies on inflammation induced by walls of gram-positive organisms use material purified from human pathogens such as S. aureus, S. pneumoniae, and a few others (Table 2). Nonpathogenic bacteria are usually not tested. The proinflammatory activities of purified walls from a few bacteria with different disease-promoting capacities were recently compared (P.A. Majcherczyk, S. Bauer, C. Decaillet, and P. Moreillon, Abstr. 43rd Intersci. Conf. Antimicrob. Agents Chemother., abstr. B1052, 2003). Nonpathogens included two *Bacillus subtilis* strains with different genetic backgrounds and one *Lactococcus lactis* strain. Pathogens included *S. aureus*, *S. pneumoniae*, and *Listeria monocytogenes*, three organisms with different disease mechanisms. Walls were purified and tested for their proinflammatory activities as described previously (52, 53, 63). The walls from *B. subtilis* and *L. monocytogenes* were ≥10 times more inflammatory than those of *S. aureus*, *S. pneumoniae*, and *L. lactis* (Majcherczyk et al., 43rd ICAAC). Thus, the inflammatory power did not correlate with the pathogenic nature of the organisms. Differences in teichoic acids were not involved either, because removal of these appendages did not alter the inflammatory power of the preparations. On the other hand, differences in the peptidoglycan amino acid content were observed. The three most inflammatory walls contained mesodiaminopimelic acid at position 3 of the stem peptide, whereas the three least inflammatory ones contained L-lysine instead.

We further searched the amino acid composition of peptidoglycan in additional gram-positive bacteria (2, 63). The organisms were sorted for the presence of mesodiaminopimelic acid or L-lysine in their stem peptides and then for their natural habitat and pathogenicity in animals. A total of 86% (19 of 22) of bacterial classes primarily colonizing animals contained lysine in their stem peptides, whereas 75% (26 of 35) of those preferentially colonizing the environment contained mesodiaminopimelic acid or ornithine instead. Likewise, 70% (18 of 26) bacterial classes responsible for infection contained lysine in their stem peptides, whereas 75% (24 of 32) of the nonpathogenic classes contained mesodiaminopimelic acid or ornithine instead. The segregation between both lysine walls and animals on the one hand and diaminopimelic acid walls and environ-ment on the other hand was statistically significant ($P < 0.0001$).

Mesodiaminopimelic acid is a precursor of L-lysine (7). Therefore, it is tempting to make the provocative speculation that gram-positive animal colonizers have evolved an L-lysine peptidoglycan in order be less well detected by innate immunity. In contrast, soil organisms and gram-negative bacteria, which contain diaminopimelic acid, might not have undergone this selective pressure because they either live in a remote niche (soil or water organisms) or are surrounded by an additional outer membrane masking the underlying peptidoglycan. Exceptions such as *Bacillus* spp. of medical importance, and other gram-positive pathogens containing diaminopimelic acid, tend to use peculiar means such as toxins or host cell invasion to infect the host (12, 62). This hypothesis would fit the model according to which successful survival at the frontier of the eukaryotes' world requires the ability of the bacteria to escape recognition and destruction by the host.

CONCLUSION AND PERSPECTIVES

Experimental evidence indicates that the peptidoglycan of gram-positive organisms can trigger inflammation mediated by innate immunity but that its most active subcomponents may be kept hidden from host recognition in its complex molecular scaffold. Conserved appendages such as teichoic acid and lipoteichoic acids are less inflammatory on their own, but they may act synergistically with peptidoglycan in some systems (47). Moreover, these molecules may interfere with other aspects of the bacterium-host response. Complement and PAF binding is one of them (14). Facilitating autolysis is another. In pneumococci, teichoic acids are decorated with choline residues that are critical for the activity of the major autolysin amidase (26). Bacterial lysis results in the release of numerous bacterial products, some of which are known as pathogenic determinants, such as pneumolysin (68), and others of which are highly inflammatory, such as peptidoglycan subcomponents and intracellular determinants (52). Pneumococcal teichoic acids were shown to undergo phase switching, being turned down during invasive infection (90, 91). Low-teichoic acid variants are less prone to autolysis and hence to host recognition. Therefore, although not directly involved in host recognition, teichoic and lipoteichoic acids act at more subtle levels to modulate bacterial survival in the eukaryotic environment.

The cohabitation of bacteria and their eukaryote hosts is important. Colonization of deep skin layers and mucosa by anaerobes that cannot invade the oxygen-rich internal milieu is thought to act as a biological barrier to oxygen-resistant pathogens, which can readily survive in the oxygenated bloodstream or tissues. Moreover, constant sensing of bioactive molecules at the level of mucosal lymphoid tissues might be critical to keep the immune system alert, modulate switches such as the Th1 and Th2 response, and maybe influence allergy. The constant inflammatory response to bacterial surface component might be profitable. As far as disease is concerned, the real problem might not be so much the ability of the host to recognize bacterial intruders as much as the capacity of pathogens to escape early recognition by innate immunity.

ACKNOWLEDGMENTS

This work was supported by the Fondation Leenards, the Fondation Santos-Suarez and grant 3200-063253.00 of the Swiss National Fund for Scientific Research.

We thank Nadine Thomas for editing the manuscript.

REFERENCES

1. **Alexander, C., and E. T. Rietschel.** 2001. Bacterial lipopolysaccharides and innate immunity. *J. Endotoxin Res.* **7:**167–202.
2. **Balows, A., H. G. Trueper, M. Dworkin, W. Harper, and K. H. Schleifer.** 1992. *The Prokaryotes. A Handbook on the Biology of Bacteria: Ecophysiology, Isolation, Identification, Applications,* 2nd ed. Springer-Verlag, New York, N.Y.
3. **Barton, G. M., and R. Medzhitov.** 2003. Toll-like receptor signaling pathways. *Science* **300:**1524–1525.
4. **Beutler, B., and E. T. Rietschel.** 2003. Innate immune sensing and its roots: the story of endotoxin. *Nat. Rev. Immunol.* **3:**169–176.
5. **Bhakdi, S., T. Klonisch, P. Nuber, and W. Fischer.** 1991. Stimulation of monokine production by lipoteichoic acids. *Infect. Immun.* **59:**4614–4620.
6. **Bochud, P. Y., F. Moser, P. Erard, F. Verdon, J. P. Studer, G. Villard, A. Cosendai, M. Cotting, F. Heim, J. Tissot, Y. Strub, M. Pazeller, L. Saghafi, A. Wenger, D. Germann, L. Matter, J. Bille, L. Pfister, and P. Francioli.** 2001. Community-acquired pneumonia. A prospective outpatient study. *Medicine* (Baltimore) **80:**75–87.
7. **Born, T. L., and J. S. Blanchard.** 1999. Structure/function studies on enzymes in the diaminopimelate pathway of bacterial cell wall biosynthesis. *Curr. Opin. Chem. Biol.* **3:**607–613.
8. **Chamaillard, M., M. Hashimoto, Y. Horie, J. Masumoto, S. Qiu, L. Saab, Y. Ogura, A. Kawasaki, K. Fukase, S. Kusumoto, M. A. Valvano, S. J. Foster, T. W. Mak, G. Nunez, and N. Inohara.** 2003. An essential role for NOD1 in host recognition of bacterial peptidoglycan containing diaminopimelic acid. *Nat. Immunol.* **4:**702–707.
9. **Chiba, H., S. Pattanajitvilai, H. Mitsuzawa, Y. Kuroki, A. Evans, and D. R. Voelker.** 2003. Pulmonary surfactant proteins A and D recognize lipid ligands on *Mycoplasma pneumoniae* and markedly augment the innate immune response to the organism. *Chest* **123:**426S.
10. **Christophides, G. K., E. Zdobnov, C. Barillas-Mury, E. Birney, S. Blandin, C. Blass, P. T. Brey, F. H. Collins, A. Danielli, G. Dimopoulos, C. Hetru, N. T. Hoa, J. A. Hoffmann, S. M. Kanzok, I. Letunic, E. A. Levashina, T. G. Loukeris, G. Lycett, S. Meister, K. Michel, L. F. Moita, H. M. Muller, M. A. Osta, S. M. Paskewitz, J. M. Reichhart, A. Rzhetsky, L. Troxler, K. D. Vernick, D. Vlachou, J. Volz, C. von Mering, J. Xu, L. Zheng, P. Bork, and F. C. Kafatos.** 2002. Immunity-related genes and gene families in Anopheles gambiae. *Science* **298:**159–165.
11. **Cossart, P., and R. Jonquieres.** 2000. Sortase, a universal target for therapeutic agents against gram-positive bacteria? *Proc. Natl. Acad. Sci. USA* **97:**5013–5015.
12. **Cossart, P., J. Pizarro-Cerda, and M. Lecuit.** 2003. Invasion of mammalian cells by *Listeria monocytogenes*: functional mimicry to subvert cellular functions. *Trends Cell Biol.* **13:**23–31.
13. **Cottagnoud, P., C. M. Gerber, P. A. Majcherczyk, F. Acosta, M. Cottagnoud, K.**

Neftel, P. Moreillon, and M. G. Tauber. 2003. The stereochemistry of the amino acid side chain influences the inflammatory potential of muramyl dipeptide in experimental meningitis. *Infect. Immun.* **71:**3663–3666.

14. Cundell, D. R., N. P. Gerard, C. Gerard, I. Idanpaan-Heikkila, and E. I. Tuomanen. 1995. *Streptococcus pneumoniae* anchor to activated human cells by the receptor for platelet-activating factor. *Nature* **377:**435–438.

15. de Gans, J., and D. van de Beek. 2002. Dexamethasone in adults with bacterial meningitis. *N. Engl. J. Med.* **347:**1549–1556.

16. Deininger, S., A. Stadelmaier, S. von Aulock, S. Morath, R. R. Schmidt, and T. Hartung. 2003. Definition of structural prerequisites for lipoteichoic acid-inducible cytokine induction by synthetic derivatives. *J. Immunol.* **170:**4134–4138.

17. Dixon, M. S., C. Golstein, C. M. Thomas, E. A. van Der Biezen, and J. D. Jones. 2000. Genetic complexity of pathogen perception by plants: the example of Rcr3, a tomato gene required specifically by Cf-2. *Proc. Natl. Acad. Sci. USA* **97:**8807–8814.

18. Dmitriev, B. A., F. V. Toukach, K. J. Schaper, O. Holst, E. T. Rietschel, and S. Ehlers. 2003. Tertiary structure of bacterial murein: the scaffold model. *J. Bacteriol.* **185:**3458–3468.

19. Dokter, W. H., A. J. Dijkstra, S. B. Koopmans, A. B. Mulder, B. K. Stulp, M. R. Halie, W. Keck, and E. Vellenga. 1994. G(AnH)MTetra, a naturally occurring 1,6-anhydro muramyl dipeptide, induces granulocyte colony-stimulating factor expression in human monocytes: a molecular analysis. *Infect. Immun.* **62:**2953–2957.

20. Dokter, W. H., A. J. Dijkstra, S. B. Koopmans, B. K. Stulp, W. Keck, M. R. Halie, and E. Vellenga. 1994. G(Anh)MTetra, a natural bacterial cell wall breakdown product, induces interleukin-1 beta and interleukin-6 expression in human monocytes. A study of the molecular mechanisms involved in inflammatory cytokine expression. *J. Biol. Chem.* **269:**4201–4206.

21. Dziarski, R., K. A. Platt, E. Gelius, H. Steiner, and D. Gupta. 2003. Defect in neutrophil killing and increased susceptibility to infection with nonpathogenic gram-positive bacteria in peptidoglycan recognition protein-S (PGRP-S)-deficient mice. *Blood* **102:**689–697.

22. Dziarski, R., R. I. Tapping, and P. S. Tobias. 1998. Binding of bacterial peptidoglycan to CD14. *J. Biol. Chem.* **273:**8680–8690.

23. Elward, K., and P. Gasque. 2003. "Eat me" and "don't eat me" signals govern the innate immune response and tissue repair in the CNS: emphasis on the critical role of the complement system. *Mol. Immunol.* **40:**85–94.

24. Fan, X., F. Stelter, R. Menzel, R. Jack, I. Spreitzer, T. Hartung, and C. Schutt. 1999. Structures in *Bacillus subtilis* are recognized by CD14 in a lipopolysaccharide binding protein-dependent reaction. *Infect. Immun.* **67:**2964–2968.

25. Fillion, I., N. Ouellet, M. Simard, Y. Bergeron, S. Sato, and M. G. Bergeron. 2001. Role of chemokines and formyl peptides in pneumococcal pneumonia-induced monocyte/macrophage recruitment. *J. Immunol.* **166:**7353–7361.

26. Fischer, W. 1997. Pneumococcal lipoteichoic and teichoic acid. *Microb. Drug Resist.* **3:**309–325.

27. Freyer, D., R. Manz, A. Ziegenhorn, M. Weih, K. Angstwurm, W. D. Docke, A. Meisel, R. R. Schumann, G. Schonfelder, U. Dirnagl, and J. R. Weber. 1999. Cerebral endothelial cells release TNF-alpha after stimulation with cell walls of *Streptococcus pneumoniae* and regulate inducible nitric oxide synthase and ICAM-1 expression via autocrine loops. *J. Immunol.* **163:**4308–4314.

28. Garcia-Bustos, J., and A. Tomasz. 1990. A biological price of antibiotic resistance: major changes in the peptidoglycan structure of penicillin-resistant pneumococci. *Proc. Natl. Acad. Sci. USA* **87:**5415–5419.

29. Garcia-Bustos, J. F., B. T. Chait, and A. Tomasz. 1987. Structure of the peptide network of pneumococcal peptidoglycan. *J. Biol. Chem.* **262:**15400–15405.

30. Girardin, S. E., I. G. Boneca, L. A. Carneiro, A. Antignac, M. Jehanno, J. Viala, K. Tedin, M. K. Taha, A. Labigne, U. Zahringer, A. J. Coyle, P. S. DiStefano, J. Bertin, P. J. Sansonetti, and D. J. Philpott. 2003. Nod1 detects a unique muropeptide from gram-negative bacterial peptidoglycan. *Science* **300:**1584–1587.

31. Girardin, S. E., I. G. Boneca, J. Viala, M. Chamaillard, A. Labigne, G. Thomas, D. J. Philpott, and P. J. Sansonetti. 2003. Nod2 is a general sensor of peptidoglycan through muramyl dipeptide (MDP) detection. *J. Biol. Chem.* **278:**8869–8872.

32. Girardin, S. E., L. H. Travassos, M. Herve, D. Blanot, I. G. Boneca, D. J. Philpott, P. J. Sansonetti, and D. Mengin-Lecreulx. 2003. Peptidoglycan molecular requirements allowing detection by Nod1 and Nod2. *J. Biol. Chem.* **278:**41702–41708.

33. Gupta, D., T. N. Kirkland, S. Viriyakosol, and R. Dziarski. 1996. CD14 is a cell-activating receptor for bacterial peptidoglycan. *J. Biol. Chem.* **271:**23310–23316.

34. Han, S. H., J. H. Kim, M. Martin, S. M. Michalek, and M. H. Nahm. 2003. Pneumococcal lipoteichoic acid (LTA) is not as potent as staphylococcal LTA in stimulating Toll-like receptor 2. *Infect. Immun.* **71:**5541–5548.

35. Hanahan, D. J. 1986. Platelet activating factor: a biologically active phosphoglyceride. *Annu. Rev. Biochem.* **55:**483–509.

36. Heine, H., and E. Lien. 2003. Toll-like receptors and their function in innate and adaptive immunity. *Int. Arch. Allergy Immunol.* **130:**180–192.

37. Heumann, D., C. Barras, A. Severin, M. P. Glauser, and A. Tomasz. 1994. Gram-positive cell walls stimulate synthesis of tumor necrosis factor alpha and interleukin-6 by human monocytes. *Infect. Immun.* **62:**2715–2721.

38. Hoffmann, J. A., F. C. Kafatos, C. A. Janeway, and R. A. Ezekowitz. 1999. Phylogenetic perspectives in innate immunity. *Science* **284:**1313–1318.

39. Honda, Z., M. Nakamura, I. Miki, M. Minami, T. Watanabe, Y. Seyama, H. Okado, H. Toh, K. Ito, T. Miyamoto, and T. Shimizu. 1991. Cloning by functional expression of platelet-activating factor receptor from guinea-pig lung. *Nature* **349:**342–346.

40. Humphries, H. E., and N. J. High. 2002. The role of *licA* phase variation in the pathogenesis of invasive disease by *Haemophilus influenzae* type b. *FEMS Immunol. Med. Microbiol.* **34:**221–230.

41. Inohara, N., Y. Ogura, F. F. Chen, A. Muto, and G. Nunez. 2001. Human Nod1 confers responsiveness to bacterial lipopolysaccharides. *J. Biol. Chem.* **276:**2551–2554.

42. Jacks-Weis, J., Y. Kim, and P. P. Cleary. 1982. Restricted deposition of C3 on M+ group A streptococci: correlation with resistance to phagocytosis. *J. Immunol.* **128:**1897–1902.

43. Jenni, R., and B. Berger-Bachi. 1998. Teichoic acid content in different lineages of *Staphylococcus aureus* NCTC8325. *Arch. Microbiol.* **170:**171–178.

44. Jersmann, H. P., C. S. Hii, G. L. Hodge, and A. Ferrante. 2001. Synthesis and surface expression of CD14 by human endothelial cells. *Infect. Immun.* **69:**479–485.

45. Jin, Y., D. Gupta, and R. Dziarski. 1998. Endothelial and epithelial cells do not respond to complexes of peptidoglycan with soluble CD14 but are activated indirectly by peptidoglycan-induced tumor necrosis factor-alpha and interleukin-1 from monocytes. *J. Infect. Dis.* **177:**1629–1638.

46. Keller, R., W. Fischer, R. Keist, and S. Bassetti. 1992. Macrophage response to bacteria: induction of marked secretory and cellular activities by lipoteichoic acids. *Infect. Immun.* **60:**3664–3672.

47. Kengatharan, K. M., S. De Kimpe, C. Robson, S. J. Foster, and C. Thiemermann. 1998. Mechanism of gram-positive shock: identification of peptidoglycan and lipoteichoic acid moieties essential in the induction of nitric oxide synthase, shock, and multiple organ failure. *J. Exp. Med.* **188:**305–315.

48. Kunz, D., N. P. Gerard, and C. Gerard. 1992. The human leukocyte platelet-activating factor receptor. cDNA cloning, cell surface expression, and construction of a novel epitope-bearing analog. *J. Biol. Chem.* **267:**9101–9106.

49. Kurt-Jones, E. A., L. Mandell, C. Whitney, A. Padgett, K. Gosselin, P. E. Newburger, and R. W. Finberg. 2002. Role of toll-like receptor 2 (TLR2) in neutrophil activation: GM-CSF enhances TLR2 expression and TLR2-mediated interleukin 8 responses in neutrophils. *Blood* **100:**1860–1868.

50. Lancefield, R. C. 1962. Current knowledge of type-specific M antigens of group A streptococci. *J. Immunol.* **89:**307–313.

51. Lancefield, R. C. 1957. Differentiation of group A streptococci with a common R antigen into three serological types, with special reference to the bactericidal test. *J. Exp. Med.* **106:**525–544.

52. Majcherczyk, P. A., H. Langen, D. Heumann, M. Fountoulakis, M. P. Glauser, and P. Moreillon. 1999. Digestion of *Streptococcus pneumoniae* cell walls with its major peptidoglycan hydrolase releases branched stem peptides carrying proinflammatory activity. *J. Biol. Chem.* **274:**12537–12543.

53. Majcherczyk, P. A., E. Rubli, D. Heumann, M. P. Glauser, and P. Moreillon. 2003. Teichoic acids are not required for *Streptococcus pneumoniae* and *Staphylococcus aureus* cell walls to trigger the release of tumor necrosis factor by peripheral blood monocytes. *Infect. Immun.* **71:**3707–3713.

54. Mathison, J. C., P. S. Tobias, E. Wolfson, and R. J. Ulevitch. 1992. Plasma lipopolysaccharide (LPS)-binding protein. A key component in macrophage recognition of gram-negative LPS. *J. Immunol.* **149:**200–206.

55. Mattsson, E., L. Verhage, J. Rollof, A. Fleer, J. Verhoef, and H. van Dijk. 1993. Peptidoglycan and teichoic acid from *Staphylococcus epidermidis* stimulate human monocytes to release tumour necrosis factor-alpha, interleukin-1 beta and interleukin-6. *FEMS Immunol. Med. Microbiol.* **7:**281–287.

56. **Means, T. K., D. T. Golenbock, and M. J. Fenton.** 2000. The biology of Toll-like receptors. *Cytokine Growth Factor Rev.* **11:**219–232.

57. **Medzhitov, R.** 2001. CpG DNA: security code for host defense. *Nat. Immunol.* **2:**15–16.

58. **Medzhitov, R.** 2001. Toll-like receptors and innate immunity. *Nat. Rev. Immunol.* **1:**135–145.

59. **Medzhitov, R., P. Preston-Hurlburt, and C. A. Janeway, Jr.** 1997. A human homologue of the Drosophila Toll protein signals activation of adaptive immunity. *Nature* **388:**394–397.

60. **Meli, D. N., S. Christen, S. L. Leib, and M. G. Tauber.** 2002. Current concepts in the pathogenesis of meningitis caused by *Streptococcus pneumoniae*. *Curr. Opin. Infect. Dis.* **15:**253–257.

61. **Michel, T., J. M. Reichhart, J. A. Hoffmann, and J. Royet.** 2001. Drosophila Toll is activated by Gram-positive bacteria through a circulating peptidoglycan recognition protein. *Nature* **414:**756–759.

62. **Mock, M., and T. Mignot.** 2003. Anthrax toxins and the host: a story of intimacy. *Cell Microbiol.* **5:**15–23.

63. **Moreillon, P., and P. A. Majcherczyk.** 2003. Proinflammatory activity of cell-wall constituents from gram-positive bacteria. *Scand. J. Infect. Dis.* **35:**632–641.

64. **Ozinsky, A., D. M. Underhill, J. D. Fontenot, A. M. Hajjar, K. D. Smith, C. B. Wilson, L. Schroeder, and A. Aderem.** 2000. The repertoire for pattern recognition of pathogens by the innate immune system is defined by cooperation between toll-like receptors. *Proc. Natl. Acad. Sci. USA* **97:**13766–13771.

65. **Prescott, S. M., G. A. Zimmerman, and T. M. McIntyre.** 1990. Platelet-activating factor. *J. Biol. Chem.* **265:**17381–17384.

66. **Rietschel, E. T., H. Brade, O. Holst, L. Brade, S. Muller-Loennies, U. Mamat, U. Zahringer, F. Beckmann, U. Seydel, K. Brandenburg, A. J. Ulmer, T. Mattern, H. Heine, J. Schletter, H. Loppnow, U. Schonbeck, H. D. Flad, S. Hauschildt, U. F. Schade, F. Di Padova, S. Kusumoto, and R. R. Schumann.** 1996. Bacterial endotoxin: chemical constitution, biological recognition, host response, and immunological detoxification. *Curr. Top. Microbiol. Immunol.* **216:**39–81.

67. **Rietschel, E. T., J. Schletter, B. Weidemann, V. El-Samalouti, T. Mattern, U. Zahringer, U. Seydel, H. Brade, H. D. Flad, S. Kusumoto, D. Gupta, R. Dziarski, and A. J. Ulmer.** 1998. Lipopolysaccharide and peptidoglycan: CD14-dependent bacterial inducers of inflammation. *Microb. Drug Resist.* **4:**37–44.

68. **Rossjohn, J., R. J. Gilbert, D. Crane, P. J. Morgan, T. J. Mitchell, A. J. Rowe, P. W. Andrew, J. C. Paton, R. K. Tweten, and M. W. Parker.** 1998. The molecular mechanism of pneumolysin, a virulence factor from *Streptococcus pneumoniae*. *J. Mol. Biol.* **284:**449–461.

69. **Sansonetti, P. J.** 2001. Rupture, invasion and inflammatory destruction of the intestinal barrier by Shigella, making sense of prokaryote-eukaryote cross-talks. *FEMS Microbiol. Rev.* **25:**3–14.

70. **Schroder, N. W., S. Morath, C. Alexander, L. Hamann, T. Hartung, U. Zahringer, U. B. Gobel, J. R. Weber, and R. R. Schumann.** 2003. Lipoteichoic acid (LTA) of *Streptococcus pneumoniae* and *Staphylococcus aureus* activates immune cells via Toll-like receptor (TLR)-2, lipopolysaccharide-binding protein (LBP), and CD14, whereas TLR-4 and MD-2 are not involved. *J. Biol. Chem.* **278:**15587–15594.

71. **Schuchat, A., K. Robinson, J. D. Wenger, L. H. Harrison, M. Farley, A. L. Reingold, L. Lefkowitz, and B. A. Perkins, and the Active Surveillance Team.** 1997. Bacterial meningitis in the United States in 1995. *N. Engl. J. Med.* **337:**970–976.

72. **Schumann, R. R., D. Pfeil, D. Freyer, W. Buerger, N. Lamping, C. J. Kirschning, U. B. Goebel, and J. R. Weber.** 1998. Lipopolysaccharide and pneumococcal cell wall components activate the mitogen activated protein kinases (MAPK) erk-1, erk-2, and p38 in astrocytes. *Glia* **22:**295–305.

73. **Schwandner, R., R. Dziarski, H. Wesche, M. Rothe, and C. J. Kirschning.** 1999. Peptidoglycan- and lipoteichoic acid-induced cell activation is mediated by toll-like receptor 2. *J. Biol. Chem.* **274:**17406–17409.

74. **Takeda, K., T. Kaisho, and S. Akira.** 2003. Toll-like receptors. *Annu. Rev. Immunol.* **21:**335–376.

75. **Takehana, A., T. Katsuyama, T. Yano, Y. Oshima, H. Takada, T. Aigaki, and S. Kurata.** 2002. Overexpression of a pattern-recognition receptor, peptidoglycan-recognition protein-LE, activates imd/relish-mediated antibacterial defense and the prophenoloxidase cascade in Drosophila larvae. *Proc. Natl. Acad. Sci. USA* **99:**13705–13710.

76. **Tauber, M. G., H. Khayam-Bashi, and M. A. Sande.** 1985. Effects of ampicillin and corticosteroids on brain water content, cerebrospinal fluid pressure, and cerebrospinal fluid lactate levels in experimental pneumococcal meningitis. *J. Infect. Dis.* **151:**528–534.

77. Tauber, M. G., A. M. Shibl, C. J. Hackbarth, J. W. Larrick, and M. A. Sande. 1987. Antibiotic therapy, endotoxin concentration in cerebrospinal fluid, and brain edema in experimental *Escherichia coli* meningitis in rabbits. *J. Infect. Dis.* **156**:456–462.

78. Timmerman, C. P., E. Mattsson, L. Martinez-Martinez, L. De Graaf, J. A. Van Strijp, H. A. Verbrugh, J. Verhoef, and A. Fleer. 1993. Induction of release of tumor necrosis factor from human monocytes by staphylococci and staphylococcal peptidoglycans. *Infect. Immun.* **61**:4167–4172.

79. Tuomanen, E., B. Hengstler, R. Rich, M. A. Bray, O. Zak, and A. Tomasz. 1987. Nonsteroidal anti-inflammatory agents in the therapy for experimental pneumococcal meningitis. *J. Infect. Dis.* **155**:985–990.

80. Tuomanen, E., H. Liu, B. Hengstler, O. Zak, and A. Tomasz. 1985. The induction of meningeal inflammation by components of the pneumococcal cell wall. *J. Infect. Dis.* **151**:859–868.

81. Tuomanen, E., A. Tomasz, B. Hengstler, and O. Zak. 1985. The relative role of bacterial cell wall and capsule in the induction of inflammation in pneumococcal meningitis. *J. Infect. Dis.* **151**:535–540.

82. van Langevelde, P., J. T. van Dissel, E. Ravensbergen, B. J. Appelmelk, I. A. Schrijver, and P. H. Groeneveld. 1998. Antibiotic-induced release of lipoteichoic acid and peptidoglycan from *Staphylococcus aureus*: quantitative measurements and biological reactivities. *Antimicrob. Agents Chemother.* **42**:3073–3078.

83. Wang, Z. M., X. Li, R. R. Cocklin, M. Wang, M. Wang, K. Fukase, S. Inamura, S. Kusumoto, D. Gupta, and R. Dziarski. 2003. Human peptidoglycan recognition protein-L is an N-acetylmuramoyl-L-alanine amidase. *J. Biol. Chem.* **278**:49044–49052.

84. Wang, Z. M., C. Liu, and R. Dziarski. 2000. Chemokines are the main proinflammatory mediators in human monocytes activated by *Staphylococcus aureus*, peptidoglycan, and endotoxin. *J. Biol. Chem.* **275**:20260–20267.

85. Watson, D. A., D. M. Musher, and J. Verhoef. 1995. Pneumococcal virulence factors and host immune responses to them. *Eur. J. Clin. Microbiol. Infect. Dis.* **14**:479–490.

86. Weber, J. R., D. Freyer, C. Alexander, N. W. Schroder, A. Reiss, C. Kuster, D. Pfeil, E. I. Tuomanen, and R. R. Schumann. 2003. Recognition of pneumococcal peptidoglycan: an expanded, pivotal role for LPS binding protein. *Immunity* **19**:269–279.

87. Weber, J. R., P. Moreillon, and E. I. Tuomanen. 2003. Innate sensors for Gram-positive bacteria. *Curr. Opin. Immunol.* **15**:408–415.

88. Weidemann, B., H. Brade, E. T. Rietschel, R. Dziarski, V. Bazil, S. Kusumoto, H. D. Flad, and A. J. Ulmer. 1994. Soluble peptidoglycan-induced monokine production can be blocked by anti-CD14 monoclonal antibodies and by lipid A partial structures. *Infect. Immun.* **62**:4709–4715.

89. Weidemann, B., J. Schletter, R. Dziarski, S. Kusumoto, F. Stelter, E. T. Rietschel, H. D. Flad, and A. J. Ulmer. 1997. Specific binding of soluble peptidoglycan and muramyldipeptide to CD14 on human monocytes. *Infect. Immun.* **65**:858–864.

90. Weiser, J. N. 1998. Phase variation in colony opacity by *Streptococcus pneumoniae*. *Microb. Drug Resist.* **4**:129–135.

91. Weiser, J. N., R. Austrian, P. K. Sreenivasan, and H. R. Masure. 1994. Phase variation in pneumococcal opacity: relationship between colonial morphology and nasopharyngeal colonization. *Infect. Immun.* **62**:2582–2589.

92. Werner, T., G. Liu, D. Kang, S. Ekengren, H. Steiner, and D. Hultmark. 2000. A family of peptidoglycan recognition proteins in the fruit fly *Drosophila melanogaster*. *Proc. Natl. Acad. Sci. USA* **97**:13772–13777.

93. Yoshimura, A., E. Lien, R. R. Ingalls, E. Tuomanen, R. Dziarski, and D. Golenbock. 1999. Cutting edge: recognition of Gram-positive bacterial cell wall components by the innate immune system occurs via Toll-like receptor 2. *J. Immunol.* **163**:1–5.

94. Yoshimura, A., H. Takada, T. Kaneko, I. Kato, D. Golenbock, and Y. Hara. 2000. Structural requirements of muramylpeptides for induction of Toll-like receptor 2-mediated NF-kappaB activation in CHO cells. *J. Endotoxin Res.* **6**:407–410.

95. Zhang, G., and S. Ghosh. 2001. Toll-like receptor-mediated NF-kappaB activation: a phylogenetically conserved paradigm in innate immunity. *J. Clin. Investig.* **107**:13–19.

INTERACTIONS OF *STREPTOCOCCUS PNEUMONIAE* WITH THE PROTEINS OF THE COMPLEMENT PATHWAYS

Margaret K. Hostetter

13

Classic experiments of a century ago provided commanding proof that serum elements were essential for the phagocytosis of virulent pneumococci (54). Thirty-three years later, the heat-stable, specific components of the opsonic interaction (antibodies) were distinguished from nonspecific, heat-labile serum proteins, now named complement (48). In the presence of antibody alone, the rate of neutrophil phagocytosis of serotype 3 *Streptococcus pneumoniae* was accelerated more than sevenfold by the addition of active complement proteins in fresh human serum. These early experiments established the essential function of complement in opsonization, phagocytosis, and killing of *S. pneumoniae*.

As shown in Fig. 1, three pathways of complement activation are now thought to trigger the opsonic deposition of C3b onto the bacterial surface:

1. the classical pathway, which can be triggered by immunoglobulin G (IgG), by IgM, and by certain non-antibody-dependent mechanisms

2. the lectin-binding pathway, which is initiated by the deposition of mannose binding lectin (MBL) or ficolin, in conjunction with MBL-associated serum proteinase 1, 2, or 3, to an array of carbohydrate groups on the bacterial surface

3. the alternative pathway, which is initiated by hydrolyzed C3 (C3·H_2O) and activated factor B (Bb)

COMPLEMENT DEFICIENCIES AND THE EPIDEMIOLOGY OF PNEUMOCOCCAL INFECTION

The propensity for any one of these pathways to be activated in defense of the host is in turn a consequence of the epidemiology of pneumococcal infection. Bacteremic illness ordinarily peaks within the first 6 to 12 months of life and is thought to reflect the absence of type-specific anticapsular antibodies from the serum of young infants. Thus, activation of a classical pathway by IgG antibodies will be impaired in infants, and "natural antibodies" or IgM is an important innate defense; in infancy, the alternative pathway assumes a contributory role.

MBL, a member of the collectin family, exhibits the collagen-like domain and the carbohydrate recognition domain of the collectin family. Three to six of these subunits comprise human MBL, which binds carbohydrates with three or four hydroxyl groups in the pyranose ring in the presence of calcium (49). Preferred

Margaret K. Hostetter, Department of Pediatrics, Yale University School of Medicine, 333 Cedar Street, LMP 4085, P.O. Box 208064, New Haven, CT 06520-8064.

The Pneumococcus. Volume Editor, Elaine I. Tuomanen,
© 2004 ASM Press, Washington, D.C.

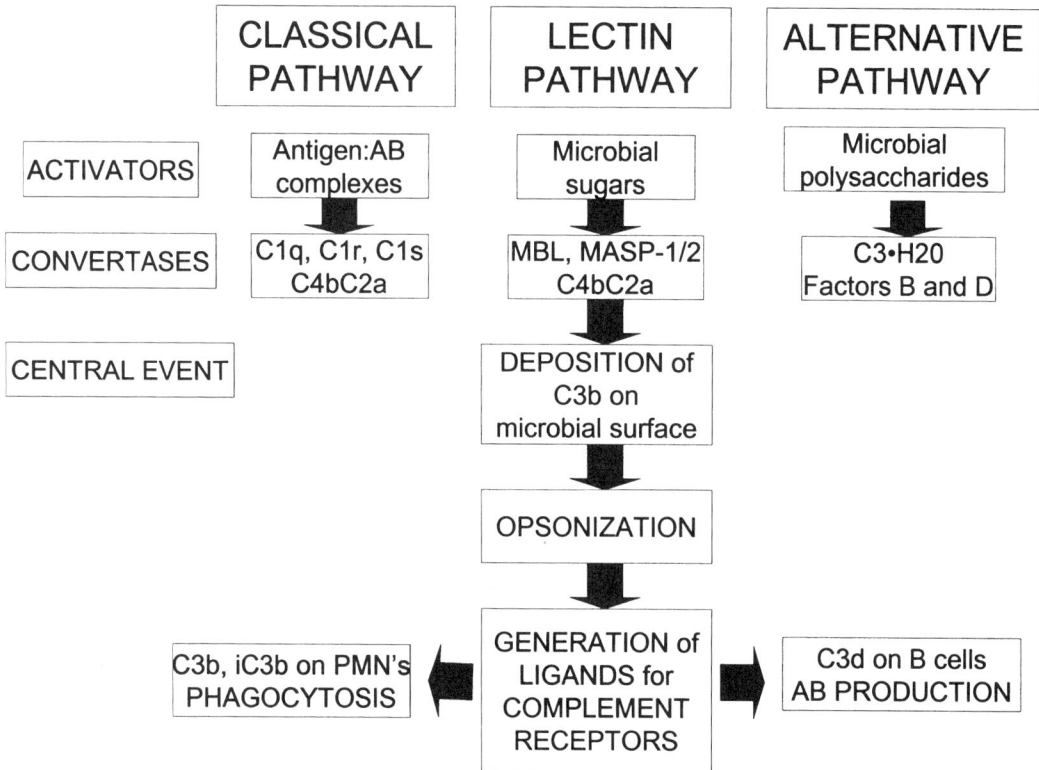

FIGURE 1 Diagram of complement-mediated opsonization and phagocytosis. AB, antibody.

ligands for MBL are mannose and N-acetyl-glucosamine (GlcNAc), whereas carbohydrates such as galactose and salicylic acid are not recognized (16). Probands homozygous for mutations of MBL have a 2.5-fold-increased risk of invasive pneumococcal infections such as pneumonia, isolated bacteremia, and meningitis (45). Unfortunately the authors of this study did not report the pneumococcal serotypes that infected these patients; based on the preferred ligands for MBL, infections with serotype 14, with a capsular polysaccharide rich in N-acetyl-glucosamine, should have been common, while infections with serotype 4 or 12, whose capsules contain substantial quanities of D-galactose, should have been unlikely.

Patients with inherited deficiencies of factor B, D, or I of the alternative complement pathway should theoretically be predisposed to in-

vasive bacterial illness; however, there seems to be no specific predisposition to pneumococcal disease. Factor I deficiency causes a permanent and unregulated activation of the alternative pathway that results in increased cleavage of C3 and consumption of factor B, factor H, and properdin. One Brazilian family with consanguineous parents had three affected children, one of whom died after sepsis, but the causative organisms covered many species (33). Other patients with factor I deficiency have suffered from pneumococcal and meningococcal infections in childhood and from recurrent erysipelas due to *Streptococcus pyogenes* as adults (4). Generation of mice deficient in factor D by gene targeting also led to a considerable slowing of the kinetics of pneumococcal opsonization (55).

At the convergence of all three pathways sits the third component of complement (C3); the central role of this protein in pneumococcal killing cannot be overemphasized. As predicted from the C3-dependent functions in Fig. 1, opsonization, bactericidal activity, and neutrophil chemotaxis are absent in affected patients. Patients genetically deficient in C3 suffer lifelong from recurrent pyogenic infections due to organisms such as *S. pneumoniae*, *S. pyogenes*, and *Staphylococcus aureus*. Infections may manifest as recurrent bacteremias after 2 years of age or as repeated episodes of erysipelas in adulthood (5, 8).

Both C3-deficient guinea pigs and dogs have recurrent bacterial infections, and C3 knockout mice generated by disrupting the murine C3 promoter have no detectable C3 protein or complement activity in serum (7, 12, 52). Challenge of these mice with *S. pneumoniae* reveals a 2,000-fold increase in bacteremia. Thus, the most compelling evidence for the role of innate immunity in defense against pneumococcal infection focuses on C3.

BIOCHEMISTRY OF COMPLEMENT-MEDIATED OPSONIZATION AND PHAGOCYTOSIS

While each of the complement pathways diagramed in Fig. 1 culminates in the cleavage of C3 and in the deposition of the opsonic form of this protein (C3b) on the pneumococcal surface, a review of the mechanisms by which C3 mediates opsonization and phagocytic killing is in order. Cleavage of C3 by the serum proteinases of the classical, lectin, or alternative complement pathway causes a conformational change that results in the exposure of a thioester bond linking Cys_{988} with Glu_{990} on the α-chain of C3 (46). The reactive carbonyl group of Glu_{990} attaches to the acceptor surface in covalent ester linkage with exposed hydroxyl groups or in covalent amide linkage with free amino groups of tyrosine residues (23, 32, 43, 44). This biochemical mechanism enables C3b to attach covalently to polysaccharides with exposed hydroxyl groups or amino sugars or to peptides exposed within the cell wall (22). If covalent attachment fails to occur within the 60-µs half-life of C3, the glu-tamyl carbonyl binds water, and the protein is rendered incapable of cellular binding.

Early experiments showed that opsonization of virulent encapsulated pneumococci (serotype 25) led to deposition of C3b on the capsule (51). Subsequent studies using ^{125}I-labeled guinea pig C3 demonstrated that C3b was bound to cell walls in both encapsulated and nonencapsulated strains, with approximately 50% of total C3b bound to teichoic acid (24). When ^{3}H-labeled purified human C3 was used to reconstitute the alternative complement pathway in agammaglobulinemic serum in which C3 and C4 had been inactivated, C3b deposition on encapsulated serotypes occurred on both the capsule and the cell wall (22). Since a considerable portion of the C3b may be deposited noncovalently, and hence improperly oriented to interact with neutrophil complement receptors (CD35 = CR1 = C3b receptor; CD11b/CD18 = CR3 = iC3b receptor), such binding assays should seek to quantitate the amount of covalently deposited C3b. When this is done, the number of C3b molecules covalently attached to an individual organism ranges about sixfold, from a low of 29,000 on a serotype 3 strain to a high of 155,000 on a serotype 4 strain (21).

C3-mediated phagocytosis is dependent upon the subsequent degradation of opsonically deposited C3b to fragments such as iC3b. C3b interacts with CD35; this receptor is unable to initiate phagocytosis because of the absence of a transmembrane domain. iC3b is the preferred ligand for the integrin CD11b/CD18, which clearly initiates phagocytosis. After opsonization in nonimmune serum, serotypes 3 and 4, resistant to phagocytosis, display roughly equal proportions of C3b, iC3b, and C3d in ester linkage to capsular polysaccharides, while serotypes 6A and 14, more susceptible to phagocytosis, display only iC3b in ester linkage (21). Techniques available at the time made it impossible to evaluate what C3 fragments may have been deposited in amide linkage.

When CD11b/CD18 on normal neutrophils was blocked by monoclonal antibodies against CD11b, phagocytosis of pneumococcal serotypes 6A and 14 was inhibited by 60 to 80%, thereby confirming the role of iC3b in mediating phagocytosis for these serotypes (17). Blockade of the neutrophil C3b receptor (CD35) failed to inhibit phagocytosis for these iC3b-bearing serotypes. For serotype 3 bearing capsule-bound C3b, iC3b, and C3d, CD11b/CD18-mediated phagocytosis accounted for only 20% of uptake; again there was no evidence for CD35-mediated phagocytosis (17). C3d does not serve as a ligand for neutrophil complement receptors but is preferentially recognized by CD21 (complement receptor type 2) on B lymphocytes and contributes to the production of anticapsular antibodies (18).

Recent work has demonstrated that the classical pathway is the dominant mechanism by which the innate immune system defends against pneumococcal infection, at least in mice (10). Mice genetically deficient in C1q ($C1q^{-/-}$) or factor B ($FB^{-/-}$) developed more rapidly progressive disease than wild-type mice after intranasal inoculation with D39, a serotype 2 organism. When the inoculum was reduced by about 60% (4×10^5 CFU), $C1q^{-/-}$ mice were less susceptible than $C3^{-/-}$ mice, although median survival differed by only 12 h. The authors concluded that both the classical and alternative pathways contributed to defense against the pneumococcus.

$C4^{-/-}$ mice, in which neither the classical nor the MBL pathway can function, were no worse than $C1q^{-/-}$ mice after intranasal inoculation, pointing to the minimal contribution of the MBL pathway in this murine model. $FB^{-/-}$ mice had an increased rate of disease progression and mortality compared with wild-type mice but were less susceptible than $C1q^{-/-}$ or $C4^{-/-}$ mice, showing that natural IgM bolsters innate immunity to S. pneumoniae in mice but is not as important as the presence of an intact classical pathway. $C1q^{-/-}$ mice had more rapidly progressive disease than $FB^{-/-}$ mice after both intranasal and intraperitoneal inoculation, although once again median survival differed by only 12 h with the D39 strain (10). These authors are to be complimented for

the breadth of their study; however, the use of the D39 strain required extraordinarily large inocula (10^6 CFU); it might have been instructive to see if these results were duplicated with the use of a highly virulent strain such as TIGR4, for which the 50% lethal dose is typically less than 10 CFU.

This brief review indicates that S. pneumoniae may interfere with the mechanisms of innate immunity by attacking components of three opsonic pathways, all of which converge on C3. Recent genetic investigations have identified a number of pneumococcal proteins that affect the activation, deposition, or cleavage of C3 so as to interfere with C3-mediated opsonization and phagocytosis. Below is a review of those pneumococcal components that target proteins of the classical or the alternative pathway. To date, no pneumococcal proteins that interfere with the lectin-binding pathway have been identified.

PNEUMOCOCCAL PROTEINS THAT INTERFERE WITH COMPLEMENT-MEDIATED FUNCTIONS

PspA: Inhibition of Complement Activation

In studies with XID mice infected with $PspA^+$ or $PspA^-$ mutants, serum C3 concentration decreased significantly within 30 min after infection with the $PspA^+$ strain, but there was markedly less activation of complement in response to the $PspA^-$ strain, as measured by serum C3 levels (47). Interestingly, more efficient activation of complement by the $PspA^-$ mutant correlated with improved clearance of the mutant. Mice deficient in C3 or in factor B failed to clear $PspA^-$ mutants, where $C5^{-/-}$ mice readily cleared the mutant strain. Analysis of serum C3 after infection purportedly showed a 7% decrease in the amount of C3b formed in the presence of the $PspA^+$ strain, which the authors considered significant in comparison to the $PspA^-$ strain (47).

Because PspAs are currently divided into six clades that comprise three families, questions arise as to their interchangeability. When a family 2 PspA from strain TIGR4 was substituted for a family 1 PspA from strain WU2, both isogenic mutants interfered with complement deposition on the pneumococcal surface and were equally virulent. Deleting the choline-binding region of the family 2 PspA generated a mutant that failed to bind human lactoferrin but retained most of its complement deposition. Virulence was moderately attenuated (39).

Similar studies were done by other groups. PspA from strain D39 (capsule type 2) was compared to family 2 PspA from WU2 (capsule type 3) (1). More C3b was detectable on D39 than on WU2, and exchange of WU2 and D39 PspAs did not change these results. Capsule type influenced C3b deposition: D39 had the most C3b, D39 expressing the type 3 capsule was intermediate, and WU2 was the lowest. The authors found that the C3 β-chain and the 46-kDa fragment of the iC3b α-chain were present on both parent strains and on PspA mutants.

Presence of the C3 α′-chain or C3d was not reported. This result might have proven helpful, since others have reported an equal distribution of C3b, iC3b, and C3d on serotype 3 organisms (21). If this distribution held true, then only one-third of covalently deposited C3 fragments would be eligible to interact with CD11b/CD18, which mediates phagocytosis of capsule type 3 (17). This observation might have explained why neither expression of PspA nor the presence of PspA antibodies correlated with protection.

CbpA: Interference with C3 Binding

Choline-binding proteins, first identified by Rosenow et al. (40), have been shown by others to mediate binding of opsonically inactive C3 (11):

1. A 90-kDa choline-binding protein called CbpA was isolated by affinity chromatography with thioester-disrupted C3.

2. CbpA identified in lysates and supernatants of exponentially growing CP1200 (unencapsulated strain) was able to bind thioester-disrupted C3 on a Western blot.

3. Purified CbpA was able to bind purified thioester-disrupted C3 with a K_d of 200 nM.

Functional CbpA has been identified in clinical isolates of serotypes 1, 3, 4, 14, and 19F and in cell lysates of PspA mutants JY119, JY118, and JY53 and their parent strain, WU2 (11).

CbpA also appears to bind secretory IgA (sIgA) (19); sIgA inhibits the binding of thioester-disrupted C3 to CbpA by approximately 50%. Scatchard analysis of C3 binding to CbpA was linear, with a K_d of 200 nM, while Scatchard analysis of the binding of sIgA was nonlinear, with a K_d of 1 μM. These results show that thioester-disrupted C3 appears to be the preferred ligand for CbpA, although there may be multiple lower-affinity sites for the binding of sIgA, one or more of which may overlap with the binding site for C3 (11).

CbpA is highly immunogenic and elicits an IgG antibody response in humans recovering from pneumococcal infection (11). Studies in the laboratory strain CP1200 demonstrate secretion of CbpA in the absence of choline. The conservation of CbpA among several encapsulated clinical isolates of *S. pneumoniae*, its production throughout pneumococcal growth, and its presence in both a cell-bound and a secreted form of 90 kDa suggest a potential role in combating opsonization and phagocytosis. Like the IgA-binding protein SpsA (19), CbpA is encoded by the same locus that encodes PspC (9).

Later studies showed that 7.2 pmol of purified CbpA was as effective as 1 U of interleukin-1α (IL-1α) in eliciting IL-8 from alveolar epithelial cells in serum-free medium (34). A *cbpA* mutant was only half as potent as a laboratory strain (CP1200). CbpA may be one of several pneumococcal proteins that contribute to the marked neutrophilic leukocytosis that has long been recognized as a hallmark of pneumococcal pneumonia or bacteremia; which of the many mediators of chemotaxis in the lung is specifically triggered by pneumococci remains unknown (31). Thus, CbpA binds C3 nonopsonically and induces IL-8, but the expression of CbpA renders incoming neutrophils incapable of phagocytosis because C3 is bound incorrectly.

PspC: Absorption of Factor H

Factor H is a negative regulator of alternative pathway activation and is composed of 20 repetitive domains termed short consensus repeats (SCRs) (57). Factor H interferes with the formation of the alternative pathway C3 convertase (C3bBb), functions as a cofactor in the degradation of C3b by factor I, and promotes the dissociation of Bb from C3 and C5 convertases, thereby impairing both the amplification loop of the alternative pathway and the generation of the membrane attack complex. Theoretically, the ability of pneumococci to bind factor H could either handicap opsonophagocytosis mediated by the alternative pathway or promote phagocytosis by enhancing cleavage of C3b to iC3b, a potent ligand for phagocytosis.

The *pspC* gene is present in 75% or more of pneumococcal strains, but the encoded proteins are highly polymorphic and are divided into 11 groups. Shared motifs include an amino-terminal signal peptide followed by an alpha-helical region encompassing 118 to 589 amino acids, a proline-rich region, and a carboxyl-terminal anchor (25). This protein is related to PspA, CbpA, and Hic, a factor H-binding inhibitor of complement (26). In type 2 pneumococci, deletion of 405 amino acids of the amino terminus ablated factor H absorption from plasma, although part of this deletion affected the proline-rich region, which is thought to be essential for efficient binding of factor H (13). Conversely, SCRs 13 to 15 of factor H are essential for binding PspC (14). The region encompassing SCRs 13 to 15 is distinct from other sites on factor H that bind bacterial proteins from *S. pyogenes*, group B streptococci, *Yersinia enterocolitica*, *Borrelia burgdorferi*, and *Neisseria gonorrhoeae* (14).

Studies of type 3 pneumococci, highly resistant to phagocytosis (48), also pointed to absorption of factor H from plasma (36) and led to the isolation of Hic. Hic is anchored to the cell wall with an LPXTG motif (26) not found in PspC groups 1 to 6, which anchor via

a choline-binding motif (25). Mutant type 3 pneumococci lacking Hic are unable to absorb factor H from human plasma and fail to bind radiolabeled factor H. The binding site for Hic on factor H is SCRs 12 to 14 (27). The presence of this protein in type 3 pneumococci may explain the cleavage of C3b to iC3b and thence to C3d in this strain (21). In fact, PspC and Hic might be more important in serotypes 6 and 14, in which degradation of C3b is halted at i3Cb.

Pla: Activation and Depletion of the Proteins of the Classical Complement Pathway

Pneumolysin (Pla), a 53-kDa intracellular protein that is highly conserved and universally expressed among virulent pneumococci, is released during autolysis (28, 30, 38). Cytolytic activities mediated by a pore-forming toxin are directed against virtually all mammalian cells, including erythrocytes, leukocytes, endothelial cells, and alveolar epithelial cells (41). Pla also activates the classical complement pathway, possibly through the binding of the Fc portion of IgG (35), and thereby leads to depletion of complement proteins and reduced opsonic activity in serum (35, 38).

A point mutation (Asp$_{385}$ → Asn) ablates complement-activating activity but retains hemolytic activity (H$^+$C$^-$ strain); a second point mutation (His$_{367}$ → Arg) retains all complement-activating activity but nullifies ~99% of the pore-forming activity that is essential for cytotoxicity (H$^-$C$^+$ strain) (3). Complement-activating activity was associated with bacterial growth in lung tissue and blood at 24 h after intratracheal challenge in mice, while cytotoxic activity best predicted enhanced bacterial replication during the first 6 h of infection and compromise of the alveolar capillary barrier. Forty-eight hours after infection, there were no differences in replication for either mutant (29).

When instilled into murine tracheas, mutants unable to activate complement led to massive neutrophil infiltration of infected bronchioles, while mutant strains in which cytolytic activity had been disrupted resulted in decreased cellular infiltrate (29). By 48 h, however, these differences had attenuated, and both H^+C^- and H^-C^+ strains were able to induce consolidation of bronchiolar spaces and associated parenchyma as well as the wild-type strain. Interestingly, strains retaining complement-activating activity (H^-C^+) and wild-type pneumococci recruited significantly greater numbers of T cells over the first 48 h after infection.

The effects of Pla have also been studied in a rat model of cirrhosis induced by intragastric installation of carbon tetrachloride (2). In cirrhotic animals, significantly lower CH50 and C3 levels are present prior to infection. Serum from cirrhotic rats infected with an H^+C^+ (wild type) strain had even lower levels of C3 as well as reduced opsonizing capacity for *S. pneumoniae*, compared to sera from noncirrhotic rats infected with the same strain (H^+C^+) or to sera from cirrhotic rats infected with H^+C^- or *ply* mutant organisms (2).

In models of meningitis, the data are conflicting. Two papers have demonstrated that Pla-deficient strains are less virulent, but a third study found that a *ply* mutant was no less virulent than the wild-type strain in murine models of meningitis (15, 50, 53). It would therefore appear that the ability of Pla to interfere with complement-mediated functions is dependent on the environment and on the pneumococcal strains tested experimentally. In a niche where complement proteins are abundant (blood or alveoli), expression of Pla contributes significantly to virulence; at a site

where complement proteins are scarce, such as the meningeal space, a major role for Pla may be more difficult to prove, especially since all mutants are derived from the D39 strain and not the more virulent TIGR4 strain.

PhpA: Degradation of C3

In vitro, exponentially growing clinical and laboratory isolates of serotypes 3, 4, and 14 have been shown to degrade first the β- and then the α-chain of purified human C3 (6). Such degradation did not occur with albumin as a substrate but was associated with heat-labile cell wall protein(s). A 79-kDa recombinant pneumococcal protein (rPhpA) encompassing a 20-kDa polypeptide with C3-degrading activity (20) proved to be a potent immunogen (56). Antibodies to this protein bound to the surface of unencapsulated *S. pneumoniae* and reduced nasopharyngeal colonization and bacteremia induced by an encapsulated type 3 strain.

Table 1 summarizes the interactions of pneumococcal capsules and proteins with components of the classical and alternative pathways of complement. In the absence of type-specific anticapsular antibodies, opsonophagocytosis proceeds by both pathways: the classical pathway presumptively triggered by natural antibodies of the IgM class and the alternative, by polysaccharides mimicking those found on bacteria (37, 42). Although a murine model has emphasized the importance of the classical pathway (10), pneumococcal infection peaks in unvaccinated children aged 6 to 12 months—a time when anticapsular antibodies acquired transplacentally have disap-

TABLE 1 Interactions of pneumococcal components with complement proteins

Pneumococcal component	C3 binding	Binding of other complement proteins	C3 degradation	Inhibition of complement activation	Cleavage of:	
					C3b to iC3b	iC3b C3d
PspA			✓	✓	?	?
PspC/Hic		✓ (factor H)				
Pla				✓		
CbpA	✓					
PhpA			✓			

peared. In these children the alternative pathway, however dilatory, is the only known defense against invasive pneumococcal disease.

Pneumococcal proteins that interact with components of the classical or alternative complement pathways are predominantly surface expressed (except for Pla), related if not completely conserved (e.g., CbpA, PspA, PspC, and Hic), and potently immunogenic. While as yet no pneumococcal protein has been found to interfere with the lectin pathway, understanding the manifold mechanisms by which *S. pneumoniae* outwits the innate immune system will almost certainly lead to improved control of the morbidity and mortality ascribable to this clever pathogen.

ACKNOWLEDGMENT

This work was completed with support from NIH grant R-01 AI 49438.

REFERENCES

1. **Abeyta, M., G. G. Hardy, and J. Yother.** 2003. Genetic alteration of capsule type but not PspA type affects accessibility of surface-bound complement and surface antigens of *Streptococcus pneumoniae*. *Infect. Immun.* **71:**218–225.
2. **Alcantara, R. B., L. C. Preheim, and M. J. Gentry-Nielsen.** 2001. Pneumolysin-induced complement depletion during experimental pneumococcal bacteremia. *Infect. Immun.* **69:**3569–3575.
3. **Alexander, J. E., A. M. Berry, J. C. Paton, J. B. Rubins, P. W. Andrew, and T. J. Mitchell.** 1998. Amino acid changes affecting the activity of pneumolysin alter the behavior of pneumococci in pneumonia. *Microb. Pathog.* **24:**167–174.
4. **Alper, C. A., N. Abramson, R. B. J. Johnston, J. H. Jandl, and F. S. Rosen.** 1970. Increased susceptibility to infection associated with abnormalities of complement-mediated functions and of the third component of complement (C3). *N. Engl. J. Med.* **282:**349–354.
5. **Alper, C. A., H. R. Colten, F. S. Rosen, A. R. Rabson, G. M. MacNab, and J. S. S. Gear.** 1972. Homozygous deficiency of C3 in a patient with repeated infections. *Lancet* **ii:**1179–1181.
6. **Angel, C. S., M. Ruzek, and M. K. Hostetter.** 1994. Degradation of C3 by *Streptococcus pneumoniae*. *J. Infect. Dis.* **170:**600–608.
7. **Auerbach, H. S., R. Burger, A. Dodds, and H. R. Colten.** 1990. Molecular basis of complement C3 deficiency in guinea pigs. *J. Clin. Investig.* **86:**96–106.
8. **Ballow, M., J. E. Shira, L. Harden, S. Y. Yang, and N. K. Day.** 1970. Complete absence of the third component of complement in man. *J. Clin. Investig.* **56:**703–710.
9. **Brooks-Walter, A., D. E. Briles, and S. K. Hollingshead.** 1999. The pspC gene of *Streptococcus pneumoniae* encodes a polymorphic protein, PspC, which elicits cross-reactive antibodies to PspA and provides immunity to pneumococcal bacteremia. *Infect. Immun.* **67:**6533–6542.
10. **Brown, J. S., T. Hussell, S. M. Gilliland, D. W. Holden, J. C. Paton, M. R. Ehrenstein, M. J. Walport, and M. Botto.** 2002. The classical pathway is the dominant complement pathway required for innate immunity to *Streptococcus pneumoniae* infection in mice. *Proc. Natl. Acad. Sci. USA* **99:**16969–16974.
11. **Cheng, Q., D. Finkel, and M. K. Hostetter.** 2000. Novel purification scheme and functions for a C3-binding protein from Streptococcus pneumoniae. *Biochemistry* **39:**5450–5457.
12. **Circolo, A., G. Garnier, W. Fukuda, X. Wang, T. Hidvegi, A. J. Szalai, D. E. Briles, J. E. Volanakis, R. A. Wetsel, and H. R. Colten.** 1999. Genetic disruption of the murine complement C3 promoter region generates deficient mice with extrahepatic expression of C3 mRNA. *Immunopharmacology* **42:**135–149.
13. **Dave, S., A. Brooks-Walter, M. K. Pangburn, and L. S. McDaniel.** 2001. PspC, a pneumococcal surface protein, binds human factor H. *Infect. Immun.* **69:**3435–3437.
14. **Duthy, T. G., R. J. Ormsby, E. Giannakis, A. D. Ogunniyi, U. H. Stroeher, J. C. Paton, and D. L. Gordon.** 2002. The human complement regulator factor H binds pneumococcal surface protein PspC via short consensus repeats 13 to 15. *Infect. Immun.* **70:**5604–5611.
15. **Friedland, I. R., M. M. Paris, S. Hickey, S. Shelton, K. Olsen, J. C. Paton, and G. H. McCracken.** 1995. The limited role of pneumolysin in the pathogenesis of pneumococcal meningitis. *J. Infect. Dis.* **172:**805–809.
16. **Fujita, T.** 2002. Evolution of the lectin-complement pathway and its role in innate immunity. *Nat. Rev. Immunol.* **2:**346–353.
17. **Gordon, D. L., G. M. Johnson, and M. K. Hostetter.** 1986. Ligand-receptor interactions in the phagocytosis of virulent Streptococcus pneumoniae by polymorphonuclear leukocytes. *J. Infect. Dis.* **154:**619–626.
18. **Griffioen, A. W., E. A. Toebes, B. J. Zegers, and G. T. Rijkers.** 1992. Role of CR2 in the human adult and neonatal in vitro antibody

response to type 4 pneumococcal polysaccharide. *Cell. Immunol.* **143:**11–22.

19. **Hammerschmidt, S., S. R. Talay, P. Brandtzaeg, and G. S. Chhatwal.** 1997. SpsA, a novel pneumococcal surface protein with specific binding to secretory immunoglobulin A and secretory component. *Mol. Microbiol.* **25:**1113–1124.

20. **Hostetter, M. K.** 1999. Opsonic and nonopsonic interactions of C3 with Streptococcus pneumoniae. *Microb. Drug Resist.* **5:**85–89.

21. **Hostetter, M. K.** 1986. Serotypic variations among virulent pneumococci in deposition and degradation of covalently bound C3b: implications for phagocytosis and antibody production. *J. Infect. Dis.* **153:**682–693.

22. **Hostetter, M. K., R. A. Krueger, and D. J. Schmeling.** 1984. The biochemistry of opsonization: central role of the reactive thiolester of the third component of complement. *J. Infect. Dis.* **150:**653–661.

23. **Hostetter, M. K., M. L. Thomas, F. S. Rosen, and B. F. Tack.** 1982. Binding of C3b proceeds by a transesterification reaction at the thiolester site. *Nature* **298:**72–75.

24. **Hummell, D. S., R. W. Berninger, and A. Tomasz.** 1981. The fixation of C3b to pneumococcal cell wall polymers as a result of activation of the alternative complement pathway. *J. Immunol.* **127:**1287–1289.

25. **Ianelli, F., M. R. Oggioni, and G. Pozzi.** 2002. Allelic variation in the highly polymorphic locus pspC of *Streptococcus pneumoniae*. *Gene* **284:**63–71.

26. **Janulczyk, R., F. Ianelli, A. G. Sjoholm, G. Pozzi, and L. Bjorck.** 2000. Hic, a novel surface protein of *Streptococcus pneumoniae* that interferes with complement function. *J. Biol. Chem.* **275:**37257–37263.

27. **Jarva, H., R. Janulczyk, J. Hellwage, P. F. Zipfel, L. Bjorck, and S. Meri.** 2002. *Streptococcus pneumoniae* evades complement attack and opsonophagocytosis by expressing the pspC locus-encoded Hic protein that binds to short consensus repeats 8–11 of factor H. *J. Immunol.* **168:**1886–1894.

28. **Johnson, M. K.** 1977. Cellular location of pneumolysin. *FEMS Microbiol. Lett.* **2:**243–245.

29. **Jounblat, R., A. Kadioglu, T. J. Mitchell, and P. W. Andrew.** 2003. Pneumococcal behavior and host responses during bronchopneumonia are affected differently by the cytolytic and complement-activating activities of pneumolysin. *Infect. Immun.* **71:**1813–1819.

30. **Kanclerski, K., and R. Mollby.** 1987. Production and purification of *Streptococcus pneumoniae* hemolysin (pneumolysin). *J. Clin. Microbiol.* **25:**222–225.

31. **Kiehl, M. G., H. Ostermann, M. L. Thomas, T. Birkfellner, and J. Kienast.** 1997. Inflammatory mediators in BAL fluid as markers of evolving pneumonia in leukocytopenic patients. *Chest* **112:**1214–1220.

32. **Law, S. A. K., T. M. Minich, and R. P. Levine.** 1981. The binding reaction between the third human complement protein and small molecules. *Biochemistry* **20:**7457–7460.

33. **Leitao, M. F., M. M. Vilela, R. Rutz, A. S. Grumach, A. Condino-Neto, and M. Kirschfink.** 1997. Complement factor I deficiency in a family with recurrent infections. *Immunopharmacology* **38:**207–213.

34. **Madsen, M., Y. Lebenthal, Q. Cheng, B. L. Smith, and M. K. Hostetter.** 2000. A pneumococcal protein that elicits interleukin-8 from pulmonary epithelial cells. *J. Infect. Dis.* **181:**1330–1336.

35. **Mitchell, T. J., P. W. Andrew, F. K. Saunders, A. N. Smith, and G. J. Boulnois.** 1991. Complement activation and antibody binding by pneumolysin via a region of the toxin homologous to a human acute-phase protein. *Mol. Microbiol.* **5:**1883–1888.

36. **Neeleman, C., S. P. M. Geelen, P. C. Aerts, M. R. Daha, T. E. Mollnes, J. J. Roord, G. Posthuma, H. van Dijk, and A. Fleer.** 1999. Resistance to both complement activation and phagocytosis in type 3 pneumococci is mediated by the binding of complement regulatory protein factor H. *Infect. Immun.* **67:**4517–4524.

37. **Pangburn, M. K.** 1989. Analysis of recognition in the alternative pathway of complement: evidence of polysaccharide size. *J. Immunol.* **142:**2766–2770.

38. **Paton, J. C., A. M. Berry, R. A. Lock, D. Hansman, and P. A. Manning.** 1986. Cloning and expression in *Escherichia coli* of the *Streptococcus pneumoniae* gene encoding pneumolysin. *Infect. Immun.* **54:**50–55.

39. **Ren, B., A. J. Szalai, O. Thomas, S. K. Hollingshead, and D. E. Briles.** 2003. Both family 1 and family 2 PspA proteins can inhibit complement deposition and confer virulence to a capsular serotype 3 strain of *Streptococcus pneumoniae*. *Infect. Immun.* **71:**75–85.

40. **Rosenow, C., P. Ryan, J. N. Weiser, S. Johnson, P. Fontan, A. Ortqvist, and H. R. Masure.** 1997. Contribution of novel choline-binding proteins to adherence, colonization and immunogenicity of Streptococcus pneumoniae. *Mol. Microbiol.* **25:**819–829.

41. **Rubins, J. B., D. Charboneau, J. C. Paton, T. J. Mitchell, P. W. Andrew, and E. F. Janoff.** 1995. Dual function of pneumolysin in

the early pathogenesis of murine pneumococcal pneumonia. *J. Clin. Investig.* **95:**142–150.

42. **Sahu, A., T. R. Kozel, and M. K. Pangburn.** 1994. Specificity of the thioester-containing reactive site of human C3 and its significance to complement activation. *Biochem. J.* **302:**429–436.

43. **Sahu, A., and M. K. Pangburn.** 1995. Tyrosine is a potential site for covalent attachment of activated complement component C3. *Mol. Immunol.* **32:**711–716.

44. **Sim, R. B., T. M. Twose, D. S. Patterson, and E. Sim.** 1981. The covalent-binding reaction of complement component C3. *Biochem. J.* **193:**115–127.

45. **Suchismita, R., K. Knox, S. Segal, D. Griffiths, C. E. Moore, K. I. Welsh, A. Smarason, N. P. Day, W. I. McPheat, D. W. Crook, and A. V. S. Hill.** 2002. MBL genotype and risk of invasive pneumococcal disease: a case-control study. *Lancet* **359:**1569–1573.

46. **Tack, B. F., R. A. Harrison, J. Janatova, M. L. Thomas, and J. W. Prahl.** 1980. Evidence for presence of an internal thiolester bond in third component of human complement. *Proc. Natl. Acad. Sci. USA* **77:**5764–5768.

47. **Tu, A.-H. T., R. L. Fulgham, M. A. McCrory, D. E. Briles, and A. J. Szalai.** 1999. Pneumococcal surface protein A inhibits complement activation by *Streptococcus pneumoniae*. *Infect. Immun.* **67:**4720–4724.

48. **Ward, H. K., and J. F. Enders.** 1933. An analysis of the opsonic and tropic action of normal and immune sera based on experiments with pneumococcus. *J. Exp. Med.* **57:**527–547.

49. **Weis, W. I., K. Drickhamer, and W. A. Hendrickson.** 1992. Structure of a C-type mannose-binding protein complexed with an oligosaccharide. *Nature* **360:**127–134.

50. **Wellmer, A., G. Zysk, J. Gerber, T. Kunst, M. Von Mering, S. Bunkowski, H. Effert, and R. Nau.** 2002. Decreased virulence of a pneumolysin-deficient strain of *Streptococcus pneumoniae* in murine meningitis. *Infect. Immun.* **70:**6504–6508.

51. **Winkelstein, J. A., A. S. Abramovitz, and A. Tomasz.** 1980. Activation of C3 via the alternative complement pathway results in fixation of C3b to the pneumococcal cell wall. *J. Immunol.* **124:**2502–2506.

52. **Winkelstein, J. A., L. C. Cork, D. E. Griffin, R. J. Adams, and D. L. Price.** 1981. Genetically determined deficiency of the third component of complement in the dog. *Science* **212:**1169–1173.

53. **Winter, A. J., S. D. Comis, M. P. Osborne, M. J. Tarlow, J. Stephen, P. W. Andrew, J. Hill, and T. J. Mitchell.** 1997. A role for pneumolysin but not neuraminidase in the hearing loss and cochlear damage induced by experimental pneumococcal meningitis in guinea pigs. *Infect. Immun.* **65:**4411–4418.

54. **Wright, A. E., and S. R. Douglas.** 1903. An experimental investigation of blood fluids in connection with phagocytosis. *Proc. R. Soc. Lond. B* **72:**357–370.

55. **Xu, Y., M. Ma, G. C. Ippolito, H. W. Schroeder, M. C. Carroll, and J. E. Volanakis.** 2001. Complement activation in factor D-deficient mice. *Proc. Natl. Acad. Sci. USA* **98:**14577–14582.

56. **Zhang, Y., A. W. Masi, V. Barniak, K. Mountzouros, M. K. Hostetter, and B. A. Green.** 2001. Recombinant PhpA protein, a unique histidine motif-containing protein from *Streptococcus pneumoniae*, protects mice against intranasal pneumococcal challenge. *Infect. Immun.* **69:**3827–3836.

57. **Zipfel, P. F., C. Skerka, J. Hellwage, S. T. Jokiranta, S. Meri, V. Brade, P. Kraiczy, M. Noris, and G. Remuzzi.** 2002. Factor H family proteins: on complement, microbes, and human disease. *Biochem. Soc. Trans.* **30:**971–978.

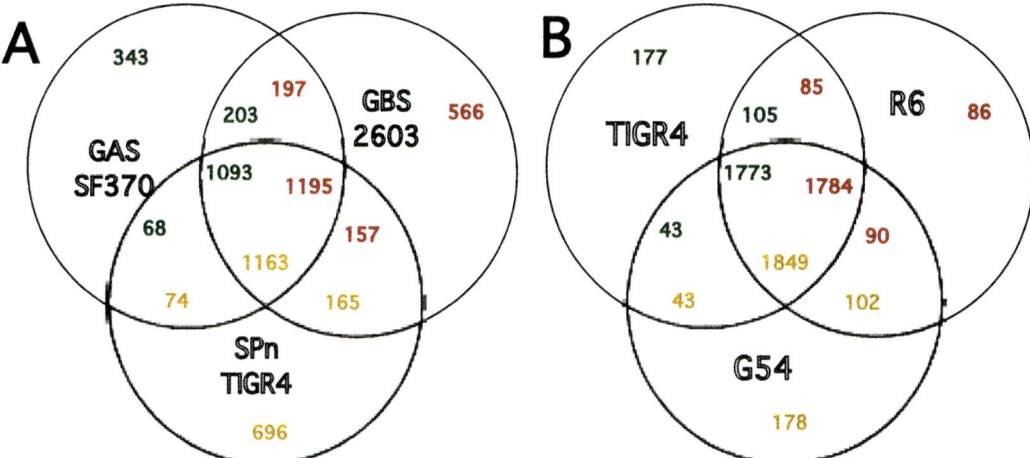

COLOR PLATE 1 [chapter 2]   Venn diagrams of BlastP comparisons of different genomes. The genomes compared in panel A are *S. pneumoniae* strain TIGR4, *S. agalactiae* strain 2603 V/R (GBS), and *S. pyogenes* strain SF370 (GAS). Numbers in the intersections represent genes shared in the respective species-to-species (A) or *S. pneumoniae* strain-to-strain (B) comparison. The colors represent the genome used as the query in each case. Numbers vary slightly depending on the query because of gene duplications in some strains or species. A BlastP e-value cutoff of $10e^{-15}$ was used in each case.

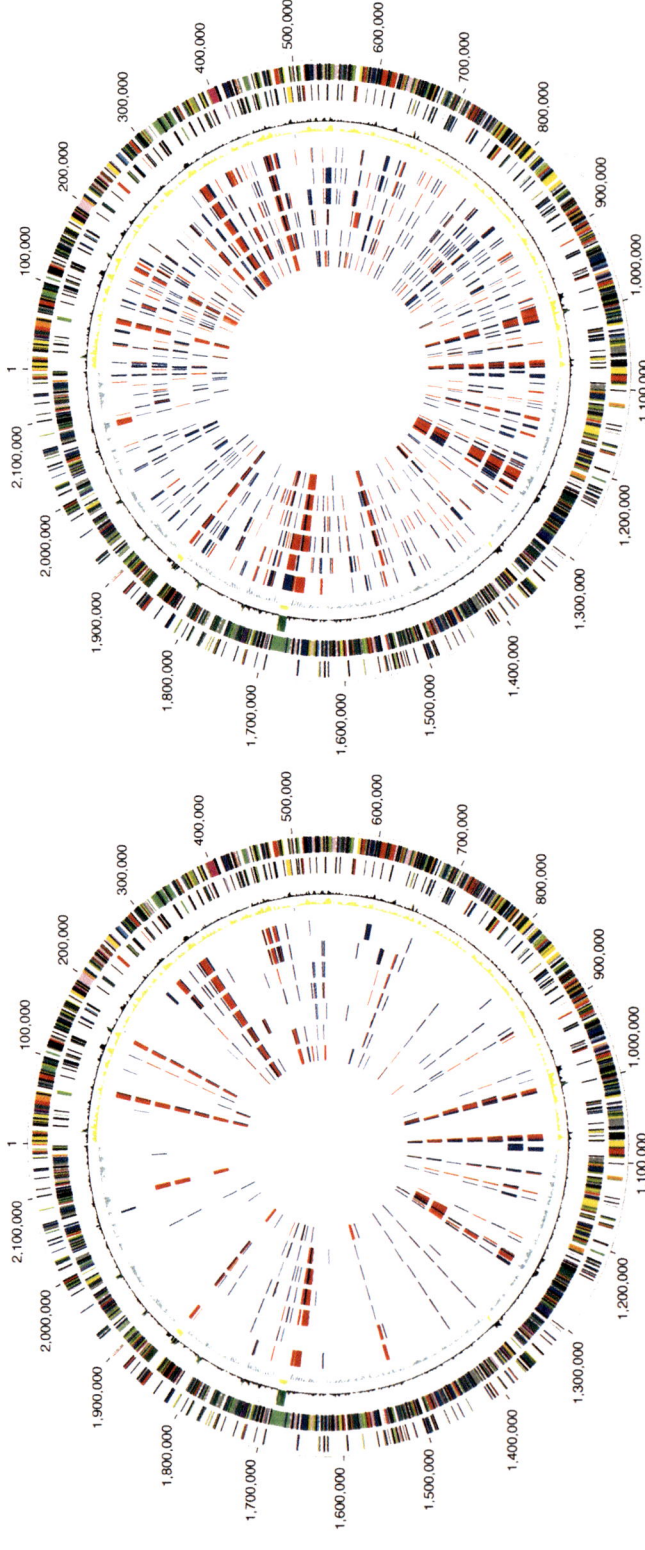

COLOR PLATE 2 [chapter 2] Circular representation of the genomic content of 13 diverse strains compared to the TIGR4 strain genome through CGH on a preliminary microarray. The cDNA microarray consisted of amplicons representing 1,954 of the 2,326 genes from the sequenced strain TIGR4. Results for the competitive hybridization between TIGR4 DNA labeled with Cy3 and the DNA from a single strain labeled with Cy5 are shown in each of the circles. Genes are either (i) present in test strain (white), (ii) absent from test strain (shown in blue for Cy3/Cy5 ratios of 5 to 10 or red for Cy3/Cy5 ratios of more than 10), or (iii) not present on the array and not analyzed (182 genes, 8% of total) (also left in white, so as to focus the discussion on those genes for which a test was performed but no orthologs were detected). The two outermost circles are the predicted genes on the forward and reverse strands, color coded by functional category. The third circle is the analysis; of atypical nucleotide composition; the most atypical regions are displayed in green. The fourth circle is the GC skew analysis; positive values are indicated in yellow, and negative values are in cyan. The next circles on the left diagram correspond to strains MA through MG (outer circle to inner circle). The circles on the right diagram correspond to strains MI through MN.

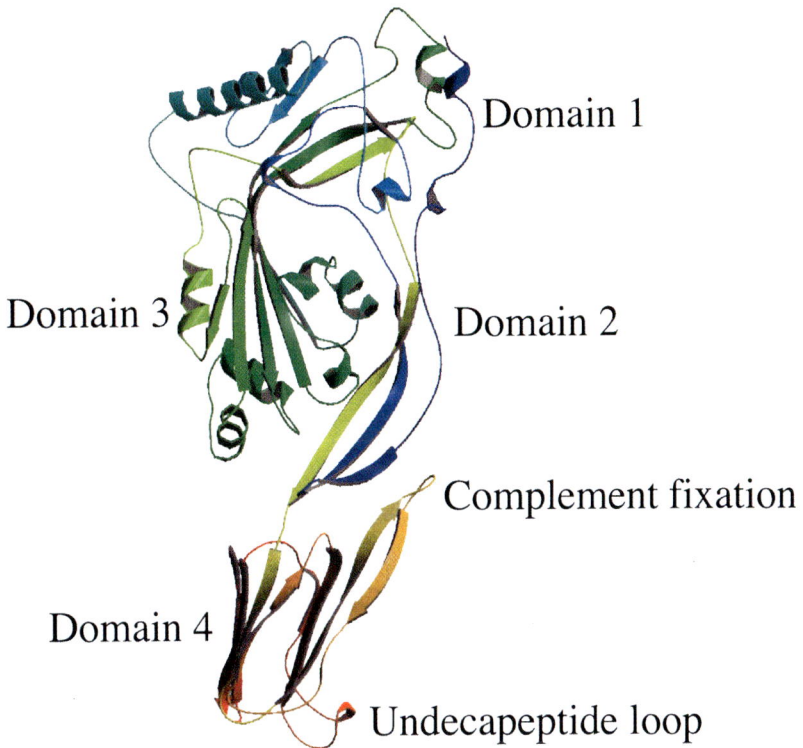

Domain 1

Domain 3

Domain 2

Complement fixation

Domain 4

Undecapeptide loop

COLOR PLATE 3 [chapter 5]   Structural model of PLY showing the four domains of the proteins and the conserved undecapeptide loop. The region involved in complement activation is marked. The model is based on reference 74.

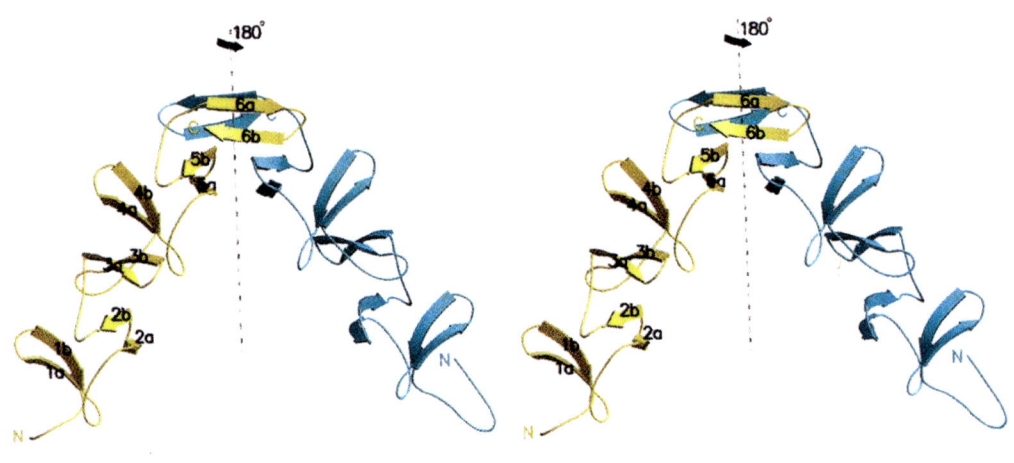

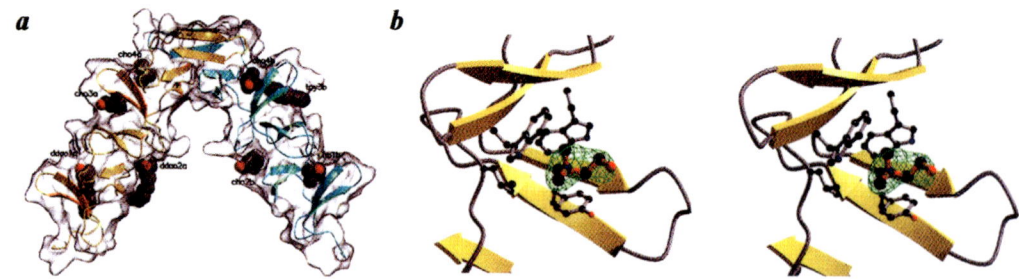

COLOR PLATE 4 [chapter 6] Predicted structural organization of LytA and its Ch-binding domain. (A) Physicochemical studies suggest the organization of a LytA dimer having four subdomains per monomer, two (N1 and N2) located N terminally and two (C1 and C2) located C terminally. Reprinted from the *Journal of Biological Chemistry* (69) with permission of the publisher. (B) Stereo ribbon diagram of the dimeric C-LytA with β-strand assignments. The two monomers are differently colored (yellow and cyan). Reprinted from the *Journal of Molecular Biology* (14) with permission of the publisher. (C) Ch-binding sites. *a*, ribbon diagram of the C-LytA dimer inscribed into the molecular surface. *b*, stereo diagram of one Ch-binding site where choline is highlighted in orange. Reprinted from *Nature Structural Biology* (15) with permission of the publisher.

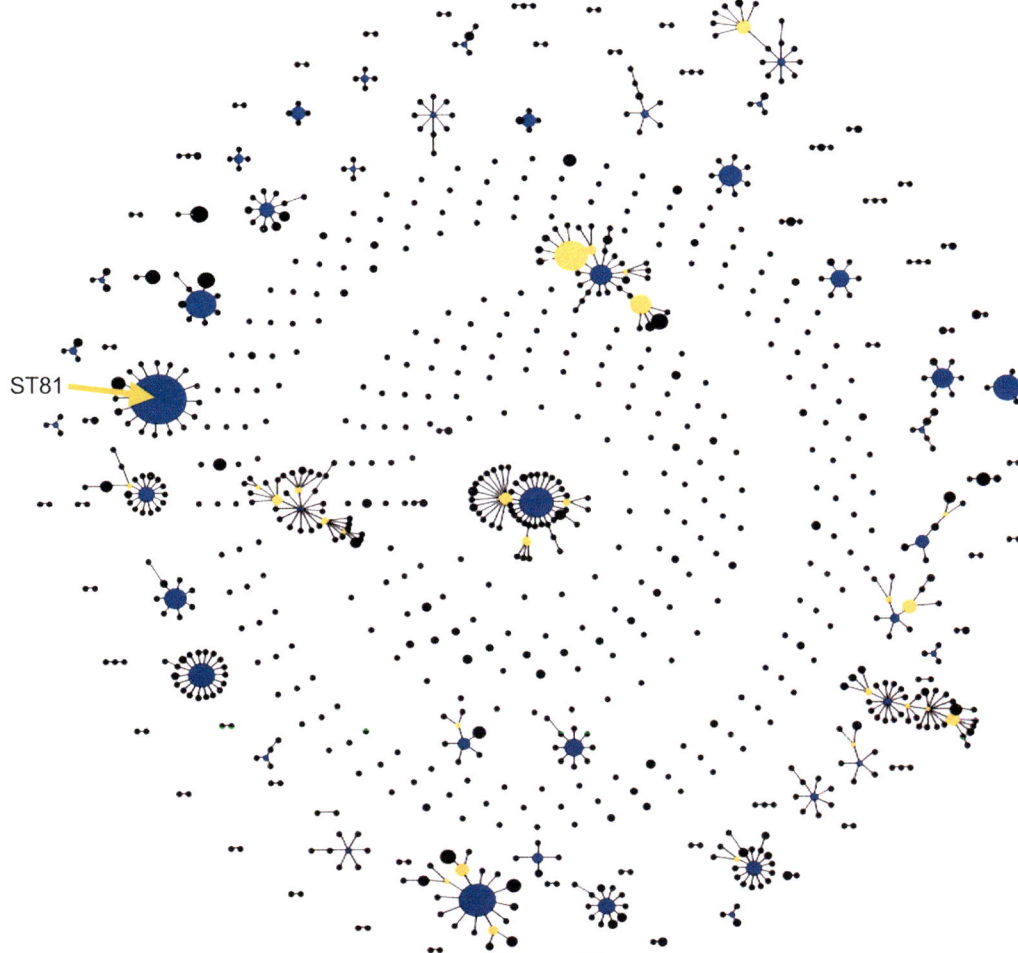

ST81

COLOR PLATE 5 [chapter 8]   Clonal complexes within the pneumococcal population represented in the MLST database. All isolates in the pneumococcal MLST database (as of September 2003) were displayed as a single eBURST diagram (a population snapshot [17a]). Clusters of linked STs correspond to clonal complexes; single STs differ from all other STs in the population at two or more loci. The areas of the circles representing the individual STs indicate the prevalence of each ST in the current database; STs in blue are the predicted founders of the clonal complex, and those in yellow are founders of subgroups within the clonal complex (17a). For clarity, ST numbers are not shown, except for ST81.

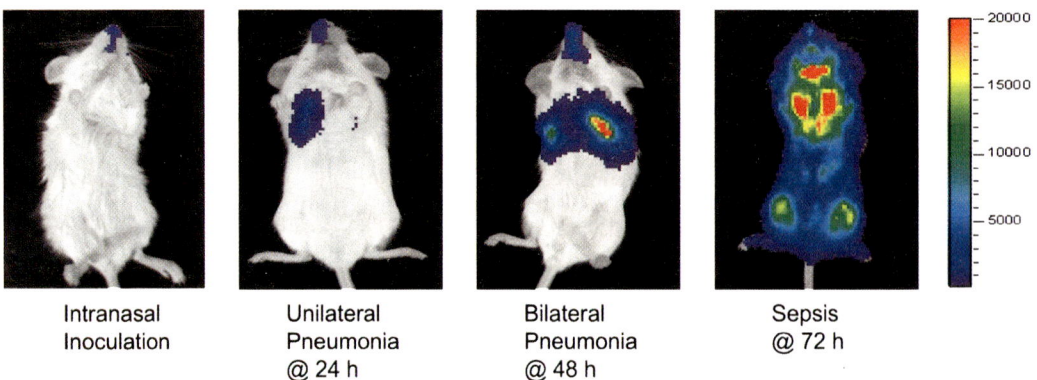

| Intranasal Inoculation | Unilateral Pneumonia @ 24 h | Bilateral Pneumonia @ 48 h | Sepsis @ 72 h |

COLOR PLATE 6 [chapter 15] Progression of pneumonia followed by in vivo imaging. Xenogen Imaging System was used to follow the spread of luciferase-bearing pneumococci introduced intranasally. Bacteria descended into the lung by 24 h, spread from one lung to the other by 48 h, and then spread to the bloodstream by 72 h.

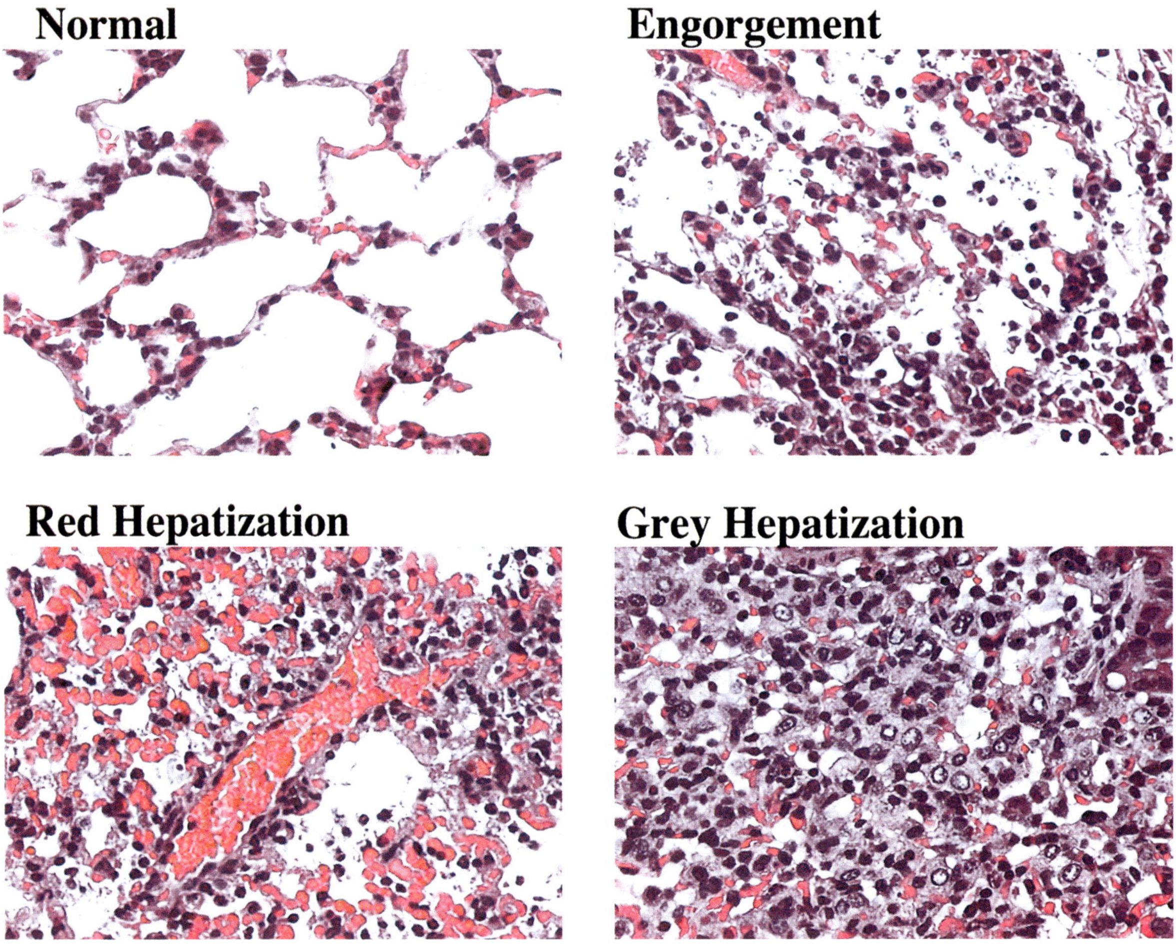

Normal

Engorgement

Red Hepatization

Grey Hepatization

COLOR PLATE 7 [chapter 15] Histopathology of three stages of consolidation. Hematoxylin and eosin staining of sections of lung representing the indicated stages of consolidation progressing from engorgement to red to grey hepatization. (Images courtesy of S. Dixon, St. Jude Children's Research Hospital.)

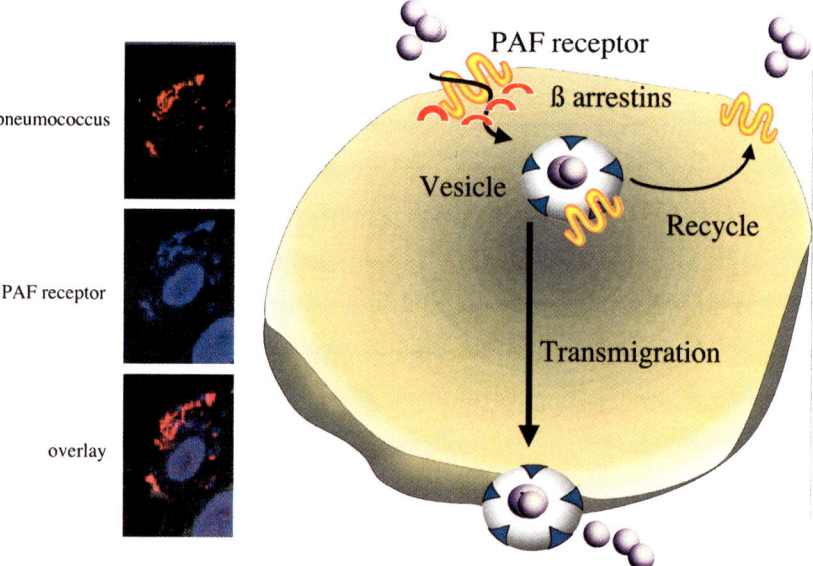

pneumococcus

PAF receptor

overlay

PAF receptor

ß arrestins

Vesicle

Recycle

Transmigration

COLOR PLATE 8 [chapter 15]   Model of pneumococcal invasion into alveolar epithelial cells. Pneumococci adhere to glycoconjugates on the surfaces of epithelial cells. A second step is required to initiate uptake and transcytosis. In the lung, the second step involves the interaction of choline on the cell wall and CbpA with PAFr. The molecular details of this event are as yet unclear but pneumococci and PAFr colocalize during invasion of type II alveolar cells in vitro. Subsequent to PAFr ligation, bacteria are taken up into a clathrin-coated vesicle that recruits the cytosolic scaffold protein β arrestin. The vesicle then moves across the cell from the apical to basolateral surface. Internalized bacteria do not escape the vesicles. In vitro, bilayers of epithelial and endothelial cells can be crossed by pneumococci in about 9 h.

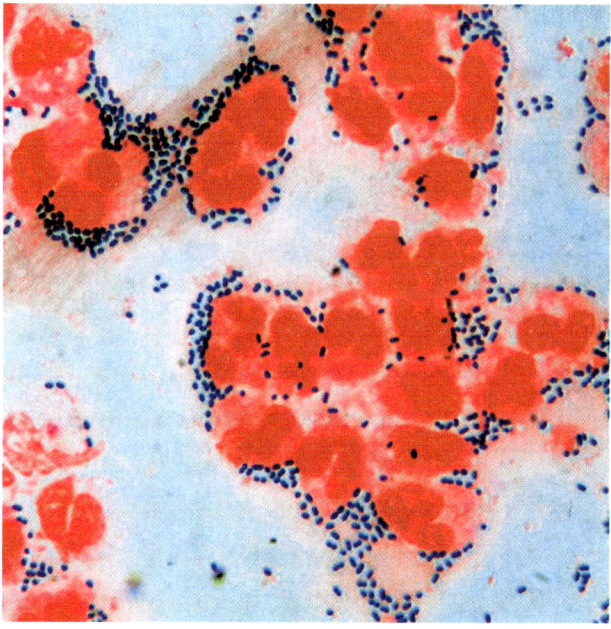

COLOR PLATE 9 [chapter 16]   Gram stain of CSF from a patient with meningitis. A large number of leukocytes surrounded by gram-positive diplococci have accumulated in the fluid. Note that in the fluid environment, few bacteria are effectively phagocytosed. (Photo courtesy of E. Halle, Charité Hospital, Berlin, Germany.)

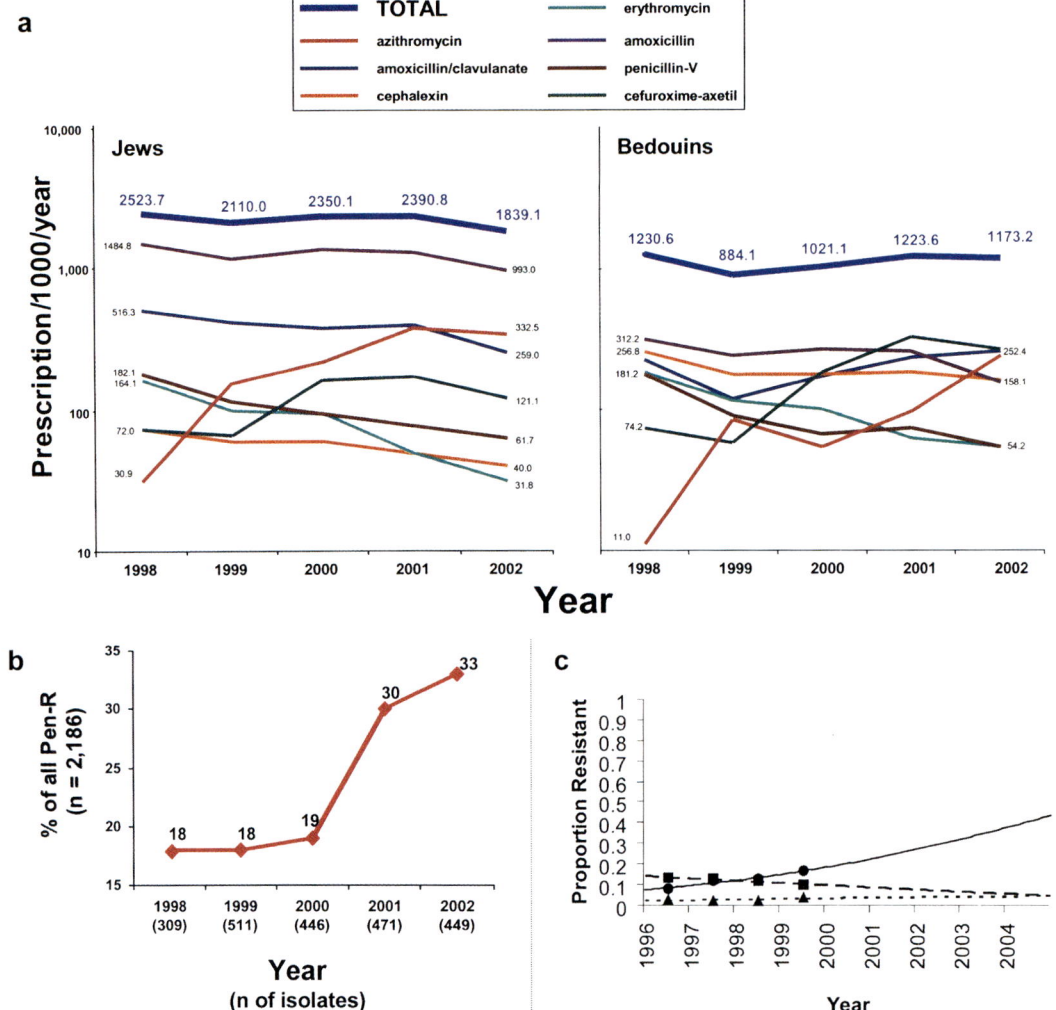

COLOR PLATE 10 [chapter 18]   (a) Antibiotic consumption among the Jewish and Bedouin children <5 years of age in southern Israel, expressed by prescription per 1,000 individuals per year. The data for the Jewish children include all prescriptions in the city of Beer-Sheva for the largest health maintenance organization covering approximately 60% of all children; for the Bedouin children, the data are derived from all prescriptions for the two largest pediatric primary care centers. (b) Proportion of all penicillin-nonsusceptible (Pen-R) *S. pneumoniae* middle ear fluid isolated obtained during the years 1998 to 2002 (from children with acute otitis media) that were dually resistant to penicillin and macrolides. (c) Actual and predicted proportions of resistance among invasive *S. pneumoniae* isolates derived from population-based surveillance in eight sites around the United States covering approximately 21 million people (81). The symbols represent the actual proportions of strains resistant to both erythromycin and penicillin (●), penicillin only (■), and erythromycin only (▲). These lines represent the predicted proportions of strains resistant to both erythromycin and penicillin (solid), penicillin alone (dashed) and erythromycin only (dotted).

A PATHOGENETIC CATEGORIZATION OF CLINICAL SYNDROMES CAUSED BY *STREPTOCOCCUS PNEUMONIAE*

Daniel M. Musher

14

As has been shown in the preceding chapters, *Streptococcus pneumoniae* has properties that enable it to colonize, to cause inflammation, and to invade (i.e., be taken up by) mammalian cells. Despite this relatively limited pathogenetic repertoire (for example, pneumococci are not motile, do not make arrays of toxins or tissue-damaging enzymes, do not sporulate, etc.), the organism is still a major cause of human disease. In accord with the scientific orientation of this textbook, the present chapter discusses clinical syndromes caused by *S. pneumoniae*, categorizing them based on pathogenesis and immune response.

Most pneumococcal disease occurs when organisms are carried from an area that is colonized (the carriage is passive, since pneumococci are, of course, nonmotile and do not "spread") into a space that is normally not colonized, such as a eustachian tube, paranasal sinus, bronchiole, alveolus, or fallopian tube (Table 1). In such spaces, if normal clearance mechanisms fail, pneumococci may proliferate, and their presence stimulates an intense inflammatory response. Ingress of polymor-

phonuclear leukocytes (PMNs) and exudation of plasma may serve to arrest the process if antibodies to surface-expressed constituents opsonize the bacteria for phagocytosis. If such antibodies are not present, the inflammatory response is likely to continue; in the case of the respiratory tract, this inflammation *is* the disease. A subset of this first category includes instances in which pneumococci reach an area because of physical disruption of an anatomic barrier, e.g., when a tear in the dura leads to invasion of the meninges or perforation of the bowel causes peritonitis or intra-abdominal abscess. A second way in which pneumococci cause disease is by invading the bloodstream, causing bacteremia. This invasion may occur at a site of colonization—for example, in the nasopharynx, causing bacteremia with no recognized focus of infection (primary bacteremia)—or in the lungs or elsewhere in the body as the result of an established infection (secondary bacteremia). In adults, most recognized cases of pneumococcal bacteremia are secondary, arising as a complication of pneumonia. The mechanisms of local invasion are discussed above.

Once bacteremia occurs, pneumococci may cause what is traditionally called hematogenous infection. Organisms may attach to up-regulated receptors on vascular endothelial cells. They may

Daniel M. Musher, Medical Service (Infectious Disease Section), Veterans Affairs Medical Center, Houston, and Departments of Medicine and Molecular Virology and Microbiology, Baylor College of Medicine, Houston, TX 77030.

TABLE 1 Classification of infections caused by *S. pneumoniae* according to pathogenetic principles

I. Colonization, followed by spread to normally sterile/near-sterile body spaces
   A. Usual conditions apply
      1. Otitis media
      2. Sinusitis
      3. Acute purulent tracheobronchitis
      4. Pneumonia (further spread—empyema)
      5. Conjunctivitis (spread from nasopharynx)
      6. Endometritis (further spread—peritonitis)
   B. Antecedent trauma
      1. Dural tear with meningitis
      2. Bowel perforation with peritonitis, abscess
II. Bacteremia
   A. "Primary," arising from invasion at the site of colonization
   B. "Secondary," arising from invasion at the site of an established infection
III. Hematogenous infection
   A. Proliferation on an endovascular surface (endocarditis)
   B. Move across vascular endothelium into tissues, causing secondary hematogenous infection
      1. Meningitis
      2. Primary peritonitis
      3. Pericarditis
      4. Septic arthritis
      5. Osteomyelitis
      6. Soft tissue infection
      7. Empyema (vs. direct extension?)

then proliferate locally, causing endocarditis (an uncommon infection) or infected mural thrombus (a decidedly rare one), or be taken up and pass through endothelial cells, causing disease at a number of sites in the body. This current understanding of hematogenous infection is based on an expanded understanding of events in which pneumococci "settle out in" or "seed" various body sites, as discussed elsewhere in this text. The resulting infections include meningitis, primary peritonitis, septic arthritis, osteomyelitis, and soft tissue infection. Empyema, which is perhaps the most common complication of pneumococcal pneumonia, and pericarditis may result from hematogenous seeding in this fashion, or by direct extension to the visceral pleura from a pulmonary focus of infection.

OTITIS MEDIA

Under normal circumstances, if small numbers of pneumococci are carried into eustachian tubes or sinuses, clearance mechanisms, chiefly ciliary action, lead to their rapid removal. Allergy, coexisting viral infection, or exposure to toxins (including cigarette smoke) may cause edema, obstructing the opening of the eustachian tube or the osteomeatal complex. Under these circumstances, pneumococci may not be cleared. In the absence of opsonizing antibody, organisms may proliferate unchecked, and clinically recognizable infection may result.

Virtually every study of acute otitis media in which material from the middle ear has been cultured has shown *S. pneumoniae* to be the most common isolate or second only to nontypeable *Haemophilus influenzae*; *Moraxella (Branhamella) catarrhalis* is a distant third (9). In these studies, which are usually carried out in children aged 6 months to 4 years, *S. pneumoniae* is implicated in about 40 to 50% of cases in which an etiologic agent is isolated or in 30 to 40% of all cases. The pneumococcus is the most prevalent pathogen in otitis media in adults as well (57). Prior infection by a respiratory virus is thought to play a major contributory role by causing congestion of the opening to the eustachian tube. Prospective longitudinal studies (13, 18, 19) have shown that when infection occurs, in most cases it follows fairly closely after colonization by a new serotype, although, of course, most instances of colonization occur without disease.

SINUSITIS

Important in the pathogenesis of acute sinusitis is congestion of the mucosal membranes caused by allergy or viral infection; resulting obstruction at the osteomeatal complex prevents clearance of bacteria. The accumulation of fluid in the paranasal sinus cavities, even during simple colds (22), provides a medium for bacterial proliferation and subsequent acute sinus infection. Not surprisingly, the bacteriology of acute maxillary sinusitis is similar to that of otitis media, with *S. pneumoniae* and/or *H. influenzae* being isolated in the great majority of cases (23).

ACUTE PURULENT TRACHEOBRONCHITIS

Concern for egregious overuse of antibiotics has led to the teaching that acute bronchitis in an otherwise healthy adult is not caused by bacteria. While this statement is probably true, in general, I do not have enough personal experience, nor does the literature give careful enough description of individual cases, to allow the conclusion that acute bronchitis is never caused by bacteria. In contrast, among persons who have chronic lung disease, damage to ciliated cells and/or increased production of mucus may prevent the clearance of inhaled or aspirated organisms. An acute bacterial infection may result, characterized by increased cough and sputum production, fever, and leukocytosis. By definition, the chest X ray shows no new infiltrate. In such patients, *S. pneumoniae* and *H. influenzae*, separately or together, are commonly implicated (6, 7). The term "acute exacerbation of chronic bronchitis" is commonly used to describe this illness, although I prefer the more specifically descriptive term "acute purulent tracheobronchitis" (39). A clinically recognizable exacerbation of chronic disease is highly associated with acquisition of a new pneumococcal strain (58).

PNEUMONIA

If potentially protective mechanisms fail to prevent both the access of pneumococci to the alveoli and their subsequent replication, pneumonia results. Bacteria proliferate in alveolar spaces and are carried along the alveolar septa; in these sites, they activate complement, generate cytokine production, and up-regulate receptors on vascular endothelial surfaces. Exudative fluid and white blood cells (WBCs) accumulate in the septa and alveoli and extend to uninvolved areas through the pores of Kohn. This filling of alveoli with microorganisms and inflammatory exudate defines the presence of pneumonia, and a clinical diagnosis is made when fluid accumulation is great enough to allow it to be seen radiographically as a nonlucent or "consolidated" area.

As mentioned earlier, prior respiratory viral infection, perhaps especially infection caused by influenza virus, appears to play a prominent role in predisposing to pneumococcal infection (11, 25, 27, 28). Up-regulation of surface receptors during viral infection may enhance pneumococcal adherence (14) and invasion. Bacteria are certainly less well cleared from the airways because of virus-induced damage. Pneumococcal disease is greatly increased in people with altered pulmonary clearance, such as those who have chronic bronchitis, asthma, or chronic obstructive pulmonary disease. Only in the past few years has a study finally supported clinical observations to associate cigarette smoking with susceptibility to pneumonia (42). It is an interesting sign of the times that Heffron's classic work (24) on the pneumococcus published in 1939 had a section on inhalation of "noxious substances" yet did not mention cigarette smoking.

Patients who have pneumococcal pneumonia, especially those who are bacteremic, very often have an underlying condition(s) that affects their ability to resist infection. For example, in a recent series of 100 cases at the Houston Veterans Affairs Medical Center, 68% of patients smoked cigarettes, 50% had chronic lung disease, and 47% were diagnosed as alcohol abusers (38). Other conditions identified in these patients and in other large series of cases are shown in Table 2. In the Houston series of cases there was no patient who was entirely free of some underlying condition, a finding that may reflect extreme care in obtaining data but also might be due to selection bias in a veteran population that seeks medical care at a veterans hospital.

The multifactorial susceptibility of aged persons to pneumococcal pneumonia is reviewed in a later chapter. The effect of alcoholism is also multifactorial and involves lifestyle (such as cold exposure and

TABLE 2 Factors predisposing to pneumococcal infection[a]

Predisposing factor	Population based, all invasive disease		Hospital based, United States, pneumonia only		
			General hospital (60), bacteremic	Veterans hospital (38)	
	Sweden (4)	Israel (50)		Bacteremic	Nonbacteremic
Age >60 yr				58	35
Cigarette smoking	40		56	69	67
Chronic lung disease	17	19	28	42	58
Alcohol abuse	32		11	58	35
Liver disease	2	6		23	21
Recent hospitalization				35	37
Previous pneumonia				21	35
Immune suppression (including HIV)		36		32	24
Neurologic disease				31	31
Diabetes	6	15	18	11	12
Congestive heart failure		35	16	27	17
Cancer	12		26	25	17
No predisposing factor				0	0

[a] Data are presented as percentages of all patients included in the series.

malnutrition), suppression of the gag reflex, and possibly deleterious effects on PMN function, although in most instances these alterations have been difficult to attribute to the effect of alcohol alone (17). A disproportionately high number of patients who have pneumococcal infection have diabetes mellitus (4, 15, 38, 50, 60), a condition in which PMN chemotaxis is reduced (35) and phagocytic function is defective (52), especially if renal insufficiency is also present. Anemia (hemoglobin less than 10 g/dl) may be present in one-third of patients with pneumococcal pneumonia (38). Many chronic diseases are associated with pneumococcal pneumonia by virtue of an association with pneumonia of whatever cause, which suggests that the predisposition is a general one rather than being specific for S. pneumoniae. Pneumococcal pneumonia follows hospitalization for all causes (31, 38) and has even been observed as a nosocomial infection (5). Other factors such as cold exposure, stress, and fatigue (24) may predispose to pneumococcal pneumonia by unknown mechanisms.

Symptoms and Physical Findings

Cough, fatigue, fever, chills, sweats, and shortness of breath are the most frequent symptoms of pneumonia; these are all more prominent in younger than in older patients (33, 37). Patients with pneumococcal pneumonia usually appear ill and have a grayish, anxious appearance that differs from that of persons with viral or mycoplasmal pneumonia. The temperature may be elevated to 102 to 103°F, the pulse to 90 to 110 beats per min, and the respiratory rate to 20 to 24 per min. Elderly patients may have only a slight temperature elevation or be afebrile but are more likely to have an increased respiratory rate (36). The absence of fever in young or middle-aged adults is associated with increased morbidity and mortality, as, especially, is hypothermia.

Radiographic Findings

In most cases of pneumococcal pneumonia, chest radiography reveals an area of infiltration involving one or more segments within a single lobe (38). Airspace consolidation is detected radiographically in most cases and is more frequent in bacteremic cases; air bronchogram, which reflects especially dense airspace consolidation, highly correlates with bacteremia (38, 43). Rarely, *S. pneumoniae* infection causes a lung abscess (62). As emphasized earlier, pneumococci do not produce highly toxic, tissue-damaging substances. Thus, abscesses do not generally occur, even at a microscopic level, and if an abscess is seen, concurrent anaerobic infection or an anatomic abnormality such as bronchial obstruction, cancer, or pulmonary infarction should be suspected; occasional cases of pneumococcal lung abscess, however, are well documented. Although pleural effusion may be found in 40% of patients with pneumococcal pneumonia by careful search, only 10% have sufficient amounts of fluid to aspirate, and in only a minority of these, perhaps 2% of the total, is empyema present (29).

Laboratory Findings

Twenty-five percent of patients with pneumococcal pneumonia have a hemoglobin of ≤10 mg/dl (38). Although the majority have leukocytosis (WBC count >12,000/mm^3), a substantial proportion may have normal WBC counts, at least at the time of admission, and 10% may have <6,000 WBC per mm^3, a finding that indicates a very poor prognosis (26). Experimental studies have suggested that this situation reflects the accumulation of all available WBCs at the infected site; more often than not, it is seen in the presence of conditions such as ethanol ingestion or malnutrition, which suppress the bone marrow (45). The low serum albumin which often is present may result from malnutrition and therefore indicate a predisposing condition or reflect catabolism and fluid shifts that are part of sepsis (38).

Diagnostic Microbiology

The etiologic role of the pneumococcus in a patient who has pneumonia is strongly suggested by microscopic demonstration of large numbers of PMNs, very few epithelial cells (PMN/epithelial cell ratio, approximately 10 to 20:1), and numerous, slightly elongated gram-positive cocci in pairs and chains in a Gram-stained sputum. If accepted terminology is strictly followed, a presumptive diagnosis of pneumococcal pneumonia is then made if *S. pneumoniae* is identified by sputum culture; the diagnosis is proven if *S. pneumoniae* is identified by blood culture. The argument that the diagnosis is never certain unless the blood culture is also positive is overly restrictive. The majority of patients with pneumococcal pneumonia do not have detectable bacteremia, and it is very uncommon for the laboratory to isolate pneumococci from sputum of a patient who does not have a clear clinical picture of pneumonia or acute purulent tracheobronchitis. Attempts to make a diagnosis based on an inadequate sputum specimen (3, 44) are largely responsible for studies claiming that microscopic examination and culture of sputum are not reliable. A recent study (40a) of patients with bacteremic pneumococcal pneumonia showed that no sputum sample was ever submitted for about 30% of patients, an inadequate sample was submitted for another 15%, and a sample after antibiotics had been given for >24 h was submitted for another 10%. For patients who had not been treated with antibiotics or who had received an antibiotic for <24 h, sputum culture was 86% sensitive. The Gram stain was more likely to be altered by antibiotic administration but was 80% sensitive for patients who had not been treated. These results also support the concept that the incidence of pneumococcal infection as a cause of community-acquired pneumonia is probably much greater than has been reported in recent years.

Other diagnostic techniques focus on the detection of antigen or antibody. In general, if the sputum is not of sufficiently good quality that the Gram stain is positive, other tests, such as co-agglutination, antigen detection, or PCR, that test for the presence of pneumococci in the sputum are not helpful because they are confounded by the same problem, namely, the amount of contaminating material relative to the number of pneumococci, as well as by the potential problem of detecting carriage rather than infection. Pneumococcal cell wall polysaccharide may be detected in urine of about two-thirds of patients with pneumococcal pneumonia versus 10 to 15% of patients with nonpneumococcal pneumonia (21); the reader may decide for him- or herself if that is likely to be a useful test. For children, the test is positive with pharyngeal colonization, and it is not useful diagnostically (10). PCR amplification may be relatively sensitive and quite specific in cerebrospinal fluid and less sensitive in blood. Antibody to pneumococcal constituents may be present in immune complexes at the time that adult patients are hospitalized for pneumonia, but tests to detect such antibody are nonspecific because they also detect serum (i.e., noncomplexed) immunoglobulin G (40), and clinical or epidemiological studies based on this kind of diagnostic technique are not reliable. Diagnosing pneumococcal infection by antibody rises is also problematic because persons who are infected may be the very ones who do not make much antibody and, for reasons that are not understood, infected persons also increase their levels of antibody to other, noninfecting serotypes, at least as detected by enzyme-linked immunosorbent assay (D. Musher, unpublished data).

Complications

Empyema, the most common complication of pneumococcal pneumonia in the preantibiotic era, occurred in about 5% of cases and remains the most common today, with an incidence of approximately 2% (29). Pleural fluid appears in a substantial proportion of cases of pneumonia but is usually an inflammatory reaction to the infection, a so-called reactive pleural effusion (29). When bacteria reach the pleural space, either hematogenously or as a result of extension of the pneumonia to the visceral pleura with spread via lymphatics, empyema results. Persistence of fever, even if "low grade," and leukocytosis after 4 to 5 days of appropriate antibiotic treatment of pneumococcal pneumonia are suggestive of empyema, and this diagnosis is even more likely if the radiograph shows persistence of pleural fluid. Under such conditions, thoracentesis is mandatory. The findings of frank pus,

bacteria by Gram stain, and fluid with pH 7.1 or less are all indications for aggressive and complete drainage with repeated needle aspiration or prompt insertion of a chest tube. If no response is seen, immediate removal of infected material by laparoscopic or open thoracotomy is indicated (1). One study of medical empyema caused by all organisms found that mortality exceeded 30% in two hospitals where the therapeutic approach was not aggressive but was less than 10% in a third hospital where it was (16).

CONJUNCTIVITIS

Isolated or epidemic conjunctivitis may occur, caused, somewhat surprisingly, by unencapsulated pneumococci (and essentially the only condition in which unencapsulated pneumococci play a role) (2, 32).

ENDOMETRITIS AND SALPINGITIS

Pneumococcal infection may involve the female reproductive organs by direct extension from the vagina (49, 53, 61). Peritonitis (12) may result from further spread into the peritoneal cavity, a syndrome that should be distinguished from primary peritonitis (see below).

PRIMARY BACTEREMIA

Bacteremia that occurs without an apparent source or focus of infection is called primary bacteremia. In a recent population-based study of adults in Israel (51), pneumonia was present in 71% of cases of pneumococcal bacteremia, meningitis was present in 8%, and otitis media or sinusitis was found in 4%; bacteremia was regarded as primary in 18%. Primary bacteremia has always been more common in children than adults; when therapy has been withheld, a focus of infection has often become apparent, usually otitis media or pneumonia.

ENDOCARDITIS

A case of pneumococcal endocarditis (47) is seen once or twice per decade at a large tertiary care hospital in the United States. A recent study of 14 cases identified in Denmark (population of 5 million) in a 10-year period (30) showed that 3 persons were alcoholic and only

1 had known of prior valvular disease; 8 had pneumonia and 4 had meningitis. I have seen a single case of an infected mural thrombus, which must be exceedingly rare.

PERICARDITIS

Purulent pericarditis (41) caused by the pneumococcus has also become exceedingly rare, whether occurring as a separate entity or together with endocarditis. The pathogenesis of this infection, just as that of empyema, might be via hematogenous spread, but it also could be via direct extension from the lungs.

MENINGITIS

Except during an epidemic of meningococcal infection, *S. pneumoniae* is the most common cause of bacterial meningitis in adults (48). In countries that have implemented effective vaccination programs for *H. influenzae* type b, the pneumococcus has become the most common sporadic cause of meningitis in children over the age of 6 months, as well, although widespread use of conjugate pneumococcal vaccine may change this situation in the future. Controversy remains over whether this syndrome results from local invasion and spread to the brain or as a result of local invasion and bacteremia with seeding of the choroid plexus and/or meninges (see reference 37 and chapter 16).

PERITONITIS

Peritonitis due to *S. pneumoniae* may be secondary to spread from female reproductive organs or from gastric or small intestinal perforation, for example, complicating gastric surgery or appendicitis (12). When pneumococci reach the peritoneal cavity from the bloodstream, often without a documented source of infection elsewhere, the condition is called primary peritonitis (12). Such cases nearly always occur in patients who have preexisting ascites which serves as a *locus minoris resistantiae* (site of diminished resistance). In children, nephrotic syndrome most commonly predisposes to this infection, whereas in adults, alcoholic cirrhosis is most often responsible.

INFECTIONS OF JOINTS, BONES, AND SOFT TISSUES

Septic arthritis (55) occurs spontaneously in a natural or prosthetic (55, 56) joint or as a complication of rheumatoid arthritis (34). When osteomyelitis occurs in adults, it tends to involve the vertebral bones (59). Epidural and brain abscesses are rarely described (20). Soft tissue infections (8, 46) occur, especially in persons who have connective tissue diseases or human immunodeficiency virus (HIV) infection. In fact, the appearance of unusual pneumococcal infections in a young adult might suggest that tests for HIV infection be undertaken (54).

CONCLUSION

S. pneumoniae, despite its somewhat limited array of tissue-damaging enzymes and toxins, remains a prominent cause of infection with a surprisingly broad array of manifestations. Disease patterns are, however, both understandable and largely predictable based on pathogenetic considerations.

REFERENCES

1. **Anstadt, M. P., C. K. Guill, E. R. Ferguson, H. S. Gordon, E. R. Soltero, A. C. Beall, Jr., and D. M. Musher.** 2003. Surgical vs. nonsurgical treatment of empyema: an outcomes analysis. *Am. J. Med. Sci.* **326:**8–14.
2. **Barker, J. H., D. M. Musher, R. Silberman, H. M. Phan, and D. A. Watson.** 1999. Genetic relatedness among nontypeable pneumococci implicated in sporadic cases of conjunctivitis. *J. Clin. Microbiol.* **37:**4039–4041.
3. **Barrett-Connor, E.** 1971. The nonvalue of sputum culture in the diagnosis of pneumococcal pneumonia. *Am. Rev. Respir. Dis.* **103:**845–848.
4. **Burman, L. A., R. Norrby, and B. Trollfors.** 1985. Invasive pneumococcal infections: incidence, predisposing factors, and prognosis. *Rev. Infect. Dis.* **7:**133–142.
5. **Chang, J. L., and J. M. Mylotte.** 1987. Pneumococcal bacteremia: updated from an adult hospital with a high rate of nosocomial cases. *J. Am. Geriatr. Soc.* **35:**747–754.
6. **Chodosh, S.** 1987. Acute bacterial exacerbations in bronchitis and asthma. *Am. J. Med.* **82:**154–163.
7. **Chodosh, S., J. McCarty, S. Farkas, M. Drehobl, R. Tosiello, M. Shan, L. Aneiro, S. Kowalsky, and The Bronchitis Study Group.** 1998. Randomized, double-blind study

of ciprofloxacin and cefuroxime axetil for treatment of acute bacterial exacerbations of chronic bronchitis. *Clin. Infect. Dis.* **27**:722–729.

8. **DiNubile, M. J., M. A. Albornoz, R. J. Stumacher, B. L. Van Uitert, S. A. Paluzzi, L. M. Bush, S. C. Nelson, and A. R. Myers.** 1991. Pneumococcal soft-tissue infections: possible association with connective tissue diseases. *J. Infect. Dis.* **163**:897–900.

9. **Dowell, S. F., J. C. Butler, G. S. Giebink, M. R. Jacobs, D. Jernigan, D. M. Musher, A. Rakowsky, and B. Schwartz.** 1999. Acute otitis media: management and surveillance in an era of pneumococcal resistance—a report from the Drug-resistant *Streptococcus pneumoniae* Therapeutic Working Group. *Pediatr. Infect. Dis. J.* **18**:1–9.

10. **Dowell, S. F., R. L. Garman, G. Liu, O. S. Levine, and Y. H. Yang.** 2001. Evaluation of Binax NOW, an assay for the detection of pneumococcal antigen in urine samples, performed among pediatric patients. *Clin. Infect. Dis.* **32**: 824–825.

11. **Dowell, S. F., C. G. Whitney, C. Wright, C. E. Rose, Jr., and A. Schuchat.** 2003. Seasonal patterns of invasive pneumococcal disease. *Emerg. Infect. Dis.* **9**:573–579.

12. **Dugi, D. D., III, D. M. Musher, J. E. Clarridge III, and R. Kimbrough.** 2001. Intraabdominal infection due to *Streptococcus pneumoniae. Medicine* (Baltimore) **80**:236–244.

13. **Faden, H., L. Duffy, R. Wasielewski, J. Wolf, D. Krystofik, and Y. Tung.** 1997. Relationship between nasopharyngeal colonization and the development of otitis media in children. Tonawanda/Williamsville Pediatrics. *J. Infect. Dis.* **175**:1440–1445.

14. **Fainstein, V., D. M. Musher, and T. R. Cate.** 1980. Bacterial adherence to pharyngeal cells during viral infection. *J. Infect. Dis.* **141**:172–176.

15. **Fang, G. D., M. Fine, J. Orloff, D. Arisumi, V. L. Yu, W. Kapoor, J. T. Grayston, S. P. Wang, R. Kohler, R. R. Muder, et al.** 1990. New and emerging etiologies for community-acquired pneumonia with implications for therapy. A prospective multicenter study of 359 cases. *Medicine* (Baltimore) **69**:307–316.

16. **Franco, M., and D. M. Musher.** 1982. Thoracic empyema: The impact of management on outcome. *Am. Rev. Respir. Dis.* **124**:82. (Abstract.)

17. **Gluckman, S. J., V. C. Dvorak, and R. R. MacGregor.** 1977. Host defenses during prolonged alcohol consumption in a controlled environment. *Arch. Intern. Med.* **137**:1539–1543.

18. **Gray, B. M., G. M. Converse III, and H. C. Dillon, Jr.** 1980. Epidemiologic studies of *Streptococcus pneumoniae* in infants: acquisition, carriage, and infection during the first 24 months of life. *J. Infect. Dis.* **142**:923–933.

19. **Gray, B. M., and H. C. Dillon, Jr.** 1988. Epidemiological studies of *Streptococcus pneumoniae* in infants: antibody to types 3, 6, 14, and 23 in the first two years of life. *J. Infect. Dis.* **158**:948–955.

20. **Grigoriadis, E., and W. L. Gold.** 1997. Pyogenic brain abscess caused by *Streptococcus pneumoniae:* case report and review. *Clin. Infect. Dis.* **25**: 1108–1112.

21. **Gutierrez, F., M. Masia, J. C. Rodriguez, A. Ayelo, B. Soldan, L. Cebrian, C. Mirete, G. Royo, and A. M. Hidalgo.** 2003. Evaluation of the immunochromatographic Binax NOW assay for detection of *Streptococcus pneumoniae* urinary antigen in a prospective study of community-acquired pneumonia in Spain. *Clin. Infect. Dis.* **36**:286–292.

22. **Gwaltney, J. M. J., C. D. Phillips, R. D. Miller, and D. K. Riker.** 1994. Computed tomographic study of the common cold. *N. Engl. J. Med.* **330**:25–30.

23. **Gwaltney, J. M. J., W. M. Scheld, M. A. Sande, and A. Sydnor.** 1992. The microbial etiology and antimicrobial therapy of adults with acute community-acquired sinusitis: a fifteen-year experience at the University of Virginia and review of other selected studies. *J. Allergy Clin. Immunol.* **90**:457–462.

24. **Heffron, R.** 1939. Pneumonia, with special reference to pneumococcus lobar pneumonia. Harvard University Press, Cambridge, Mass. (Reprint, 1979.)

25. **Hodges, R., and C. MacLeod.** 1946. Epidemic pneumococcal pneumonia. II. The influence of population characteristics and the environment. *Am. J. Hyg.* **44**:193–206.

26. **Hook, E. W. I., C. A. Horton, and D. R. Schaberg.** 1983. Failure of intensive care unit support to influence mortality from pneumococcal bacteremia. *JAMA* **249**:1055–1057.

27. **Jones, E. E., P. L. Alford, A. L. Reingold, et al.** 1998. Predisposition to invasive pneumococcal illness following parainfluenza type 3 virus infection in chimpanzees. *J. Am. Vet. Med. Assoc.* **185**:1351–1353.

28. **Kim, P. E., D. M. Musher, W. P. Glezen, M. C. Rodriguez-Barradas, W. K. Nahm, and C. E. Wright.** 1996. Association of invasive pneumococcal disease with season, atmospheric conditions, air pollution, and the isolation of respiratory viruses. *Clin. Infect. Dis.* **22**:100–106.

29. **Light, R. W., W. M. Girard, S. G. Jenkinson, and R. B. George.** 1980. Parapneumonic effusions. *Am. J. Med.* **69**:507–512.

30. **Lindberg, J., J. Prag, and H. C. Schonheyder.** 1998. Pneumococcal endocarditis is not just a disease of the past: an analysis of 16 cases diagnosed in Denmark 1986–1997. *Scand. J. Infect. Dis.* **30**:469–472.

31. Lipsky, B. A., E. J. Boyko, T. S. Inui, and T. D. Koepsell. 1986. Risk factors for acquiring pneumococcal infections. *Arch. Intern. Med.* **146:** 2179–2185.

32. Martin, M., J. H. Turco, M. E. Zegans, R. R. Facklam, S. Sodha, J. A. Elliott, J. H. Pryor, B. Beall, D. D. Erdman, Y. Y. Baumgartner, P. A. Sanchez, J. D. Schwartzman, J. Montero, A. Schuchat, and C. G. Whitney. 2003. An outbreak of conjunctivitis due to atypical *Streptococcus pneumoniae*. *N. Engl. J. Med.* **348:**1112–1121.

33. Metlay, J. P., R. Schulz, Y.-H. Li, D. E. Singer, T. J. Marrie, C. M. Coley, L. J. Hough, D. S. Obrosky, W. N. Kapoor, and M. J. Fine. 1997. Influence of age on symptoms at presentation in patients with community-acquired pneumonia. *Arch. Intern. Med.* **157:**1453– 1459.

34. Morley, P. K., R. G. Hull, and M. A. Hall. 1987. Pneumococcal septic arthritis in rheumatoid arthritis. *Ann. Rheum. Dis.* **46:**482–484.

35. Mowat, A. G., and J. Baum. 1980. Chemotaxis of polymorphonuclear leukocytes from patients with diabetes mellitus. *N. Engl. J. Med.* **142:** 869–875.

36. Murphy, T. F., and B. C. Fine. 1984. Bacteremic pneumococcal pneumonia in the elderly. *Am. J. Med. Sci.* **288:**114–118.

37. Musher, D. M. *Streptococcus pneumoniae. In* G. L. Mandell, J. E. Bennett, and R. Dolin (ed.), *Mandell, Douglass, and Bennett's Principles and Practice of Infectious Diseases,* 6th ed., in press. John Wiley & Sons, New York, N. Y.

38. Musher, D. M., I. Alexandraki, E. A. Graviss, N. Yanbeiy, A. Eid, L. A. Inderias, H. M. Phan, and E. Solomon. 2000. Bacteremic and nonbacteremic pneumococcal pneumonia. A prospective study. *Medicine* (Baltimore) **79:**210–221.

39. Musher, D. M., K. R. Kubitschek, J. Crennan, and R. E. Baughn. 1983. Pneumonia and acute febrile tracheobronchitis due to *Haemophilus influenzae. Ann. Intern. Med.* **99:**444–450.

40. Musher, D. M., R. Mediwala, H. M. Phan, G. Chen, and R. E. Baughn. 2001. Nonspecificity of assaying for IgG antibody to pneumolysin in circulating immune complexes as a means to diagnose pneumococcal pneumonia. *Clin. Infect. Dis.* **32:**534–538.

40a. Musher, D. M., R. Montoya, and A. Wanahita. Reliability of microscopic examination of gram-stained sputum and sputum culture in patients with bacteremic pneumococcal pneumonia. *Clin. Infec. Dis.,* in press.

41. New England Journal of Medicine. 1990. Case records of the Massachusetts General Hospital. Weekly clinicopathological exercises. Case 49–1990. A 47-year-old Cape Verdean man with pericardial disease. *N. Engl. J. Med.* **323:**1614– 1624.

42. Nuorti, J. P., J. C. Butler, M. M. Farley, L. H. Harrison, A. McGeer, M. S. Kolczak, R. F. Breiman, and Active Bacterial Core Surveillance Team. 2000. Cigarette smoking and invasive pneumococcal disease. *N. Engl. J. Med.* **342:**681–689.

43. Ort, S., J. L. Ryan, G. Barden, and N. D'Espopo. 1983. Pneumococcal pneumonia in hospitalized patients. Clinical and radiological presentations. *JAMA* **249:**214–218.

44. Perlino, C. A. 1984. Laboratory diagnosis of pneumonia due to *Streptococcus pneumoniae. J. Infect. Dis.* **150:**139–144.

45. Perlino, C. A., and D. Rimland. 1985. Alcoholism, leukopenia, and pneumococcal sepsis. *Am. Rev. Respir. Dis.* **132:**757–760.

46. Peters, N. S., S. J. Eykyn, and A. G. Rudd. 1989. Pneumococcal cellulitis: a rare manifestation of pneumococcaemia in adults. *J. Infect.* **19:**57–59.

47. Powderly, W. G., S. L. Stanley, Jr., and G. Medoff. 1986. Pneumococcal endocarditis: report of a series and review of the literature. *Rev. Infect. Dis.* **8:**786–791.

48. Quagliarello, V. J., and W. M. Scheld. 1997. Treatment of bacterial meningitis. *N. Engl. J. Med.* **336:**708–716.

49. Rahav, G., L. Ben-David, and E. Persitz. 1991. Postmenopausal pneumococcal tubo-ovarian abscess. *Rev. Infect. Dis.* **13:**896–897.

50. Rahav, G., Y. Toledano, D. Engelhard, A. Simhon, A. E. Moses, T. Sacks, and M. Shapiro. 1997. Invasive pneumococcal infections: a comparison between adults and children. *Medicine* (Baltimore) **76:**295–303.

51. Raz, R., G. Elhanan, Z. Shimoni, R. Kitzes, C. Rudnicki, Y. Igra, A. Yinnon, and the Israeli Adult Pneumococcal Bacteremia Group. 1997. Pneumococcal bacteremia in hospitalized Israeli adults: epidemiology and resistance to penicillin. *Clin. Infect. Dis.* **24:**1164–1168.

52. Repine, J. E., C. C. Clawson, and F. C. Goetz. 1980. Bactericidal function of neutrophils from patients with acute bacterial infections and from diabetics. *J. Infect. Dis.* **6:**869–875.

53. Robinson, E. N. J. 1990. Pneumococcal endometritis and neonatal sepsis. *Rev. Infect. Dis.* **12:** 416–422.

54. Rodriguez Barradas, M. C., D. M. Musher, R. J. Hamill, M. Dowell, J. T. Bagwell, and C. V. Sanders. 1992. Unusual manifestations of pneumococcal infection in human immunodeficiency virus-infected individuals: the past revisited. *Clin. Infect. Dis.* **14:**192–199.

55. Ross, J. J., C. L. Saltzman, P. Carling, and D. S. Shapiro. 2003. Pneumococcal septic

arthritis: review of 190 cases. *Clin. Infect. Dis.* **36:** 319–327.

56. **Ryczak, M., M. Sands, R. B. Brown, and J. H. Sklar.** 1987. Pneumococcal arthritis in a prosthetic knee. A case report and review of the literature. *Clin. Orthop.* **1987:**224–227.

57. **Schwartz, L. E., and R. B. Brown.** 1992. Purulent otitis media in adults. *Arch. Intern. Med.* **152:**2301–2304.

58. **Sethi, S., N. Evans, B. J. Grant, and T. F. Murphy.** 2002. New strains of bacteria and exacerbations of chronic obstructive pulmonary disease. *N. Engl. J. Med.* **347:**465–471.

59. **Turner, D. P., V. C. Weston, and P. Ispahani.** 1999. *Streptococcus pneumoniae* spinal infection in Nottingham, United Kingdom: not a rare event. *Clin. Infect. Dis.* **28:**873–881.

60. **Watanakunakorn, C., and T. A. Bailey.** 1997. Adult bacteremic pneumococcal pneumonia in a community teaching hospital, 1992–1996. A detailed analysis of 108 cases. *Arch. Intern. Med.* **157:**1965–1971.

61. **Westh, H., L. Skibsted, and B. Korner.** 1990. *Streptococcus pneumoniae* infections of the female genital tract and in the newborn child. *Rev. Infect. Dis.* **12:**416–422.

62. **Yangco, B. G., and S. C. Deresinski.** 1980. Necrotizing or cavitating pneumonia due to *Streptococcus pneumoniae:* report of four cases and review of the literature. *Medicine* (Baltimore) **59:**449–457.

ATTACHMENT AND INVASION OF THE RESPIRATORY TRACT

Elaine I. Tuomanen

15

EPIDEMIOLOGY

Respiratory infections are a leading cause of death in children in both the developed and the developing worlds (102). Although the morbidity rate has remained high, improved access to medical care and the discovery of antibiotics have decreased the mortality rate in the United States from over 70/100,000 in 1940 to less than 5/100,000 since 1980 (27). In ambulatory children, the etiology of pneumonia is identified in ~40% of cases (79). Pneumococci cause approximately one-quarter of the cases of community-acquired pneumonia in both children and the elderly (34), yielding an attack rate of 1 to 4 out of 1,000 individuals per annum. The proportion of hospitalized children with complicated pneumonia has risen steadily over the last decade from 23 to 53% (16, 127). The case fatality rate is 6% in the younger group and 15% in very elderly patients (34), with antibiotic resistance increasing this value to 25% (98).

Factors associated with a greater risk of pneumococcal disease include human immunodeficiency virus infection (30), sickle cell disease (59), hemolytic-uremic syndrome (10),

and dialysis (130). A particularly dramatic predisposing factor for pneumococcal pneumonia is infection with influenza A virus (81).

CLINICAL FEATURES

The description of the patient with lobar pneumococcal pneumonia is considered one of the classics of medicine (97). The five cardinal signs are sudden-onset pleuritic chest pain (>75% of cases), a shaking chill (lasting ~30 min in 50% of cases), and cough, followed by fever and production of rusty sputum (55). Tachypnea ensues, often accompanied by retractions and grunting in the infant and abdominal pain in the toddler (37). The patient appears flushed, with cyanotic lips. Decreased air entry, crepitant rales indicating fluid in the alveoli, and tactile fremitus as fluid induces vibration in alveolar air are commonly highly localized to the involved segment. These findings may disappear with advanced consolidation only to reappear during recovery (redux crepitus). Lobar pneumonia is more commonly associated with serotypes 1, 5, 6, 7, and 14 in children and 1, 2, 3, 5, 7, and 8 in adults. Bronchopneumonia is frequently associated with serotypes 6, 18, and 19 in children and 3, 7, 8, 10, 18, and 20 in adults (55).

Despite intense inflammation, the course of pneumococcal pneumonia is characteristically surprisingly uncomplicated. However, up to

Elaine I. Tuomanen, Department of Infectious Diseases, St. Jude Children's Research Hospital, Memphis, TN 38105.

The Pneumococcus. Volume Editor, Elaine I. Tuomanen,
© 2004 ASM Press, Washington, D.C.

30% of children develop pleural effusions and 14% have empyema (14, 127). In most cases, these complications do not require intervention and resolve completely with medical management. Lung abscess and necrotizing pneumonia are exceedingly rare for pneumococcal disease.

DIAGNOSIS

The history and physical findings usually establish the diagnosis of pneumonia. Lobar consolidation on chest X ray or computed tomography (CT) scan is suggestive but not diagnostic of bacterial etiology (79) (Fig. 1A and B). If sputum or pleural fluid is available, Gram stain looking for organisms, a high number of leukocytes (>25/high-power field) and a low number of epithelial cells can support the diagnosis (Fig. 1C). Increased levels of C-reactive protein and interleukin-6 (IL-6) in serum may indicate bacterial etiology (96). Culture of sputum, blood, or pleural fluid serves to identify the etiology of pneumonia, but only 10 to 15% of patients have positive cultures. Sputum is particularly difficult for children to produce, and many patients have received prior antibiotics that make bacterial recovery unreliable (6). Thus, most patients are treated empirically.

Because of these difficulties in diagnosis, the application of molecular diagnostic techniques has been a high priority. Unfortunately, for

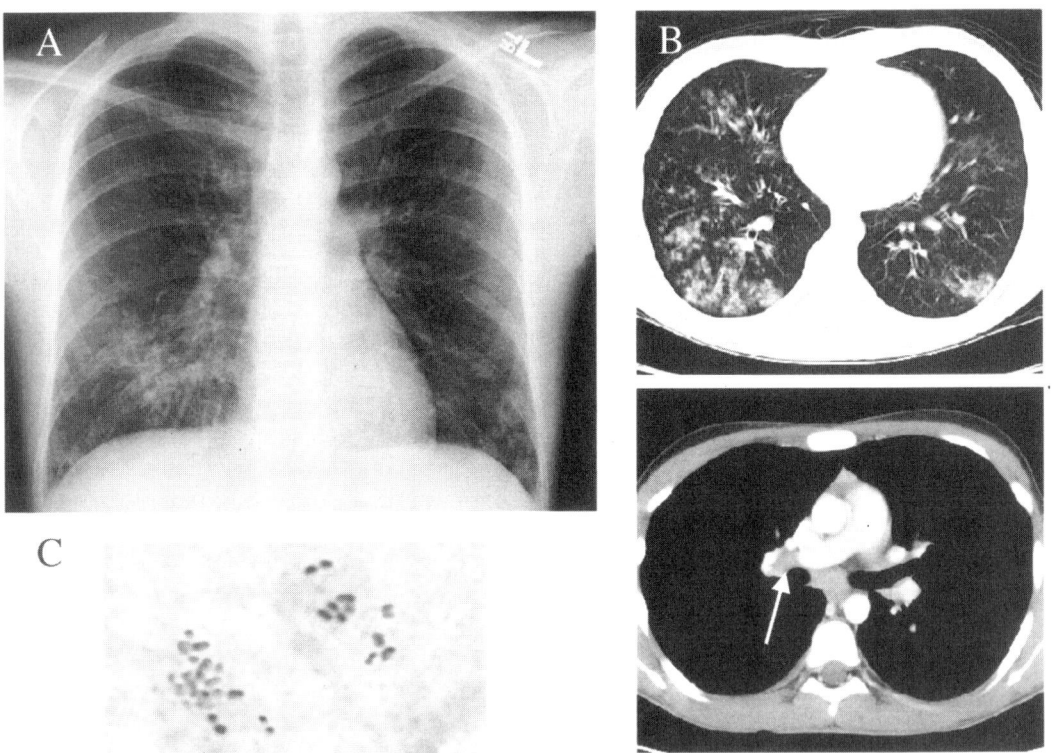

FIGURE 1 Diagnosis of pneumonia. (A) Chest X ray. Bilateral lower lobe (R > L) airspace disease in a 14-year-old-boy with sudden onset of fever and cough. (B) Top, axial contrast-enhanced CT through the lung bases and filmed with lung windows demonstrates patchy bilateral lower lobe air space disease (R > L). There are no associated pleural effusions. Bottom, axial contrast-enhanced CT through the mid-chest and filmed with mediastinal windows demonstrates a right hilar lymph node (arrow). (Images courtesy of S. Kaste, St. Jude Children's Research Hospital.) (C) Sputum Gram stain. Gram-positive cocci are visible in meshwork of sputum and cellular debris. (Image courtesy of R. Hayden, St. Jude Children's Research Hospital.)

each study showing promise there are studies showing lack of efficacy. Circulating immune complexes in serum proved to be unreliable markers (93), and immunochromatography to detect antigens in blood, urine, or sputum showed mixed results (26, 92). PCR assays to detect the autolysin LytA, the transporter PsaA, or the toxin pneumolysin have each been found to be sensitive but not specific, especially for young children (15, 31, 80, 90, 91, 120).

EXPERIMENTAL MODELS OF DISEASE

Even before 1940, hundreds of studies had been published examining the course of pneumococcal pneumonia using aerosol, intranasal, intratracheal, intraperitoneal, subcutaneous, intradermal, or intravenous inoculation of mice, rabbits, guinea pigs, monkeys, cats, dogs, rats, or pigeons (94). Reliable infection of the lungs of rabbits was achieved only with inoculation via either the nose (10 µl dropped into the anterior (naris) or trachea (~50 µl instilled through a catheter inserted into the mainstem bronchus via tracheostomy). Intravenous inoculation has not reproducibly produced pneumonia even in recent models, indicating that direct entry to the lung via the airways is a requirement for lung disease (103, 116). Major advances in pneumonia models in the 1930s and 1940s included the ability to reliably produce lethality by adding 5% hog gastric mucin to the inoculum (94) and the use of logarithmically growing bacteria to increase virulence (150). More recently, small inocula of pneumococci have been shown to cause pneumonia and bacteremia if passed in animals (116).

Animal imaging has added a new dimension to the ability to map the course of infection. The use of bioluminescent pneumococci stably transformed with the gram-positive *lux* transposon cassette and thereby detectable within the living mouse by a sensitive charge-coupled device camera (Xenogen Imaging System) is one such advance (38) (Color Plate 6). For example, mice infected intranasally with strain A66.1 developed pneumonia only,

those challenged with D39 experienced high-grade sepsis, while TIGR4 infection resulted in low-grade pneumonia and bacteremia ultimately progressing to meningitis (95). Such bacterial strain specific differences in virulence have been shown in many model systems (11) and reflect similar differences in humans (3). Organ-specific differences in virulence were suggested by early studies in which intranasal virulence did not correlate with intraperitoneal virulence (94). The major effect of different genetic backgrounds of animals on the course of infection has been particularly apparent in mice (95). Enhanced susceptibility to disease in CBA/Ca mice has been shown to be due to impaired tumor necrosis factor (TNF) secretion (65). With the recognized differences in virulence among strains and in susceptibility among animals, the comparison of virulence data among labs requires use of the same bacterium-animal pairing.

PATHOGENESIS

It has been argued that the pneumococcus should be considered a commensal of the nasopharynx since each of >90 serotypes is carried asymptomatically for up to 6 weeks at a time in up to half of healthy individuals (2). Invasive disease usually arises during the first few days of colonization by a new serotype (45, 46). Infection of mice intranasally results in the appearance of bacteria in the trachea, heart, lymphatics, and lungs within a minute (103). Positive blood cultures can be obtained for 30% of these mice at 10 min, but bacteremia is quickly cleared. The role of the lymphatics in pneumococcal dissemination is less clear. Pneumococcal replication in cervical and pulmonary lymphatics has been shown to serve as a reservoir available to seed the bloodstream (35).

Pneumococci are presumably aerosolized from the nasopharynx to the alveolus, bypassing the ciliated epithelium, to which they are not able to attach (132). Bronchopneumonia, which characteristically involves airways more than alveoli, is promoted by an antecedent injury to the ciliated mucosa which provides conditions for pneumococcal attachment.

Such a tropism of pneumococci for damaged epithelium was demonstrated in autopsy material from patients dying of pneumonia during the influenza pandemic of 1957 to 1958 (75). Damaged epithelium can also arise by smoking or infection with other respiratory pathogens. Exposure of the endobronchial submucosa allows pneumococcal attachment to basement membrane components such as fibronectin and collagen (69, 136). Once localized to the airway, release of pneumolysin exacerbates the mucosal damage since epithelial cells are particularly sensitive to toxin-induced cytolysis and inhibition of ciliary beat frequency (33, 104, 113, 115).

Localization to Alveoli

Alveolar epithelial cells provide excellent attachment sites for pneumococci (133). Pathology of the very early pneumonic lesion shows bacteria floating freely in serous fluid coating the surfaces of alveolar type 2 cells (103, 150). The presence of pneumococci in the alveolus need not lead to inflammation unless bacteria maintain localization to this site and multiply (68, 126). The pneumococcus adheres to sialylated cell surface glycoconjugates as a first step in associating loosely with human cells (5, 20, 61). It has been suggested that the multiple neuraminidases of *Streptococcus pneumoniae* may then cleave terminal sialic acid, exposing cryptic receptors and enhancing closer adherence (99). One such receptor may be GlcNAc(β1–3)Gal (70), a determinant recognized by many respiratory pathogens. This two-step sequence accords with evidence from the chinchilla model of otitis media that shows progressive loss of sialic acid on the surface of the eustachian tube as the pneumococcus tracks up to the middle ear (73, 131). Viruses with neuraminidase activity, such as influenza and parainfluenza viruses, may act synergistically with pneumococcus in this fashion, priming the respiratory epithelium for adherence (52, 83). This notion is supported by in vitro, in vivo, and clinical studies showing greater adherence of pneumococci to virus-infected cells (47, 52, 101, 139). This effect is critically dependent on the viral neuraminidase and can be prevented by neuraminidase inhibitors (82).

Progression to Pneumonia

The landmark description of the evolution of pulmonary consolidation was published by Laennec in 1932 (71) and later refined by Loosli (74). The pneumonic lesion progresses through three stages: engorgement, red hepatization, and grey hepatization (Color Plate 7). During the early phase of engorgement a serous transudate accumulates in the alveoli as pneumococci multiply and pass from alveolus to alveolus within a lobe through the pores of Kohn (48). The edema fluid derives from leakage from the systemic circulation as evidenced by staining of the lesion, but not surrounding healthy lung, in animals bearing intravascular dye (68). The lung appears heavy yet crepitant, with a violet discoloration. An incision into the lung yields a frothy fluid, and pathologic examination shows engorged capillaries and alveolar epithelial cells. This early stage of pneumonia is essentially asymptomatic and short-lived, probably lasting only several hours (55). However, these changes continue to be found at the leading edge of the spreading pneumonic lesion throughout the course of disease.

Sutliff and Friedemann (124) and Rich and McKee (106) showed that a "heat stable and nonprotein" component of the pneumococcus could produce profound alveolar edema independent of leukocytes. Later studies identified the edema factor as the peptidoglycan-teichoic acid cell wall complex. Mixtures of pneumococcal cell wall components recreate many features of the symptom complex of pneumococcal pneumonia (135). After cell wall challenge, bronchoalveolar lavage fluid is flooded with serum components, neutrophils, and fibrin. The teichoic acid-peptidoglycan network induces a variety of mammalian cells to produce TNF-α, IL-1, IL-6, IL-10, IL-12, nitric oxide, and vascular endothelial growth factor and promotes a procoagulant state (39, 41, 56, 57, 105, 137, 142) (Table 1). Many of these effects are mediated through cell wall binding to toll-like receptor 2 (118, 142,

TABLE 1 Activities of pneumococcal cell wall in the lung

Binding to PAFr
Induction of cytokines
   TNF
   IL-1, IL-6, IL-10, IL-12
   Vascular endothelial growth factor
   Nitric oxide
Signaling via:
   Activation of ERK, JNK, and p38 MAP kinases
   Translocation of NF-κB to the nucleus
Induction of leukocyte migration via increased expression of:
   Endothelial ICAM-1
   Leukocyte integrin Mac-1
Fixation of complement
Induction of procoagulant state

---

151), an innate immune system receptor present on the surfaces of respiratory epithelial cells (Fig. 2).

An important clinical correlate of cell wall-induced inflammation occurs in conjunction with antibiotic therapy. Penicillin triggers the rapid release of cell wall components, with a secondary increase in inflammation immediately after antibiotics are initiated. Steroids and nonsteroidal anti-inflammatory agents (134, 135) decrease edema in animal models of pneumonia. Hydrocortisone given simultaneously with penicillin has been shown to suppress clinical toxemia in patients with pneumococcal pneumonia (140).

ADHESIN CbpA

Pneumococci that are free-floating in the alveolus eventually adhere to alveolar type II cells (20). CbpA (also designated PspC, SpsA, and Hic) is the most abundant of the choline-binding proteins and functions as an adhesin in the upper and lower respiratory tract (112). Attachment of CbpA also induces chemokine and ICAM production from the cell (77, 89). At least 75% of pneumococcal isolates produce CbpA. Strains lacking CbpA cannot multiply in the lung, although if injected directly into the blood they show no loss of virulence (4).

CbpA has been shown to interact with several different human receptor proteins, including the polymeric immunoglobulin receptor (pIgR) (12, 49, 153), which traffics secretory IgA for secretion, platelet-activating factor receptor (PAFr) (19), inhibitor of complement factor H (23, 29, 62), and complement component C3 (122). Of these, PAFr and epithelium-derived C3 play essential roles in pneumococcal interactions with cells in the lung that lead to progression from pneumonia to bacteremia and meningitis (4, 13, 110).

A schematic domain structure of CbpA in its most common form is shown in Fig. 3 (112). CbpA (110 kDa, 693 amino acids for the TIGR4 strain) has 10 choline-binding motifs at the C terminus that are responsible for binding choline groups on the bacterial cell wall. A variant of CbpA, termed Hic, is anchored covalently to the cell wall by an LPXTGE motif rather than via choline (62). The presence of one or the other cell wall binding domain divides clinical isolates into two major groups (60). The N terminus (~440 amino acids) is comprised of several helical domains that each contain multiple repeats of the leucine zipper motif commonly involved in protein self-association. Domain A is ~25 amino acids in length, while R1 and R2 are nearly identical and each contain 110 amino acids (112). These domains are variably present in CbpA from clinical isolates, leading to a division of the 2 major groups into 11 smaller groups (6 with the choline-binding domain and 5 with LPXTGE) (13, 60). A YRNYPT sequence in the helical domain is known to be important for CbpA binding to the secretory component of IgA (50), but, as pIgR is not expressed in the lung, this activity of CbpA is relevant to colonization only (12). Sequences required to bind to other receptors, such as PAFr expressed in lung, have not been determined.

PULMONARY VIRULENCE DETERMINANTS

In vivo genetic screens have identified many virulence determinants important in the lung (Table 2). These determinants have been iden-

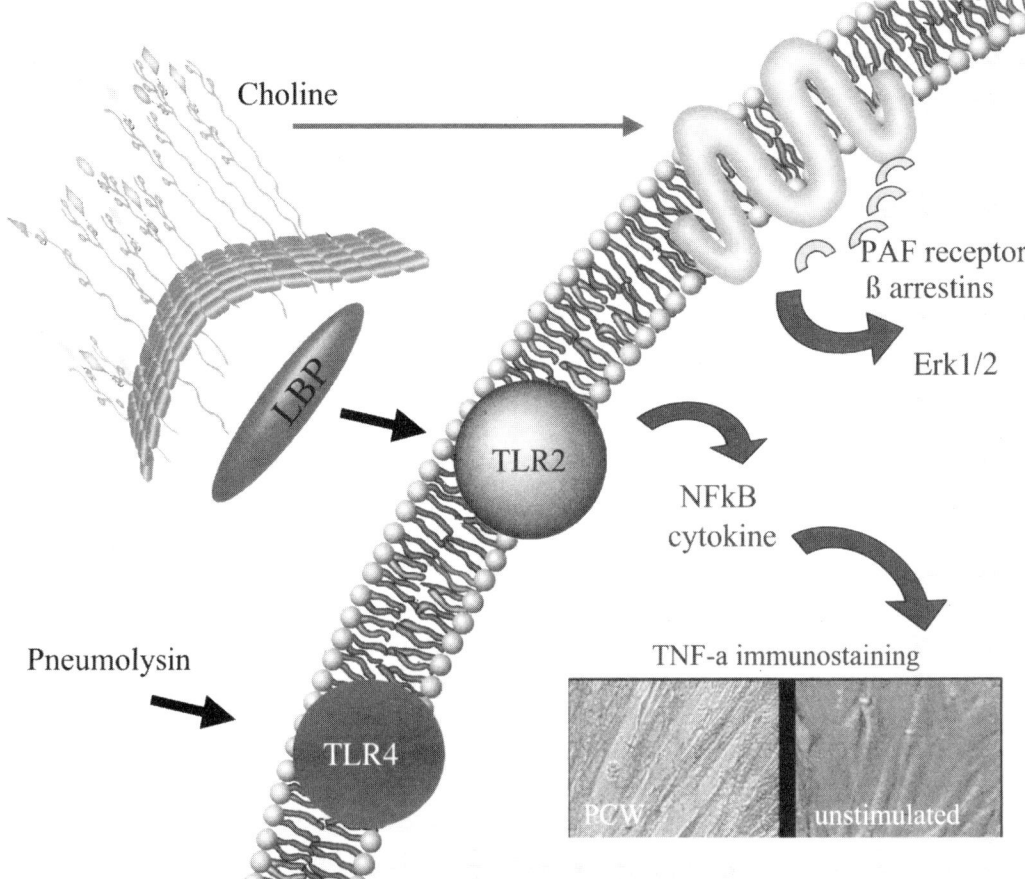

FIGURE 2 Model for the inflammatory response to cell wall in the lung. In the alveolar space, pneumococcal cell wall (PCW) (shown as layers of peptidoglycan decorated by projecting choline-containing teichoic acids) interacts with two different signaling pathways. The choline on the teichoic acid binds to PAFr that initiates bacterial uptake into vacuoles supported by the scaffold protein β arrestin and activation of Erk kinases (19, 119). The peptidoglycan portion bound to a soluble peptidoglycan recognition protein (for example, lipopolysaccharide-binding protein [LBP] [142]) is presented to toll-like receptor 2 (TLR2). TLR2 also binds lipoteichoic acid (118), and TLR4 has been found to bind pneumolysin (78). Engagement of TLRs elicits production of TNF (shown in inset by immunostaining) and IL-1, resulting in separation of epithelial cells and accumulation of a serous exudate in alveoli.

tified individually and as coregulated elements that contribute to a phenotype suitable for mucosal adherence and invasion. One marker of this phase-variable, mucosal phenotype is the transparent colonial morphology (21). Transparent pneumococci have greater expression of CbpA, LytA, and choline on their surfaces, consistent with the role of these elements in adherence (66). The two-component systems Cia, rr01, and Zmp have been shown to play a

role in regulating genes whose products promote mucosal interactions (53, 121, 128). Hava and Camilli (53) challenged animals intranasally with a library of mutants and harvested the lungs in search of genetic determinants of colonization and pneumonia. This approach was validated by finding known virulence factors such as HtrA (serine protease), IgA proteases, CbpA, LytC (an autolysin), CbpD (unknown function [43]), PspA (lactoferrin-binding pro-

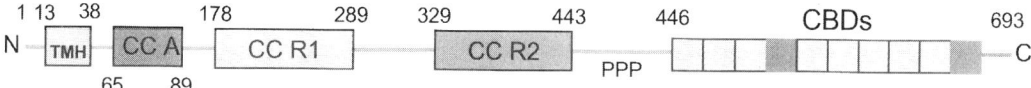

FIGURE 3 Schematic domain structure of CbpA. Domains labeled A, R1, and R2 contain multiple repeats of the classical leucine zipper heptad repeat that is found in coiled-coil proteins. PPP is a proline-rich region, and TMH is a putative transmembrane α-helix that serves as a signal sequence for secretion. Ten repeats of the choline-binding motif are found within the C-terminal choline-binding domain (CBD). (Annotated from the TIGR4 sequence.)

tein), and competence factors D and E. The screen identified a new regulatory element, RrgA, that controls transcription of several genes in a pathogenicity islet required for colonization of the nasopharynx and lung but not operative in the bloodstream (54). The precise function of these genes is not clear, although three encode sortases that covalently link proteins to the cell wall.

Damage to the Alveolar Epithelium

The second phase of consolidation involves leakage of erythrocytes into the alveoli leading to the appearance of red hepatization. The

TABLE 2 Pulmonary virulence determinants[a]

Pathogenic factors
  Cell wall hydrolases
  IgA proteases
  Choline-binding proteins: CbpA, PspA, CbpC, CbpD
  Toxins: pneumolysin, superoxide dismutase, pyruvate oxidase
  Serine proteases: HtrA, PrtA
  Metalloproteases

Metabolic pathways
  Amino acid biosynthesis
  Vitamin biosynthesis
  DNA metabolism
  Energy metabolism
  Fatty acid metabolism
  Permeases and transporters

Signaling elements
  Transcription factors
  Two-component systems

[a] Summarized from reference 53.

lung is no longer crepitant, and a clotted exudate fills the alveolus, giving the surface of the lung a granular appearance. Maximum consolidation, usually at day 3, leads to a lung so dense that it does not collapse with intentional pneumothorax (9). Fibrin deposition is pronounced in this phase and progressively decreases perfusion of the lesion (74). The fibrin mesh in the alveolus contains erythrocytes, bacteria, some leukocytes, and desquamated epithelium. Fibrin strands pass through the pores of Kohn from one alveolus to another. Lymphatics are dilated and filled with cells and fibrin. Pneumococci in the lesion are still alive (76) and if left to multiply will continue to provoke edema and fibrin formation sufficient to kill the experimental animal (149). Mice become bacteremic during this phase, i.e., at ~36 h after respiratory tract challenge, coincident with the peak of bacterial multiplication and well before the peak of inflammation, at >80 h (22). In leukopenic animals, the bacterial load in the lung increases but there is no decrease in the incidence of bacteremia, indicating that events in this stage of consolidation are sufficient to lead to invasion without the effects of leukocytes (141).

BACTERIAL TOXINS AND ALVEOLAR DAMAGE

Damage to the pulmonary epithelium at this stage of pneumonia is greater in leukopenic animals (141), indicating prominent direct toxicity of bacterial elements. Two toxins that fulfill criteria to mediate direct pulmonary damage are pneumolysin and H_2O_2 (8, 123). Pneumolysin is a thiol-activated cytolysin with both pore-forming activity and complement-activating activity (99). Mutants defective in production of pneumolysin are greatly attenuated in experimental pneumococcal pneumonia models (8, 64), and the hemolytic activity of the toxin contributes most significantly to damage to epithelial cells during the stage of red hepatization (113). In vitro, pneumolysin directly kills alveolar epithelial cells, pulmonary endothelial cells, and, to a lesser extent, macrophages (58, 114, 115). It enhances inflammation by inducing IL-6 and MIP2 and by directly activating complement (63, 109).

H_2O_2 is a reactive oxygen species that is produced in particularly large amounts by the pneumococcus because it lacks catalase (100). H_2O_2 can diffuse freely into cells and oxidizes mitochondria and DNA mediating direct toxicity to most cells, including those in the alveolus (28). A mutant lacking pyruvate oxidase (the major enzyme activity producing H_2O_2) shows greatly reduced virulence in animal models for pneumonia (123). The effects of pneumolysin and H_2O_2 are additive and include not only killing of epithelial cells but also arresting ciliary beating (32). This suggests that substantial reduction in lung cell cytotoxicity during pneumonia would likely require inactivation of both H_2O_2 and pneumolysin.

Host Response

In the final stage of consolidation, leukocytes migrate into the lesion and the lung takes on an appearance known as grey hepatization (48). Alveoli are densely packed with leukocytes and lysed erythrocytes. Capillaries are compressed and filled with thrombi, but despite very poor perfusion, there is no necrosis of the lung parenchyma. During the phase of grey hepatization, the patient appears toxic (97), but the degree of toxemia does not correlate with the extent of pulmonary consolidation for reasons that are still not known.

The influx of leukocytes leads to trapping of the pneumococci against the alveolar wall in a process called surface phagocytosis (150). The presence of the opsonin C-reactive protein (CRP) in pulmonary secretions is an important innate defense component. CRP binds choline on the surfaces of pneumococci and many other respiratory pathogens (144, 145) and acts by promoting bacterial engulfment by neutrophils (87, 88) and by blocking bacterial adherence to the PAF on epithelial cells (44). Killing of bacteria by either engulfment or neutrophil-released products may be successful for a low number of bacteria (76). However, the presence of inadequate leukocytes often makes the inflammation worse. For example, release of neutrophil defensins increases adherence of respiratory pathogens, thereby counteracting the benefit of their antimicrobial activity (42). Invading leukocytes are also a major source of elastase, a proteolytic activity that is particularly high in lavage fluid from pneumococcal infection (84).

CHEMOKINES AND LEUKOCYTE MIGRATION

The pneumonic lesion evolves during grey hepatization to marshal a wide array of host defenses. Chemokine production from the epithelium leads to the recruitment of leukocytes. Chemoattracts include bacterial factors, such as formylmethionyl-leucyl-phenylalanine, and at least eight chemokines, the genes for which are up-regulated upon exposure of epithelial cells to pneumococci: RANTES, MIP-1a, MIP-1β, MIP-2, IP-10, MCP-1, T-cell activation factor 3, and KC (36). The timing of KC and MIP-2 expression coincides with neutrophil recruitment (Fig. 4). Even leukopenic animals exhibit increases in IL-1, IL-6, MIP-1, MIP-2, and MCP (141), indicating a dominant role for the epithelium in generating early inflammatory molecules. The process of CbpA binding to the epithelium induces chemokine and ICAM production (89), and the binding of CbpA to C3 increases IL-8 production (77). Complement fixation to the cell wall results in production of the chemokine C5a (129, 148) and complement-mediated opsonization of bacteria is an important mechanism of clearance (18). Pneumolysin also activates complement, and loss of this activity greatly attenuates virulence in the lung at this stage, as opposed to the

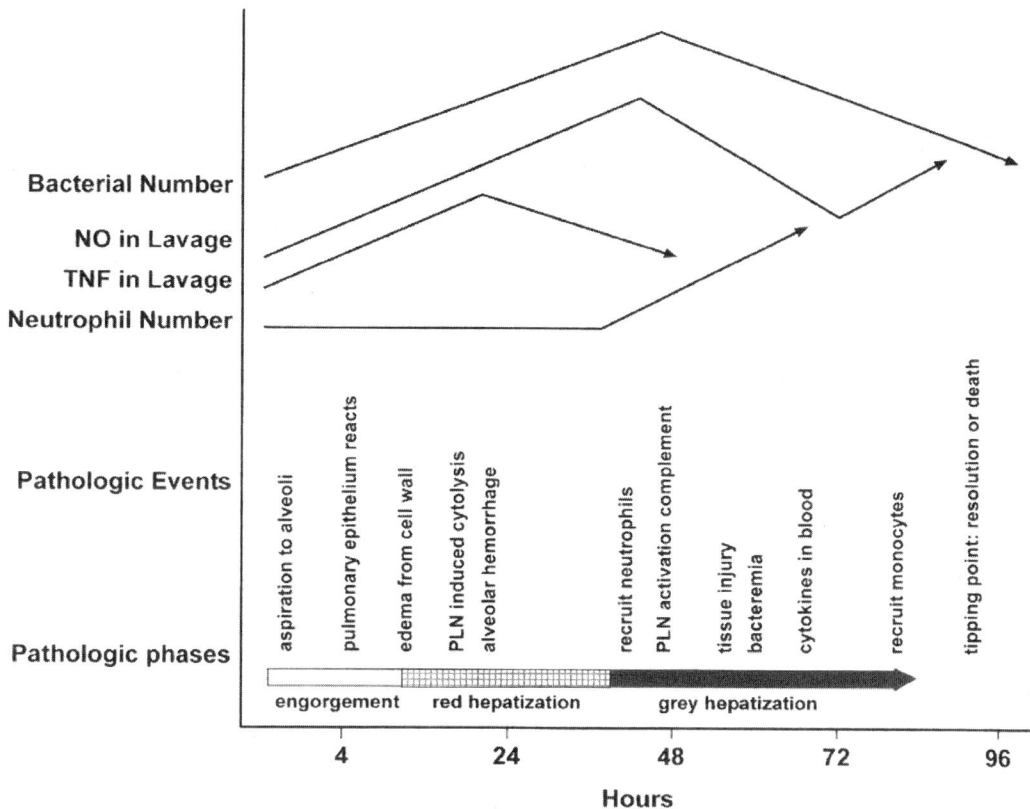

FIGURE 4 Kinetics of events during progression of pneumococcal pneumonia. Bacterial multiplication proceeds unimpeded during the stages of engorgement and red hepatization, peaking at 36 h in the stage of grey hepatization. Bacteremia is a result of pneumococcal adherence to and invasion of alveolar cells. The edema characteristic of engorgement arises from cell-wall induced signaling in epithelial cells and activation of the alternative pathway of the complement cascade by cell wall. Cytokines begin to appear in bronchoalveolar lavage fluid in the first few hours of engorgement but do not reach a maximum until the phase of red hepatization (18 to 24 h). At this stage, the activated endothelium expresses tissue factor forming a platform for procoagulant activity, and the cytolytic activity of pneumolysin is prominent. During the stage of grey hepatization, polymorphonuclear leukocytes (PLN) are recruited and begin to control pneumococcal multiplication. Complement activation by pneumolysin aids in this clearance. The outcome of the infection depends at least in part on the ability of the host to withstand the inflammation associated with bacterial death (i.e., tipping point).

greater role of its cytotoxic activity earlier in the disease (1, 113).

The migration of leukocytes into the alveolar spaces has several features unique to the pneumococcus and the lung. Pneumococci induce an increased expression of the β2 integrin Mac-1, thereby enhancing leukocyte stickiness (147). Half of the leukocytes use this Mac-1-dependent mechanism of entry, as evidenced by a 50% reduction in leukocyte density in bronchoalveolar lavage fluid of animals treated with anti-CD18 antibody (17). The additional leukocytes are recruited by a mechanism unique to the pneumococcus and thus far observed only in the lung (25, 85). This mechanism involves PAFr, since it is inhibited by PAFr antagonists (17). It also appears to involve expression of galectin-3 in lung epithelia and endothelia. Galectin-3 is a β-galactoside binding lectin that promotes neutrophil adhesion (117).

CYTOKINES

Both the inflamed epithelium and the incoming leukocytes produce a variety of cytokines (7) (Fig. 4), and the pattern has different kinetics in response to different strains of pneumococci (86). TNF and IL-1β are inflammatory signals critical to eventual survival (125). Either TNF or IL-1 is required for survival of the experimental animal, as demonstrated by the ability of increased TNF to compensate for loss of IL-1 in mice defective in IL-1 receptor (107). IL-18, but not IL-12, is also protective in the context of pneumococcal pneumonia (72). Interestingly, gamma interferon has no apparent role, at least in the early acute stages of pneumonia (108). Perhaps this is due to the appearance of mononuclear cells only late in the course of inflammation, with IgG being detected as late as day 8 (51).

The ability of the inflammatory response to control bacterial multiplication without release of inflammatory debris by extensive bacterial lysis is required for recovery and represents a tipping point between life and death in the course of disease (22). Recovery is heralded by natural crisis, a phenomenon characterized by rapid defervescence. During resolution, the lung becomes translucent and jelly-like, with a creamy, purulent exudate. Eventually macrophages replace the neutrophils and absorb the debris (111). This process usually proceeds to complete resolution, but delay leads to deposition of granulation tissue and fibrosis (67).

Invasion

A requirement for the evolution of pneumonia to invasive disease includes the up-regulation of receptors on alveolar cells that participate in bacterial uptake. In vitro studies have demonstrated that invasion of alveolar cells involves recognition of PAFr by pneumococcus. PAFr is a G protein-coupled, seven transmembrane-spanning receptor that binds the chemokine PAF and augments inflammation. It is recognized as a portal for pneumococcal endocytosis by a G protein-independent mechanism (Color Plate 8) (19). Although pneumococci bind to PAFr, they do not induce signal transduction, nor do they interfere with PAF-associated signaling (40). Using COS cells transfected with components of the PAFr, it has been shown that the presence of PAFr is necessary for pneumococcal invasion (19). More recently it has been appreciated that PAFr must be supported by β1 arrestin, which is the scaffold protein linking PAFr to the endocytotic machinery (J. Radin, personal communication). The binding of the natural ligand, PAF, to PAFr is dependent on phosphorylcholine, and this determinant is an integral part of the pneumococcal cell wall. This choline-dependent invasion mechanism appears to apply to a large number of respiratory pathogens. A surface protein of mycoplasmas (24), lipopolysaccharide of *Haemophilus influenzae* (145, 146), *Pseudomonas aeruginosa* surface protein, and pili of *Neisseria* spp. (144) can be decorated with a phosphorylcholine adduct. In fact, pneumococci and *Haemophilus* share very similar genes for adding phosphorylcholine to surface molecules (146, 152). The presence of the adduct is subject to phase variation such that it is present when bacteria are on the mucosal surface and need to interact with PAFr but is removed for optimal bacterial survival in the bloodstream (143, 145, 146). In the case of the pneumococcus, the adduct is removed by Pce, a choline-binding protein which is a phosphorylcholine esterase (138). The benefit of removing the phosphorylcholine is the ability to escape the host innate defense element CRP that binds to choline, blocks attachment, and opsonizes these pathogens (44, 87, 88, 145). Thus, choline is useful in attaching to the lung but detrimental to the organism in the bloodstream.

CONCLUSION

Pneumococcal pneumonia is very common and causes a dramatic clinical picture due to intense inflammation in the lung. Innate and antibody-mediated defenses can control the infection in some cases and the severe lobar pneumonia will resolve such that the original pulmonary architecture is preserved. However, antibiotics are increasingly required to control this infection. Given the exceptionally high incidence of pneumonia, use of the pneumococcal vaccine to prevent infection is both prudent and cost-effective.

REFERENCES

1. **Alexander, J., A. Berry, J. Paton, J. Rubins, P. Andrew, and T. Mitchell.** 1998. Amino acid changes affecting the activity of pneumolysin alter the behaviour of pneumococci in pneumonia. *Microb. Pathog.* **24:**167–174.
2. **Austrian, R.** 1986. Some aspects of the pneumococcal carrier state. *J. Antimicrob. Chemother.* **18**(Suppl. A):35–45.
3. **Austrian, R., and J. Gold.** 1964. Pneumococcal bacteremia and especial reference to bacteremic pneumococcal pneumonia. *Ann. Intern. Med.* **60:**759–776.
4. **Balachandran, P., A. Brooks-Walter, A. Virolainen-Julkunen, S. Hollingshead, and D. Briles.** 2002. Role of pneumococcal surface protein C in nasopharyngeal carriage and pneumonia and its ability to elicit protection against carriage of *Streptococcus pneumoniae*. *Infect. Immun.* **70:**2526–2534.
5. **Barthelson, R., A. Mobasseri, D. Zopf, and P. Simon.** 1998. Adherence of *Streptococcus pneumoniae* to respiratory epithelial cells is inhibited by sialylated oligosaccharides. *Infect. Immun.* **66:**1439–1444.
6. **Bartlett, J., and L. Mundy.** 1995. Community acquired pneumonia. *N. Engl. J. Med.* **333:**1618–1623.
7. **Bergeron, Y., N. Ouellet, A. Deslauriers, M. Simard, M. Olivier, and M. Bergeron.** 1998. Cytokine kinetics and other host factors in response to pneumococcal pulmonary infection in mice. *Infect. Immun.* **66:**912–922.
8. **Berry, A., J. Yother, D. Briles, D. Hansman, and J. Paton.** 1989. Reduced virulence of a defined pneumolysin-negative mutant of *Streptococcus pneumoniae*. *Infect. Immun.* **57:**2037–2042.
9. **Blake, F., M. Howard, and W. Hull.** 1935. Artificial pneumothorax in the treatment of lobar pneumonia. *JAMA* **105:**1489–1503.
10. **Brandt, J., C. Wong, S. Mihm, J. Robers, J. Smith, E. Brewer, R. Thiagarajan, and B. Warady.** 2002. Invasive pneumococcal disease and hemolytic uremic syndrome. *Pediatrics* **110:**371–376.
11. **Briles, D., M. Crain, B. Gray, C. Forman, and J. Yother.** 1992. A strong association between capsular type and mouse virulence among human isolates of *Streptococcus pneumoniae*. *Infect. Immun.* **60:**111–116.
12. **Brock, S., P. McGraw, P. Wright, and J. J. Crowe.** 2002. The human polymeric immunoglobulin receptor facilitates invasion of epithelial cells by *Streptococcus pneumoniae* in a strain-specific and cell type-specific manner. *Infect. Immun.* **70:**5091–5095.
13. **Brooks-Walter, A., D. Briles, and S. Hollingshead.** 1999. The *pspC* gene of *Streptococcus pneumoniae* encodes a polymorphic protein, PspC, which elicits cross-reative antibodies to PspA and provides immunity to pneumococcal bacteremia. *Infect. Immun.* **67:**6533–6542.
14. **Buckingham, S., M. King, and M. Miller.** 2003. Incidence and etiologies of complicated parapneumonic effusions in children, 1996–2001. *Pediatr. Infect. Dis. J.* **22:**499–504.
15. **Butler, J., S. Bosshardt, M. Phelan, S. Moroney, M. Tondella, M. Farley, A. Schuchat, and B. Fields.** 2003. Classical and latent class analysis evaluation of sputum PCR and urine antigen testing for diagnosis of pneumococcal pneumonia in adults. *J. Infect. Dis.* **187:**1416–1423.
16. **Byington, C., L. Spencer, and T. Johnson.** 2002. An epidemiological investigation of a sustained high rate of pediatric parapneumonic empyema: risk factors and microbiological associations. *Clin. Infect. Dis.* **34:**434–440.
17. **Cabellos, C., D. E. MacIntyre, M. Forrest, M. Burroughs, S. Prasad, and E. Tuomanen.** 1992. Differing roles of platelet-activating factor during inflammation of the lung and subarachnoid space. *J. Clin. Investig.* **90:**612–618.
18. **Coonrod, J., and K. Yoneda.** 1982. Complement and opsonins in alveolar secretions and serum of rats with pneumonia due to *Streptococcus pneumoniae*. *Rev. Infect. Dis.* **3:**310–322.
19. **Cundell, D., N. Gerard, C. Gerard, I. Idanpaan-Heikkila, and E. Tuomanen.** 1995. *Streptococcus pneumoniae* anchors to activated eukaryotic cells by the receptor for platelet activating factor. *Nature* **377:**435–438.
20. **Cundell, D., and E. Tuomanen.** 1994. Receptor specificity of adherence of *Streptococcus pneumoniae* to human type II pneumocytes and vascular endothelial cells in vitro. *Microb. Pathog.* **17:**361–374.
21. **Cundell, D., J. Weiser, J. Shen, A. Young, and E. Tuomanen.** 1995. Relationship between colonial morphology and adherence of *Streptococcus pneumoniae*. *Infect. Immun.* **63:**757–761.
22. **Dallaire, F., N. Ouellet, Y. Bergeron, V. Turmel, M. Gauthier, M. Simard, and M.**

Bergeron. 2001. Microbiological and inflammatory factors associated wtih the development of pneumococcal pneumonia. *J. Infect. Dis.* **184:** 292–300.

23. **Dave, S., A. Brooks-Walter, M. Pangburn, and L. McDaniel.** 2001. PspC, a pneumococcal surface protein, binds human factor H. *Infect. Immun.* **69:**3435–3437.

24. **Deutsch, J., M. Salman, and S. Rottem.** 1995. An unusual polar lipid from the cell membrane of Mycoplasma fermentans. *Eur. J. Biochem.* **227:**897–902.

25. **Doerschuk, C. M., R. K. Winn, H. O. Coxson, and J. M. Harlan.** 1990. CD18-dependent and -independent mechanisms of neutrophil emigration in the pulmonary and systemic microcirculation of rabbits. *J. Immunol.* **144:** 2327–2333.

26. **Dominguez, J., S. Blanco, C. Rodrigo, M. ZAzuara, N. Galí, A. Mainou, A. Esteve, A. Castellví, C. Prat, L. Matas, and V. Ausina.** 2003. Usefulness of urinary antigen detection by an immunochomatographic test for diagnosis of pneumococcal pneumonia in children. *J. Clin. Microbiol.* **41:**2161–2163.

27. **Dowell, S., B. Kupronis, E. R. Zell, and D. Shay.** 2000. Mortality from pneumonia in children in the United States, 1939 through 1996. *N. Engl. J. Med.* **342:**1399–1407.

28. **Duane, P., J. Rubins, H. Weisel, and E. Janoff.** 1993. Identification of hydrogen peroxide as a *Streptococcus pneumoniae* toxin for rat alveolar epithelial cells. *Infect. Immun.* **61:**4392–4397.

29. **Duthy, T. G., R. J. Ormsby, E. Giannakis, A. D. Ogunniyi, U. H. Stroeher, J. C. Paton, and D. L. Gordon.** 2002. The human complement regulator factor H binds pneumococcal surface protein PspC via short consensus repeats 13 to 15. *Infect. Immun.* **70:**5604–5611.

30. **Dworkin, M. S., J. W. Ward, D. L. Hanson, J. L. Jones, J. E. Kaplan, and Adolescent Spectrum of HIV Disease Project.** 2001. Pneumococcal disease among human immunodeficiency virus-infected persons: incidence, risk factors, and impact of vaccination. *Clin. Infect. Dis.* **32:**794–800.

31. **Falguera, M., A. Lopez, A. Nogues, J. Porcel, and M. Rubio-Caballero.** 2002. Evaluation of the PCR method for detection of *Streptococcus pneumoniae* DNA in pleural fluid samples. *Chest* **122:**2212–2216.

32. **Feldman, C., R. Anderson, R. Cockeran, T. Mitchell, P. Cole, and R. Wilson.** 2002. The effects of pneumolysin and hydrogen peroxide, alone and in combination, on human ciliated epithelium in vitro. *Respir. Med.* **96:**580–585.

33. **Feldman, C., T. J. Mitchell, P. W. Andrew, G. J. Boulnois, R. C. Read, H. C. Todd, P. J. Cole, and R. Wilson.** 1990. The effect of *Streptococcus pneumoniae* pneumolysin on human respiratory epithelium in vitro. *Microb. Pathog.* **9:**275–284.

34. **Fernandez-Sabe, N., R. Carratala, B. Roson, J. Dorca, R. Verdaguer, F. Manresa, and F. Gudiol.** 2003. Community acquired pneumonia in very elderly patients: causative organisms, clinical characteristics, and outcomes. *Medicine* (Baltimore) **82:**159–169.

35. **Field, M., M. Shaffer, J. Enders, and C. Drinker.** 1937. The distribution in the blood and lymph of pneumococcus type III injected intravenously in rabbits, and the effect of treatment with specific antiserum on the infection of the lymph. *J. Exp. Med.* **65:**469–485.

36. **Fillion, I., N. Ouellet, M. Simard, Y. Bergeron, S. Sato, and M. Bergeron.** 2001. Role of chemokines and formyl peptides in pneumococcal pneumonia-induced monocyte/macrophage recruitment. *J. Immunol.* **166:**7353–7361.

37. **Fine, M., M. Smith, and C. Carson.** 1994. Efficacy of pneumococcal vaccination in adults: a meta-analysis of randomized clinical controlled trials. *Arch. Intern. Med.* **154:**2666–2677.

38. **Francis, K. P., J. Yu, C. Bellinger-Kawahara, D. Joh, M. J. Hawkinson, G. Xiao, T. F. Purchio, M. G. Caparon, M. Lipsitch, and P. R. Contag.** 2001. Visualizing pneumococcal infections in the lungs of live mice using bioluminescent *Streptococcus pneumoniae* transformed with a novel gram-positive *lux* transposon. *Infect. Immun.* **69:**3350–3358.

39. **Freyer, D., R. Manz, A. Ziegenhorn, M. Weih, K. Angstwurm, W. Docke, A. Meisel, R. Schumann, G. Schonfelder, U. Dirnagl, and J. Weber.** 1999. Cerebral endothelial cells release TNFa after stimulation with cell walls of *Streptococcus pneumoniae* and regulate iNOS and ICAM-1 expression via autocrine loops. *J. Immunol.* **163:**4308–4314.

40. **Garcia Rodriguez, C., D. R. Cundell, E. I. Tuomanen, L. F. Kolakowski, Jr., C. Gerard, and N. P. Gerard.** 1995. The role of N-glycosylation for functional expression of the human platelet-activating factor receptor. Glycosylation is required for efficient membrane trafficking. *J. Biol. Chem.* **270:**25178–25184.

41. **Geelen, S., C. Bhattacharyya, and E. Tuomanen.** 1992. Induction of procoagulant activity on human endothelial cells by *Streptococcus pneumoniae*. *Infect. Immun.* **60:**4179–4183.

42. **Gorter, A., P. Hiemstra, S. de Bentzmann, S. van Wetering, J. Dankert, and L. van**

Alphen. 2000. Stimulation of bacterial adherence by neutrophil defensins varies among bacterial species but not among host cell types. *FEMS Immunol. Med. Microbiol.* **28:**105–111.

43. **Gosink, K., E. Mann, C. Guglielmo, E. Tuomanen, and R. Masure.** 2000. Role of novel choline binding proteins in virulence of *Streptococcus pneumoniae. Infect. Immun.* **68:** 5690–5695.

44. **Gould, J., and J. Weiser.** 2002. The inhibitory effect of C-reactive protein on bacterial phosphorylcholine platelet-activating factor receptor-mediated adherence is blocked by surfactant. *J. Infect. Dis.* **186:**361–371.

45. **Gray, B., G. Converse, and H. Killon.** 1980. Epidemiologic studies of *Streptococcus pneumoniae* in infants: acquisition, carriage and infection during the first 24 months of life. *J. Infect. Dis.* **142:**923–933.

46. **Gray, B. M., and H. C. Dillon, Jr.** 1986. Clinical and epidemiologic studies of pneumococcal infection in children. *Pediatr. Infect. Dis.* **5:**201–207.

47. **Hakansson, A., A. Kidd, G. Wadell, H. Sabharwal, and C. Svanborg.** 1994. Adenovirus infection enhances in vitro adherence of *Streptococcus pneumoniae. Infect. Immun.* **62:**2707–2714.

48. **Hamburger, M., and O. Robertson.** 1940. Studies of the pathogenesis of experimental pneumococcus pneumonia in the dog. *J. Exp. Med.* **72:**261–274.

49. **Hammerschmidt, S., S. Talay, P. Brandtzaeg, and G. Chhatwal.** 1997. SpsA, a novel pneumococcal surface protein with specific binding to secretory immunoglobulin A and secretory component. *Mol. Microbiol.* **25:** 1113–1124.

50. **Hammerschmidt, S., M. Tillig, S. Wolff, J. Vaerman, and G. Chhatwal.** 2000. Species-specific binding of human secretory component to SpsA protein of *Streptococcus pneumoniae* via a hexapeptide motif. *Mol. Microbiol.* **36:**726–736.

51. **Hand, W., and J. Cantey.** 1974. Antibacterial mechanisms of the lower respiratory tract. I. Immunoglobulin synthesis and secretion. *J. Clin. Investig.* **53:**354–362.

52. **Harford, C., V. Leidler, and M. Hara.** 1948. Effect of the lesion due to influenza virus on the resistance of mice to inhaled pneumococci. *J. Exp. Med.* **86:**53–68.

53. **Hava, D., and A. Camilli.** 2002. Large-scale identification of serotype 4 *Streptococcus pneumoniae* virulence factors. *Mol. Microbiol.* **45:**1389–1405.

54. **Hava, D., C. Hemsley, and A. Camilli.** 2003. Transcriptional regulation in the *Streptococcus pneumoniae rlrA* pathogenicity islet by RlrA. *J. Bacteriol.* **185:**413–421.

55. **Heffron, R.** 1939. *Pneumonia.* Commonwealth Fund, New York, N.Y.

56. **Hermann, C., I. Spreitzer, N. Schroeder, S. Morath, M. Lehner, W. Fischer, C. Schutt, R. Schumann, and T. Hartung.** 2002. Cytokine induction by purified lipoteichoic acids from various bacterial species—role of LBP, sCD14, CD14 and failure to induce IL-12 and subsequent IFN release. *Eur. J. Immunol.* **32:** 541–551.

57. **Heumann, D., C. Barras, A. Severin, M. P. Glauser, and A. Tomasz.** 1994. Gram-positive cell walls stimulate synthesis of tumor necrosis factor alpha and interleukin-6 by human monocytes. *Infect. Immun.* **62:**2715–2721.

58. **Hirst, R., H. Yesilkaya, E. Clitheroe, A. Rutman, N. Dufty, T. Mitchell, C. O'-Callaghan, and P. Andrew.** 2002. Sensitivities of human monocytes and epithelial cells to pneumolysin are different. *Infect. Immun.* **70:** 1017–1022.

59. **Hord, J., R. Byrd, L. Stowe, B. Windsor, and K. Smith-Whitley.** 2002. *Streptococcus pneumoniae* sepsis and meningitis during the penicillin prophylaxis era in children with sickle cell disease. *J. Pediatr. Hematol. Oncol.* **24:**470–472.

60. **Iannelli, F., M. Oggioni, and G. Pozzi.** 2002. Allelic variation in the highly polymorphic locus *pspC* of *Streptococcus pneumoniae. Gene* **284:**63–71.

61. **Idanpaan-Heikkila, I., P. Simon, C. Cahill, K. Sokol, and E. Tuomanen.** 1997. Oligosaccharides interfere with the establishment and progression of experimental pneumococcal pneumonia. *J. Infect. Dis.* **176:**704–712.

62. **Jarva, H., R. Janulczyk, J. Hellwage, P. Zipfel, L. Bjorck, and S. Meri.** 2002. *Streptococcus pneumoniae* evades complement attack and opsonophagocytosis by expressing the pspC locus-encoded Hic protein that binds short consensus repeats 8–11 of factor H. *J. Immunol.* **168:**1886–1894.

63. **Jounblat, R., A. Kadioglu, T. J. Mitchell, and P. W. Andrew.** 2003. Pneumococcal behavior and host responses during bronchopneumonia are affected differently by the cytolytic and complement-activating activities of pneumolysin. *Infect. Immun.* **71:**1813–1819.

64. **Kadioglu, A., N. Gingles, K. Grattan, A. Kerr, T. Mitchell, and P. Andrew.** 2000. Host cellular immune response to pneumococcal lung infection in mice. *Infect. Immun.* **68:**492–501.

65. **Kerr, A., J. Irvine, J. Search, N. Gingles, A. Kadioglu, P. Andrew, W. McPheat, C.**

Booth, and T. Mitchell. 2002. Role of inflammatory mediators in resistance and susceptibility to pneumococcal infection. *Infect. Immun.* **70:**1547–1557.

66. Kim, J., and J. Weiser. 1998. Association of intrastrain phase variation in quantity of capsular polysaccharide and teichoic acid with the virulence of *Streptococcus pneumoniae. J. Infect. Dis.* **177:**368–377.

67. Kline, B. 1917. Experimental study of organization in lobar pneumonia. *J. Exp. Med.* **26:**239–248.

68. Kline, B., and M. Winternitz. 1915. Studies on experimental pneumonia in rabbits. VIII. Intra vitam staining in experimental pneumonia, and the circulation in the pneumonic lung. *J. Exp. Med.* **21:**311–319.

69. Kostrzynska, M., and T. Wadstrom. 1992. Binding of laminin, type IV collagen, and vitronectin by *Streptococcus pneumoniae. Zentbl. Bakteriol.* **277:**80–83.

70. Krivan, H. C., D. D. Roberts, and V. Ginsburg. 1988. Many pulmonary pathogenic bacteria bind specifically to the carbohydrate sequence GalNacB1–4Gal found in some glycolipids. *Proc. Natl. Acad. Sci. USA* **85:**6157–6161.

71. Laennec, R. 1932. *A Treatise on the Diseases of the Chest and on Mediate Auscultation.* SS & Wm Wood, New York, N.Y.

72. Lauw, F., J. Branger, S. Florquin, P. Speelman, S. Van Deventer, S. Akira, and T. van der Poll. 2002. IL-18 improves the early antimicrobial host response to pneumococcal pneumonia. *J. Immunol.* **168:**372–378.

73. Linder, T., R. Dandiles, D. Lime, and T. DeMaria. 1994. Effect of intranasal inoculation of *Streptococcus pneumoniae* on the structure of the surface carbohydrates of the chinchilla eustachian tube and middle ear mucosa. *Microb. Pathog.* **16:**435–441.

74. Loosli, C. 1940. Pathogenesis and pathology of lobar pneumonia. *Lancet* **i:**49–54.

75. Louria, D., H. Blumenfeld, J. Ellis, E. Kilbourne, and D. Rogers. 1959. Studies on influenza in the pandemic of 1957–58. II. Pulmonary complications of influenza. *J. Clin. Investig.* **38:**213–265.

76. MacCallum, W. 1925. *Textbook of Pathology.* W. B. Saunders, Philadelphia, Pa.

77. Madsen, M., Y. Lebenthal, Q. Cheng, B. Smith, and M. Hostetter. 2000. A pneumococcal protein that elicits interleukin-8 from pulmonary epithelial cells. *J. Infect. Dis.* **181:**1330–1336.

78. Malley, R., P. Henneke, S. Morse, M. Cieslewicz, M. Lipsitch, C. Thompson, E.

Kurt-Jones, J. Paton, M. Wessels, and D. Golenbock. 2003. Recognition of pneumolysin by Toll-like receptor 4 confers resistance to pneumococcal infection. *Proc. Natl. Acad. Sci. USA* **100:**1966–1971.

79. Marrie, T., H. Durant, and L. Yates. 1989. Community-acquired pneumonia requiring hospitalization. *Rev. Infect. Dis.* **11:**568.

80. McAvin, J. C., P. A. Reilly, R. M. Roudabush, W. J. Barnes, A. Salmen, G. W. Jackson, K. Beninga, A. Astorga, F. K. McCleskey, W. B. Huff, D. Niemeyer, and K. L. Lohman. 2001. Sensitive and specific method for rapid identification of *Streptococcus pneumoniae* using real-time fluorescence PCR. *J. Clin. Microbiol.* **39:**3446–3451.

81. McCullers, J., and E. Tuomanen. 2001. Molecular pathogenesis of pneumococcal pneumonia. *Front. Biosci.* **6:**877–889.

82. McCullers, J., and K. Bartmess. 2003. Role of neuraminidase in lethal synergism between influenza virus and *Streptococcus pneumoniae. J. Infect. Dis.* **187:**1000–1009.

83. McCullers, J., and J. Rehg. 2002. Lethal synergism between influenza virus and *Streptococcus pneumoniae*: characterization of a mouse model and the role of platelet-activating factor receptor. *J. Infect. Dis.* **186:**341–350.

84. Melby, K., G. Toews, and A. Pierce. 1985. Pulmonary elastase activity in response to *Streptococcus pneumoniae* and *Pseudomonas aeruginosa. Am. Rev. Respir. Dis.* **131:**559–563.

85. Mileski, W., J. Harlan, C. Rice, and R. Winn. 1990. *Streptococcus pneumoniae*-stimulated macrophages induce neutrophils to emigrate by a CD18-independent mechanism of adherence. *Circ. Shock* **31:**259–267.

86. Mohler, J., E. Azoulay-Dupuis, C. Amory-Rivier, J. Mazoit, J. Bedos, V. Rieux, and P. Moine. 2003. *Streptococcus pneumoniae* strain-dependent lung inflammatory responses in a murine model of pneumococcal pneumonia. *Intensive Care Med.* **29:**808–816.

87. Mold, C., K. Edwards, and H. Gewura. 1982. Binding of C-reactive protein to bacteria. *Infect. Immun.* **38:**392–395.

88. Mold, C., B. Rodic-Polic, and T. Du Clos. 2002. Protection from *Streptococcus pneumoniae* infection by C reactive protein and natural antibody requires complement but not Fc gamma receptors. *J. Immunol.* **168:**6375–6381.

89. Murdoch, C., R. Read, Q. Zhang, and A. Finn. 2002. Choline binding protein A of *Streptococcus pneumoniae* elicits chemokine production and expression of intercellular adhesion molecule 1 (CD54) by human alveolar epithelial cells. *J. Infect. Dis.* **186:**1253–1260.

90. **Murdoch, D.** 2003. Nucleic acid amplification tests for the diagnosis of pneumonia. *Clin. Infect. Dis.* **36:**1162–1170.

91. **Murdoch, D., T. Anderson, K. Beynon, A. Chua, A. Fleming, R. Laing, G. Town, G. Mills, S. Chambers, and L. Jennings.** 2003. Evaluation of a PCR assay for detection of *Streptococcus pneumoniae* in respiratory and nonrespiratory samples from adults with community-acquired pneumonia. *J. Clin. Microbiol.* **41:**63–66.

92. **Murdoch, D., R. Laing, G. Mills, N. Karalus, G. Town, S. Mirrett, and L. Reller.** 2001. Evaluation of a rapid immunochromatographic test for detection of *Streptococcus pneumoniae* antigen in urine samples from adults with community-acquired pneumonia. *J. Clin. Microbiol.* **39:**3495–3498.

93. **Musher, D., R. Mediwala, H. Phan, G. Chen, and R. Baughn.** 2001. Nonspecificity of assaying for IgG antibody to pneumolysin in circulating immune complexes as a means to diagnose pneumococcal pneumonia. *Clin. Infect. Dis.* **32:**534–538.

94. **Nungester, W., and L. Jourdonais.** 1936. Mucin as an aid in the experimental production of lobal pneumonia. *J. Infect. Dis.* **59:**258–265.

95. **Orihuela, C. J., G. Gao, M. McGee, J. Yu, K. P. Francis, and E. Tuomanen.** 2003. Organ-specific models of *Streptococcus pneumoniae* disease. *Scand. J. Infect. Dis.* **35:**647–652.

96. **Ortqvist, A., J. Hedlund, B. Wretlind, A. Carlstrom, and M. Kalin.** 1995. Diagnostic and prognostic value of interleukin-6 and C-reactive protein in community acquired pneumonia. *Scand. J. Infect. Dis.* **27:**457–462.

97. **Osler, W.** 1897. On certain features in the prognosis of pneumonia. *Am. J. Med. Sci.* **113:**1–10.

98. **Pallares, R., J. Linares, M. Vadillo, C. Cabellos, F. Manresa, P. Viladrich, R. Martin, and F. Gudiol.** 1995. Resistance to penicillin and cephalosporins and mortality from severe pneumococcal pneumonia in Barcelona, Spain. *N. Engl. J. Med.* **333:**474–480.

99. **Paton, J., A. Berry, and R. Lock.** 1997. Molecular analysis of putative pneumococcal virulence proteins. *Microb. Drug Resist.* **3:**1–10.

100. **Pericone, C. D., K. Overweg, P. M. W. Hermans, and J. N. Weiser.** 2000. Inhibitory and bactericidal effects of hydrogen peroxide production by *Streptococcus pneumoniae* on other inhabitants of the upper respiratory tract. *Infect. Immun.* **68:**3990–3997.

101. **Plotkowski, M. C., E. Puchelle, G. Beck, J. Jacquot, and C. Hannoun.** 1986. Adherence of type 1 Streptococcus pneumoniae to tracheal epithelium of mice infected with influenza A/PR8 virus. *Am. Rev. Respir. Dis.* **134:**1040–1044.

102. **Programming, N.C. Office of Statistics.** 1999. Deaths and death rates for the 10 leading causes of death in specified age groups: United States, 1997. *Natl. Vital Stat. Rep.* **47:**27–37.

103. **Rake, G.** 1936. Pathogenesis of pneumococcus infection in mice following intranasal instillation. *J. Exp. Med.* **63:**17–37.

104. **Rayner, C. F., A. D. Jackson, A. Rutman, A. Dewar, T. J. Mitchell, P. W. Andrew, P. J. Cole, and R. Wilson.** 1995. Interaction of pneumolysin-sufficient and -deficient isogenic variants of *Streptococcus pneumoniae* with human respiratory mucosa. *Infect. Immun.* **63:**442–447.

105. **Reisenfeld-Orn, I., S. Wolpe, J. Garcia-Bustos, M. Hoffmann, and E. Tuomanen.** 1989. Production of interleukin-1 but not tumor necrosis factor by human monocytes stimulated with pneumococcal cell surface components. *Infect. Immun.* **57:**1890–1893.

106. **Rich, A., and C. McKee.** 1939. The pathogenicity of avirulent pneumococci for animals deprived of leukocytes. *Bull. Johns Hopkins Hosp.* **64:**434–446.

107. **Rijneveld, A., S. Florquin, J. Branger, P. Speelman, S. Van Deventer, and T. van der Poll.** 2001. TNF-alpha compensates for the impaired host defense of IL-1 type 1 receptor-deficient mice during pneumococcal pneumonia. *J. Immunol.* **167:**5240–5246.

108. **Rijneveld, A., F. Lauw, M. Schultz, S. Florquin, A. Te Velde, P. Speelman, S. Van Deventer, and T. van der Poll.** 2002. The role of interferon-gamma in murine pneumococcal pneumonia. *J. Infect. Dis.* **185:**91–97.

109. **Rijneveld, A., G. van den Dobbelsteen, S. Florquin, T. Standiford, P. Speelman, L. van Alphen, and T. van der Poll.** 2002. Roles of interleukin-6 and macrophage inflammatory protein-2 in pneumolysin-induced lung inflammation in mice. *J. Infect. Dis.* **185:**123–126.

110. **Ring, A., J. Weiser, and E. Tuomanen.** 1998. Pneumococcal penetration of the blood brain barrier: molecular analysis of a novel re-entry path. *J. Clin. Investig.* **102:**347–360.

111. **Robertson, O., S. Woo, S. Cheer, and L. King.** 1928. Study of the mechanism of recovery from experimental pneumococcus infection. *J. Exp. Med.* **47:**317–333.

112. **Rosenow, C., P. Ryan, J. Weiser, S. Johnson, P. Fontan, A. Ortqvist, and H. Masure.** 1997. Contribution of a novel choline

binding protein to adherence, colonization, and immunogenicity of *Streptococcus pneumoniae*. *Mol. Microbiol.* **25**:819–829.

113. **Rubins, J., D. Charboneau, C. Fasching, A. Berry, J. Paton, J. Alexander, P. Andrew, T. Mitchell, and E. Janoff.** 1996. Distinct role for pneumolysin's cytotoxic and complement activities in the pathogenesis of pneumococcal pneumonia. *Am. J. Respir. Crit. Care Med.* **153**:1339–1346.

114. **Rubins, J., P. Duane, D. Charboneau, and E. Janoff.** 1992. Toxicity of pneumolysin to pulmonary endothelial cells in vitro. *Infect. Immun.* **60**:1740–1746.

115. **Rubins, J. B., P. G. Duane, D. Clawson, D. Charboneau, J. Young, and D. E. Niewoehner.** 1993. Toxicity of pneumolysin to pulmonary alveolar epithelial cells. *Infect. Immun.* **61**:1352–1358.

116. **Saladino, R., A. Stack, G. Fleisher, C. Thompson, D. Briles, L. Kobzik, and G. Siber.** 1997. Development of a model of low-inoculum *Streptococcus pneumoniae* intrapulmonary infection in infant rats. *Infect. Immun.* **65**:4701–4704.

117. **Sato, S., N. Ouellet, I. Pelletier, M. Simard, A. Rancourt, and M. Bergeron.** 2002. Role of galectin-3 as an adhesion molecule for neutrophil extravasation during streptococcal pneumonia. *J. Immunol.* **168**:1813–1822.

118. **Schroeder, N., S. Morath, C. Alexander, L. Hamann, I. Spreitzer, T. Hartung, U. Zahringer, U. Goebel, J. Weber, and R. Schumann.** 2003. Lipoteichoic acid of *S. pneumoniae* and *S. aureus* activate immune cells via toll-like receptor (TLR)-2, and not TLR-4 and MD-2. *J. Biol. Chem.* **278**:15587–15594.

119. **Schumann, R., D. Pfeil, D. Freyer, W. Buerger, N. Lamping, C. Kirschning, U. Goebel, and J. Weber.** 1998. Lipopolysaccharide and pneumococcal cell wall components activate the mitogen activated protein kinases (MAPK) ERK-1, ERK-2, and p38 in astrocytes. *Glia* **22**:295–305.

120. **Scott, J. A. G., E. L., Marston, A. J. Hall, and K. Marsh.** 2003. Diagnosis of pneumococcal pneumonia by *psaA* PCR analysis of lung aspirates from adult patients in Kenya. *J. Clin. Microbiol.* **41**:2554–2559.

121. **Sebert, M. H., L. M. Palmer, M. Rosenberg, and J. N. Weiser.** 2002. Microarray-based identification of *htrA*, a *Streptococcus pneumoniae* gene that is regulated by the CiaRH two-component system and contributes to nasopharyngeal colonization. *Infect. Immun.* **70**:4059–4067.

122. **Smith, B., and M. Hostetter.** 2000. C3 as substrate for adhesion of *Streptococcus pneumoniae*. *J. Infect. Dis.* **182**:497–508.

123. **Spellerberg, B., D. Cundell, J. Sandros, B. Pearce, I. Idänpään-Heikkilä, C. Rosenow, and H. Masure.** 1996. Pyruvate oxidase as a determinant of virulence in *Streptococcus pneumoniae*. *Mol. Microbiol.* **19**:803–813.

124. **Sutliff, W., and T. Friedemann.** 1938. A soluble edema-producing substance from the pneumococcus. *J. Immunol.* **34**:455–467.

125. **Takashima, K., K. Tateda, T. Matsumoto, Y. Iizawa, M. Nakao, and K. Yamaguchi.** 1997. Role of tumor necrosis factor alpha in pathogenesis of pneumococcal pneumonia in mice. *Infect. Immun.* **65**:257–260.

126. **Talbot, U., A. Paton, and J. Paton.** 1996. Uptake of *Streptococcus pneumoniae* by respiratory epithelial cells. *Infect. Immun.* **64**:3772–3777.

127. **Tan, T., E. J. Mason, E. Wald, W. Varson, G. Schutze, J. Bradley, L. Givner, R. Yogev, K. Kim, and S. Kaplan.** 2002. Clinical characteristics of children with complicated pneumonia caused by *Streptococcus pneumoniae*. *Pediatrics* **110**:1–6.

128. **Throup, J., K. Koretke, A. Bryant, K. Ingraham, A. Chalker, Y. Ge, A. Marra, N. Wallis, J. Brown, D. Holmes, M. Rosenberg, and M. Burnham.** 2000. A genomic analysis of two-component signal transduction in *Streptococcus pneumoniae*. *Mol. Micriobiol.* **35**:566–576.

129. **Toews, G., and W. Vial.** 1984. The role of C5 in the polymorphonuclear leukocyte recruitment in response to *Streptococcus pneumoniae*. *Am. Rev. Respir. Dis.* **129**:82–86.

130. **Tokars, J., M. Frank, M. Alter, and M. Arduino.** 2002. National surveillance of dialysis-associated diseases in the United States, 2000. *Semin. Dial.* **15**:162–171.

131. **Tong, H., M. McIver, L. Fisher, and T. DeMaria.** 1999. Effect of lacto-N-neotetraose, asialoganglioside-GM1 and neuraminidase on adherence of otitis media associated serotypes of *Streptococcus pneumoniae* to chinchilla tracheal epithelium. *Microb. Pathog.* **26**:111–119.

132. **Tuomanen, E.** 1986. Piracy of adhesins: attachment of superinfecting pathogens to respiratory cilia by secreted adhesins of *Bordetella pertussis*. *Infect. Immun.* **54**:905–908.

133. **Tuomanen, E., R. Austrian, and H. Masure.** 1995. The pathogenesis of pneumococcal infection. *N. Engl. J. Med.* **332**:1280–1284.

134. **Tuomanen, E., B. Hengstler, R. Rich, M. Bray, O. Zak, and A. Tomasz.** 1987. Nonsteroidal anti-inflammatory agents in the therapy of experimental pneumococcal meningitis. *J. Infect. Dis.* **155**:985–990.

135. **Tuomanen, E., R. Rich, and O. Zak.** 1987. Induction of pulmonary inflammation by components of the pneumococcal cell surface. *Am. Rev. Respir. Dis.* **135:**869–874.

136. **van der Flier, M., N. Chhun, T. Wizemann, J. Min, J. McCarthy, and E. Tuomanen.** 1995. Adherence of *Streptococcus pneumoniae* to immobilized fibronectin. *Infect. Immun.* **63:**4317–4322.

137. **van der Flier, M., F. Coenjaerts, J. Kimpen, A. Hoepelman, and S. Geelen.** 2000. *Streptococcus pneumoniae* induces secretion of vascular endothelial growth factor by human neutrophils. *Infect. Immun.* **68:**4792–4794.

138. **Vollmer, W., and A. Tomasz.** 2001. Identification of the teichoic acid phosphorylcholine esterase of *Streptococcus pneumoniae. Mol. Microbiol.* **39:**1610–1622.

139. **Wadowsky, R., S. Mietzner, D. Skoner, W. Doyle, and P. Fireman.** 1995. Effect of experimental influenza A virus infection on isolation of *Streptococcus pneumoniae* and other aerobic bacteria from the oropharynges of allergic and nonallergic adult subjects. *Infect. Immun.* **63:**1153–1157.

140. **Wagner, H., I. Bennett, L. Lasagna, L. Cluff, M. Rosenthal, and G. Mirick.** 1956. The effect of hydrocortisone upon the course of pneumococcal pneumonia treated with penicillin. *Bull. Johns Hopkins Hosp.* **98:**197–215.

141. **Wang, E., M. Simard, N. Ouellet, Y. Bergeron, D. Beauchamp, and M. Bergeron.** 2002. Pathogenesis of pneumococcal pneumonia in cyclophosphamide-induced leukopenia in mice. *Infect. Immun.* **70:**4226–4238.

142. **Weber, J. R., D. Freyer, C. Alexander, N. W. Schroder, A. Reiss, C. Kuster, D. Pfeil, E. I. Tuomanen, and R. R. Schumann.** 2003. Recognition of pneumococcal peptidoglycan, an expanded, pivotal role for LPS binding protein. *Immunity* **19:**269–279.

143. **Weiser, J., R. Austrian, P. Sreenivasan, and H. Masure.** 1994. Phase variation in pneumococcal opacity: relationship between colonial morphology and nasopharyngeal colonization. *Infect. Immun.* **62:**2582–2589.

144. **Weiser, J., J. B. Goldberg, N. Pan, L. Wilson, and M. Virji.** 1998. The phosphorylcholine epitope undergoes phase variation on a 43-kilodalton protein in *Pseudomonas aeruginosa* and on pili of *Neisseria meningitidis* and *Neisseria gonorrhoeae. Infect. Immun.* **66:**4263–4267.

145. **Weiser, J., N. Pan, K. McGowan, D. Musher, A. Martin, and J. Richards.** 1998. Phosphorylcholine on the lipopolysaccharide of *Haemophilus influenzae* contributes to persistence in the respiratory tract and sensitivity to serum killing mediated by C-reactive protein. *J. Exp. Med.* **187:**631–640.

146. **Weiser, J., M. Shchepetov, and S. Chong.** 1997. Decoration of lipopolysaccharide with phosphorylcholine: a phase-variable characteristic of *Haemophilus influenzae. Infect. Immun.* **65:**943–950.

147. **Williams, J. J., M. Pahl, D. Kwong, J. Zhang, D. Hatakeyama, K. Ahmed, M. Naderi, M. Kim, and N. Vazii.** 2003. Modulation of neutrophil complement receptor 3 expression by pneumococci. *Clin. Sci.* **104:**615–625.

148. **Winkelstein, J., and A. Tomasz.** 1978. Activation of the alternative complement pathway by pneumococcal cell wall teichoic acid. *J. Immunol.* **120:**174–178.

149. **Winternitz, M., and A. Hirschfelder.** 1913. Studies upon experimental pneumonia in rabbits. *J. Exp. Med.* **17:**657–665.

150. **Wood, W. J., R. Smith, and B. Watson.** 1946. Studies on the mechanism of recovery in pneumococcal pneumonia. IV. The mechanism of phagocytosis in the absence of antibody. *J. Exp. Med.* **84:**387–401.

151. **Yoshimura, A., E. Lien, R. Ingalls, E. Tuomanen, R. Dziarski, and D. Golenbock.** 1999. Recognition of gram positive bacterial cell wall components by the innate immune system occurs via toll-like receptor 2. *J. Immunol.* **63:**1–5.

152. **Zhang, J., I. Idanpaan-Heikkila, W. Fischer, and E. Tuomanen.** 1999. The pneumococcal *lic D2* is involved in phosphorylcholine metabolism. *Mol. Microbiol.* **31:**1477–1488.

153. **Zhang, J.-R., K. Mostov, M. Lamm, M. Nanno, S. Shimida, M. Ohwaki, and E. Tuomanen.** 2000. The polymeric immunoglobulin receptor translocates pneumococci across human nasopharyngeal epithelial cells. *Cell* **102:**827–837.

PATHOGENESIS OF
PNEUMOCOCCAL MENINGITIS

Joerg R. Weber

16

PNEUMOCOCCAL MENINGITIS: MOST IMPORTANT, MOST DISASTROUS

Pneumococcal meningitis is still an unresolved problem in clinical medicine. Although highly effective antibiotics kill the bacteria efficiently, mortality rates of up to 34% are reported, which are unacceptably high (18). Additionally, up to 50% of the survivors suffer from long-term sequelae (5, 15, 73). Recently, a landmark study of the efficacy of dexamethasone as an adjunctive treatment to antibiotics showed that both mortality and morbidity can be reduced in pneumococcal meningitis (15). It also became clear, however, that only an understanding of the molecular basis of the disease would allow further improvement of the outcome, particularly in view of upcoming challenges such as increasing resistance to currently used antibiotics (16, 66, 96). Different animal models as well as cell culture systems of relevant brain structures, in combination with new molecular techniques, provide an exciting area of research which will hopefully translate into a general understanding of how pneumococci interact with and damage our brain and create new opportunities for neuroprotection.

Epidemiology

During the last 20 years, the epidemiology of bacterial meningitis has dramatically changed. *Haemophilus influenzae*, formerly a major cause of meningitis, virtually disappeared in developed countries and serves as a remarkable example of a successful vaccination campaign. Nowadays, pneumococci are the most important cause of bacterial meningitis in children and adults in the United States as well as in Europe. The incidence of the disease varies from 1.1 to 1.2 per 100,000 per year in the United States (94) and western Europe (3) and up to 12 per 100,000 per year in Africa (57). The risk of disease is highest under the age of 5 and over age 60 years. Some predisposing factors, such as splenectomy, malnutrition, or sickle cell disease, are known (21, 34). The use of conjugate pneumococcal vaccines has led to a significant decline in invasive pneumococcal disease, including meningitis, in those regions promoting this approach (95). An emerging problem is the spread of pneumococci resistant to beta-lactam antibiotics (79). Prolonged persistence of pneumococci in the cerebrospinal fluid (CSF) may result in higher mortality as well as in pronounced neurological damage in survivors (20, 51). Longer persistence of living bacteria urges us to understand in detail the effects of pneumococcal toxins and released cell

Joerg R. Weber, Department of Neurology, Charité, Universitätsmedizin Berlin, D-10098 Berlin, Germany.

The Pneumococcus. Volume Editor, Elaine I. Tuomanen.
© 2004 ASM Press, Washington, D.C.

wall components and their contribution to neuronal damage.

Clinical Features and Prognosis

Early clinical features of bacterial meningitis are nonspecific and include fever, headache and, later on, meningismus (neck stiffness). These clinical signs indicate inflammatory activation of the trigeminal sensory nerve fibers in the meninges and can be blocked experimentally by 5-HT receptor agonists (31). Approximately 20% of patients develop focal neurologic signs, such as epileptic seizures or paresis of a limb, and 20% present with impaired consciousness. The key to confirm the diagnosis is proof of bacteria in CSF by Gram-staining or, more specifically, a positive pneumococcal culture and by influx of leukocytes in the CSF (Color Plate 9). There is an ongoing controversy about the necessity of a cranial computed tomography scan (CT) before lumbar puncture. A cranial CT provides information concerning brain edema, hydrocephalus, and intracranial complications, such as infarcts, and local infections, such as mastoiditis. Those patients who present with focal neurologic deficits or those who have disturbed consciousness must have a cranial CT before lumbar puncture if possible (27). Limited access to CT should never delay antibiotic treatment in patients with suspected bacterial meningitis.

A number of recent studies demonstrate the unfavorable outcome of pneumococcal meningitis, reporting mortality rates up to 40%. More alarming is the surprisingly high rate of sequelae of up to 50% in survivors of the disease. Hearing loss arising from bacterial invasion of the cochlea is one of the major disabilities after meningitis and occurs in up to 21% of survivors (4, 15). Major clinical complications are raised intracranial pressure as a consequence of brain edema, hydrocephalus, vascular complications, and sepsis, all contributing to neuronal injury (18, 34). Dexamethasone is able to reduce the mortality and unfavorable outcome by about 50% and decreases hearing loss by about 30% (15). Dexamethasone should be given before or, at the latest, together with the antibiotic therapy (four doses of 10 mg/day for 4 days or 0.15 mg/kg of bodyweight for 4 days) based on the idea of preventing an overshooting inflammatory response initiated by highly inflammatory bacterial components released by antibiotic lysis. Another study performed in Malawi showed no effect of dexamethasone treatment in children. A different antibiotic treatment regimen and the high number of human immunodeficiency virus patients in this trial may explain this difference (54). In summary, dexamethasone treatment should be recommended in areas of the world where beta-lactam antibiotics are the first line of antibiotic treatment.

WHAT CAN WE LEARN FROM ANIMAL MODELS?

A major impetus for meningitis research has been the introduction of reliable animal models in the last 30 years. Most models are based on the injection of pneumococci or their inflammatory components into the subarachnoidal space, followed by the hallmarks of meningitis: the influx of leukocytes into CSF, development of brain edema, intracranial pressure, and typical histological changes. An additional advantage is that depending on the experimental animal, CSF can be withdrawn and analyzed. Early models used monkeys (35) and rabbits (13, 82). Later, the rat model provided the advantage of exact physiological monitoring and allowed advanced techniques, such as cranial windows for studying online leukocyte-endothelium interactions and cerebral blood flow (62, 65, 88). Recently, the spectrum of meningitis models has been expanded to include the mouse, thereby connecting meningitis research to transgenic technology and modern immunology (24, 31). All these animal models share a similar approach of inducing meningitis by injecting the pneumococci somewhere into the subarachnoidal space, resulting in predictable and reproducible levels of inflammation. However, they also share the same limitation in that they do not really model the most common natural course of disease that proceeds from the mucosa to the blood and then into the CSF. However, these models are widely used and simulate the less common form of bacterial invasion from local infections such as mastoiditis. Promising new approaches infect animals intranasally (50, 103) and study the invasion of pneumococci into the central nervous system (CNS) as visualized with bioluminescent bacteria detected by a charge-coupled device (Xenogen Imaging System). Beside their classic role in meningitis research these models are used to examine more general aspects of innate immunity (91) and are excellent research tools.

HOW DO PNEUMOCOCCI ENTER THE BRAIN?

The brain is completely separated from the blood by tight blood-CSF and blood-brain barriers. These barriers keep larger molecules, as well as bacteria and viruses, out of the immune-privileged CNS. Direct access to CNS through dural defects, e.g. after neurosurgery or trauma or due to local infections, is rare but potentially possible. More commonly, the successful invasive pathogen carries effective tools in order to invade and replicate within the CNS. For meningitis, invasion occurs specifically into the CSF space rather than the brain parenchyma. Colonization of the nasopharynx, translocation into the bloodstream, and subsequent high-grade bacteremia are thought to be necessary to invade successfully into the subarachnoidal space (14).

In principle, there are three potential entry sites into CSF (Fig. 1). (i) The blood-CSF barrier formed by the choroid plexus is composed of a fenestrated endothelial layer and the choroidal epithelium that generates the barrier to the CSF by forming tight junctions (80). (ii) The blood-CSF barrier in the subarachnoidal space is represented by vessels that are mainly composed of endothelial cells forming tight junctions (75). (iii) The real blood-brain barrier consists of endothelial cells of cerebral vessels that closely interact

with pericytes as well as astrocytes to contribute to the barrier function (25). Until now the anatomic location of bacterial entry has been controversial. Some evidence generated in the monkey model suggests that the choroid plexus may be the site of invasion (14). Meningococci have been found in the plexus as well as in the meninges (64). Animals developing pneumococcal meningitis after intraperitoneal or intranasal challenge show inflammatory infiltration mostly around the leptomeningeal blood vessels supporting subarachnoidal and arachnoidal vessels as potential entry sites (68, 103). Taking these few data from different models together, the anatomic site of bacterial entry into the CSF is still unknown.

Pneumococci adhere to and finally invade brain microvascular endothelial cells (BMEC) by transcytosis and reach the basal surface of the endothelial cell layer (67). Initial adherence is dependent on pneumococcal choline-binding protein A (CpbA) displayed on the pneumococcal cell wall (69). Invasion of pneumococci is mediated by cell wall phosphorylcholine interacting with the platelet-activating factor (PAF) receptor on eukaryotic cells (12), including human cerebral endothelial cells (67). Interestingly, pneumococcal phase variation is important, and only transparent pneumococci are successful invaders. The transparent phenotype carries significantly more CbpA and choline on the teichoic acid of its cell wall (36). Bacterial invasion only happens after stimulation of the endothelial monolayer, for instance, with tumor necrosis factor α (TNF-α). Strikingly, the pneumococcus itself, in particular its cell wall, induces TNF-α production, and therefore the bacterium carries the tools necessary for activation of endothelial cells (23). In summary, the cell wall may serve as an activator of the endothelial barrier and sets up ideal conditions for transparent pneumococci to interact with the PAF receptor and enter the CNS successfully (Fig. 2). Once the bacteria have entered the subarachnoidal space, they take advantage of the immune privilege of the CNS. At the

A.

B. Choroid plexus   **C. Cerebral capillaries**

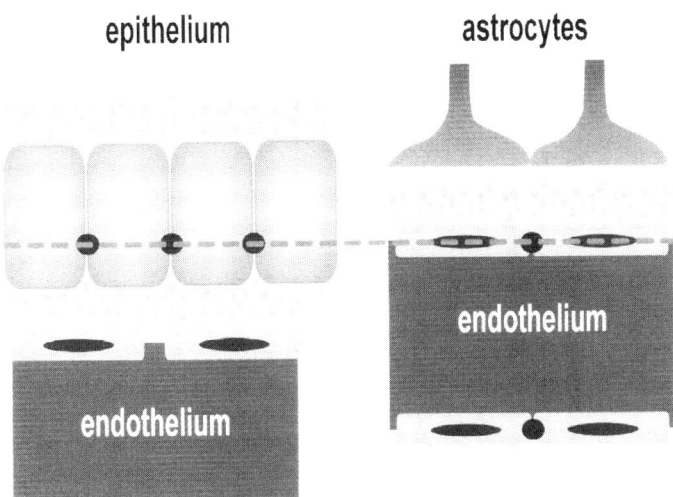

FIGURE 1 Schematic anatomy of the blood-brain barrier. (A) Cerebral vessels mark the blood-brain barrier as formed by tight junctions and supporting cells. (B) In detail, the barrier at the choroid plexus is formed at the layer of epithelial CSF secreting cells that overlay fenestrated endothelial cells. Translocated bacteria are indicated as small ovals. (C) The barrier at the cerebral capillary is formed by tight junctions between endothelial cells that are supported by astrocytes. Translocated bacteria are indicated as small ovals.

beginning of the inflammatory cascade, leukocytes, complement, and immunoglobulins are virtually absent in the CSF, and the pneumococcus can replicate and induce inflammation (76).

ONCE IN THE CSF, HOW DO PNEUMOCOCCI CAUSE INFLAMMATION?
In addition to activating the cerebral endothelium, different pneumococcal surface components and toxins induce production of proinflammatory cytokines in most cells of the CNS. Meningeal and perivascular macrophages may also play a crucial role in early meningeal inflammation (63). Heat-killed unencapsulated pneumococci, purified pneumococcal cell wall and lipoteichoic acid cause meningitis indistinguishable from that caused by living pneumococci (82, 92). Different pattern recognition molecules capable of sensing pneumococci have been postulated. Surprisingly, lipopolysaccharide (LPS)-binding protein (LBP), a molecule historically attributed to LPS signaling, binds to the pneumococcal cell wall and augments activation of BMEC and macrophages. The biological activity of the cell wall critically relies on the intact peptidoglycan

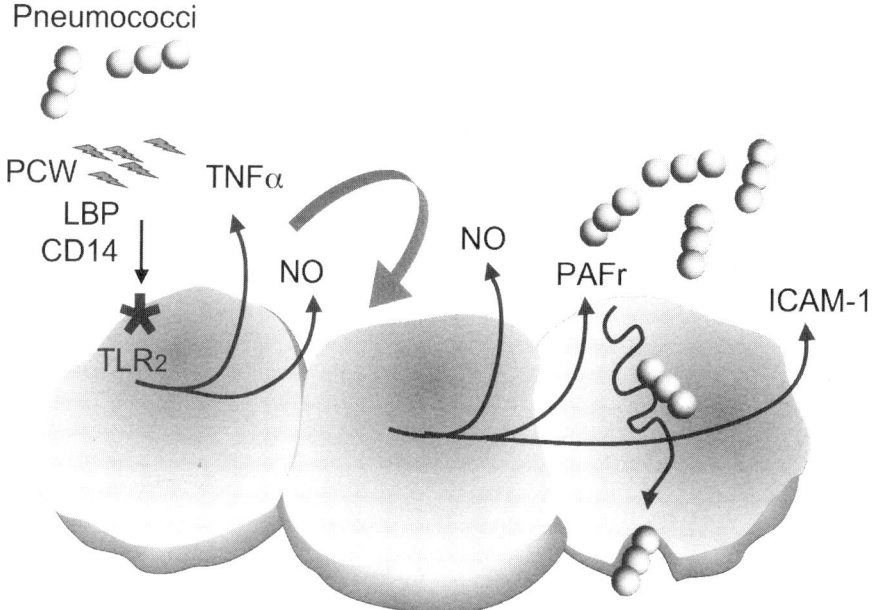

FIGURE 2 Autocrine loop enhancing transmigration of pneumococci across the blood-brain barrier. Pneumococcal cell wall (PCW) induces endothelial cells to produce TNF and NO. These in turn further activate the neighboring cells to produce more NO and up-regulate the PAF receptor (PAFr) and ICAM-1. PAF receptor enables bacterial translocation into the CSF, while ICAM-1 recruits leukocytes.

backbone and in part on the presence of stem peptides. Anionic groups in the stem peptides or the glycan backbone may serve as an iterative binding motif for LBP. Translation of this in vitro observation to experimental meningitis in LBP deficient mice resulted in a significant decrease of invading leukocytes into the CSF following pneumococcal cell wall or living bacteria (91).

CD14, another pattern recognition molecule, may play a role in pneumococcal immune activation in meningitis (9). For heat-killed pneumococci, lipoteichoic acid, and cell wall, CD14 dependent and independent signaling pathways seem to exist (10, 11, 74). Heat-killed pneumococci, peptidoglycan, and lipoteichoic acid signal through the TLR-2 receptor and activate NF-κB in a reporter gene assay (72, 91, 99). Interestingly, pneumolysin, a pore forming toxin and critical virulence factor of pneumococci, signals through the TLR-4 receptor (49). Experiments in

TLR-2 and TLR-4 double-knockout peritoneal macrophages provide some evidence for other signaling receptors that may be involved (37). Although there is clear evidence that major surface components of pneumococci initiate inflammation in a TLR-2-dependent fashion, meningitis in TLR-2-deficient mice challenged with living pneumococci is associated with higher mortality and pronounced bacterial growth and inflammation compared to that in wild-type animals (19). On balance, TLR-2 seems to be the most important receptor for the early inflammatory response caused by pneumococci and contributes to host-driven bacterial control.

HOW DO PNEUMOCOCCI SIGNAL?
Engagement of TLR-2, TLR-4, PAF receptor, and other receptors by pneumococcal ligands drives a series of major intracellular downstream signaling events, including activation of erk1/2, p38 mitogen-activated protein

kinase, and NF-κB. Purified pneumococcal cell wall induces the phosphorylation of erk1/2 and p38 kinases (74) and inhibition of these pathways in microglia is associated with significantly reduced cytokine and chemokine release (26). Inhibition of the erk1/2 pathway reduces TNF-α release but has no effect on nitric oxide (NO) production, whereas inhibition of p38 decreases NO accumulation but not TNF-α (55). Upstream events that might explain such differences are currently unknown. The involvement of the NF-κB pathway in pneumococcal signaling was first described for human and murine monocytes (78). In a rat model of pneumococcal meningitis, NF-κB activity was increased and pharmacological inhibition resulted in lower leukocyte influx and decreased interleukin-6 (IL-6) levels (38). NF-κB activation has been connected to the TLR-2 receptor (91, 99).

Pro- and anti-inflammatory cytokines are important regulators of the immune response to bacteria and other pathogens. TNF-α, IL-1β, and IL-6 have been detected in the CSF of patients suffering from pneumococcal meningitis (81, 85) as well as in several animal models. Inhibition of TNF-α results in decreased leukocyte invasion in pneumococcal meningitis (71). Although recombinant species-specific TNF-α injected intracisternally causes only mild inflammation, it greatly augments the effect of otherwise inactive, low concentrations of pneumococcal cell wall (1). Studies of meningitis in TNF-α-deficient mice have not really clarified the role of this potent cytokine. TNF-α-deficient mice die earlier than controls after intracerebral infection, but no effect on leukocyte recruitment or bacterial density in the CSF has been reported. Additionally, neuronal damage is not affected by TNF-α or TNF receptor deficiency (93).

Antibodies directed against IL-1β reduce the influx of neutrophils into the CSF and point to a relevant role for this cytokine (71). Caspase-1 is required to generate mature IL-1β, and caspase-1 depletion is associated with a reduction of intracranial complications in experimental meningitis (42). A recent study of IL-1 receptor type 1-deficient mice in a blood-borne-meningitis model produced conflicting results, with higher bacterial titers in the CSF and increased mortality, suggesting that IL-1 contributes to control of bacterial growth (101).

Deficiency in IL-18, another cytokine cleaved by caspase-1, results in reduced inflammation and better survival in experimental pneumococcal meningitis (100). IL-6 seems to play a regulatory role, since its depletion augments inflammation but partially prevents vascular permeability (59). IL-10, a cytokine with anti-inflammatory properties, decreases inflammation in pneumococcal meningitis and reduces cytokine and chemokine production but does not affect bacterial clearance (39, 102). Transforming growth factor β (TGF-β), another anti-inflammatory cytokine, is also able to prevent intracranial complications in early pneumococcal rat meningitis (60). Taken together, proinflammatory and anti-inflammatory cytokines regulate a complex immune response in a time dependent fashion. Attenuation of intracranial inflammatory changes leads to significant beneficial effect on outcome only if control of bacterial growth is not compromised.

Chemokines are divided into four subfamilies (for review, see reference 70), are produced by microglia stimulated with pneumococcal cell wall (28), and are up-regulated in animal models of pneumococcal meningitis. Interference with the inflammatory cascade at several levels reduces chemokine concentrations in brain tissue of infected animals (42). Chemokines are elevated in the CSF of meningitis patients regardless of the bacterial agent (77). IL-8 has been shown to play a major role in leukocyte chemotaxis in *H. influenzae* type b and LPS-induced meningitis (17). It is assumed but not proven that a similar role is likely in pneumococcal disease.

LEUKOCYTES: GOOD OR BAD?

Bacterial meningitis is characterized by two processes targeting the blood-brain barrier. The invasion of pneumococci is followed by the recruitment of highly activated leukocytes. Leukocyte migration from blood into CSF is a multistep process beginning with rolling, firm adhesion and finally migration. In pneumococcal meningitis, all these steps happen in the arachnoidal microvessels (48a, 90) and the triggers for leukocyte-endothelium interactions include histamine (89) and cytokines. The early steps of the leukocyte-endothelium interaction are mediated by selectins and their ligands. Inhibition of leukocyte rolling by interfering with the function of P- and L-selectin decreases CSF leukocytosis (90). Firm adhesion is mostly mediated by ICAM-1 on the endothelial side and CD18 adhesion molecules on the leukocytes. The pneumococcal cell wall induces ICAM-1 up-regulation in BMEC through a TNF-dependent autocrine loop (23) (Fig. 2). Clearly, TNF from other sources may add to this effect. Following firm adhesion, the leukocytes migrate between endothelial cells, loosening tight junctions by as yet-uncharacterized interactions with junctional components.

Blocking antibodies directed against CD18 on the leukocyte or ICAM-1 on the endothelial cell decrease leukocyte-induced brain edema, intracranial pressure, and blood flow alterations (83, 87). This translates to improved survival of infected animals and less neuronal apoptosis in the hippocampus (6). Of clinical importance is the observation that lower numbers of leukocytes in the CSF are not necessarily associated with increased bacterial growth, in part because phagocytosis in a fluid environment is not effective regardless of the number of cells (83). Thus, dexamethasone and other antileukocyte agents represent useful strategies to improve the outcome of disease. A further challenge and still unresolved problem is the detailed understanding of leukocyte functions so as to discriminate harmful and beneficial effects. Matrix metalloproteases (MMPs) are zinc-dependent endopeptidases that degrade extracellular matrix components and may represent an example of a potentially harmful leukocyte product (98). Neutrophils appear to be the major source of MMPs and potentially use these proteases to degrade the basal lamina on their way into the CSF. In an infant rat model of pneumococcal meningitis, MMP inhibitors decreased inflammatory alterations, including neuronal damage (43, 46). MMPs are elevated in the CSF of meningitis patients, and an increased concentration of MMP-9 is correlated to a poor clinical outcome of bacterial meningitis (47). Pneumococci produce zinc metalloproteinase capable of activation of MMP-9. However, the role of this virulence factor in meningitis is still unknown (58).

OXIDATIVE STRESS

As in other injury models of the CNS, oxidative mechanisms have been proposed to play a crucial role in tissue damage in bacterial meningitis. Reactive oxygen species and reactive nitrogen intermediates are both major effector molecules involved in bacterial clearance but also result in bystander host cell injury. Different antioxidative strategies interfering with generation of reactive oxygen species attenuate intracranial complications, including neuronal damage in pneumococcal meningitis in the adult as well as in the infant animal model (2, 61, 62).

Inhibition of inducible NO synthase (NOS) ameliorates inflammation in pneumococcal meningitis. This observation accords with less inflammation in inducible-NOS knockout mice (40, 97). In contrast, genetic deficiency of endothelial NOS augments blood-brain barrier damage and inflammatory complications (41). Oxygen and nitrogen species react to form the oxidant peroxynitrite, a metabolite of known neurotoxicity. Peroxynitrite is detected as nitrotyrosine in the CSF of patients and experimental animals and has been correlated with an unfavorable outcome (32). Peroxynitrite is found in leukocytes and the meninges and along penetrating blood vessels (33). One of the challenges of the future will be to identify cellular sources of radicals and consider the substantial ability of pneumococci to produce radicals themselves.

CELLULAR (NEURONAL) DAMAGE

Although survivors of bacterial meningitis suffer from pronounced memory and learning deficits (52, 84), meningitis is technically a disease of the meninges and the blood-brain barrier, not the brain parenchyma. Thus, neuronal injury arises indirectly from the bacteria and an excessive immune response by the host. In the infant rat model of experimental meningitis, neuronal injury occurs in the hippocampus and the cortex and is both necrotic and apoptotic (45, 48). In the rabbit model, apoptosis is a predominant mechanism of neuronal damage and focuses on the dentate gyrus (6, 104). Hippocampal neurons of patients who have died from pneumococcal meningitis show the morphological criteria of apoptosis and active caspase-3 (56). Additionally, hippocampal atrophy is frequently observed in survivors of bacterial meningitis (22).

Several factors contribute in a complex fashion to neuronal damage in pneumococcal meningitis. One trigger is the host inflammatory response, since blocking leukocyte invasion into the CSF prevents about 50% of the damage. In experimental meningitis, concentrations of several caspases are significantly elevated and active caspase-3 is present in the dentate gyrus (86). Therefore, it is surprising that intrathecal application of z-VAD-fmk, a broad-spectrum caspase inhibitor, rescues only ~50% of neurons (6). These results suggest that while half the damage is caspase dependent, a portion of the damage is independent of host inflammation and the classical caspase apoptotic pathway. In vitro, pneumococci induce cell death in different cell types, like microglia and neurons. Treatment of those cells with z-VAD-fmk does not prevent pneumococcus-induced cell death, providing an additional strong argument supporting a mixture of caspase-dependent and -independent execution of cell death. Activity measurements of several caspases, including the key executioner caspase-3, are negative in cell cultures undergoing apoptosis in response to pneumococci. A potential mechanism of the caspase-independent cell death is rapid damage of mitochondria that is followed by the release of cytochrome c and apoptosis-inducing factor (AIF) (7). AIF drives chromatin condensation into large 50-kb fragments, and cells incubated with living pneumococci do not generate classical, oligonucleosomal DNA fragments but, rather, show large 50-kb fragments

on pulsed field gel electrophoresis. In addition, microinjected antibodies directed against AIF partially prevent cell death (7). At first look, these results appear in sharp contrast to the beneficial effect of caspase inhibitors in experimental meningitis. However, synthesis of the available data indicates two independent major contributors to neuronal damage. Neurons and glia appear to respond to pneumococci and the host inflammatory response by initiating both caspase-dependent and -independent cell death pathways.

Until recently the question of whether bacterial metabolites and toxins contribute to neuronal damage in meningitis remained unsettled. Two major pneumococcal toxins were implicated in neuronal damage in studies using mutant bacteria. Incubation of brain cells with strains defective in production of pneumolysin, a pore-forming, proinflammatory cytolysin (53), and H_2O_2 resulted in significant but not complete reduction of apoptosis. Application of these toxins in purified form induced apoptotic cell death in neurons and microglia in a dose-dependent fashion, and the process appeared to be caspase independent. Injection of double-knockout bacteria intrathecally in rabbits caused significantly less apoptosis of neurons, proving the in vivo role of these toxins (8). Thus, the host response appears to drive a caspase-dependent neuronal cell death, while the bacterial toxins initiate AIF-dependent apoptosis (Fig. 3).

In experimental meningitis, cellular damage has been shown in neurons and, interestingly, in the ependyma. Localized loss of cilia, a decrease of the ciliary beat frequency, and destruction of the ependymal ultrastructure result from pneumococcal meningitis (29, 30). Dysfunction of these cells may contribute to increased intracranial pressure and increased outflow resistance and may expose underlying neurons to host and bacterial cytotoxins.

Translating these concepts to the clinical situation suggests the basis for several treatment modifications to decrease sequelae arising from neuronal loss. Not only must patients be treated as early as possible with sufficient antibiotics but also the host response should be down

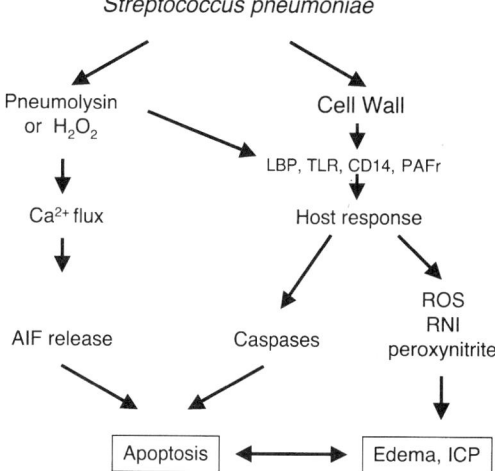

FIGURE 3 Schematic of two pathways of neuronal injury in pneumococcal meningitis. ICP, intracranial pressure; ROS, reactive oxygen species; RNI, reactive nitrogen intermediates.

modulated to prevent excess inflammation. This is currently achieved by the use of dexamethasone together with the first doses of antibiotic. Although the beneficial effect of dexamethasone has been shown in clinical trials, it should be considered that, at least in experimental models, this drug augments apoptosis in the dentate gyrus (44, 104). Although the overall beneficial effect of dexamethasone seems to outweigh this proapoptotic effect, it suggests that better interventions should be sought.

REFERENCES

1. **Angstwurm, K., D. Freyer, U. Dirnagl, U. K. Hanisch, R. R. Schumann, K. M. Einhäupl, and J. R. Weber**. 1998. Tumour necrosis factor alpha induces only minor inflammatory changes in the central nervous system, but augments experimental meningitis. *Neuroscience* **86**:627–634.
2. **Auer, M., L. A. Pfister, D. Leppert, M. G. Tauber, and S. L. Leib**. 2000. Effects of clinically used antioxidants in experimental pneumococcal meningitis. *J. Infect. Dis.* **182**:347–350.
3. **Berg, S., B. Trollfors, B. A. Claesson, K. Alestig, L. Gothefors, S. Hugosson, L. Lindquist, P. Olcen, V. Romanus, and K. Strangert**. 1996. Incidence and prognosis of meningitis due to Haemophilus influenzae, Strep-

tococcus pneumoniae and Neisseria meningitidis in Sweden. *Scand. J. Infect. Dis.* **28**:247–252.
4. **Bhatt, S. M., A. Lauretano, C. Cabellos, C. Halpin, R. A. Levine, W. Z. Xu, J. B. Nadol, Jr., and E. Tuomanen**. 1993. Progression of hearing loss in experimental pneumococcal meningitis: correlation with cerebrospinal fluid cytochemistry. *J. Infect. Dis.* **167**:675–683.
5. **Bohr, V., O. B. Paulson, and N. Rasmussen**. 1984. Pneumococcal meningitis. Late neurologic sequelae and features of prognostic impact. *Arch. Neurol.* **41**:1045–1049.
6. **Braun, J. S., R. Novak, K. H. Herzog, S. M. Bodner, J. L. Cleveland, and E. I. Tuomanen**. 1999. Neuroprotection by a caspase inhibitor in acute bacterial meningitis. *Nat. Med.* **5**:298–302.
7. **Braun, J. S., R. Novak, P. J. Murray, C. M. Eischen, S. A. Susin, G. Kroemer, A. Halle, J. R. Weber, E. I. Tuomanen, and J. L. Cleveland**. 2001. Apoptosis-inducing factor mediates microglial and neuronal apoptosis caused by pneumococcus. *J. Infect. Dis.* **184:**1300–1309.
8. **Braun, J. S., J. E. Sublett, D. Freyer, T. J. Mitchell, J. L. Cleveland, E. I. Tuomanen, and J. R. Weber**. 2002. Pneumococcal pneumolysin and H₂O₂ mediate brain cell apoptosis during meningitis. *J. Clin. Investig.* **109:**19–27.
9. **Cauwels, A., K. Frei, S. Sansano, C. Fearns, R. Ulevitch, W. Zimmerli, and R. Landmann**. 1999. The origin and function of soluble CD14 in experimental bacterial meningitis. *J. Immunol.* **162**:4762–4772.
10. **Cauwels, A., E. Wan, M. Leismann, and E. Tuomanen**. 1997. Coexistence of CD14-dependent and independent pathways for stimulation of human monocytes by gram-positive bacteria. *Infect. Immun.* **65**:3255–3260.
11. **Cleveland, M. G., J. D. Gorham, T. L. Murphy, E. Tuomanen, and K. M. Murphy**. 1996. Lipoteichoic acid preparations of gram-positive bacteria induce interleukin-12 through a CD14-dependent pathway. *Infect. Immun.* **64:**1906–1912.
12. **Cundell, D. R., N. P. Gerard, C. Gerard, I. Idanpaan-Heikkila, and E. I. Tuomanen**. 1995. Streptococcus pneumoniae anchor to activated human cells by the receptor for platelet-activating factor. *Nature* **377**:435–438.
13. **Dacey, R. G., and M. A. Sande**. 1974. Effect of probenecid on cerebrospinal fluid concentrations of penicillin and cephalosporin derivatives. *Antimicrob. Agents Chemother.* **6**:437–441.
14. **Daum, R. S., D. W. Scheifele, V. P. Syriopoulou, D. Averill, and A. L. Smith**.

1978. Ventricular involvement in experimental Hemophilus influenzae meningitis. *J. Pediatr.* **93**:927–930.

15. **de Gans, J., and D. van de Beek**. 2002. Dexamethasone in adults with bacterial meningitis. *N. Engl. J. Med.* **347**:1549–1556.

16. **Doern, G. V., K. P. Heilmann, H. K. Huynh, P. R. Rhomberg, S. L. Coffman, and A. B. Brueggemann**. 2001. Antimicrobial resistance among clinical isolates of Streptococcus pneumoniae in the United States during 1999–2000, including a comparison of resistance rates since 1994–1995. *Antimicrob. Agents Chemother.* **45**:1721–1729.

17. **Dumont, R. A., B. D. Car, N. N. Voitenok, U. Junker, B. Moser, O. Zak, and T. O'Reilly**. 2000. Systemic neutralization of interleukin-8 markedly reduces neutrophilic pleocytosis during experimental lipopolysaccharide-induced meningitis in rabbits. *Infect. Immun.* **68**:5756–5763.

18. **Durand, M. L., S. B. Calderwood, D. J. Weber, S. I. Miller, F. S. Southwick, V. S. Caviness, Jr., and M. N. Swartz**. 1993. Acute bacterial meningitis in adults. A review of 493 episodes. *N. Engl. J. Med.* **328**:21–28.

19. **Echchannaoui, H., K. Frei, C. Schnell, S. L. Leib, W. Zimmerli, and R. Landmann**. 2002. Toll-like receptor 2-deficient mice are highly susceptible to Streptococcus pneumoniae meningitis because of reduced bacterial clearing and enhanced inflammation. *J. Infect. Dis.* **186**:798–806.

20. **Fiore, A. E., J. F. Moroney, M. M. Farley, L. H. Harrison, J. E. Patterson, J. H. Jorgensen, M. Cetron, M. S. Kolczak, R. F. Breiman, and A. Schuchat**. 2000. Clinical outcomes of meningitis caused by Streptococcus pneumoniae in the era of antibiotic resistance. *Clin. Infect. Dis.* **30**:71–77.

21. **Fraser, D. W., C. P. Darby, R. E. Koehler, C. F. Jacobs, and R. A. Feldman**. 1973. Risk factors in bacterial meningitis: Charleston County, South Carolina. *J. Infect. Dis.* **127**:271–277.

22. **Free, S. L., L. M. Li, D. R. Fish, S. D. Shorvon, and J. M. Stevens**. 1996. Bilateral hippocampal volume loss in patients with a history of encephalitis or meningitis. *Epilepsia* **37**:400–405.

23. **Freyer, D., R. Manz, A. Ziegenhorn, M. Weih, K. Angstwurm, W. D. Docke, A. Meisel, R. R. Schumann, G. Schonfelder, U. Dirnagl, and J. R. Weber**. 1999. Cerebral endothelial cells release TNF-alpha after stimulation with cell walls of Streptococcus pneumoniae and regulate inducible nitric oxide synthase and ICAM-1 expression via autocrine loops. *J. Immunol.* **163**:4308–4314.

24. **Gerber, J., G. Raivich, A. Wellmer, C. Noeske, T. Kunst, A. Werner, W. Bruck, and R. Nau**. 2001. A mouse model of Streptococcus pneumoniae meningitis mimicking several features of human disease. *Acta Neuropathol.* **101**:499–508.

25. **Gloor, S. M., M. Wachtel, M. F. Bolliger, H. Ishihara, R. Landmann, and K. Frei**. 2001. Molecular and cellular permeability control at the blood-brain barrier. *Brain Res. Rev.* **36**:258–264.

26. **Hanisch, U. K., M. Prinz, K. Angstwurm, K. G. Hausler, O. Kann, H. Kettenmann, and J. R. Weber**. 2001. The protein tyrosine kinase inhibitor AG126 prevents the massive microglial cytokine induction by pneumococcal cell walls. *Eur. J. Immunol.* **31**:2104–2115.

27. **Hasbun, R., J. Abrahams, J. Jekel, and V. J. Quagliarello**. 2001. Computed tomography of the head before lumbar puncture in adults with suspected meningitis. *N. Engl. J. Med.* **345**:1727–1733.

28. **Hausler, K. G., M. Prinz, C. Nolte, J. R. Weber, R. R. Schumann, H. Kettenmann, and U. K. Hanisch**. 2002. Interferon-gamma differentially modulates the release of cytokines and chemokines in lipopolysaccharide- and pneumococcal cell wall-stimulated mouse microglia and macrophages. *Eur. J. Neurosci.* **16**:2113–2122.

29. **Hirst, R. A., B. Gosai, A. Rutman, P. W. Andrew, and C. O'Callaghan**. 2003. Streptococcus pneumoniae damages the ciliated ependyma of the brain during meningitis. *Infect. Immun.* **71**:6095–6100.

30. **Hirst, R. A., K. S. Sikand, A. Rutman, T. J. Mitchell, P. W. Andrew, and C. O'Callaghan**. 2000. Relative roles of pneumolysin and hydrogen peroxide from Streptococcus pneumoniae in inhibition of ependymal ciliary beat frequency. *Infect. Immun.* **68**:1557–1562.

31. **Hoffmann, O., N. Keilwerth, M. B. Bille, U. Reuter, K. Angstwurm, R. R. Schumann, U. Dirnagl, and J. R. Weber**. 2002. Triptans reduce the inflammatory response in bacterial meningitis. *J. Cereb. Blood Flow Metab.* **22**:988–996.

32. **Kastenbauer, S., U. Koedel, B. F. Becker, and H. W. Pfister**. 2002. Oxidative stress in bacterial meningitis in humans. *Neurology* **58**:186–191.

33. **Kastenbauer, S., U. Koedel, and H. W. Pfister**. 1999. Role of peroxynitrite as a mediator of pathophysiological alterations in experimental pneumococcal meningitis. *J. Infect. Dis.* **180**:1164–1170.

34. **Kastenbauer, S., and H. W. Pfister**. 2003. Pneumococcal meningitis in adults: spectrum of complications and prognostic factors in a series of 87 cases. *Brain* **126:**1015–1025.

35. **Kaufmann, A. F., and K. D. Quist**. 1969. Pneumococcal meningitis and peritonitis in rhesus monkeys. *J. Am. Vet. Med. Assoc.* **155:** 1158–1162.

36. **Kim, J. O., and J. N. Weiser**. 1998. Association of intrastrain phase variation in quantity of capsular polysaccharide and teichoic acid with the virulence of Streptococcus pneumoniae. *J. Infect. Dis.* **177:**368–377.

37. **Koedel, U., B. Angele, T. Rupprecht, H. Wagner, A. Roggenkamp, H. W. Pfister, and C. J. Kirschning**. 2003. Toll-like receptor 2 participates in mediation of immune response in experimental pneumococcal meningitis. *J. Immunol.* **170:**438–444.

38. **Koedel, U., I. Bayerlein, R. Paul, B. Sporer, and H. W. Pfister**. 2000. Pharmacologic interference with NF-kappaB activation attenuates central nervous system complications in experimental pneumococcal meningitis. *J. Infect. Dis.* **182:**1437–1445.

39. **Koedel, U., A. Bernatowicz, K. Frei, A. Fontana, and H. W. Pfister**. 1996. Systemically (but not intrathecally) administered IL-10 attenuates pathophysiologic alterations in experimental pneumococcal meningitis. *J. Immunol.* **157:** 5185–5191.

40. **Koedel, U., A. Bernatowitz, R. Paul, K. Frei, A. Fontana, and H. W. Pfister**. 1995. Experimental pneumococcal meningitis: cerebrovascular alterations, brain edema, and meningeal inflammation are linked to the production of nitric oxide. *Ann. Neurol.* **37:**313–323.

41. **Koedel, U., R. Paul, F. Winkler, S. Kastenbauer, P. L. Huang, and H. W. Pfister**. 2001. Lack of endothelial nitric oxide synthase aggravates murine pneumococcal meningitis. *J. Neuropathol. Exp. Neurol.* **60:**1041–1050.

42. **Koedel, U., F. Winkler, B. Angele, A. Fontana, R. A. Flavell, and H. W. Pfister**. 2002. Role of caspase-1 in experimental pneumococcal meningitis: evidence from pharmacologic caspase inhibition and caspase-1-deficient mice. *Ann. Neurol.* **51:**319–329.

43. **Leib, S. L., J. M. Clements, R. L. Lindberg, C. Heimgartner, J. M. Loeffler, L. A. Pfister, M. G. Tauber, and D. Leppert**. 2001. Inhibition of matrix metalloproteinases and tumour necrosis factor alpha converting enzyme as adjuvant therapy in pneumococcal meningitis. *Brain* **124:**1734–1742.

44. **Leib, S. L., C. Heimgartner, Y. D. Bifrare, J. M. Loeffler, and M. G. Tauber**. 2003. Dexamethasone aggravates hippocampal apoptosis and learning deficiency in pneumococcal meningitis in infant rats. *Pediatr. Res.* **54:**353–357.

45. **Leib, S. L., Y. S. Kim, L. L. Chow, R. A. Sheldon, and M. G. Tauber**. 1996. Reactive oxygen intermediates contribute to necrotic and apoptotic neuronal injury in an infant rat model of bacterial meningitis due to group B streptococci. *J. Clin. Investig.* **98:**2632–2639.

46. **Leib, S. L., D. Leppert, J. Clements, and M. G. Tauber**. 2000. Matrix metalloproteinases contribute to brain damage in experimental pneumococcal meningitis. *Infect. Immun.* **68:**615–620.

47. **Leppert, D., S. L. Leib, C. Grygar, K. M. Miller, U. B. Schaad, and G. A. Hollander**. 2000. Matrix metalloproteinase (MMP)-8 and MMP-9 in cerebrospinal fluid during bacterial meningitis: association with blood-brain barrier damage and neurological sequelae. *Clin. Infect. Dis.* **31:**80–84.

48. **Loeffler, J. M., R. Ringer, M. Hablutzel, M. G. Tauber, and S. L. Leib**. 2001. The free radical scavenger alpha-phenyl-tert-butyl nitrone aggravates hippocampal apoptosis and learning deficits in experimental pneumococcal meningitis. *J. Infect. Dis.* **183:**247–252.

48a. **Lorenzl, S., U. Koedel, U. Dirnagl, G. Ruckdeschel, and H. W. Pfisten**. 1993. Imaging of leukocyte-endothelium interaction using in vivo confocal laser scanning microscopy during the early phase of experimental pneumococcal meningitis. *J. Infect. Dis.* **168:**927–933.

49. **Malley, R., P. Henneke, S. C. Morse, M. J. Cieslewicz, M. Lipsitch, C. M. Thompson, E. Kurt-Jones, J. C. Paton, M. R. Wessels, and D. T. Golenbock**. 2003. Recognition of pneumolysin by Toll-like receptor 4 confers resistance to pneumococcal infection. *Proc. Natl. Acad. Sci. USA* **100:**1966–1971.

50. **Marra, A., and D. Brigham**. 2001. Streptococcus pneumoniae causes experimental meningitis following intranasal and otitis media infections via a nonhematogenous route. *Infect. Immun.* **69:** 7318–7325.

51. **McCullers, J. A., B. K. English, and R. Novak**. 2000. Isolation and characterization of vancomycin-tolerant Streptococcus pneumoniae from the cerebrospinal fluid of a patient who developed recrudescent meningitis. *J. Infect. Dis.* **181:**369–373.

52. **Merkelbach, S., H. Sittinger, I. Schweizer, and M. Müller**. 2000. Cognitive outcome after

bacterial meningitis. *Acta Neurol. Scand.* **102:** 118–123.

53. **Mitchell, T. J.** 2000. Virulence factors and the pathogenesis of disease caused by Streptococcus pneumoniae. *Res. Microbiol.* **151:**413–419.

54. **Molyneux, E. M., A. L. Walsh, H. Forsyth, M. Tembo, J. Mwenechanya, K. Kayira, L. Bwanaisa, A. Njobvu, S. Rogerson, and G. Malenga.** 2002. Dexamethasone treatment in childhood bacterial meningitis in Malawi: a randomised controlled trial. *Lancet* **360:**211–218.

55. **Monier, R. M., K. L. Orman, E. A. Meals, and B. K. English.** 2002. Differential effects of p38- and extracellular signal-regulated kinase mitogen-activated protein kinase inhibitors on inducible nitric oxide synthase and tumor necrosis factor production in murine macrophages stimulated with Streptococcus pneumoniae. *J. Infect. Dis.* **185:**921–926.

56. **Nau, R., A. Soto, and W. Bruck.** 1999. Apoptosis of neurons in the dentate gyrus in humans suffering from bacterial meningitis. *J. Neuropathol. Exp. Neurol.* **58:**265–274.

57. **O'Dempsey, T. J., T. F. McArdle, N. Lloyd-Evans, I. Baldeh, B. E. Lawrence, O. Secka, and B. Greenwood.** 1996. Pneumococcal disease among children in a rural area of west Africa. *Pediatr. Infect. Dis. J.* **15:**431–437.

58. **Oggioni, M. R., G. Memmi, T. Maggi, D. Chiavolini, F. Iannelli, and G. Pozzi.** 2003. Pneumococcal zinc metalloproteinase ZmpC cleaves human matrix metalloproteinase 9 and is a virulence factor in experimental pneumonia. *Mol. Microbiol.* **49:**795–805.

59. **Paul, R., U. Koedel, F. Winkler, B. C. Kieseier, A. Fontana, M. Kopf, H. P. Hartung, and H. W. Pfister.** 2003. Lack of IL-6 augments inflammatory response but decreases vascular permeability in bacterial meningitis. *Brain* **126:** 1873–1882.

60. **Pfister, H. W., K. Frei, B. Ottnad, U. Koedel, A. Tomasz, and A. Fontana.** 1992. Transforming growth factor beta 2 inhibits cerebrovascular changes and brain edema formation in the tumor necrosis factor alpha-independent early phase of experimental pneumococcal meningitis. *J. Exp. Med.* **176:**265–268.

61. **Pfister, H. W., U. Ködel, U. Dirnagl, R. L. Haberl, G. Ruckdeschel, and K. M. Einhäupl.** 1992. Effect of catalase on regional cerebral blood flow and brain edema during the early phase of experimental pneumococcal meningitis. *J. Infect. Dis.* **166:**1442–1445.

62. **Pfister, H. W., U. Koedel, R. L. Haberl, U. Dirnagl, W. Feiden, G. Ruckdeschel, and K. Einhäupl.** 1990. Microvascular changes during

the early phase of experimental bacterial meningitis. *J. Cereb. Blood Flow Metab.* **10:**914–922.

63. **Polfliet, M. M., P. J. Zwijnenburg, A. M. van Furth, T. van der Poll, E. A. Dopp, C. Renardel de Lavalette, E. M. van Kesteren-Hendrikx, N. van Rooijen, C. D. Dijkstra, and T. K. van den Berg.** 2001. Meningeal and perivascular macrophages of the central nervous system play a protective role during bacterial meningitis. *J. Immunol.* **167:**4644–4650.

64. **Pron, B., M. K. Taha, C. Rambaud, J. C. Fournet, N. Pattey, J. P. Monnet, M. Musilek, J. L. Beretti, and X. Nassif.** 1997. Interaction of Neisseria meningitidis with the components of the blood-brain barrier correlates with an increased expression of PilC. *J. Infect. Dis.* **176:**1285–1292.

65. **Quagliarello, V. J., W. J. Long, and W. M. Scheld.** 1986. Morphologic alterations of the blood-brain barrier with experimental meningitis in the rat. Temporal sequence and role of encapsulation. *J. Clin. Investig.* **77:**1084–1095.

66. **Richter, S. S., K. P. Heilmann, S. L. Coffman, H. K. Huynh, A. B. Brueggemann, M. A. Pfaller, and G. V. Doern.** 2002. The molecular epidemiology of penicillin-resistant Streptococcus pneumoniae in the United States, 1994–2000. *Clin. Infect. Dis.* **34:**330–339.

67. **Ring, A., J. N. Weiser, and E. I. Tuomanen.** 1998. Pneumococcal trafficking across the blood-brain barrier. Molecular analysis of a novel bidirectional pathway. *J. Clin. Investig.* **102:** 347–360.

68. **Rodriguez, A. F., S. L. Kaplan, E. P. Hawkins, and E. O. Mason, Jr.** 1991. Hematogenous pneumococcal meningitis in the infant rat: description of a model. *J. Infect. Dis.* **164:**1207–1209.

69. **Rosenow, C., P. Ryan, J. N. Weiser, S. Johnson, P. Fontan, A. Ortqvist, and H. R. Masure.** 1997. Contribution of novel choline-binding proteins to adherence, colonization and immunogenicity of Streptococcus pneumoniae. *Mol. Microbiol.* **25:**819–829.

70. **Rossi, D., and A. Zlotnik.** 2000. The biology of chemokines and their receptors. *Annu. Rev. Immunol.* **18:**217–242.

71. **Saukkonen, K., S. Sande, C. Cioffe, S. Wolpe, B. Sherry, A. Cerami, and E. Tuomanen.** 1990. The role of cytokines in the generation of inflammation and tissue damage in experimental gram-positive meningitis. *J. Exp. Med.* **171:**439–448.

72. **Schroder, N. W., S. Morath, C. Alexander, L. Hamann, T. Hartung, U. Zahringer, U. B.**

Gobel, J. R. Weber, and R. R. Schumann. 2003. Lipoteichoic acid (LTA) of Streptococcus pneumoniae and Staphylococcus aureus activates immune cells via Toll-like receptor (TLR)-2, lipopolysaccharide-binding protein (LBP), and CD14, whereas TLR-4 and MD-2 are not involved. *J. Biol. Chem.* **278:**15587–15594.

73. **Schuchat, A., K. Robinson, J. D. Wenger, L. H. Harrison, M. Farley, A. L. Reingold, L. Lefkowitz, and B. A. Perkins**. 1997. Bacterial meningitis in the United States in 1995. Active Surveillance Team. *N. Engl. J. Med.* **337:** 970–976.

74. **Schumann, R. R., D. Pfeil, W. Bürger, N. Lamping, C. J. Kirschning, U. Goebel, and J. R. Weber**. 1998. Lipopolysaccharide and pneumococcal cell wall components activate the mitogen activated protein kinases (MAPK) erk-1, erk-2, and p 38 in astrocytes. *Glia* **22:**295–305.

75. **Segal, M. B.** 2000. The choroid plexuses and the barriers between the blood and the cerebrospinal fluid. *Cell. Mol. Neurobiol.* **20:**183–196.

76. **Simberkoff, M. S., N. H. Moldover, and J. Rahal, Jr.** 1980. Absence of detectable bactericidal and opsonic activities in normal and infected human cerebrospinal fluids. A regional host defense deficiency. *J. Lab. Clin. Med.* **95:** 362–372.

77. **Spanaus, K. S., D. Nadal, H. W. Pfister, J. Seebach, U. Widmer, K. Frei, S. Gloor, and A. Fontana.** 1997. C-X-C and C-C chemokines are expressed in the cerebrospinal fluid in bacterial meningitis and mediate chemotactic activity on peripheral blood-derived polymorphonuclear and mononuclear cells in vitro. *J. Immunol.* **158:** 1956–1964.

78. **Spellerberg, B., C. Rosenow, W. Sha, and E. I. Tuomanen.** 1996. Pneumococcal cell wall activates NF-kappa B in human monocytes: aspects distinct from endotoxin. *Microb. Pathog.* **20:**309–317.

79. **Stanek, R. J., and M. A. Mufson.** 1999. A 20-year epidemiological study of pneumococcal meningitis. *Clin. Infect. Dis.* **28:**1265–1272.

80. **Strazielle, N., and J. F. Ghersi-Egea.** 2000. Choroid plexus in the central nervous system: biology and physiopathology. *J. Neuropathol. Exp. Neurol.* **59:**561–574.

81. **Tauber, M. G., and B. Moser.** 1999. Cytokines and chemokines in meningeal inflammation: biology and clinical implications. *Clin. Infect. Dis.* **28:**1–11.

82. **Tuomanen, E., H. Liu, B. Hengstler, O. Zak, and A. Tomasz.** 1985. The induction of meningeal inflammation by components of the pneumococcal cell wall. *J. Infect. Dis.* **151:** 859–868.

83. **Tuomanen, E. I., K. Saukkonen, S. Sande, C. Cioffe, and S. D. Wright.** 1989. Reduction of inflammation, tissue damage, and mortality in bacterial meningitis in rabbits treated with monoclonal antibodies against adhesion-promoting receptors of leukocytes. *J. Exp. Med.* **170:**959–969.

84. **van de Beek, D., B. Schmand, J. de Gans, M. Weisfelt, H. Vaessen, J. Dankert, and M. Vermeulen.** 2002. Cognitive impairment in adults with good recovery after bacterial meningitis. *J. Infect. Dis.* **186:**1047–1052.

85. **van Furth, A. M., J. J. Roord, and R. van Furth.** 1996. Roles of proinflammatory and antiinflammatory cytokines in pathophysiology of bacterial meningitis and effect of adjunctive therapy. *Infect. Immun.* **64:**4883–4890.

86. **von Mering, M., A. Wellmer, U. Michel, S. Bunkowski, A. Tlustochowska, W. Bruck, U. Kuhnt, and R. Nau.** 2001. Transcriptional regulation of caspases in experimental pneumococcal meningitis. *Brain Pathol.* **11:**282–295.

87. **Weber, J. R., K. Angstwurm, W. Bürger, K. M. Einhäupl, and U. Dirnagl.** 1995. Anti ICAM-1 (CD54) monoclonal antibody reduces inflammatory changes in experimental bacterial meningitis. *J. Neuroimmunol.* **63:**63–68.

88. **Weber, J. R., K. Angstwurm, D. Freyer, M. Weih, C. Busch, W. Bürger, and U. Dirnagl.** 1996. Heparansulfate and heparin attenuate regional cerebral blood flow (rCBF) increase in experimental bacterial meningitis. *Soc. Neurosci. Abstr.* **22:**677.

89. **Weber, J. R., K. Angstwurm, T. Rosenkranz, U. Lindauer, W. Bürger, K. M. Einhäupl, and U. Dirnagl.** 1997. Histamine (H1) receptor antagonist inhibits leukocyte rolling in pial vessels in the early phase of bacterial meningitis in rats. *Neurosci. Lett.* **226:**17–20.

90. **Weber, J. R., K. Angstwurm, T. Rosenkranz, U. Lindauer, D. Freyer, W. Burger, C. Busch, K. M. Einhaupl, and U. Dirnagl.** 1997. Heparin inhibits leukocyte rolling in pial vessels and attenuates inflammatory changes in a rat model of experimental bacterial meningitis. *J. Cereb. Blood Flow Metab.* **17:**1221–1229.

91. **Weber, J. R., D. Freyer, C. Alexander, N. W. Schroder, A. Reiss, C. Kuster, D. Pfeil, E. I. Tuomanen, and R. R. Schumann.** 2003. Recognition of pneumococcal peptidoglycan: an expanded, pivotal role for LPS binding protein. *Immunity* **19:**269–279.

92. **Weber, J. R., P. Moreillon, and E. I. Tuomanen.** 2003. Innate sensors for Gram-positive bacteria. *Curr. Opin. Immunol.* **15:**408–415.

93. Wellmer, A., J. Gerber, J. Ragheb, G. Zysk, T. Kunst, A. Smirnov, W. Bruck, and R. Nau. 2001. Effect of deficiency of tumor necrosis factor alpha or both of its receptors on *Streptococcus pneumoniae* central nervous system infection and peritonitis. *Infect. Immun.* **69:**6881–6886.

94. Wenger, J. D., A. W. Hightower, R. R. Facklam, S. Gaventa, C. V. Broome, and The Bacterial Meningitis Study Group. 1990. Bacterial meningitis in the United States, 1986: report of a multistate surveillance study. *J. Infect. Dis.* **162:**1316–1323.

95. Whitney, C. G., M. M. Farley, J. Hadler, L. H. Harrison, N. M. Bennett, R. Lynfield, A. Reingold, P. R. Cieslak, T. Pilishvili, D. Jackson, R. R. Facklam, J. H. Jorgensen, and A. Schuchat. 2003. Decline in invasive pneumococcal disease after the introduction of protein-polysaccharide conjugate vaccine. *N. Engl. J. Med.* **348:**1737–1746.

96. Whitney, C. G., M. M. Farley, J. Hadler, L. H. Harrison, C. Lexau, A. Reingold, L. Lefkowitz, P. R. Cieslak, M. Cetron, E. R. Zell, J. H. Jorgensen, and A. Schuchat. 2000. Increasing prevalence of multidrug-resistant Streptococcus pneumoniae in the United States. *N. Engl. J. Med.* **343:**1917–1924.

97. Winkler, F., U. Koedel, S. Kastenbauer, and H. W. Pfister. 2001. Differential expression of nitric oxide synthases in bacterial meningitis: role of the inducible isoform for blood-brain barrier breakdown. *J. Infect. Dis.* **183:**1749–1759.

98. Yong, V. W., C. Power, P. Forsyth, and D. R. Edwards. 2001. Metalloproteinases in biology and pathology of the nervous system. *Nat. Rev. Neurosci.* **2:**502–511.

99. Yoshimura, A., E. Lien, R. R. Ingalls, E. Tuomanen, R. Dziarski, and D. Golenbock. 1999. Cutting edge: recognition of Gram-positive bacterial cell wall components by the innate immune system occurs via Toll-like receptor 2. *J. Immunol.* **163:**1–5.

100. Zwijnenburg, P. J., T. van der Poll, S. Florquin, S. Akira, K. Takeda, J. J. Roord, and A. M. van Furth. 2003. Interleukin-18 gene-deficient mice show enhanced defense and reduced inflammation during pneumococcal meningitis. *J. Neuroimmunol.* **138:**31–37.

101. Zwijnenburg, P. J., T. van der Poll, S. Florquin, J. J. Roord, and A. M. van Furth. 2003. IL-1 receptor type 1 gene-deficient mice demonstrate an impaired host defense against pneumococcal meningitis. *J. Immunol.* **170:**4724–4730.

102. Zwijnenburg, P. J., T. van der Poll, S. Florquin, J. J. Roord, and A. M. van Furth. 2003. Interleukin-10 negatively regulates local cytokine and chemokine production but does not influence antibacterial host defense during murine pneumococcal meningitis. *Infect. Immun.* **71:**2276–2279.

103. Zwijnenburg, P. J., T. van der Poll, S. Florquin, S. J. van Deventer, J. J. Roord, and A. M. van Furth. 2001. Experimental pneumococcal meningitis in mice: a model of intranasal infection. *J. Infect. Dis.* **183:**1143–1146.

104. Zysk, G., W. Bruck, J. Gerber, Y. Bruck, H. W. Prange, and R. Nau. 1996. Anti-inflammatory treatment influences neuronal apoptotic cell death in the dentate gyrus in experimental pneumococcal meningitis. *J. Neuropathol. Exp. Neurol.* **55:**722–728.

IMMUNODEFICIENCY AND INVASIVE PNEUMOCOCCAL DISEASE

Edward N. Janoff and Jeffrey B. Rubins

17

INTRODUCTION

In both children and adults, *Streptococcus pneumoniae* infection most often begins with successful colonization of the nasopharynx. Asymptomatic infections with these encapsulated pathogens are typically transient but occur sequentially, and the organism is considered part of the normal upper respiratory flora in humans. Mucosal diseases are the most common complications. These diseases range in severity from otitis media in the auditory canal (45) to sinusitis in the upper respiratory tract to bronchitis in the large upper airways and pneumonia in the small lower airways (11). Invasive pneumococcal disease, such as bacteremia and meningitis, and recovery of *S. pneumoniae* from normally sterile sites (e.g., pleural fluid, joint, peritoneal fluid) comprise a smaller subset of these infections. Although invasive disease occurs in all populations, rates of these serious infections are substantially higher among persons with some degree of immune compromise. However, "immunod-

eficiency" is a relative term. Immune competence implies that we are able to adapt successfully and resist injury from the microbial challenges of our individual environments. Immunodeficiency suggests that, per exposure to a pathogen, symptomatic infection is more likely to occur than in the healthy general population. The immunological, genetic, and environmental factors that contribute to this predisposition or increased risk of disease are many and of varying impact.

IMPACT OF IMMUNODEFICIENCY ON INVASIVE PNEUMOCOCCAL DISEASE

Invasive pneumococcal disease affects persons of selected races and ages, in certain social conditions, and with specific underlying diseases at rates that greatly exceed those of the general population (Fig. 1). In the United States, rates of pneumococcal bacteremia, a reliable if insensitive marker of the burden of this bacterial infection, are remarkably consistent, at approximately 19 to 23 per 100,000 per year in most heterogeneous populations (21, 33, 177). However, even within these populations, rates vary by race, with a threefold higher incidence among black Americans than among Caucasians, differences that have not been explained entirely by differences in socioeco-

Edward N. Janoff, Mucosal and Vaccine Research Center and Infectious Disease Section, Veterans Affairs Medical Center, University of Minnesota School of Medicine, Minneapolis, MN 55417. *Jeffrey B. Rubins*, Mucosal and Vaccine Research Center and Pulmonary Section, Veterans Affairs Medical Center, University of Minnesota School of Medicine, Minneapolis, MN 55417.

The Pneumococcus. Volume Editor, Elaine I. Tuomanen,
© 2004 ASM Press, Washington, D.C.

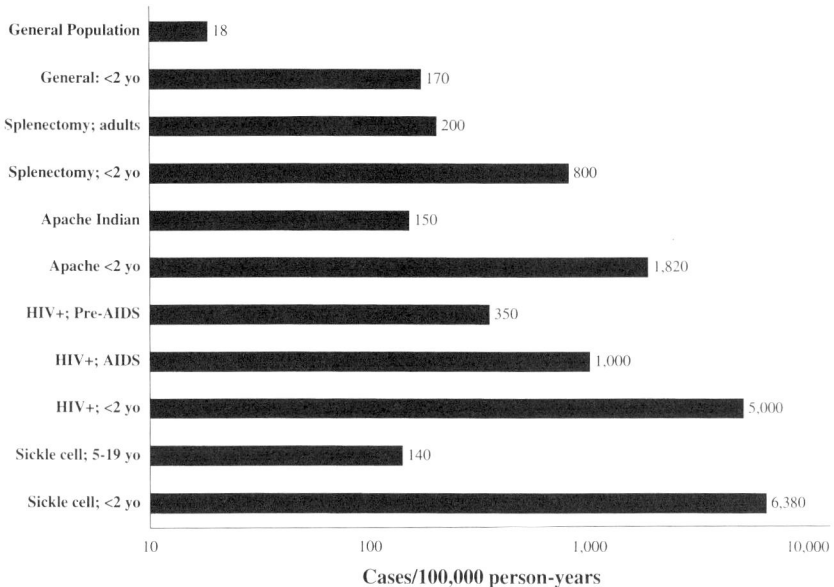

FIGURE 1 Rates of invasive pneumococcal disease by risk group and age (all ages versus those in children <2 years of age [yo]). Some rates are approximated from several studies.

nomic factors. Similarly, Native American populations in geographically divergent areas from Alaska to Arizona have a 5- to 10-fold higher annual incidence of pneumococcal bacteremia than that of most U.S. populations (74 to 207 per 100,000) (56, 60, 61). Independent of race, children under 2 years experience yet another 10-fold increase in the incidence of bacteremia compared with matched adult populations (18, 27, 56, 58, 61, 69). Adults with chronic exposure to inhaled pollutants, such as tobacco smokers (160) and African gold miners (12, 55), are prominent among high-risk groups. Thus, specific communities within the general population, including those living with poverty, crowding, pollution, and extreme stress (rigorous military training), are preferentially susceptible to invasive pneumococcal disease (39, 107, 124; M. Reichler, R. Reynolds, B. Schwartz, D. Musher, D. Pratt, G. Hohenhaus, J. Struewing, B. Plikaytis, J. Elliott, and R. Breiman, Program Abstr. 31st Intersci. Conf. Antimicrob. Agents Chemother., abstr. 49, 1991).

In addition to these ethnicity-, exposure-, and age-related groups, persons with selected underlying medical illnesses and immune defects comprise a substantial proportion of cases of invasive pneumococcal disease. Patients with approved indications for pneumococcal vaccine account for about half of all cases, particularly among adults (47, 177). Included in this group are those with underlying medical illnesses (chronic obstructive pulmonary disease, diabetes, and renal, liver, and heart failure), who make up about a quarter of all cases. Conditions such as asplenia or sickle cell disease, although not widely distributed in the population (the latter affects ≈1/600 African Americans), are disproportionately represented in series of pneumococcal bacteremia (177). Such disproportionate representations among patients with *S. pneumoniae* are more difficult to document when the prevalence of the underlying defect is quite rare, such as with selected defects in complement production (59, 150, 183) or congenital antibody defects (190, 208). In these settings, however, the strong associations are revealed by retrospective and prospective surveillance of these patients and summarized case reports.

More obvious is the remarkable prominence of human immunodeficiency virus type 1 (HIV-1) infection in determining the burden of serious pneumococcal disease. The prevalence of HIV-1 is quite high (up to 0.5% of the adult population in the United States and up to 35% of selected populations in Sub-Saharan Africa), and the incidence of this bacterial infection is particularly high in this population (up to 1 to 1.6% per year in affected adults in United States and Africa and 5 to 10% in children) (Fig. 1) (83, 144, 173). Moreover, pneumonia is more often complicated by bacteremia, and recurrence rates are increased by five- to sixfold to 15 to 25% with HIV-1 disease (118, 144, 149, 173). As a result of these factors, the proportion of patients with invasive pneumococcal disease who were coinfected with HIV-1 ranged from almost 15 to 20% of patients in U.S. cities (149, 177) to 40% in nonelderly U.S. men (105) to over two-thirds of cases in African adults and children (96, 118, 122, 144). Based on these data, HIV-1 and other causes of immune compromise contribute significantly to the rates of serious pneumococcal infections in communities across the world.

In addition to an increased incidence of invasive pneumococcal disease, case-fatality rates may also be higher among persons with underlying diseases and specific immune defects. Depending on age, death occurred almost two to four times as often among patients with any approved indication for pneumococcal vaccination than among those who did not meet these criteria during multistate surveillance in the United States (47, 177) (Fig. 2). In a related study, case-fatality rates by risk factor were increased by over four- to fivefold among patients with organ failure (liver, heart, kidney) and invasive pneumococcal disease and by two- to threefold with a variety of other conditions (75) (Fig. 3). Increased age, liver and kidney disease, and neutropenia have been confirmed as

risk factors for death in other studies from Europe (16).

Findings that advanced HIV-1 infection and AIDS are associated with an increased risk of related mortality are inconsistent (75, 118), but most data suggest that early stages of HIV infection convey no increased risk. The threefold-higher HIV-1-associated fatality rates observed among South African children with bacteremic pneumonia may involve differences in age and nutrition status, as well as HIV-1-associated immune defects. Moreover, despite an increased incidence, mortality does not appear to be increased among black compared with white Americans (75).

The relationship between risk of disease and outcome is inconsistent. Some conditions, such as sickle cell disease and asplenia, predispose to both an increased incidence of invasive pneumococcal disease and a higher mortality rate. To a less obvious extent, the incidence of this infection among patients with diabetes and organ failure is likely increased (160), as is attributable mortality (177). In contrast, children under 2 years of

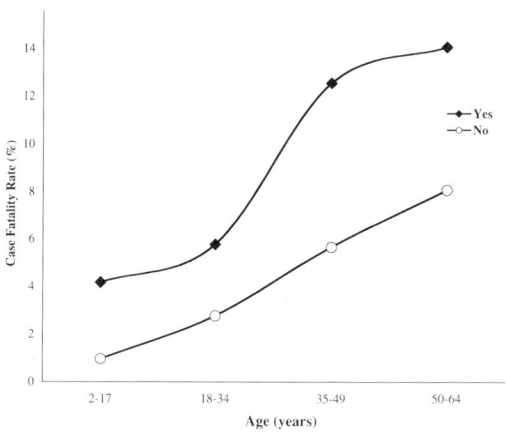

FIGURE 2 Association of case-fatality rates among 15,860 patients (ages 2 to 104 years) with invasive pneumococcal disease by age with and without U.S. Advisory Committee of Immunization Practices indications for pneumococcal vaccine in nine areas of the United States, 1995 to 1998. Overall case-fatality rates were 11.7% among those with indications ("Yes") and 4.0% among those without indications ("No") (177).

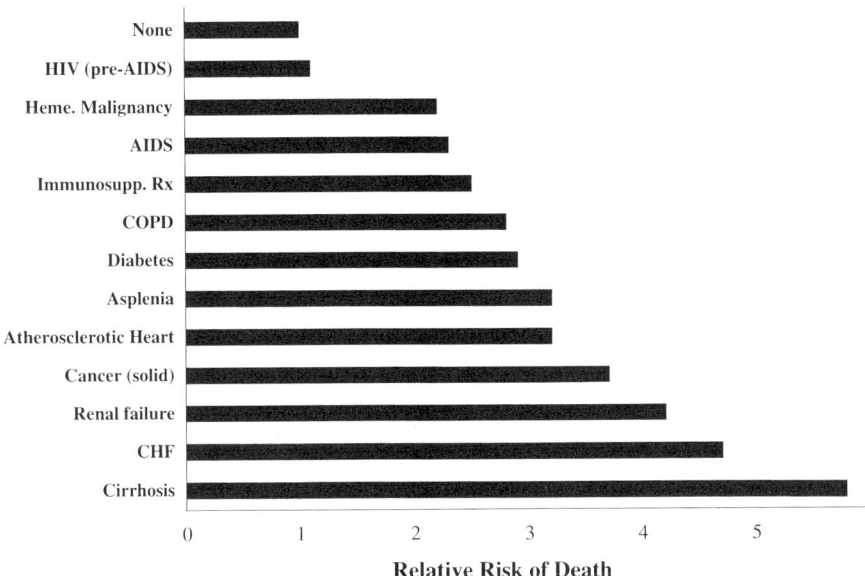

FIGURE 3   Relative risk of death from invasive pneumococcal disease among 3,049 hospitalized patients with bacteremic pneumococcal pneumonia and specific underlying illnesses in nine cities in the United States and Canada. Overall case-fatality rates were 12% (range, 6 to 27%) (75). Heme., hematologic; immunosupp., immunosuppressive; CHF, congestive heart failure.

age, particularly infants, experience many times the incidence of pneumococcal bacteremia but also a much lower mortality rate than those that occur in adults, particularly the elderly. Similarly, invasive pneumococcal disease occurs 50- to 100-fold more often among patients with AIDS, but relative mortality rates in adults are typically similar to those of, or at most increased by twofold compared with those of, seronegative patients with bacteremia (85, 117, 118, 177). Whether smoking, which is reported to confer a substantial risk among nonimmunocompromised persons (160), is also associated with increased mortality is unknown.

The final observation that reveals that the immune defects commonly predispose to initial episodes of invasive pneumococcal disease is the rate of recurrent disease. The presence of two or more episodes of invasive pneumococcal infection (usually reinfections rather than relapses) is highly predictive of underlying immunodeficiency (53, 125, 149, 179).

Indeed, the rate of recurrence is up to 50 times higher than the rate pre-event in the general population, suggesting an identification and selection of susceptible hosts by circulating organisms (125). HIV-1 infection, particularly advanced disease, accounts for up to 50% of recurrences in children and adults. Other prominent susceptibilities include sickle cell disease in children and chronic liver disease and hematologic malignancies (multiple myeloma, lymphoma, chemotherapy with neutropenia) in adults.

In this chapter, we focus on the rates and clinical manifestations of invasive pneumococcal disease in patients at high risk, on proposed mechanisms of impaired defense, as well as on strategies for prevention of invasive pneumococcal disease in these groups. These syndromes include splenectomy, sickle cell disease, malignancy, underlying medical illnesses, and concomitant infection with HIV-1. We introduce the functional relationship of host defenses acting in concert to protect against pneumococcal disease. Other instructive articles in this volume on *S. pneumoniae* highlight the pathogenic mechanisms utilized by the organism to evade these host defenses. We describe how individual host defects may each contribute independently to the increased risk of invasive disease, and we observe that the potential number of immunological risk factors is greatest and rates of invasive pneumococcal disease are highest among HIV type 1 (HIV-1)-infected patients. In this context, the reader is encouraged to consider what factors convey natural immunity against initial pneumococcal infection and what are the critical components of protection against lethal invasive disease (e.g., immune versus nonimmune factors, anatomic defenses, inflammation, and systemic versus mucosal immunity). Identifying these factors may stimulate innovative approaches to prevent invasive pneumococcal disease in the immunocompromised host.

CHILDREN

The extremely high rates of invasive pneumococcal disease in apparently otherwise healthy children (Fig. 1) suggest the presence of specific defects related to this class of organisms. Indeed, rates of other invasive mucosal pathogens with polysaccharide capsules, such as *Haemophilus influenzae* type b and *Neisseria meningitidis,* are also similarly increased. Anatomic imbalances, such as alignment of the eustachian tube, high rates of preceding viral and allergic episodes, and close contact with other children likely contribute to this risk. However, humoral defects also figure prominently. A physiological problem is related to the lack of prior experience with the antigen and a lack of prior specific immunity and memory. Another may be the transient hypogammaglobulinemia that occurs in the second year of life. Finally, young children show a marked inability to generate antibody responses to capsular polysaccharide antigens. This humoral defect, likely under T-cell control, may derive in

part from compromised production of the immunoglobulin G2 (IgG2) subclass, which comprises an appreciable proportion of capsule-specific IgG (7, 52, 204). The enhanced ability of capsular polysaccharides to stimulate specific IgG when conjugated to protein antigens was the basis for the outstanding clinical efficacy of newer vaccines directed to prevent invasive *H. influenzae* disease in children (68). This success derived in part from the shift to production from IgG2 to predominantly IgG1 subclass anticapsular antibodies, which are readily produced by young children. Such conjugate vaccines have recently been shown to be remarkably effective in preventing invasive pneumococcal disease in children (28, 29).

The risks for invasive pneumococcal disease in children may decline with age due to maturation of the immune system, to an expanded repertoire of immunological memory, and to the use of this new generation of vaccines. Nevertheless, some patients with specific genetic immune defects remain at risk for life, whereas other previously healthy persons develop into high-risk patients due to the acquisition of hematologic, oncological, or medical problems. Each specific immune defect of antibody production, complement activation, and phagocytosis, as well as virulence factors of the organism (109), likely contributes to increased rates of pneumococcal disease in children and adults (Table 1 and Fig. 4).

ANTIBODY DEFECTS

Specific humoral responses to pneumococcal capsular polysaccharides have long been recognized as critical to effective opsonophagocytosis and killing of the organism (36, 92, 221). Defective humoral responses to *S. pneumoniae* contribute to the increased rates of invasive infection not only in patients with primary antibody deficiencies but also in those with defects in complement and splenic function. Although specific rates are not available, patients with hypogammaglobulinemia are susceptible to invasive *S. pneumoniae* infection (182). Specifi-

TABLE 1 Factors which may be associated with increased rates or severity of invasive *S. pneumoniae* infection in the compromised host

Acquired immune defects
  Anatomic
    Hypoasplenia or asplenia (functional, congenital, surgical)
    Poor eustachian tube drainage (children)
    Cranial cerebrospinal fluid leak (congenital, traumatic)
    Cochlear implants
    Chronic lung disease and remodeling (?)
  Humoral
    Panhypogammaglobulinemia
    Selective IgG2 deficiency [and non-G2m(n) allotypes] (respiratory infections, ± selective IgA deficiency)
    Selective unresponsiveness to polysaccharides
    Hypocomplementemia (congenital C3, nephrotic syndrome, liver failure)
  Cellular
    Neutropenia (especially with cancer chemotherapy)
    Neutrophil dysfunction (diabetes?)
    Transplantation
    Lymphoid malignancies
  Combined
    HIV/AIDS
    Sickle cell disease
Environmental
  Environmental smoke
  Smoking (tobacco)
  Crowding
  Stress (e.g., military training)
Genetic
  Mannose-binding protein deficiency/polymorphisms
  FcγRIIa polymorphisms
  Tumor necrosis factor α receptor polymorphisms
  C-reactive protein polymorphisms
  IRAK-4 deficiency
Extremes of age
  <2 yr
  >65 yr

cally, IgG2 appears to be the predominant antibody class for protective humoral responses to *S. pneumoniae* (143, 189). IgG2 deficiency probably underlies, in part, the increased rates of pneumococcal bacteremia in hypogammaglobulinemia and in some patients with IgA deficiency (141). Although patients with selective IgA deficiency are not generally suscepti-

ble to higher rates of pneumococcal infection, the subset who have concomitantly impaired IgG2 responses to pneumococcal capsular polysaccharides do appear to be at increased risk (141). As described below, hematologic malignancies which impair polyclonal antibody production, such as chronic lymphocytic leukemia (CLL), in which B-cell development

Obstacles to
Killing of *S. pneumoniae*

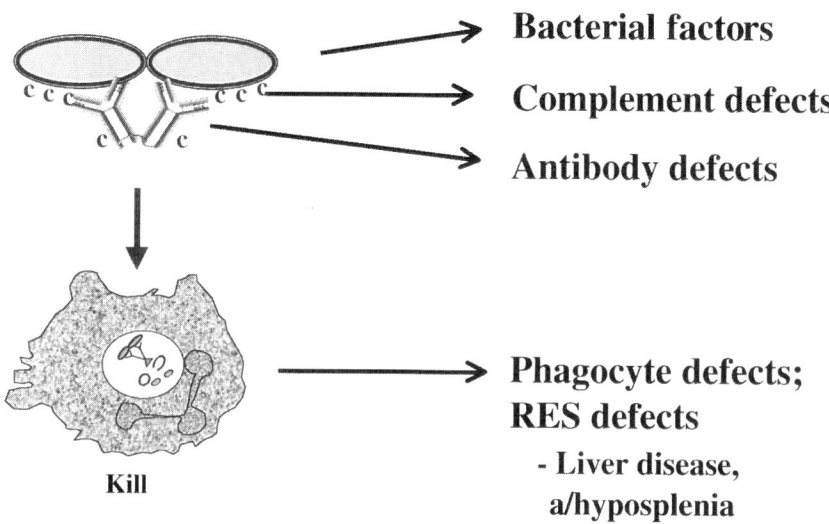

Bacterial factors

Complement defects

Antibody defects

Kill

**Phagocyte defects;
RES defects**
- **Liver disease,
a/hyposplenia**

FIGURE 4   Mechanisms which may impair killing of *S. pneumoniae*.

is arrested at an early stage, and multiple myeloma, in which the monoclonal arrest is at the well-differentiated plasma cell stage, provide a common background for a high incidence and recurrence rate of invasive pneumococcal disease (50, 51, 53, 112, 133, 179, 217). Thus, both acquired and genetic antibody defects may increase the risk of pneumococcal disease.

However, even subsets of otherwise healthy adults (11 to 20%) with normal total IgG levels may fail to respond to pneumococcal polysaccharide antigens in the 23-valent vaccine (156, 185). Musher et al. demonstrated that vaccine responder and nonresponder potential was hereditable in a large family in a non-HLA-related "mixed, codominant fashion" (156). A genetic component of pneumococcal polysaccharide responsiveness was also suggested by a close relationship among monozygotic twins between antibody G2m(n) allotype and total IgG2 levels, between the paired avidities of capsule-specific antibodies (132), as well as

between levels of vaccine-specific antibodies in twins (129). However, specific clonal responses to the polysaccharides differed between these twins, implying that immunoglobulin gene utilization for the antigen-binding V_H region, mutation, and clonal selection are not genetically determined in this context (131).

Although the majority of total IgG in serum is of the IgG1 subclass, as noted, an appreciable proportion of pneumococcal capsule-specific IgG is of the IgG2 subclass (41, 52, 204). Allotypes are antigenic hereditary allelic polymorphisms on immunoglobulin molecules. The presence of the IgG2 allotype G2m(n), also known as G2m(23), has been associated with increased levels of specific antibody after pneumococcal immunization (7, 193). Moreover, Caucasian children without the G2m(n) allotype were represented with increased frequency among those with serious infections with another encapsulated organism, *H. influenzae* type b (7). Similarly, persons with selective IgG2 subclass deficiency are at increased risk of poor vaccine responses and pneumococcal mucosal disease, particularly in the presence of concomitant IgA deficiency. In contrast to G2m(n), expression of the kappa light chain allotype Km(1), which is associated with decreased capsule-specific kappa chain responses to *H. influenzae* type b and *N. meningitidis* group C, has no such association with the magnitude of responses to immunization with pneumococcal polysaccharides (6).

Other genetic elements that may affect antibody responses to polysaccharides include polymorphisms in immunoglobulin V_H3 genes (3, 194), which dominate capsule-specific antibody responses (17, 49), and blood group expression. Indeed, responses to pneumococcal capsular polysaccharides may be higher among persons with types O and B, and pneumococcal disease may be less frequent among those with type B blood antigen (174). Therefore, genetic variations may introduce elements of immune hyporesponsiveness in individuals within the population which may limit responsiveness to vaccines and increase susceptibility to infection with *S. pneumoniae*.

COMPLEMENT DEFECTS

Complement, a primitive component of the immune system, facilitates both killing of *S. pneumoniae* and induction of capsule-specific antibody responses (42, 62, 72, 73). Complement activation is considered essential for efficient opsonophagocytosis and killing of *S. pneumoniae*. This process involves activation of classical, alternative, or lectin pathways (220) for activation and binding of early complement components (C1, C4, C2, and C3, especially C3b of the classical pathway) to the bacterial surface. Activation of the membrane attack complex by terminal complement components (C5 to C9) does not appear to be required in this process because the thick peptidoglycan cell wall resists the direct bacterial lysis which characterizes defense against many gram-negative pathogens. Nevertheless, C5a and, to a lesser extent, C3a serve as anaphylotoxins to recruit and activate phagocytes to the site of infection, and C5a and its receptor may

facilitate both mucosal and systemic bacterial clearance (108, 186). Complement can bind to *S. pneumoniae* and mediate phagocytosis by both antibody-dependent and -independent mechanisms (110). Highlighting the importance of complement in control of the organisms, *S. pneumoniae* is the most common cause of invasive bacterial infection among patients with congenital or acquired deficiencies of early components of the classical pathway (C1, C4, and C2) and the alternative pathway (factor I, factor H, and factor B); C3 is common to both pathways.

Factor C2 deficiency is associated with an increased incidence of invasive *S. pneumoniae* infection. C2 deficiency, which may be related to defects in translation and synthesis, as well as secretion of the protein, is the most common complement deficiency among persons of European origin (120). Pneumococcal disease may be more common when additional defects in alternate pathway components (e.g., factor B) are present (159). Even isolated C2 deficiencies are associated with an increased incidence of *S. pneumoniae* infection, presumably because C2 is more rapidly fixed by the classical complement pathway (110). Similarly, deficiency of C4b, rather than C4a, is most closely associated with meningitis and invasive disease (25, 184). Of note, C2, C4, and factor B are each encoded on chromosome 6 between the HLA-A and HLA-D loci.

Although less common, C3 deficiency is associated with increased rates of pyogenic infections with encapsulated organisms such as *S. pneumoniae* (227) and with recurrent pneumococcal infections (77). C3 deficiency affects the efficiency of both complement activation and induction of antibody responses. In addition to their inability to fix C3b to the bacterial surface for opsonophagocytosis, these patients have almost total deficiency of antibodies to pneumococcal capsular polysaccharides (104).

Such complement-related antibody defects likely relate to observations that C3 is also required to induce optimal humoral responses to polysaccharide antigens (62, 98, 99, 164).

Complement receptors, specifically CD21 (CR2) and CD35 (CR1), are proposed to link innate and adaptive humoral immune responses by linking C3d/g-antigen complexes with the B-cell receptor (surface Ig) and CD19 (42, 62, 102). Such linkage enhances the immunogenicity of antigens by several log units, which may be of particular relevance when antigen levels are low, such as with asymptomatic exposure or colonization. Based on these data, complement components, and each complement pathway (34), may well be instrumental in the induction of "natural immunity" to *S. pneumoniae*, in the acute response to infection to support killing (192), and in the ability to induce protective antibody responses. Additionally, CR3 and CR4 have also been proposed to support complement-independent antibody-mediated phagocytosis of *Cryptococcus neoformans*, which also expresses a polysaccharide capsule (213).

SPLENECTOMY AND SPLENIC DYSFUNCTION

Impaired splenic function may result from congenital, traumatic, or surgical loss or as a consequence of chronic hemolytic disease, particularly sickle cell disease. Surgical removal may occur in association with staging or treatment of lymphoid malignancies or with concomitant chemotherapy and/or radiation therapy, so the infectious consequences of splenectomy may be multifactorial. Nevertheless, asplenia and splenic dysfunction are associated with substantially increased rates of invasive *S. pneumoniae* infection (Fig. 1) (88). Although frequently described, consistent determinations of splenectomy-related incidence data and contributory factors are somewhat limited. The annual incidence of *S. pneumoniae* bacteremia in splenectomized adults is estimated to be significantly increased at 92 to ≈300 per 100,000 patients (5- to 15-fold that in the general population) (163, 198) and higher yet in children (15, 209). These complications may be highest during the first 10 years after splenectomy, but the risk of serious infection continues for decades. The lifetime

risk of a sepsis-related death in a child undergoing splenectomy may approach 1 in 5, and that of an adult may approach 1 in 16 (209).

S. pneumoniae comprises at least half of the cases of overwhelming postsplenectomy sepsis (OPSS), a syndrome in this population with high mortality (20 to 70%) (136, 218, 233). Other organisms associated with this syndrome include *H. influenzae* type b, *N. meningitidis*, group B streptococci, and, less commonly, gram-negative bacilli (*Escherichia coli*, *Capnocytophaga canimorsus*, *Pseudomonas aeruginosa*, and *Bacteroides fragilis*) or the protozoan *Babesia*. OPSS can be characterized by a deceptively insidious onset with low-grade fever, malaise, and sore throat followed by the rapid development of overwhelming infection, hypotension, mental status changes, and disseminated intravascular coagulation. Primary bacteremias with high numbers of organisms per milliliter of blood but no obvious source are common (>50% of cases), as is meningitis, although a pulmonary source can also be identified in some patients. Mortality and chronic sequelae are highest in those with meningitis (197). Dramatic consequences of OPSS may include adrenal insufficiency (Waterhouse-Friderichsen syndrome) and peripheral symmetric gangrene with necrosis of multiple digits and limbs.

Most of these organisms associated with invasive splenectomy-related infections possess polysaccharide capsules that serve as mechanisms of pathogenicity and to evade host recognition. Specific antibodies to the capsule, in collaboration with complement and intact reticuloendothelial function, are typically required for clearance of these intravascular organisms. The principal defect associated with splenic loss and dysfunction appears to be decreased clearance of circulating organisms, mediated by phagocytes in the sinuses of the splenic red pulp. The "pit count," the residual proportion of abnormal (pocked or pitted) circulating erythrocytes which would normally be cleared by the spleen, is significantly increased in asplenic patients (14 to 58% versus <1% in controls) (37), as is the time required to clear antibody-coated red cells (216). Both complement and antibody contribute to clearance of *S. pneumoniae* by the liver and spleen (30, 227). Regarding antibodies, levels of total IgM in serum are often decreased in the splenectomized patients (64), but total IgG and IgA are normal. Moreover, levels of capsule-specific IgG also appear to be normal prior to immunization in these patients.

The ability of splenectomized patients to generate effective humoral immune responses to pneumococcal polysaccharide vaccine is controversial. Several reports suggest that such responses are impaired in this population (63, 100, 126, 176). Optimal induction of primary immune responses to polysaccharide antigens likely involves a functionally intact spleen containing B cells expressing a high density of surface CD21 and C3d receptors (CD21) (147, 166). However, these structures and cells are also present in regional lymph nodes where specific B cells can also be induced and selected prior to transport to, e.g., the marrow for terminal differentiation and antibody production. In addition, many studies demonstrate that patients with splenectomy generate responses of a magnitude and duration comparable to those among control subjects (89, 130, 162). These data suggest that defects other than those related to capsule-specific antibody production, such as the loss of reticuloendothelial cell activity described above, contribute most prominently to their higher rates of invasive *S. pneumoniae* infection (Fig. 4).

Clinically, effective prevention of invasive pneumococcal infections after splenectomy requires proactive education and intervention. If possible at the time of splenectomy, reticuloendothelial function may be best preserved if only partial splenectomy, splenic repair, or autotransplantation can be accomplished (216). However, the presence of residual splenic tissue does appear to enhance decrements in mitogen-stimulated antibody production in vitro (64). Pneumococcal vaccines, including the earlier 14-valent and current 23-valent polysaccharide vaccines, may provide substantial protection against invasive pneumococcal disease in this high-risk population. In sequen-

tial 8- to 9-year periods during which pneumococcal vaccine was not or was given to children undergoing splenectomy in Denmark, pneumococcal bacteremia developed in 15 of 384 children (3.9%) versus 0 of 280 (0%), respectively (130). Use of penicillin prophylaxis or early therapy was not addressed. Vaccine recommendations are summarized in Table 2. Other approaches are logical, if not rigorously supported by prospective randomized data. Penicillin prophylaxis is usually recommended for at least several years after splenectomy, particularly in children. Infections that have occurred on such prophylaxis have not been highly associated with the development of antimicrobial resistance. Compliance with prophylaxis may be problematic and should be stressed. Independent of the use of prophylaxis, most patients are advised to have ready access to appropriate antimicrobials should related symptoms of impending infection develop. Finally, a medical-alert bracelet may help physicians rapidly identify splenectomized patients who present with serious infections and limited mental status.

SICKLE CELL DISEASE

Sickle cell anemia produces functional asplenia resulting in an extraordinary susceptibility to pneumococcal bacteremia. Prior to the introduction of the pneumococcal vaccine in 1978, the age-specific annual incidence of *S. pneumoniae* bacteremia in patients with sickle cell disease varied from 5,700 to 42,100 per 100,000 for children under 2 years old and was approximately 1,100 per 100,000 for those over 4 years old, with a total case-fatality rate of 26.8% (230, 232). Children with sickle cell anemia may have reduced serum opsonization of *S. pneumoniae* via the classical complement pathway secondary to low levels of IgG and IgM antibodies to capsular polysaccharides (26). B-cell maturation appears to be arrested at a step dependent on the B-cell-activating cytokine interleukin-4, an abnormality associated with decreased numbers of antibody-producing cells in response to *S. pneumoniae* polysaccharides and mitogens (172). Despite

TABLE 2 Immunocompromised persons for whom pneumococcal vaccine is recommended

23-valent polysaccharide vaccine[a]
    Adults with chronic illness
        Cardiovascular disease
        Pulmonary disease
        Diabetes mellitus
        Alcoholism
        Cirrhosis
        Cerebrospinal fluid leaks
    Immunocompromised adults
        HIV infection
        Persons with splenic dysfunction or anatomic asplenia
        Hodgkin's disease
        Leukemia, lymphoma
        Multiple myeloma
        Chronic renal failure
        Nephrotic syndrome
        Other immunosuppression (e.g., organ transplantation)
7-valent protein-polysaccharide conjugate vaccine[b]
    Children <5 yr of age with specific chronic illnesses
        Anatomic or functional asplenia
        Sickle cell disease
        Nephrotic syndrome
        Cerebrospinal fluid leaks
        HIV infection

[a] Strongly consider revaccination ≥6 years after the first dose for those patients at highest risk of fatal pneumococcal infection (e.g., asplenic patients) or for those at highest risk of rapid decline in antibody levels (e.g., those with chronic renal failure, nephrotic syndrome, or transplanted organs).
[b] Consider revaccination after 3 to 5 years if ≤10 years old at revaccination (44, 46–48).

these humoral defects, older children with sickle cell disease can generate adequate antibody responses to pneumococcal vaccine, especially after a booster immunization (121, 171). Use of the vaccine has decreased the overall incidence of *S. pneumoniae* infection in these children (230). However, as noted, the vaccine is substantially less effective in all children under 2 years (26) and has not reduced the high rates of invasive infection in this age range (230). Additional defects in unidentified classical and alternative complement cofactors (27), as well as in neutrophil oxidative activity (111), which have not as yet been fully characterized, have been proposed to occur in these children.

CANCER AND TRANSPLANTATION

Despite high levels of total immunoglobulin, usually IgG, patients with multiple myeloma experience high rates of pneumococcal infection, recurrence, and mortality (179). These patients are often functionally akin to those with hypogammaglobulinemia, because their high antibody levels are typically clonal, representing only a single antibody of irrelevant specificity, and other clones and isotypes are very low. These conditions likely underlie observations that C3b binding to *S. pneumoniae* is significantly decreased with sera from myeloma patients with recurrent pneumococcal infections even though levels of C3, total hemolytic complement, and C-reactive protein are normal (50). As discussed earlier, unlike myeloma, in which the malignant B-cell clone is terminally differentiated, the predisposition for invasive pneumococcal disease among adults with CLL derives from their arrested early B-cell development and associated hypogammaglobulinemia (51, 112, 133, 217). Infections are most common among patients with CLL and decreased levels of IgG2 and IgG4, and perhaps IgA, in serum (5). However, the efficacy of chronic infusions of pooled intravenous immunoglobulins from healthy donors to prevent pneumococcal and other infections with CLL has been variable (54, 101, 224), perhaps because levels of relevant specific antibodies in these preparations are also variable.

Patients with solid tumors, such as lymphomas, also experience increased rates of *S. pneumoniae* infection. The risk of fulminant bacterial sepsis, most often with *S. pneumoniae*, among patients with Hodgkin's disease is approximately 330 cases per 100,000 patients (84). Their risks include a high prevalence of splenectomy and decreased responsiveness to *S. pneumoniae* capsular polysaccharides, a feature shared by patients with acute lymphocytic lymphoma (76, 81). Furthermore, bone marrow transplant therapy for hematologic malignancies can cause increased rates of pneumococcal infection in both children and adults (197, 228). These infections most often occur from >3 months to many years after engraftment and are more common among those with chronic graft-versus-host disease, poor CD4$^+$ T-cell recovery, and persistent hypogammaglobulinemia (especially IgG2) (139). Consequently, specific antibody responses to pneumococcal polysaccharide vaccine are not optimal, especially in the presence of chronic graft-versus-host reactions (14, 90, 103, 197, 207). These pneumococcal infections, usually primary bacteremias but also meningitis and other foci, present with symptoms similar to those without immune compromise. Recurrent disease, usually reinfection, is common (15 to 20%), and mortality ranges from 6 to 33%.

Rates and associated findings of pneumococcal disease among patients with solid-organ transplants (especially kidney but also liver and heart) are similar to those in patients with marrow transplants. The additive risks of immunosuppressive medications and splenectomy, which were quite common among renal transplant infections in the past, resulted in rates of pneumococcal disease of ≈3% per year (2,800 to 3,600 cases/100,000 patient years) (142), although these rates have declined (197). Responses to the polysaccharide vaccine are substantially decreased in magnitude, function, and duration (148). The 7-valent conjugate vaccine does not consistently enhance capsule-specific antibody levels, nor their opsonic activity, in convalescent-phase sera compared with those in response to the 23-valent polysaccharide vaccine (140).

UNDERLYING MEDICAL CONDITIONS

In addition to these conditions in which specific defects in humoral, complement, or splenic defense functions can be identified, a number of other potentially immunocompromising diseases have been associated with increased susceptibility to invasive *S. pneumoniae* infection (39). Diseases such as diabetes mellitus, alcoholism, cirrhosis, and nonhematologic malignancies appear to confer a two- to fivefold increased risk of pneumococcal bacteremia in case-control series; true population-based incidences are not known (21, 35, 39, 97, 119, 134, 138, 145, 146, 164, 165, 223). The predisposition of such patients to pneumococcal bacteremia is often multifactorial, with defects in mechanical defenses (e.g., weakened gag reflex and cough) and suppression of neutrophil phagocytosis and killing activities contributing to potential deficiencies in humoral responses and reticuloendothelial clearance mechanisms.

Particularly during exacerbations of their renal disease, patients with nephrotic syndrome are extremely susceptible to pneumococcal peritonitis and bacteremia. The marked decrease in serum opsonizing activity in these patients is likely related to decreased levels of complement factor B in serum (molecular weight, 80,000) secondary to urinary losses (227). In addition, serum IgG antibodies to *S. pneumoniae* polysaccharides are decreased, partly due to urinary losses but also due in part to decreased antibody production (86, 206).

STRESS

Outbreaks of pneumococcal pneumonia and bacteremia have been well reported among younger men in areas of crowding, such as prisons (107), and in the setting of rigorous and highly demanding military training (57). Among military recruits, increased rates, including outbreaks of respiratory infections with adenoviruses and *N. meningitidis,* have been addressed by increasing the living area per person and vaccination. Similarly, crowding may well promote exposure to *S. pneumoniae,* colonization, and disease in these settings. Another predisposing factor may be the physiological consequences of emotional stress. Such changes may result from chronic or acute stress. Responses to influenza virus vaccine, both antibody and cellular, were impaired among long-term caregivers of patients with Alzheimer's disease (123, 219), as were antibody responses to pneumococcal polysaccharide vaccine (94). Similarly, the magnitude of responses to hepatitis B vaccine was associated with the relative ability or inability to adapt to examination-related stress among medical students (93). Investigators have proposed that the β-adrenergic responses to stress trigger the sympathetic-adrenal medullary and hypothalamic-pituitary adrenal axes (reviewed in reference 95). Catecholamines and cortisol may then modulate a range of immune activity. Despite these changes, in an animal model, mice which overexpress corticotropin-releasing hormone, a central mediator of the stress response, had protection from lethal challenge with *S. pneumoniae* after immunization equal to that of control mice (154). Future research in this area should include evaluation of humoral responses to pneumococcal vaccine and its clinical efficacy among military trainees, who experience high rates of stress and pneumococcal disease, as well as the multifaceted physiological consequences of and adaptations to severe stress in this group.

GENETICS

Genetic factors serve to modulate the impact of environmental exposure on rates and severity of disease. These factors can affect the rapidity, character, and magnitude of the inflammatory and specific immune response to the pathogen. Genetic variants of antibodies and complement production have been implicated in the predisposition to invasive pneumococcal infections, as discussed above, as have variants of innate immune factors and phagocyte receptors. Mannose binding lectin (MBL) has been characterized as a pattern recognition protein which binds to canonical structures containing mannose or *N*-acetylglucosamine on a range of viral, fungal, and bacterial pathogens, including *S. pneumoniae*. MBL activates complement on the surface of organisms by the MBL pathway.

Genetic polymorphisms and decreased levels of the protein in serum have been associated with opsonic defects in vitro and a predisposition to infection, including respiratory infections among children 6 to 17 months of age, febrile neutropenic episodes in children with malignancies, and severe or unusual infections in adults (87, 128, 157, 167, 210–212). These observations may indicate that MBL deficiency may reveal, modulate, or accentuate other underlying immune or inflammatory abnormalities or processes and thus becomes penetrant in combination. Of two case control studies, one from Great Britain revealed an increased prevalence of deficiency-associated MBL polymorphisms in a broad age range of 229 patients with invasive pneumococcal disease compared with that in 353 non-age-matched control subjects (12 versus 5%, respectively). The homozygous variant group was also overrepresented among patients with type 14 *S. pneumoniae* (32 versus 13%). The target carbohydrates are present in the capsules of some pneumococci, and type 14 does contain *N*-acetylglu-cosamine, but MBL did not bind to any of five isolates tested to date (158). In the second study, no preferential expression of variant MBL alleles was evident among 140 adults with pneumococcal bacteremia in Denmark compared with that in 250 control subjects. Whether or not MBL is found to have an independent effect on the incidence of pneumococcal infection when other clinical, immunological, and genetic factors are included, these defects will only contribute a relatively small proportion of the overall risk in the susceptible population.

A similar situation may pertain with polymorphisms in Fcγ receptor II (FcγRIIa), which binds IgG2 and has two codominant alleles. One allele, FcγRIIa-H131, binds IgG2 with high affinity and likely contributes to phagocytosis of organisms, whereas the second, FcγRIIa-R131, binds the antibody less efficiently with low affinity. The prevalence of the low-affinity FcγRIIa-R131 was increased among 42 adults with bacteremic pneumococcal pneumonia compared with rates in 28 patients with nonbacteremic pneumonia or 136 uninfected control subjects (50 versus 28 versus 29%; $P < 0.05$) (231). These results were not consistent with those in HIV-infected children with and without invasive pneumococcal disease, among whom receptor allelic frequencies were similar (2). The relative abilities of the two receptors to mediate phagocytosis of IgG2-opsonized organisms have been confirmed in vitro (178). Whether FcγR polymorphisms also affect antibody responses to vaccine and clinical outcomes with invasive disease is under consideration (191, 231).

Most recently, children with an uncommon inherited deficiency of interleukin-associated kinase (IRAK-4) have been shown to experience selective serious and recurrent infections with extracellular pyogenic bacteria, including *S. pneumoniae* (168). The response-modifying gene is required for activation of NF-κB and mitogen-activated protein kinase, which typically convey signals for cytokine production upon surface engagement of Toll-like receptors and interleukin-1 receptor by their intracytoplasmic Toll–interleukin-1 receptor domain. In vitro activation confirmed the lack of cytokine production by monocytes from these patients. The risk of infection was greatest in the three children earlier in life, but by ages 6, 7, and 11 years, they had resolved. This observation suggests that complementary acquired defense mechanisms may be sufficient later in life to compensate for apparently severe defects that manifest in more immunologically naïve children.

HIV-1 INFECTION

HIV-1 infection is among the most prominent risks for invasive pneumococcal infection in children and nonelderly adults worldwide. As described in detail above, rates of disease are increased by up to 100-fold compared with age-matched populations, and recurrence rates are substantially increased as well. Predisposing factors for these infections in patients with HIV-1 disease are many but none clearly predominates. The incidence of bacterial pneumonia is increased five- to 10-fold during HIV-1 infection (5.5 versus 0.9 per 100 person-years in seronegative persons) (106), particularly in drug users (31, 200). Although *S. pneumoniae* and *H. influenzae* are common causes in this population (106, 137, 152, 169, 200), pneumonia with *H. influenzae* is uncommonly bacteremic, whereas rates of bacteremic pneumococcal pneumonia are high (60 to 80%, compared with 10 to 20% with typical community-acquired pneumococcal pneumonia) (13, 24, 33, 114, 135, 137, 169, 173, 180, 196, 199, 200, 229). As a result, *S. pneumoniae* is a leading cause of bacteremia during HIV-1-infection, accounting for 13 to 31% of all bacteremias in these adult patients (114). Invasive pneumococcal disease (bacteremia or meningitis) affects 5 to 12% of HIV-1-infected children per year in the United States (10, 70), compared with approximately 1% of seropositive adults (Fig. 1). With higher numbers of persons infected with HIV-1 and higher rates of this bacterial infection (1.6%/year in adults) (83), the impact of HIV-1-associated pneumococcal disease in Sub-Saharan Africa is far greater than in industrialized nations. Detailed epidemiological and clinical features have been recently reviewed (85, 118).

HIV-1-associated immunodeficiency, rather than behavioral (e.g., smoking), medical (e.g., splenectomy, liver disease), environmental (e.g., seasonality), or bacteriological features (serotypes or colonization), appears to underlie the remarkable rates of disease in this population (118, 173). Increased rates of disease among intravenous drug users are independent of HIV-1 infection (31, 106, 199, 200). Pneumococcal infections occur at all stages of HIV-1 infection, although rates are highest and most, but not all, cases occur among patients with <200 CD4$^+$ T cells/μl (80, 91, 117, 196). However, numbers of CD4$^+$ T cells are at best only markers of immune status. More relevant are the specific mechanisms of host defense against *S. pneumoniae*.

Natural antibodies, those found in sera of healthy persons in the absence of overt previous infection or immunization, may be generated in response to asymptomatic colonization (155), exposure to cross-reactive antigens, or the presence of polyreactive low-affinity antibodies (43). Baseline levels of IgG to pneumococcal polysaccharides were significantly lower in both asymptomatic HIV-1-infected patients and AIDS patients than in seronegative homosexual and heterosexual subjects (9, 115, 116), but these results are not consistently reported (41, 225). Moreover, among HIV-1-infected patients with acute pneumococcal bacteremia in the United States and Kenya, levels of capsule-specific IgG were often significantly lower than those in healthy seronegative control subjects (117, 173), although reported levels vary (82, 116).

Levels of IgG2 in serum may also be decreased during HIV-1 infection (41, 151, 175), but more important than the levels or structure of antibodies to *S. pneumoniae* is their ability to mediate clearance of the organism. Levels of

capsule-specific antibodies are directly related to the functional ability of serum to support phagocytic killing of the organism (92, 114). Such activity in vitro has been decreased among HIV-1-infected patients both before and after bacteremic infection in the United States and Africa (116, 117). Thus, for selected serotypes, levels of functionally active antibody to the pneumococcal capsule, a critical virulence factor, may be decreased in association with high rates of invasive disease during HIV-1 infection. No consistent data confirm that antibody avidity, a measure of the strength of antibody-antigen binding and, often, functional activity, may be decreased during HIV-1 infections. In addition, selective decrements in both total and capsule-specific antibodies encoded by the VH3 family of variable-region genes have been proposed to underlie the humoral immune defects associated with HIV infection, particularly those related to responses to *S. pneumoniae* (1, 22, 23, 195).

Antibodies in serum to the major cytotoxin produced by *S. pneumoniae* (38), pneumolysin, which modulates both tissue and immune integrity, were significantly decreased in HIV-1-infected patients in the United States compared with those in seronegative control subjects (8). Similarly, HIV-1-infected patients in Kenya who later developed pneumococcal bacteremia also had significantly lower levels of antipneumolysin antibody levels at baseline than did seronegative control subjects. Thus, lower levels of antibodies to pneumolysin, a toxin which promotes tissue invasion, were associated with the higher incidence of bacteremic pneumococcal infections among HIV-1-infected patients.

Regarding other facets of the phagocytic process, complement activity (CH$_{50}$, C3, C4) may be normal or low (18, 215), and products of abnormal complement activation, which may inhibit clearance of opsonized particles, are increased (170) during HIV-1 infection. Levels of C-reactive protein, a prototypic hepatic acute-phase reactant, were normal in sera and commensurate in those with and without HIV-1 infection both prior to and during acute pneumococcal infection (Janoff et al., unpublished data). Polymorphonuclear leukocyte-mediated, antibody-dependent cellular cytotoxicity may be normal (201) or depressed (153) and chemotaxis may be impaired (67) in HIV-1-infected patients. Moreover, no association has been found between FcγRIIa allotype and pneumococcal disease during HIV-1 infection (2, 178, 222). Nevertheless, polymorphonuclear leukocyte dysfunction and other defects in phagocyte function, such as delayed and decreased clearance of opsonized particles by macrophages in the liver and spleen (18, 19, 92), may contribute to the high rates of pneumococcal infection in these patients.

In summary, despite the many fold-increased rates of pneumonia and, particularly, invasive pneumococcal disease during HIV-1 infection, no specific overriding risk or target for intervention has been confirmed. Risks are most likely multifaceted and include mild to moderate impairment of antibody production and function, neutrophil responses, and clearance by the reticuloendothelial cells in the liver and spleen. Independent of these vagaries, use of highly active antiretroviral therapy appears to have decreased rates of invasive disease by about half; prophylactic antibiotics used for other indications may lower rates, as may immunization, the most accessible, safe, and inexpensive intervention available.

EFFICACY OF PNEUMOCOCCAL VACCINE

The protection provided by both the 23-valent polysaccharide vaccine in adults and the 7-valent conjugate vaccine in children is most apparent and consistent against invasive pneumococcal disease. The burden of morbidity and, particularly, mortality of invasive pneumococcal disease falls predominantly on the older adult population (21, 78, 177). In five of six case control studies among older adults, the vaccine provided significant protection (44 to 81%) (71, 79, 113, 202, 203, 214). However, in three of these studies, the vaccine was not effective in the ≈20% of patients who were

severely immunocompromised (about 20%, with confidence intervals [CI] overlapping 0) but was effective (54 to 70%) in noncompromised older adults (Fig. 5) (113, 202, 203). Other proposed mechanisms of vaccine failure among the elderly include infection with non-vaccine serotypes, waning of antibody responses, inadequate or ineffective capsule-specific antibody responses, or compromised nonhumoral defense (187). The prominence of underlying medical conditions in bacteremic patients in this population supports the latter.

Rubins et al. have shown that baseline levels, avidity, and killing activity of specific IgG are relatively normal in noncompromised older adults, although a subset of these subjects show consistently impaired responses to the majority of vaccine serotypes (185, 188). This subset may be greater for patients with rheumatologic disease and therapy (66), and opsonophagocytic activity may be depressed in debilitated older adults (181). Thus, a variety of medical and age-related complications, particularly solid and hematologic malignancies, rather than aging itself (40), may limit vaccine efficacy in the expanding aging population. Use of protein-polysaccharide conjugate vaccines has not shown enhanced immunogenicity in this population to date, although immunization of young children may reduce exposure and disease in older adults (226).

The ability of the 23-valent pneumococcal vaccine to protect patients with HIV-1 infection against invasive disease has been controversial. The only large prospective, randomized, placebo-controlled trial of this vaccine showed no evidence of protection against invasive pneumococcal disease and showed an increased rate of all-cause pneumonia among HIV-1-infected adults in Uganda (83). However, a study of similar design with a 9-valent conjugate vaccine showed 53% (95% CI, 21 to 73) vaccine efficacy against all invasive pneumococcal disease among HIV-1-infected children in South Africa (127). The attributable effect was confirmed by a 65% efficacy against

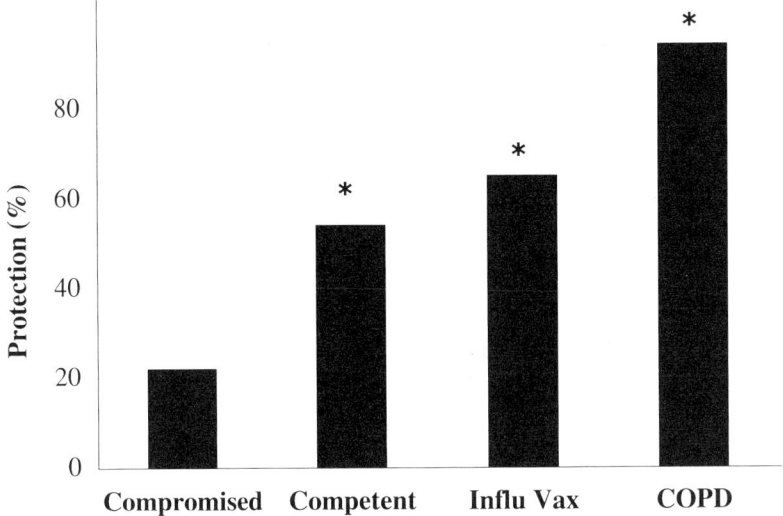

FIGURE 5 Effectiveness of 23-valent pneumococcal polysaccharide vaccine against invasive pneumococcal disease in adults >65 years of age (Seattle, Wash.) in a retrospective cohort study monitoring 127,180 patient years (84,203 after immunization), including 61 cases of invasive disease. Compromised patients (those with cancer, immunosuppressive therapy, chronic renal failure, and liver failure) comprised 19% of subjects and 26% of patients with invasive disease. Values for immunocompetent patients with chronic lung disease were unadjusted for influenza virus vaccine and other factors. $P \leq 0.06$ for protection compared with unimmunized adults in the same group (116). COPD, chronic obstructive pulmonary disease.

vaccine serotypes but none against nonvaccine types. Results of two efficacy studies among adults with HIV-1 disease in the United States suggest some level of protection in selected groups. In a retrospective case control study (176 cases and 327 control subjects), adjusted vaccine efficacy against invasive disease was 76% among whites, but the vaccine was ineffective among black subjects (32). In a multisite observational cohort study in the United States, efficacy (RR, 0.5; 95% CI, 0.3 to 0.9) against severe pneumococcal disease was also proposed for patients with >500 CD4$^+$ T cells (65). Efforts to foster specific responses to vaccine have shown no consistently enhanced immunogenicity with a single dose of conjugate vaccine in HIV-1-infected adults compared with the polysaccharide alone (4). However, including a conjugate in a two-dose regimen may augment capsule-specific antibody levels and function in this population (74). Reimmunization with the

23-valent polysaccharide vaccine elicited feeble responses (214), but conjugate vaccines are now being tested in this context.

The value of development of new vaccines and their use in high-risk populations has recently been demonstrated among Native Americans, who experience extremely high rates of invasive pneumococcal disease (Fig. 1). Although the 23-valent polysaccharide vaccine was not effective in preventing these infections in a case control study of Navajo adults (20), a 7-valent protein-polysaccharide conjugate vaccine (PnCRM7) was very effective ($\approx$80%) in protecting against invasive pneumococcal disease in a randomized placebo-controlled trial among 8,292 Navajo and White Mountain Apache children, who experience even higher rates of disease than adults (161). Thus, newer vaccines may overcome selected obstacles to host defense against serious pneumococcal infections.

SUMMARY

Specific immune defects can be identified for particular groups at higher risk of invasive pneumococcal infection, whereas more generalized deficiencies in immune and nonimmune defenses underlie bacteremia in others. These defects involve each component of the protective host response—antibodies, complement, and phagocytes—and each component is susceptible to evasive mechanisms employed by the organism. The basis for increased rates and immune defects can be environmental, behavioral, genetic, or acquired. Pneumococcal vaccine is effective against invasive disease in the majority of patients at risk, but its efficacy is limited among those with the most severe immune compromise, who often represent those at highest risk of disease and mortality. Our understanding of this dynamic and complex host-pathogen interaction facilitates our ability to provide the most effective protective strategies, be they active or passive immunization, behavioral change (e.g., smoking), or antimicrobial prophylaxis. The development of newer vaccines which are more universally immunogenic and which combat mucosal disease more effectively is an important goal in preventing pneumococcal disease in the immunocompromised host.

ACKNOWLEDGMENTS

We thank Ann Emery for excellent secretarial assistance.

This work was supported by the National Institutes of Health (AI39445, AI34051, and AI48796), the Veterans Affairs Research Service, and the Mucosal and Vaccine Research Center.

REFERENCES

1. **Abadi, J., J. Friedman, R. A. Mageed, R. Jefferis, M. C. Rodriguez-Barradas, and L. Pirofski.** 1998. Human antibodies elicited by a pneumococcal vaccine express idiotypic determinants indicative of V_H3 gene segment usage. *J. Infect. Dis.* **178:**707–716.

2. **Abadi, J., Z. Zhong, J. Dobroszycki, and L.-A. Pirofski.** 1997. FcγRIIa polymorphism in human immunodeficiency virus–infected children with invasive pneumococcal disease. *Pediatr. Res.* **42:**259–262.

3. **Adderson, E. E., F. H. Azmi, P. M. Wilson, P. G. Shackelford, and W. L. Carroll.** 1993. The human VH3b gene subfamily is highly polymorphic. *J. Immunol.* **151:**800–809.

4. **Ahmed, F., M. C. Steinhoff, M. C. Rodriguez-Barradas, R. G. Hamilton, D. M. Musher, and K. E. Nelson.** 1996. Effect of human immunodeficiency virus type 1 infection on the antibody response to a glycoprotein conjugate pneumococcal vaccine: results from a randomized trial. *J. Infect. Dis.* **173:**83–90.

5. **Aittoniemi, J., A. Miettinen, S. Laine, M. Sinisalo, P. Laippala, L. Vilpo, and J. Vilpo.** 1999. Opsonising immunoglobulins and mannan-binding lectin in chronic lymphocytic leukemia. *Leuk. Lymphoma* **34:**381–385.

6. **Ambrosino, D. M., V. A. Barrus, G. G. DeLange, and G. R. Siber.** 1986. Correlation of the Km(1) immunoglobulin allotype with anti-polysaccharide antibodies in Caucasian adults. *J. Clin. Investig.* **78:**361–365.

7. **Ambrosino, D. M., G. Schiffman, E. C. Gotschlich, P. H. Schur, G. A. Rosenberg, G. G. DeLange, E. vanLoghem, and G. R. Siber.** 1985. Correlation between G2M(n) immunoglobulin allotype and human antibody response and susceptibility to polysaccharide encapsulated bacteria. *J. Clin. Investig.* **75:**1935–1942.

8. **Amdahl, B. M., J. B. Rubins, C. L. Daley, C. F. Gilks, P. C. Hopewell, and E. N. Janoff.** 1995. Impaired natural immunity to pneumolysin during human immunodeficiency virus infection in the United States and Africa. *Am. J. Respir. Crit. Care Med.* **152:**2000–2004.

9. **Amman, A. J., G. Schiffman, D. Abrams, P. Volberding, J. Ziegler, and M. Conant.** 1984. B-cell immunodeficiency in acquired immune deficiency syndrome. *JAMA* **251:**1447–1449.

10. **Andiman, W. A., J. Mezger, and E. Shapiro.** 1994. Invasive bacterial infections in children born to women infected with human immunodeficiency virus type 1. *J. Pediatr.* **124:**846–852.

11. **Austrian, R.** 1986. Pneumococcal pneumonia: diagnostic, epidemiologic, therapeutic and prophylactic considerations. *Chest* **90:**738–743.

12. **Austrian, R.** 1981. Pneumococcus: the first hundred years. *Rev. Infect. Dis.* **3:**183–189.

13. **Austrian, R., and J. Gold.** 1964. Pneumococcal bacteremia with especial reference to bacteremic pneumococcal pneumonia. *Ann. Intern. Med.* **60:**759–776.

14. **Avanzini, M. A., A. M. Carra, R. Maccario, M. Zecca, P. Pignatti, M. Marconi, P. Comoli, F. Bonetti, P. DeStefano, and F. Locatelli.** 1995. Antibody response to pneumococcal vaccine in children receiving bone marrow transplantation. *J. Clin. Immunol.* **15:**137–144.

15. **Baccarani, M., M. Fiacchini, P. Galieni, F. Gherlinzoni, R. Fanin, G. Fasola, P. Mazza, and S. Tura.** 1986. Meningitis and septicaemia in adults splenectomized for Hodgkin's disease. *Scand. J. Haematol.* **36:**492–498.

16. **Balakrishnan, I., P. Crook, R. Morris, and S. H. Gillespie.** 2000. Early predictors of mortality in pneumococcal bacteraemia. *J. Infect.* **40:**256–261.

17. **Baxendale, H. E., Z. Davis, H. N. White, M. B. Spellerberg, F. K. Stevenson, and D. Goldblatt.** 2000. Immunogenetic analysis of the immune response to pneumococcal polysaccharide. *Eur. J. Immunol.* **30:**1214–1223.

18. **Bender, B. S., J. F. Bohnsack, S. H. Sourlis, M. M. Frank, and T. C. Quinn.** 1987. Demonstration of defective C3-receptor-mediated clearance by the reticuloendothelial system in patients with acquired immunodeficiency syndrome. *J. Clin. Investig.* **79:**715–720.

19. **Bender, B. S., M. M. Frank, T. J. Lawley, W. J. Smith, C. M. Brickman, and T. C. Quinn.** 1985. Defective reticuloendothelial system Fc-receptor function in patients with the acquired immunodeficiency syndrome. *J. Infect. Dis.* **152:**409–412.

20. **Benin, A. L., K. L. O'Brien, J. P. Watt, R. Reid, E. R. Zell, S. Katz, C. Donaldson, A. Parkinson, A. Schuchat, M. Santosham, and C. G. Whitney.** 2003. Effectiveness of the 23-valent polysaccharide vaccine against invasive pneumococcal disease in Navajo adults. *J. Infect. Dis.* **188:**81–89.

21. **Bennett, N. M., J. Buffington, and F. M. La Force.** 1992. Pneumococcal bacteremia in Monroe County, New York. *Am. J. Public Health* **82:**1513–1516.

22. **Berberian, L., J. Shukla, R. Jefferis, and J. Braun.** 1994. Effects of HIV infection on VH3 (D12 idiotope) B cells in vivo. *J. Acquir. Immune Defic. Syndr.* **7:**641–646.

23. **Berberian, L., Y. Valles-Ayoub, N. Sun, O. Martinez-Maza, and J. Braun.** 1991. A V$_H$ clonal deficit in human immunodeficiency virus-positive individuals reflects a B-cell maturational arrest. *Blood* **78:**175–179.

24. **Berstein, L. J., B. Z. Krieger, B. Novick, M. J. Sicklick, and A. Rubinstein.** 1985. Bacterial infection in the acquired immunodeficiency syndrome of children. *Pediatr. Infect. Dis.* **4:**472–475.

25. **Bishof, N. A., T. R. Welch, and L. S. Beischel.** 1990. C4B deficiency: a risk factor for bacteremia with encapsulated organisms. *J. Infect. Dis.* **162:**248–250.

26. **Bjornson, A. B., and J. S. Lobel.** 1987. Direct evidence that decreased serum opsonization of Streptococcus pneumoniae via the alternative complement pathway in sickle cell disease is related to antibody deficiency. *J. Clin. Investig.* **79:**388–398.

27. **Bjornson, A. B., J. S. Lobel, and K. S. Harr.** 1985. Relation between serum opsonic activity for *Streptococcus pneumoniae* and complement function in sickle cell disease. *J. Infect. Dis.* **152:** 701–709.

28. **Black, S., H. Shinefield, B. Fireman, E. Lewis, P. Ray, J. R. Hansen, L. Elvin, K. M. Ensor, J. Hackell, G. Siber, F. Malinoski, D. Madore, I. Chang, R. Kohberger, W. Watson, R. Austrian, K. Edwards, and the Northern California Kaiser Permanente Vaccine Study Center Group.** 2000. Efficacy, safety and immunogenicity of heptavalent pneumococcal conjugate vaccine in children. *Pediatr. Infect. Dis. J.* **19:**187–195.

29. **Black, S. B., H. R. Shinefield, S. Ling, J. Hansen, B. Fireman, D. Spring, J. Noyes, E. Lewis, P. Ray, J. Lee, and J. Hackell.** 2002. Effectiveness of heptavalent pneumococcal conjugate vaccine in children younger than five years of age for prevention of pneumonia. *Pediatr. Infect. Dis. J.* **21:**810–815.

30. **Bohnsack, J. F., and E. J. Brown.** 1986. The role of the spleen in resistance to infection. *Annu. Rev. Med.* **37:**49–59.

31. **Boschini, A., C. Smacchia, M. Di Fine, A. Schiesari, P. Ballarini, M. Arlotti, C. Gabrielli, G. Castellani, M. Genova, P. Pantani, A. Cozzi Lepri, and G. Rezza.** 1996. Community-acquired pneumonia in a cohort of former injection drug users with and without human immunodeficiency virus infection: incidence, etiologies, and clinical aspects. *Clin. Infect. Dis.* **23:**107–113.

32. **Breiman, R. F., D. W. Keller, M. A. Phelan, and D. H. Sniadack.** 2000. Evaluation of effectiveness of the 23-valent pneumococcal capsular polysaccharide vaccine for HIV-infected patients. *Arch. Intern. Med.* **160:**2633–2638.

33. **Breiman, R. F., J. S. Spika, V. J. Navarro, P. M. Darden, and C. P. Darby.** 1990. Pneumococcal bacteremia in Charleston County, South Carolina: a decade later. *Arch. Intern. Med.* **150:**1401–1405.

34. **Brown, J. S., T. Hussell, S. M. Gilliland, D. W. Holden, J. C. Paton, M. R. Ehren-**

stein, M. J. Walport, and M. Botto. 2002. The classical pathway is the dominant complement pathway required for innate immunity to Streptococcus pneumoniae infection in mice. *Proc. Natl. Acad. Sci. USA* **99:**16969–16974.

35. **Bruyn, G. A. W., J. W. M. van der Meer, J. Hermans, and W. Knoppert.** 1988. Pneumococcal bacteremia in adults over a 10-year period at University Hospital, Leiden. *Rev. Infect. Dis.* **10:**446–450.

36. **Bruyn, G. A. W., B. J. M. Zegers, and R. van Furth.** 1992. Mechanisms of host defense against infection with *Streptococcus pneumoniae*. *Clin. Infect. Dis.* **14:**251–262.

37. **Buchanan, G. R., S. J. Smith, C. A. Holtkamp, and J. P. Fuseler.** 1983. Bacterial infection and splenic reticuloendothelial function in children with hemoglobin SC disease. *Pediatrics* **72:**93–98.

38. **Bulnois, G. J., J. C. Paton, T. J. Mitchell, and P. W. Andrew.** 1991. Structure and function of pneumolysin, the multifunctional thiol-activated toxin of *Streptococcus pneumoniae*. *Mol. Microbiol.* **5:**2611–2616.

39. **Burman, L. Å., R. Norrby, and B. Trollfors.** 1985. Invasive pneumococcal infections: incidence, predisposing factors, and prognosis. *Rev. Infect. Dis.* **7:**133–142.

40. **Carson, P. J., K. L. Nichol, J. O'Brien, P. Hilo, and E. N. Janoff.** 2000. Immune function and vaccine responses in the healthy advanced elderly. *Arch. Intern. Med.* **160:**2017–2024.

41. **Carson, P. J., R. L. Schut, M. L. Simpson, J. O'Brien, and E. N. Janoff.** 1995. Antibody class and subclass responses to pneumococcal polysaccharides following immunization of human immunodeficiency virus-infected patients. *J. Infect. Dis.* **172:**340–345.

42. **Carter, R. H., and D. T. Fearon.** 1992. CD19: lowering the threshold for antigen receptor stimulation of B lymphocytes. *Science* **256:**105–106.

43. **Casali, P., and A. L. Notkins.** 1989. CD5+ lymphocytes and the human B cell repertoire. *Immunol. Today* **10:**364–368.

44. **Centers for Disease Control.** 1993. FDA approval of use of a new *Haemophilus* b conjugate vaccine and a combined diphtheria-tetanus-pertussis and *Haemophilus* b conjugate vaccine for infants and children. *Morb. Mortal. Wkly. Rep.* **42:**296–298.

45. **Centers for Disease Control.** 1984. Update: pneumococcal polysaccharide vaccine usage—United States. *Ann. Intern. Med.* **101:**348–350.

46. **Centers for Disease Control and Prevention.** 2000. Preventing pneumococcal disease among infants and young children. *Morb. Mortal. Wkly. Rep.* **49:**1–35.

47. **Centers for Disease Control and Prevention.** 1997. Prevention of pneumococcal disease: recommendations of the Advisory Committee on Immunization Practices (ACIP). *Morb. Mortal. Wkly. Rep.* **46**(RR-8)**:**1–24.

48. **Centers for Disease Control and Prevention.** 1999. USPHS/IDSA guidelines for the prevention of opportunistic infections in persons infected with human immunodeficiency virus. *Morb. Mortal. Wkly. Rep.* **48**(RR-10)**:**1–66.

49. **Chang, Q., J. Abadi, P. Alpert, and L. Pirofski.** 2000. A pneumococcal capsular polysaccharide vaccine induces a repertoire shift with increased VH3 expression in peripheral B cells from human immunodeficiency virus (HIV)-uninfected but not HIV-infected persons. *J. Infect. Dis.* **181:**1313–1321.

50. **Cheson, B. D., H. S. Walker, M. E. Health, R. J. Gobel, and J. Janatova.** 1984. Defective binding of the third component of complement (C3) to *Streptococcus pneumoniae* in multiple myeloma. *Blood* **63:**949–957.

51. **Chou, M., A. E. Brown, A. Blevins, and D. Armstrong.** 1983. Severe pneumococcal infection in patients with neoplastic disease. *Cancer* **51:**1546–1550.

52. **Chudwin, D. S., S. G. Artrip, and G. Schiffman.** 1987. Immunoglobulin G class and subclass antibodies to pneumococcal capsular polysaccharides. *Clin. Immunol. Immunopathol.* **44:**114–121.

53. **Coccia, M. R., R. R. Facklam, L. D. Saravolatz, and O. Manzor.** 1998. Recurrent pneumococcal bacteremia: 34 episodes in 15 patients. *Clin. Infect. Dis.* **26:**982–985.

54. **Cooperative Group for the Study of Immunoglobulin in Chronic Lymphocytic Leukemia.** 1988. Intravenous Immunoglobulin for the prevention of infection in chronic lymphocytic leukemia: a randomized, controlled trial. *N. Engl. J. Med.* **319:**902–907.

55. **Corbett, E. L., G. J. Churchyard, S. Charalambos, B. Samb, V. Moloi, T. C. Clayton, A. D. Grant, J. Murray, R. J. Hayes, and K. M. De Cock.** 2002. Morbidity and mortality in South African gold miners: impact of untreated disease due to human immunodeficiency virus. *Clin. Infect. Dis.* **34:**1251–1258.

56. **Cortese, M. M., M. Wolff, J. Almeido-Hill, R. Reid, J. Ketcham, and M. Santosham.** 1992. High incidence rates of invasive pneumococcal disease in the White Mountain Apache population. *Arch. Intern. Med.* **152:**2277–2282.

57. **Crum, N. F., M. R. Wallace, C. R. Lamb, A. M. Conlin, D. E. Amundson, P. E. Ol-**

son, M. A. Ryan, T. J. Robinson, G. C. Gray, and K. C. Earhart. 2003. Halting a pneumococcal pneumonia outbreak among United States Marine Corps trainees. *Am. J. Prev. Med.* **25**:107–111.

58. **Dagan, R., D. Engelhard, and E. Piccard.** 1992. Epidemiology of invasive childhood pneumococcal infections in Israel: the Israeli Pediatric Bacteremia and Meningitis Group. *JAMA* **268**:3328–3332.

59. **Dalmasso, A. P.** 1986. Complement in the pathophysiology and diagnosis of human diseases. *Crit. Rev. Clin. Lab. Sci.* **24**:123–183.

60. **Davidson, M., A. J. Parkinson, L. R. Bulkow, M. A. Fitzgerald, H. V. Peters, and D. J. Parks.** 1994. The epidemiology of invasive pneumococcal disease in Alaska, 1986–90: ethnic differences and opportunities for prevention. *J. Infect. Dis.* **170**:368–376.

61. **Davidson, M., C. D. Schraer, A. J. Parkinson, J. F. Campbell, R. R. Facklam, R. B. Wainwright, A. P. Lanier, and W. L. Heyward.** 1989. Invasive pneumococcal disease in an Alaska native population, 1980 through 1986. *JAMA* **261**:715–718.

62. **Dempsey, P. W., M. E. Allison, S. Akkaraju, C. C. Goodnow, and D. T. Fearon.** 1996. C3d of complement as a molecular adjuvant: bridging innate and acquired immunity. *Science* **271**:348–350.

63. **DiPadova, F., M. Durig, F. Harder, C. DiPadova, and C. Zanussi.** 1985. Impaired antipneumococcal antibody production in patients without spleens. *Br. Med. J. Clin. Res. Educ.* **290**:14–16.

64. **Drew, P. A., G. K. Kiroff, A. Ferrante, and R. C. Cohen.** 1984. Alterations in immunoglobulin synthesis by peripheral blood mononuclear cells from splenectomized patients with and without splenic regrowth. *J. Immunol.* **132**:191–196.

65. **Dworkin, M. S., J. W. Ward, D. L. Hanson, J. L. Jones, and J. E. Kaplan.** 2001. Pneumococcal disease among human immunodeficiency virus-infected persons: incidence, risk factors, and impact of vaccination. *Clin. Infect. Dis.* **32**:794–800.

66. **Elkayam, O., D. Paran, D. Caspi, I. Litinsky, M. Yaron, D. Charboneau, and J. B. Rubins.** 2002. Immunogenicity and safety of pneumococcal vaccination in patients with rheumatoid arthritis or systemic lupus erythematosus. *Clin. Infect. Dis.* **34**:147–153.

67. **Ellis, M., S. Gupta, S. Galant, S. Hakim, C. VandeVen, C. Toy, and M. S. Cairo.** 1988. Impaired neutrophil function in patients with AIDS or AIDS-related complex: a comprehensive evaluation. *J. Infect. Dis.* **158**:1268–1276.

68. **Eskola, J., H. Käyhty, A. K. Takala, H. Peltola, P.-R. Rönnberg, E. Kela, E. Pekkanen, P. H. McVerry, and P. H. Mäkelä.** 1990. A randomized, prospective field trial of a conjugate vaccine in the protection of infants and young children against invasive *Haemophilus influenzae* type b disease. *N. Engl. J. Med.* **323**:1381–1387.

69. **Eskola, J., A. K. Takala, E. Kela, E. Pekkanen, R. Kalliokoski, and M. Leinonen.** 1992. Epidemiology of invasive pneumococcal infections in children in Finland. *JAMA* **268**:3323–3327.

70. **Farley, J. J., J. C. King, Jr., P. Nair, S. E. Hines, R. L. Tressler, and P. E. Vink.** 1994. Invasive pneumococcal disease among infected and uninfected children of mothers with human immunodeficiency virus infection. *J. Pediatr.* **124**:853–858.

71. **Farr, B. M., L. Johnston, D. K. Cobb, M. J. Fisch, T. P. Germanson, K. A. Adal, and A. M. Anglim.** 1995. Preventing pneumococcal bacteremia in patients at risk. *Arch. Intern. Med.* **155**:2336–2340.

72. **Fearon, D. T.** 1997. Seeking wisdom in innate immunity. *Nature* **388**:323–324.

73. **Fearon, D. T., and R. M. Locksley.** 1996. The instructive role of innate immunity in the acquired immune response. *Science* **272**:50–53.

74. **Feikin, D. R., C. M. Elie, M. B. Goetz, J. L. Lennox, G. M. Carlone, S. Romero-Steiner, P. F. Holder, W. A. O'Brien, C. G. Whitney, J. C. Butler, and R. F. Breiman.** 2001. Randomized trial of the quantitative and functional antibody responses to a 7-valent pneumococcal conjugate vaccine and/or 23-valent polysaccharide vaccine among HIV-infected adults. *Vaccine* **20**:545–553.

75. **Feikin, D. R., A. Schuchat, M. Kolczak, N. L. Barrett, L. H. Harrison, L. Lefkowitz, A. McGeer, M. M. Farley, D. J. Vugia, C. Lexau, K. R. Stefonek, J. E. Patterson, and J. H. Jorgensen.** 2000. Mortality from invasive pneumococcal pneumonia in the era of antibiotic resistance, 1995–1997. *Am. J. Public Health* **90**:223–229.

76. **Feldman, S., W. Malone, R. Wilbur, and G. Schiffman.** 1985. Pneumococcal vaccination in children with acute lymphocytic leukemia. *Med. Pediatr. Oncol.* **13**:69–72.

77. **Figueroa, J. E., and P. Densen.** 1991. Infectious diseases associated with complement deficiencies. *Clin. Microbiol. Rev.* **4**:359–395.

78. **Filice, G. A., C. P. Darby, and D. W. Fraser.** 1980. Pneumococcal bacteremia in

Charleston County, South Carolina. *Am. J. Epidemiol.* **112**:828–835.

79. **Forrester, H. L., D. W. Jahnigen, and F. M. LaForce.** 1987. Inefficacy of pneumococcal vaccine in a high-risk population. *Am. J. Med.* **83**:425–430.

80. **Frankel, R. E., M. Virata, C. Hardalo, F. L. Altice, and G. Friedland.** 1996. Invasive pneumococcal disease: clinical features, serotypes, and antimicrobial resistance patterns in cases involving patients with and without human immunodeficiency virus infection. *Clin. Infect. Dis.* **23**:577–584.

81. **Frederiksen, B., L. Specht, J. Henrichsen, F. K. Pedersen, and J. Pedersen-Bjergaard.** 1989. Antibody response to pneumococcal vaccine in patients with early stage Hodgkin's disease. *Eur. J. Haematol.* **43**:45–49.

82. **French, N., C. F. Gilks, A. Mujugira, C. Fasching, J. O'Brien, and E. N. Janoff.** 1998. Pneumococcal vaccination in HIV-1-infected adults in Uganda: humoral response and two vaccine failures. *AIDS* **12**:1683–1689.

83. **French, N., J. Nakiyingi, L. M. Carpenter, E. Lugada, C. Watera, K. Moi, M. Moore, D. Antvelink, D. Mulder, E. N. Janoff, J. Whitworth, and C. F. Gilks.** 2000. 23-valent pneumococcal polysaccharide vaccine in HIV-1-infected Ugandan adults: double-blind, randomised and placebo controlled trial. *Lancet* **355**:2106–2111.

84. **Frezzato, M., G. Castaman, and F. Rodeghiero.** 1993. Fulminant sepsis in adults splenectomized for Hodgkin's disease. *Haematologica* **78**(6 Suppl. 2):73–77.

85. **Fry, A. M., R. R. Facklam, C. G. Whitney, B. D. Plikaytis, and A. Schuchat.** 2003. Multistate evaluation of invasive pneumococcal diseases in adults with human immunodeficiency virus infection: serotype and antimicrobial resistance patterns in the United States. *J. Infect. Dis.* **188**:643–652.

86. **Garin, E. H., and D. J. Barrett.** 1988. Pneumococcal polysaccharide immunization in patients with active nephrotic syndrome. *Nephron* **50**:383–388.

87. **Garred, P., J. Strom, L. Quist, E. Taaning, and H. O. Madsen.** 2003. Association of mannose binding lectin polymorphisms with sepsis and fatal outcome, in patients with systemic inflammatory response syndrome. *J. Infect. Dis.* **188**:1394–1403.

88. **Gaston, M. H., J. I. Verter, G. Woods, C. Pegelow, J. Kelleher, G. Presbury, H. Zarkowsky, E. Vichinsky, R. Iyer, J. S. Lobel, S. Diamond, C. T. Holbrook, F. M. Gill, K. Ritchey, and J. M. Falletta.** 1986. Prophylaxis with oral penicillin in children with sickle cell anemia: a randomized trial. *N. Engl. J. Med.* **314**:1593–1599.

89. **Giebink, G. S., J. E. Foker, Y. Kim, and G. Schiffman.** 1980. Serum antibody and opsonic responses to vaccination with pneumococcal capsular polysaccharide in normal and splenectomized children. *J. Infect. Dis.* **141**:404–412.

90. **Giebink, G. S., P. I. Warkentin, N. K. Ramsay, and J. H. Kersey.** 1986. Titers of antibody to pneumococci in allogeneic bone marrow transplant recipients before and after vaccination with pneumococcal vaccine. *J. Infect. Dis.* **154**:590–596.

91. **Gilks, C. F., L. S. Otieno, R. J. Brindle, R. S. Newnham, G. N. Lule, J. B. O. Were, P. M. Simani, S. M. Bhatt, G. B. A. Okelo, P. G. Waiyaki, and D. A. Warrell.** 1992. The presentation and outcome of HIV-related disease in Nairobi. *Quarterly J. Med.* **297**:25–32.

92. **Gillespie, S. H.** 1989. Aspects of pneumococcal infection including bacterial virulence, host response and vaccination. *J. Med. Microbiol.* **28**:237–248.

93. **Glaser, R., J. K. Kiecolt-Glaser, W. B. Malarkey, and J. F. Sheridan.** 1998. The influence of psychological stress on the immune response to vaccines. *Ann. N. Y. Acad. Sci.* **840**:649–655.

94. **Glaser, R., J. Sheridan, W. B. Malarkey, R. C. MacCallum, and J. K. Kiecolt-Glaser.** 2000. Chronic stress modulates the immune response to a pneumococcal pneumonia vaccine. *Psychosom. Med.* **62**:804–807.

95. **Glass, R. I., A.-M. Svennerholm, B. J. Stoll, M. R. Khan, K. M. Hossain, M. Imdadul Huq, and J. Holmgren.** 1983. Protection against cholera in breast-fed children by antibodies in breast milk. *N. Engl. J. Med.* **308**:1389–1392.

96. **Gordon, S. B., M. Chaponda, A. L. Walsh, C. J. Whitty, M. A. Gordon, C. E. Machili, C. F. Gilks, M. J. Boeree, S. Kampondeni, R. C. Read, and M. E. Molyneux.** 2002. Pneumococcal disease in HIV-infected Malawian adults: acute mortality and long-term survival. *AIDS* **16**:1409–1417.

97. **Gransden, W. R., S. J. Eykyn, and I. Phillips.** 1985. Pneumococcal bacteremia: 325 episodes at St. Thomas's hospital. *Br. Med. J.* **290**:505–508.

98. **Griffioen, A. W., G. T. Rijkers, P. Janssens-Korpela, and B. J. M. Zegers.** 1991. Pneumococcal polysaccharides complexed with C3d bind to human B lymphocytes via complement receptor type 2. *Infect. Immun.* **59**:1839–1845.

99. **Griffioen, A. W., E. A. Toebes, B. J. Zegers, and G. T. Rijkers.** 1992. Role of CR2 in the human adult and neonatal in vitro antibody response to type 4 pneumococcal polysaccharide. *Cell. Immunol.* **143:**11–22.

100. **Grimfors, G., M. Bjorkholm, L. Hammarstrom, J. Askergren, C. I. Smith, and G. Holm.** 1989. Type-specific anti-pneumococcal antibody subclass response to vaccination after splenectomy with special reference to lymphoma patients. *Eur. J. Haematol.* **43:**404–410.

101. **Gurrieri, C., P. McGuire, H. Zan, X. J. Yan, A. Cerutti, E. Albesiano, S. L. Allen, V. Vinciguerra, K. R. Rai, M. Ferrarini, P. Casali, and N. Chiorazzi.** 2002. Chronic lymphocytic leukemia B cells can undergo somatic hypermutation and intraclonal immunoglobulin V(H)DJ(H) gene diversification. *J. Exp. Med.* **196:**629–639.

102. **Haas, K. M., M. Hasegawa, D. A. Steeber, J. C. Poe, M. D. Zabel, C. B. Bock, D. R. Karp, D. E. Briles, J. H. Weis, and T. F. Tedder.** 2002. Complement receptors CD21/35 link innate and protective immunity during *Streptococcus pneumoniae* infection by regulating IgG3 antibody responses. *Immunity* **17:**713–723.

103. **Hammarström, V., K. Pauksen, J. Azinge, G. Oberg, and P. Ljungman.** 1993. Pneumococcal immunity and response to immunization with pneumococcal vaccine in bone marrow transplant patients: the influence of graft versus host reaction. *Support Care Cancer* **1:**195–199.

104. **Hazelwood, M. A., D. S. Kumararatne, A. D. Webster, M. Goodall, P. Bird, and M. Daha.** 1992. An association between homozygous C3 deficiency and low levels of anti-pneumococcal capsular polysaccharide antibodies. *Clin. Exp. Immunol.* **87:**404–409.

105. **Hibbs, J. R., J. M. Douglas, Jr., F. N. Judson, W. L. McGill, C. A. M. Rietmeijer, and E. N. Janoff.** 1997. Prevalence of human immunodeficiency virus infection, mortality rate, and serogroup distribution among patients with pneumococcal bacteremia at Denver General Hospital, 1984–1994. *Clin. Infect. Dis* **25:**195–199.

106. **Hirschtick, R. E., J. Glassroth, M. C. Jordan, T. C. Wilcosky, J. M. Wallace, P. A. Kvale, N. Markowitz, M. J. Rosen, B. T. Mangura, and P. C. Hopewell.** 1995. Bacterial pneumonia in persons infected with the human immunodeficiency virus. *N. Engl. J. Med.* **333:**845–851.

107. **Hoge, C. W., M. R. Reichler, E. A. Dominguez, J. C. Bremer, T. D. Mastro, K. A. Hendricks, D. M. Musher, J. A. Elliott, R. R. Facklam, and R. F. Breiman.** 1994. An epidemic of pneumococcal disease in an overcrowded, inadequately ventilated jail. *N. Engl. J. Med.* **331:**643–648.

108. **Hopken, U. E., B. Lu, N. P. Gerard, and C. Gerard.** 1996. The C5a chemoattractant receptor mediates mucosal defence to infection. *Nature* **383:**86–89.

109. **Hornef, M. W., M. J. Wick, M. Rhen, and S. Normark.** 2002. Bacterial strategies for overcoming host innate and adaptive immune responses. *Nat. Immunol.* **3:**1033–1040.

110. **Hostetter, M. K.** 1986. Serotypic variations among virulent pneumococci in deposition and degradation of covalently bound C3b: implications for phagocytosis and antibody production. *J. Infect. Dis.* **153:**682–693.

111. **Humbert, J. R., E. L. Winsor, J. M. Githens, and J. B. Schmitz.** 1990. Neutrophil dysfunctions in sickle cell disease. *Biomed. Pharmacother.* **44:**153–158.

112. **Itala, M., H. Helenius, J. Nikoskelainen, and K. Remes.** 1992. Infections and serum IgG levels in patients with chronic lymphocytic leukemia. *Eur. J. Haematol.* **48:**266–270.

113. **Jackson, L. A., K. M. Neuzil, O. Yu, P. Benson, W. E. Barlow, A. L. Adams, C. A. Hanson, L. D. Mahoney, D. K. Shay, W. W. Thompson, and the Vaccine Safety Datalink.** 2003. Effectiveness of pneumococcal polysaccharide vaccine in older adults. *N. Engl. J. Med.* **348:**1747–1755.

114. **Janoff, E. N., R. F. Breiman, C. L. Daley, and P. C. Hopewell.** 1992. Pneumococcal disease during HIV infection. Epidemiologic, clinical, and immunologic perspectives. *Ann. Intern. Med.* **117:**314–324.

115. **Janoff, E. N., J. M. Douglas, M. Gabriel, M. J. Blaser, A. J. Davidson, D. L. Cohn, and F. N. Judson.** 1988. Class-specific antibody response to pneumococcal capsular polysaccharides in men infected with human immunodeficiency virus type 1. *J. Infect. Dis.* **158:**983–990.

116. **Janoff, E. N., C. Fasching, J. C. Ojoo, J. O'Brien, and C. F. Gilks.** 1997. Responsiveness of human immunodeficiency virus type 1-infected Kenyan women with or without prior pneumococcal disease to pneumococcal vaccine. *J. Infect. Dis.* **175:**975–978.

117. **Janoff, E. N., J. O'Brien, P. Thompson, J. Ehret, G. Meiklejohn, G. Duvall, and J. M. Douglas, Jr.** 1993. *Streptococcus pneumoniae* colonization, bacteremia, and immune response among persons with human immunodeficiency virus infection. *J. Infect. Dis.* **167:**49–56.

118. **Janoff, E. N., and J. B. Rubins.** 2000. *Invasive Pneumococcal Disease in Immunocompromised Patients.* Mary Ann Liebert, Inc., Larchmont, N.Y.

119. **Jetté, L. P., and F. Lamothe.** 1989. Surveillance of invasive *Streptococcus pneumoniae* infection in Quebec, Canada, from 1984 to 1986: serotype distribution, antimicrobial susceptibility, and clinical characteristics. *J. Clin. Microbiol.* **27:**1–5.

120. **Johnson, C. A., P. Densen, R. A. Wetsel, F. S. Cole, N. E. Goeken, and H. R. Colten.** 1992. Molecular heterogeneity of C2 deficiency. *N. Engl. J. Med.* **326:**871–874.

121. **Kaplan, J., S. Sarnaik, and G. Schiffman.** 1986. Revaccination with polyvalent pneumococcal vaccine in children with sickle cell anemia. *Am. J. Pediatr.* **8:**80–82.

122. **Karstaedt, A. S., M. Khoosal, and H. H. Crewe-Brown.** 2001. Pneumococcal bacteremia in adults in Soweto, South Africa, during the course of a decade. *Clin. Infect. Dis.* **33:**610–614.

123. **Kiecolt-Glaser, J. K., R. Glaser, S. Gravenstein, W. B. Malarkey, and J. Sheridan.** 1996. Chronic stress alters the immune response to influenza virus vaccine in older adults. *Proc. Natl. Acad. Sci. USA* **93:**3043–3047.

124. **Kim, P. E., D. M. Musher, W. P. Glezen, M. C. Rodriguez-Barradas, W. K. Nahm, and C. E. Wright.** 1996. Association of invasive pneumococcal disease with season, atmospheric conditions, air pollution, and the isolation of respiratory viruses. *Clin. Infect. Dis.* **22:**100–106.

125. **King, M. D., C. G. Whitney, F. Parekh, M. M. Farley, and Active Bacterial Core Surveillance Team/Emerging Infections Program Network.** 2003. Recurrent invasive pneumococcal disease: a population-based assessment. *Clin. Infect. Dis.* **37:**1029–1036.

126. **Kiroff, G. K., A. N. Hodgen, P. A. Drew, and G. G. Jamieson.** 1985. Lack of effect of splenic regrowth on the reduced antibody responses to pneumococcal polysaccharides in splenectomized patients. *Clin. Exp. Immunol.* **62:**48–56.

127. **Klugman, K. P., S. A. Madhi, R. E. Huebner, R. Kohberger, N. Mbelle, N. Pierce, and the Vaccine Trialists Group.** 2003. A trial of a 9-valent pneumococcal conjugate vaccine in children with and those without HIV infection. *N. Engl. J. Med.* **349:**1341–1348.

128. **Koch, G., B. D. Lok, A. van Oudenaren, and R. Benner.** 1982. The capacity and mechanism of bone marrow antibody formation by thymus-independent antigens. *J. Immunol.* **128:**1497–1501.

129. **Kohler, P. F., V. J. Rivera, E. D. Eckert, T. J. Bouchard, Jr., and L. L. Heston.** 1985. Genetic regulation of immunoglobulin and specific antibody levels in twins reared apart. *J. Clin. Investig.* **75:**883–888.

130. **Konradsen, H. B., and J. Henrichsen.** 1991. Pneumococcal infections in splenectomized children are preventable. *Acta Paediatr. Scand.* **80:**423–427.

131. **Konradsen, H. B., J. Henrichsen, H. Wachmann, and N. Holm.** 1993. The influence of genetic factors on the immune response as judged by pneumococcal vaccination of mono- and dizygotic Caucasian twins. *Clin. Exp. Immunol.* **92:**532–536.

132. **Konradsen, H. B., V. A. Oxelius, M. Hahn-Zoric, and L. A. Hanson.** 1994. The importance of G1m and 2 allotypes for the IgG2 antibody levels and avidity against pneumococcal polysaccharide type 1 within mono- and dizygotic twin-pairs. *Scand. J. Immunol.* **40:**251–256.

133. **Kontoyianis, D. P., E. J. Anaissie, and G. P. Bodey.** 1993. *Infection in Chronic Lymphocytic Leukemia: a Reappraisal in Chronic Lymphocytic Leukemia Scientific Advances and Clinical Developments.* Marcel Dekker, Inc., New York, N.Y.

134. **Koziel, H., and M. J. Koziel.** 1995. Pulmonary complications of diabetes mellitus. Pneumonia. *Infect. Dis. Clin. N. Am.* **9:**65–96.

135. **Krasinski, K., W. Borkowsky, S. Bonk, R. Lawrence, and S. Chandwani.** 1988. Bacterial infections in human immunodeficiency virus-infected children. *Pediatr. Infect. Dis. J.* **7:**323–328.

136. **Krivit, W.** 1977. Overwhelming postsplenectomy infection. *Am. J. Hematol.* **2:**193–201.

137. **Krumholz, H. M., M. A. Sande, and B. Lo.** 1989. Community-acquired bacteremia in patients with acquired immunodeficiency syndrome: clinical presentation, bacteriology and outcome. *Am. J. Med.* **86:**776–779.

138. **Kuikka, A., J. Syrjanen, O. V. Renkonen, and V. V. Valtonen.** 1992. Pneumococcal bacteraemia during a recent decade. *J. Infect.* **24:**157–168.

139. **Kulkarni, S., R. Powles, J. Treleaven, U. Riley, S. Singhal, C. Horton, B. Sirohi, N. Bhagwati, S. Meller, R. Saso, and J. Mehta.** 2000. Chronic graft versus host disease is associated with long-term risk for pneumococcal infections in recipients of bone marrow transplants. *Blood* **95:**3683–3686.

140. **Kumar, D., C. Rotstein, G. Miyata, D. Arlen, and A. Humar.** 2003. Randomized, double-blind, controlled trial of pneumococcal

vaccination in renal transplant recipients. *J. Infect. Dis.* **187**:1639–1645.

141. **Lane, P. J. L., and I. C. M. Maclennan.** 1986. Impaired IgG2 anti-pneumococcal antibody responses in patients with recurrent infection and normal IgG2 levels but no IgA. *Clin. Exp. Immunol.* **65**:427–433.

142. **Linnemann, C. C., Jr., and M. R. First.** 1979. Risk of pneumococcal infections in renal transplant patients. *JAMA* **241**:2619–2621.

143. **Lortan, J. E., A. S. Kaniuk, and M. A. Monteil.** 1993. Relationship of in vitro phagocytosis of serotype 14 *Streptococcus pneumoniae* to specific class and IgG subclass antibody levels in healthy adults. *Clin. Exp. Immunol.* **91**:54–57.

144. **Madhi, S. A., K. Petersen, A. Madhi, M. Khoosal, and K. P. Klugman.** 2000. Increased disease burden and antibiotic resistance of bacteria causing severe community-acquired lower respiratory tract infections in human immunodeficiency virus type 1-infected children. *Clin. Infect. Dis.* **31**:170–176.

145. **Marfin, A. A., J. Sporrer, P. S. Moore, and A. D. Siefkin.** 1995. Risk factors for adverse outcome in persons with pneumococcal pneumonia. *Chest* **107**:457–462.

146. **Marrie, T. J.** 1994. Pneumonia and carcinoma of the lung. *J. Infect.* **29**:45–52.

147. **Martin, F., A. M. Oliver, and J. F. Kearney.** 2001. Marginal zone and B1 B cells unite in the early response against T-independent blood-borne particulate antigens. *Immunity* **14**:617–629.

148. **McCashland, T. M., L. C. Preheim, and M. J. Gentry.** 2000. Pneumococcal vaccine response in cirrhosis and liver transplantation. *J. Infect. Dis.* **181**:757–760.

149. **McEllistrem, M. C., A. B. Mendelsohn, M. A. Pass, J. A. Elliott, C. G. Whitney, J. A. Kolano, and L. H. Harrison.** 2002. Recurrent invasive pneumococcal disease in individuals with human immunodeficiency virus infection. *J. Infect. Dis.* **185**:1364–1368.

150. **Morgan, B. P., and M. J. Walport.** 1991. Complement deficiency and disease. *Immunol. Today* **12**:301–306.

151. **Müller, F., S. S. Frøland, and P. Brandtzaeg.** 1989. Altered IgG-subclass distribution in lymph node cells and serum of adults infected with human immunodeficiency virus (HIV). *Clin. Exp. Immunol.* **78**:153–158.

152. **Mundy, L. M., P. G. Auwaerter, D. Oldach, M. L. Warner, A. Burton, E. Vance, C. A. Gaydos, J. M. Joseph, R. Gopalan, R. D. Moore, T. C. Quinn, P. Charache, and J. G. Bartlett.** 1995. Community-acquired pneumonia: impact of immune

status. *Am. J. Respir. Crit. Care Med.* **152**:1309–1315.

153. **Murphy, P. M., H. C. Lane, A. S. Fauci, and J. I. Gallin.** 1988. Impairment of neutrophil bactericidal capacity in patients with AIDS. *J. Infect. Dis.* **158**:627–630.

154. **Murray, S. E., H. R. Lallman, A. D. Heard, M. B. Rittenberg, and M. P. Stenzel-Poore.** 2001. A genetic model of stress displays decreased lymphocytes and impaired antibody responses without altered susceptibility to *Streptococcus pneumoniae*. *J. Immunol.* **167**:691–698.

155. **Musher, D. M., J. E. Groover, M. R. Reichler, F. X. Riedo, B. Schwartz, D. A. Watson, R. E. Baughn, and R. F. Breiman.** 1997. Emergence of antibody to capsular polysaccharides of *Streptococcus pneumoniae* during outbreaks of pneumonia: association with nasopharyngeal colonization. *Clin. Infect. Dis.* **24**:441–446.

156. **Musher, D. M., J. E. Groover, D. A. Watson, J. P. Pandey, M. C. Rodriguez-Barradas, R. E. Baughn, M. S. Pollack, E. A. Graviss, M. de Andrade, and C. I. Amos.** 1997. Genetic regulation of the capacity to make immunoglobulin G to pneumococcal capsular polysaccharides. *J. Investig. Med.* **45**:57–68.

157. **Neth, O., I. Hann, M. W. Turner, and N. J. Klein.** 2001. Deficiency of mannose-binding lectin and burden of infection in children with malignancy: a prospective study. *Lancet* **358**:614–618.

158. **Neth, O., D. L. Jack, A. W. Dodds, H. Holzel, N. J. Klein, and M. W. Turner.** 2000. Mannose-binding lectin binds to a range of clinically relevant microorganisms and promotes complement deposition. *Infect. Immun.* **68**:688–693.

159. **Newman, S. L., L. B. Vogler, R. D. Feigin, and R. B. Johnston, Jr.** 1979. Recurrent septicemia associated with congenital deficiency of C2 and partial deficiency of factor B and the alternate complement pathway. *N. Engl. J. Med.* **299**:290–292.

160. **Nuorti, J. P., J. C. Butler, M. M. Farley, L. H. Harrison, A. McGeer, M. S. Kolczak, R. F. Breiman, and Active Bacterial Core Surveillance Team.** 2000. Cigarette smoking and invasive pneumococcal disease. *N. Engl. J. Med.* **342**:681–689.

161. **O'Brien, K. L., L. H. Moulton, R. Reid, R. Weatherholtz, J. Oski, L. Brown, G. Kumar, A. Parkinson, D. Hu, J. Hackell, I. Chang, R. Kohberger, G. Siber, and M. Santosham.** 2003. Efficacy and safety of seven-

valent conjugate pneumococcal vaccine in American Indian children: group randomised trial. *Lancet* **362**:355–361.

162. **Oldfield, S., S. Jenkins, H. Yeoman, D. Gray, and I. C. M. MacLennan.** 1985. Class and subclass anti-pneumococcal antibody responses in splenectomized patients. *Clin. Exp. Immunol.* **61**:664–673.

163. **O'Neal, B. J., and J. C. McDonald.** 1981. The risk of sepsis in the asplenic adult. *Ann. Surg.* **194**:775–778.

164. **Ortqvist, A., M. Kalin, I. Julander, and M. A. Mufson.** 1993. Deaths in bacteremic pneumococcal pneumonia. A comparison of two populations—Huntington, WV and Stockholm, Sweden. *Chest* **103**:710–716.

165. **Perlino, C. A., and D. Rimland.** 1985. Alcoholism, leukopenia, and pneumococcal sepsis. *Am. Rev. Respir. Dis.* **132**:757–760.

166. **Peset-Llopis, M. J., G. Harms, M. J. Hardonk, and W. Timens.** 1996. Human immune response to pneumococcal polysaccharides: complement-mediated localization preferentially on CD21-positive splenic marginal zone B cells and follicular dendritic cells. *J. Allergy Clin. Immunol.* **97**:1015–1024.

167. **Peterslund, N. A., C. Koch, J. C. Jensenius, and S. Thiel.** 2001. Association between deficiency of mannose-binding lectin and severe infections after chemotherapy. *Lancet* **358**:637–638.

168. **Picard, C., A. Puel, M. Bonnet, C. L. Ku, J. Bustamante, K. Yang, C. Soudais, S. Dupuis, J. Feinberg, C. Fieschi, C. Elbim, R. Hitchcock, D. Lammas, G. Davies, A. Al-Ghonaium, H. Al-Rayes, S. Al-Jumaah, S. Al-Hajjar, I. Z. Al-Mohsen, H. H. Frayha, R. Rucker, T. R. Hawn, A. Aderem, H. Tufenkeji, S. Haraguchi, N. K. Day, R. A. Good, M. A. Gougerot-Pocidalo, A. Ozinsky, and J. L. Casanova.** 2003. Pyogenic bacterial infections in humans with IRAK-4 deficiency. *Science* **299**:2076–2079.

169. **Polsky, B., J. W. M. Gold, E. Wimbey, J. Dryjansky, A. E. Brown, G. Schiffman, D. Armstrong, and T. C. Quinn.** 1986. Bacterial pneumonia in patients with the acquired immune deficiency syndrome. *Ann. Intern. Med.* **104**:38–41.

170. **Puppo, F., R. Ruzzenenti, S. Brenci, L. Lanza, M. Scudeletti, and F. Indiveri.** 1991. Major histocompatibility gene products and human immunodeficiency virus infection. *J. Lab. Clin. Med.* **117**:91–100.

171. **Rao, S. P., K. Rajkumar, G. Schiffman, N. Desai, C. Unger, and S. T. Miller.** 1995.

Anti-pneumococcal antibody levels three to seven years after first booster immunization in children with sickle cell disease, and after a second booster. *J. Pediatr.* **127**:590–592.

172. **Rautonen, N., N. L. Martin, J. Rautonen, Y. Rooks, W. C. Mentzer, and D. W. Wara.** 1992. Low number of antibody producing cells in patients with sickle cell anemia. *Immunol. Lett.* **34**:201–211.

173. **Redd, S. C., G. W. Rutherford, M. A. Sande, A. R. Lifson, W. K. Hadley, R. R. Facklam, and J. S. Spika.** 1990. The role of human immunodeficiency virus in pneumococcal bacteremia in San Francisco residents. *J. Infect. Dis.* **162**:1012–1017.

174. **Reed, W. P., G. W. Drach, and R. C. Williams, Jr.** 1974. Antigens common to human and bacterial cells. IV. Studies of human pneumococcal disease. *J. Lab. Clin. Med.* **83**:599–610.

175. **Reimer, C. B., C. M. Black, R. C. Holman, T. W. Wells, R. M. Ramirez, J. A. Sa-Ferreira, J. K. Nicholson, and J. S. McDougal.** 1988. Hypergammaglobulinemia associated with human immunodeficiency virus infection. *Monogr. Allergy* **23**:83–96.

176. **Reinert, R. R., A. Kaufhold, O. Kuhnemund, and R. Lutticken.** 1994. Serum antibody responses to vaccination with 23-valent pneumococcal vaccine in splenectomized patients. *Int. J. Med. Microbiol. Virol. Parasitol. Infect. Dis.* **281**:481–490.

177. **Robinson, K. A., W. Baughman, G. Rothrock, N. L. Barrett, M. Pass, C. Lexau, B. Damaske, K. Stefonek, B. Barnes, J. Patterson, E. R. Zell, A. Schuchat, and C. G. Whitney.** 2001. Epidemiology of invasive *Streptococcus pneumoniae* infections in the United States, 1995–1998: opportunities for prevention in the conjugate vaccine era. *JAMA* **285**:1729–1735.

178. **Rodriguez, M. E., W.-L. van der Pol, L. A. M. Sanders, and J. G. J. van de Winkel.** 1999. Crucial role of FcγRIIa (CD32) in assessment of functional anti-*Streptococcus pneumoniae* antibody activity in human sera. *J. Infect. Dis.* **179**:423–433.

179. **Rodriguez-Creixems, M., P. Munoz, E. Miranda, T. Pelaez, and R. Alonso.** 1996. Recurrent pneumococcal bacteremia. A warning of immunodeficiency. *Arch. Intern. Med.* **156**:1429–1434.

180. **Rodriquez-Barradas, M. C., D. M. Musher, R. J. Hamill, M. Dowell, J. T. Bagwell, and C. V. Sanders.** 1992. Unusual manifestations of pneumococcal infection in human immunodeficiency virus-infected individuals: the past revisited. *Clin. Infect. Dis.* **14**:192–199.

181. Romero-Steiner, S., D. M. Musher, M. S. Cetron, L. B. Pais, J. E. Groover, A. E. Fiore, B. D. Plikaytis, and G. M. Carlone. 1999. Reduction in functional antibody activity against *Streptococcus pneumoniae* in vaccinated elderly individuals highly correlates with decreased IgG antibody avidity. *Clin. Infect. Dis.* **29:**281–288.

182. Rosen, F. S., and C. A. Janeway. 1966. The gamma globulins. III: The antibody deficiency syndromes. *N. Engl. J. Med.* **275:** 709–715.

183. Ross, S. C., and P. Densen. 1984. Complement deficiency states and infection: epidemiology, pathogenesis and consequences of neisserial and other infections in an immune deficiency. *Medicine* (Baltimore) **63:**243–273.

184. Rowe, P. C., R. H. McLean, R. A. Wood, R. J. Leggiadro, and J. A. Winkelstein. 1989. Association of homozygous C4B deficiency with bacterial meningitis. *J. Infect. Dis.* **160:**448–451.

185. Rubins, J. B., M. Alter, J. Loch, and E. N. Janoff. 1999. Determination of antibody responses of elderly adults to all 23 capsular polysaccharides after pneumococcal vaccination. *Infect. Immun.* **67:**5979–5984.

186. Rubins, J. B., D. Charboneau, J. C. Paton, T. J. Mitchell, P. W. Andrew, and E. N. Janoff. 1995. Dual function of pneumolysin in the early pathogenesis of murine pneumococcal pneumonia. *J. Clin. Investig.* **95:**142–150.

187. Rubins, J. B., and E. N. Janoff. 2001. Pneumococcal disease in the elderly: what is preventing vaccine efficacy? *Drugs Aging* **18:**305–311.

188. Rubins, J. B., A. K. Puri, J. Loch, D. Charboneau, R. MacDonald, N. Opstad, and E. N. Janoff. 1998. Magnitude, duration, quality, and function of pneumococcal vaccine responses in elderly adults. *J. Infect. Dis.* **178:**431–440.

189. Rynnel-Dagoo, B., A. Freijd, L. Hammarström, V. Oxelius, and M. A. Persson. 1986. Pneumococcal antibodies of different immunoglobulin subclasses in normal and IgG subclass deficient individuals of various ages. *Acta Oto-Laryngol.* **101:**146–151.

190. Ryser, O., A. Morell, and W. H. Hitzig. 1988. Primary immunodeficiencies in Switzerland: first report of the national registry in adults and children. *J. Clin. Immunol.* **8:**479–485.

191. Saeland, E., J. H. Leusen, G. Vidarsson, W. Kuis, E. A. Sanders, I. Jonsdottir, and J. G. van de Winkel. 2003. Role of leukocyte immunoglobulin G receptors in vaccine-induced immunity to Streptococcus pneumoniae. *J. Infect. Dis.* **187:**1686–1693.

192. Saeland, E., G. Vidarsson, J. H. Leusen, E. Van Garderen, M. H. Nahm, H. Vile-Weekhout, V. Walraven, A. M. Stemerding, J. S. Verbeek, G. T. Rijkers, W. Kuis, E. A. Sanders, and J. G. Van De Winkel. 2003. Central role of complement in passive protection by human IgG1 and IgG2 anti-pneumococcal antibodies in mice. *J. Immunol.* **170:** 6158–6164.

193. Sarvas, H., N. Rautonen, S. Sipinen, and O. Mäkelä. 1989. IgG subclasses of pneumococcal antibodies—effect of allotype G2m(n). *Scand. J. Immunol.* **29:**229–237.

194. Sasso, E. H., J. H. Buckner, and L. A. Suzuki. 1995. Ethnic differences of polymorphism of an immunoglobulin VH3 gene. *J. Clin. Investig.* **96:**1591–1600.

195. Scamurra, R. W., D. J. Miller, L. Dahl, M. Abrahamsen, V. Kapur, S. M. Wahl, E. C. Milner, and E. N. Janoff. 2000. Impact of HIV-1 infection on VH3 gene repertoire of naive human B cells. *J. Immunol.* **164:**5482–5491.

196. Schuchat, A., C. V. Broome, A. Hightower, S. J. Costa, and W. Parkin. 1991. Use of surveillance for invasive pneumococcal disease to estimate the size of the immunosuppressed HIV-infected population. *JAMA* **265:**3275–3279.

197. Schutze, G. E., E. O. Mason, Jr., E. R. Wald, W. J. Barson, J. S. Bradley, T. Q. Tan, K. S. Kim, L. B. Givner, R. Yogev, and S. L. Kaplan. 2001. Pneumococcal infections in children after transplantation. *Clin. Infect. Dis.* **33:**16–21.

198. Selby, C., S. Hart, P. Ispahani, and P. J. Toghill. 1987. Bacteremia in adults after splenectomy or splenic irradiation. *Q. J. Med.* **63:**523–530.

199. Selwyn, P. A., P. Alcabes, D. Hartel, D. Buono, E. E. Schoenbaum, R. S. Klein, K. Davenny, and G. H. Friedland. 1992. Clinical manifestations and predictors of disease progression in drug users with human immunodeficiency virus infection. *N. Engl. J. Med.* **327:** 1697–1703.

200. Selwyn, P. A., A. R. Feingold, D. Hartel, E. E. Schoenbaum, M. H. Alderman, R. S. Klein, and G. H. Friedland. 1988. Increased risk of bacterial pneumonia in HIV-infected intravenous drug users without AIDS. *AIDS* **2:**267–272.

201. Shah, T. P., and F. R. Sattler. 1987. Polymorphonuclear leukocyte-mediated, antibody-dependent cellular cytotoxicity in patients with AIDS. *J. Infect. Dis.* **155:**594–595. (Letter.)

202. Shapiro, E. D., A. T. Berg, R. Austrian, D. Schroeder, V. Parcells, A. Margolis, R. K. Adair, and J. D. Clemens. 1991. The protec-

tive efficacy of polyvalent pneumococcal polysaccharide vaccine. *N. Engl. J. Med.* **325**:1453–1460.

203. **Shapiro, E. D., and J. D. Clemens.** 1984. A controlled evaluation of the protective efficacy of pneumococcal vaccine for patients at high risk of serious pneumococcal infections. *Ann. Intern. Med.* **101**:325–330.

204. **Siber, G. R., P. H. Schur, A. C. Aisenberg, S. A. Weitzman, and G. Schiffman.** 1980. Correlation between serum IgG2 concentrations and the antibody response to bacterial polysaccharide antigens. *N. Engl. J. Med.* **303**:178–182.

205. **Sims, R. V., W. C. Steinmann, J. H. McConville, L. R. King, W. C. Zwick, and J. S. Schwartz.** 1988. The clinical effectiveness of pneumococcal vaccine in the elderly. *Ann. Intern. Med.* **108**:653–657.

206. **Spika, J. S., N. A. Halsey, C. T. Le, A. J. Fish, G. M. Lum, B. A. Lauer, G. Schiffman, and G. S. Giebink.** 1986. Decline of vaccine-induced antipneumococcal antibody in children with nephrotic syndrome. *Am. J. Kidney Dis.* **7**:466–470.

207. **Spoulou, V., P. Victoratos, J. P. Ioannidis, and S. Grafakos.** 2000. Kinetics of antibody concentration and avidity for the assessment of immune response to pneumococcal vaccine among children with bone marrow transplants. *J. Infect. Dis.* **182**:965–969.

208. **Stiehm, E., T. Chin, A. Haas, and A. Peerless.** 1986. Infectious complications of the primary immunodeficiencies. *Clin. Immunol. Immunopathol.* **40**:69–86.

209. **Styrt, B.** 1990. Infection associated with asplenia: risks, mechanisms, and prevention. *Am. J. Med.* **88**:33N–42N.

210. **Sumiya, M., M. Super, P. Tabona, R. J. Levinsky, T. Arai, M. W. Turner, and J. A. Summerfield.** 1991. Molecular basis of opsonic defect in immunodeficient children. *Lancet* **337**:1569–1570.

211. **Summerfield, J. A., S. Ryder, M. Sumiya, M. Thursz, A. Gorchein, M. A. Monteil, and M. W. Turner.** 1995. Mannose binding protein gene mutations associated with unusual and severe infections in adults. *Lancet* **345**:886–889.

212. **Super, M., S. Thiel, J. Lu, R. J. Levinsky, and M. W. Turner.** 1989. Association of low levels of mannan-binding protein with a common defect of opsonisation. *Lancet* **ii**:1236–1239.

213. **Taborda, C. P., and A. Casadevall.** 2002. CR3 (CD11b/CD18) and CR4 (CD11c/CD18) are involved in complement-independent antibody-mediated phagocytosis of Cryptococcus neoformans. *Immunity* **16**:791–802.

214. **Tasker, S. A., M. R. Wallace, J. B. Rubins, W. B. Paxton, J. O'Brien, and E. N. Janoff.** 2002. Reimmunization with 23-valent pneumococcal vaccine for patients infected with human immunodeficiency virus type 1: clinical, immunologic, and virologic responses. *Clin. Infect. Dis.* **34**:813–821.

215. **Tausk, F. A., A. McCutchan, P. Spechko, R. D. Schreiber, and I. Gigli.** 1986. Altered erythrocyte C3b receptor expression, immune complexes, and complement activation in homosexual men in varying risk groups for acquired immune deficiency syndrome. *J. Clin. Investig.* **78**:977–982.

216. **Traub, A., G. S. Giebink, C. Smith, C. C. Kuni, M. L. Brekke, D. Edlund, and J. F. Perry.** 1987. Splenic reticuloendothelial function after splenectomy, spleen repair, and spleen autotransplantation. *N. Engl. J. Med.* **317**:1559–1564.

217. **Twomey, J. T.** 1973. Infections complicating multiple myeloma and chronic lymphocytic leukemia. *Arch. Intern. Med.* **132**:562–565.

218. **Van Wyck, D. B.** 1983. Overwhelming postsplenectomy infection (OPSI): the clinical syndrome. *Lymphology* **16**:107–114.

219. **Vedhara, K., N. K. Cox, G. K. Wilcock, P. Perks, M. Hunt, S. Anderson, S. L. Lightman, and N. M. Shanks.** 1999. Chronic stress in elderly carers of dementia patients and antibody response to influenza vaccination. *Lancet* **353**:627–631.

220. **Walport, M. J.** 2001. Complement. Second of two parts. *N. Engl. J. Med.* **344**:1140–1144.

221. **Ward, H. K., and J. F. Enders.** 1933. An analysis of the opsonic and tropic action of normal and immune sera based on experiments with the pneumococcus. *J. Exp. Med.* **1933:** 527–547.

222. **Warmerdam, P. A. M., J. G. J. van de Winkel, A. Vlug, N. A. C. Westerdaal, and P. J. A. Capel.** 1991. A single amino acid in the second Ig-like domain of the human Fcγ receptor II is critical for human IgG2 binding. *J. Immunol.* **147**:1338–1343.

223. **Watanakunakorn, C., A. Greifenstein, K. Stroh, D. G. Jarjoura, D. Blend, A. Cugino, and A. J. Ognibene.** 1993. Pneumococcal bacteremia in three community teaching hospitals from 1980 to 1989. *Chest* **103**:1152–1156.

224. **Weeks, J. C., M. R. Tierney, and M. C. Weinstein.** 1991. Cost effectiveness of prophylactic intravenous immune globulin in chronic lymphocytic leukemia. *N. Engl. J. Med.* **325**:81–86.

225. **Weiss, P. J., M. R. Wallace, E. C. Oldfield III, J. O'Brien, and E. N. Janoff.** 1995.

Response of recent HIV-seroconverters to the pneumococcal polysaccharide vaccine and *Haemophilus influenzae* type b conjugate vaccine. *J. Infect. Dis.* **171:**1217–1222.

226. **Whitney, C. G., M. M. Farley, J. Hadler, L. H. Harrison, N. M. Bennett, R. Lynfield, A. Reingold, P. R. Cieslak, T. Pilishvili, D. Jackson, R. R. Facklam, J. H. Jorgensen, A. Schuchat, and the Active Bacterial Core Surveillance of the Emerging Infections Program Network.** 2003. Decline in invasive pneumococcal disease after the introduction of protein-polysaccharide conjugate vaccine. *N. Engl. J. Med.* **348:**1737–1746.

227. **Winkelstein, J. A.** 1984. Complement and the host's defense against the pneumococcus. *Crit. Rev. Microbiol.* **11:**187–208.

228. **Winston, D. J., G. Schiffman, D. C. Wang, S. A. Feig, C.-H. Lin, E. L. Marso, W. G. Ho, L. S. Young, and R. P. Gale.** 1979. Pneumococcal infections after human bone-marrow transplantation. *Ann. Intern. Med.* **91:**835–841.

229. **Witt, D. J., D. E. Craven, and W. R. McCabe.** 1987. Bacterial infections in adult patients with the acquired immune deficiency syndrome (AIDS) and AIDS-related complex. *Am. J. Med.* **82:**900–906.

230. **Wong, W. Y., D. R. Powars, L. Chan, A. Hiti, C. Johnson, and G. Overturf.** 1991. Polysaccharide encapsulated bacterial infection in sickle cell anemia: a thirty year epidemiologic experience. *Am. J. Hematol.* **39:**176–182.

231. **Yee, A. M., H. M. Phan, R. Zuniga, J. E. Salmon, and D. M. Musher.** 2000. Association between FcγRIIa-R131 allotype and bacteremic pneumococcal pneumonia. *Clin. Infect. Dis.* **30:**25–28.

232. **Zarkowsky, H. S., D. Gallagher, F. M. Gill, W. C. Wang, J. M. Falletta, W. M. Lande, P. S. Levy, J. I. Verter, and D. Wethers.** 1986. Bacteremia in sickle hemoglobinopathies. *J. Pediatr.* **109:**579–585.

233. **Zarrabi, M. H., and F. Rosner.** 1984. Serious infections in adults following splenectomy for trauma. *Arch. Intern. Med.* **144:**1421–1424.

TREATMENT AND PREVENTION

CHANGING THE ECOLOGY OF PNEUMOCOCCI WITH ANTIBIOTICS AND VACCINES

Ron Dagan and Marc Lipsitch

18

INTRODUCTION

The upper respiratory tract (URT) of typical, healthy humans harbors a number of bacterial species, including *Streptococcus pneumoniae*, *Staphylococcus* spp., *Haemophilus influenzae*, *Moraxella catarrhalis*, and *Neisseria meningitidis*. These species, which are normally commensal but can become pathogenic when they enter normally sterile sites, are accompanied in the URT by a number of nonpathogenic species (4, 73). Multiple strains of a single species may also be present, as demonstrated for *S. pneumoniae* in classic carriage studies that found up to five coexisting pneumococcal serotypes in the URT of healthy children (52).

With such diversity in a confined environment, there are ample opportunities for interactions among components of microbial flora. Attention has focused on inhibitory interactions, mediated by production of soluble inhibitory factors (83) and possibly by competition for nutrients, receptors, or physical space (70, 83). Additionally, stimulation of nonspe-

cific or cross-reactive immune responses by one bacterial population may result in inhibition of other populations (76). On the other hand, bacteria resident in the URT also secrete a number of chemicals, including those that inhibit host immune responses (111) and antibiotic action (16), whose action could plausibly result in enhancement of one strain or species by another, if the secreted products of the resident species were to detoxify the environment for the other (68).

In a microbial community structured by such interactions, disturbances that inhibit the growth of a subset of the bacteria present will have consequences not only for the susceptible bacteria but also for those whose abundance is regulated, in part, by the presence of the susceptible bacteria (78). In particular, antimicrobial treatment kills or inhibits nasopharyngeal bacteria susceptible to the antimicrobial at the concentration achieved in the nasopharynx, and vaccines inhibit acquisition of pneumococci against which the vaccine antigens are protective. To the extent that antimicrobial-susceptible pneumococci compete with resistant pneumococci, antimicrobial treatment that inhibits susceptible strains provides a selective advantage to resistant strains or species. Likewise, vaccine-induced immune responses that reduce carriage

Ron Dagan, Pediatric Infectious Disease Unit, Soroka University Medical Center, and Faculty of Health Sciences, Ben-Gurion University of the Negev, P.O. Box 151, Beer-Sheva 84101, Israel. *Marc Lipsitch*, Department of Epidemiology, Harvard School of Public Health, 677 Huntington Avenue, Boston, MA 02115.

The Pneumococcus. Volume Editor, Elaine I. Tuomanen,
© 2004 ASM Press, Washington, D.C.

of certain pneumococci will provide an advantage to their competitors, whether these are pneumococcal strains that do not carry epitopes against which the vaccine is directed or other species. This chapter describes the mechanisms and consequences for the pneumococcal population of the selective pressures imposed by antibiotics and vaccines. Although we suspect that inhibiting pneumococci will also have effects on other species of URT bacteria, we focus our attention on the selective effects of vaccines and antibiotics on pneumococci.

THE EFFECT OF ANTIBIOTICS ON *S. PNEUMONIAE* ECOLOGY

The direct effects of antibiotics on *S. pneumoniae* in the nasopharynx of the treated patient depend on the pharmacokinetics and pharmacodynamics of the agent. These direct effects then lead, through the interactions of bacterial populations, to effects on the overall ecology of the nasopharynx of the treated person, and then to effects on the superpopulation of pneumococci in the community as a whole.

Pharmacokinetic and Pharmacodynamic Principles To Predict Bacterial Eradication in the Nasopharynx

To achieve eradication of a microorganism from a body site by an anti-infective agent, three conditions must be fulfilled. (i) The anti-infective agent must be active against the organism. Activity is measured in vitro by the agent's MIC for the organism. (ii) The agent needs to reach the site where the organism is found in concentrations that exceed the MIC. (iii) It must be present at the site for a time period that is sufficient for this action.

The pharmacology of antimicrobial chemotherapy in acute otitis media can be divided into (i) the pharmacokinetic component—i.e., the dosing regimen, the absorption of the drug, and its distribution and elimination, which determine the time course of the drug concentrations in serum and in tissues such as the mucosal tissues that sur-

round the nasopharynx—and (ii) the pharmacodynamic component, which deals with the association between concentrations of the drug at the site of infection (or colonization) and its antimicrobial effect (24). There has been an increasing interest in the use of these pharmacokinetic and pharmacodynamic parameters in the prediction of bacteriological and clinical efficacy (37).

For drugs such as beta-lactams, clindamycin, and trimethoprim-sulfamethoxazole (TS), the main determinant for predicting bacteriological eradication is achieving concentrations of free, unbound drug that are above the MIC and last at least 40 to 50% of the dosing interval (23). For fluoroquinolones, aminoglycosides, and most macrolides (including azalides and ketolides), the ratio of peak concentration to MIC is a more important determinant than time over MIC, because of the prolonged postantibiotic effect (23, 35, 88) and more rapid killing at higher concentrations.

For any antimicrobial agent, the presence of adequate concentrations in the lumen of the nasopharynx will not only eliminate or reduce the concentration of susceptible pneumococci at that site but also reduce the probability of new colonization by susceptible pneumococci.

For organisms, such as *S. pneumoniae*, that are carried and multiply mainly as extracellular organisms, the important measurement for the pharmacokinetic and pharmacodynamic calculation is the extracellular concentrations of the unbound antibacterial agent at the site of the infection (or colonization). Drugs that concentrate mainly within the cells, such as azithromycin, often achieve only very low extracellular concentrations despite high tissue levels, with the resulting effect only on organisms for which the drug has very low MIC. Similarly, a drug that is strongly bound to protein, such as ceftriaxone, may not strongly affect the organisms carried in the nasopharynx, because of its limited ability to reach the nasopharynx with high unbound concentrations (53).

Another important determinant of a drug's ecological impact is its half-life, since the prolonged presence of antibiotics, even at concentrations below the MIC, may inhibit susceptible strains and promote the establishment or outgrowth of resistant ones. An illustrative example is the study by Kastner and Guggenbichler (61), in which five macrolides were compared for the ability to select for nasopharyngeal flora that is macrolide resistant for 6 weeks after administration. Treatment was given for 7 days (except for azithromycin, given for 3 days). One week after initiation of treatment, 42 to 92% of all children carried macrolide-resistant flora. After 4 and 6 weeks, the proportions of children treated with azithromycin carrying macrolide-resistant microorganisms in the throat were 87 and 85%, respectively, versus ≤42 and ≤25%, respectively, of the children in all other macrolide groups ($P < 0.005$) (Fig. 1). Azithromycin has a very long elimination half-life (up to 72 h), while the elimination half-life of the other macrolides is <8 h (87).

S. pneumoniae resistance to the beta-lactam antibiotic class evolved mainly by complex restructuring of the targets of the beta-lactams, the penicillin-binding proteins (PBPs) (54).

The PBP targets in penicillin-nonsusceptible S. pneumoniae are modified to low-binding affinity versions of the native molecule, and the eventual level of resistance is determined by how many and to what extent targets are modified (19). Since restructuring of the PBPs is mediated by stepwise alterations in PBPs, the resulting increase in penicillin MIC is gradual.

The resulting distribution of levels of resistance to beta-lactam drugs means that there is no sharp cutoff between resistant and susceptible strains for any given beta-lactam and that a strain that is completely resistant to achievable concentrations of one beta-lactam may be quite susceptible to clinically achieved concentrations of another. Thus, in the community, S. pneumoniae remains mostly susceptible to amoxicillin (and often the injectable expanded-spectrum cephalosporins, such as ceftriaxone), even when resistant to other classes or to oral penicillin or cephalosporins (59, 60). This phenomenon explains the differential effect observed on S. pneumoniae nasopharyngeal carriage after treatment with high-dose amoxicillin (in formulations with or without clavulanate) in comparison with other drugs.

For antibiotic classes other than the beta-lactams, most resistance is expressed as high-

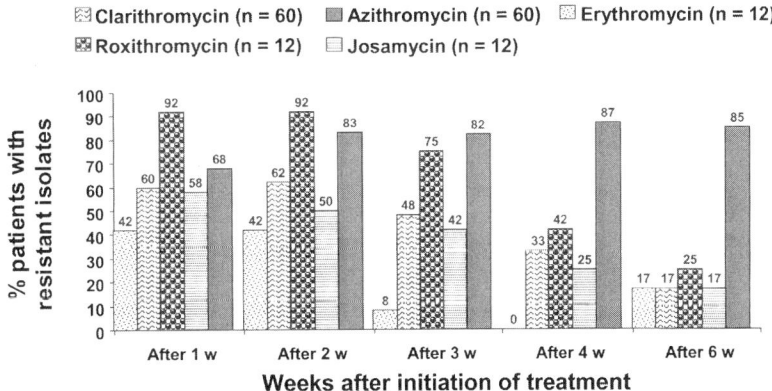

FIGURE 1 Effect of treatment with various macrolides on the carriage of macrolide-resistant microorganisms in the throat 1 to 6 weeks after initiation of treatment (61). All children were treated for 7 days (except the azithromycin group, which was treated for 3 days). The median age of the children was 4.2 years, without a significant difference in age distribution between the group. $P < 0.005$ for azithromycin versus all other groups.

level resistance sharply dividing organisms between those that are fully susceptible and those that are fully resistant.

Population-Biological Considerations

The direct effect of antibiotic treatment is to inhibit or kill those bacteria that are susceptible to the antibiotic at the concentration achieved. To the extent that strains of pneumococci compete to colonize the nasopharynx, antibiotic-mediated inhibition of susceptible strains will promote the growth of resistant strains in the treated individual, if they are present (70). Even if no resistant strain is present in the treated individual, inhibition of susceptible strains in an antibiotic-treated individual may promote the spread of resistant strains in the community by reducing transmission of the susceptible strains with which the resistant strains compete (72).

To understand the relationship between antibiotic use and antibiotic resistance in the pneumococci colonizing the nasopharynx, several questions should be answered. Among the most important are (i) is antibiotic use by an individual associated with an increased risk that that individual will carry a resistant strain of pneumococcus? and (ii) how is the overall level of use of antibiotics in a community associated with the prevalence of resistance in pneumococci in that community?

The first question can most simply be answered by assessing whether carriage of resistant strains is higher among individuals who have recently received antibiotics than among those who have not. Such comparisons are often reported as odds ratios, and we term the odds ratio associated with this question the "simple" odds ratio, OR_S. If a study finds that OR_S is >1, this implies that the treated individuals are themselves at higher risk of resistant carriage. However, if treatment kills susceptible strains but does not promote acquisition or outgrowth of resistant strains in the individual, the OR_S will remain less than or equal to one. Many studies of the association between antibiotic use and pneumococcal resistance have asked a slightly different question: among those

individuals who carry pneumococci, are those who have been treated more likely to have a resistant strain than those who have not been treated? Such studies calculate what can be called a "conditional" odds ratio (OR_C), so called because it is calculated conditional on having some pneumococcal strain. A treatment that kills susceptible strains but does not increase the treated individual's risk of acquiring a resistant strain will result in an elevated OR_C (though it does not result in an elevated OR_S). Put another way, a finding that the OR_C is >1 indicates that treatment eliminates susceptible strains but may not mean that treated individuals are at higher risk of having a resistant strain.

The second question concerns the association between use and resistance at the community level: do communities with higher levels of antimicrobial use have higher levels (or faster-increasing levels) of resistance? Any antibiotic that inhibits susceptible strains more than it inhibits resistant (or less susceptible) strains will exert natural selection in favor of the strains against which it is less active. The degree of selection exerted depends on the relative activities against the resistant and susceptible strains. The OR_C measured in individual-level studies summarizes the relative effect of treatment on resistant versus susceptible strains and is therefore a good summary, measured at the individual level, of the selective effect that treatment can have at the community-wide level. Since the OR_C can be greater than one even when the antibiotic has activity against resistant strains, it follows that even a drug that reduces an individual's risk of resistance (measured by OR_S) will promote resistance as long as it is more active against susceptible than against resistant strains (70).

More directly, the association at the community level between antibiotic use and resistance may be measured by comparing these variables for whole communities. Population genetic theory suggests that the rate of increase in antibiotic resistance should be directly related to the rate of antibiotic use (71); however, as suggested above, a detailed understanding of pharmacology suggests that some drugs should promote resistance more effectively than others.

Efforts to control antimicrobial use have been undertaken in hopes of slowing, or even reversing, the increase in antibiotic resistance in pneumococci and other organisms. An important factor in determining the size and timescale of effects from such efforts is the extent (if any) to which drug-resistant strains are handicapped in competition with drug-susceptible strains when they colonize hosts who are not taking the antibiotic. In particular, the possibility that reductions in antibiotic use might result in an absolute decline in antibiotic resistance (rather than simply a slowing of the increase in resistance) depends on the existence of such a "fitness cost"; if there is no disadvantage to being resistant, then there would be no selective pressure to drive down the frequency of resistant strains, even if antimicrobial use were somehow reduced to zero (71). In other pathogens, although such fitness costs often exist, they can rapidly be reduced or eliminated by compensatory evolution, in which mutations appear that counteract the deleterious effects of resistance (1, 71). To date, few studies have addressed this issue for clinically important resistance mechanisms in pneumococci (47, 100), though it is an area of active research. Perhaps more important in the case of pneumococci is the observation that a number of multiply resistant clones have spread globally, suggesting that if such fitness costs exist, they may be offset in particular clones by association with other genetic characteristics that are fitness enhancing. Taken together, these considerations suggest that while reducing use may slow the increase in resistance, it may be difficult to reduce the absolute level of resistance simply by controlling antibiotic use (71).

Most pneumococcal resistance to most clinically important classes of antibiotics is high-level resistance, meaning that the MIC for resistant strains is well above clinically achievable concentrations. However, resistance to penicillins is a major exception; a pneumococcal strain for which the penicillin MIC is ≥ 0.125 µg/ml is considered nonsusceptible to penicillin, and a strain for which the MIC is 1 µg/ml is considered fully resistant, while many beta-lactams can reach concentrations considerably higher than these cutoffs. As a result, it is possible for beta-lactams to inhibit strains that are defined as resistant.

The selective effects of antibiotics on pneumococci are complicated by the fact that most strains that are resistant to one clinically important drug are also resistant to others (41, 81, 100). When such associations exist, use of one drug may promote not only resistance to that drug but also resistance to other, associated classes of drugs. Such effects do not require biochemical similarity or physical genetic linkage of the resistance mechanisms; rather, resistance determinants to the two drugs tend to occur together in the pneumococcal population (linkage disequilibrium).

Evidence for the Role of Antibiotic Drugs in the Ecology of *S. pneumoniae*

Combining these population biological and pharmacokinetic and pharmacodynamic considerations, one can predict both how antimicrobial treatment will affect individual risk of carrying both resistant and susceptible organisms and how community-wide levels of antimicrobial use should affect population prevalence resistance. For antimicrobial classes such as macrolides, trimethoprim-sulfamethoxazole, and most oral cephalosporins, one expects treatment to reduce carriage of susceptible strains and to leave resistant carriage unaffected, or perhaps increase it if resistant subpopulations are able to outgrow when susceptible strains are killed; in terms of the "Pharmacokinetic and pharmacodynamic principles to predict bacterial eradication in the nasopharynx," above, such drugs are expected to have OR_Ss greater than or equal to one (increased risk of individual carriage of resistant organisms) and OR_Cs greater than one (greater effect on susceptible than resistant strains). For those penicillins that are active even against "nonsusceptible" strains, we expect carriage of susceptible strains to be reduced, but also (in the short term) that carriage of nonsusceptible/resistant strains will be reduced in treated individuals, albeit less than carriage of susceptible strains is reduced. Thus, we expect an OR_S less than one but an OR_C greater than one. At the community level, we expect that all of these drugs will select for increases over time in the prevalence of resistance to themselves and to other drug classes for which resistance is associated. A more detailed rationale for these predictions is given in references 70, 71, and 81. Here we use these predictions as a framework to evaluate the empirical evidence bearing on these questions.

EVIDENCE FOR THE EFFECT OF ANTIBIOTICS ON S. PNEUMONIAE CARRIAGE DURING TREATMENT AND DURING THE SHORT TO INTERMEDIATE TERM FOLLOW-UP (WEEKS TO A FEW MONTHS)

Tables 1 and 2 summarize the results of studies that have assessed the relationship between antimicrobial use and carriage of antimicrobial-resistant pneumococci. In Table 1, we list studies that were observational and in most cases retrospective: individuals were swabbed to assess pneumococcal carriage, and the results were tested for association with a history of antibiotic use (2, 3, 12, 17, 40, 51, 63, 75, 85, 96, 98, 99, 101, 107–109, 114). Table 2 lists prospective intervention studies in which carriage of pneumococci was assessed before and after treatment with an antimicrobial agent (21, 22, 25, 29, 40, 45, 46, 53, 86, 103; R. Dagan, D. Greenberg, A. Leiberman, N. Peled, and E. Leibovitz, 43rd Intersci. Conf. Antimicrob. Agents Chemother., abstr. G-1856, 2003; R. Dagan, E. Leibovitz, L. Piglansky, and P. Yagupsky, 39th Intersci. Conf. Antimicrob. Agents Chemother., abstr. 1028, 1999). An advantage of the intervention studies is that they remove the influence of many confounding variables whose effects may be present in the observational studies, especially, but not only, when the observational studies are analyzed for univariate associations. An advantage of the observational studies is that they typically follow the association between use and resistance for several months, as opposed to a time frame of weeks in most intervention studies.

Effects of Treatment on Individual Risk of Carrying Resistant Pneumococci (OR$_S$).

The studies summarized in Tables 1 and 2 show several consistent themes. For those studies that consider the effects of macrolides, trimethoprim-sulfamethoxazole, and in some cases oral cephalosporins, an individual is usually more likely to carry a resistant strain of pneumococcus after treatment than without treatment ($OR_S > 1$). For amoxicillin, with or without clavulanate, the prospective trials largely show a reduction in carriage of resistant strains, as predicted above ($OR_S < 1$), especially with short treatments and high doses. These findings are consistent with the expectations described above. However, several retrospective studies (51, 84) and one arm of a prospective trial (103) show that amoxicillin or unspecified beta-lactams are associated with an increased risk of carrying a resistant strain ($OR_S > 1$). There are multiple factors that may explain these differences: increased risk tended to be found when treatment was for long periods or with low doses (51), in the setting of an outbreak of a resistant strain (84), and in studies assessing relatively longer periods between use and resistance (84). The possibility that confounding played a role—since children who have many infections with resistant strains may be likely to receive more antibiotics—cannot be ruled out in the case of the observational studies.

Effects of Treatment on the Proportion of Resistant Pneumococci Carried.

As noted above, antimicrobial treatment can promote resistance in the community even in the cases (such as high-dose amoxicillin, described above) in which it reduces an individual's risk for carrying the resistant strain. The condition for it to do so is that it must reduce carriage of susceptible strains more than it reduces carriage of resistant strains; the measure of this effect is the OR_C. In many intervention

TABLE 1 Observation studies of the association between antibiotic used and carriage of antibiotic-nonsusceptible *S. pneumoniae*[a]

No.	Ref	Pub year	Geographic location	Study type	No. of patients	Time[b]	Risk factor	OR[S]	OR[C]
1	96	1981	Denver, Colo.	DCC, P, RO	62	2 mo	A	+[c]U	NA
2	101	1984	South Africa	Hosp, P	81	1 mo	β	NA	+U
3	99	1992	Cleveland, Ohio	DCC, P, RO	192	3 mo	≥3A	+[c]U	NA
4	2	1996	Iceland	DCC, P	448	12 mo	B ≥ 1	NA	+[c]M
							TS ≥ 1	NA	+[c]M
							ERY ≥ 1	NA	+[c]M
5	3	1996	Memphis, Tenn.	Ped pract, P	100	1 mo	β	NA	+[c]U
						2 mo	β	NA	+U
						3 mo	β	NA	+U
6	12	1996	Nebraska	DCC, P	12	2 mo	A	NA	+U
7	17	1996	United States	Prophylaxis	20	6 mo	β	+[c,d]U	NA
8	109	1996	Yukon-Kuskokwim delta	P, carriage + clinical isolates	95	3 mo	A	NA	=[c]U
9	51	1998	France	School, P	941	1 mo	Low β[f]	+[c]U	NA
							Low β[f]	+[c]U, M	NA
10	75	1998	Kentucky	DCC, P	104	2 wk	A	NA	=U
11	85	1998	Sweden	DCC, P, RO	1,029	6 mo	TS	+[c]U, M	NA
							AMP/AMOX	+U	NA
12	114	1998	Israel	DCC, LF, P	48	1 mo	A	+[c]M	NA
13	63	1999	Toronto, Canada	DCC, P	586	CAU	A	NA	+[c]U, M
						1 mo	A	N	+[c]U, M
14	108	1999	Greece	DCC, P	315	1 mo	A	+[c]U	+[c]U
						3 mo	A	+U	+U
15	40	2000	Malawi	Rural children	251	1 mo	TS	+[c]U	NA
							SP	+[e,g]U	NA
16	107	2000	Greece	Hosp, ped pract, P	2,448	3 mo	MAC	+[e,h]U	NA
17	98	2003	Israel	Ped pract, P	404	3 mo	A	+[c]M	+[c]M

[a] A, antibiotics; AMOX, amoxicillin; AMP, ampicillin; B, before-after treatment; β, β-lactams; CAU, current antibiotic use (at time of sampling); DCC, day care center(s); ERY, erythromycin; Hosp, hospital; LF, longitudinal follow-up; M, multivariate analysis; MAC, macrolides; NA, not available or not applicable; P, prevalence study; ped pract, pediatric practices; RO, study during outbreak of a resistant strain; SP, sulfadoxine-pyrimethamine; TS, trimethoprim-sulfamethoxazole; U, univariate analysis (for observational studies); +, increased rate; =, no change in rate.

[b] Time prior to culture during which antibiotic risk factor is measured.

[c] Statistically significant at a P of <0.05; otherwise, not statistically significant.

[d] Penicillin nonsusceptible.

[e] No numbers given, but paper states association was not significant.

[f] Significant associations found only with low or long doses.

[g] TS-resistant S. pneumoniae.

[h] Macrolide-resistant S. pneumoniae.

TABLE 2 Prospective studies of the effect of antibiotics on carriage of antibiotic-resistant *S. pneumoniae* comparing carriage before treatment to carriage during and immediately after treatment[a]

No.	Pub year	Ref	Geographic location	Study type	Risk factor	Pretreatment			During treatment			Immediately after treatment				Posttreatment			
						No.	% R_I	% R_{SP}	No.	OR_S	OR_C	Time after completion of Rx	No.	OR_S	OR_C	Time after completion of Rx	No.	OR_S	OR_C
1	1998	29	Israel	RTT	CU	42	33[b]	50[b]	42	(−)	(+)[b]	NA[b]	NA	NA	NA	NA	NA	NA	NA
					Cef	58	29[b]	41[b]	58	(−)	(=)[b]	NA[b]	NA	NA	NA	NA	NA	NA	NA
2	1999	22	France	RTT	CT	247	32[b]	55[b]	NA	NA	NA	2–4 d	230	(−)	(+)	NA	NA	NA	NA
					AC (HD)	250	32[b]	53[b]	NA	NA	NA	2–4 d	235	(−)[c]	(+)[d]	NA	NA	NA	NA
3	1998	25	France	RTT	CX	214	20[b]	39[b]	NA	NA	NA	1–5 d	214	(+)	(+)[d]	20–30 d	214	(=)	(=)
					AC (HD)	212	20[b]	39[b]	NA	NA	NA	1–5 d	212	(−)[c]	(+)[c]	20–30 d	212	(−)	(−)
4	1997	21	France	RTT	AC (HD)	185	24[b]	43[b]	NA	NA	NA	2–6 d	185	(−)[c]	(+)[c]	NA	NA	NA	NA
					CP	179	22	40	NA	NA	NA	2–6 d	179	(−)[c]	(+)[c]	NA	NA	NA	NA
5	2000	86	Texas	NRTT	AZI	146	2	4	NA	NA	NA	12 d	144	(+)	(+)[c]	4.4 mo	137	(+)[c]	(+)[c]
6	2000	40	Malawi	NRTT	TS	284	42	NA	NA	NA	NA	2 d	250	(+)[c]	NA	3.5 wk	163	(=)	NA
					SP	365	38	NA	NA	NA	NA	6 d	302	(=)	NA	4 wk	202	(+)[c]	NA
7	2000	45	Texas	RTT[f]	AC	64	13[c]	28[c]	NA	NA	NA	4 d	64	(−)	(−)[c]	2 mo	64	(−)	(+)
					AC	65	16[f]	34[f]	NA	NA	NA	4 d	64	(−)	(+)[f]	2 mo	64	(−)	(+)
					AZI	68	21	40	NA	NA	NA	10 d	68	(=)	(+)[c]	2 mo	68	(−)[c]	(+)[c]
8	2002	46	Texas	RTT[i]	AC (HD)	55	15[c]	29[c]	NA	NA	NA	4 d	55	(−)[c,e]	(−)[c]	2 mo	55	(−)[c,e]	(−)[c]
					AC (HD)	55	20[f]	38[f]	NA	NA	NA	4 d	55	(−)[c,f]	(+)[f]	2 mo	55	(−)[c,f]	(+)
					AZI	60	23	52	NA	NA	NA	10 d	60	(−)	(+)	2 mo	60	(−)	(+)

9	2001	103	Dominican Republic	RTT	AMOX	392	26[b]	36[b]	365	(−)[c]	(+)[c]	NA	NA	NA	NA	18 d	346	(+)	(+)[c]
				RTT	AMOX (HD) (short course)	391	27[b]	36[b]	375	(−)[c]	(+)[c]	NA	NA	NA	NA	23 d	355	(−)	(+)[c]
10	2003	Dagan[j]	Israel	RTT	AZI	70	19[g]	27[g]	65	(+)[g]	(+)[g]	2–6 d	48	(+)[g]	(+)[c,g]	11–20 d	45	(+)[c,g]	(+)[c,g]
					AC (HD)	73	45[b]	62[b]	65	(−)[b,c]	(+)[b,c]	2–6 d	52	(−)[b,c]	(+)[b]	11–20 d	53	(−)[b]	(+)[b]
11	1999	Dagan[k]	Israel	NRTT	TS	71	34[b]	45[b]	57	(+)[b]	(+)[b]	2–4 d	47	(+)[b]	(+)[b]	11–30 d	37	(+)[b]	(+)[b]
12	2002	53	Israel	NRTT	CT (1 day)	83	59	91[b]	83	(−)[b]	(+)[b]	1–3 d	69	(=)[b]	(=)[b]	18–30 d	45	(=)[b]	(−)[b]
				NRTT	CT (3 days)	87	56	89[b]	87	(−)[b,c]	(+)[b]	1–3 d	73	(=)[b]	(=)[b]	16–28 d	50	(=)[b]	(−)[b]

[a] AC, amoxicillin–clavulanate; AC (HD), amoxicillin/clavulanate at ≥90 mg/kg/day; AMOX, amoxicillin; AMOX (HD), amoxicillin at ≥80 mg/kg/day; AZI, azithromycin; Cef, cefaclor; CP, cefpodoxime–proxetil; CT, ceftriaxone; CU, cefuroxime; CX, cefixime; d, days; NA, not available or not applicable; NRTT, nonrandomized clinical trial; % R_I, percentage of individuals carrying resistant S. pneumoniae; % R_{sp}, percentage of all S. pneumoniae isolates that were resistant; RTT, randomized treatment trial; Rx, treatment; SP, sulfadoxine–pyrimethamine; TS, trimethoprim–sulfamethoxazole; (+) increased rate; (−), decreased rate; (=), no change in rate.

[b] Penicillin nonsusceptible.

[c] Statistically significant at a P of <0.05; otherwise, not statistically significant.

[d] Before calculations based on colonized individuals only in both groups.

[e] Penicillin-intermediately susceptible S. pneumoniae.

[f] Penicillin-fully resistant S. pneumoniae.

[g] Macrolide-resistant S. pneumoniae.

[h] TS-resistant S. pneumoniae.

[i] Children who received additional antibiotic treatment between end of treatment and end of 2 months of follow-up were not excluded.

[j] Dagan et al., 43rd ICAAC, abstr. G-1856.

[k] Dagan et al., 39th ICAAC, abstr. 1028.

studies, OR_C could be calculated during treatment, and in nearly all studies for which OR_C during treatment could be calculated, it was greater than 1.0, although not always statistically significant (29, 103; Dagan et al., 43rd ICAAC, abstr. G-1856; Dagan et al., 39th ICAAC). The one exception was in the study using cefaclor, which has poor activity against pneumococci (29). The OR_C was still >1.0 in most of the intervention studies during the immediate posttreatment period (1 to 10 days posttreatment) (21, 22, 25, 29, 40, 45, 46, 53, 86, 103; Dagan et al., 43rd ICAAC, abstr. G-1856; Dagan et al., 39th ICAAC). This was less obvious when a period of >10 days posttreatment was considered, but a tendency toward an effect lasting for up to several months was seen (25, 40, 45, 46, 53, 86, 103; Dagan et al., 43rd ICAAC, abstr. G-1856; Dagan et al., 39th ICAAC). This finding is consistent with the observational studies, in which the OR_C was elevated in all but two cases, again only sometimes with statistical significance. Overall, there is compelling evidence that regardless of antimicrobial classes, treated individuals carrying pneumococci are more likely to have resistant strains than untreated individuals carrying pneumococci. This is consistent with our expectations for individual effects and lends strength to the prediction (evaluated in "Evidence for ecological effect at the level of society" below) that ecological associations should be found between antimicrobial use and resistance at a population level.

The Effect of a Drug on Carriage of *S. pneumoniae* Resistant to Another Antibiotic Class in the Era of MDR.

Strains resistant to only one drug class are gradually being replaced in many regions by strains that are dually resistant or multidrug resistant (MDR) (resistant to ≥ 3 antibiotic classes) (81). Consequently, use of one class of antibiotics may promote carriage of *S. pneumoniae* resistant to one or more other classes. One of the first reports dealing with such cross-selection was from Iceland, where multiresistance (essentially to penicillin, macrolides, and TS) was

highly prevalent (2). In this study, any antibiotic treatment in the last 12 months was highly associated with nasopharyngeal carriage of penicillin-nonsusceptible *S. pneumoniae*. While previous treatment with one or two courses of beta-lactam drugs increased the odds of carrying penicillin-nonsusceptible *S. pneumoniae* by 6.8 (95% confidence interval, 1.8 to 25.5), the respective figures for TS and erythromycin use were 7.2 (95% confidence interval, 1.7 to 30.1) and 8.6 (95% confidence interval, 1.1 to 64.0). If the child received ≥ 3 courses of a beta-lactam drug in the last 12 months, the risk was 6.0 (95% confidence interval, 1.4 to 24.8) compared to that for those not receiving antibiotics, while the risk of carrying penicillin-nonsusceptible *S. pneumoniae* was 13.1 (95% confidence interval, 3.2 to 54.9) and 12.2 (95% confidence interval, 2.0 to 75.4) for 3 or more courses of TS and erythromycin, respectively. Similarly, during an outbreak of penicillin-resistant *S. pneumoniae* in day care centers in Sweden, TS treatment during the last 6 months was the strongest risk factor for carrying penicillin-resistant *S. pneumoniae* (85).

Several prospective studies looking at the effect of treatment with one class on carriage of *S. pneumoniae* resistant to another class were conducted recently in southern Israel, where dual resistance and multiresistance are highly prevalent among *S. pneumoniae* strains and are still increasing in frequency (26). The first study, conducted in 1998, investigated the effect of TS on nasopharyngeal carriage of *S. pneumoniae* in relation to both TS resistance and penicillin nonsusceptibility (Dagan et al., 39th ICAAC). In this study, 71 children were treated for acute otitis media with TS for 10 days. Pneumococcal nasopharyngeal cultures were obtained on day 1 (pretreatment) and on days 4 to 6, 12 to 14, and 21 to 40 (Fig. 2). Carriage of TS-resistant *S. pneumoniae* was observed in 34% of the children before initiation of treatment but increased to 51% on days 4 to 6 (during treatment). This proportion persisted until days 21 to 40. At the same time, the na-

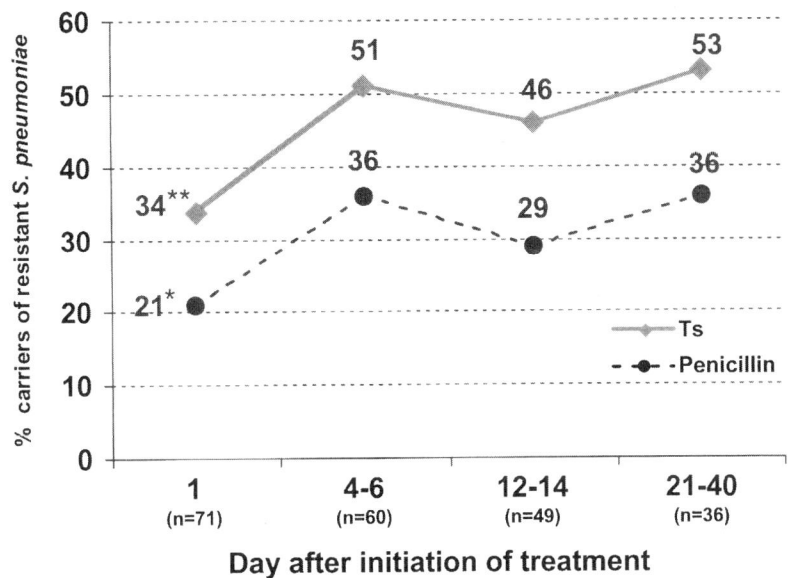

FIGURE 2 Nasopharyngeal carriage of TS-resistant and penicillin-nonsusceptible *S. pneumoniae* in children with acute otitis media treated with TS for 10 days (Dagan et al., 39th ICAAC, abstr. 1028). *, $P = 0.038$ day 1 versus all other days combined. **, $P = 0.1$ day 1 versus all other days combined.

sopharyngeal carriage of penicillin-nonsusceptible *S. pneumoniae* increased from 21 to 36%. This proportion persisted to days 21 to 40. The promotion of penicillin-nonsusceptible *S. pneumoniae* carriage by TS treatment can be explained by the close association between Ts and penicillin nonsusceptibility: for 92 and 67% of all TS-resistant *S. pneumoniae* isolates, the penicillin MICs were ≥0.1 and ≥1.0 μg/ml, respectively.

A study with a similar design was conducted in the years 2001 to 2002 in children with acute otitis media treated for 5 days with azithromycin (Dagan et al., 43rd ICAAC, abstr. G-1856). In this study, the proportion of children who carried macrolide-resistant *S. pneumoniae* doubled within 4 to 6 days from initiation of treatment. In parallel, the proportion of MDR *S. pneumoniae* carriage also doubled (Fig. 3). This effect persisted at least until 2 to 3 weeks posttreatment. The exact duration of this effect was not studied, but at an additional visit on days 70 to 120, carriage rates of both macrolide-resistant and MDR *S. pneumoniae* were again similar to the pretreatment samples.

The antibiotic resistance patterns of middle ear fluid isolates in 2001 and 2002 in the community in southern Israel showed macrolide resistance in 24 and 25% of pneumococci, respectively, and MDR proportions of 23 and 25%, respectively (unpublished data). In contrast, in 1998 the macrolide and MDR proportions for similar specimens were only 13% each. Since a similar study with identical entry criteria and design (including the azithromycin regimen) was conducted in 1998, the effect of azithromycin on the carriage rate of macrolide-resistant and MDR *S. pneumoniae* could be compared between the two periods (R. Dagan, D. Greenberg, A. Leiberman, N. Peled, and E. Leibovitz, 43rd Intersci. Conf. Antimicrob. Agents Chemother., abstr. G-898, 2003) (Fig. 4). The results suggest that the ability of a drug such as azithromycin to promote carriage of macrolide resistance and MDR depends on prevalence of resistance in the community. The effect of azithromycin on the promotion of carriage of resistant *S. pneumoniae* was much less apparent in 1998 than in 2001 to 2002, when the rate of both macrolide resistance and MDR was twofold higher than that in 1998.

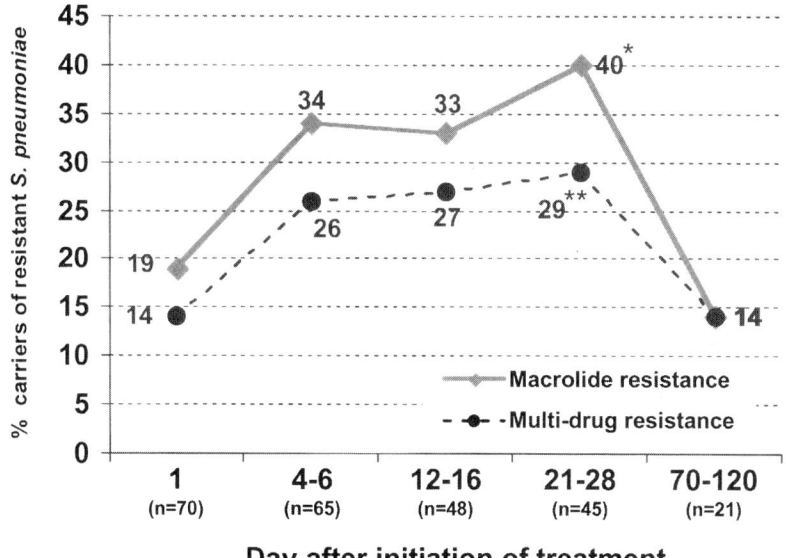

FIGURE 3 Nasopharyngeal carriage of macrolide-resistant or MDR *S. pneumoniae* in children with acute otitis media treated with azithromycin for 5 days (Dagan et al., 43rd ICAAC, abstr. G-1856). *, *P* < 0.05 compared to day 1. **, *P* = 0.095 compared to day 1.

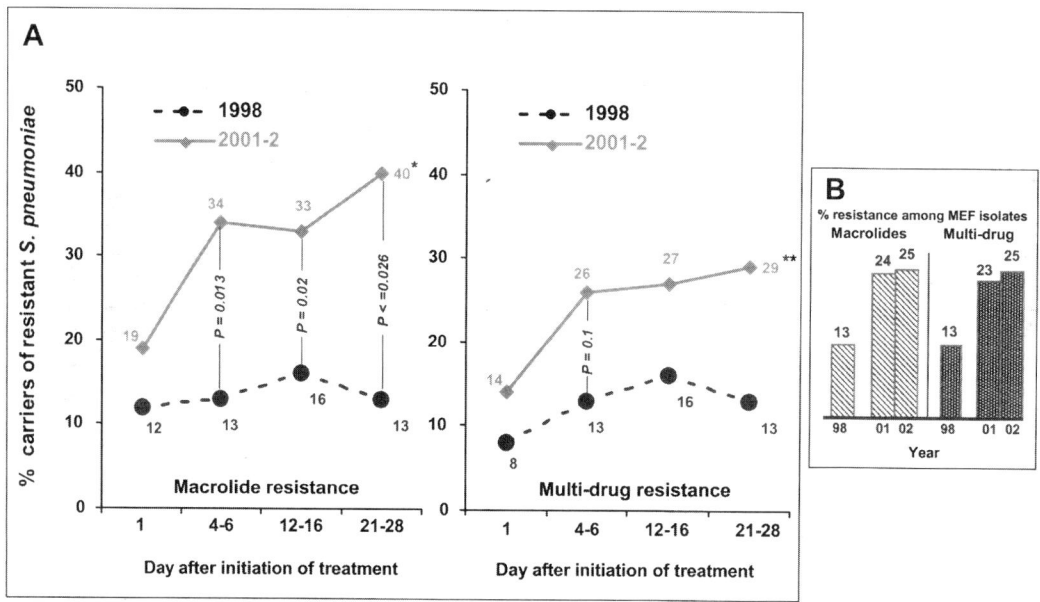

FIGURE 4 Nasopharyngeal carriage of macrolide-resistant and MDR *S. pneumoniae* in children with acute otitis media treated with azithromycin for 5 days in two periods: 1998 and 2001 to 2002 (Dagan et al., 43rd ICAAC, abstr. G-898). (A) Comparison of the effect of azithromycin on carriage of macrolide-resistant and MDR *S. pneumoniae* between 1998 and 2001 to 2002; (B) relation of macrolide resistance and MDR in *S. pneumoniae* isolated from middle ear fluid (MEF) in southern Israel in 1998 (*n* = 563), 2001 (*n* = 672), and 2002 (*n* = 639) (unpublished data).

These results demonstrate that the absence of an effect or a minimum effect of a drug on carriage of nonsusceptible *S. pneumoniae* in the absence of widespread resistance can be misleading. Even if no increase in resistance is observed in individuals who receive a drug when community-wide prevalence of resistance is low (8), there are still good reasons to expect that such an effect may appear once the community-wide prevalence increases.

In contrast to the effect of azithromycin on macrolide-resistant and MDR *S. pneumoniae* carriage, treatment with high-dose amoxicillin reduces the risk of carrying penicillin-nonsusceptible and MDR *S. pneumoniae*. This was clearly exemplified in a study conducted in the years 2001 to 2002 in southern Israel, where approximately 24% of all *S. pneumoniae* isolates

were macrolide resistant and MDR was also observed in the same order of magnitude (Dagan et al., 43rd ICAAC, abstr. G-1856). In this study, children aged 6 to 30 months with acute otitis media were randomized to receive either high-dose oral amoxicillin-clavulanate (amoxicillin dose of 90 mg/kg of body weight/day) for 10 days or oral azithromycin (10 mg/kg on the first day followed by 5 mg/kg daily for an additional 4 days). Nasopharyngeal cultures were obtained before treatment, 3 to 5 days after initiation of treatment, immediately posttreatment, 2 to 3 weeks after completion of treatment, and 2.5 to 4.0 months after completion of treatment (Fig. 5).

The graphs presented in Fig. 5 demonstrate the following. (i) Both drugs reduced profoundly the carriage of both penicillin-suscep-

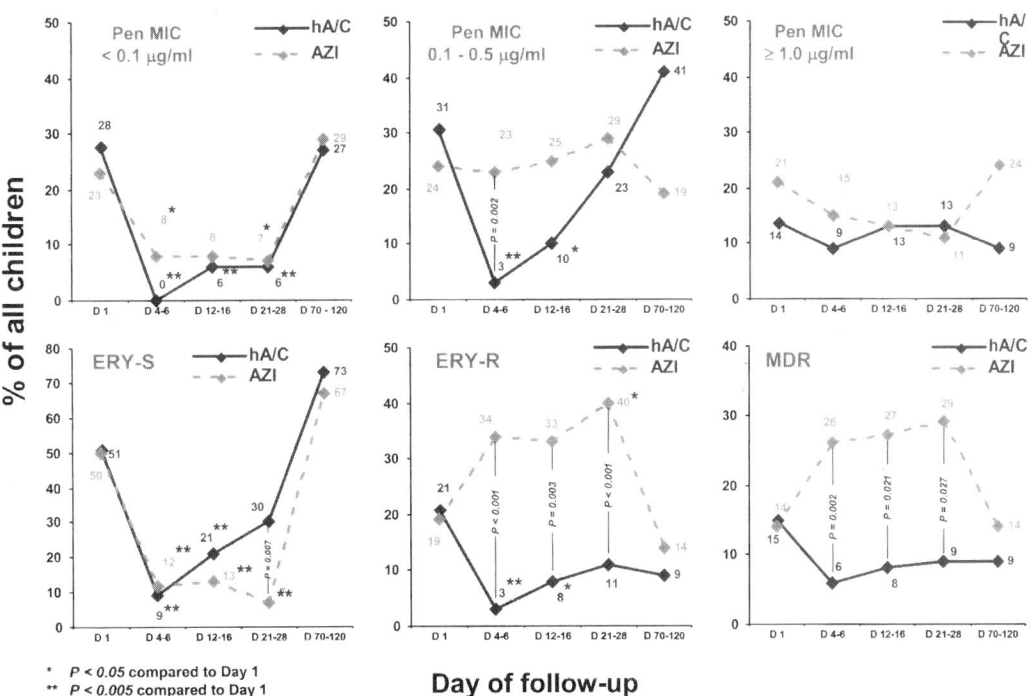

FIGURE 5 *S. pneumoniae* nasopharyngeal carriage in children treated with high-dose amoxicillin-clavulanate for 10 days versus 5 days of azithromycin for acute otitis media, by susceptibility to penicillin, susceptibility to erythromycin, and MDR. Pen, penicillin; ERY, erythromycin; hA/C, high-dose amoxicillin-clavulanate; AZI, azithromycin; D, day; S, susceptible; R, resistant.

tible and macrolide-susceptible *S. pneumoniae*. (ii) Full recolonization of susceptible *S. pneumoniae* was not observed within the first month of follow-up, but recolonization of macrolide-susceptible *S. pneumoniae* was significantly slower after treatment with azithromycin than with high-dose amoxicillin-clavulanate. (iii) In contrast to that of penicillin-susceptible *S. pneumoniae*, carriage of penicillin-nonsusceptible *S. pneumoniae* was neither reduced nor increased when treated with azithromycin; when treated with high-dose amoxicillin-clavulanate, penicillin-nonsusceptible *S. pneumoniae* for which the MIC was <1.0 μg/ml was markedly reduced during treatment, with a gradual recolonization after completion of the treatment. Little effect on carriage of *S. pneumoniae* for which the penicillin MIC was ≥1.0 μg/ml was observed with either of the drugs. (iv) Both carriage of macrolide-resistant *S. pneumoniae* (measured by erythromycin resistance) and carriage of MDR *S. pneumoniae* were *reduced* by high-dose amoxicillin-clavulanate, while a dramatic *increase* in carriage of both macrolide-resistant and MDR *S. pneumoniae* was seen in the group treated with azithromycin. This differential effect lasted for more than 1 month.

This study showed clearly that, in light of the dual resistance to penicillin and macrolides in regions where penicillin resistance, macrolide resistance, and MDR are prevalent in *S. pneumoniae*, high-dose amoxicillin (or amoxicillin-clavulanate) is still ecologically safer in terms of promotion of drug-resistant and MDR *S. pneumoniae* than azithromycin.

The examples presented above show clearly that (i) antibiotic treatment alters the ecology of *S. pneumoniae* in the individual's flora by promoting the carriage of antibiotic-resistant *S. pneumoniae* at the expense of antibiotic-susceptible *S. pneumoniae*; (ii) the effect often lasts beyond the treatment period; (iii) some drugs may have a more profound and more prolonged effect than others; and (iv) in the era of MDR, the effect of a given drug may extend beyond its own class and promote carriage of *S. pneumoniae* resistant to other classes.

EVIDENCE FOR ECOLOGICAL EFFECT AT THE LEVEL OF SOCIETY

The proportions of resistance to various classes of antimicrobial agents vary substantially not only between countries (6, 15, 113) but also among regions within countries (R. J. Davidson, C. C. K. Chan, G. Doern, and G. G. Zhanel, 13th Eur. Congr. Clin. Microbiol. Infect. Dis., abstr. P1031, 2003) (44, 81, 84) or even groups such as day care center attendees within one region (34, 48). Given the effect of antibiotics on the antibiotic-resistant *S. pneumoniae* carriage in the individual reviewed above, it is plausible that variation in the use of antibiotics among communities will be associated with the differences in prevalent resistance patterns in disease. Indeed, several studies have demonstrated a clear association between the use of specific antibiotics or antibiotic classes and respective antibiotic resistance in *S. pneumoniae* both in carriage and in disease. These studies either measured the association of increase in the consumption of specific antibiotics with time and respective class resistance or conducted cross-sectional ecological studies, testing the association between antibiotic consumption and the respective resistance in each community. These studies demonstrated a class association between consumption of beta-lactam (34, 36, 44, 84, 95, 102), macrolide (Davidson et al., 13th ECCMID; H. Goosens, M. Elseviers, M. Ferech, E. Hendrickx, R. Vanderstichele, A. Bryskier, and ESAC National Representatives, 43rd Intersci. Conf. Antimicrob. Agents Chemother., abstr. C2–67, 2003) (5, 44, 58, 93), TS (93), or fluoroquinolones (20) and antibiotic resistance to the respective classes among *S. pneumoniae* isolates in the community. Interestingly, the use of a compound active against malaria (sulfa-doxaline-pyrimethamine [Fansidar]) was associated with an increase in pneumococcal resistance to TS in the community assumingly due to cross-resistance (40).

As reviewed above, the predicted effect of the consumption of antibiotics on *S. pneumoniae* antibiotic resistance is not expected to be equal. Given the higher potential of the oral cephalosporins, TS, and macrolides in promoting carriage of drug-resistant and MDR *S. pneumoniae*, it is expected that the effect of these drugs will be significantly more marked than that of the penicillins. Furthermore, because of the extremely long half-life of azithromycin, the drug is expected to have a driving role in drug resistance and MDR among *S. pneumoniae* strains in the community. Indeed, these differential effects could be observed in various studies. Several studies observed that macrolide consumption was an important drive for increasing penicillin resistance in a given community (15, 44).

Oral cephalosporin use was also found to be a more important promoter of penicillin resistance among *S. pneumoniae* strains in the society than penicillin (44, 102). It is therefore not surprising to observe that in various communities, the proportion of *S. pneumoniae* strains that are resistant to two or more drugs (most often beta-lactam–TS or beta-lactam–TS–macrolide) is increasing at the expense of single-drug resistance. McCormick et al. analyzed temporal trends in the proportion of singly and dually resistant organisms and found that pneumococcal strains resistant to both penicillin and erythromycin were increasing faster than strains singly resistant to either in the United States during the years 1996 to 1999 (81). Using a mathematical transmission model that assumed a logistic trend in resistance as expected with a constant level of antimicrobial use, it was predicted that by July of 2004, in the absence of vaccine, 41% of pneumococcal isolates in the surveillance area used for the collection of their data in the various studies will be dually resistant, with 5% resistant to penicillin only and 5% resistant to erythromycin only. The predictions in this model assumed that antimicrobial use stayed constant; if current trends toward increased use of azithromycin and reduced use of oral penicillins (58, 80, 106) continue, the reasoning advanced here suggests that resistance will climb faster.

The pattern of increase in azithromycin consumption in the presence of a decrease in amoxicillin consumption was found in Israel during the years 1998 to 2002 (Color Plate 10a). This was seen in parallel with an increase in macrolide resistance and MDR rate in middle ear fluid isolates (Color Plate 10b). During this period, the proportion of penicillin-nonsusceptible *S. pneumoniae* strains that became dually resistant to penicillin and macrolides rose rapidly (Color Plate 10c).

To summarize the effect of antibiotics on *S. pneumoniae* ecology, it is clear that antibiotic treatment in any community profoundly affects *S. pneumoniae* ecology. With the incredible ability of *S. pneumoniae* to rapidly become nonsusceptible to a multitude of drugs and at the same time to keep its fitness, it is unlikely that a reversal of the trend of increasing antibiotic resistance and MDR will be fully reversed by antibiotic restriction as the sole measure. However, reducing antibiotic use and at the same time emphasizing the restriction of specific drugs with the most marked effect on ecology, while preferring drugs that are ecologically safer such as the penicillins, is likely to result in the slowing down of the process of increasing resistance among *S. pneumoniae* in the community.

THE EFFECT OF PNEUMOCOCCAL CONJUGATE VACCINES ON *S. PNEUMONIAE* ECOLOGY

Mechanisms of the Vaccines' Effect on Carriage

The control of nasopharyngeal carriage is important to managing pneumococcal disease and essential to reducing person-to-person spread of *S. pneumoniae*. It is therefore important to understand the potential effect of the pneumococcal vaccine on pneumococcal carriage.

Immunization with conjugate pneumococcal vaccines reduces nasopharyngeal carriage of *S. pneumoniae* of serotypes included in the vaccine (VT), and occasionally those immunologically related to VT, in infants and toddlers (27, 28, 31, 32, 79, 90, 110, 115; R. Dagan, O. Zamir, N. Tirosh, I. Belmaker, and L. Guy, 40th Intersci. Conf. Antimicrob.

Agents Chemother., abstr. 47, 2000; K. M. Edwards, G. Wandling, P. Palmer, and M. D. Decker, 37th Annu. Meet. Infect. Dis. Soc. Am., abstr. 34, 1999; T. M. Kilpi, A. Palmu, M. Leinonen, H. Käyhty, and P. H. Mäkelä, 41st Intersci. Conf. Antimicrob. Agents Chemother., abstr. G-2036, 2001; K. Kristinsson, S. T. Sigurdardottir, T. Gudnason, S. Kjartansson, K. Davidsdottir, O. Leroy, and I. Jansdottir, 37th Intersci. Conf. Antimicrob. Agents Chemother., abstr. G-5, 1997; K. L. O'Brien, M. Bronsdon, and G. M. Carlone, Pediatr. Acad. Soc. Annu. Meet. Pediatr. Res., vol. 49, abstr. 1463, 2001).

Conjugate vaccines reduce the prevalence of carriage by preventing new acquisition and not by shortening the duration of already existing colonization. This is true for both *H. influenzae* type b (7) and *S. pneumoniae* (28). This phenomenon, although well established, is not completely understood, since the attachment of *S. pneumoniae* to mucosal surfaces of mammalian tissues is not mediated by any known mechanism involving the polysaccharide capsule (97, 104). One frequently cited speculation for the mechanism of this phenomenon is that, when in sufficient amount, the antibodies bind to the polysaccharide capsule and create a steric inhibition of the interaction of pneumococcal surface proteins with binding sites on the mammalian epithelial cell surface.

Previous studies have suggested that the determining factor for this vaccine effect is the presence in saliva of anticapsular immunoglobulin G (IgG) leaking from the serum rather than secretory anticapsular IgA (38). Furthermore, saliva anti-pneumococcal polysaccharide IgG concentrations are related to serum concentrations, which supports the idea that salivary anti-polysaccharide specific IgG derives from serum (62, 67, 89). Secretory IgA plays only a minor role in protection from colonization, possibly because *S. pneumoniae* has specific proteins that cleave secretory IgA but not IgG (64; H. Kayhty, S. Rapola, B. Simell, and T. Kilpi, 2nd Int. Symp. Pneumococci Pneumococcal Dis., abstr. 018, 2000).

The importance of systemic IgG in preventing colonization has also been demonstrated experimentally in infant rats, in which passively transferred, systemic anticapsular IgG reduced intralitter transmission of *S. pneumoniae* (77). Similar results were obtained by direct inoculation of pretreated animals versus controls. The few infected rats that became colonized despite pretreatment were less heavily colonized than controls, suggesting that the effect of circulating IgG on carriage is not an "all or none" phenomenon, but probably a continuous phenomenon.

If this speculation reflects the real chain of events, the inhibition of colonization should depend on the concentrations of capsule-specific IgG in serum. Indeed, in a study conducted with toddlers attending day care centers is southern Israel, using a 9-valent pneumococcal conjugate vaccine (conjugated to CRM_{197} [PnCRM9]), it was clearly shown that the rate of acquisition of new *S. pneumoniae* strains was inversely related to the concentration of serotype-specific antipolysaccharide serum IgG (R. Dagan, N. Givon-Lavi, M. Sikuler-Cohen, D. Zamir, D. Fraser, R. Kohberger, and G. Siber, 41st Intersci. Conf. Antimicrob. Agents Chemother., abstr. G-2040, 2001). Furthermore, in the study, prevention of acquisition of serotype 6A was inversely associated with the postvaccination concentration of serum anti-6B IgG (the PnCRM9 contained the polysaccharide of serotype 6B but not 6A). This is proof of the concept that the concentration of circulating serotype-specific IgG concentration not only determines acquisition (and thereby carriage) of the specific serotype against which the antibodies are directed but also can influence the acquisition and carriage of serotypes that are antigenically related (belonging to the same serogroup but not having the same serotype).

While existing conjugate vaccines provide some cross-reactive protection against carriage of pneumococci in the same serogroup as the antigens included in the vaccine, they provide no protection against other serogroups that are not in the vaccine. Therefore, insofar as pneumococcal strains compete with one another, the use of such a vaccine is expected to result in increased carriage of non-vaccine serotype (non-VT) pneumococci. This response is closely analogous to the way in which antimicrobial use inhibits susceptible strains and thereby provides an advantage to resistant strains.

The mechanisms and consequences of such increases in the carriage of non-VT pneumococci—a phenomenon often called "serotype replacement"—are complicated and not yet fully understood. Replacement may occur by any or all of the following mechanisms: (i) increases in carriage of existing strains that carry non-VT; (ii) appearance of new strains of non-VT that were previously absent or very rare; and (iii) acquisition of new, non-VT capsules by existing VT strains via transformation with DNA encoding biosynthesis genes for non-VT capsules (69). The consequences of replacement depend largely on the extent to which the non-VT strains that increase in frequency are virulent (capable of causing disease) (105). It is clear that the capsular serotype itself is a major determinant of virulence (18), but it is equally clear that several serotypes excluded from the current vaccines, and especially from the 7-valent vaccine used in the United States, are capable of causing disease under the right conditions (42, 55, 56). Serotypes whose present contribution to invasive disease is small may be capable of causing more disease in the absence of their competitors. On the other hand, if the replacing serotypes turn out to be largely avirulent, then the phenomenon of replacement may be beneficial, as the increased prevalence of non-VT pneumococci competitively inhibits VT strains. Other bacterial species, including *H. influenzae* and *Staphylococcus aureus*, are inhibited in vitro by pneumococci, most importantly via secreted hydrogen peroxide (83, 92). There is thus the possibility that these species will also increase in frequency in colonizing the human URT as pneumococcal vaccines become more widespread.

Evidence for the Role of Vaccines in the Carriage of *S. pneumoniae*

The nonconjugate pneumococcal polysaccharide vaccines do not have a significant effect on carriage of *S. pneumoniae* in children and adults (30, 74). In contrast, a marked effect of the conjugate vaccines on carriage was observed in many studies (27, 28, 30, 32, 79, 90, 110, 115; Dagan et al., 40th ICAAC; Edwards et al., 37th Annu. Meet. IDSA; Kilpi et al., 41st ICAAC; Kristinsson et al., 37th ICAAC; O'Brien et al., Pediatr. Acad. Soc. Annu. Meet. Pediatr. Res., 2001). Table 3 presents the studies conducted and published thus far to document the effect of pneumococcal vaccines on carriage of *S. pneumoniae* (27, 28, 30, 32, 79, 90, 110, 115; Dagan et al., 40th ICAAC; Edwards et al., 37th Annu. Meet. IDSA; Kilpi et al., 41st ICAAC; Kristinsson et al., 37th ICAAC; O'Brien et al., Pediatr. Acad. Soc. Annu. Meet. Pediatr. Res., 2001). Despite variations in the nature of the conjugate vaccines, populations, and the age at which vaccines were administered, a significant reduction in carriage of the serotypes included in the vaccine was clearly observed in all studies. However, in most of the studies, a "replacement" phenomenon occurred: an increase in the carriage of *S. pneumoniae* serotypes not included in the vaccine was observed in conjunction with a decrease in the carriage of serotypes included in the vaccine.

The effect of reduction of carriage of the serotypes included in the vaccine and the increase of the non-VT pneumococci after administration with conjugate pneumococcal vaccines may last for several years. In a study conducted in the Philippines (A. A. Palmu, T. Kaijalainen, J. Verho, E. Herva, P. H. Mäkelä and T. M. Kilpi, 3rd Int. Symp. Pneumococci Pneumococcal Dis., p. 24, 2002), Palmu et al. vaccinated 754 infants at ages 2, 4, 6, and 12 months with either a 7-valent CRM_{197} pneumococcal conjugate vaccine (PnCRM7) or a control vaccine. These children were tested at the age of 4.5 years for *S. pneumoniae* carriage. The rates of carriage of serotypes included in the vaccine were 8.5 and 13.6% in the vaccine

TABLE 3 Studies on the effect of conjugate pneumococcal vaccine on carriage of *S. pneumoniae* and antibiotic-resistant *S. pneumoniae*

Authors	Reference	Conjugate vaccine (valence)[a]	Site	Age (mo) at vaccination	Reduction in serotypes included in the vaccine	Reduction in resistant pneumococci	Increase in non-VT
Dagan et al.	30	PnOMPC7	Israel	12–18	Yes	Yes	No
Dagan et al.	32	PnT; PncD (4-valent)	Israel	2, 4, 6	Yes	Yes	±[b]
Obaro et al.	90	PnCRM5	The Gambia	2, 3, 4	Yes	ND[c]	Yes
Kristinsson et al.	37th ICAAC, abstr. G-5	PnT; PncD (8-valent)	Iceland	3, 4, 6	Yes	ND	Yes
Mbelle et al.	79	PnCRM9	South Africa	1.5, 2.5, 3.5	Yes	Yes	Yes
Edwards et al.	IDSA 1999, abstr. 34	PnCRM9	United States	2, 4, 6, 12	Yes	ND	Yes
Dagan et al.	40th ICAAC, abstr. 47	PnT/D (11-valent)	Israel	2, 4, 6, 12	Yes	Yes	No
Dagan et al.	27, 28	PnCRM9	Israel	12–35	Yes	Yes	Yes
O'Brien et al.	Pediatr. Res. 2001, abstr. 1463	PnCRM7	United States (Native American)	2, 4, 6, 12–15	Yes	ND	Yes
Kilpi et al.[d]	19th Annu. Meet. ESPID, p. 4, 2001	PnCRM7	Finland	2, 4, 6, 12	Yes	ND	ND
Veenhoven et al.	110	PnCRM7 + PS23	The Netherlands	12–72	Yes	ND	Yes
Yeh et al.	115	PnOMPC7	United States	2, 4, 6, 12	Yes[e]	ND	Yes[f]

[a] PnOMPC = 7-valent pneumococcal conjugate vaccine, conjugated to outer membrane complex of *N. meningitidis* B; PnT, pneumococcal conjugate vaccine, conjugated to tetanus toxoid; PnD, pneumococcal conjugate vaccine, conjugated to diphtheria toxoid; PnT/D, pneumococcal conjugate vaccine, conjugated to a mixture of tetanus and diphtheria toxoids; PnCRM, pneumococcal conjugate vaccine, conjugated to CRM₁₉₇ protein (PnCRM5, 5-valent; PnCRM7, 7-valent; PnCRM9, 9 valent); PS23, pneumococcal polysaccharide vaccine (23-valent).

[b] ±, not conclusive.

[c] ND, not done.

[d] T. M. Kilpi, R. Syrjanen, A. Palnu, A. Herva, J. Eskola, P. H. Makela, and Fin 5, OM Group.

[e] Small sample size permitted to observe reduction in only serotype 23F.

[f] Small sample size did not permit statistically significant difference.

and control groups, respectively (relative risk, 0.62; 95% confidence interval, 0.41 to 0.95). The respective carriage rate of the non-VT was 20.2 versus 16% (representing a relative risk of 1.25 for carrying a non-VT *S. pneumoniae* strain) among vaccinees. In another study conducted among Navajo and White Mountain Apache in the United States (E. V. Millar, K. L. O'Brien, J. P. Watt, J. Dallas, S. Katz, M. A. Bronsdon, J. Elliot, R. Reid, and M. Santosham, 3rd ISPPD, p. 54, 2002). A total of 238 infants and toddlers were randomized by community to receive either PnCRM7 or control vaccine (98% received at least three doses). The children were cultured at a mean age of 3.2 years and 2.6 years after last dose. The rates of carriage of vaccine serotypes in vaccinees and controls were 11.5 and 31.0% (P = 0.05), and those for the non-VT pneumococci were 88.5 and 68.9%, respectively (P = 0.09).

Although the replacement phenomenon is remarkable, its clinical significance is not fully elucidated. Two controlled, double-blind efficacy studies were conducted with the PnCRM7 vaccine (10, 91), and one controlled, double-blind efficacy study was conducted with the PnCRM9 vaccine (66). In all three studies the primary outcome was invasive disease. In all three studies, excellent efficacy in preventing disease caused by the serotypes included in the vaccine was demonstrated, but these studies did not suggest a significant increase in disease with the serotypes not included in the vaccine. Surveys in the United States after introduction of the vaccine suggested a slight increase in the incidence of invasive infections caused by the serotypes that were both not included in the vaccine and not cross-reactive with the VT, but this did not reach statistical significance. Furthermore, the extent was negligible compared to the decrease in morbidity seen with the VT and cross-reactive serotypes (112).

The situation is completely different when mucosal infections are considered. Studies on the efficacy of pneumococcal conjugate vaccines in the prevention of acute otitis media

exemplify the extent of the replacement phenomenon in mucosal infections. The pathogenesis of most cases of acute otitis media infections starts with a viral upper respiratory infection that, at the same time, facilitates colonization of bacteria on the pharyngeal mucosa and creates a negative pressure in the middle ear cavity. This may result in aspiration of pharyngeal content, including pathogens, to the middle ear cavity as the first line in acute otitis media (57). Thus, it is to be expected that if a replacement of VT *S. pneumoniae* by non-VT strains occurs in the nasopharynx, increased chance of aspiration of non-VT may result in increased rate of acute otitis media caused by non-VT *S. pneumoniae* after vaccination. This was indeed found in two prospective randomized double-blind studies conducted with PnCRM7 (39) and a 7-valent pneumococcal conjugate vaccine conjugated to outer membrane protein of *N. meningitidis* B (PnOMPC7) (65) (Fig. 6).

In both studies, the efficacy against VT *S. pneumoniae* was 56 to 57%. In one study (PnCRM7), the efficacy was >50% even for the VT-related serotypes. However, for the serotypes that were unrelated to the VT, there was a negative efficacy, with a 27 to 33% increased rate of acute otitis media episodes. Furthermore, an increase in acute otitis media caused by *S. pneumoniae* pathogens (*H. influenzae* and *M. catarrhalis*) was observed. The overall reduction in these two studies, being the balance of the excellent efficacy against VT *S. pneumoniae*, the moderate efficacy against VT-related *S. pneumoniae*, and the negative efficacy against all others, was very modest at best: for PnCRM7 it was 6% (95% confidence interval, −4 to 16%), and for PnOMPC7 it was −1% (95% confidence interval, −12 to 10%).

Postlicensure data are confined to the United States and are limited, but they suggest a strong effect of replacement on acute otitis media episodes. In rural Kentucky a significant reduction in the proportion of acute otitis media caused by VT *S. pneumoniae*, and a parallel increase in *H. influenzae* and *M. catarrhalis*, was observed in 2001 to 2002 com-

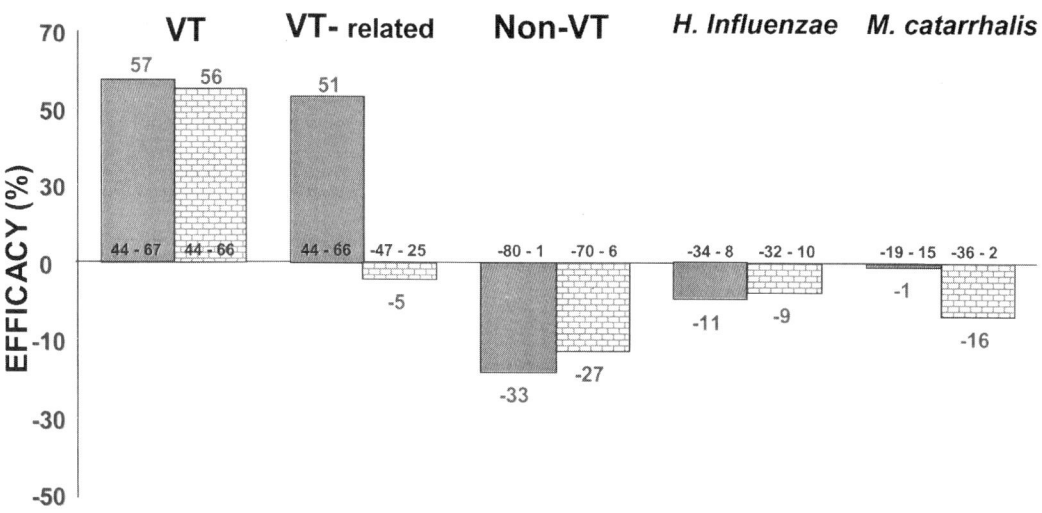

FIGURE 6 Effect of PnCRM7 (39) and PnOMPC7 (65) in prevention of acute otitis media caused by VT *S. pneumoniae*, VT-related *S. pneumoniae*, non-VT *S. pneumoniae*, *H. influenzae*, and *M. catarrhalis*. Positive values represent a reduction in episodes; negative values represent an increase in cases.

pared with the prevaccination era (S. Block, Pediatr. Acad. Soc. Annu. Meet., abstr. LB3). In a study on children who underwent tympanocentesis in Pittsburgh, Pa., an absolute decrease of cases caused by VT *S. pneumoniae* was seen, with a parallel increase in non-VT, compared with the prevaccination era (M. McEllistrem, J. M. Adams, E. O. Mason, and E. R. Wald, 43rd Intersci. Conf. Antimicrob. Agents Chemother., abstr. G-2047, 2003) (Fig. 7).

Thus, for acute otitis media, as a representative of a mucosal infection, the replacement phenomenon observed in the nasopharynx was clearly associated with a replacement disease. Data do not exist for the other respiratory infections. For pneumonia, the overall efficacy in reducing alveolar pneumonia was modest, around 20%, in children <3 years of age (11, 66). Since in these studies no bacteriological results are provided, the role of the replacement phenomenon in the overall efficacy of pneumonia cannot be assessed.

Herd Protection by Conjugate Pneumococcal Vaccines

The reduction in nasopharyngeal carriage of VT pneumococci is important because it will reduce the spread of these serotypes. Two prospective double-blind studies in different settings clearly have shown the existence of this phenomenon. In one double-blind comparative study conducted in southern Israel, toddlers attending day care centers were vaccinated with a PnCRM9 or control vaccine (49). In this study, attendees at day care centers and their younger siblings who stayed at home were monitored after vaccination of the day care attendees. A marked reduction in vaccine-type pneumococcal carriage was seen in the younger siblings of those who were vaccinated with the PnCRM9 vaccine compared with siblings of the control (Fig. 8). In parallel, a clear trend of increase in carriage of non-VT pneumococci was observed in the younger siblings of the vaccinated day care toddlers, compared to younger siblings of the nonvaccinated day care center attendees.

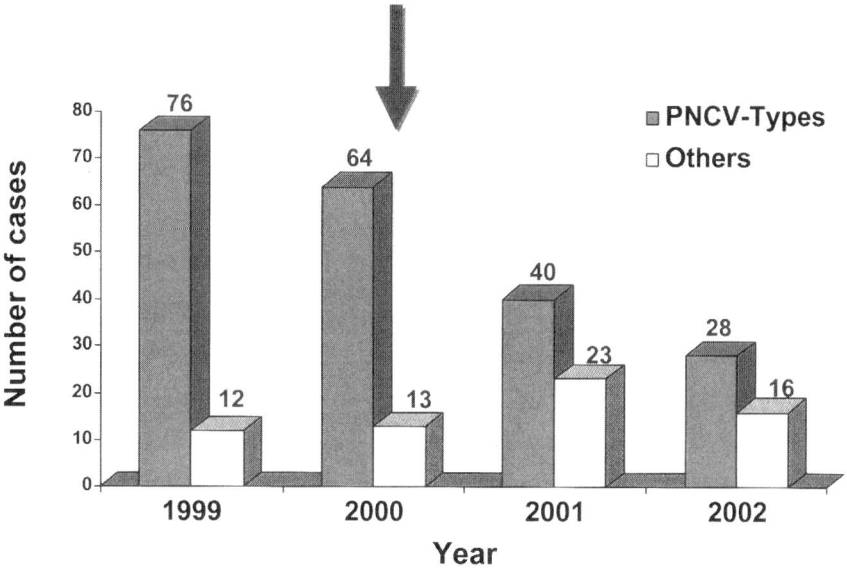

FIGURE 7 Number of pneumococcal acute otitis media episodes caused by VT (PNCV) and non-VT *S. pneumoniae* at the Children's Hospital of Pittsburgh from 1 January 1999 to 31 December 2002 (McEllistrem et al., 43rd ICAAC, abstr. G-2047). The arrow indicates the start of widespread vaccination with PnCRM7 in the United States.

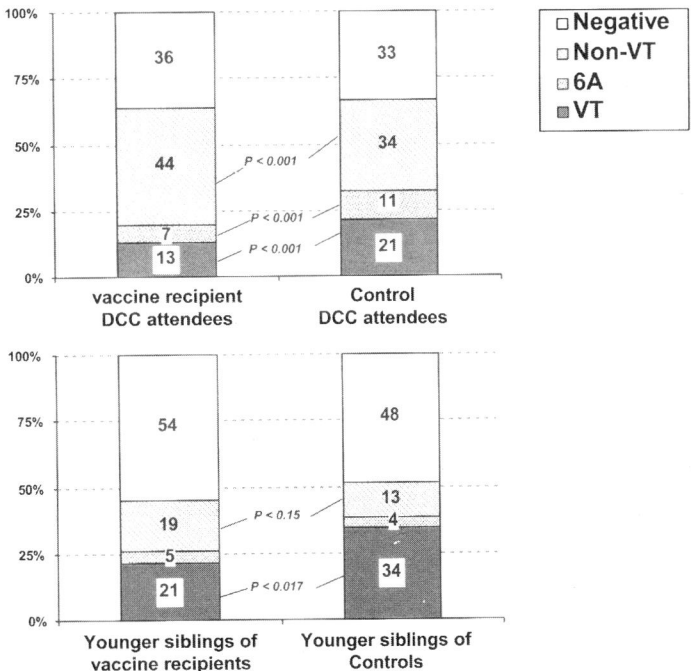

FIGURE 8 Distribution of VT (serotypes 1, 4, 5, 6B, 9V, 14, 18C, 19F and 23F), serotype 6A, non-VT pneumococci (all other serotypes), and no growth in 149 cultures from 23 younger siblings of day care center (DCC) attendees receiving PnCRM9 and 157 cultures from 23 younger siblings of the control group. This was compared with 1,886 cultures from the 132 older DCC attendee recipients of the PnCRM9 vaccine and 1,862 cultures from 130 DCC attendees receiving control vaccines.

In a second study, Navajo and White Mountain Apache children younger than 2 years of age were randomized according to their community of residence to receive either a PnCRM7 or meningococcus C conjugate vaccine. Nasopharyngeal swabs were cultured for *S. pneumoniae* from 598 nonimmunized infants residing in both vaccinated (by PnCRM7 vaccine) and control communities (K. L. O'Brien, M. A. Bronsdon, J. Becenti, G. M. Carlone, R. R. Facklam, S. Kvamme, R. Abboud, B. Schwartz, and R. Reid, 41st Intersci. Conf. Antimicrob. Agents Chemother., abstr. G-S2032, 2001). A 24% reduction in carriage of vaccine-associated serotypes was noted in unvaccinated infants residing in the PnCRM7 communities versus those living in control vaccine communities. The reduction was found both in infants who lived with a PnCRM7-vaccinated sibling and in those who did not have direct contact with a PnCRM7-vaccinated child. This study demonstrated the indirect protective effect of pneumococcal vaccination on those in the community.

Postlicensure data from the United States confirm the prelicensure observations on herd protection. The introduction of infant and toddler vaccination with PnCRM7 in the United States was associated within less than 2 years with a reduction of 32% in invasive pneumococcal infections in adults aged 20 to 39 years (the age group most likely to represent the parents), a reduction of 8% in the age group of 40 to 64 years, and a reduction of 18% in the age group of ≥65 years (most likely to represent the grandparents) (112).

Future Prospects

As use of capsular polysaccharide-conjugate vaccines increases, continued surveillance for increases in otitis and especially invasive disease from non-VT is critical. While results to date have been optimistic overall, there is considerable historic precedent for changes in serotype distribution (42) and for the appearance and rapid global spread of particularly successful pneumococcal clones under the right selective conditions (82). In particular, mathematical models of serotype replacement suggest that the extent of replacement will increase as vaccine coverage increases (69) and that nearly undetectable changes in the setting of a clinical trial may be consistent with much more dramatic changes when large fractions of the transmitting population are vaccinated.

The advent of next-generation vaccines against conserved protein antigens (13, 14) may provide a solution to the problem of serotype replacement, since these vaccines should be effective against all pneumococcal serotypes. On the other hand, since the extent of the serotype replacement problem remains to be seen, this advantage must be weighed against the possibility of inducing increases in other pathogenic species, as well as against other considerations of cost, coverage, and efficacy that will become clearer as such vaccines are developed.

INTERACTIONS BETWEEN SELECTIVE EFFECTS OF ANTIBIOTIC USE AND VACCINES

Mechanisms of Interaction

While the major effect of antibiotic use is to select for resistant strains, and the major effect of vaccination is to select against VT strains, there are several potential interactions between selection by host immunity (natural or vaccine enhanced) and by antibiotics. The sequencing of the pneumococcal genome revealed that the genetic locus encoding capsule biosynthetic enzymes, which determines the capsular serotype, is closely linked to the loci encoding penicillin and cephalosporin resistance. It has recently been shown that drug-susceptible, serotype 4 pneumococci exposed to DNA from a penicillin-resistant, type 19F strain and selected with penicillin or cefotaxime can acquire not only resistant alleles of the PBPs but also the new serotype, due to cotransformation of the linked *cps* locus with one or both *pbp* genes (K. P. Trzcinski, C. M. Thompson, and M. Lipsitch, 43rd Intersci. Conf. Antimicrob. Agents Chemother., abstr. C-1113, 2003). These experiments showed that selection by antimicrobial agents can result in "hitchhiking" selection of new capsular type and suggested that selection by host immunity might also result in changes in penicillin resistance profiles.

At the population level, the serotypes responsible for invasive disease have shifted gradually over the years. Isolates collected from Boston City Hospital over a 40-year period (42) show a shift from a predominance of low-numbered serotypes in the preantibiotic era to a recent predominance of such previously rare serogroups as 6, 14, 19, and 23, which now account for a high proportion of pneumococcal infections in many parts of the world, especially in the developed world (55, 56). Because the serotypes that have increased in frequency in recent decades include the serotypes with the highest proportion of antibiotic resistance, it is tempting to speculate that antibiotic selection may have been responsible for some of this shift in serotype composition. On the other hand, a recent study of geographic variation in the pneumococcal resistance to penicillin and erythromycin within the United States revealed that regions with a high prevalence of resistance did not have a higher proportion of those serotypes most associated with resistance; instead, serotype composition was quite similar across regions, and regions with high resistance prevalence had a higher prevalence of resistance within each of the serotypes (81).

Of immediate interest is the extent to which use of the pneumococcal conjugate vaccine will reduce the prevalence of resistant pneumococci. Use of the vaccine should reduce the incidence of resistant infections (and carriage of resistant strains) because of the particularly high prevalence of resistance among VT pneumococci. To the extent that serotype replacement occurs, it will further reduce the proportion of resistant isolates among pneumococci causing disease. Moreover, by reducing overall pneumococcal disease incidence, use of the vaccine should reduce the need for antibiotic use, thereby further reducing selection pressure for antibiotic resistance. The durability of these benefits will depend on whether drug resistance remains largely confined to the serogroups included in the vaccine; if resistant strains acquire new, nonvaccine capsular types, or if non-VT strains acquire resistance determinants, these benefits will decline over time.

Evidence of the Effect of Conjugate Vaccines on Reduction of Antibiotic Resistance

As expected, in all studies that investigated the effect of conjugate vaccines on nasopharyngeal carriage of antibiotic-resistant *S. pneumoniae*, a reduction in carriage of such strains was observed (27, 30, 32, 79; Dagan et al., 40th ICAAC, abstr. 47) (Table 3). Furthermore, the use of conjugate vaccines reduced the use of antibiotics in two prospective double-blind studies. In a study conducted in northern California, administration of the PnCRM7 vaccine reduced the use of antibiotics by 5.3% in patients given vaccines versus controls; the reduction was 5.0% for drugs generally used as first-line agents (such as amoxicillin and ampicillin) and 11.2% for those often used as second-line agents (cephalosporins, amoxicillin-clavulanate, and azithromycin) (S. Black, oral presentation, 19th Annu. Meet. Eur. Soc. Paediatr. Infect. Dis.). From first dose to age 3.5 years, PnCRM7 prevented a total of three antibiotic prescriptions per 100 children vaccinated per protocol (43). In another controlled, double-blind study performed with toddlers attending day care centers in southern Israel, the administration of PnCRM9 reduced the use of antibiotics in vaccine recipients versus controls by 17% (33), in parallel with the reduction in carriage of antibiotic-resistant *S. pneumoniae* and respiratory diseases (27, 33).

Furthermore, a significant reduction of carriage of antibiotic-resistant and MDR *S. pneumoniae* in the younger siblings of the day care attendees as the result of vaccination of their older siblings was observed (49).

In a large-scale double-blind randomized efficacy study conducted in South Africa, immunization of infants with PnCRM9 resulted in a significant reduction in invasive pneumococcal disease caused by antibiotic-resistant *S. pneumoniae* in both human immunodeficiency virus-positive and -negative subjects (66). The overall reductions for penicillin-resistant, TS-resistant, and any resistant *S. pneumoniae* invasive infections were 67% (95% confidence interval, 19 to 78%), 56% (95% confidence interval, 16 to 78%), and 56% (95% confidence interval, 21 to 77%), respectively.

Preliminary postlicensure data from the Centers for Disease Control and Prevention indicate that in the United States, a dramatic reduction in invasive infection caused by antibiotic-resistant pneumococci occurred in 2001 and 2002 in comparison to the years 1999 to 2000. This dramatic reduction was seen in invasive infection caused by penicillin-nonsusceptible, macrolide-resistant pneumococci as well as those caused by *S. pneumoniae* dually resistant to penicillin and macrolides (M. H. Kyaw, M. M. Farley, J. Hadler, L. H. Harrison, N. M. Bennett, R. Lynfield, A.

Reingold, P. R. Cieslak, R. R. Facklam, J. Jorgensen, and C. G. Whitneys, 43rd Intersci. Conf. Antimicrob. Agents Chemother., abstr. G-2045, 2003) (Fig. 9A). In children aged <2 years (mostly vaccinated), reductions of 88, 81, and 86% were seen in the incidence of invasive infections caused by penicillin-nonsusceptible, macrolide-nonsusceptible and penicillin-plus-macrolide-nonsusceptible *S. pneumoniae*, respectively, in 2002 compared with 1999.

Furthermore, as expected, a herd protection effect was clearly seen in older age groups that were not vaccinated (Fig. 9B). For invasive infections caused by penicillin-nonsusceptible *S. pneumoniae*, reductions of 71, 47, and 37% were observed in children aged 2 to 4 years (mostly unvaccinated) and adults (who were not vaccinated with PnCRM7) aged 20 to 39 years and ≥65 years, respectively.

Data regarding postvaccination reduction in antibiotic resistance in acute otitis media are scarce, but an effect is suggested by a report

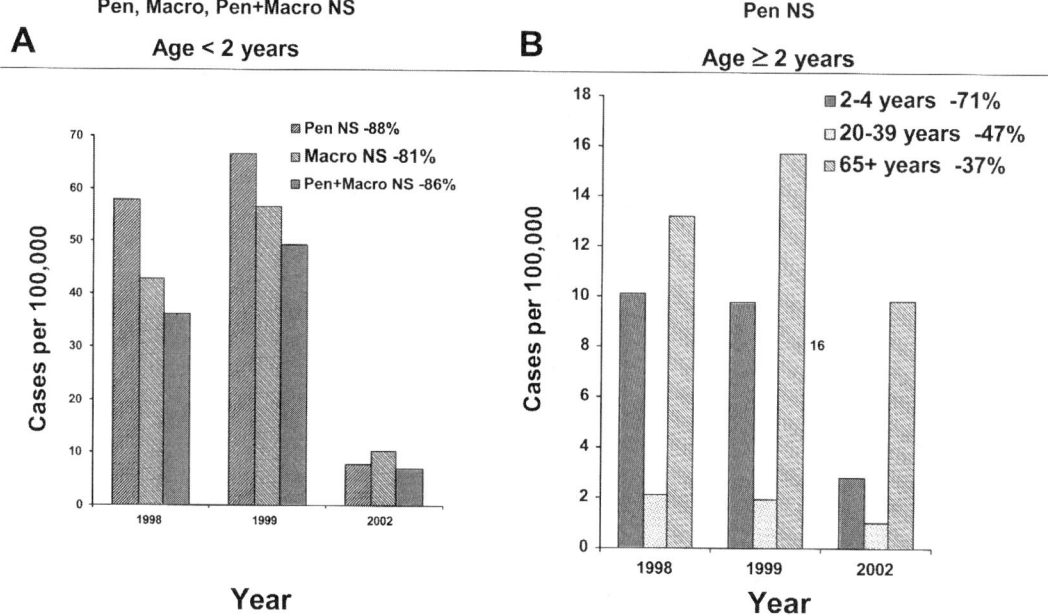

FIGURE 9 National trends in antibiotic-resistant invasive pneumococcal disease in the conjugate-vaccine era in the United States after the introduction of PnCRM7 to infants and toddlers in 2000 (Kyaw et al., 43rd ICAAC, abstr. G-2045). The reduction figures are for comparison of 2002 to 1999. (A) Penicillin, macrolide, and penicillin plus macrolide nonsusceptibility in children 2 years of age; (B) penicillin nonsusceptibility in children aged 2 to 4 years and adults aged 20 to 39 years and ≥65 years. Pen, penicillin; Macro, macrolides; NS, nonsusceptible.

from rural Kentucky on the reduction of acute otitis media caused by penicillin-resistant *S. pneumoniae* in parallel with an increased proportion of cases by non-VT *S. pneumoniae* and non-*S. pneumoniae* pathogens (S. Block, Pediatr. Acad. Soc. Annu. Meet., Abstr. LB3, 2003).

The dramatic effect of the pneumococcal conjugate vaccines on disease caused by antibiotic-resistant *S. pneumoniae* may be hampered if a replacement by non-VT that are resistant to antibiotic occurs. Nowadays, most antibiotic resistance, especially high-level resistance and MDR, is mainly confined to VT or VT-related *S. pneumoniae*, but non-VT *S. pneumoniae* persistent clones that are nonsusceptible to penicillin and cause disease were already reported. Such examples are the *S. pneumoniae* serotype 35B clone causing invasive disease in the United States (9) and serotype 15B/C, 21, 33F, and 35B clones causing acute otitis media in southern Israel (94). These findings are even more intriguing after the recent findings that some penicillin-nonsusceptible *S. pneumoniae* clones are in fact derived from capsular switch of known vaccine-type serotypes. An example for this is a recent finding of a 15B/C penicillin-nonsusceptible clone causing acute otitis media in southern Israel which is the same clone as a penicillin-nonsusceptible 19F *S. pneumoniae* clone found in Costa Rica and a penicillin- and TS-nonsusceptible serotype 11A *S. pneumoniae* clone causing acute otitis media in southern Israel that is of the same clone as the international *S. pneumoniae* Spain[9V]-3 clone (N. Porat and R. Dagan, unpublished data).

Thus, reduction of disease caused by antibiotic-resistant *S. pneumoniae* observed with the use of pneumococcal conjugate vaccine, although dramatic, may change direction if antibiotic pressure is not reduced in the community.

CONCLUSION

We have emphasized throughout this chapter that the effects of vaccines and antibiotics on the ecology of pneumococci take place against a background of a genetically diverse population structured by direct and indirect ecological interactions between strains. The importance of understanding pneumococcal population structure has been appreciated since the earliest days (50). We have argued that this understanding can be extended to reconcile the disparate results of clinical trials of antibiotics and vaccines, to evaluate the relative selective effect of different antibiotics for resistant strains, and to project the effects of vaccines and antibiotics on future patterns in the prevalence of serotypes and resistant strains. Continued improvements in our understanding of the ongoing evolution of pneumococcal populations and our effects on it will require ongoing surveillance of isolates from pneumococcal carriage and disease, typed using improved, molecular typing methods, and combined with further mechanistic studies of the ways in which vaccines, natural immunity, and antibiotics alter the flora in treated individuals.

REFERENCES

1. **Andersson, D. I.** 2003. Persistence of antibiotic resistant bacteria. *Curr. Opin. Microbiol.* **6:** 452–456.

2. **Arason, V. A., K. G. Kristinsson, J. A. Sigurdsson, G. Stefansdottir, S. Molstad, and S. Gudmundsson.** 1996. Do antimicrobials increase the carriage rate of penicillin resistant pneumococci in children? Cross sectional prevalence study. *BMJ* **313:**387–391.

3. **Arnold, K. E., R. J. Leggiadro, R. F. Breiman, H. B. Lipman, B. Schwartz, M. A. Appleton, K. O. Cleveland, H. C. Szeto, B. C. Hill, F. C. Tenover, J. A. Elliott, and R. R. Facklam.** 1996. Risk factors for carriage of drug-resistant Streptococcus pneumoniae among children in Memphis, Tennessee. *J. Pediatr.* **128:**757–764.

4. **Austrian, R.** 1986. Some aspects of the pneumococcal carrier state. *J. Antimicrob. Chemother.* **18**(Suppl. A)**:**35–45.

5. **Baquero, F.** 1999. Evolving resistance patterns of Streptococcus pneumoniae: a link with long-acting macrolide consumption? *J. Chemother.* **11**(Suppl. 1)**:**35–43.

6. **Baquero, F.** 1995. Pneumococcal resistance to beta-lactam antibiotics: a global geographic overview. *Microb. Drug Resist.* **1:**115–120.

7. **Barbour, M. L., R. T. Mayon-White, C. Coles, D. W. Crook, and E. R. Moxon.** 1995. The impact of conjugate vaccine on carriage of Haemophilus influenzae type b. *J. Infect. Dis.* **171**:93–98.

8. **Batt, S. L., B. M. Charalambous, A. W. Solomon, C. Knirsch, P. A. Massae, S. Safari, N. E. Sam, D. Everett, D. C. Mabey, and S. H. Gillespie.** 2003. Impact of azithromycin administration for trachoma control on the carriage of antibiotic-resistant *Streptococcus pneumoniae*. *Antimicrob. Agents Chemother.* **47**: 2765–2769.

9. **Beall, B., M. C. McEllistrem, R. E. Gertz, Jr., D. J. Boxrud, J. M. Besser, L. H. Harrison, J. H. Jorgensen, and C. G. Whitney.** 2002. Emergence of a novel penicillin-nonsusceptible, invasive serotype 35B clone of Streptococcus pneumoniae within the United States. *J. Infect. Dis.* **186**:118–122.

10. **Black, S., H. Shinefield, B. Fireman, E. Lewis, P. Ray, J. R. Hansen, L. Elvin, K. M. Ensor, J. Hackell, G. Siber, F. Malinoski, D. Madore, I. Chang, R. Kohberger, W. Watson, R. Austrian, and K. Edwards.** 2000. Efficacy, safety and immunogenicity of heptavalent pneumococcal conjugate vaccine in children. *Pediatr. Infect. Dis. J.* **19**:187–195.

11. **Black, S. B., H. R. Shinefield, S. Ling, J. Hansen, B. Fireman, D. Spring, J. Noyes, E. Lewis, P. Ray, J. Lee, and J. Hackell.** 2002. Effectiveness of heptavalent pneumococcal conjugate vaccine in children younger than five years of age for prevention of pneumonia. *Pediatr. Infect. Dis. J.* **21**:810–815.

12. **Boken, D. J., S. A. Chartrand, E. S. Moland, and R. V. Goering.** 1996. Colonization with penicillin-nonsusceptible Streptococcus pneumoniae in urban and rural child-care centers. *Pediatr. Infect. Dis. J.* **15**:667–672.

13. **Briles, D. E., E. Ades, J. C. Paton, J. S. Sampson, G. M. Carlone, R. C. Huebner, A. Virolainen, E. Swiatlo, and S. K. Hollingshead.** 2000. Intranasal immunization of mice with a mixture of the pneumococcal proteins PsaA and PspA is highly protective against nasopharyngeal carriage of *Streptococcus pneumoniae*. *Infect. Immun.* **68**:796–800.

14. **Briles, D. E., S. K. Hollingshead, G. S. Nabors, J. C. Paton, and A. Brooks-Walter.** 2000. The potential for using protein vaccines to protect against otitis media caused by Streptococcus pneumoniae. *Vaccine* **19**(Suppl. 1):S87–S95.

15. **Bronzwaer, S. L., O. Cars, U. Buchholz, S. Molstad, W. Goettsch, I. K. Veldhuijzen, J. L. Kool, M. J. Sprenger, and J. E. Degener.** 2002. A European study on the relationship between antimicrobial use and antimicrobial resistance. *Emerg. Infect. Dis.* **8**:278–282.

16. **Brook, I.** 2001. The role of beta-lactamase producing bacteria and bacterial interference in streptococcal tonsillitis. *Int. J. Antimicrob. Agents* **17**: 439–442.

17. **Brook, I., and A. E. Gober.** 1996. Prophylaxis with amoxicillin or sulfisoxazole for otitis media: effect on the recovery of penicillin-resistant bacteria from children. *Clin. Infect. Dis.* **22**:143–145.

18. **Brueggemann, A. B., D. T. Griffiths, E. Meats, T. Peto, D. W. Crook, and B. G. Spratt.** 2003. Clonal relationships between invasive and carriage Streptococcus pneumoniae and serotype- and clone-specific differences in invasive disease potential. *J. Infect. Dis.* **187**:1424–1432.

19. **Chambers, H. F.** 1999. Penicillin-binding protein-mediated resistance in pneumococci and staphylococci. *J. Infect. Dis.* **179**(Suppl. 2):S353–S359.

20. **Chen, D. K., A. McGeer, J. C. de Azavedo, D. E. Low, and Canadian Bacterial Surveillance Network.** 1999. Decreased susceptibility of Streptococcus pneumoniae to fluoroquinolones in Canada. *N. Engl. J. Med.* **341**:233–239.

21. **Cohen, R., E. Bingen, E. Varon, F. de La Rocque, N. Brahimi, C. Levy, M. Boucherat, J. Langue, and P. Geslin.** 1997. Change in nasopharyngeal carriage of Streptococcus pneumoniae resulting from antibiotic therapy for acute otitis media in children. *Pediatr. Infect. Dis. J.* **16**:555–560.

22. **Cohen, R., M. Navel, J. Grunberg, M. Boucherat, P. Geslin, M. Derriennic, F. Pichon, and J. M. Goehrs.** 1999. One dose ceftriaxone vs. ten days of amoxicillin/clavulanate therapy for acute otitis media: clinical efficacy and change in nasopharyngeal flora. *Pediatr. Infect. Dis. J.* **18**:403–409.

23. **Craig, W. A.** 1998. Pharmacokinetic/pharmacodynamic parameters: rationale for antibacterial dosing of mice and men. *Clin. Infect. Dis.* **26**:1–10.

24. **Craig, W. A., and D. Andes.** 1996. Pharmacokinetics and pharmacodynamics of antibiotics in otitis media. *Pediatr. Infect. Dis. J.* **15**:255–259.

25. **Dabernat, H., P. Geslin, F. Megraud, P. Begue, J. Boulesteix, C. Dubreuil, F. de La Roque, A. Trinh, and A. Scheimberg.** 1998. Effects of cefixime or co-amoxiclav treatment on nasopharyngeal carriage of Streptococcus pneumoniae and Haemophilus influenzae in children with acute otitis media. *J. Antimicrob. Chemother.* **41**:253–258.

26. **Dagan, R., N. Givon-Lavi, L. Shkolnik, P. Yagupsky, and D. Fraser.** 2000. Acute otitis media caused by antibiotic-resistant Streptococcus

pneumoniae in southern Israel: implication for immunizing with conjugate vaccines. *J. Infect. Dis.* **181:**1322–1329.

27. **Dagan, R., N. Givon-Lavi, O. Zamir, and D. Fraser.** 2003. Effect of a nonavalent conjugate vaccine on carriage of antibiotic-resistant Streptococcus pneumoniae in day-care centers. *Pediatr. Infect. Dis. J.* **22:**532–540.

28. **Dagan, R., N. Givon-Lavi, O. Zamir, M. Sikuler-Cohen, L. Guy, J. Janco, P. Yagupsky, and D. Fraser.** 2002. Reduction of nasopharyngeal carriage of Streptococcus pneumoniae after administration of a 9-valent pneumococcal conjugate vaccine to toddlers attending day care centers. *J. Infect. Dis.* **185:**927–936.

29. **Dagan, R., E. Leibovitz, D. Greenberg, P. Yagupsky, D. M. Fliss, and A. Leiberman.** 1998. Dynamics of pneumococcal nasopharyngeal colonization during the first days of antibiotic treatment in pediatric patients. *Pediatr. Infect. Dis. J.* **17:**880–885.

30. **Dagan, R., R. Melamed, M. Muallem, L. Piglansky, D. Greenberg, O. Abramson, P. M. Mendelman, N. Bohidar, and P. Yagupsky.** 1996. Reduction of nasopharyngeal carriage of pneumococci during the second year of life by a heptavalent conjugate pneumococcal vaccine. *J. Infect. Dis.* **174:**1271–1278.

31. **Dagan, R., R. Melamed, M. Muallem, L. Piglansky, and P. Yagupsky.** 1996. Nasopharyngeal colonization in southern Israel with antibiotic-resistant pneumococci during the first 2 years of life: relation to serotypes likely to be included in pneumococcal conjugate vaccines. *J. Infect. Dis.* **174:**1352–1355.

32. **Dagan, R., M. Muallem, R. Melamed, O. Leroy, and P. Yagupsky.** 1997. Reduction of pneumococcal nasopharyngeal carriage in early infancy after immunization with tetravalent pneumococcal vaccines conjugated to either tetanus toxoid or diphtheria toxoid. *Pediatr. Infect. Dis. J.* **16:**1060–1064.

33. **Dagan, R., M. Sikuler-Cohen, O. Zamir, J. Janco, N. Givon-Lavi, and D. Fraser.** 2001. Effect of a conjugate pneumococcal vaccine on the occurrence of respiratory infections and antibiotic use in day-care center attendees. *Pediatr. Infect. Dis. J.* **20:**951–958.

34. **De Lencastre, H., K. G. Kristinsson, A. Brito-Avo, I. S. Sanches, R. Sa-Leao, J. Saldanha, E. Sigvaldadottir, S. Karlsson, D. Oliveira, R. Mato, M. Aires de Sousa, and A. Tomasz.** 1999. Carriage of respiratory tract pathogens and molecular epidemiology of Streptococcus pneumoniae colonization in healthy children attending day care centers in Lisbon, Portugal. *Microb. Drug Resist.* **5:**19–29.

35. **den Hollander, J. G., J. D. Knudsen, J. W. Mouton, K. Fuursted, N. Frimodt-Moller, H. A. Verbrugh, and F. Espersen.** 1998. Comparison of pharmacodynamics of azithromycin and erythromycin in vitro and in vivo. *Antimicrob. Agents Chemother.* **42:**377–382.

36. **Diekema, D. J., A. B. Brueggemann, and G. V. Doern.** 2000. Antimicrobial-drug use and changes in resistance in Streptococcus pneumoniae. *Emerg. Infect. Dis.* **6:**552–556.

37. **Drusano, G. L., and W. A. Craig.** 1997. Relevance of pharmacokinetics and pharmacodynamics in the selection of antibiotics for respiratory tract infections. *J. Chemother.* **9**(Suppl. 3)**:**38–44.

38. **Eskola, J., S. Black, and H. Shinefield.** 2003. Pneumococcal conjugate vaccines, p. 589–624. *In* S. A. Plotkin and M. D. Orenstein (ed.), *Vaccines,* 4th ed. W. B. Saunders Company, Philadelphia, Pa.

39. **Eskola, J., T. Kilpi, A. Palmu, J. Jokinen, J. Haapakoski, E. Herva, A. Takala, H. Kayhty, P. Karma, R. Kohberger, G. Siber, and P. H. Makela.** 2001. Efficacy of a pneumococcal conjugate vaccine against acute otitis media. *N. Engl. J. Med.* **344:**403–409.

40. **Feikin, D. R., S. F. Dowell, O. C. Nwanyanwu, K. P. Klugman, P. N. Kazembe, L. M. Barat, C. Graf, P. B. Bloland, C. Ziba, R. E. Huebner, and B. Schwartz.** 2000. Increased carriage of trimethoprim/sulfamethoxazole-resistant Streptococcus pneumoniae in Malawian children after treatment for malaria with sulfadoxine/pyrimethamine. *J. Infect. Dis.* **181:**1501–1505.

41. **Fenoll, A., I. Jado, D. Vicioso, A. Perez, and J. Casal.** 1998. Evolution of *Streptococcus pneumoniae* serotypes and antibiotic resistance in Spain: update (1990 to 1996). *J. Clin. Microbiol.* **36:**3447–3454.

42. **Finland, M., and M. W. Barnes.** 1977. Changes in occurrence of capsular serotypes of *Streptococcus pneumoniae* at Boston City Hospital during selected years between 1935 and 1974. *J. Clin. Microbiol.* **5:**154–166.

43. **Fireman, B., S. B. Black, H. R. Shinefield, J. Lee, E. Lewis, and P. Ray.** 2003. Impact of the pneumococcal conjugate vaccine on otitis media. *Pediatr. Infect. Dis. J.* **22:**10–16.

44. **Garcia-Rey, C., L. Aguilar, F. Baquero, J. Casal, and R. Dal-Re.** 2002. Importance of local variations in antibiotic consumption and geographical differences of erythromycin and penicillin resistance in *Streptococcus pneumoniae. J. Clin. Microbiol.* **40:**159–164.

45. **Ghaffar, F., L. S. Muniz, K. Katz, J. Reynolds, J. L. Smith, P. Davis, I. R. Friedland, and G. H. McCracken, Jr.** 2000. Effects

of amoxicillin/clavulanate or azithromycin on nasopharyngeal carriage of Streptococcus pneumoniae and Haemophilus influenzae in children with acute otitis media. *Clin. Infect. Dis.* **31**:875–880.

46. **Ghaffar, F., L. S. Muniz, K. Katz, J. L. Smith, T. Shouse, P. Davis, and G. H. Mc-Cracken, Jr.** 2002. Effects of large dosages of amoxicillin/clavulanate or azithromycin on nasopharyngeal carriage of Streptococcus pneumoniae, Haemophilus influenzae, nonpneumococcal alpha-hemolytic streptococci, and Staphylococcus aureus in children with acute otitis media. *Clin. Infect. Dis.* **34**:1301–1309.

47. **Gillespie, S. H., L. L. Voelker, and A. Dickens.** 2002. Evolutionary barriers to quinolone resistance in Streptococcus pneumoniae. *Microb. Drug Resist.* **8**:79–84.

48. **Givon-Lavi, N., R. Dagan, D. Fraser, P. Yagupsky, and N. Porat.** 1999. Marked differences in pneumococcal carriage and resistance patterns between day care centers located within a small area. *Clin. Infect. Dis.* **29**:1274–1280.

49. **Givon-Lavi, N., D. Fraser, and R. Dagan.** 2003. Vaccination of day-care center attendees reduces carriage of Streptococcus pneumoniae among their younger siblings. *Pediatr. Infect. Dis. J.* **22**:524–532.

50. **Griffith, F.** 1928. The significance of pneumococcal types. *J. Hyg.* **27**:8–159.

51. **Guillemot, D., C. Carbon, B. Balkau, P. Geslin, H. Lecoeur, F. Vauzelle-Kervroedan, G. Bouvenot, and E. Eschwege.** 1998. Low dosage and long treatment duration of beta-lactam: risk factors for carriage of penicillin-resistant Streptococcus pneumoniae. *JAMA* **279**:365–370.

52. **Gundel, M., and G. Okura.** 1933. Untersuchungen uber das gleichzeitige Vorkommen mehrerer Pneumokokkentypen bei Gesunden und ihre Bedeutung für die Epidemiologie. *Z. Hyg. Infektionskr.* **114**:678–704.

53. **Haiman, T., E. Leibovitz, L. Piglansky, J. Press, P. Yagupsky, A. Leiberman, and R. Dagan.** 2002. Dynamics of pneumococcal nasopharyngeal carriage in children with nonresponsive acute otitis media treated with two regimens of intramuscular ceftriaxone. *Pediatr. Infect. Dis. J.* **21**:642–647.

54. **Hakenbeck, R., T. Grebe, D. Zahner, and J. B. Stock.** 1999. Beta-lactam resistance in Streptococcus pneumoniae: penicillin-binding proteins and non-penicillin-binding proteins. *Mol. Microbiol.* **33**:673–678.

55. **Hausdorff, W. P., J. Bryant, C. Kloek, P. R. Paradiso, and G. R. Siber.** 2000. The contribution of specific pneumococcal serogroups to different disease manifestations: implications for conjugate vaccine formulation and use, part II. *Clin. Infect. Dis.* **30**:122–140.

56. **Hausdorff, W. P., J. Bryant, P. R. Paradiso, and G. R. Siber.** 2000. Which pneumococcal serogroups cause the most invasive disease: implications for conjugate vaccine formulation and use, part I. *Clin. Infect. Dis.* **30**:100–121.

57. **Heikkinen, T.** 2000. Role of viruses in the pathogenesis of acute otitis media. *Pediatr. Infect. Dis. J.* **19**:S17–S22.

58. **Hyde, T. B., K. Gay, D. S. Stephens, D. J. Vugia, M. Pass, S. Johnson, N. L. Barrett, W. Schaffner, P. R. Cieslak, P. S. Maupin, E. R. Zell, J. H. Jorgensen, R. R. Facklam, and C. G. Whitney.** 2001. Macrolide resistance among invasive Streptococcus pneumoniae isolates. *JAMA* **286**:1857–1862.

59. **Jacobs, M. R., D. Felmingham, P. C. Appelbaum, and R. N. Gruneberg.** 2003. The Alexander Project 1998–2000: susceptibility of pathogens isolated from community-acquired respiratory tract infection to commonly used antimicrobial agents. *J. Antimicrob. Chemother.* **52**:229–246.

60. **Jones, R. N., A. H. Mutnick, and D. J. Varnam.** 2002. Impact of modified nonmeningeal *Streptococcus pneumoniae* interpretive criteria (NC-CLS M100-S12) on the susceptibility patterns of five parenteral cephalosporins: report from the SENTRY Antimicrobial Surveillance Program (1997 to 2001). *J. Clin. Microbiol.* **40**:4332–4333.

61. **Kastner, U., and J. P. Guggenbichler.** 2001. Influence of macrolide antibiotics on promotion of resistance in the oral flora of children. *Infection* **29**:251–256.

62. **Kauppi, M., J. Eskola, and H. Kayhty.** 1995. Anti-capsular polysaccharide antibody concentrations in saliva after immunization with Haemophilus influenzae type b conjugate vaccines. *Pediatr. Infect. Dis. J.* **14**:286–294.

63. **Kellner, J. D., E. L. Ford-Jones, and Toronto Child Care Centre Study Group.** 1999. Streptococcus pneumoniae carriage in children attending 59 Canadian child care centers. *Arch. Pediatr. Adolesc. Med.* **153**:495–502.

64. **Kilian, M., J. Reinholdt, H. Lomholt, K. Poulsen, and E. V. Frandsen.** 1996. Biological significance of IgA1 proteases in bacterial colonization and pathogenesis: critical evaluation of experimental evidence. *APMIS* **104**:321–338.

65. **Kilpi, T., H. Ahman, J. Jokinen, K. S. Lankinen, A. Palmu, H. Savolainen, M. Gronholm, M. Leinonen, T. Hovi, J. Eskola, H. Kayhty, N. Bohidar, J. C. Sadoff, and P. H. Makela.** 2003. Protective efficacy of a second pneumococcal conjugate vaccine against pneumococcal acute otitis media in infants and chil-

dren: randomized, controlled trial of a 7-valent pneumococcal polysaccharide-meningococcal outer membrane protein complex conjugate vaccine in 1666 children. *Clin. Infect. Dis.* **37:**1155–1164.

66. **Klugman, K. P., S. A. Madhi, R. E. Huebner, R. Kohberger, N. Mbelle, and N. Pierce.** 2003. A trial of a 9-valent pneumococcal conjugate vaccine in children with and those without HIV infection. *N. Engl. J. Med.* **349:**1341–1348.

67. **Korkeila, M., H. Lehtonen, H. Ahman, O. Leroy, J. Eskola, and H. Kayhty.** 2000. Salivary anti-capsular antibodies in infants and children immunised with Streptococcus pneumoniae capsular polysaccharides conjugated to diphtheria or tetanus toxoid. *Vaccine* **18:**1218–1226.

68. **Lenski, R. E., and S. E. Hattingh.** 1986. Coexistence of two competitors on one resource and one inhibitor: a chemostat model based on bacteria and antibiotics. *J. Theor. Biol.* **122:**83–93.

69. **Lipsitch, M.** 1999. Bacterial vaccines and serotype replacement: lessons from Haemophilus influenzae and prospects for Streptococcus pneumoniae. *Emerg. Infect. Dis.* **5:**336–345.

70. **Lipsitch, M.** 2001. Measuring and interpreting associations between antibiotic use and penicillin resistance in Streptococcus pneumoniae. *Clin. Infect. Dis.* **32:**1044–1054.

71. **Lipsitch, M.** 2001. The rise and fall of antimicrobial resistance. *Trends Microbiol.* **9:**438–444.

72. **Lipsitch, M., and M. H. Samore.** 2002. Antimicrobial use and antimicrobial resistance: a population perspective. *Emerg. Infect. Dis.* **8:**347–354.

73. **Mackowiak, P. A.** 1982. The normal microbial flora. *N. Engl. J. Med.* **307:**83–93.

74. **MacLeod, C. M., R. G. Hodges, M. Heidelberger, and W. G. Bernhard.** 1945. Prevention of pneumococcal pneumonia by immunization with specific capsular polysaccharides. *J. Exp. Med.* **82:**445–465.

75. **Mainous, A. G., III, M. E. Evans, W. J. Hueston, W. B. Titlow, and L. J. McCown.** 1998. Patterns of antibiotic-resistant Streptococcus pneumoniae in children in a day-care setting. *J. Fam. Pract.* **46:**142–146.

76. **Malley, R., M. Lipsitch, A. Stack, R. Saladino, G. Fleisher, S. Pelton, C. Thompson, D. Briles, and P. Anderson.** 2001. Intranasal immunization with killed unencapsulated whole cells prevents colonization and invasive disease by capsulated pneumococci. *Infect. Immun.* **69:**4870–4873.

77. **Malley, R., A. M. Stack, M. L. Ferretti, C. M. Thompson, and R. A. Saladino.** 1998. Anticapsular polysaccharide antibodies and nasopharyngeal colonization with *Streptococcus pneumoniae* in infant rats. *J. Infect. Dis.* **178:**878–882.

78. **May, R. M.** 1973. *Stability and Complexity in Model Ecosystems.* Princeton University Press, Princeton, N.J.

79. **Mbelle, N., R. E. Huebner, A. D. Wasas, A. Kimura, I. Chang, and K. P. Klugman.** 1999. Immunogenicity and impact on nasopharyngeal carriage of a nonavalent pneumococcal conjugate vaccine. *J. Infect. Dis.* **180:**1171–1176.

80. **McCaig, L. F., R. E. Besser, and J. M. Hughes.** 2003. Antimicrobial drug prescription in ambulatory care settings, United States, 1992–2000. *Emerg. Infect. Dis.* **9:**432–437.

81. **McCormick, A. W., C. G. Whitney, M. M. Farley, R. Lynfield, L. H. Harrison, N. M. Bennett, W. Schaffner, A. Reingold, J. Hadler, P. Cieslak, M. H. Samore, and M. Lipsitch.** 2003. Geographic diversity and temporal trends of antimicrobial resistance in Streptococcus pneumoniae in the United States. *Nat. Med.* **9:**424–430.

82. **McGee, L., L. McDougal, J. Zhou, B. G. Spratt, F. C. Tenover, R. George, R. Hakenbeck, W. Hryniewicz, J. C. Lefévre, A. Tomasz, and K. P. Klugman.** 2001. Nomenclature of major antimicrobial-resistant clones of *Streptococcus pneumoniae* defined by the Pneumococcal Molecular Epidemiology Network. *J. Clin. Microbiol.* **39:**2565–2571.

83. **McLeod, J., and J. Gordon.** 1922. Production of hydrogen peroxide by bacteria. *Biochem. J.* **16:**499–506.

84. **Melander, E., K. Ekdahl, G. Jonsson, and S. Molstad.** 2000. Frequency of penicillin-resistant pneumococci in children is correlated to community utilization of antibiotics. *Pediatr. Infect. Dis. J.* **19:**1172–1177.

85. **Melander, E., S. Molstad, K. Persson, H. B. Hansson, M. Soderstrom, and K. Ekdahl.** 1998. Previous antibiotic consumption and other risk factors for carriage of penicillin-resistant Streptococcus pneumoniae in children. *Eur. J. Clin. Microbiol. Infect. Dis.* **17:**834–838.

86. **Morita, J. Y., E. Kahn, T. Thompson, L. Laclaire, B. Beall, G. Gherardi, K. L. O'Brien, and B. Schwartz.** 2000. Impact of azithromycin on oropharyngeal carriage of group A Streptococcus and nasopharyngeal carriage of macrolide-resistant Streptococcus pneumoniae. *Pediatr. Infect. Dis. J.* **19:**41–46.

87. **Nightingale, C. H.** 1997. Pharmacokinetics and pharmacodynamics of newer macrolides. *Pediatr. Infect. Dis. J.* **16:**438–443.

88. **Novelli, Λ., S. Fallani, M. I. Cassetta, S. Arrigucci, and T. Mazzei.** 2002. In vivo pharma-

codynamic evaluation of clarithromycin in comparison to erythromycin. *J. Chemother.* **14**:584–590.

89. **Nurkka, A., H. Ahman, M. Korkeila, V. Jantti, H. Kayhty, and J. Eskola.** 2001. Serum and salivary anti-capsular antibodies in infants and children immunized with the heptavalent pneumococcal conjugate vaccine. *Pediatr. Infect. Dis. J.* **20**:25–33.

90. **Obaro, S. K., R. A. Adegbola, W. A. Banya, and B. M. Greenwood.** 1996. Carriage of pneumococci after pneumococcal vaccination. *Lancet* **348**:271–272.

91. **O'Brien, K. L., L. H. Moulton, R. Reid, R. Weatherholtz, J. Oski, L. Brown, G. Kumar, A. Parkinson, D. Hu, J. Hackell, I. Chang, R. Kohberger, G. Siber, and M. Santosham.** 2003. Efficacy and safety of sevenvalent conjugate pneumococcal vaccine in American Indian children: group randomised trial. *Lancet* **362**:355–361.

92. **Pericone, C. D., K. Overweg, P. W. Hermans, and J. N. Weiser.** 2000. Inhibitory and bactericidal effects of hydrogen peroxide production by *Streptococcus pneumoniae* on other inhabitants of the upper respiratory tract. *Infect. Immun.* **68**:3990–3997.

93. **Pihlajamaki, M., P. Kotilainen, T. Kaurila, T. Klaukka, E. Palva, and P. Huovinen.** 2001. Macrolide-resistant Streptococcus pneumoniae and use of antimicrobial agents. *Clin. Infect. Dis.* **33**:483–488.

94. **Porat, N., G. Barkai, M. R. Jacobs, R. Trefler, and R. Dagan.** 2004. Four antibiotic resistant S. pneumoniae clones unrelated to the pneumococcal conjugate vaccine serotypes causing acute otitis media in southern Israel. *J. Infect. Dis.* **189**:385–392.

95. **Pradier, C., B. Dunais, H. Carsenti-Etesse, and P. Dellamonica.** 1997. Pneumococcal resistance patterns in Europe. *Eur. J. Clin. Microbiol. Infect. Dis.* **16**:644–647.

96. **Radetsky, M. S., G. R. Istre, T. L. Johansen, S. W. Parmelee, B. A. Lauer, A. M. Wiesenthal, and M. P. Glode.** 1981. Multiply resistant pneumococcus causing meningitis: its epidemiology within a day-care centre. *Lancet* **ii**:771–773.

97. **Rapola, S., V. Jantti, R. Haikala, R. Syrjanen, G. M. Carlone, J. S. Sampson, D. E. Briles, J. C. Paton, A. K. Takala, T. M. Kilpi, and H. Kayhty.** 2000. Natural development of antibodies to pneumococcal surface protein A, pneumococcal surface adhesin A, and pneumolysin in relation to pneumococcal carriage and acute otitis media. *J. Infect. Dis.* **182**:1146–1152.

98. **Regev-Yochay, G., M. Raz, B. Shainberg, R. Dagan, M. Varon, M. Dushenat, and E. Rubinstein.** 2003. Independent risk factors for carriage of penicillin-non-susceptible Streptococcus pneumoniae. *Scand. J. Infect. Dis.* **35**:219–222.

99. **Reichler, M. R., A. A. Allphin, R. F. Breiman, J. R. Schreiber, J. E. Arnold, L. K. McDougal, R. R. Facklam, B. Boxerbaum, D. May, R. O. Walton, and M. R. Jacobs.** 1992. The spread of multiply resistant Streptococcus pneumoniae at a day care center in Ohio. *J. Infect. Dis.* **166**:1346–1353.

100. **Rieux, V., C. Carbon, and E. Azoulay-Dupuis.** 2001. Complex relationship between acquisition of beta-lactam resistance and loss of virulence in Streptococcus pneumoniae. *J. Infect. Dis.* **184**:66–72.

101. **Robins-Browne, R. M., A. B. Kharsany, and H. J. Koornhof.** 1984. Antibiotic-resistant pneumococci in hospitalized children. *J. Hyg.* **93**:9–16.

102. **Samore, M. H., M. K. Magill, S. C. Alder, E. Severina, L. Morrison-De Boer, J. L. Lyon, K. Carroll, J. Leary, M. B. Stone, D. Bradford, J. Reading, A. Tomasz, and M. A. Sande.** 2001. High rates of multiple antibiotic resistance in Streptococcus pneumoniae from healthy children living in isolated rural communities: association with cephalosporin use and intrafamilial transmission. *Pediatrics* **108**:856–865.

103. **Schrag, S. J., C. Pena, J. Fernandez, J. Sanchez, V. Gomez, E. Perez, J. M. Feris, and R. E. Besser.** 2001. Effect of short-course, high-dose amoxicillin therapy on resistant pneumococcal carriage: a randomized trial. *JAMA* **286**:49–56.

104. **Simell, B., M. Korkeila, H. Pursiainen, T. M. Kilpi, and H. Kayhty.** 2001. Pneumococcal carriage and otitis media induce salivary antibodies to pneumococcal surface adhesin a, pneumolysin, and pneumococcal surface protein a in children. *J. Infect. Dis.* **183**:887–896.

105. **Spratt, B. G., and B. M. Greenwood.** 2000. Prevention of pneumococcal disease by vaccination: does serotype replacement matter? *Lancet* **356**:1210–1211.

106. **Steinman, M. A., R. Gonzales, J. A. Linder, and C. S. Landefeld.** 2003. Changing use of antibiotics in community-based outpatient practice, 1991–1999. *Ann. Intern. Med.* **138**:525–533.

107. **Syrogiannopoulos, G. A., I. N. Grivea, T. A. Davies, G. D. Katopodis, P. C. Appelbaum, and N. G. Beratis.** 2000. Antimicrobial use and colonization with erythromycin-re-

sistant Streptococcus pneumoniae in Greece during the first 2 years of life. *Clin. Infect. Dis.* **31:**887–893.

108. **Tsolia, M., G. Kouppari, A. Zaphiropoulou, S. Gavrili, M. Tsirepa, D. Kafetzis, and T. Karpathios.** 1999. Prevalence and patterns of resistance of Streptococcus pneumoniae strains isolated from carriers attending day care centers in the area of Athens. *Microb. Drug Resist.* **5:**271–278.

109. **Ussery, X. T., B. D. Gessner, H. Lipman, J. A. Elliott, M. J. Crain, P. C. Tien, A. J. Parkinson, M. Davidson, R. R. Facklam, and R. F. Breiman.** 1996. Risk factors for nasopharyngeal carriage of resistant Streptococcus pneumoniae and detection of a multiply resistant clone among children living in the Yukon-Kuskokwim Delta region of Alaska. *Pediatr. Infect. Dis. J.* **15:**986–992.

110. **Veenhoven, R., D. Bogaert, C. Uiterwaal, C. Brouwer, H. Kiezebrink, J. Bruin, E. IJzerman, P. Hermans, R. de Groot, B. Zegers, W. Kuis, G. Rijkers, A. Schilder, and E. Sanders.** 2003. Effect of conjugate pneumococcal vaccine followed by polysaccharide pneumococcal vaccine on recurrent acute otitis media: a randomized study. *Lancet* **361:**2189–2195.

111. **Weiser, J. N., D. Bae, C. Fasching, R. W. Scamurra, A. J. Ratner, and E. N. Janoff.** 2003. Antibody-enhanced pneumococcal adherence requires IgA1 protease. *Proc. Natl. Acad. Sci. USA* **100:**4215–4220.

112. **Whitney, C. G., M. M. Farley, J. Hadler, L. H. Harrison, N. M. Bennett, R. Lynfield, A. Reingold, P. R. Cieslak, T. Pilishvili, D. Jackson, R. R. Facklam, J. H. Jorgensen, and A. Schuchat.** 2003. Decline in invasive pneumococcal disease after the introduction of protein-polysaccharide conjugate vaccine. *N. Engl. J. Med.* **348:**1737–1746.

113. **Whitney, C. G., M. M. Farley, J. Hadler, L. H. Harrison, C. Lexau, A. Reingold, L. Lefkowitz, P. R. Cieslak, M. Cetron, E. R. Zell, J. H. Jorgensen, and A. Schuchat.** 2000. Increasing prevalence of multidrug-resistant Streptococcus pneumoniae in the United States. *N. Engl. J. Med.* **343:**1917–1924.

114. **Yagupsky, P., N. Porat, D. Fraser, F. Prajgrod, M. Merires, L. McGee, K. P. Klugman, and R. Dagan.** 1998. Acquisition, carriage, and transmission of pneumococci with decreased antibiotic susceptibility in young children attending a day care facility in southern Israel. *J. Infect. Dis.* **177:**1003–1012.

115. **Yeh, S. H., K. M. Zangwill, H. Lee, S. J. Chang, V. I. Wong, D. P. Greenberg, and J. I. Ward.** 2003. Heptavalent pneumococcal vaccine conjugated to outer membrane protein of Neisseria meningitidis serogroup b and nasopharyngeal carriage of Streptococcus pneumoniae in infants. *Vaccine* **21:**2627–2631.

PNEUMOCOCCAL INFECTIONS: THERAPEUTIC STRATEGIES AND PITFALLS

Kathryn M. Edwards

19

PNEUMOCOCCAL INFECTIONS: THERAPEUTIC STRATEGIES AND PITFALLS

Streptococcus pneumoniae is one of the most important bacterial causes of respiratory infection and invasive disease in children and adults. Its impact is particularly severe in the developing world, where over 2 million cases of pneumonia occur annually and meningitis and sepsis are not uncommon. Data on the burden of *S. pneumoniae* disease in the United States from the Centers for Disease Control and Prevention (CDC) are depicted in Fig. 1. While these rates are less than those seen in developing countries, the burden is still substantial.

For many years penicillin or ampicillin was the drug of choice for all pneumococcal infections because the organism was sensitive to antibiotic concentrations easily achievable by either the oral or the parenteral route. However, over the last decade the rapid emergence of antibiotic-resistant pneumococcal organisms has complicated the therapy of pneumococcal infections in both adults and children (20, 59).

In children the rates of antibiotic resistance exceed those in adults because children frequently harbor *S. pneumoniae* in the nasopharynx and receive antibiotics more often than adults, thus promoting selection of antibiotic-resistant strains.

The goal of this chapter is to review how increased rates of antibiotic-resistant *S. pneumoniae* have influenced the morbidity and mortality associated with pneumococcal disease and to present optimal therapeutic approaches for management of these infections in children and adults.

OTITIS MEDIA IN CHILDREN

Numerous pediatric studies have used tympanocentesis to definitively determine the bacterial etiology of acute otitis media in children and to measure the therapeutic effect of antibiotics on sterilization of the middle ear fluid. Most drugs used for the therapy of acute otitis media exert their effect by time-dependent killing, in which concentrations of antibiotics in the middle ear effusions exceed the MICs for the pathogens for at least 40 to 50% of the dosing interval (15). The efficacy of various antibiotics for the treatment of acute otitis media can be accurately assessed using the "double-tap" tympanocentesis methodology first described by Howie and Ploussard (31). In

Kathryn M. Edwards, Division of Infectious Diseases, Department of Pediatrics, CCC-5323 Medical Center North, Vanderbilt University School of Medicine, Nashville, TN 37232.

The Pneumococcus. Volume Editor, Elaine I. Tuomanen,
© 2004 ASM Press, Washington, D.C.

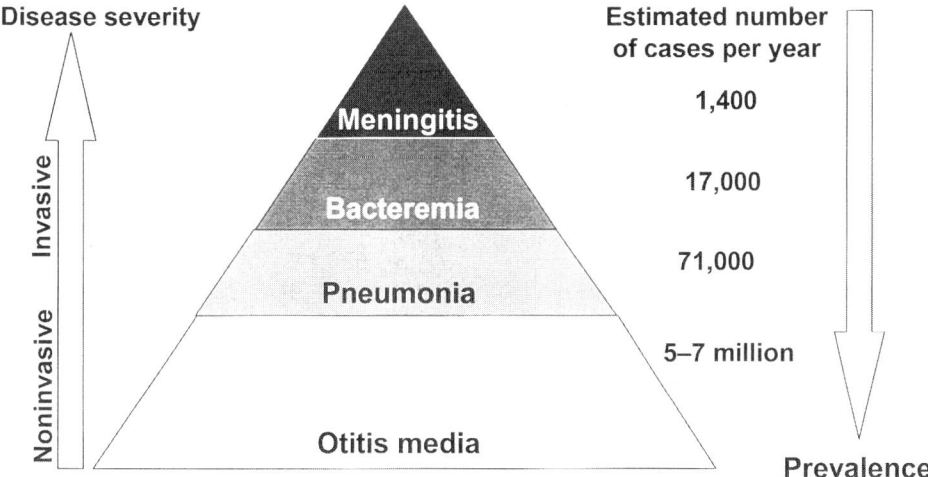

FIGURE 1 *S. pneumoniae*: pathogen with significant disease burden in children in the United States. Available from http://www.pneumo.com; data obtained from *Morb. Mortal. Wkly. Rep.* **46:**1–24.

these double-tap studies, tympanocentesis fluid is obtained prior to the start of antibiotic therapy, antibiotic therapy is initiated, and then a repeat tympanocentesis is performed after 3 to 4 days of treatment. By comparing bacterial growth of the tympanocentesis fluid before and after therapy, the efficacy of the specific antibiotic agent can be determined with relatively few studied subjects (31). Double-tap studies comparing multiple different antimicrobials and their dosages have been reported and provide the best evidence for effective therapy.

Although the exact distribution of bacterial organisms causing acute otitis media differs among the various reports, most studies suggest that *S. pneumoniae* makes up nearly half of all cases (10, 27). For many years *S. pneumoniae* organisms were highly sensitive to penicillin (MIC < 0.1 μg/ml), and oral amoxicillin therapy at a dose of 30 to 40 mg/kg of body weight/day was the standard of care for the treatment of acute otitis media. However, in the past decade increasing numbers of penicillin-intermediate (MIC = 0.1 to 1.0 μg/ml) or nonsusceptible (MIC > 1.0 μg/ml) *S. pneumoniae* strains have been reported (29, 60). Increasing reports of refractory acute otitis media have also been linked to *S. pneumoniae* intermediate and nonsusceptible strains (15, 62).

Using the double tap methodology, Dagan in Israel compared many different treatments for acute otitis media due to *S. pneumoniae* and recently summarized his findings (Fig. 2) (15). Of the 324 patients with bacteriologically proven pneumococcal acute otitis media, the failure rate was 7% among those infected with susceptible strains and 37% among those infected with organisms with altered susceptibility. A subsequent report by Dagan et al. demonstrated the efficacy of high-dose amoxicillin-clavulanate (80 to 90 mg/kg/day of amoxicillin) for the therapy of pneumococcal acute otitis media (16). In this double tap study, 122 of 125 *S. pneumoniae* isolates (98%) were eradicated from the middle ear, including 31 of 34 (91%) nonsusceptible isolates (MIC > 2 μg/ml), using high-dose amoxicillin-clavulanate. In addition, symptoms and otoscopic signs of acute inflammation improved or completely resolved in the majority of subjects. The authors concluded that "high dose amoxicillin-clavulanate was highly efficacious in children with AOM, including children <24 months of age and those with nonsusceptible Spn organisms." Cefdinir, a new oral extended-spectrum agent comparable to cefuroxime, has also been studied in 125 children with culture-confirmed acute otitis media at a dose of 7 mg/kg twice daily for 5 days in an open-label, double tap evaluation. When only the pneumococcal isolates were studied, eradication rates at the end of therapy were 73% for penicillin-intermediate isolates and 50% for penicillin-resistant strains. The authors concluded that "cefdinir should be considered a suitable second line antibiotic agent" (11). However, these data highlight once again that resistant organisms are more difficult to treat.

In an attempt to reduce the inappropriate use of antibiotics, a consensus CDC panel convened several years ago to formulate guidelines for antibiotic therapy of acute otitis media in "an era of pneumococcal resistance"

(21). The panel's conclusions were that "oral amoxicillin should remain the first line antimicrobial agent for treating acute otitis media. In view of the increasing prevalence of drug resistant *Streptococcus pneumoniae*, the safety of amoxicillin at higher than standard dosages and evidence that higher dosages can achieve effective middle ear fluid concentrations, an increase in the dosage used for empiric treatment from 40–45 mg/kg/day to 80–90 mg/kg/day is recommended. For patients with clinically defined treatment failure, useful alternative agents include oral amoxicillin-clavulanate, cefuroxime axetil, and intramuscular ceftriaxone." The American Academy of Pediatrics will soon release guidelines for the therapy of acute otitis media in children. Although these guidelines have not been published, indications are that they will be similar to those proposed by the CDC panel (21).

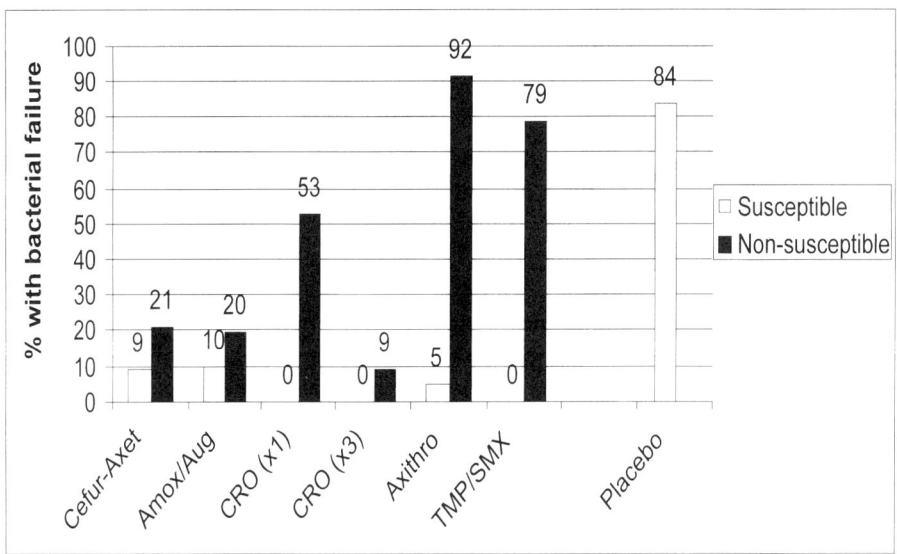

FIGURE 2 Bacteriological response to antibiotics in 324 patients with pneumococcal acute otitis media studied by the double tympanocentesis method. Agents used include cefuroxime axetil (Cefur-Axet) (30 mg/kg/day), amoxicillin (Amox) (40 mg/kg/day), augmentin (Aug) (45 mg/kg/day), intramuscular ceftriaxone at one dose of 50 mg/kg [CRO (×1)], intramuscular ceftriaxone at three daily doses of 50 mg/kg [CRO (×3)], oral azithromycin (Axithro) at 10 mg/kg/day for 3 days or 10 mg/kg/day for 1 day and 5 mg/kg/day for the following 4 days, and oral trimethoprim-sulfamethoxazole (TMP/SMX) at 8/40 mg/kg. (Chart modified from reference 15.)

ALTERNATIVE STRATEGIES FOR THE PREVENTION OF ACUTE OTITIS MEDIA IN CHILDREN

In the late 1990s Uhari and colleagues in Finland published several articles that suggested that xylitol sugar was effective in the prevention of acute otitis media in children. Xylitol, a sugar alcohol that occurs naturally in many different types of berries, inhibits the growth of *S. pneumoniae* in vitro (but not other bacteria associated with acute otitis media). In the initial Uhari study, 306 children with a history of recurrent otitis media were randomized to receive either xylitol chewing gum or sucrose chewing gum for 2 months (54). During the 2-month follow-up period, at least one episode of acute otitis media was seen in 21% of the control group and in 12% of the xylitol-treated group ($P = 0.04$). In addition, significantly fewer antibiotics were administered to the xylitol-treated group than to the control group. In a subsequent study from the same group, 857 children were recruited from Finnish day-care centers and randomized to one of five treatment groups: control syrup, xylitol syrup, control chewing gum, xylitol chewing gum, or xylitol lozenges (55). During the 3-month follow-up period after randomization, the occurrence of acute otitis media and the use of antibiotic therapy were significantly lower in those children receiving xylitol syrup or gum than in children in the control group. Finally, in a third randomized controlled trial, 270 children with tympanostomy tubes placed for chronic effusion or recurrent otitis were randomized to receive either xylitol or placebo (56). In this trial no differences were noted in the rates of otitis media between the control and xylitol-treated groups. Further large-scale studies of this agent have not been conducted, but it remains an interesting but unproven prevention strategy.

BACTEREMIA IN CHILDREN

In the mid-1970s it was recognized that a significant number of children between the ages of 3 and 36 months with high fever but no focal findings on examination had occult pneumococcal bacteremia (6, 42, 44, 53). With the elimination of *Haemophilus influenzae* type b (Hib) disease after the licensure of the Hib conjugate vaccine in 1990, the rate of bacteremia in febrile children 3 to 36 months of age was reported to be approximately 2%, with *S. pneumoniae* responsible for 83 to 92% of the cases (2, 38). The impact of delayed antibiotic therapy for children with occult bacteremia has been studied; meta-analyses have estimated that children with occult pneumococcal bacteremia who are not treated with antibiotics at the initial evaluation have a 2.7 to 5.8% risk of subsequently developing meningitis (7, 47).

Between 1984 and 1995, reports of occult bacteremia due to intermediate-resistant *S. pneumoniae* strains appeared. However, since therapy generally consisted of parenteral extended- or broad-spectrum cephalosporins, treatment failures were uncommon (33). Then in 1998 Buckingham et al. reported "breakthrough" meningitis in a previously healthy toddler during treatment with parenteral cephalosporins for bacteremia (12). When the susceptibility of that organism was evaluated, cefotaxime and cefuroxime MICs were 2.0 and 8.0 µg/ml, respectively. Raising further concerns about antibiotic resistance and bacteremia, a case of breakthrough pneumonia was subsequently reported for a previously healthy toddler with occult pneumococcal bacteremia; an organism for which the cefuroxime MIC was 8 µg/ml was isolated from this child (22). In an eight-center children's hospital surveillance study of pneumococcal infections, Kaplan et al. reported the outcome of 100 children with *S. pneumoniae* bacteremia secondary to penicillin- and cephalosporin-nonsusceptible infections (34). All but one child, subsequently demonstrated to have severe combined immunodeficiency, were successfully treated with oral or parenteral beta-lactam antibiotics. In another case series, Silverstein et al., reviewing 922 cases of pneumococcal bacteremia, found 56 due to penicillin-nonsusceptible organisms (49). They concluded that "reduced susceptibility to penicillin and cephalosporins did not affect the outcome of pneumococcal bacteremia and with current practices, intermediate penicillin resistance was of little clinical significance in nonmeningitic systemic pneumococcal infections."

Based on these informative studies, an immunocompetent child between the ages of 3 and 36 months with culture-proven pneumococcal bacteremia, without meningitis, due to a nonsusceptible isolate can be effectively treated with a parenteral broad-spectrum cephalosporin as an appropriate initial therapy. Prior to beginning this therapy, repeat blood cultures should be obtained to document persistent bacteremia. If the repeat culture is subsequently found to be negative prior to the start of antibiotic therapy, the clinician may choose to use an oral (amoxicillin at 80 to 90 mg/kg/day) or parenteral (cefotaxime at 100 mg/kg/day or ceftriaxone at 75 mg/kg/day) agent, depending on the clinical status of the child.

BACTEREMIA SINCE LICENSURE OF THE 7-VALENT PNEUMOCOCCAL CONJUGATE VACCINE

In February 2000, a 7-valent capsular polysaccharide-protein conjugate vaccine was licensed for use in infants and children in the United States. Later that same year the vaccine was recommended by public- and private-sector advisory groups for all children <2 years of age and for children between the ages of 24 and 59 months with immunocompromising conditions or chronic illnesses putting them at high risk for *S. pneumoniae* disease (1, 3). In the pivotal efficacy trials conducted at Northern California Kaiser Permanente leading to vaccine licensure, the pneumococcal conjugate vaccine had a 94% efficacy for the prevention of invasive pneumococcal disease due to vaccine serotypes and an 89% efficacy for prevention of invasive disease due to all pneumococcal serotypes (9).

The efficacy trial data are confirmed by a recent publication of population-based data from the Active Bacterial Core Surveillance of the CDC, which documents a striking reduction in invasive *S. pneumoniae* disease in children <2 years of age since vaccine licensure (58). Rates of invasive *S. pneumoniae* disease in this age group fell from 188 cases/100,000 prior to licensure to 59 cases/100,000 after licensure, a 69% decline. Invasive *S. pneumoniae* disease due to the vaccine and vaccine-related serotypes declined by 78 and 50%, respectively, both highly statistically significant. Since most of the serotypes associated with antibiotic nonsusceptibility are included in the 7-valent conjugate vaccine, rates of disease with antibiotic-resistant *S. pneumoniae* organisms have also declined. Finally, a recent cost analysis using invasive disease figures derived since vaccine licensure suggests that when "the rate of occult bacteremia falls below 0.5% with the widespread use of the conjugate pneumococcal vaccine, then strategies that use empirical diagnostic testing and antibiotic treatment should be eliminated" (37). Disease rates are approaching these levels. However, one must acknowledge that some cases of pneumococcal bacteremia will persist. Some cases will be due to nonvaccine serotypes, some will result from vaccine failure in immunocompromised or healthy hosts, and some will be seen in unimmunized or incompletely immunized children. Since it is estimated that high fever without a source accounts for nearly 14% of pediatric office visits, and since cases of pneumococcal bacteremia are now relatively uncommon, it is unlikely that the total number of office visits for fever without source will decrease as a result of the routine administration of the conjugate vaccine (35).

So what is a practitioner to do? For infants less than 2 months of age with fever, appropriate blood, urine, and cerebrospinal fluid cultures should be obtained; hospitalization and parenteral antibiotic therapy should be started. Initial empiric therapy with ampicillin and gentamicin or ampicillin and cefotaxime appears appropriate. This recommendation is supported by recent data on the rates of neonatal pneumococcal bacteremia from the surveillance of eight children's hospitals in the United States from 1993 until the present time (30). Out of a total of 4,428 cases of pneumococcal bacteremia, only 29 cases were seen in infants <30 days of age. The mean age of presentation was 18 days;

infants were likely to be full-term and to present with pneumonia, meningitis, or acute otitis media. A recent case report of neonatal pneumococcal sepsis in association with fatal maternal pneumococcal sepsis also appeared (32).

If a child has documented pneumococcal bacteremia after vaccination, it should be reported to the CDC via their website http://www.cdc.gov/nip. Reporting should document the vaccination history, including dates of vaccination and the source of vaccine and lot number. One should also obtain the pneumococcal isolate for serotyping and susceptibility testing, and acute- and convalescent-phase sera for measurement of the immune responses to vaccine and disease.

In summary, given the current rate of decline in invasive pneumococcal disease in children 3 to 36 months of age, it is time to reconsider replacing routine diagnostic studies and empirical use of parenteral antimicrobials in febrile children without focal findings with careful observation and clinical judgment.

BACTEREMIA IN ADULTS

The story for adults is much different; occult bacteremia is rarely described. A large prospective international observational study of 844 consecutive adult patients hospitalized with blood cultures positive for *S. pneumoniae* from December 1998 to January 2001 was recently reported (61). In this study the mean age of the study patients was 52.1 years, with a range of 15 to 97 years. One or more underlying chronic medical conditions were present in 48% of the subjects, and 20% were seriously ill and admitted to intensive care units. Seven hundred ninety-three of the 844 pneumococcal isolates were available for in vitro susceptibility testing. Of these isolates, 15% had intermediate susceptibility to penicillin (MIC = 0.12 to 1 μg/ml), 9.6% were resistant to penicillin (MIC > 2 μg/ml), and 13 isolates were highly resistant (MIC ≥ 3 μg/ml). A striking difference was noted in the prevalence of penicillin-nonsusceptible pneumococci by region. The percentages of nonsusceptible *S. pneumoniae* strains in the various countries are as follows: Taiwan had

57%, Hong Kong had 53%, the United States had 31%, France had 29%, Spain had 28%, South Africa had 24%, Brazil had 19%, New Zealand had 19%, Sweden had 7%, and Argentina had 3%. Using multivariate analyses, underlying disease and prior receipt of at least one antibiotic from 1 to 70 days prior to the episode of bacteremia were found to be significantly associated with antibiotic resistance. Although penicillin resistance was not a risk factor for mortality, patient age, severity of illness on presentation, and an underlying medical condition associated with immunosuppression were all significantly associated with mortality.

Another major finding of this report was the impact of pneumococcal antimicrobial susceptibility on patient outcome. Concordant antibiotic therapy (receipt of a single antibiotic with in vitro activity against *S. pneumoniae* within the first 2 days of therapy) versus discordant therapy (inactive in vitro) was assessed at 14 days. Discordant therapy with penicillin, cefotaxime, and ceftriaxone did not result in a higher mortality rate. In contrast, cefuroxime therapy of pneumococcal bacteremia in subjects in which the cefuroxime MIC was >8 μg/ml was associated with higher mortality, and this association was most apparent in subjects who were not critically ill. Discordant therapy with cefuroxime also showed a trend for slower defervescence, although statistical significance was not attained (Table 1). The conclusions of this large study are that

TABLE 1 Factors evaluated for mortality in 360 patients receiving antibiotic monotherapy on multivariate analyses[a]

Factor	*P* value
Critical illness	0.0001
Discordant therapy	
Cefuroxime	0.0175
Penicillin	NS
Ceftriaxone or cefotaxime	NS

[a] On multivariate analysis, severity of illness and cefuroxime-associated discordant therapy were significantly associated with higher mortality rate, but discordant therapy involving penicillin or ceftriaxone or cefotaxime was not. NS, not significant. Modified from reference 61.

the use of penicillin and broad-spectrum cephalosporins such as ceftriaxone or cefotaxime remains effective for the treatment of pneumococcal bacteremia and that cefuroxime should be used with caution when the MIC is >8 µg/ml.

PNEUMONIA IN CHILDREN

Although *S. pneumoniae* is widely regarded as the most common cause of bacterial pneumonia in children, definitive confirmation of the organism from blood or pleural fluid occurs less than 50% of the time, thus making the impact of increasing antibiotic resistance on clinical outcome difficult to discern (50). One of the first studies to compare the clinical characteristics and outcomes in children with pneumonia attributable to penicillin-susceptible and -nonsusceptible strains of *S. pneumoniae* was reported by Tan et al. (51).

They evaluated 254 children diagnosed with pneumococcal pneumonia from 1 September 1993 to 31 August 1996 who were enrolled in the eight-children's-hospital United States Pediatric Multicenter Pneumococcal Surveillance Study. Most of these children were hospitalized and treated with extended- or broad-spectrum cephalosporins. No significant differences were found in the clinical presentation or outcome of therapy between children with penicillin-susceptible and -nonsusceptible pneumococcal organisms. Conclusions of this study were that "hospitalized patients were more likely to have underlying illnesses, multiple lobe involvement, and the presence of pleural effusions. . . and in otherwise normal patients with pneumonia attributable to penicillin-resistant pneumococcal isolates, therapy with standard beta-lactam agents was effective." More recently the same group conducted a retrospective study of 368 children hospitalized with confirmed pneumococcal pneumonia from 1 September 1993 to 31 January 2000 (52). Of the 368 isolates obtained from the study children, 47 (12.8%) were intermediate and 37 (10.1%) were resistant to penicillin; 18 (5%) were intermediate to ceftriaxone, and 9 (2.5%) were resistant to ceftriaxone. Over the study period the proportion of hospitalized patients with complicated pneumococcal pneumonia increased from 22.6% in 1994 to 53% in 1999. Despite this increase, 98% of the patients recovered from their pneumonia. Moreover, antibiotic-resistant organisms were not seen more often in complicated disease.

Hardie et al. also evaluated 64 cases of complicated parapneumonic effusions in children during a 4-year period at a large children's hospital (26). Of the 23 cases of *S. pneumoniae* empyema, 17 were penicillin susceptible and 6 were nonsusceptible. Patients with effusions growing nonsusceptible organisms had a higher incidence of bacteremia than those with penicillin-susceptible organisms, but no significant differences in duration of chest tube drainage, febrile days, oxygen use, or hospital stay were noted between the susceptible and nonsusceptible groups. In a recent publication summarizing international trials evaluating the clinical efficacy of high-dose penicillin or ampicillin against pneumococcal pneumonia caused by susceptible, intermediate, and resistant strains, there were no significant differences in the morbidity or mortality of patients infected with the different strains (17).

In summary, therapy of pneumococcal pneumonia with penicillin, ampicillin, or cefuroxime is adequate for children hospitalized with pneumococcal pneumonia due to isolates for which the penicillin MIC is ≤2 µg/ml (50). Outpatient pneumonia can be treated with amoxicillin (80 to 90 mg/kg/day divided twice a day), amoxicillin-clavulanate (80 to 90 mg/kg/day divided twice a day), cefuroxime axetil, or cefdinir. Quinolones are used often in adults but have a limited role in children because of concerns over cartilage damage in growing children and because other antimicrobials are available for the therapy of pneumonia. However, when the penicillin or the cephalosporin MIC is ≥4 µg/ml, alternative agents like clindamycin or vancomycin should be considered.

PNEUMONIA IN ADULTS

Pneumonia is the sixth most common cause of death in the United States. Annually 2 to 3 million cases of community-acquired pneumonia result in approximately 10 million physician visits, 500,000 hospitalizations, and 45,000 deaths in the United States (41). The incidence of pneumococcal pneumonia of all serotypes is estimated to be 258 cases per 100,000 population in all ages and 962 cases per 100,000 in persons aged ≥65 years. Mortality among hospitalized adults averages about 14%, compared to <1% for those not hospitalized (24). The Infectious Disease Society of America (IDSA) has published guidelines for the management of community-acquired pneumonia in adults and has recently updated them (8, 40). These two articles provide evidence for each therapeutic guideline and an exhaustive list of references.

The IDSA guidelines recommend that two blood cultures and a Gram stain and culture of the expectorated sputum be obtained prior to antibiotic therapy in hospitalized patients. The rationale for this approach is to improve the care of the individual patient by identifying the pathogen, determining its susceptibility to the antibiotics that have been started empirically, and allowing the therapy to be narrowed when the agent is identified. For the initial empirical therapy of outpatient pneumonia, a macrolide, doxycycline, or a fluoroquinolone with enhanced activity against *S. pneumoniae* is recommended. For hospitalized patients with pneumonia, an extended-spectrum cephalosporin (cefotaxime or ceftriaxone) combined with a macrolide or a fluoroquinolone alone is recommended. For adult patients in the intensive care unit, ceftriaxone, cefotaxime, ampicillin–sulbactam, or piperacillin-tazobactam in combination with a fluoroquinolone or macrolide is recommended by the IDSA guidelines. Parenteral antibiotics can be switched to oral agents when the patient is clinically improving, which is usually accomplished in 3 to 5 days.

S. pneumoniae is the most commonly identified cause of community-acquired pneumonia in adults, accounting for between 9 and 55% of all cases requiring hospitalization (28). The IDSA guidelines specifically address pneumococcal pneumonia since it accounts for about two-thirds of bacteremic pneumonia and is the most frequently lethal community-acquired pneumonia. When it has been determined that pneumococcal pneumonia is present and the organism is sensitive to penicillin, penicillin or amoxicillin is the agent of choice. Cefotaxime or ceftriaxone is the preferred parenteral agent for the treatment of pneumococcal pneumonia without meningitis for strains for which the MIC is <2 μg/ml. Amoxicillin is the preferred antibiotic for oral treatment of pneumococcal pneumonia involving susceptible strains (40). Since the publication of the IDSA guidelines in 2000, fluoroquinolone use has increased, resistance to these agents has been reported, and clinical failures have been described (18). Vancomycin, linezolid, and quinupristin-dalfopristin are the only drugs with proven in vitro activity against strains for which the MIC is >4 μg/ml.

Regarding the outcomes of adults with pneumococcal pneumonia with susceptible and nonsusceptible strains, a number of studies have recently appeared that suggest there are no differences in morbidity and mortality in adults (14, 57). In contrast, one CDC study found the mortality associated with nonsusceptible *S. pneumoniae* to be increased threefold when the organisms were resistant to penicillin and increased sevenfold when due to ceftriaxone-resistant strains (23). However, this study was criticized for not determining the nature of the treatment in each case (36).

In summary, the IDSA guidelines for adults suggest that proven pneumococcal pneumonia, even in critically ill patients, should be treated with cefotaxime (1 g every 6 to 8 h) or ceftriaxone (1 g every 12 to 24 h). Many patients have received ampicillin (1 to 2 g every 6 h) with good response. Although vancomycin would provide good coverage for all resistant strains, this agent should be used very carefully and its use restricted to highly resistant strains and for the therapy of meningitis as will be outlined below. Linezolid and quinupristin-dalfopristin are other therapies that could be considered.

MENINGITIS

S. pneumoniae is the most common cause of meningitis since the licensure of the Hib conjugate vaccine and is the bacterial organism associated with the highest case-fatality rate (21%). Nonetheless, the rate of pneumococcal meningitis remains rather low, with approximately 1.1 cases/100,000 children and adults combined (48). Antibiotic-resistant pneumococcal meningitis, not unexpectedly, has become a concern, with over 36% of cases due to nonsusceptible strains. Since the licensure of the Hib conjugate vaccine, bacterial meningitis in the United States has become a disease predominantly of adults rather than of infants and young children. Currently it is estimated that there are over 3,000 cases of pneumococcal meningitis diagnosed each year in the United States. Because the blood-brain barrier is the most difficult barrier to circumvent, the therapy of pneumococcal meningitis in the era of increasing antibiotic resistance has had to be modified.

THERAPY FOR MENINGITIS IN CHILDREN

As a result of the increasing rates of antibiotic resistance, the Committee on Infectious Diseases of the American Academy of Pediatrics established new guidelines for the treatment of children with meningitis. For infants and children older than 1 month with suspected bacterial meningitis, the combination of high-dose cefotaxime (225 to 300 mg/kg/day in three or four divided doses) or ceftriaxone (100 mg/kg/day in one or two divided doses) plus vancomycin (60 mg/kg/day in four divided doses) has been recommended (4). After the susceptibility of the isolate is determined, then the therapy can be tailored to the specific pathogen.

Several studies have compared the morbidity and mortality of pneumococcal meningitis in children with susceptible and nonsusceptible strains. Arditi and colleagues reported on 180 children with pneumococcal meningitis enrolled in the United States Pediatric Multicenter Pneumococcal Surveillance Study from 1 September 1993 to 31 August 1996 (5). Although 14 (7.7%) died, no deaths were due to documented treatment failure with a resistant strain and only one child had bacteriological failure with a penicillin-resistant (MIC = 2 μg/ml) organism. Follow-up of the survivors demonstrated that 41 (25%) had motor deficits and 48 (32%) had unilateral ($n = 26$) or bilateral ($n = 22$) moderate to severe hearing loss. Overall, 12.7 and 6.6% of the pneumococcal isolates were intermediate and resistant to penicillin and 4.4 and 2.8% were intermediate and resistant to ceftriaxone, respectively. Clinical presentation, cerebrospinal fluid indices on admission, hospital course, morbidity rates, and mortality rates were similar for patients infected with penicillin- or ceftriaxone-susceptible versus -nonsusceptible organisms. However, the relatively small numbers of nonsusceptible isolates and the inclusion of vancomycin in the treatment regimen for the majority of the patients limited the power of this study to detect significant differences between the groups. Nonetheless, the results showed that nonsusceptible strains were not intrinsically more virulent than susceptible ones.

In another review by Buckingham et al., 86 cases of culture-confirmed pneumococcal meningitis in children were reported; 34 isolates were nonsusceptible to penicillin (12 resistant), and 17 were nonsusceptible to cefotaxime (12 resistant) (13). Interestingly, antibiotic susceptibility was not significantly associated with death, intensive care unit admission, mechanical ventilation, focal neurologic deficits, seizures, secondary fever, abnormal neuroimaging studies, or hospital days. However, children with penicillin-resistant isolates had significantly higher median blood leukocyte counts (24,100 versus 15,700/μl; $P = 0.03$), lower median cerebrospinal fluid protein concentrations (85 versus 219 mg/dl;

$P = 0.04$), and higher cerebrospinal fluid glucose concentrations ($\geq$50 mg/dl in 7 of 11 versus 15 of 68; $P = 0.009$) and had lower rates of sensorineural hearing loss (1 of 8 versus 25 of 40; $P = 0.02$) than children with isolates that were susceptible to penicillin. These data also support the findings of Arditi et al. that nonsusceptible strains are not more virulent than susceptible ones.

Finally, pneumococcal isolates tolerant to vancomycin have been reported in cases of meningitis associated with poor therapeutic responses. Vancomycin tolerance and its role in treatment failures will need to be carefully monitored (43).

USE OF STEROIDS IN CHILDREN WITH MENINGITIS

In 1991 the results of a placebo-controlled, double-blind trial of dexamethasone therapy conducted with 101 infants and children with culture-proven bacterial meningitis or clinical signs and cerebrospinal fluid findings of meningitis were reported (46). Subjects were randomized to receive either dexamethasone and cefotaxime ($n = 52$) or cefotaxime plus placebo ($n = 49$). In the group receiving dexamethasone (0.15 mg/kg), it was given 15 to 20 min before the first dose of cefotaxime and was repeated every 6 h for 4 days. By 12 h after the beginning of therapy, the mean opening cerebrospinal pressure and the estimated cerebral perfusion pressure had improved significantly in the dexamethasone-treated children but worsened in the children treated only with cefotaxime; the meningeal inflammation and the concentrations of tumor necrosis factor alpha and platelet-activating factor in the cerebrospinal fluid had decreased in the dexamethasone-treated children and increased in the control group; and after 24 h the clinical condition and mean prognostic score were significantly better in the dexamethasone treated group than among the controls. At follow-up examination after a mean of 15 months, 14% of the dexamethasone-treated children and 38% of the surviving controls had one or more

neurologic or audiological sequelae ($P = 0.007$). The relative risk of sequelae for a child receiving placebo compared with a dexamethasone-treated child was 3.8 (95% confidence interval [CI], 1.3 to 11.5). Since this study was published before the widespread use of Hib immunization, the majority of cases in this report were secondary to Hib meningitis. This complicates the interpretation of the data for cases of pneumococcal meningitis.

In contrast to these findings, in the study by Arditi et al. mentioned earlier, 40 children treated with dexamethasone ($\geq$8 doses) initiated before or within 1 h after the first dose of antibiotics had significantly greater moderate or severe hearing loss (46%) than children not receiving dexamethasone (23%) (5). The incidence of any neurologic deficits also was significantly higher in the dexamethasone-treated group (55 versus 33%). However, children in the dexamethasone group more frequently required mechanical ventilation and had lower initial concentrations of glucose in the cerebrospinal fluid than children who did not receive any dexamethasone. Thus, when severity of illness and the incidence of any deafness and/or any neurologic sequelae were controlled, there were no longer significant differences between the dexamethasone and control groups.

In 1997, McIntyre et al. reported the results of a meta-analysis to evaluate the effectiveness of dexamethasone in bacterial meningitis in the subcategories of causative organism and timing and nature of antibiotic therapy (45). Randomized, concurrently controlled trials published from 1988 to November 1996 were selected. Of 16 studies identified, 5 were excluded. In Hib meningitis, dexamethasone reduced severe hearing loss overall (combined odds ratio [OR], 0.31; 95% CI, 0.14 to 0.69). In pneumococcal meningitis, only studies in which dexamethasone was given early suggested protection, which was significant for severe hearing loss (combined OR, 0.09; 95% CI, 0.0 to 0.71) and approached significance for any neurologic or hearing deficit (combined OR, 0.23; 95% CI, 0.04 to 1.05). For all organisms combined, the OR suggested protection against neurologic deficits other than hearing loss but was not significant (OR, 0.59; 95% CI, 0.34 to 1.02). Outcomes were similar in studies that used 2 versus more than 2 days of dexamethasone therapy. Adverse effects were not significantly increased with dexamethasone except for secondary fever. The incidence of gastrointestinal tract bleeding increased with longer duration of dexamethasone treatment (0.5% in controls, 0.8% with 2 days of treatment, 3.0% with 4 days of treatment) (45).

Many have recommended a repeat lumbar puncture in pneumococcal meningitis to ensure sterilization of the cerebrospinal fluid after 36 to 48 h of antibiotic therapy when the MIC of cefotaxime or ceftriaxone for the organism is >2 μg/ml. This appears particularly important if dexamethasone therapy is being used concomitantly (33). Whether dexamethasone use should become routine practice in cases of pediatric meningitis remains controversial. Some recent data in experimental models of pneumococcal meningitis demonstrate increased apoptosis of cells in the dentate gyrus when animals are given dexamethasone (39). This has raised concern whether similar processes might occur in children. As stated by the American Academy of Pediatrics in the *Red Book*, "experts vary in recommending the use of corticosteroids in pneumococcal meningitis; data are not sufficient to demonstrate a clear benefit in children" (4).

MENINGITIS IN ADULTS

As mentioned earlier, the review of meningitis by Schuchat et al. demonstrated that the case-fatality rate in adults was 24%, much higher than the fatality rates reported for children (48). In addition, the morbidity in adults was also considerable. However, limited data are available on clinical outcomes of meningitis due to nonsusceptible *S. pneumoniae* strains in adults. Fiore et al. analyzed population-based active-surveillance data from 109 cases of pneumococcal meningitis from November 1994 to April 1996 in both children and adults. Pneumococcal isolates were resistant to cefotaxime in 9% of cases and had intermediate susceptibility in 11% (25). Although children were more likely to have cephalosporin-nonsusceptible pneumococcal meningitis, the mortality rate was significantly higher among adults aged 18 to 64 years. Upon admission vancomycin was given to 29% of patients, and within 48 h of admission it was given to 52%. Nonsusceptibility to cefotaxime was not associated with increased mortality, prolonged length of hospital or intensive care unit stay, requirement of intubation or oxygen, intensive care unit care, discharge to another medical or long-term-care facility, or neurologic deficit. The empirical use of vancomycin likely influenced these findings.

Recommended empirical therapy for adults with suspected bacterial meningitis includes administration of both vancomycin and a broad-spectrum cephalosporin (ceftriaxone or cefotaxime) intravenously. When the organism is found to be a pneumococcus susceptible to penicillin, monotherapy with penicillin can be used. If the pneumococcal organism is resistant to penicillin but susceptible to broad-spectrum cephalosporins, the latter should be used. If the organism is resistant to penicillin and the broad-spectrum cephalosporins, then vancomycin or rifampin plus a broad-spectrum cephalosporin is appropriate.

USE OF STEROIDS IN ADULTS WITH MENINGITIS

A large prospective randomized double-blind multicenter trial of the use of dexamethasone in bacterial meningitis in adults was recently completed by de Gans and van de Beek (19). In this trial adults with bacterial meningitis were randomized to receive either placebo or dexamethasone (10 mg) administered 15 to 20 min before or with the first dose of antibiotics and given every 6 h for the first 4 days. Of the 301 subjects enrolled, 157 received dexamethasone and 144 received placebo; baseline characteristics of the two groups were similar. Dexamethasone therapy was associated with a statistically significant reduction in the risk of an unfavorable outcome (relative risk [RR] = 0.59 [95% CI, 0.37 to 0.94]; P = 0.03) and was also associated with reduced mortality (RR = 0.48 [95% CI, 0.24 to 0.96]; P = 0.04). In patients with culture-proven pneumococcal meningitis, there was also improved outcome noted in the dexamethasone-treated group (RR = 0.50 [95% CI, 0.30 to 0.83]; P = 0.006). The authors concluded that early treatment with dexamethasone improved the outcome of acute bacterial meningitis in adults.

SUMMARY OF THE TREATMENT OF PNEUMOCOCCAL INFECTIONS IN CHILDREN AND ADULTS

Tables 2 and 3 summarize therapeutic options for pneumococcal infections in children and adults. Whether the increased use of conjugate

TABLE 2 Antibiotic therapy for pneumococcal infections in children

Infection	Agent	Dosage[a]
Acute otitis media		
First-line therapy	Oral amoxicillin	80–90 mg/kg/day ÷ BID
For clinical defined treatment failures at 3 days (second-line therapy)	Oral amoxicillin-clavulanate	80–90 mg of the amoxicillin component/kg/day ÷ BID
	Oral cefuroxime axetil	30 mg/kg/day ÷ BID
	Oral cefdinir	14 mg/kg/day ÷ BID
	Intramuscular ceftriaxone	50 mg/kg/day (1–3 days)
Pneumonia		
Outpatient pneumonia	Oral amoxicillin	80–90 mg/kg/day ÷ BID
	Oral amoxicillin-clavulanate	80–90 mg of amoxicillin/kg/day ÷ BID
	Oral cefuroxime axetil	30 mg/kg/day ÷ BID
	Oral cefdinir	14 mg/kg twice daily ÷ BID
Inpatient pneumonia (initial empirical therapy)	Intravenous penicillin *or* Intravenous cefuroxime	250,000–400,000 U/kg/day ÷ q4–6h 200–240 mg/kg/day ÷ q8h
Pneumococcal-susceptible to penicillin (MIC ≤ 2 μg/ml)	Intravenous penicillin	250,000–400,000 U/kg/day ÷ q4–6h
Pneumococcal, nonsusceptible to penicillin, susceptible to cephalosporin	Intravenous cefuroxime *or* Intravenous ceftriaxone *or* Intravenous cefotaxime	200–240 mg/kg/day ÷ q8h 100 mg/kg/day ÷ q12h 225–300 mg/kg/day ÷ q8h
Pneumococcal, nonsusceptible to penicillin, nonsusceptible to cephalosporin	Intravenous vancomycin	40–60 mg/kg/day ÷ q6h
Meningitis		
Initial empirical therapy	Vancomycin *and* Ceftriaxone *or* Cefotaxime	60 mg/kg/day ÷ q6h 100 mg/kg/day ÷ q12h 225–300 mg/kg/day ÷ q8h
Pneumococcal, susceptible to penicillin	Penicillin	250,000–400,000 U/kg/day ÷ q4–6h
Pneumococcal, resistant to penicillin, susceptible to cephalosporins	Ceftriaxone *or* Cefotaxime	100 mg/kg/day ÷ q12h 225–300 mg/kg/day ÷ q8h
Resistant to penicillin and cephalosporins	Vancomycin *and* Ceftriaxone *or* Cefotaxime Rifampin may also be added to vancomycin	60 mg/kg/day ÷ q6h 100 mg/kg/day ÷ q12h 225–300 mg/kg/day ÷ q8h 20 mg/kg/day ÷ q12h

[a] BID, twice a day; q4–6h, every 4 to 6 h.

TABLE 3 Antibiotic therapy for pneumococcal infections in adults

Infection	Agent	Dosage[a]
Pneumonia		
Outpatient pneumonia	Macrolides	
	Erythromycin	500 mg q6h
	Azithromycin	500 mg q.d.
	Clarithromycin	500 mg BID
	or	
	Doxycycline	
	or	
	Antipneumococcal fluoroquinolone	
	Moxifloxacin	400 mg q.d.
	Gatifloxacin	400 mg q.d.
	Levofloxacin	500 mg q.d.
Inpatient pneumonia (initial empirical therapy)	Intravenous cefotaxime	1 g q8h
	or	
	Ceftriaxone *plus*	2 g q.d.
	Macrolides	
	Erythromycin	500 mg q6h
	Azithromycin	500 mg q.d.
	Clarithromycin	500 mg BID
	or	
	Antipneumococcal fluoroquinolone alone	
	Moxifloxacin	400 mg q.d.
	Gatifloxacin	400 mg q.d.
	Levofloxacin	500 mg q.d.
Pneumococcal, susceptible to penicillin (MIC ≤ 2 μg/ml)	Intravenous penicillin	1–2 million U
Pneumococcal, nonsusceptible to penicillin, susceptible to cephalosporin	Intravenous ceftriaxone	2 g q.d.
	Intravenous cefotaxime	1 g q8h
Pneumococcal, nonsusceptible to penicillin, nonsusceptible to cephalosporin	Intravenous vancomycin	1 g q12h
Meningitis		
Initial empirical therapy	Vancomycin	15 mg/kg q6h
	Ceftriaxone *or*	2g q12h
	Cefotaxime	2g q4–6h
Pneumococcal, susceptible to penicillin	Penicillin	3–4 million U q4h
	Ampicillin	2 g q4h
Pneumococcal, resistant to penicillin, susceptible to cephalosporins	Ceftriaxone *or*	2 g q12h
	Cefotaxime	2 g q4–6h
Resistant to penicillin and cephalosporins	Vancomycin *and*	15 mg/kg q6h
	Ceftriaxone *or*	2 g q12h
	Cefotaxime	2 g q4–6h
	Rifampin may also be added to vancomycin	600 mg q24h

[a] q6h, every 6 h; q.d., once a day; BID, twice a day.

vaccines and the reduced rates of inappropriate antibiotic use will lead to decreased antibiotic resistance to the pneumococcus in the future remains to be determined. One can only hope that this will be the case.

REFERENCES

1. **Advisory Committee on Immunization Practices.** 2000. Preventing pneumococcal disease among infants and young children. Recommendations of the Advisory Committee on Immunization Practices (ACIP). *Morb. Mortal. Wkly. Rep.* **49:**1–35.

2. **Alpern, E. R., E. A. Alessandrini, L. M. Bell, K. N. Shaw, and K. L. McGowan.** 2000. Occult bacteremia from a pediatric emergency department: current prevalence, time to detection, and outcome. *Pediatrics* **106:**505–511.

3. **American Academy of Pediatrics.** 2000. American Academy of Pediatrics, Committee on Infectious Diseases. Policy statement: recommendations for the prevention of pneumococcal infections, including the use of pneumococcal conjugate vaccine (Prevnar), pneumococcal polysaccharide vaccine, and antibiotic prophylaxis. *Pediatrics* **106:**362–366.

4. **American Academy of Pediatrics, Committee on Infectious Diseases.** 2003. *Red Book: Report of the Committee on Infectious Diseases.* American Academy of Pediatrics, Elk Grove Village, Ill.

5. **Arditi, M., E. O. Mason, Jr., J. S. Bradley, T. Q. Tan, W. J. Barson, G. E. Schutze, E. R. Wald, L. B. Givner, K. S. Kim, R. Yogev, and S. L. Kaplan.** 1998. Three-year multicenter surveillance of pneumococcal meningitis in children: clinical characteristics, and outcome related to penicillin susceptibility and dexamethasone use. *Pediatrics* **102:**1087–1097.

6. **Baraff, L. J., J. W. Bass, G. R. Fleisher, J. O. Klein, G. H. McCracken, Jr., K. R. Powell, and D. L. Schriger.** 1993. Practice guideline for the management of infants and children 0 to 36 months of age with fever without source. Agency for Health Care Policy and Research. *Ann. Emerg. Med.* **22:**1198–1210.

7. **Baraff, L. J., S. I. Lee, and D. L. Schriger.** 1993. Outcomes of bacterial meningitis in children: a meta-analysis. *Pediatr. Infect. Dis. J.* **12:**389–394.

8. **Bartlett, J. G., S. F. Dowell, L. A. Mandell, T. M. File, Jr., D. M. Musher, and M. J. Fine for the Infectious Diseases Society of America.** 2000. Practice guidelines for the management of community-acquired pneumonia in adults. *Clin. Infect. Dis.* **31:**347–382.

9. **Black, S., H. Shinefield, B. Fireman, E. Lewis, P. Ray, J. R. Hansen, L. Elvin, K. M. Ensor, J. Hackell, G. Siber, F. Malinoski, D. Madore, I. Chang, R. Kohberger, W. Watson, R. Austrian, K. Edwards, and Northern California Kaiser Permanente Vaccine Study Center Group.** 2000. Efficacy, safety and immunogenicity of heptavalent pneumococcal conjugate vaccine in children. *Pediatr. Infect. Dis. J.* **19:**187–195.

10. **Block, S. L., J. Hedrick, C. J. Harrison, R. Tyler, A. Smith, R. Findlay, and E. Keegan.** 2002. Pneumococcal serotypes from acute otitis media in rural Kentucky. *Pediatr. Infect. Dis. J.* **21:**859–865.

11. **Block, S. L., J. A. Hedrick, J. Kratzer, M. A. Nemeth, and K. J. Tack.** 2000. Five-day twice daily cefdinir therapy for acute otitis media: microbiologic and clinical efficacy. *Pediatr. Infect. Dis. J.* **19:**S153–S158.

12. **Buckingham, S. C., S. P. Brown, and V. H. Joaquin.** 1998. Breakthrough bacteremia and meningitis during treatment with cephalosporins parenterally for pneumococcal pneumonia. *J. Pediatr.* **132:**174–176.

13. **Buckingham, S. C., J. A. McCullers, J. Lujan-Zilbermann, K. M. Knapp, K. L. Orman, and B. K. English.** 2001. Pneumococcal meningitis in children: relationship of antibiotic resistance to clinical characteristics and outcomes. *Pediatr. Infect. Dis. J.* **20:**837–843.

14. **Castillo, E. M., L. S. Rickman, S. K. Brodine, E. K. Ledbetter, and C. Kelly.** 2000. Streptococcus pneumoniae: bacteremia in an era of penicillin resistance. *Am. J. Infect. Control* **28:**239–243.

15. **Dagan, R.** 2001. Treatment of acute otitis media—challenges in the era of antibiotic resistance. *Vaccine* **19**(Suppl. 1):S9–S16.

16. **Dagan, R., A. Hoberman, C. Johnson, E. L. Leibovitz, A. Arguedas, F. V. Rose, B. R. Wynne, and M. R. Jacobs.** 2001. Bacteriologic and clinical efficacy of high dose amoxicillin/clavulanate in children with acute otitis media. *Pediatr. Infect. Dis. J.* **20:**829–837.

17. **Dagan, R., K. P. Klugman, W. A. Craig, and F. Baquero.** 2001. Evidence to support the rationale that bacterial eradication in respiratory tract infection is an important aim of antimicrobial therapy. *J. Antimicrob. Chemother.* **47:**129–140.

18. **Davidson, R., R. Cavalcanti, J. L. Brunton, D. J. Bast, J. C. de Azavedo, P. Kibsey, C. Fleming, and D. E. Low.** 2002. Resistance to levofloxacin and failure of treatment of pneumococcal pneumonia. *N. Engl. J. Med.* **346:**747–750.

19. **de Gans, J., and D. van de Beek.** 2002. Dexamethasone in adults with bacterial meningitis. *N. Engl. J. Med.* **347:**1549–1556.

20. **Doern, G. V., A. B. Brueggemann, H. Huynh, and E. Wingert.** 1999. Antimicrobial resistance with Streptococcus pneumoniae in the United States, 1997–98. *Emerg. Infect. Dis.* **5:** 757–765.

21. **Dowell, S. F., J. C. Butler, G. S. Giebink, M. R. Jacobs, D. Jernigan, D. M. Musher, A. Rakowsky, and B. Schwartz.** 1999. Acute otitis media: management and surveillance in an era of pneumococcal resistance—a report from the Drug-Resistant Streptococcus pneumoniae Therapeutic Working Group. *Pediatr. Infect. Dis. J.* **18:**1–9.

22. **Dowell, S. F., T. Smith, K. Leversedge, and J. Snitzer.** 1999. Failure of treatment of pneumonia associated with highly resistant pneumococci in a child. *Clin. Infect. Dis.* **29:**462–463.

23. **Feikin, D. R., A. Schuchat, M. Kolczak, N. L. Barrett, L. H. Harrison, L. Lefkowitz, A. McGeer, M. M. Farley, D. J. Vugia, C. Lexau, K. R. Stefonek, J. E. Patterson, and J. H. Jorgensen.** 2000. Mortality from invasive pneumococcal pneumonia in the era of antibiotic resistance, 1995–1997. *Am. J. Public Health* **90:** 223–229.

24. **Fine, M. J., M. A. Smith, C. A. Carson, S. S. Mutha, S. S. Sankey, L. A. Weissfeld, and W. N. Kapoor.** 1996. Prognosis and outcomes of patients with community-acquired pneumonia. A meta-analysis. *JAMA* **275:**134–141.

25. **Fiore, A. E., J. F. Moroney, M. M. Farley, L. H. Harrison, J. E. Patterson, J. H. Jorgensen, M. Cetron, M. S. Kolczak, R. F. Breiman, and A. Schuchat.** 2000. Clinical outcomes of meningitis caused by Streptococcus pneumoniae in the era of antibiotic resistance. *Clin. Infect. Dis.* **30:**71–77.

26. **Hardie, W. D., N. E. Roberts, S. F. Reising, and C. D. Christie.** 1998. Complicated parapneumonic effusions in children caused by penicillin-nonsusceptible Streptococcus pneumoniae. *Pediatrics* **101:**388–392.

27. **Hausdorff, W. P., G. Yothers, R. Dagan, T. Kilpi, S. I. Pelton, R. Cohen, M. R. Jacobs, S. L. Kaplan, C. Levy, E. L. Lopez, E. O. Mason, Jr., V. Syriopoulou, B. Wynne, and J. Bryant.** 2002. Multinational study of pneumococcal serotypes causing acute otitis media in children. *Pediatr. Infect. Dis. J.* **21:**1008–1016.

28. **Heffelfinger, J. D., S. F. Dowell, J. H. Jorgensen, K. P. Klugman, L. R. Mabry, D. M. Musher, J. F. Plouffe, A. Rakowsky, A. Schuchat, and C. G. Whitney.** 2000. Management of community-acquired pneumonia in the era of pneumococcal resistance: a report from the Drug-Resistant Streptococcus pneumoniae Therapeutic Working Group. *Arch. Intern. Med.* **160:**1399–1408.

29. **Henderson, F. W., P. H. Gilligan, K. Wait, and D. A. Goff.** 1988. Nasopharyngeal carriage of antibiotic-resistant pneumococci by children in group day care. *J. Infect. Dis.* **157:**256–263.

30. **Hoffman, J. A., E. O. Mason, G. E. Schutze, T. Q. Tan, W. J. Barson, L. B. Givner, E. R. Wald, J. S. Bradley, R. Yogev, and S. L. Kaplan.** 2003. Streptococcus pneumoniae infections in the neonate. *Pediatrics* **112:**1095–1102.

31. **Howie, V. M., and J. H. Ploussard.** 1969. The "in vivo sensitivity test"—bacteriology of middle ear exudate, during antimicrobial therapy in otitis media. *Pediatrics* **44:**940–944.

32. **Hughes, B. R., J. L. Mercer, and L. B. Gosbel.** 2001. Neonatal pneumococcal sepsis in association with fatal maternal pneumococcal sepsis. *Aust. N. Z. J. Obstet. Gynaecol.* **41:**457–458.

33. **Kaplan, S. L., and E. O. Mason, Jr.** 2002. Mechanisms of pneumococcal antibiotic resistance and treatment of pneumococcal infections in 2002. *Pediatr. Ann.* **31:**250–260.

34. **Kaplan, S. L., E. O. Mason, Jr., W. J. Barson, T. Q. Tan, G. E. Schutze, J. S. Bradley, L. B. Givner, K. S. Kim, R. Yogev, and E. R. Wald.** 2001. Outcome of invasive infections outside the central nervous system caused by Streptococcus pneumoniae isolates nonsusceptible to ceftriazone in children treated with beta-lactam antibiotics. *Pediatr. Infect. Dis. J.* **20:**392–396.

35. **Klein, J. O.** 2002. Management of the febrile child without a focus of infection in the era of universal pneumococcal immunization. *Pediatr. Infect. Dis. J.* **21:**584–588. (Discussion, **21:**613–614.)

36. **Klugman, K. P., and C. Feldman.** 2001. Streptococcus pneumoniae respiratory tract infections. *Curr. Opin. Infect. Dis.* **14:**173–179.

37. **Lee, G. M., G. R. Fleisher, and M. B. Harper.** 2001. Management of febrile children in the age of the conjugate pneumococcal vaccine: a cost-effectiveness analysis. *Pediatrics* **108:**835–844.

38. **Lee, G. M., and M. B. Harper.** 1998. Risk of bacteremia for febrile young children in the post-Haemophilus influenzae type b era. *Arch. Pediatr. Adolesc. Med.* **152:**624–628.

39. **Leib, S. L., C. Heimgartner, Y. D. Bifrare, J. M. Loeffler, and M. G. Taauber.** 2003. Dexamethasone aggravates hippocampal apoptosis and learning deficiency in pneumococcal meningitis in infant rats. *Pediatr. Res.* **54:** 353–357.

40. **Mandell, L. A., J. G. Bartlett, S. F. Dowell, T. M. File, Jr., D. M. Musher, and C. Whitney.** 2003. Update of practice guidelines for the

management of community-acquired pneumonia in immunocompetent adults. *Clin. Infect. Dis.* **37**:1405–1433.

41. **Marston, B. J., J. F. Plouffe, T. M. File, Jr., B. A. Hackman, S. J. Salstrom, H. B. Lipman, M. S. Kolczak, R. F. Breiman, and The Community-Based Pneumonia Incidence Study Group.** 1997. Incidence of community-acquired pneumonia requiring hospitalization. Results of a population-based active surveillance study in Ohio. *Arch. Intern. Med.* **157**:1709–1718.

42. **McCarthy, P. L., G. W. Grundy, S. Z. Spiesel, and T. F. Dolan, Jr.** 1976. Bacteremia in children: an outpatient clinical review. *Pediatrics* **57**:861–868.

43. **McCullers, J. A., B. K. English, and R. Novak.** 2000. Isolation and characterization of vancomycin-tolerant Streptococcus pneumoniae from the cerebrospinal fluid of a patient who developed recrudescent meningitis. *J. Infect. Dis.* **181**:369–373.

44. **McGowan, J. E., Jr., L. Bratton, J. O. Klein, and M. Finland.** 1973. Bacteremia in febrile children seen in a "walk-in" pediatric clinic. *N. Engl. J. Med.* **288**:1309–1312.

45. **McIntyre, P. B., C. S. Berkey, S. M. King, U. B. Schaad, T. Kilpi, G. Y. Kanra, and C. M. Perez.** 1997. Dexamethasone as adjunctive therapy in bacterial meningitis. A meta-analysis of randomized clinical trials since 1988. *JAMA* **278**:925–931.

46. **Odio, C. M., I. Faingezicht, M. Paris, M. Nassar, A. Baltodano, J. Rogers, X. Saez-Llorens, K. D. Olsen, and G. H. McCracken, Jr.** 1991. The beneficial effects of early dexamethasone administration in infants and children with bacterial meningitis. *N. Engl. J. Med.* **324**:1525–1531.

47. **Rothrock, S. G., M. B. Harper, S. M. Green, M. C. Clark, R. Bachur, D. P. McIlmail, P. A. Giordano, and J. L. Falk.** 1997. Do oral antibiotics prevent meningitis and serious bacterial infections in children with Streptococcus pneumoniae occult bacteremia? A meta-analysis. *Pediatrics* **99**:438–444.

48. **Schuchat, A., K. Robinson, J. D. Wenger, L. H. Harrison, M. Farley, A. L. Reingold, L. Lefkowitz, B. A. Perkins, and Active Surveillance Team.** 1997. Bacterial meningitis in the United States in 1995. *N. Engl. J. Med.* **337**:970–976.

49. **Silverstein, M., R. Bachur, and M. B. Harper.** 1999. Clinical implications of penicillin and ceftriaxone resistance among children with pneumococcal bacteremia. *Pediatr. Infect. Dis. J.* **18**:35–41.

50. **Tan, T. Q.** 2003. Antibiotic resistant infections due to Streptococcus pneumoniae: impact on therapeutic options and clinical outcome. *Curr. Opin. Infect. Dis.* **16**:271–277.

51. **Tan, T. Q., E. O. Mason, Jr., W. J. Barson, E. R. Wald, G. E. Schutze, J. S. Bradley, M. Arditi, L. B. Givner, R. Yogev, K. S. Kim, and S. L. Kaplan.** 1998. Clinical characteristics and outcome of children with pneumonia attributable to penicillin-susceptible and penicillin-nonsusceptible Streptococcus pneumoniae. *Pediatrics* **102**:1369–1375.

52. **Tan, T. Q., E. O. Mason, Jr., E. R. Wald, W. J. Barson, G. E. Schutze, J. S. Bradley, L. B. Givner, R. Yogev, K. S. Kim, and S. L. Kaplan.** 2002. Clinical characteristics of children with complicated pneumonia caused by Streptococcus pneumoniae. *Pediatrics* **110**:1–6.

53. **Teele, D. W., S. I. Pelton, M. J. Grant, J. Herskowitz, D. J. Rosen, C. E. Allen, R. S. Wimmer, and J. O. Klein.** 1975. Bacteremia in febrile children under 2 years of age: results of cultures of blood of 600 consecutive febrile children seen in a "walk-in" clinic. *J. Pediatr.* **87**:227–230.

54. **Uhari, M., T. Kontiokari, M. Koskela, and M. Niemela.** 1996. Xylitol chewing gum in prevention of acute otitis media: double blind randomised trial. *BMJ* **313**:1180–1184.

55. **Uhari, M., T. Kontiokari, and M. Niemela.** 1998. A novel use of xylitol sugar in preventing acute otitis media. *Pediatrics* **102**:879–884.

56. **Uhari, M., T. Tapiainen, and T. Kontiokari.** 2000. Xylitol in preventing acute otitis media. *Vaccine* **19**(Suppl. 1):S144–147.

57. **Watanabe, H., S. Sato, K. Kawakami, K. Watanabe, K. Oishi, N. Rikitomi, T. Ii, H. Ikeda, A. Sato, and T. Nagatake.** 2000. A comparative clinical study of pneumonia by penicillin-resistant and -sensitive Streptococcus pneumoniae in a community hospital. *Respirology* **5**:59–64.

58. **Whitney, C. G., M. M. Farley, J. Hadler, L. H. Harrison, N. M. Bennett, R. Lynfield, A. Reingold, P. R. Cieslak, T. Pilishvili, D. Jackson, R. R. Facklam, J. H. Jorgensen, and A. Schuchat.** 2003. Decline in invasive pneumococcal disease after the introduction of protein-polysaccharide conjugate vaccine. *N. Engl. J. Med.* **348**:1737–1746.

59. **Whitney, C. G., M. M. Farley, J. Hadler, L. H. Harrison, C. Lexau, A. Reingold, L. Lefkowitz, P. R. Cieslak, M. Cetron, E. R. Zell, J. H. Jorgensen, and A. Schuchat.** 2000. Increasing prevalence of multidrug-resistant Streptococcus pneumoniae in the United States. *N. Engl. J. Med.* **343**:1917–1924.

60. **Yagupsky, P., N. Porat, D. Fraser, F. Prajgrod, M. Merires, L. McGee, K. P. Klugman, and R. Dagan.** 1998. Acquisition, carriage, and transmission of pneumococci with decreased antibiotic susceptibility in young children attending a day care facility in southern Israel. *J. Infect. Dis.* **177:**1003–1012.

61. **Yu, V. L., C. C. Chiou, C. Feldman, A. Ortqvist, J. Rello, A. J. Morris, L. M. Baddour, C. M. Luna, D. R. Snydman, M. Ip, W. C. Ko, M. B. Chedid, A. Andremont, and K. P. Klugman.** 2003. An international prospective study of pneumococcal bacteremia: correlation with in vitro resistance, antibiotics administered, and clinical outcome. *Clin. Infect. Dis.* **37:**230–237.

62. **Zenni, M. K., S. H. Cheatham, J. M. Thompson, G. W. Reed, A. B. Batson, P. S. Palmer, K. L. Holland, and K. M. Edwards.** 1995. Streptococcus pneumoniae colonization in the young child: association with otitis media and resistance to penicillin. *J. Pediatr.* **127:**533–537.

CLINICAL RELEVANCE OF ANTIBIOTIC RESISTANCE IN PNEUMOCOCCAL INFECTIONS

Keith P. Klugman

20

INTRODUCTION

While the global emergence of antimicrobial resistance in the pneumococcus has been documented in a variety of surveillance networks and individual research studies, the clinical relevance of that resistance can best be measured by the use of pharmacodynamics. Antimicrobial susceptibility (measured by MIC) is compared to the achievable unbound fraction of antibiotic at the site of the infection. The measures derived from the relationship between drug levels and MIC describe the ability of a given drug to kill or inhibit bacterial growth, and they provide tools to measure the impact of resistance on bacterial eradication and clinical outcome (10).

Pharmacodynamic principles predict success when beta-lactam, or trimethoprim-sulfamethoxazole, drug levels exceed the MIC for 40 to 50% of the dosing interval. The pharmacodynamic principles which predict the clinical impact of resistance to most other

classes of antimicrobials, including the azalides and fluoroquinolones, suggest that the ratio of the area under the curve of the 24-h drug concentration divided by the MIC should exceed 25 for the treatment of pneumococcal infections in immunocompetent adults and children (33) (Table 1). The clinical relevance of antibiotic resistance in the treatment of pneumococcal pneumonia has recently been reviewed (37). This chapter updates that review and expands it to the consideration of other pneumococcal diseases such as meningitis, otitis media, sinusitis, exacerbations of chronic bronchitis, and the limited literature on other types of infection such as infections of the pleura and endocarditis.

PNEUMONIA

Beta-Lactam Agents

Pharmacodynamics predict that high doses of intravenous penicillin remain useful for the treatment of pneumococcal pneumonia up to MICs of 4 µg/ml (5, 33). Bacteriologically confirmed failures of high-dose intravenous penicillin therapy of pneumococcal pneumonia have not been reported to date, although some patients will die despite likely bacterial eradication if they are at high risk of mortal-

Keith P. Klugman, Department of International Health, Rollins School of Public Health and Division of Infectious Diseases, School of Medicine, Emory University, 1518 Clifton Rd. NE, Room 764, Atlanta, GA 30322, and University of the Witwatersrand/Medical Research Council/National Institute for Communicable Diseases, Respiratory and Meningeal Pathogens Research Unit, Johannesburg, South Africa.

The Pneumococcus, Volume Editor, Elaine I. Tuomanen,
© 2004 ASM Press, Washington, D.C.

TABLE 1 Pharmacodynamic breakpoints above which antibiotics used to treat *Streptococcus pneumoniae* infections outside of the meninges are predicted to fail[a]

Drug	Route of administration	Pharmacodynamic breakpoint (μg/ml)
Amoxicillin	Oral	2
Cefaclor	Oral	0.5
Cefuroxime	Oral	1
Cefprozil	Oral	1
Loracarbef	Oral	0.5
Cefixime	Oral	0.5
Erythromycin	Oral	0.25
Clarithromycin	Oral	0.25
Azithromycin	Oral	0.12
Ciprofloxacin	Oral	1
Ofloxacin	Oral	2
Levofloxacin	Oral	1
Gatifloxacin	Oral	1
Moxifloxacin	Oral	1
Penicillin G	i.v.	4
Ampicillin	i.v.	2
Cefuroxime	i.v.	4
Cefotaxime	i.v.	2
Ceftriaxone	i.v.	2
Cefepime	i.v.	4
Ceftazidime	i.v.	8
Meropenem	i.v.	1

[a] Drugs given at usual therapeutic doses. Values are the highest free (unbound) drug concentrations achieved for 40% of the dosing interval for beta-lactams other than cephalosporins and 50% of the dosing interval for cephalosporins and macrolides; the values derived for the azalide, azithromycin, and the fluoroquinolones are 24-h free drug areas under the drug concentration curve (AUC) divided by 25, thought to predict successful treatment of immunocompetent persons. Immunocompromised persons may only be able to be cured of infections caused by strains for which the azalide or fluoroquinolone MICs are five times lower (AUC/125). i.v., intravenous. Table reproduced from reference 33.

ity, regardless of the susceptibility of the organism (3, 23, 24). A large number of studies with both children (9, 16, 26, 27, 34, 58) and adults (20, 49, 51, 53, 54, 59, 68) have failed to document an association between outcome and resistance when patients have been stratified by severity of disease and the beta-lactam drugs given were intravenous penicillin, ampicillin, amoxicillin, cefotaxime, or ceftriaxone. The only two studies that have found such an association either could not stratify for severity of disease (22) or did not document the susceptibility of the pneumococci to the drugs that failed (22, 63). Pharmacodynamic breakpoints do, however, suggest that beta-lactams less active against pneumococci may lead to failure. Examples of such failures are recorded for cefazolin (60), cefuroxime (6, 19, 68), and ticarcillin (54). Few data are available on the failure of oral beta-lactam therapy in pneumonia caused by penicillin-resistant strains leading to persistence of the strain and development of bacteremia. Such failures may be anticipated given the observation that bacteremia with susceptible strains may follow inadequate dosing of oral amoxicillin (66) and the observed failures of low-dose amoxicillin for the prophylaxis of penicillin-resistant pneumococcal bacteremia in patients with sickle cell disease (1).

Macrolides

There are now many case reports documenting the failure of macrolides against resistant strains leading to strain persistence and the development of pneumococcal bacteremia (25, 36, 40, 45, 61, 67). These data are consistent with two recent series in which concordance of macrolide resistance in blood isolates and previous macrolide therapy was sought (44, 66). Of particular concern is the recent observation that macrolide resistance due to mutational changes in rRNA genes and/or the genes of ribosomal proteins may lead to the emergence of resistance during therapy, associated with clinical failure (8, 35, 52, 57).

Fluoroquinolones

The pharmacodynamics of ciprofloxacin and levofloxacin do not support their use for the management of pneumococcal pneumonia (62), and clinical failures have been documented by many authors especially in the face of emerging resistance to these agents (15, 21, 31, 56, 57, 65). Resistance emerges during therapy, and patients who are immunocompromised may be at particular risk, perhaps because they may be colonized with pneumococci for prolonged periods (2).

Trimethoprim-Sulfamethoxazole

Bacteremic pneumonia caused by a resistant strain has been described following trimethoprim-sulfamethoxazole therapy in a child (40) and an adult (43) and following prophylaxis with this agent (46), suggesting that the MICs of the agent for resistant strains probably exceed the levels achievable by oral dosing.

Tetracycline

The failure of tetracycline therapy, associated with recurrent sputum isolation of resistant strains, has been recorded in the setting of pneumonia (30, 55, 64).

Streptogramins and Ketolides

The failure of pristinamycin with growth of the resistant strain from blood on therapy has been documented (7). Although neither ketolides nor streptogramins were used to treat this case of pneumococcal pneumonia, a patient in whom macrolide use was continued despite the presence of a macrolide-resistant organism was found to be infected with pneumococci resistant to both streptogramins and ketolides (57).

EMPYEMA

The development of empyema caused by beta-lactam-resistant strains has been described (49). Although a causal relationship with beta-lactam resistance was suggested in that study, more data are needed to confirm the risk of empyema due to failure of beta-lactams to eradicate beta-lactam-resistant pneumococcal strains from patients with empyema. In contrast, the emergence during therapy of a macrolide-resistant pneumococcus from the pus of an empyema associated with clinical failure of the erythromycin therapy has been reported (17).

EXACERBATIONS OF CHRONIC BRONCHITIS

The emergence of tetracycline resistance (4, 30, 55, 64) is well described and is associated with clinical failure in patients previously exposed to this class of antibiotics during the treatment of exacerbations of chronic bronchitis. Persistence of resistant strains in patients treated with trimethoprim has been reported (50), and more recently the emergence of resistance and treatment failure in an exacerbation of chronic bronchitis have been reported with levofloxacin (15).

SINUSITIS

There are few data confirming the bacteriological failure of sinusitis therapy associated with resistant pneumococci. One such case is the development of sinusitis and bacteremia during cotrimoxazole prophylactic therapy in a 17-year-old receiving azothioprine and prednisone (47).

OTITIS MEDIA

The relationship between adverse clinical outcome and failure to sterilize the middle ear fluid during therapy of acute otitis media has been established (14). Double-tympanocentesis studies have clearly demonstrated the relevance of pharmacodynamic principles for the prediction of bacterial eradication from the middle ear (10, 11). These types of studies have defined the persistence of bacterial pathogens, including the pneumococcus, in the middle ear following placebo (32), which allows comparison of the eradication of resistant strains to the placebo effect. The efficacy of cefaclor approaches that of placebo (12), as the time above MIC for pneumococci with intermediate-level penicillin resistance (MIC, 0.1 to 1 μg/ml) achieved by this agent is zero (33). In contrast, there is an intermediate level of eradication (21%) (12) for cefuroxime which achieves 35% time above the MIC for 90% of penicillin-intermediately resistant strains (33). A more active beta-lactam, such as high-dose (90 mg/kg of body weight/day) amoxicillin given with clavulanate, achieves well in excess of 50% time above MIC for these strains and achieved an eradication of 100% (13). Against penicillin-resistant pneumococci, the only oral beta-lactam achieving more than 40% time above the MIC is high-dose amoxicillin, which was associated with a 9% bacteriological failure against such strains in a double-tap tympanocentesis study (13). A regimen of 3 days of ceftriaxone given intramuscularly once daily has been shown to be more successful in the eradication of penicillin-nonsusceptible strains (3% failure) than has a one-day regimen (48% failure) (42). Azithromycin given for 3 or 5 days is similar to placebo in the eradication of macrolide-resistant pneumococci (11), as is trimethoprim-sulfamethoxazole for the eradication of strains resistant to that agent (41).

ENDOCARDITIS

There has been one large study of penicillin-resistant pneumococcal endocarditis reported (48). The authors conclude that intermediately penicillin-resistant strains respond to high-dose penicillin but that cephalosporin-resistant strains should be treated with a combination of cefotaxime or ceftriaxone plus vancomycin.

MENINGITIS

As antimicrobial penetration into the cerebrospinal fluid is limited by the blood-brain barrier, lower levels of resistance are associated with clinical failure, which has been shown to occur even with intermediately beta-lactam-resistant strains. Clinical failure of penicillin therapy for the treatment of penicillin-intermediately resistant pneumococcal meningitis, associated with increased neurologic sequelae and death, was shown in a prospective study (29). Similarly, cefotaxime and ceftriaxone have failed when used as monotherapy for the treatment of pneumococcal meningitis caused by strains either intermediately or fully resistant to these drugs (defined as MIC of 1 or ≥2 μg/ml, respectively) (38). An innovative approach to the assessment of combination therapy has been the assessment of the bactericidal activity against cephalosporin-resistant pneumococci in cerebrospinal fluid taken from patients receiving monotherapy or combination therapy. These studies suggest the superiority of ceftriaxone combination with vancomycin or rifampin (39), as well as the combination of cefotaxime plus vancomycin (18), over the use of ceftriaxone or cefotaxime alone (28) for the treatment of cephalosporin-resistant pneumococcal meningitis.

FAILURE OF PENICILLIN PROPHYLAXIS

One of the consequences of emerging penicillin resistance is the potential failure of penicillin for the prophylaxis of bacteremia caused by penicillin-resistant strains. There is evidence that low doses of amoxicillin given as prophylaxis to patients with sickle cell disease (125 mg twice daily for children <2 years of age and 250 mg twice daily for older children) may lead to higher proportions of penicillin-resistant pneumococcal bacteremias in patients compliant with that program. Children compliant with prophylaxis were at 2.2 times increased risk of infection with penicillin-intermediately resistant strains than susceptible strains and at 2.5 times increased risk for penicillin-resistant pneumococcal strains (1). These data are consistent with the inability of low-dose amoxicillin to achieve levels in blood sufficient to prevent disease caused by resistant strains.

CONCLUSIONS

Pharmacodynamic principles explain the clinical failures observed with the emergence of resistance in some classes of antibiotics and also explain the successful continued use of the more active drugs despite the emergence of resistance. They thus allow the development of rational guidelines for the treatment of infections caused by antibiotic-resistant pneumococci.

REFERENCES

1. Adamkiewicz, T. V., S. Sarnaik, G. R. Buchanan, R. V. Iyer, S. T. Miller, C. H. Pegelow, Z. R. Rogers, E. Vichinsky, J. Elliott, R. R. Facklam, K. L. O'Brien, B. Schwartz, C. A. Van Beneden, M. J. Cannon, J. R. Eckman, H. Keyserling, K. Sullivan, W. Y. Wong, and W. C. Wang. 2003. Invasive pneumococcal infections in children with sickle cell disease in the era of penicillin prophylaxis, antibiotic resistance, and 23-valent pneumococcal polysaccharide vaccination. *J. Pediatr.* 143:438–444.

2. Anderson, K. B., J. S. Tan, T. M. File, Jr., J. R. DiPersio, B. M. Willey, and D. E. Low. 2003. Emergence of levofloxacin-resistant pneumococci in immunocompromised adults after therapy for community-acquired pneumonia. *Clin. Infect. Dis.* 37:376–381.

3. Austrian, R., and J. Gold. 1964. Pneumococcal bacteremia with special reference to bacteremic pneumococcal pneumonia. *Ann. Intern. Med.* 60:759–776.

4. Bizzozero, O. J., Jr., and V. T. Andriole. 1969. Tetracycline-resistant pneumococcal infection. Incidence, clinical presentation, and laboratory evaluation. *Arch. Intern. Med.* 123:388–393.

5. Bryan, C. S., R. Talwani, and M. S. Stinson. 1997. Penicillin dosing for pneumococcal pneumonia. *Chest* 112:1657–1664.

6. Buckingham, S. C., S. P. Brown, and V. H. Joaquin. 1998. Breakthrough bacteremia and meningitis during treatment with cephalosporins parenterally for pneumococcal pneumonia. *J. Pediatr.* 132:174–176.

7. Burucoa, C., T. Pasdeloup, C. Chapon, J. L. Fauchere, and R. Robert. 1995. Failure of pristinamycin treatment in a case of pneumococcal pneumonia. *Eur. J. Clin. Microbiol. Infect. Dis.* 14:341–342.

8. Butler, J. C., J. L. Lennox, L. K. McDougal, J. A. Sutcliffe, A. Tait-Kamradt, and F. C. Tenover. 2003. Macrolide-resistant pneumococcal endocarditis and epidural abscess that develop during erythromycin therapy. *Clin. Infect. Dis.* 36:e19–e25.

9. Choi, E. H., and H. J. Lee. 1998. Clinical outcome of invasive infections by penicillin-resistant *Streptococcus pneumoniae* in Korean children. *Clin. Infect. Dis.* 26:1346–1354.

10. Craig, W. A. 1998. Pharmacokinetic/pharmacodynamic parameters: rationale for antibacterial dosing of mice and men. *Clin. Infect. Dis.* 26:1–10; quiz, 11–12.

11. Dagan, R. 2003. Achieving bacterial eradication using pharmacokinetic/pharmacodynamic principles. *Int. J. Infect. Dis.* 7(Suppl. 1):S21–S26.

12. Dagan, R., O. Abramson, E. Leibovitz, R. Lang, S. Goshen, D. Greenberg, P. Yagupsky, A. Leiberman, and D. M. Fliss. 1996. Impaired bacteriologic response to oral cephalosporins in acute otitis media caused by pneumococci with intermediate resistance to penicillin. *Pediatr. Infect. Dis. J.* 15:980–985.

13. Dagan, R., A. Hoberman, C. Johnson, E. L. Leibovitz, A. Arguedas, F. V. Rose, B. R. Wynne, and M. R. Jacobs. 2001. Bacteriologic and clinical efficacy of high dose amoxicillin/clavulanate in children with acute otitis media. *Pediatr. Infect. Dis. J.* 20:829–837.

14. Dagan, R., E. Leibovitz, D. Greenberg, P. Yagupsky, D. M. Fliss, and A. Leiberman. 1998. Early eradication of pathogens from middle ear fluid during antibiotic treatment of acute otitis media is associated with improved clinical outcome. *Pediatr. Infect. Dis. J.* 17:776–782.

15. Davidson, R., R. Cavalcanti, J. L. Brunton, D. J. Bast, J. C. de Azavedo, P. Kibsey, C. Fleming, and D. E. Low. 2002. Resistance to levofloxacin and failure of treatment of pneumococcal pneumonia. *N. Engl. J. Med.* 346:747–750.

16. Deeks, S. L., R. Palacio, R. Ruvinsky, D. A. Kertesz, M. Hortal, A. Rossi, J. S. Spika, J. L. Di Fabio, and The *Streptococcus pneumoniae* Working Group. 1999. Risk factors and course of illness among children with invasive penicillin-resistant *Streptococcus pneumoniae*. *Pediatrics* 103:409–413.

17. Dixon, J. M. S. 1967. Pneumococcus resistant to erythromycin and lincomycin. *Lancet* i:573.

18. Doit, C., J. Barre, R. Cohen, S. Bonacorsi, A. Bourrillon, and E. H. Bingen. 1997.

Bactericidal activity against intermediately cephalosporin-resistant *Streptococcus pneumoniae* in cerebrospinal fluid of children with bacterial meningitis treated with high doses of cefotaxime and vancomycin. *Antimicrob. Agents Chemother.* **41:**2050–2052.

19. **Dowell, S. F., T. Smith, K. Leversedge, and J. Snitzer.** 1999. Failure of treatment of pneumonia associated with highly resistant pneumococci in a child. *Clin. Infect. Dis.* **29:**462–463.

20. **Einarsson, S., M. Kristjansson, K. G. Kristinsson, G. Kjartansson, and S. Jonsson.** 1998. Pneumonia caused by penicillin-non-susceptible and penicillin-susceptible pneumococci in adults: a case-control study. *Scand. J. Infect. Dis.* **30:**253–256.

21. **Empey, P. E., H. R. Jennings, A. C. Thornton, R. P. Rapp, and M. E. Evans.** 2001. Levofloxacin failure in a patient with pneumococcal pneumonia. *Ann. Pharmacother.* **35:**687–690.

22. **Feikin, D. R., A. Schuchat, M. Kolczak, N. L. Barrett, L. H. Harrison, L. Lefkowitz, A. McGeer, M. M. Farley, D. J. Vugia, C. Lexau, K. R. Stefonek, J. E. Patterson, and J. H. Jorgensen.** 2000. Mortality from invasive pneumococcal pneumonia in the era of antibiotic resistance, 1995–1997. *Am. J. Public Health* **90:**223–229.

23. **Feldman, C., J. M. Kallenbach, S. D. Miller, J. R. Thorburn, and H. J. Koornhof.** 1985. Community-acquired pneumonia due to penicillin-resistant pneumococci. *N. Engl. J. Med.* **313:**615–617.

24. **Fine, M. J., M. A. Smith, C. A. Carson, S. S. Mutha, S. S. Sankey, L. A. Weissfeld, and W. N. Kapoor.** 1996. Prognosis and outcomes of patients with community-acquired pneumonia. A meta-analysis. *JAMA* **275:**134–141.

25. **Fogarty, C., R. Goldschmidt, and K. Bush.** 2000. Bacteremic pneumonia due to multidrug-resistant pneumococci in 3 patients treated unsuccessfully with azithromycin and successfully with levofloxacin. *Clin. Infect. Dis.* **31:**613–615.

26. **Friedland, I. R.** 1995. Comparison of the response to antimicrobial therapy of penicillin-resistant and penicillin-susceptible pneumococcal disease. *Pediatr. Infect. Dis. J.* **14:**885–890.

27. **Friedland, I. R., and K. P. Klugman.** 1992. Antibiotic-resistant pneumococcal disease in South African children. *Am. J. Dis. Child.* **146:**920–923.

28. **Friedland, I. R., and K. P. Klugman.** 1997. Cerebrospinal fluid bactericidal activity against cephalosporin-resistant *Streptococcus pneumoniae* in children with meningitis treated with high-dosage cefotaxime. *Antimicrob. Agents Chemother.* **41:**1888–1891.

29. **Friedland, I. R., and K. P. Klugman.** 1992. Failure of chloramphenicol therapy in penicillin-resistant pneumococcal meningitis. *Lancet* **339:**405–408.

30. **Hansman, D., and G. Andrews.** 1967. Hospital infection with pneumococci resistant to tetracycline. *Med. J. Aust.* **i:**498–501.

31. **Ho, P. L., W. S. Tse, K. W. Tsang, T. K. Kwok, T. K. Ng, V. C. Cheng, and R. M. Chan.** 2001. Risk factors for acquisition of levofloxacin-resistant *Streptococcus pneumoniae*: a case-control study. *Clin. Infect. Dis.* **32:**701–707.

32. **Howie, V. M., and J. H. Ploussard.** 1972. Efficacy of fixed combination antibiotics versus separate components in otitis media. Effectiveness of erythromycin estrolate, triple sulfonamide, ampicillin, erythromycin estolate-triple sulfonamide, and placebo in 280 patients with acute otitis media under two and one-half years of age. *Clin. Pediatr.* **11:**205–214.

33. **Jacobs, M. R.** 2001. Optimisation of antimicrobial therapy using pharmacokinetic and pharmacodynamic parameters. *Clin. Microbiol. Infect.* **7:**589–596.

34. **Kaplan, S. L., E. O. Mason, Jr., W. J. Barson, T. Q. Tan, G. E. Schutze, J. S. Bradley, L. B. Givner, K. S. Kim, R. Yogev, and E. R. Wald.** 2001. Outcome of invasive infections outside the central nervous system caused by *Streptococcus pneumoniae* isolates nonsusceptible to ceftriaxone in children treated with beta-lactam antibiotics. *Pediatr. Infect. Dis. J.* **20:**392–396.

35. **Kays, M. B., M. F. Wack, D. W. Smith, and G. A. Denys.** 2002. Azithromycin treatment failure in community-acquired pneumonia caused by *Streptococcus pneumoniae* resistant to macrolides by a 23S rRNA mutation. *Diagn. Microbiol. Infect. Dis.* **43:**163–165.

36. **Kelley, M. A., D. J. Weber, P. Gilligan, and M. S. Cohen.** 2000. Breakthrough pneumococcal bacteremia in patients being treated with azithromycin and clarithromycin. *Clin. Infect. Dis.* **31:**1008–1011.

37. **Klugman, K. P.** 2002. Bacteriological evidence of antibiotic failure in pneumococcal lower respiratory tract infections. *Eur. Respir. J. Suppl.* **36:**3s–8s.

38. **Klugman, K. P.** 1994. Pneumococcal resistance to the third-generation cephalosporins: clinical, laboratory and molecular aspects. *Int. J. Antimicrob. Agents* **4:**63–67.

39. **Klugman, K. P., I. R. Friedland, and J. S. Bradley.** 1995. Bactericidal activity against cephalosporin-resistant *Streptococcus pneumoniae* in cerebrospinal fluid of children with acute bacterial meningitis. *Antimicrob. Agents Chemother.* **39:**1988–1992.

40. Klugman, K. P., H. J. Koornhof, V. Kuhnle, S. D. Miller, P. J. Ginsburg, and A. C. Mauff. 1986. Meningitis and pneumonia due to novel multiply resistant pneumococci. *Br. Med. J.* **292:**730.

41. Leiberman, A., E. Leibovitz, L. Piglansky, S. Raiz, J. Press, P. Yagupsky, and R. Dagan. 2001. Bacteriologic and clinical efficacy of trimethoprim-sulfamethoxazole for treatment of acute otitis media. *Pediatr. Infect. Dis. J.* **20:**260–264.

42. Leibovitz, E., L. Piglansky, S. Raiz, J. Press, A. Leiberman, and R. Dagan. 2000. Bacteriologic and clinical efficacy of one day vs. three day intramuscular ceftriaxone for treatment of nonresponsive acute otitis media in children. *Pediatr. Infect. Dis. J.* **19:**1040–1045.

43. Linares, J., J. L. Perez, J. Garau, and R. Martin. 1983. Cotrimoxazole resistance in pneumococci. *Eur. J. Clin. Microbiol.* **2:**473–474.

44. Lonks, J. R., J. Garau, L. Gomez, M. Xercavins, A. Ochoa de Echaguen, I. F. Gareen, P. T. Reiss, and A. A. Medeiros. 2002. Failure of macrolide antibiotic treatment in patients with bacteremia due to erythromycin-resistant *Streptococcus pneumoniae*. *Clin. Infect. Dis.* **35:**556–564.

45. Lonks, J. R., and A. A. Medeiros. 1993. High rate of erythromycin and clarithromycin resistance among *Streptococcus pneumoniae* isolates from blood cultures from Providence, R.I. *Antimicrob. Agents Chemother.* **37:**1742–1745.

46. Madhi, S. A., K. Petersen, A. Madhi, A. Wasas, and K. P. Klugman. 2000. Impact of human immunodeficiency virus type 1 on the disease spectrum of *Streptococcus pneumoniae* in South African children. *Pediatr. Infect. Dis. J.* **19:**1141–1147.

47. Markman, M., J. Mannisi, J. D. Dick, B. Filburn, G. W. Santos, and R. Saral. 1982. Sulfamethoxazole-trimethoprim-resistant pneumococcal sepsis. *JAMA* **248:**3011–3012.

48. Martinez, E., J. M. Miro, B. Almirante, J. M. Aguado, P. Fernandez-Viladrich, M. L. Fernandez-Guerrero, J. L. Villanueva, F. Dronda, A. Moreno-Torrico, M. Montejo, P. Llinares, and J. M. Gatell. 2002. Effect of penicillin resistance of Streptococcus pneumoniae on the presentation, prognosis, and treatment of pneumococcal endocarditis in adults. *Clin. Infect. Dis.* **35:**130–139.

49. Metlay, J. P., J. Hofmann, M. S. Cetron, M. J. Fine, M. M. Farley, C. Whitney, and R. F. Breiman. 2000. Impact of penicillin susceptibility on medical outcomes for adult patients with bacteremic pneumococcal pneumonia. *Clin. Infect. Dis.* **30:**520–528.

50. Moore, E. P., and E. W. Williams. 1988. Hospital transmission of multiply antibiotic-resistant *Streptococcus pneumoniae*. *J. Infect.* **16:**199–200.

51. Moroney, J. F., A. E. Fiore, L. H. Harrison, J. E. Patterson, M. M. Farley, J. H. Jorgensen, M. Phelan, R. R. Facklam, M. S. Cetron, R. F. Breiman, M. Kolczak, and A. Schuchat. 2001. Clinical outcomes of bacteremic pneumococcal pneumonia in the era of antibiotic resistance. *Clin. Infect. Dis.* **33:**797–805.

52. Musher, D. M., M. E. Dowell, V. D. Shortridge, R. K. Flamm, J. H. Jorgensen, P. Le Magueres, and K. L. Krause. 2002. Emergence of macrolide resistance during treatment of pneumococcal pneumonia. *N. Engl. J. Med.* **346:**630–631.

53. Pallares, R., O. Capdevila, J. Linares, I. Grau, H. Onaga, F. Tubau, M. H. Schulze, P. Hohl, and F. Gudiol. 2002. The effect of cephalosporin resistance on mortality in adult patients with nonmeningeal systemic pneumococcal infections. *Am. J. Med.* **113:**120–126.

54. Pallares, R., J. Linares, M. Vadillo, C. Cabellos, F. Manresa, P. F. Viladrich, R. Martin, and F. Gudiol. 1995. Resistance to penicillin and cephalosporin and mortality from severe pneumococcal pneumonia in Barcelona, Spain. *N. Engl. J. Med.* **333:**474–480.

55. Percival, A., E. C. Armstrong, and G. C. Turner. 1969. Increased incidence of tetracycline-resistant pneumococci in Liverpool in 1968. *Lancet* **i:**998–1000.

56. Perez-Trallero, E., J. M. Garcia-Arenzana, J. A. Jimenez, and A. Peris. 1990. Therapeutic failure and selection of resistance to quinolones in a case of pneumococcal pneumonia treated with ciprofloxacin. *Eur. J. Clin. Microbiol. Infect. Dis.* **9:**905–906.

57. Perez-Trallero, E., J. M. Marimon, L. Iglesias, and J. Larruskain. 2003. Fluoroquinolone and macrolide treatment failure in pneumococcal pneumonia and selection of multidrug-resistant isolates. *Emerg. Infect. Dis.* **9:**1159–1162.

58. Quach, C., K. Weiss, D. Moore, E. Rubin, A. McGeer, and D. E. Low. 2002. Clinical aspects and cost of invasive Streptococcus pneumoniae infections in children: resistant vs. susceptible strains. *Int. J. Antimicrob. Agents* **20:**113–118.

59. Roson, B., J. Carratala, F. Tubau, J. Dorca, J. Linares, R. Pallares, F. Manresa, and F. Gudiol. 2001. Usefulness of beta-lactam therapy for community-acquired pneumonia in the era of drug-resistant Streptococcus pneumoniae: a randomized study of amoxicillin-clavulanate and ceftriaxone. *Microb. Drug Resist.* **7:**85–96.

60. Sacho, H., K. P. Klugman, H. J. Koornhof, and P. Ruff. 1987. Community-acquired pneu-

monia in an adult due to a multiply-resistant pneumococcus. *J. Infect.* **14:**188–189.

61. **Sanchez, C., R. Armengol, J. Lite, I. Mir, and J. Garau.** 1992. Penicillin-resistant pneumococci and community-acquired pneumonia. *Lancet* **339:**988.

62. **Tillotson, G., X. Zhao, and K. Drlica.** 2001. Fluoroquinolones as pneumococcal therapy: closing the barn door before the horse escapes. *Lancet Infect. Dis.* **1:**145–146.

63. **Turett, G. S., S. Blum, B. A. Fazal, J. E. Justman, and E. E. Telzak.** 1999. Penicillin resistance and other predictors of mortality in pneumococcal bacteremia in a population with high human immunodeficiency virus seroprevalence. *Clin. Infect. Dis.* **29:**321–327.

64. **Turner, G. C.** 1963. Tetracycline-resistant pneumococci in a general hospital. *Lancet* **ii:**1292–1295.

65. **Urban, C., N. Rahman, X. Zhao, N. Mariano, S. Segal-Maurer, K. Drlica, and J. J. Rahal.** 2001. Fluoroquinolone-resistant *Streptococcus pneumoniae* associated with levofloxacin therapy. *J. Infect. Dis.* **184:**794–798.

66. **Van Kerkhoven, D., W. E. Peetermans, L. Verbist, and J. Verhaegen.** 2003. Breakthrough pneumococcal bacteraemia in patients treated with clarithromycin or oral beta-lactams. *J. Antimicrob. Chemother.* **51:**691–696.

67. **Waterer, G. W., R. G. Wunderink, and C. B. Jones.** 2000. Fatal pneumococcal pneumonia attributed to macrolide resistance and azithromycin monotherapy. *Chest* **118:**1839–1840.

68. **Yu, V. L., C. C. Chiou, C. Feldman, A. Ortqvist, J. Rello, A. J. Morris, L. M. Baddour, C. M. Luna, D. R. Snydman, M. Ip, W. C. Ko, M. B. Chedid, A. Andremont, and K. P. Klugman.** 2003. An international prospective study of pneumococcal bacteremia: correlation with in vitro resistance, antibiotics administered, and clinical outcome. *Clin. Infect. Dis.* **37:**230–237.

MECHANISMS FOR PENICILLIN RESISTANCE IN *STREPTOCOCCUS PNEUMONIAE*: PENICILLIN-BINDING PROTEINS, GENE TRANSFER, AND CELL WALL METABOLISM

Carina Bergmann, Fang Chi, Shwan Rachid,[†]
and Regine Hakenbeck

21

Penicillin resistance in *Streptococcus pneumoniae* has been studied in the laboratory since the early 1940s, long before it became an actual problem in clinical isolates in the late 1970s. Beta-lactamase production has never been observed in this organism; instead, resistance is based on a complex mutational pathway that involves multiple alterations in several penicillin target proteins, the penicillin-binding proteins (PBPs). Observations made in laboratory mutants are distinct from evolutionary mechanisms that occur outside. In laboratory mutants, non-PBP genes also contribute to resistance. Intraspecies gene transfer of PBP variants between commensal streptococci and the pathogen *S. pneumoniae* appears to be responsible for the emergence of clinical isolates and guarantees efficient spread between clones of the pathogen.

PBPs IN *S. PNEUMONIAE* AND THEIR ROLE IN BETA-LACTAM RESISTANCE

Each bacterial species contains a characteristic set of PBPs. These membrane proteins catalyze late steps of murein biosynthesis outside the cytoplasmic membrane. According to the domain structure and conserved amino acid motifs in the N-terminal domain, high-molecular-weight (HMW) class A and class B, and low-molecular-weight (LMW) PBPs are distinguished (for review, see reference 18). The penicillin-binding domain common to all PBPs is believed to perform the crucial penicillin-sensitive reaction: a transpeptidation between two muropeptides resulting in the cross-linked structure of the murein network. It contains three characteristic amino acid sequences: SXXK with the active-site serine and the SXN and the KS(T)G boxes. It is this domain, and only this domain, where mutations crucial for the development of resistance against beta-lactams have been located. The mutations result in a decreased affinity to beta-lactams; hence, a higher concentration of the antibiotic is needed for inhibition of the low-affinity PBP variant and consequently for growth inhibition and lysis in vivo. The N-terminal domain of class A HMW PBPs can act as a transglycosylase, a reaction that can be inhibited by moenomycin. The function of C-terminal domains is not known. The penicillin-binding domain of the LMW PBPs is associated with D,D-carboxypeptidase activity.

PBPs interact with beta-lactam antibiotics by forming a covalent complex via the active-

Carina Bergmann, Fang Chi, Shwan Rachid, and Regine Haken-beck, Department of Microbiology, University of Kaiserslautern, D-67663 Kaiserslautern, Germany.
†Present address: German Research Centre for Biotechnology (GBF), D-38124 Braunschweig, Germany.

The Pneumococcus. Volume Editor, Elaine I. Tuomanen,
© 2004 ASM Press, Washington, D.C.

site serine. Using radioactive penicillin derivatives, the penicilloyl-PBP complexes can be visualized by fluorography (for a review, see reference 23). Recently, fluorescent beta-lactam derivatives have also been used for PBP labeling (8, 59), and antisera as well as monoclonal antibodies against beta-lactams can be used to detect some (but not all) penicilloyl-PBP complexes on Western blots (4, 24). These methods are particularly useful for the detection of low-affinity PBP variants in resistant strains since the antibiotics can be used in high concentrations sufficient to achieve labeling of the PBPs.

S. pneumoniae contains six PBPs: the three HMW class A PBP1a, -1b, and -2a; two HMW class B PBPs, PBP-2x and -2b; and the LMW PBP3 (25). All six PBPs have been found as low-affinity variants associated with beta-lactam resistance as specified below. Low-affinity forms of either PBP2x or PBP2b confer resistance alone; i.e., they are primary resistance determinants (20) (Fig. 1). Both class B HMW PBPs, and only these two, are apparently essential, and attempts to delete them have not been successful (32). They are believed to represent the crucial transpeptidases in the pneumococcus, but no in vitro test system with natural substrates is yet available. Thioester substrates can be used to study the hydrolysis and transfer reactions of the proteins, and this has been proven particularly useful for the characterization of enzymatic activities associated with low-affinity PBPs from resistant strains (1).

The class A HMW PBP1a is responsible for high-level resistance in resistant clinical isolates, provided that the strains contain already altered PBP2b and/or PBP2x (44). PBP2a plays a role as a resistance determinant in laboratory mutants (36). Although it occasionally appears as a low-affinity variant in some clinical isolates and resistant transformants (27), molecular evidence of its function in the resistance process has not yet been obtained (15). PBP1b has been implicated in resistance so far only in *S. pneumoniae* transformants using donor DNA from a high-level resistant *Streptococcus mitis* organism (27). Each of the class A HMW PBPs can be deleted individually, but attempts to isolate PBP1a and PBP2a double mutants have failed, suggesting that they may be able to substitute functionally for each other (30, 46). These studies also suggested that PBP2a functions as the main transglycosylase in *S. pneumoniae*. PBP2a has been shown recently to contain transglycosylase activity in vitro, and it is likely that PBP1a can also perform this function (10).

A low-affinity PBP3 was found to confer cefotaxime resistance in a laboratory mutant (34). PBP3, a D,D-carboxypeptidase (26), can be deleted, but the mutants, which contain a thickened cell wall and a biochemically altered murein, cannot divide regularly (49, 50).

Methicillin-resistant *Staphylococcus aureus* organisms become completely susceptible to beta-lactams upon disruption of the *fem* genes (for factor essential for methicillin resistance). The *S. aureus* FemABX peptidyltransferases are responsible for the formation of the murein interpeptide linkage (for review, see reference 48), and *S. pneumoniae* with mutations in the *fem* homologs *fibAB* produce a cross-link-defective murein (57). Similarly, almost the entire beta-lactam resistance in *S. pneumoniae* collapses when the *femAB* homologs *fibAB*, also called *murMN*, are inactivated (17, 57), and in the susceptible R6 strain, disruption of *fibA* has an impact on penicillin susceptibility (57). It has been shown that MurM is required for high-level penicillin and cefotaxime resistance in a clinical Hungarian isolate (56). On the other hand, high-level oxacillin and cefotaxime resistance associated with alterations in all five HMW PBPs was achieved by transformation of *S. pneumoniae* with chromosomal *S. mitis* DNA without involving the *fibAB* genes, suggesting that they are not necessarily required for high resistance levels (27).

TOLERANCE IS ASSOCIATED WITH BETA-LACTAM-RESISTANT STRAINS

It was observed early that many penicillin-resistant clinical isolates also exhibited a tolerant phenotype: treatment with antibiotic concen-

A

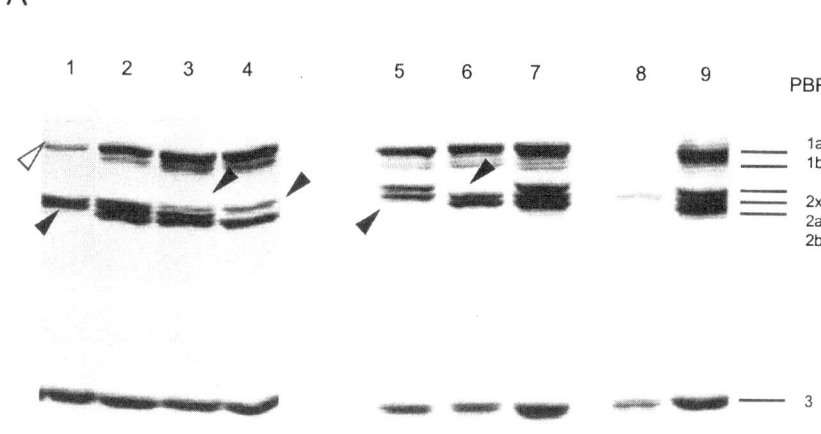

B

Strain	Mutations in PBP	selection steps	MIC (µg/ml) for selective antibiotic	
			cefotaxime	piperacillin
R6	none	none	0.02	0.04
1	2b	5		0.64
3	2x	1	0.16	
4	2x,2a	4	0.32	
5	2b	1		0.01
6	2x	1	0.2	
8	1a,1b,2x,2a,2b	4	30	4

FIGURE 1 PBPs in beta-lactam-resistant *S. pneumoniae*. (A) Cell lysates were incubated with ^{3}[H]propionylampicillin, proteins were separated by sodium dodecyl sulfate-polyacrylamide gel electrophoresis, and PBP–beta-lactam complexes were visualized after fluorography. The six PBPs present in the sensitive parental strain R6 (lanes 2, 7, and 9) are marked on the right. The arrowheads indicate low-affinity PBP variants. Strains were laboratory mutants selected with piperacillin (lane 1) and cefotaxime (lanes 3 and 4) and beta-lactam-resistant transformants of the R6 strain, obtained with chromosomal DNA of a high-level beta-lactam-resistant *S. mitis* strain and selection with piperacillin (lane 5), cefotaxime (lane 6), and four rounds of transformation and selection with cefotaxime and benzylpenicillin (lane 8). (B) MICs and altered PBPs of the strains shown in panel A. No mutation in PBP1a occurs in the mutant shown in lane 1, although the apparent amount of PBP1a is reduced (white arrowhead).

trations above the MIC did not lead to cellular lysis and killing (39). It was then shown that in contrast to the highly lytic penicillins, extended-spectrum cephalosporins such as cefotaxime induce a tolerant response, and this property was associated with the fact that these drugs do not interact with PBP2b (29). It was only a logical conclusion that the low-affinity PBP2b found in most resistant clinical isolates mimics the cefotaxime effect: here the beta-lactams cannot interact with PBP2b, and hence the cells exhibit a tolerant phenotype. Indeed, strains that contain a low-affinity PBP2b as the only resistance determinant exhibit a tolerant phenotype upon beta-lactam treatment, and this is true for laboratory mutants as well as mutants with highly altered *pbp2b* genes from clinical isolates (20, 47).

This scenario has two implications. First, the use of extended-spectrum cephalosporins for treating infections caused by penicillin-resistant *S. pneumoniae* selects easily for cefotaxime resistance since there is one target less compared to penicillins: only *pbp2x* and *pbp1a* play a role in cefotaxime resistance, whereas *pbp2b* is needed for oxacillin and penicillin resistance (3, 14, 27, 44). Indeed, high-level cefotaxime-resistant strains where only *pbp2x* and *pbp1a* were altered have been isolated in the United States, one of them containing a mutation in PBP2x responsible for extremely high cefotaxime resistance levels that was known to be easily selectable with cefotaxime in the laboratory: the T550A change close to the K547SG box (6, 42). In addition, a clone with cephalosporin resistance and penicillin susceptibility carrying altered *pbp1a* and *pbp2x* genes but unaltered *pbp2b* genes has been reported in South Africa (53). Furthermore, the fact that a low-affinity PBP2b mediates tolerance, i.e., a vast increase in the potential to survive drug treatment, but confers only a slight decrease in susceptibility, raises the question of the actual selection pressure for this phenotype. Is it the antibiotic treatment or is it the better survival strategy associated with the mutation that is sufficiently selective? The question as to whether there is a price to be paid for resistance development is a central one. The context described above suggests that there is no such price, rather, a low-affinity PBP2b means double benefit: it is the prerequisite for high-level penicillin resistance, and it allows better survival. In view of the fact that PBP2b is encoded by an essential chromosomal gene, it may prove difficult to dilute it out of the bacterial population even without the selective pressure of beta-lactam therapy.

GENE TRANSFER AND THE EVOLUTION OF MOSAIC GENES IN CLINICAL ISOLATES

In contrast to the point mutations that can be identified in *pbp* genes of resistant laboratory mutants, *pbp* genes of resistant clinical isolates encoding low-affinity forms of PBP2x, PBP2b, and PBP1a contain sequence blocks with up to over 20% divergence in their DNA sequence compared to those in sensitive genes, resulting in approximately 10% amino acid difference (12, 38, 40). Two implications arise immediately: such sequence blocks cannot have evolved by mutations, since there are too many alterations and most of them do not result in amino acid alterations, and the amino acids that are relevant for resistance cannot be immediately deduced by sequence comparison. It was also recognized that such mosaic blocks are shared between resistant *S. pneumoniae* and commensal streptococci (13, 47). In the following text, the primary resistance determinant *pbp2x* is used as an example for the evolution of such mosaic genes.

Mosaic genes are the result of gene transfer events followed by recombination into the chromosome. In order to find genes that are potential ancestors of the mosaic blocks, *pbp2x* genes in closely related commensal streptococci were investigated. Indeed, two penicillin-sensitive *S. mitis* strains contained such *pbp2x* genes: the British NCTC10712 strain and the South African isolate M3 (51). Similarly, *pbp2b* genes closely related to the sequence blocks found in resistant *S. pneumoniae* were detected in sensitive commensal streptococci (11). The commensal flora such as *S. mitis* has to endure every antibiotic therapy of its host, and it is therefore likely that resistant strains appear soon and are likely to remain undetected since antibiograms of, e.g., *S. mitis* isolates are not determined routinely in the hospital. Once a powerful resistance determinant has arisen in the commensal species, *S. pneumoniae*, a naturally genetically competent organism, can incorporate the potent gene candidate and selection of such transformants is guaranteed through the next antibiotic treatment. Surprisingly, it was possible to transform *pbp2x* genes even of the sensitive *S. mitis* into a sensitive *S. pneumoniae* acceptor strain, and such transformants were selectable although only slightly more resistant than both the donor and the acceptor strain. This indicates that the presence of a *pbp2x* homolog in a different genetic background can be sufficient to induce a change the susceptibility to the antibiotic.

Sequence comparison of a large number of *pbp2x* genes from resistant *S. pneumoniae* and *S. mitis* clearly shows that many different *pbp2x* homologs must be present in the gene pool that is accessible to *S. mitis* and *S. pneumoniae*. One mosaic block common to many different clones can be recognized (47), and its distribution in major antimicrobial-resistant clones isolated worldwide is schematically shown in Fig. 2. The length of this block is different in each of the clones except for the two Spanish serotype 23F and 9V clones, although the

DNA sequence is nearly identical. There are at least another five sequence types present in the *pbp2x* genes shown in Fig. 2, all of which are again approximately 20% different from each other. Several more sequence variants are found in the data bank of *pbp2x* genes constructed in our laboratory, which will be published in detail elsewhere. The large number of *pbp2x* variants is surprising, and it is still not clear whether this reflects the number of different streptococcal species that participate in this game, whether *pbp2x* is a variable gene in at least one of these species, and whether the transformation event by itself contributes to the variability in some way. Some *pbp2x* variants are identical in genetically different clones, and in some cases, such as in the two main Spanish clones of serotype 23F and 9V, apparently all three genes *pbp2x*, *pbp2b*, and *pbp1a* are of identical mosaic structure (7).

The borders of the mosaic blocks within the *pbp* genes are not distributed arbitrarily; rather,

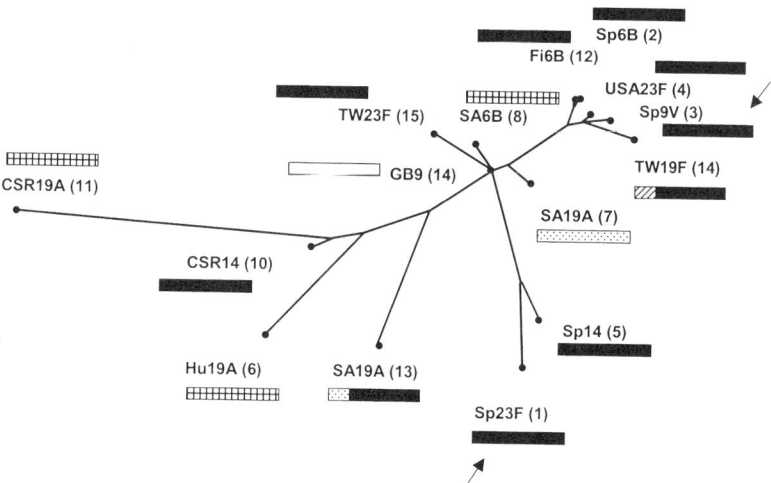

FIGURE 2 Mosaic structure of *pbp2x* genes in *S. pneumoniae* clones. The genetic relatedness of *S. pneumoniae* clones was calculated on the basis of their multilocus sequence type data using the split-tree program. The patterns in the *pbp2x* genes indicate the relationship of altered sequence blocks; the blocks themselves are not shown. Genes marked in black contain one highly related mosaic block that covers the penicillin-binding domain and various portions of the gene encoding the N- and C-terminal parts of the protein; it is identical in size and sequence in the two clones marked by the arrows. The genes with the checkered pattern are all identical in sequence. The *pbp2x* gene of the one penicillin-sensitive clone has no mosaic structure and is shown in white.

they frequently coincide with the domain structure of the protein (51). Adjacent genes are also frequently part of the recombination events and appear variable in resistant strains: the *ddl* gene adjacent to *pbp2b* (16) and the *ftsL* gene upstream of *pbp2x* (our own unpublished results). In addition, the fact that serotype switching occurs in some clones could be associated with large recombinational replacements of the *cps-pbp1a* region (5). Using chromosomal DNA of high-level resistant isolates, one round of transformation is sufficient to produce alterations in two PBP genes that are located far away on the chromosome: *pbp2b* and *pbp2x*, as well as *pbp2a* and *pbp2x* (27). It is not known how many and what other chromosomal genes are cotransferred during the *pbp*-associated resistance development, and whether they are relevant for the resistant phenotype.

MUTATIONS IN PBP GENES ASSOCIATED WITH RESISTANCE

Examination of laboratory mutants demonstrated the complexity of resistance development with respect to PBP mutations. Two different beta-lactam antibiotics were used to select for spontaneous, independent mutant families that consisted of members with increasing resistance levels: piperacillin, a highly lytic penicillin that interacts with all PBPs at low concentrations, and cefotaxime, which does not interact with PBP2b. Each mutant obtained after four to six selection steps was affected in between one and three PBPs; each contained an individual set of up to four mutations in one PBP; the order of genes affected during the selection procedure varied, and depending on the beta-lactam used for selection, different PBPs were targeted (36). The localization of the PBP2x mutations—13 distinct sites were recognized in the mutants—did not map anywhere near the active sites except for two cases: the T550A change, close to the K547SG box, and the H394Y change (35, 37). The Q552E mutation can also be selected in laboratory mutants (52).

The identification of amino acid changes corresponding to mosaic genes that are important for the resistance phenotype is obvious if such a mutation has been selected as a single point mutation in the laboratory. Indeed, the T550A change has been found in a high-level cefotaxime-resistant strain from the United States, the Q552E change is present in many clinical isolates, and the H394Y change close to the S395SN box also occurs (6, 20, 35, 45, 52). Other mutations include those close to the active site S337TMK at T338 (A, P, or G), and high-level cephalosporin resistance appeared to be related to the M339F change (2, 43, 45, 47). Figure 3 shows the distribution of these changes in a variety of mosaic genes.

This relatively small number of positions in PBP2x implicated in resistance in the large number of clinical isolates investigated so far is in contrast to the many distinct mutations found in the few laboratory mutants obtained under identical selection conditions. Apart from the fact that certainly not all relevant substitutions have been identified in the clinical isolates, it could be possible that some of the mutations selected in the laboratory are associated with growth defects in vivo or are not sufficiently cross-protective within the beta-lactam antibiotic family. A high-resolution structure of PBP2x from a sensitive as well as a resistant strain is now available (9, 19), and the impact of some mutations has been discussed (43). However, the effects of mutations distant to the active site remain to be clarified.

In the other PBPs, mutations close to active-site motifs have also been implicated in resistance: the T446A change in PBP2b close to S443SN, which has been associated with tolerance, is present in most isolates (11, 20). PBP1a frequently contains a T371A or T371S change close to the active-site S370 (2, 54). The same positions were affected in the PBPs of a high-level-resistant *S. mitis*: T371S in PBP1a, T337A or T337G in PBP2x, and T446A in PBP2b, and even PBP2a that was associated with resistance in transformation experiments had the same T-to-A change di-

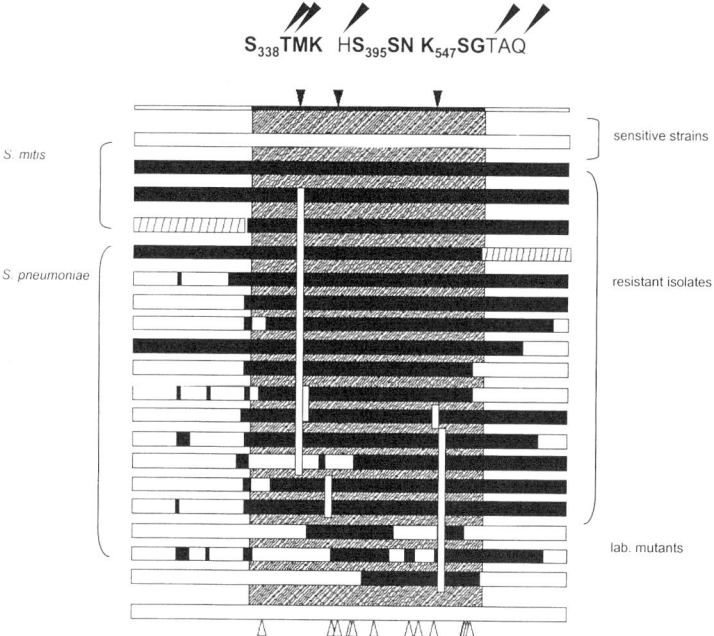

FIGURE 3 Mosaic block structure of *pbp2x* genes and mutations associated with resistance. White indicates sequences of sensitive strains. Divergent sequences representing mosaic blocks are shown in black independent of their relatedness. The positions of the active-site motifs are shown at the top. The arrows above the conserved motifs indicate amino acid substitutions that have been shown to be associated with resistance phenotype (T338, M339, H394, T550, Q552). Altered positions in the mosaic genes are marked by white bars across the gene. The positions of mutations identified in cefotaxime-resistant laboratory mutants are indicated at the bottom. The region covering the central penicillin-binding domain is marked by grey shading.

rectly adjacent to the active-site serine (27, 47). Moreover, PBP3 in a cefotaxime-resistant laboratory mutant had the change T242I close to the K239TG box (34), and the C-terminal end of the penicillin-binding domain has been implicated in laboratory mutants of PBP2x and PBP2b as well as in PBP2b of clinical isolates (12, 28, 31).

The MICs of penicillin G and cefotaxime for susceptible strains are around 0.02 μg/ml. One should remember that the increase in MIC mediated by mutations in either PBP2x or PBP2b alone, the two PBPs which are primarily affected during the resistance process, is hardly clinically significant. Independent of whether there is one mutation in PBP2b or several in a highly altered mosaic variant, it

barely confers a twofold increase. Single mutations in PBP2x mediate between 2- and 30-fold increase in MIC, depending on the mutation, and again, with *pbp* variants of clinical isolates similar values were obtained. As mentioned above, high-level penicillin resistance requires the introduction of at least one HMW PBP. Isolates for which the penicillin MICs are 1 μg/ml and higher generally contain altered *pbp2b*, *pbp2x*, and *pbp1a*; high-level cefotaxime-resistant strains for which the MICs are up to 30 μg/ml, such as in the isolates from the United States, contained only altered *pbp2x* and *pbp1a* genes (but some were hypersusceptible to penicillin due the T550A mutation [6]), and the high cefotaxime and penicillin MICs for an *S. mitis* isolate (128 and 64 μg/ml, respectively) required low-affinity variants of five HMW PBPs (33).

PBP-INDEPENDENT MECHANISMS OF BETA-LACTAM RESISTANCE IN LABORATORY MUTANTS

The number of PBP mutations identified in the laboratory mutants did not match the number of selection steps, and in some cases even the first-step mutant did not contain any detectable PBP alteration. Even more astounding was the discovery that independent of the selective beta-lactam, all mutants were defective in competence development. Two genes associated with this phenotype were identified, and similar to the situation with the PBP genes, a different gene locus was affected in the piperacillin-resistant mutants versus the cefotaxime-resistant mutants. Two of the three piperacillin-resistant mutants contained a mutation in the gene CpoA, a putative glycosyltransferase, that was suggested to be involved in teichoic acid biosynthesis (21). Mutations in *cpoA* also affected the apparent amount of PBP1a. Both *cpoA* alleles conferred low-level resistance to piperacillin as well as a defect in genetic transformation, but the molecular mechanism of these phenotypes is not understood.

All cefotaxime-resistant mutants contained mutations in a histidine protein kinase gene, *ciaH*; again, each mutant had one or two distinct *ciaH* mutations, and the selection step at which the mutation occurred differed also (22, 58). *ciaH* is part of a two-component system of the classical OmpR/EnvZ type. Meanwhile, putative target genes of the CiaR response regulator have been identified by characterization of DNA fragments that interact specifically with the protein, and *cia*-dependent regulation has been verified by transcriptome analysis for some of the genes (41). Several of the putative *cia* target genes are related to the biosynthesis of cell wall polysaccharides. These results in combination with phenotypic analyses of the mutants suggested that the *cia*-system can monitor the integrity of the cell wall and that antibiotic-induced damage can be counteracted to some degree by activation of cell wall biosynthesis. The *cia* system is part of a complex regulatory network; it is able to inhibit the expression of competence-related genes under conditions in which its primary target genes are activated (41). No mutations in the *cia* system have been found in resistant clinical isolates, and it is likely that they can be tolerated by the cells only under laboratory conditions. A non-PBP-mediated mechanism has been suggested for a high-level resistant Hungarian clone, but its molecular nature remains to be identified (55).

Although main principles of the evolution of resistance to beta-lactam antibiotics are understood, there are still many questions unsolved. For several resistant laboratory mutants the resistance determinants are not known yet. The molecular mechanism of protection against cell wall antibiotics mediated by the non-PBP genes *ciaH* and *cpoA* is not clear, and what is the nature of the non-PBP genes discussed in clinical isolates? More point mutations in the PBPs will certainly be identified, but can we deduce the evolutionary pathway in detail in retrospective studies? What is the actual in vivo function of the PBPs—their penicillin-sensitive activities, as well as those associated with the non-penicillin-binding domains? And is there a price for resistance?

ACKNOWLEDGMENTS

This work was supported by the DFG (Ha 1011/8-4, 9-1), the BMBF (031U213B and 13N8349), the EU (QRLT-1999-31020 and QRLT-2000-00873), and the Schwerpunkt Biotechnologie der Universität Kaiserslautern.

REFERENCES

1. **Adam, M., C. Damblon, M. Jamin, W. Zorzi, V. Dusart, M. Galleni, A. El Kharroubi, G. Piras, B. G. Spratt, W. Keck, J. Coyette, J.-M. Ghuysen, M. Nguyen-Distèche, and J.-M. Frère.** 1991. Acyltransferase activities of the high-molecular-mass essential penicillin-binding proteins. *Biochem. J.* **279**:601– 604.
2. **Asahi, Y., Y. Takeuchi, and K. Ubukata.** 1999. Diversity of substitutions within or adjacent to conserved amino acid motifs of penicillin-binding protein 2X in cephalosporin-resistant *Streptococcus pneumoniae* isolates. *Antimicrob. Agents Chemother.* **43**:1252–1255.

3. Barcus, V. A., K. Ghanekar, M. Yeo, T. J. Coffey, and C. G. Dowson. 1995. Genetics of high level penicillin resistance in clinical isolates of *Streptococcus pneumoniae*. *FEMS Microbiol. Lett.* **126:**299–303.

4. Briese, T., H. Ellerbrok, H.-M. Schier, and R. Hakenbeck. 1988. Reactivity of anti-β-lactam antibodies with β-lactam-penicillin-binding protein complexes, p. 404–409. *In* P. Actor, L. Daneo-Moore, M. L. Higgins, M. R. J. Salton, and G. D. Shockman (ed.), *Antibiotic Inhibition of Bacterial Cell Surface Assembly and Function.* American Society for Microbiology, Washington, D.C.

5. Coffey, T. J., M. Daniels, M. C. Enright, and B. G. Spratt. 1999. Serotype 14 variants of the Spanish penicillin-resistant serotype 9V clone of *Streptococcus pneumoniae* arose by large recombinational replacements of the *cpsA-pbp1a* region. *Microbiology* **145:**2023–2031.

6. Coffey, T. J., M. Daniels, L. K. McDougal, C. G. Dowson, F. C. Tenover, and B. G. Spratt. 1995. Genetic analysis of clinical isolates of *Streptococcus pneumoniae* with high-level resistance to expanded-spectrum cephalosporins. *Antimicrob. Agents Chemother.* **39:**1306–1313.

7. Coffey, T. J., C. G. Dowson, M. Daniels, J. Zhou, C. Martin, B. G. Spratt, and J. M. Musser. 1991. Horizontal transfer of multiple penicillin-binding protein genes, and capsular biosynthetic genes, in natural populations of *Streptococcus pneumoniae*. *Mol. Microbiol.* **5:**2255–2260.

8. Dargis, M., and F. Malouin. 1994. Use of biotinylated β-lactams and chemiluminescence for study and purification of penicillin-binding proteins in bacteria. *Antimicrob. Agents Chemother.* **38:** 973–980.

9. Dessen, A., N. Mouz, E. Gordon, J. Hopkins, and O. Dideberg. 2001. Crystal structure of PBP2x from a highly penicillin-resistant *Streptococcus pneumoniae* clinical isolate: a mosaic framework containing 83 mutations. *J. Biol. Chem.* **276:**45105–45112.

10. Di Guilmi, A. M., A. Dessen, O. Dideberg, and T. Vernet. 2003. The glycosyltransferase domain of penicillin-binding protein 2a from *Streptococcus pneumoniae* catalyzes the polymerization of murein glycan chains. *J. Bacteriol.* **185:** 4418–4423.

11. Dowson, C. G., T. J. Coffey, C. Kell, and R. A. Whiley. 1993. Evolution of penicillin resistance in *Streptococcus pneumoniae*; the role of *Streptococcus mitis* in the formation of a low affinity PBP2B in *S. pneumoniae*. *Mol. Microbiol.* **9:**635–643.

12. Dowson, C. G., A. Hutchison, J. A. Brannigan, R. C. George, D. Hansman, J. Liñares, A. Tomasz, J. M. Smith, and B. G. Spratt. 1989. Horizontal transfer of penicillin-binding protein genes in penicillin-resistant clinical isolates of *Streptococcus pneumoniae*. *Proc. Natl. Acad. Sci. USA* **86:**8842–8846.

13. Dowson, C. G., A. Hutchison, N. Woodford, A. P. Johnson, R. C. George, and B. G. Spratt. 1990. Penicillin-resistant viridans streptococci have obtained altered penicillin-binding protein genes from penicillin-resistant strains of *Streptococcus pneumoniae*. *Proc. Natl. Acad. Sci. USA* **87:**5858–5862.

14. Dowson, C. G., A. P. Johnson, E. Cercenado, and R. C. George. 1994. Genetics of oxacillin resistance in clinical isolates of *Streptococcus pneumoniae* that are oxacillin resistant and penicillin susceptible. *Antimicrob. Agents Chemother.* **38:**49–53.

15. du Plessis, M., A. M. Smith, and K. P. Klugman. 2000. Analysis of penicillin-binding protein lb and 2a genes from *Streptococcus pneumoniae*. *Microb. Drug Resist.* **6:**127–131.

16. Enright, M. C., and B. G. Spratt. 2004. Extensive variation in the *ddl* gene of penicillin-resistant *Streptococcus pneumoniae* results from a hitchhiking effect driven by the penicillin-binding protein 2b gene. *Mol. Biol. Evol.* **16:**1687–1695.

17. Filipe, S. R., and A. Tomasz. 2000. Inhibition of the expression of penicillin-resistance in *Streptococcus pneumoniae* by inactivation of cell wall muropeptide branching genes. *Proc. Natl. Acad. Sci. USA* **97:**4891–4896.

18. Goffin, C., and J.-M. Ghuysen. 2002. Biochemistry and comparative genomics of SxxK superfamily acyltransferases offer a clue to the mycobacterial paradox: presence of penicillin-susceptible target proteins versus lack of efficiency of penicillin as therapeutic agent. *Microbiol. Mol. Biol. Rev.* **66:**706–738.

19. Gordon, E., N. Mouz, E. Duée, and O. Dideberg. 2000. The crystal structure of the penicillin-binding protein 2x from *Streptococcus pneumoniae* and its acyl-enzyme form: implication in drug resistance. *J. Mol. Biol.* **299:**477–485.

20. Grebe, T., and R. Hakenbeck. 1996. Penicillin-binding proteins 2b and 2x of *Streptococcus pneumoniae* are primary resistance determinants for different classes of β-lactam antibiotics. *Antimicrob. Agents Chemother.* **40:**829–834.

21. Grebe, T., J. Paik, and R. Hakenbeck. 1997. A novel resistance mechanism for β-lactams in *Streptococcus pneumoniae* involves CpoA, a putative glycosyltransferase. *J. Bacteriol.* **179:**3342–3349.

22. Guenzi, E., A. M. Gasc, M. A. Sicard, and R. Hakenbeck. 1994. A two-component signal-transducing system is involved in competence and penicillin susceptibility in laboratory mutants of *Streptococcus pneumoniae*. *Mol. Microbiol.* **12:**505–515.

23. **Hakenbeck, R.** 2000. Detection of low affinity penicillin-binding protein variants in *Streptococcus pneumoniae*, p. 265–271. *In* S. H. Gillespie (ed.), *Antibiotic Resistance Methods and Protocols*. Humana Press, Totowa, N.J.

24. **Hakenbeck, R., T. Briese, and H. Ellerbrok.** 1986. Antibodies against the benzylpenicilloyl moiety as a probe for penicillin-binding proteins. *Eur. J. Biochem.* **157:**101–106.

25. **Hakenbeck, R., K. Kaminski, A. König, M. van der Linden, J. Paik, P. Reichmann, and D. Zähner.** 1999. Penicillin-binding proteins in β-lactam-resistant *Streptococcus pneumoniae*. *Microb. Drug Resist.* **5:**91–99.

26. **Hakenbeck, R., and M. Kohiyama.** 1982. Purification of penicillin-binding protein 3 from *Streptococcus pneumoniae*. *Eur. J. Biochem.* **127:**231–236.

27. **Hakenbeck, R., A. König, I. Kern, M. van der Linden, W. Keck, D. Billot-Klein, R. Legrand, B. Schoot, and L. Gutmann.** 1998. Acquisition of five high-M_r penicillin-binding protein variants during transfer of high-level β-lactam resistance from *Streptococcus mitis* to *Streptococcus pneumoniae*. *J. Bacteriol.* **180:**1831–1840.

28. **Hakenbeck, R., C. Martin, C. Dowson, and T. Grebe.** 1994. Penicillin-binding protein 2b of *Streptococcus pneumoniae* in piperacillin-resistant laboratory mutants. *J. Bacteriol.* **176:**5574–5577.

29. **Hakenbeck, R., S. Tornette, and N. F. Adkinson.** 1987. Interaction of non-lytic β-lactams with penicillin-binding proteins in *Streptococcus pneumoniae*. *J. Gen. Microbiol.* **133:**755–760.

30. **Hoskins, J., P. Matsushima, D. L. Mullen, J. Tang, G. Zhao, T. I. Meier, T. I. Nicas, and S. R. Jaskunas.** 1999. Gene disruption studies of penicillin-binding proteins 1a, 1b and 2a in *Streptococcus pneumoniae*. *J. Bacteriol.* **181:**6552–6555.

31. **James, P. A.** 1990. Comparison of four methods for the determination of MIC and MBC of penicillin for viridans streptococci and the implications for penicillin tolerance. *J. Antimicrob. Chemother.* **25:**209–216.

32. **Kell, C. M., U. K. Sharma, C. G. Dowson, C. Town, T. S. Balganesh, and B. G. Spratt.** 1993. Deletion analysis of the essentiality of penicillin-binding proteins 1A, 2B and 2X of *Streptococcus pneumoniae*. *FEMS Microbiol. Lett.* **106:**171–175.

33. **König, A., R. R. Reinert, and R. Hakenbeck.** 1998. *Streptococcus mitis* with unusual high level resistance to β-lactam antibiotics. *Microb. Drug Resist.* **4:**45–49.

34. **Krauß, J., and R. Hakenbeck.** 1997. A mutation in the D,D-carboxypeptidase penicillin-binding protein 3 of *Streptococcus pneumoniae* contributes to cefotaxime resistance of the laboratory mutant C604. *Antimicrob. Agents Chemother.* **41:** 936–942.

35. **Krauß, J., M. van der Linden, T. Grebe, and R. Hakenbeck.** 1996. Penicillin-binding proteins 2x and 2b as primary PBP-targets in *Streptococcus pneumoniae*. *Microb. Drug Resist.* **2:**183–186.

36. **Laible, G., and R. Hakenbeck.** 1987. Penicillin-binding proteins in β-lactam-resistant laboratory mutants of *Streptococcus pneumoniae*. *Mol. Microbiol.* **1:**355–363.

37. **Laible, G., and R. Hakenbeck.** 1991. Five independent combinations of mutations can result in low-affinity penicillin-binding protein 2x of *Streptococcus pneumoniae*. *J. Bacteriol.* **173:**6986–6990.

38. **Laible, G., B. G. Spratt, and R. Hakenbeck.** 1991. Inter-species recombinational events during the evolution of altered PBP 2x genes in penicillin-resistant clinical isolates of *Streptococcus pneumoniae*. *Mol. Microbiol.* **5:**1993–2002.

39. **Liu, H. H., and A. Tomasz.** 1985. Penicillin tolerance in multiply drug-resistant natural isolates of *Streptococcus pneumoniae*. *J. Infect. Dis.* **152:**365–372.

40. **Martin, C., C. Sibold, and R. Hakenbeck.** 1992. Relatedness of penicillin-binding protein 1a genes from different clones of penicillin-resistant *Streptococcus pneumoniae* isolated in South Africa and Spain. *EMBO J.* **11:**3831–3836.

41. **Mascher, T., M. Merai, N. Balmelle, A. de Saizieu, and R. Hakenbeck.** 2003. The *Streptococcus pneumoniae* cia regulon: CiaR target sites and transcription profile analysis. *J. Bacteriol.* **185:** 60–70.

42. **McDougal, L. K., J. K. Rasheed, J. W. Biddle, and F. C. Tenover.** 1995. Identification of multiple clones of extended-spectrum cephalosporin-resistant *Streptococcus pneumoniae* isolates in the United States. *Antimicrob. Agents Chemother.* **39:**2282–2288.

43. **Mouz, N., A. M. Di Guilmi, E. Gordon, R. Hakenbeck, O. Dideberg, and T. Vernet.** 1999. Mutations in the active site of penicillin-binding protein PBP2x from *Streptococcus pneumoniae*. Role in the specificity for β-lactam antibiotics. *J. Biol. Chem.* **274:**19175–19180.

44. **Muñóz, R., C. G. Dowson, M. Daniels, T. J. Coffey, C. Martin, R. Hakenbeck, and B. G. Spratt.** 1992. Genetics of resistance to third-generation cephalosporins in clinical isolates of *Streptococcus pneumoniae*. *Mol. Microbiol.* **6:**2461–2465.

45. **Nagai, K., T. A. Davies, M. R. Jacobs, and P. C. Appelbaum.** 2002. Effects of amino acid alterations in penicillin-binding proteins (PBPs) 1a, 2b, and 2x on PBP affinities of penicillin, ampicillin, amoxicillin, cefditoren, cefuroxime, cefprozil, and cefaclor in 18 clinical isolates of penicillin-susceptible, -intermediate, and -resistant pneumococci. *Antimicrob. Agents Chemother.* **46:**1273–1280.

46. **Paik, J., I. Kern, R. Lurz, and R. Hakenbeck.** 1999. Mutational analysis of the *Streptococcus pneumoniae* bimodular class A penicillin-binding proteins. *J. Bacteriol.* **181:**3852–3856.

47. **Reichmann, P., A. König, J. Liñares, F. Alcaide, F. C. Tenover, L. McDougal, S. Swidsinski, and R. Hakenbeck.** 1997. A global gene pool for high-level cephalosporin resistance in commensal *Streptococcus* spp. and *Streptococcus pneumoniae*. *J. Infect. Dis.* **176:**1001–1012.

48. **Rohrer, S., and B. Berger-Bächi.** 2003. FemABX peptidyl transferases: a link between branched-chain cell wall peptide formation and beta-lactam resistance in gram-positive cocci. *Antimicrob. Agents Chemother.* **47:**837–846.

49. **Schuster, C., B. Dobrinski, and R. Hakenbeck.** 1990. Unusual septum formation in *Streptococcus pneumoniae* mutants with an alteration in the D,D-carboxypeptidase penicillin-binding protein 3. *J. Bacteriol.* **172:**6499–6505.

50. **Severin, A., C. Schuster, R. Hakenbeck, and A. Tomasz.** 1992. Altered murein composition in a D,D-carboxypeptidase mutant of *Streptococcus pneumoniae*. *J. Bacteriol.* **174:**5152–5155.

51. **Sibold, C., J. Henrichsen, A. König, C. Martin, L. Chalkley, and R. Hakenbeck.** 1994. Mosaic *pbpX* genes of major clones of penicillin-resistant *Streptococcus pneumoniae* have evolved from *pbpX* genes of a penicillin-sensitive *Streptococcus oralis*. *Mol. Microbiol.* **12:**1013–1023.

52. **Sifaoui, F., M. D. Kitzis, and L. Gutmann.** 1996. In vitro selection of one-step mutants of *Streptococcus pneumoniae* resistant to different oral β-lactam antibiotics is associated with alterations of PBP2x. *Antimicrob. Agents Chemother.* **40:**152–156.

53. **Smith, A. M., R. F. Botha, H. J. Koornhof, and K. P. Klugman.** 2001. Emergence of a pneumococcal clone with cephalosporin resistance and penicillin susceptibility. *Antimicrob. Agents Chemother.* **45:**2648–2650.

54. **Smith, A. M., and K. P. Klugman.** 1998. Alterations in PBP1A essential for high-level penicillin resistance in *Streptococcus pneumoniae*. *Antimicrob. Agents Chemother.* **42:**1329–1333.

55. **Smith, A. M., and K. P. Klugman.** 2000. Non-penicillin-binding protein mediated high-level penicillin and cephalosporin resistance in a Hungarian clone of *Streptococcus pneumoniae*. *Microb. Drug Resist.* **6:**105–110.

56. **Smith, A. M., and K. P. Klugman.** 2001. Alterations in MurM, a cell wall muropeptide branching enzyme, increase high-level penicillin and cephalosporin resistance in *Streptococcus pneumoniae*. *Antimicrob. Agents Chemother.* **45:**2393–2396.

57. **Weber, B., K. Ehlert, A. Diehl, P. Reichmann, H. Labischinski, and R. Hakenbeck.** 2000. The *fib* locus in *Streptococcus pneumoniae* is required for peptidoglycan crosslinking and PBP-mediated beta-lactam resistance. *FEMS Microbiol. Lett.* **188:**81–85.

58. **Zähner, D., K. Kaminski, M. van der Linden, T. Mascher, M. Merai, and R. Hakenbeck.** 2002. The *ciaR/ciaH* regulatory network of *Streptococcus pneumoniae*. *J. Mol. Microbiol. Biotechnol.* **4:**211–216.

59. **Zhao, G., T. I. Meir, S. D. Kahl, K. R. Gee, and L. C. Blaszczak.** 1999. Bocillin FL, a sensitive and commercially available reagent for detection of penicillin-binding proteins. *Antimicrob. Agents Chemother.* **43:**1124–1128.

MACROLIDE, QUINOLONE, AND OTHER NON-β-LACTAM ANTIBIOTIC RESISTANCE IN STREPTOCOCCUS PNEUMONIAE

Karita Ambrose and David S. Stephens

22

INTRODUCTION

The treatment of infections due to *Streptococcus pneumoniae* has become a complicated global problem due to antibiotic resistance. The emergence of resistance to penicillin and other beta-lactam antibiotics in pneumococci in the 1980s and 1990s led to the increased use of macrolides, fluoroquinolones, and other non-beta-lactam antibiotics for pneumococcal infections. While these antibiotics provided effective alternative therapy for *S. pneumoniae* infections, the increased prevalence of resistance to these drugs has paralleled their increased use. In the United States, antimicrobial resistance of *S. pneumoniae* rose rapidly, and by the late 1990s, 34% of invasive pneumococcal isolates collected nationwide were resistant to penicillin (25, 113). Rates of resistance of *S. pneumoniae* to non-beta-lactams were as follows: macrolides, 25%; clindamycin, 9%; tetracycline, 16%; chloramphenicol, 8%; and trimethoprim-sulfamethoxazole, 30% (26). In these reports, 9 to 14% of pneumococcal isolates were resistant to three or more drug classes, which included increased resistance to both erythromycin and trimethoprim-sulfamethoxazole (116).

Selective pressure associated with the widespread use of antibiotics, in general, has resulted in resistance to various classes of antimicrobial agents (24). In *S. pneumoniae*, levels of macrolide resistance are correlated directly with increased usage (42). In one study, 24% of patients had breakthrough bacteremia with an erythromycin-resistant pneumococcal strain while taking a macrolide, whereas individuals not taking a macrolide had erythromycin-susceptible pneumococcal bacteremia (56). Likewise, the prevalence of pneumococci with reduced fluoroquinolone susceptibility has been correlated with the annual number of fluoroquinolone prescriptions in both Canada and the United States (13, 14). Clonal expansion and spread of multiresistant *S. pneumoniae* have also contributed significantly to the dissemination of antimicrobial resistance. The total population of antibiotic-resistant pneumococci increasingly is dominated by a small number of highly successful clones (49). Understanding the genetic basis of resistance in these clones, and how resistance is acquired and spread, may provide insights that lead to effective strategies to reduce transmission of multiply resistant pneumococci. This chapter focuses on the

Karita Ambrose, Department of Medicine, Emory University School of Medicine, Atlanta, GA 30322. *David S. Stephens*, Departments of Medicine and Microbiology and Immunology, Emory University School of Medicine, Atlanta, GA 30322, and VA Medical Center, Decatur, GA 30033.

TABLE 1 Molecular mechanisms of antibiotic resistance in *S. pneumoniae*

Macrolides	Erm (23S rRNA methyltransferases)
	Mutations in 23S rRNA or L4 and L22 ribosomal protein genes
	MefE or MefA/Mel-mediated efflux
Quinolones	Mutations in DNA gyrase (GyrA) or topoisomerase (ParC) genes
	PmrR-mediated efflux
Tetracycline	Tet(M) or Tet(O) ribosomal protection proteins
Chloramphenicol	CAT catalyzes conversion of chloramphenicol to inactive metabolites
Trimethoprim-sulfamethoxazole	Mutations in DHFR gene
	Mutations in DHPS gene
Rifampin	Mutations in RNA polymerase (RpoB) gene

molecular mechanisms of non-beta-lactam resistance in *S. pneumoniae*, summarized in Table 1. Resistance to beta-lactam antibiotics is discussed in chapter 21.

MOLECULAR MECHANISMS OF RESISTANCE

Molecular Basis for Resistance to Macrolides

Macrolides are composed of amino and/or neutral sugars attached to 14-membered (erythromycin and clarithromycin), 15-membered (azithromycin), or 16-membered (josamycin, spiramycin, midecamycin, miocamycin, and tylosin) lactone rings. Most 16-membered macrolides have limited use or are used primarily for veterinary purposes. Macrolides are bacteriostatic and block the elongation step of protein synthesis by binding to the 50S ribosomal subunit, stimulating dissociation of the peptidyl-tRNA molecule from the ribosome and resulting in premature release of the peptide chain (113). Resistance to macrolides can occur by enzymatic inactivation, modification of the target by methylation or mutation, and active efflux. For *S. pneumoniae*, resistance by enzymatic inactivation of the antibiotic has not been described.

The first mechanism of macrolide resistance described for pneumococci was posttranscriptional modification of the 23S rRNA by adenine-N^6 methyltransferase (113). In *Escherichia coli*, these enzymes add either one or two methyl groups to a single adenine residue (A2058) within domain V, the peptidyl transferase center, of the 23S rRNA. The secondary structure of this region is conserved, consisting of a loop formed at the junction of five helices. The binding sites of lincosamides and streptogramin B antibiotics overlap with those of macrolides; thus, methylation confers cross-resistance (MLS_B resistance phenotype) and high MICs, ≥ 64 μg ml^{-1}, of macrolides are noted. Genes encoding these methylases have been designated *erm* (erythromycin ribosome methylase) and are found in numerous gram-positive bacteria. The structural basis of the interaction of the ribosome and erythromycin A was recently elucidated (92). Interestingly, the antibiotic binding sites were composed exclusively of segments of 23S rRNA at the peptidyl transferase cavity and did not involve any interaction of the drugs with ribosomal proteins.

Acquisition of *erm* in *S. pneumoniae* is due primarily to conjugative transposons related to Tn*1545*, Tn*1545*-like elements, or a Tn*917*-like element (see "Diversity of resistance determinants in pneumococci," below). Based on a new nomenclature devised for *erm* genes, the *erm*(B) class predominates in pneumococci (85). More recently, *erm*(A) has been reported to occur in a pneumococcal isolate from Greece (105) and in an isolate which also possessed *erm*(B) (7). Expression of *erm*(B) is dependent on the conformation of newly synthesized mRNA and can result in either constitutive or inducible expression of the methylase (114). Methylase activity in pneumococci can be either inducible or constitutive; isolates that express *erm* constitutively were found to have a large deletion or DNA duplication in the upstream regulatory regions (86).

An emerging mechanism of resistance to macrolides in pneumococci has been mutation in rRNA (23S rRNA) and the ribosomal protein L4 and L22 genes. Based on surveillance from central and eastern European countries, 15% of the erythromycin-resistant isolates harbored mutations that altered ribosomal protein L4 (68). There are four copies of 23S rRNA in *S. pneumoniae*, and macrolide resistance phenotypes occur when at least two of the copies carry a mutation (108). Various point mutations in the 23S rRNA and ribosomal protein L4 and L22 genes have been reported that confer macrolide resistance in pneumococci (12, 53, 68, 82). Additionally, unusual macrolide resistance phenotypes can result from mutations in the 23S rRNA or ribosomal protein L4 gene. A macrolide-lincosamide (ML) phenotype was due to A2059G mutations, whereas an A2062C mutation conferred resistance to 16-membered macrolides and streptogramin ($M_{16}S$) (23, 107). An unusual macrolide phenotype was also reported where a three-amino-acid substitution in ribosomal protein L4 produced an MS phenotype and a six-amino-acid insertion caused further resistance to ketolides (107). Pneumococcal isolates with multiple mutations in the 23S rRNA, as well as combinations of ribosomal protein (both L4 and L22) alterations and 23S rRNA mutations, have also been reported that confer resistance to macrolides (82).

The second mechanism of macrolide resistance in *S. pneumoniae* is due to expression of the Mef efflux pump. The *mef* gene provides lower levels of resistance to erythromycin and other macrolides than can be achieved by alterations of the ribosome. Thus, the MICs for pneumococci expressing *mef* typically are 4 to 32 μg ml^{-1}. Further, pneumococcal strains have been identified that carry both *mef* and *erm*(B) genes and bear the MLS$_B$ phenotype (52, 57).

In the United States, macrolide resistance due to the presence of Mef rapidly increased in the 1980s. This trend has also been observed in other countries, such as South Africa and Canada (35, 43, 103, 118). The *mef* gene encodes a 44-kDa hydrophobic protein with 12 putative membrane-spanning domains. Mef belongs to the major facilitator superfamily class of efflux pumps and drives efflux via the proton motive force (16). The pump is specific for 14- and 15-membered macrolides, and resistance is designated the M phenotype (106). Two variants, *mefE* and *mefA*, are found in isolates of *S. pneumoniae* and *Streptococcus pyogenes*, respectively. However, *mefA* has been shown to be present in some Italian isolates of *S. pneumoniae* (22, 71).

The genetic elements harboring both *mefA* and *mefE* in *S. pneumoniae* have been described. The *mefE* gene is present on the 5.4- or 5.5-kb macrolide efflux genetic assembly (mega) element and is part of an operon that also includes the downstream *mel* gene, which is a homolog of *msr*, and encodes a 488-amino-acid protein (36). In staphylococci, *msrA* encodes an ATP-binding cassette (ABC) transporter protein which results in energy-dependent efflux of erythromycin (88). Proteins of the ABC transporter superfamily are organized such that two ATP-binding domains located cytoplasmically interact with two hydrophobic domains consisting of six to eight transmembrane domains. Both MsrA and Mel contain ATP-binding domains characteristic of the ABC transporters; however, they lack the hydrophobic segments encoding the transmembrane domains. MsrA is predicted to interact with unidentified chromosomally encoded transmembrane complexes to allow efflux of erythromycin (87). Convergent to the *mefE/mel* operon are three open reading frames (ORFs) with homologies to stress response genes of Tn*5252*. Because *mefE* and *mel* are cotranscribed as an operon, they are predicted to encode a dual efflux pump.

The *mefA*-containing element, Tn*1207.1*, is a 7.2-kb defective transposon containing *mefA* and four downstream ORFs with homologies and a gene order similar to those of the corresponding ORFs in mega and three additional ORFs at the 5′ end with homologies to recombinase and integrase genes (90).

Other mechanisms that confer macrolide resistance in gram-positive bacteria have been identified, such as additional efflux systems and proteins that modify the antibiotic rather than the target. In staphylococci, resistance to macrolides includes active efflux by *msrA*, the *mel* homolog in *S. pneumoniae*; target modification by *erm*; and inactivation of antibiotic by esterases which can elicit resistance to macrolides, lincosamides, or streptogramin B antibiotics. MsrA is plasmid borne, confers resistance to 14-membered macrolides and streptogramin B (MS_B phenotype), and is capable of inducing macrolide resistance independently when cloned in a susceptible staphylococcal host (88). The esterase activity identified in *Staphylococcus* hydrolyzed 14- and 16-membered macrolides. The virginiamycin family of gene products modifies streptogramins by hydrolysis (streptogramin B-*vgb*) or by adding an acetyl group to the antibiotic (streptogramin A-*vat*). Most of these resistance determinants are plasmid borne (85). Recently, a plasmid in *Staphylococcus aureus* was identified that harbored three resistance determinants for macrolides (61).

Mechanisms of Quinolone Resistance

Quinolones are a potent class of antimicrobial agents with broad-spectrum activity that target type II topoisomerases: DNA gyrase and topoisomerase IV. Both DNA gyrase and topoisomerase IV are heterotetramers composed of two subunits each of GyrA and GyrB for the former and ParC and ParE for the latter. The main function of DNA gyrase is to maintain the correct level of DNA supercoiling, which facilitates DNA unwinding during replication and transcription, whereas topoisomerase IV is responsible for the unlinking of daughter chromosomes following DNA synthesis, a process known as decatenation. Quinolones interact with the enzyme-DNA complex to stimulate DNA cleavage and inhibit religation of the resulting double-stranded breaks, leading to cell death (3).

Fluoroquinolones are used for the treatment of pneumococcal infections when resistance to other antimicrobials is suspected, and they have become widely employed for empirical coverage of bacterial pneumonia. In the United States, pneumococcal resistance to fluoroquinolones remains rare compared to the increasing trend of resistance reported in Canada (14, 84). Fluoroquinolone resistance in pneumococci is mediated by mutations in the quinolone resistance-determining regions (QRDR) of *gyrA* and/or *parC* or by active efflux of the antibiotic from the cell. There is a stepwise mechanism of fluoroquinolone resistance in pneumococci that is also dependent on the antibiotic used to select resistance. In ciprofloxacin-resistant pneumococci, a Ser_{79} substitution to Tyr ($Ser_{79} \rightarrow Tyr$) in ParC confers low-level resistance, whereas secondary mutations in *gyrA* result in high-level resistance (93). In contrast, mutations occur first with *gyrA* when selection is with sparfloxacin and correspond to those seen in *E. coli*, resulting in a Ser_{83} substitution to either Phe or Tyr. High-level resistance to sparfloxacin results from subsequent mutations in the QRDR of *parC* (73, 118). The primary targets of other quinolones have been determined; trovafloxacin, levofloxacin, and norfloxacin target topoisomerase IV (ParC), whereas the primary target of gatifloxacin is DNA gyrase (GyrA) (34).

Active efflux of fluoroquinolones has been demonstrated in some resistant isolates of *S. pneumoniae* (10, 37, 122). Gill et al. described a putative efflux pump for fluoroquinolones, PmrA, that has 24% amino acid identity to the multidrug efflux pumps NorA and Bmr (37).

PmrA has 12 transmembrane segments comparable to efflux proteins of the major facilitator superfamily proton-dependent pumps. The pump provides low-level resistance to the fluoroquinolone norfloxacin and is a less prevalent mechanism of resistance in clinical isolates than alterations of the topoisomerases. Other studies also suggest the presence of a non-PmrA pump in S. pneumoniae that may confer resistance to a broader range of substrates, including ciprofloxacin and other fluoroquinolones (77, 78).

Other Non-Beta-Lactam Antibiotic Resistance Mechanisms

TETRACYCLINE

Tetracyclines are broad-spectrum antimicrobial agents that inhibit protein synthesis by blocking attachment of aminoacyl-tRNA to the acceptor site of the ribosome. Because tetracycline is relatively inexpensive, and is safe for use in older children and adults, it has been used in the treatment of pneumococcal infections. The emergence of tetracycline-resistant pneumococci has limited use of this antibiotic for pneumococcal infections. In the United States, resistance to tetracyclines doubled from 8% in 1994 to 1995 to 17% in 1999 to 2000 (26). Active efflux, enzymatic alteration, and ribosomal protection have been described as mechanisms for tetracycline resistance in bacteria. In S. pneumoniae, ribosomal protection is the only mechanism of tetracycline resistance thus far described and results primarily by acquisition of tet(M) and occasionally tet(O) gene products (47, 58, 94, 120); Tet(M) and Tet(O) share approximately 77% amino acid similarity (119). While it is unclear how Tet(M) and Tet(O) function, they are homologous to elongation factors EF–Tu and EF-G and have ribosome-dependent GTPase activity (15).

CHLORAMPHENICOL

Resistance to chloramphenicol is found in pneumococcal clinical isolates (49, 84). However, with the exception of Spain, where about 30 to 50% of isolates are resistant, levels of chloramphenicol resistance of pneumococci in most other parts of the world have remained low (47, 54, 75). In a recent United States study, chloramphenicol resistance was found in 8% of invasive isolates, which represented an increase of 4% from 1995 to 2000 (26). Chloramphenicol inhibits protein synthesis by targeting peptidyl transferase, the enzyme responsible for the catalysis of peptide bond formation during translation (64, 65, 100). Clinical isolates that are resistant to chloramphenicol encode an enzyme, chloramphenicol acetyltransferase (CAT), that catalyzes the conversion of chloramphenicol to its 1-acetoxy, 3-acetoxy, and 1,3-diacetoxy derivatives. These derivatives are no longer able to bind the 50S subunit and inhibit peptidyl transferase activity. In S. pneumoniae, the cat gene is carried on the conjugative transposon Tn5253 and on other similar conjugative or composite transposons (112). Maintenance of cat in countries where chloramphenicol is very rarely used is probably due to selection for resistance to other agents in strains that are multiresistant (49).

TRIMETHOPRIM-SULFAMETHOXAZOLE

Trimethoprim-sulfamethoxazole or co-trimoxazole has been widely used for treatment of upper respiratory and urinary tract infections, and as antibiotic prophylaxis in patients with AIDS, because of its broad coverage, synergistic effects, and low cost. The rate of pneumococcal resistance to co-trimoxazole increased drastically in the United States from 0.6% between 1979 and 1987 to 10% between 1991 and 1992 and was 29% in 1998 (9, 102, 116). In South Africa, the rate of resistance to trimethoprim-sulfamethoxazole among 259 clinical isolates was recently reported to be 64% (2). The efficacy of using trimethoprim versus trimethoprim-sulfamethoxazole in combination for pneumococcal infections has been debated. Reports suggest that there is little difference between the two alternatives and that trimethoprim alone at a higher dose may be effective and have fewer side effects than the combination (2, 11).

Trimethoprim functions by inhibiting dihydrofolate reductase (DHFR). This blocks the reduction of dihydrofolate to tetrahydrofolate, a precursor for the synthesis of purines, thymidylate, and certain amino acids. Likewise, sulfamethoxazole inhibits dihydropteroate synthase (DHPS) and generates the same effect as trimethoprim. Alteration of the DHFR gene either by chromosomal mutation or through acquisition of an exogenous DHFR gene can lead to trimethoprim resistance. In pneumococci, trimethoprim resistance is usually mediated by a point mutation in the DHFR gene leading to an Ile_{100}-to-Leu substitution, although resistance may be enhanced by other mutations in the gene (60). Similarly, altered chromosomal DHPS genes can lead to resistance to sulfonamides. Clinical isolates resistant to sulfonamides were shown to have a 3- or 6-bp duplication resulting in the addition of one or two amino acids. These mutations were sufficient in conferring resistance on susceptible strains upon transformation (2, 59). Isolates that are resistant to trimethoprim are rarely sensitive to trimethoprim-sulfamethoxazole (2). Thus, trimethoprim-sulfamethoxazole resistance correlates more strongly with trimethoprim resistance than sulfamethoxazole resistance.

RIFAMPIN

Rifampin can be used in combination with vancomycin and extended-spectrum cephalosporins for treatment of penicillin-resistant pneumococcal meningitis. Pneumococcal resistance to rifampin has remained low globally compared to resistance to other antibiotics. In the United States, the rate of resistance to rifampin was 0.5% in 1995 (25). This resistance rate is comparable to the 0.1 and 1.5% rates identified in Spain and Brazil, respectively, but resistance can emerge rapidly on therapy (55, 98).

Rifampin functions by binding to the β-subunit of RNA polymerase and causes premature termination during DNA transcription. Resistance to rifampin occurs via point mutations in the RNA polymerase gene, rpoB (30, 72, 111). Mutations in two different regions of rpoB have been shown to confer resistance to rifampin (30, 72). Additionally, there are suggestions that pneumococcal resistance to rifampin may emerge from interspecies recombination events (30).

GENETIC BASES FOR ACQUISITION OF ANTIBIOTIC RESISTANCE DETERMINANTS

Chromosomal mutations that reduce the sensitivity of the antibiotic target(s) are the most common mechanism of resistance to several non-beta-lactam antimicrobials, including quinolones, trimethoprim, and rifampin, and the horizontal spread of resistance to these antibiotics into different pneumococcal strains is considered to be due to genetic transformation. However, the resistance determinants for the macrolides, tetracycline, and chloramphenicol are typically borne on conjugative transposons which mediate their spread into other strains (Table 2).

Conjugative transposons are discrete DNA elements that encode gene products necessary for movement between replicative DNA

TABLE 2 Genetic bases for antibiotic resistance in pneumococci

Conjugative or composite transposons (Tn916-Tn1545, Tn3951, Tn3701, Tn3872 (Tn917), Tn5251–Tn5253)
  cat, tet(M), erm(B)
Elements related to conjugative transposons
  Mega (carries mefE/mel)
  Tn1207.1 (carries mefA/mel)
Plasmid integration
  pC194 into Tn5252
Point mutations and/or homologous recombination with related species
  pbp
  tet(M)
  QRDR of type II topoisomerase
  dhfr
  dhps
  rpoB

species without using the host recombination machinery (83, 89). The primary step in transfer of the transposon involves excision and formation of a nonreplicative covalently closed circular intermediate that can be transferred either intra- or intercellularly (17, 19, 89, 96). Conjugative transposons differ from transposons in that they lack terminal repeats and do not generate a duplicate target sequence when they insert (69). Conjugal transfer of genetic elements between pneumococcal isolates is believed to account for much of the rapid increase in resistance to some non-beta-lactam antibiotics.

The Tn916-Tn1545 family of transposons has a broad host range and has been introduced or found naturally in more than 50 different species of bacteria (18). This family of transposons accounts for the vast majority of erythromycin and tetracycline resistance in pneumococci. In one study, all pneumococcal isolates resistant to erythromycin had sequences homologous to the integrase of Tn1545 (79). In addition to transposing intercellularly, genetic elements of this family can transpose intracellularly to plasmids and other conjugative elements, a mechanism allowing dissemination of multiple resistance determinants (4, 18). An example of this is illustrated in the pneumococcal composite transposon Tn5253, formerly Ω(cat tet) BM6001 (4, 112). The transposon is 65.5 kb and is a composite of two transposons, Tn5251 [containing tet(M)] inserted into Tn5252 (containing the cat determinant) (4). Tn5251 and Tn5252 are both capable of independent conjugal transfer; however, this has not been observed when they are associated in Tn5253 (4). Additionally, the composite transposon, like Tn5252, shows site-specific integration into the pneumococcal genome, in contrast to the more random integrations of Tn5251 (4, 112). A second composite transposon, Tn3872, found in pneumococcal isolates of serotypes 23F and 6B, had erm(B) present. A Tn917-like element inserted at orf9 into Tn916 or Tn916 modified elements generated Tn3872 (63).

While S. pneumoniae is capable of harboring and maintaining plasmids, plasmid-mediated antibiotic resistance in pneumococci is rare. Transformation of the staphylococcal cat-containing plasmid pC194 into S. pneumoniae resulted in loss of the plasmid at a rate of 2% per generation when grown in the absence of chloramphenicol (5, 41). Sequence analysis of cat in Tn5252 identified homologies with sequences of pC194 and insertion of the entire staphylococcal plasmid in some isolates. Thus, integration of pC194 into a conjugative element provides an example of an alternate mechanism by which resistance markers can be acquired and disseminated in S. pneumoniae (117). This mechanism of dissemination has also been illustrated for Tet(M) in tetracycline-resistant clinical isolates of pathogenic Neisseria in which truncated Tn916-like elements were inserted into conjugative plasmids (104).

Homologous recombination involving genes encoding antibiotic resistance determinants can result in novel alleles with a mosaic phenotype. The best example is illustrated in the emergence of penicillin-resistant penicillin-binding proteins. Resistance to penicillin has been attributed to the replacement of regions of the pneumococcal pbp genes with the corresponding regions from the homologous pbp genes of closely related viridans group streptococcal species, leading to hybrid (mosaic) pbp genes (28). Mosaic genes have also been identified with non-beta-lactam antibiotic resistance determinants. Homologous recombination has contributed to the evolution and heterogeneity of the tetracycline resistance determinant, tet(M) (27, 70). A mosaic structure was detected among eight different tet(M) genes from gram-negative and gram-positive bacteria, including pneumococci, and could be traced to two separate alleles. The two alleles differed in their GC content and varied by 8% (70). Horizontal genetic exchange has been shown to disrupt clonal structure by reassorting alleles among lineages. Diversity of tet(M) alleles was also shown in both inter- and intraclonal lineages of multidrug-resistant clones (27). Mosaic structures of *tet*(M) are believed to be generated after integration of conjugative transposons and other mobile genetic elements into the chromosome. This reasoning is supported by studies demonstrating that the presence of one element does not prevent the acquisition of a second element, thus providing the opportunity for homologous recombination between alleles (69).

Interspecies recombination between pneumococci and viridans group streptococci has also been reported for the QRDR of type II topoisomerase genes of fluoroquinolone-resistant pneumococci, although this appears to be a relatively rare mechanism of acquiring resistance compared to acquisition by point mutation (6).

Antibiotic-resistant pneumococcal clones may be selected and amplified by antimicrobial therapy (49). An important factor in the emergence of antibiotic resistance in pneumococci has been selective pressure created by oral antimicrobial usage in the management of respiratory tract infections (24). In most cases disease caused by antibiotic-resistant pneumococci probably occurs as a result of prior carriage of a resistant strain, but emergence of an L22 ribosomal protein mutant during treatment with azithromycin of a patient with pneumococcal pneumonia provides strong evidence that selective pressure arising from the use of macrolides may rapidly result in clinically important resistance (67).

The use of fluoroquinolones to treat pneumococcal infections is the most important factor in the emergence of fluoroquinolone resistance in *S. pneumoniae* (24). In Canada, the percentage of pneumococci with reduced susceptibility to fluoroquinolones increased from 0% of isolates in 1993 to 1.7% of isolates in 1997 to 1998. This increase correlated directly with the number of annual fluoroquinolone prescriptions, from 0.8 to 5.5 prescriptions per 100 persons a year from 1988 to 1997 (38). Surveillance studies in the United States have also reported an increase from 3.1 to 4.6 prescriptions per 100 persons a year from 1993 to 1998 (13). Fluoroquinolone-resistant pneumococcal isolates increased from 2.6% in 1995 to 3.8% of isolates in 1997. Further, mutations in the topoisomerases are able to spread by transformation into multiresistant isolates, allowing resistance to disseminate clonally or horizontally (24, 38, 48).

In pneumococci, resistance to erythromycin is primarily inducible. Constitutive expression of the methylase due to deletions, duplications, or point mutations in the attenuator sequence does not occur often in pneumococci (53, 114). When expression is constitutive, the *erm*(B) mRNA is active, leading to constitutive methylation of the 23S rRNA. Inducible expression allows the mRNA to be made in an inactive conformation that becomes active by the presence of an inducing macrolide. While induction of *erm*(B) has not been studied extensively, a model of *erm*(C) induction has been proposed for staphylococci which is dependent on the size of the inducing macrolide. Bacteria in which resistance is inducible are proposed to regulate their expression through the use of leader peptide sequences. Macrolides capable of inducing can partially block the ribosome-peptidyl-tRNA complex, allowing some nascent peptides to slip around the antibiotic and synthesis not to be inhibited (114). Additionally, MefE efflux resistance is induced with the presence of erythromycin (53).

DIVERSITY OF RESISTANCE DETERMINANTS IN PNEUMOCOCCI

Conjugative Transposons (Tn*916*-Tn*1545*, Tn*3951*, Tn*3701*, Tn*5251* to Tn*5253*)

Conjugative transposons in pneumococci belong mostly to two classes of elements: the Tn*916*-Tn*1545*-type elements and the Tn*5252* composite elements that include Tn*3701* and Tn*3951* identified in other streptococcal species (4, 21, 33, 50, 51, 112). These elements range in size from 18 to 60 kb and carry single or multiple antibiotic resistance determinants. The Tn*916*-Tn*1545* family of conjugative transposons is ubiquitous and has more than 10 members. Tn*916* was one of the first transposons found on the chromosome of a multidrug-resistant *Enterococcus faecalis* strain (33). Tn*916* is an 18-kb element that harbors the Tet(M) gene and 23 other ORFs that play a role in conjugative transposition (18, 31). Based on sequence homologies, the putative functions of many of the ORFs have been assigned (18, 31). Regions both upstream and downstream of *tet*(M) encode products important in transfer, transposition, or both. ORFs 1 and 2 encode gene products related to the excisionase and integrase of lambdoid bacteriophage, while ORF 5 is homologous to *traA*, which is necessary for conjugative transfer. Tet(M) and the leader sequence that controls expression of *tet*(M) are encoded by ORFs 11 and 12, respectively. ORFs 14 to 24 encode gene products required for intercellular and intracellular transposition.

In pneumococci, *tet*(M) was first identified on a 25.3-kb transposon, Tn*1545*, that contained the additional resistance determinants *erm*(B) and *aphA-3*. Sequence analyses have shown that the termini of Tn*916* and Tn*1545* are identical for at least 200 bp. Tn*1545* has a broad host range, mediating conjugation with both gram-negative and gram-positive bacteria. The mechanism of transposition is proposed to be similar to that of Tn*916*.

The composite transposon Tn*5253*, composed of Tn*5251* and Tn*5252*, is larger and more complex (4, 112). There were no significant homologies observed between Tn*5251* and Tn*5252* based on DNA hybridization. The 18-kb Tn*5251* component of Tn*5253* belongs to the Tn*916*-Tn*1545* family of conjugative transposons. Sequence analysis of a >4,000-bp region showed little difference between Tn*5251* and Tn*916*, and most differences were clustered in a 0.68-kb segment of *tet*(M) (81). Tn*5252* is 47 kb and represents a different class of conjugative transposons due to its size, conjugal properties, and narrower host range (80, 101). Sequence analyses of the latter transposon revealed ORFs with homologies to genes for integrases, excisionases, DNA relaxases, and transcriptional regulators. Interruption of genes with homologies to integrase

and relaxase genes disrupted transfer functions of the element (45). Also identified in Tn*5252* was an operon-like region with homologs of *umuDC* that encode SOS repair functions. These gene products were able to restore UV-inducible mutagenic repair of chromosomal DNA in cells defective in error-prone repair in *E. coli*, *S. pneumoniae*, and *E. faecalis* (66).

Elements Related to Conjugative Transposons (Mega and Tn*1207.1*)

The *mefE*-carrying mega and *mefA*-carrying Tn*1207.1* nonconjugative transposons appear to be related to conjugative transposons. ORFs 3, 4, and 5 of mega and ORFs 6, 7, and 8 of Tn*1207.1*, respectively, are homologs and share similarity to ORFs 11, 12, and 13, respectively, of the Tn*5252* conjugative transposon. Additionally, three ORFs present upstream of *mefA* in Tn*1207.1* with homologies to recombinase and integrase genes may allow for conjugal transfer in strains where the element is not truncated (22). Differences in the chromosomal insertion sites of mega and Tn*1207.1* were also found. While Tn*1207.1* was integrated at a single specific chromosomal site (*celB*) in all the strains examined, mega was found inserted at four different sites (22, 36). The class I insertion site was within a putative phosphomethylpyrimidine kinase gene, class II insertions were within a DNA-3-methyladenine glycosylase gene, the class III insertion was within a putative capsule biosynthesis gene located outside of the capsule biosynthesis region, and the class IV insertion site was in an RNA methyltransferase gene (36). Insertion of Tn*1207.1* in *celB* impaired competence in those pneumococcal isolates. Most of the insertion types identified in pneumococcal isolates harboring mega were either class I or class II, and these strains remained naturally competent.

Nonconjugative Transposons (Tn*917*)

Tn*917* is an enterococcus-derived 5.3-kb nonconjugative transposable element that provides erythromycin resistance [*erm*(B)] but not tetracycline or kanamycin resistance (99, 109, 110). Tn*917* and Tn*917*-like elements are prevalent in *Enterococcus* and have been identified in pneumococci as part of the composite transposon Tn*3872*, a member of the Tn*3* family of transposons with >95% homology to the staphylococcal transposon Tn*551* (76, 121). Tn*917* generates a 5-bp duplication upon insertion and exhibits a relatively low frequency of excision (76). Transposition of Tn*917* is inducible at concentrations of erythromycin as low as 0.001 μg/ml and was the first example where the antibiotic to which a transposon provides resistance could enhance its ability to transpose (99, 109).

Once conjugative elements enter the cell they may lose genes required for conjugal transfer but can still be spread horizontally by genetic transformation. The elements mega and Tn*1207.1* are related to conjugative transposons and appear to have initially entered pneumococci via illegitimate, site-specific recombination events (22, 36, 90). However, they have been disseminated in the pneumococcal population horizontally by transformation and the expansion of successful clones (35, 36).

In the United States, population-based surveillance for *S. pneumoniae* invasive disease showed that the rapid increase in macrolide resistance was correlated with the spread of mega in pneumococci. Mega was identified in genetically unrelated isolates of pneumococci at identical insertion sites and it appeared in new serotypes throughout a 6-year surveillance, suggesting that transformation and homologous recombination rapidly spread mega in *S. pneumoniae* (35). Some clinical isolates possessing Tn*1207.1* were capable of transferring the element by conjugation. These may represent a variant transpositionally active form of this element or complementation of the transposition defect of Tn*1207.1* by the transposition functions encoded by related transposons in these isolates (22).

NEW ANTIBIOTICS AND EFFECTS OF CONJUGATE VACCINES ON RESISTANCE

Resistance genes are commonly encountered in pneumococci, whether they are modified forms of chromosomal genes that alter the targets of antibiotic action or increase efflux or are novel genes, such as those for inactivating enzymes, that are carried on mobile genetic elements. Horizontal spread of antibiotic resistance genes occurs in pneumococci both by transformation and by conjugative transposition, and resistance may also be introduced from related species. There now exists a large pool of resistance determinants in the pneumococcal population, and particularly within the childhood serotypes these are frequently carried in the nasopharynges of young children and are most commonly exposed to antibiotics.

The ease with which resistance can spread horizontally, and the plethora of antibiotic resistance genes and mechanisms, has led to the emergence and subsequent global dissemination of successful pneumococcal strains (clones) that are resistant to multiple classes of antibiotics. A number of new antibiotics have been introduced in recent years, such as the ketolides, Synercid (quinupristin-dalfopristin), and linezolid, and these provide new therapeutic options for the treatment of pneumococcal infections and particularly for infections caused by these multiresistant strains. Novel classes of antibiotics are also being developed, such as those that inhibit aminoacyl tRNA synthetases (46, 91).

Ketolides are semisynthetic derivatives of the 14-membered macrolides that differ from erythromycin via modification with a 3-keto group. Telithromycin (HMR 3647) is the first ketolide to be used clinically and showed a high level of clinical efficacy against *S. pneumoniae* isolates with resistance to erythromycin and penicillin in Japan (32). Clinical cure rates were 92% for those with resistance to penicillin G and 86% for those with resistance to erythromycin. However, acquisition of *erm* alone or in combination with a mutation in ribosomal protein L4 has been reported to confer high-level ketolide resistance on two clinical isolates of *S. pneumoniae* (1).

Synercid (quinupristin-dalfopristin) is a semisynthetic combination of streptogramin A and streptogramin B that is bactericidal for a number of gram-positive cocci, including penicillin-resistant pneumococci. In vitro studies have shown high rates of susceptibility for pneumococci and other streptococcal species (29, 40). A five- or six-amino-acid tandem duplication in the L22 ribosomal protein has been reported to confer low-level resistance to quinupristin-dalfopristin at a rate of 0.02% (44).

Other strategies to reduce the rate of antimicrobial resistance include reducing antibiotic use and modification of antibiotic regimens. Studies have shown that primary care sites receiving household and office-based patient educational materials for treatment of uncomplicated acute bronchitis in adults had a substantial decline in the antibiotic prescriptions rates from 74 to 48%, which was not seen at control sites (39). From 1992 to 1996, a decrease in the prevalence of erythromycin-resistant *S. pyogenes* isolates in Finland was observed which correlated with a decline in total macrolide consumption in outpatient therapy. A reduction of 2.4 daily doses per 1,000 inhabitants in 1991 to a level of 1.4 in 1992 resulted in an approximately 8% decrease of erythromycin-resistant isolates from 1992 to 1996. The difference was more prominent 2 years following the imposition of national guidelines (97). While results of this magnitude have not been observed with pneumococci, it is likely that the decreased antibiotic consumption could reverse the trend and incidence of resistant pneumococcal isolates (95).

Exciting, recent data indicate that the total burden of antimicrobial resistance can be reduced by the use of pneumococcal conjugate vaccines that reduce colonization and transmission of the vaccine serotypes (20, 62, 115). In the United States, penicillin resistance has been reduced 35% in invasive pneumococcal isolates since the conjugate vaccine was introduced (115). Similar reductions are also found in younger siblings of vaccinated children compared with those of nonvaccinated children (8, 74). Much of this impact is due to the high levels of antibiotic resistance in the serotypes commonly carried by children which correspond to those covered by the conjugate vaccines. Despite these encouraging trends in the new era of conjugate vaccines, antibiotic resistance in pneumococci will remain a continuing challenge.

ACKNOWLEDGMENTS

We thank Fred Tenover, Brian Spratt, and surveillance personnel of Georgia Emerging Infections Program, Active Bacterial Core Surveillance, for helpful assistance.

This work was supported by a Fellowships in Research Science and Teaching (FIRST) Award, Emory University, and Atlanta VA Merit Award.

REFERENCES

1. **Ackermann, G., and A. C. Rodloff.** 2003. Drugs of the 21st century: telithromycin (HMR 3647)—the first ketolide. *J. Antimicrob. Chemother.* **51:**497–511.
2. **Adrian, P. V., and K. P. Klugman.** 1997. Mutations in the dihydrofolate reductase gene of trimethoprim-resistant isolates of *Streptococcus pneumoniae. Antimicrob. Agents Chemother.* **41:** 2406–2413.
3. **Anderson, V. E., T. D. Gootz, and N. Osheroff.** 1998. Topoisomerase IV catalysis and the mechanism of quinolone action. *J. Biol. Chem.* **273:**17879–17885.
4. **Ayoubi, P., A. O. Kilic, and M. N. Vijayakumar.** 1991. Tn*5253*, the pneumococcal omega (*cat tet*) BM6001 element, is a composite structure of two conjugative transposons, Tn *5251* and Tn *5252. J. Bacteriol.* **173:**1617–1622.
5. **Ballester, S., P. Lopez, J. C. Alonso, M. Espinosa, and S. A. Lacks.** 1986. Selective advantage of deletions enhancing chloramphenicol acetyltransferase gene expression in *Streptococcus pneumoniae* plasmids. *Gene* **41:**153–163.
6. **Bast, D. J., J. C. de Azavedo, T. Y. Tam, L. Kilburn, C. Duncan, L. A. Mandell, R. J. Davidson, and D. E. Low.** 2001. Interspecies recombination contributes minimally to fluoroquinolone resistance in *Streptococcus pneumoniae. Antimicrob. Agents Chemother.* **45:**2631–2634.
7. **Betriu, C., M. Redondo, M. L. Palau, A. Sanchez, M. Gomez, E. Culebras, A. Boloix, and J. J. Picazo.** 2000. Comparative in vitro activities of linezolid, quinupristin-dalfopristin, moxifloxacin, and trovafloxacin against erythromycin-susceptible and -resistant streptococci. *Antimicrob. Agents Chemother.* **44:**1838–1841.
8. **Black, S., and H. Shinefield.** 1997. Issues and challenges: pneumococcal vaccination in pediatrics. *Pediatr. Ann.* **26:**355–60.

9. **Breiman, R. F., J. C. Butler, F. C. Ten-over, J. A. Elliott, and R. R. Facklam.** 1994. Emergence of drug-resistant pneumococcal infections in the United States. *JAMA* **271:**1831–1835.

10. **Brenwald, N. P., M. J. Gill, and R. Wise.** 1998. Prevalence of a putative efflux mechanism among fluoroquinolone-resistant clinical isolates of *Streptococcus pneumoniae*. *Antimicrob. Agents Chemother.* **42:**2032–2035.

11. **Brumfitt, W., and J. M. Hamilton-Miller.** 1993. Reassessment of the rationale for the combinations of sulphonamides with diaminopyrimidines. *J. Chemother.* **5:**465–469.

12. **Canu, A., B. Malbruny, M. Coquemont, T. A. Davies, P. C. Appelbaum, and R. Leclercq.** 2002. Diversity of ribosomal mutations conferring resistance to macrolides, clindamycin, streptogramin, and telithromycin in *Streptococcus pneumoniae*. *Antimicrob. Agents Chemother.* **46:**125–131.

13. **Centers for Disease Control and Prevention.** 2001. Resistance of *Streptococcus pneumoniae* to fluoroquinolones—United States, 1995–1999. *Morb. Mortal. Wkly. Rep.* **50:**800–804.

14. **Chen, D. K., A. McGeer, J. C. de Azavedo, D. E. Low, and Canadian Bacterial Surveillance Network.** 1999. Decreased susceptibility of *Streptococcus pneumoniae* to fluoroquinolones in Canada. *N. Engl. J. Med.* **341:**233–239.

15. **Chopra, I., and M. Roberts.** 2001. Tetracycline antibiotics: mode of action, applications, molecular biology, and epidemiology of bacterial resistance. *Microbiol. Mol. Biol. Rev.* **65:**232–260.

16. **Clancy, J., J. Petitpas, F. Dib-Hajj, W. Yuan, M. Cronan, A. V. Kamath, J. Bergeron, and J. A. Retsema.** 1996. Molecular cloning and functional analysis of a novel macrolide-resistance determinant, *mefA*, from *Streptococcus pyogenes*. *Mol. Microbiol.* **22:**867–879.

17. **Clewell, D. B., S. E. Flannagan, Y. Ike, J. M. Jones, and C. Gawron-Burke.** 1988. Sequence analysis of termini of conjugative transposon Tn916. *J. Bacteriol.* **170:**3046–3052.

18. **Clewell, D. B., S. E. Flannagan, and D. D. Jaworski.** 1995. Unconstrained bacterial promiscuity: the Tn916-Tn1545 family of conjugative transposons. *Trends Microbiol.* **3:**229–236.

19. **Clewell, D. B., and C. Gawron-Burke.** 1986. Conjugative transposons and the dissemination of antibiotic resistance in streptococci. *Annu. Rev. Microbiol.* **40:**635–659.

20. **Dagan, R., R. Melamed, M. Muallem, L. Piglansky, D. Greenberg, O. Abramson, P. M. Mendelman, N. Bohidar, and P. Yagupsky.** 1996. Reduction of nasopharyngeal carriage of pneumococci during the second year of life by a heptavalent conjugate pneumococcal vaccine. *J. Infect. Dis.* **174:**1271–1278.

21. **David, F., G. de Cespedes, F. Delbos, and T. Horaud.** 1993. Diversity of chromosomal genetic elements and gene identification in antibiotic-resistant strains of *Streptococcus pneumoniae* and *Streptococcus bovis*. *Plasmid* **29:**147–153.

22. **Del Grosso, M., F. Iannelli, C. Messina, M. Santagati, N. Petrosillo, S. Stefani, G. Pozzi, and A. Pantosti.** 2002. Macrolide efflux genes *mef* (A) and *mef* (E) are carried by different genetic elements in *Streptococcus pneumoniae*. *J. Clin. Microbiol.* **40:**774–778.

23. **Depardieu, F., and P. Courvalin.** 2001. Mutation in 23S rRNA responsible for resistance to 16-membered macrolides and streptogramins in *Streptococcus pneumoniae*. *Antimicrob. Agents Chemother.* **45:**319–323.

24. **Doern, G. V.** 2001. Antimicrobial use and the emergence of antimicrobial resistance with *Streptococcus pneumoniae* in the United States. *Clin. Infect. Dis.* **33**(Suppl. 3)**:**S187–S192.

25. **Doern, G. V., A. Brueggemann, H. P. Holley, Jr., and A. M. Rauch.** 1996. Antimicrobial resistance of *Streptococcus pneumoniae* recovered from outpatients in the United States during the winter months of 1994 to 1995: results of a 30-center national surveillance study. *Antimicrob. Agents Chemother.* **40:**1208–1213.

26. **Doern, G. V., K. P. Heilmann, H. K. Huynh, P. R. Rhomberg, S. L. Coffman, and A. B. Brueggemann.** 2001. Antimicrobial resistance among clinical isolates of *Streptococcus pneumoniae* in the United States during 1999–2000, including a comparison of resistance rates since 1994–1995. *Antimicrob. Agents Chemother.* **45:**1721–1729.

27. **Doherty, N., K. Trzcinski, P. Pickerill, P. Zawadzki, and C. G. Dowson.** 2000. Genetic diversity of the *tet* (M) gene in tetracycline-resistant clonal lineages of *Streptococcus pneumoniae*. *Antimicrob. Agents Chemother.* **44:**2979–2984.

28. **Dowson, C. G., T. J. Coffey, and B. G. Spratt.** 1994. Origin and molecular epidemiology of penicillin-binding-protein-mediated resistance to beta-lactam antibiotics. *Trends Microbiol.* **2:**361–366.

29. **Eliopoulos, G. M.** 2003. Quinupristin-dalfopristin and linezolid: evidence and opinion. *Clin. Infect. Dis.* **36:**473–481.

30. **Enright, M., P. Zawadski, P. Pickerill, and C. G. Dowson.** 1998. Molecular evolution of rifampicin resistance in *Streptococcus pneumoniae*. *Microb. Drug Resist.* **4:**65–70.

31. **Flannagan, S. E., L. A. Zitzow, Y. A. Su, and D. B. Clewell.** 1994. Nucleotide sequence

of the 18-kb conjugative transposon Tn*916* from *Enterococcus faecalis*. *Plasmid* **32**:350–354.

32. **Fogarty, C. M., S. Kohno, P. Buchanan, M. Aubier, and M. Baz.** 2003. Community-acquired respiratory tract infections caused by resistant pneumococci: clinical and bacteriological efficacy of the ketolide telithromycin. *J. Antimicrob. Chemother.* **51**:947–955.

33. **Franke, A. E., and D. B. Clewell.** 1981. Evidence for a chromosome-borne resistance transposon (Tn*916*) in *Streptococcus faecalis* that is capable of "conjugal" transfer in the absence of a conjugative plasmid. *J. Bacteriol.* **145**:494–502.

34. **Fukuda, H., and K. Hiramatsu.** 1999. Primary targets of fluoroquinolones in *Streptococcus pneumoniae*. *Antimicrob. Agents Chemother.* **43:** 410–412.

35. **Gay, K., W. Baughman, Y. Miller, D. Jackson, C. G. Whitney, A. Schuchat, M. M. Farley, F. Tenover, and D. S. Stephens.** 2000. The emergence of *Streptococcus pneumoniae* resistant to macrolide antimicrobial agents: a 6-year population-based assessment. *J. Infect. Dis.* **182**:1417–1424.

36. **Gay, K., and D. S. Stephens.** 2001. Structure and dissemination of a chromosomal insertion element encoding macrolide efflux in *Streptococcus pneumoniae*. *J. Infect. Dis.* **184**:56–65.

37. **Gill, M. J., N. P. Brenwald, and R. Wise.** 1999. Identification of an efflux pump gene, *pmrA*, associated with fluoroquinolone resistance in *Streptococcus pneumoniae*. *Antimicrob. Agents Chemother.* **43**:187–189.

38. **Goldstein, E. J., and S. M. Garabedian-Ruffalo.** 2002. Widespread use of fluoroquinolones versus emerging resistance in pneumococci. *Clin. Infect. Dis.* **35**:1505–1511.

39. **Gonzales, R., J. F. Steiner, A. Lum, and P. H. Barrett, Jr.** 1999. Decreasing antibiotic use in ambulatory practice: impact of a multidimensional intervention on the treatment of uncomplicated acute bronchitis in adults. *JAMA* **281**:1512–1519.

40. **Gordon, K. A., M. L. Beach, D. J. Biedenbach, R. N. Jones, P. R. Rhomberg, and A. H. Mutnick.** 2002. Antimicrobial susceptibility patterns of beta-hemolytic and viridans group streptococci: report from the SENTRY Antimicrobial Surveillance Program (1997–2000). *Diagn. Microbiol. Infect. Dis.* **43**:157–162.

41. **Horinouchi, S., and B. Weisblum.** 1982. Nucleotide sequence and functional map of pC194, a plasmid that specifies inducible chloramphenicol resistance. *J. Bacteriol.* **150**:815–825.

42. **Hyde, T. B., K. Gay, D. S. Stephens, D. J. Vugia, M. Pass, S. Johnson, N. L. Barrett, W. Schaffner, P. R. Cieslak, P. S. Maupin,** E. R. Zell, J. H. Jorgensen, R. R. Facklam, and C. G. Whitney. 2001. Macrolide resistance among invasive Streptococcus pneumoniae isolates. *JAMA* **286**:1857–1862.

43. **Johnston, N. J., J. C. De Azavedo, J. D. Kellner, and D. E. Low.** 1998. Prevalence and characterization of the mechanisms of macrolide, lincosamide, and streptogramin resistance in isolates of *Streptococcus pneumoniae*. *Antimicrob. Agents Chemother.* **42**:2425–2426.

44. **Jones, R. N., D. J. Farrell, and I. Morrissey.** 2003. Quinupristin-dalfopristin resistance in *Streptococcus pneumoniae*: novel L22 ribosomal protein mutation in two clinical isolates from the SENTRY Antimicrobial Surveillance Program. *Antimicrob. Agents Chemother.* **47**:2696–2698.

45. **Kilic, A. O., M. N. Vijayakumar, and S. F. al-Khaldi.** 1994. Identification and nucleotide sequence analysis of a transfer-related region in the streptococcal conjugative transposon Tn*5252*. *J. Bacteriol.* **176**:5145–5150.

46. **Kim, S., S. W. Lee, E. C. Choi, and S. Y. Choi.** 2003. Aminoacyl-tRNA synthetases and their inhibitors as a novel family of antibiotics. *Appl. Microbiol. Biotechnol.* **61**:278–288.

47. **Klugman, K. P.** 1990. Pneumococcal resistance to antibiotics. *Clin. Microbiol. Rev.* **3**:171–196.

48. **Klugman, K. P.** 2003. The role of clonality in the global spread of fluoroquinolone-resistant bacteria. *Clin. Infect. Dis.* **36**:783–785.

49. **Klugman, K. P.** 2002. The successful clone: the vector of dissemination of resistance in *Streptococcus pneumoniae*. *J. Antimicrob. Chemother.* **50**(Suppl. S2):1–5.

50. **Le Bouguenec, C., G. de Cespedes, and T. Horaud.** 1988. Molecular analysis of a composite chromosomal conjugative element (Tn*3701*) of *Streptococcus pyogenes*. *J. Bacteriol.* **170**:3930–3936.

51. **Le Bouguenec, C., G. de Cespedes, and T. Horaud.** 1990. Presence of chromosomal elements resembling the composite structure Tn*3701* in streptococci. *J. Bacteriol.* **172**:727–734.

52. **Leclercq, R., and P. Courvalin.** 1991. Bacterial resistance to macrolide, lincosamide, and streptogramin antibiotics by target modification. *Antimicrob. Agents Chemother.* **35**:1267–1272.

53. **Leclercq, R., and P. Courvalin.** 2002. Resistance to macrolides and related antibiotics in *Streptococcus pneumoniae*. *Antimicrob. Agents Chemother.* **46**:2727–2734.

54. **Linares, J., J. Garau, C. Dominguez, and J. L. Perez.** 1983. Antibiotic resistance and serotypes of *Streptococcus pneumoniae* from patients with community-acquired pneumococcal disease. *Antimicrob. Agents Chemother.* **23**:545–547.

55. **Linares, J., R. Pallares, T. Alonso, J. L. Perez, J. Ayats, F. Gudiol, P. F. Viladrich, and R. Martin.** 1992. Trends in antimicrobial resistance of clinical isolates of *Streptococcus pneumoniae* in Bellvitge Hospital, Barcelona, Spain (1979–1990). *Clin. Infect. Dis.* **15:**99–105.

56. **Lonks, J. R., J. Garau, and A. A. Medeiros.** 2002. Implications of antimicrobial resistance in the empirical treatment of community-acquired respiratory tract infections: the case of macrolides. *J. Antimicrob. Chemother.* **50**(Suppl. S2):87–92.

57. **Luna, V. A., P. Coates, E. A. Eady, J. H. Cove, T. T. Nguyen, and M. C. Roberts.** 1999. A variety of gram-positive bacteria carry mobile *mef* genes. *J. Antimicrob. Chemother.* **44:**9–25.

58. **Luna, V. A., and M. C. Roberts.** 1998. The presence of the *tetO* gene in a variety of tetracycline-resistant *Streptococcus pneumoniae* serotypes from Washington State. *J. Antimicrob. Chemother.* **42:**613–619.

59. **Maskell, J. P., A. M. Sefton, and L. M. Hall.** 1997. Mechanism of sulfonamide resistance in clinical isolates of *Streptococcus pneumoniae. Antimicrob. Agents Chemother.* **41:**2121–2126.

60. **Maskell, J. P., A. M. Sefton, and L. M. Hall.** 2001. Multiple mutations modulate the function of dihydrofolate reductase in trimethoprim-resistant *Streptococcus pneumoniae. Antimicrob. Agents Chemother.* **45:**1104–1108.

61. **Matsuoka, M., K. Endou, H. Kobayashi, M. Inoue, and Y. Nakajima.** 1998. A plasmid that encodes three genes for resistance to macrolide antibiotics in *Staphylococcus aureus. FEMS Microbiol. Lett.* **167:**221–227.

62. **Mbelle, N., R. E. Huebner, A. D. Wasas, A. Kimura, I. Chang, and K. P. Klugman.** 1999. Immunogenicity and impact on nasopharyngeal carriage of a nonavalent pneumococcal conjugate vaccine. *J. Infect. Dis.* **180:**1171–1176.

63. **McDougal, L. K., F. C. Tenover, L. N. Lee, J. K. Rasheed, J. E. Patterson, J. H. Jorgensen, and D. J. LeBlanc.** 1998. Detection of Tn*917*-like sequences within a Tn*916*-like conjugative transposon (Tn*3872*) in erythromycin-resistant isolates of *Streptococcus pneumoniae. Antimicrob. Agents Chemother.* **42:**2312–2318.

64. **Monro, R. E., and K. A. Marcker.** 1967. Ribosome-catalysed reaction of puromycin with a formylmethionine-containing oligonucleotide. *J. Mol. Biol.* **25:**347–350.

65. **Monro, R. E., and D. Vazquez.** 1967. Ribosome-catalysed peptidyl transfer: effects of some inhibitors of protein synthesis. *J. Mol. Biol.* **28:**161–165.

66. **Munoz-Najar, U., and M. N. Vijayakumar.** 1999. An operon that confers UV resistance by evoking the SOS mutagenic response in streptococcal conjugative transposon Tn*5252. J. Bacteriol.* **181:**2782–2788.

67. **Musher, D. M., M. E. Dowell, V. D. Shortridge, R. K. Flamm, J. H. Jorgensen, P. Le Magueres, and K. L. Krause.** 2002. Emergence of macrolide resistance during treatment of pneumococcal pneumonia. *N. Engl. J. Med.* **346:**630–631.

68. **Nagai, K., P. C. Appelbaum, T. A. Davies, L. M. Kelly, D. B. Hoellman, A. T. Andrasevic, L. Drukalska, W. Hryniewicz, M. R. Jacobs, J. Kolman, J. Miciuleviciene, M. Pana, L. Setchanova, M. K. Thege, H. Hupkova, J. Trupl, and P. Urbaskova.** 2002. Susceptibilities to telithromycin and six other agents and prevalence of macrolide resistance due to L4 ribosomal protein mutation among 992 pneumococci from 10 Central and Eastern European countries. *Antimicrob. Agents Chemother.* **46:**371–377.

69. **Norgren, M., and J. R. Scott.** 1991. The presence of conjugative transposon Tn*916* in the recipient strain does not impede transfer of a second copy of the element. *J. Bacteriol.* **173:**319–324.

70. **Oggioni, M. R., C. G. Dowson, J. M. Smith, R. Provvedi, and G. Pozzi.** 1996. The tetracycline resistance gene *tet*(M) exhibits mosaic structure. *Plasmid* **35:**156–163.

71. **Oster, P., A. Zanchi, S. Cresti, M. Lattanzi, F. Montagnani, C. Cellesi, and G. M. Rossolini.** 1999. Patterns of macrolide resistance determinants among community-acquired *Streptococcus pneumoniae* isolates over a 5-year period of decreased macrolide susceptibility rates. *Antimicrob. Agents Chemother.* **43:**2510–2512.

72. **Padayachee, T., and K. P. Klugman.** 1999. Molecular basis of rifampin resistance in *Streptococcus pneumoniae. Antimicrob. Agents Chemother.* **43:**2361–2365.

73. **Pan, X. S., and L. M. Fisher.** 1997. Targeting of DNA gyrase in *Streptococcus pneumoniae* by sparfloxacin: selective targeting of gyrase or topoisomerase IV by quinolones. *Antimicrob. Agents Chemother.* **41:**471–474.

74. **Pelton, S. I., R. Dagan, B. M. Gaines, K. P. Klugman, D. Laufer, K. O'Brien, and H. J. Schmitt.** 2003. Pneumococcal conjugate vaccines: proceedings from an Interactive Symposium at the 41st Interscience Conference on Antimicrobial Agents and Chemotherapy. *Vaccine* **21:**1562–1571.

75. **Perez, J. L., J. Linares, J. Bosch, M. J. Lopez de Goicoechea, and R. Martin.** 1987. Antibiotic resistance of *Streptococcus pneumoniae* in childhood carriers. *J. Antimicrob. Chemother.* **19:**278–280.

76. **Perkins, J. B., and P. J. Youngman.** 1984. A physical and functional analysis of Tn917, a *Streptococcus* transposon in the Tn3 family that functions in *Bacillus. Plasmid* **12**:119–138.

77. **Pestova, E., J. J. Millichap, F. Siddiqui, G. A. Noskin, and L. R. Peterson.** 2002. Non-PmrA-mediated multidrug resistance in *Streptococcus pneumoniae. J. Antimicrob. Chemother.* **49**:553–556.

78. **Piddock, L. J., and M. M. Johnson.** 2002. Accumulation of 10 fluoroquinolones by wild-type or efflux mutant *Streptococcus pneumoniae. Antimicrob. Agents Chemother.* **46**:813–820.

79. **Poyart-Salmeron, C., P. Trieu-Cuot, C. Carlier, and P. Courvalin.** 1991. Nucleotide sequences specific for Tn1545-like conjugative transposons in pneumococci and staphylococci resistant to tetracycline. *Antimicrob. Agents Chemother.* **35**:1657–1660.

80. **Pozzi, G., R. A. Musmanno, E. A. Renzoni, M. R. Oggioni, and M. G. Cusi.** 1988. Host-vector system for integration of recombinant DNA into chromosomes of transformable and nontransformable streptococci. *J. Bacteriol.* **170**:1969–1972.

81. **Provvedi, R., R. Manganelli, and G. Pozzi.** 1996. Characterization of conjugative transposon Tn5251 of *Streptococcus pneumoniae. FEMS Microbiol. Lett.* **135**:231–236.

82. **Reinert, R. R., A. Wild, P. Appelbaum, R. Lutticken, M. Y. Cil, and A. Al-Lahham.** 2003. Ribosomal mutations conferring resistance to macrolides in *Streptococcus pneumoniae* clinical strains isolated in Germany. *Antimicrob. Agents Chemother.* **47**:2319–2322.

83. **Rice, L. B.** 2000. Bacterial monopolists: the bundling and dissemination of antimicrobial resistance genes in gram-positive bacteria. *Clin. Infect. Dis.* **31**:762–769.

84. **Richter, S. S., K. P. Heilmann, S. L. Coffman, H. K. Huynh, A. B. Brueggemann, M. A. Pfaller, and G. V. Doern.** 2002. The molecular epidemiology of penicillin-resistant *Streptococcus pneumoniae* in the United States, 1994–2000. *Clin. Infect. Dis.* **34**:330–339.

85. **Roberts, M. C., J. Sutcliffe, P. Courvalin, L. B. Jensen, J. Rood, and H. Seppala.** 1999. Nomenclature for macrolide and macrolide-lincosamide-streptogramin B resistance determinants. *Antimicrob. Agents Chemother.* **43**:2823–2830.

86. **Rosato, A., H. Vicarini, and R. Leclercq.** 1999. Inducible or constitutive expression of resistance in clinical isolates of streptococci and enterococci cross-resistant to erythromycin and lincomycin. *J. Antimicrob. Chemother.* **43**:559–562.

87. **Ross, J. I., E. A. Eady, J. H. Cove, and S. Baumberg.** 1995. Identification of a chromosomally encoded ABC-transport system with which the staphylococcal erythromycin exporter MsrA may interact. *Gene* **153**:93–98.

88. **Ross, J. I., E. A. Eady, J. H. Cove, W. J. Cunliffe, S. Baumberg, and J. C. Wootton.** 1990. Inducible erythromycin resistance in staphylococci is encoded by a member of the ATP-binding transport super-gene family. *Mol. Microbiol.* **4**:1207–1214.

89. **Salyers, A. A., N. B. Shoemaker, A. M. Stevens, and L. Y. Li.** 1995. Conjugative transposons: an unusual and diverse set of integrated gene transfer elements. *Microbiol. Rev.* **59**:579–590.

90. **Santagati, M., F. Iannelli, M. R. Oggioni, S. Stefani, and G. Pozzi.** 2000. Characterization of a genetic element carrying the macrolide efflux gene *mef*(A) in *Streptococcus pneumoniae. Antimicrob. Agents Chemother.* **44**:2585–2587.

91. **Schimmel, P., J. Tao, and J. Hill.** 1998. Aminoacyl tRNA synthetases as targets for new anti-infectives. *FASEB J.* **12**:1599–1609.

92. **Schlunzen, F., R. Zarivach, J. Harms, A. Bashan, A. Tocilj, R. Albrecht, A. Yonath, and F. Franceschi.** 2001. Structural basis for the interaction of antibiotics with the peptidyl transferase centre in eubacteria. *Nature* **413**:814–821.

93. **Schmitz, F. J., P. G. Higgins, S. Mayer, A. C. Fluit, and A. Dalhoff.** 2002. Activity of quinolones against gram-positive cocci: mechanisms of drug action and bacterial resistance. *Eur. J. Clin. Microbiol. Infect. Dis.* **21**:647–659.

94. **Schmitz, F. J., M. Perdikouli, A. Beeck, J. Verhoef, and A. C. Fluit.** 2001. Molecular surveillance of macrolide, tetracycline and quinolone resistance mechanisms in 1191 clinical European *Streptococcus pneumoniae* isolates. *Int. J. Antimicrob. Agents* **18**:433–436.

95. **Schrag, S. J., B. Beall, and S. F. Dowell.** 2000. Limiting the spread of resistant pneumococci: biological and epidemiologic evidence for the effectiveness of alternative interventions. *Clin. Microbiol. Rev.* **13**:588–601.

96. **Scott, J. R., P. A. Kirchman, and M. G. Caparon.** 1988. An intermediate in transposition of the conjugative transposon Tn916. *Proc. Natl. Acad. Sci. USA* **85**:4809–4813.

97. **Seppala, H., T. Klaukka, J. Vuopio-Varkila, A. Muotiala, H. Helenius, K. Lager, P. Huovinen, and The Finnish Study Group for Antimicrobial Resistance.** 1997. The effect of changes in the consumption of macrolide antibiotics on erythromycin resistance in group A streptococci in Finland. *N. Engl. J. Med.* **337**:441–446.

98. **Sessegolo, J. F., A. S. Levin, C. E. Levy, M. Asensi, R. R. Facklam, and L. M. Teixeira.** 1994. Distribution of serotypes and antimicrobial resistance of *Streptococcus pneumoniae* strains isolated in Brazil from 1988 to 1992. *J. Clin. Microbiol.* **32:**906–911.

99. **Shaw, J. H., and D. B. Clewell.** 1985. Complete nucleotide sequence of macrolide-lincosamide-streptogramin B-resistance transposon Tn917 in *Streptococcus faecalis. J. Bacteriol.* **164:** 782–796.

100. **Shaw, W. V., and A. G. Leslie.** 1991. Chloramphenicol acetyltransferase. *Annu. Rev. Biophys. Biophys. Chem.* **20:**363–386.

101. **Shoemaker, N. B., M. D. Smith, and W. R. Guild.** 1980. DNase-resistant transfer of chromosomal *cat* and *tet* insertions by filter mating in Pneumococcus. *Plasmid* **3:**80–87.

102. **Spika, J. S., R. R. Facklam, B. D. Plikaytis, M. J. Oxtoby, and The Pneumococcal Surveillance Working Group.** 1991. Antimicrobial resistance of *Streptococcus pneumoniae* in the United States, 1979–1987. *J. Infect. Dis.* **163:**1273–1278.

103. **Sutcliffe, J., A. Tait-Kamradt, and L. Wondrack.** 1996. *Streptococcus pneumoniae* and *Streptococcus pyogenes* resistant to macrolides but sensitive to clindamycin: a common resistance pattern mediated by an efflux system. *Antimicrob. Agents Chemother.* **40:**1817–1824.

104. **Swartley, J. S., C. F. McAllister, R. A. Hajjeh, D. W. Heinrich, and D. S. Stephens.** 1993. Deletions of Tn916-like transposons are implicated in *tetM*-mediated resistance in pathogenic *Neisseria. Mol. Microbiol.* **10:**299–310.

105. **Syrogiannopoulos, G. A., I. N. Grivea, A. Tait-Kamradt, G. D. Katopodis, N. G. Beratis, J. Sutcliffe, P. C. Appelbaum, and T. A. Davies.** 2001. Identification of an *erm*(A) erythromycin resistance methylase gene in *Streptococcus pneumoniae* isolated in Greece. *Antimicrob. Agents Chemother.* **45:**342–344.

106. **Tait-Kamradt, A., J. Clancy, M. Cronan, F. Dib-Hajj, L. Wondrack, W. Yuan, and J. Sutcliffe.** 1997. *mefE* is necessary for the erythromycin-resistant M phenotype in *Streptococcus pneumoniae. Antimicrob. Agents Chemother.* **41:** 2251–2255.

107. **Tait-Kamradt, A., T. Davies, P. C. Appelbaum, F. Depardieu, P. Courvalin, J. Petitpas, L. Wondrack, A. Walker, M. R. Jacobs, and J. Sutcliffe.** 2000. Two new mechanisms of macrolide resistance in clinical strains of *Streptococcus pneumoniae* from Eastern Europe and North America. *Antimicrob. Agents Chemother.* **44:**3395–3401.

108. **Tait-Kamradt, A., T. Davies, M. Cronan, M. R. Jacobs, P. C. Appelbaum, and J. Sutcliffe.** 2000. Mutations in 23S rRNA and ribosomal protein L4 account for resistance in pneumococcal strains selected in vitro by macrolide passage. *Antimicrob. Agents Chemother.* **44:**2118–2125.

109. **Tomich, P. K., F. Y. An, and D. B. Clewell.** 1980. Properties of erythromycin-inducible transposon Tn917 in *Streptococcus faecalis. J. Bacteriol.* **141:**1366–1374.

110. **Tomich, P. K., F. Y. An, and D. B. Clewell.** 1979. A transposon (Tn917) in *Streptococcus faecalis* that exhibits enhanced transposition during induction of drug resistance. *Cold Spring Harbor Symp. Quant. Biol.* **43**(Pt. 2):1217–1221.

111. **van Tilburg, P. M., D. Bogaert, M. Sluijter, A. R. Jansz, R. de Groot, and P. W. Hermans.** 2001. Emergence of rifampin-resistant *Streptococcus pneumoniae* as a result of antimicrobial therapy for penicillin-resistant strains. *Clin. Infect. Dis.* **33:**e93–e96.

112. **Vijayakumar, M. N., S. D. Priebe, and W. R. Guild.** 1986. Structure of a conjugative element in *Streptococcus pneumoniae. J. Bacteriol.* **166:** 978–984.

113. **Weisblum, B.** 1995. Erythromycin resistance by ribosome modification. *Antimicrob. Agents Chemother.* **39:**577–585.

114. **Weisblum, B.** 1995. Insights into erythromycin action from studies of its activity as inducer of resistance. *Antimicrob. Agents Chemother.* **39:**797–805.

115. **Whitney, C. G., M. M. Farley, J. Hadler, L. H. Harrison, N. M. Bennett, R. Lynfield, A. Reingold, P. R. Cieslak, T. Pilishvili, D. Jackson, R. R. Facklam, J. H. Jorgensen, and A. Schuchat.** 2003. Decline in invasive pneumococcal disease after the introduction of protein-polysaccharide conjugate vaccine. *N. Engl. J. Med.* **348:**1737–1746.

116. **Whitney, C. G., M. M. Farley, J. Hadler, L. H. Harrison, C. Lexau, A. Reingold, L. Lefkowitz, P. R. Cieslak, M. Cetron, E. R. Zell, J. H. Jorgensen, and A. Schuchat.** 2000. Increasing prevalence of multidrug-resistant *Streptococcus pneumoniae* in the United States. *N. Engl. J. Med* **343:**1917–1924.

117. **Widdowson, C. A., P. V. Adrian, and K. P. Klugman.** 2000. Acquisition of chloramphenicol resistance by the linearization and integration of the entire staphylococcal plasmid pC194 into the chromosome of *Streptococcus pneumoniae. Antimicrob. Agents Chemother.* **44:** 393–395.

118. **Widdowson, C. A., and K. P. Klugman.** 1999. Molecular mechanisms of resistance to

commonly used non-betalactam drugs in *Streptococcus pneumoniae*. *Semin. Respir. Infect.* **14:**255–268.

119. **Widdowson, C. A., and K. P. Klugman.** 1998. The molecular mechanisms of tetracycline resistance in the pneumococcus. *Microb. Drug Resist.* **4:**79–84.

120. **Widdowson, C. A., K. P. Klugman, and D. Hanslo.** 1996. Identification of the tetracycline resistance gene, *tet*(O), in *Streptococcus pneumoniae*. *Antimicrob. Agents Chemother.* **40:**2891–2893.

121. **Wu, S. W., H. de Lencastre, and A. Tomasz.** 1999. The *Staphylococcus aureus* transposon Tn*551*: complete nucleotide sequence and transcriptional analysis of the expression of the erythromycin resistance gene. *Microb. Drug Resist.* **5:**1–7.

122. **Zeller, V., C. Janoir, M. D. Kitzis, L. Gutmann, and N. J. Moreau.** 1997. Active efflux as a mechanism of resistance to ciprofloxacin in *Streptococcus pneumoniae*. *Antimicrob. Agents Chemother.* **41:**1973–1978.

THE HUMORAL IMMUNE RESPONSE TO *STREPTOCOCCUS PNEUMONIAE*

Clifford M. Snapper, Jesus Colino, Abdul Q. Khan, and Zheng Qi Wu

23

ADAPTIVE IMMUNITY IS MEDIATED BY ANTIBODY

Mechanisms of Protection

Adaptive immunity to extracellular bacteria, such as *Streptococcus pneumoniae*, is largely conferred by antibody. Antibody specific for both bacterial polysaccharide and protein antigens has been shown to protect the host from infection with otherwise lethal *S. pneumoniae* strains (1). Antibody binding to the bacterial surface can activate complement either through the classical (immunoglobulin M [IgM] and IgG) or alternative (IgA) pathways. Subsequent binding of antibody and C3b to Fc and C3b receptors, respectively, expressed by phagocytic cells (neutrophils and macrophages) results in opsonophagocytosis and rapid killing of ingested bacteria. The associated generation of other complement fragments can indirectly mediate protection by enhancing other aspects of the innate response, as well as adaptive immunity (65, 66). In particular the generation of C3d, the ligand for complement receptor type 2 (CR2

[CD21]), may boost antigen-specific Ig responses both through enhanced signaling via B-cell Ig receptor (BCR)-CD21 coligation on the B cell surface and through antigen retention on CD21-expressing FDC in germinal centers (12).

Ig Isotypes

Distinct Ig isotypes possess overlapping and unique effector functions on the basis of the particular Fc region expressed (76). The isotype of the expressed Ig can determine its relative ability to activate complement, bind to distinct Fc receptors on multiple cell types, cross the placenta, penetrate into tissue or transcytose through epithelial cells, and resist the action of proteases. Ig isotype also can determine antibody half-life in serum and the antigen valency of Ig through formation of pentamers (IgM) or dimers (IgA) or through self-aggregation (mouse IgG3). Thus, the pattern of Ig isotypes elicited during a bacterial infection or after vaccination, in addition to the epitope specificity and affinity of the Ig, can impact the level of protection afforded by such antibody. Switching from the initial expression of IgM to particular downstream Ig isotypes by an activated B cell is a directed process dependent on specific cytokines signaling in concert with activation stimuli mediated through BCR, CD40, or Toll-like

Clifford M. Snapper, Jesus Colino, and Abdul Q. Khan, Department of Pathology, The Uniformed Services University of the Health Sciences, 4301 Jones Bridge Road, Bethesda, MD 20814. *Zheng Qi Wu,* Autoimmunity Branch, NIAMS, National Institutes of Health, 9000 Rockville Pike, Bethesda, MD 20892.

receptors (TLR). Cytokines will also promote proliferation and maturation to antibody secretion of the isotype switched, as well as the IgM$^+$ B cell.

DIFFERENTIAL REGULATION OF ANTIPOLYSACCHARIDE AND ANTIPROTEIN IG RESPONSES

As mentioned above, antibody specific for both bacterial polysaccharide and protein can be protective against extracellular bacteria, including S. pneumoniae. Much of our current knowledge of the comparative regulation of antipolysaccharide and antiprotein responses has come from studies using purified, soluble polysaccharides and proteins, with or without haptenation and/or inclusion of adjuvant (61). These studies revealed that antipolysaccharide responses, in contrast to those specific for protein, are more rapid, fail to induce germinal center reactions (with some exceptions), show little generation of memory or affinity maturation, and have a more restricted Ig isotype profile (i.e., largely IgM and IgG3 in the mouse [74], IgM and IgG2c in the rat [19], and IgM and IgG2 in the human [5]). Proteins, in distinct contrast to polysaccharides, are enzymatically processed within endosomes to generate surface major histocompatibility complex (MHC) class II peptide complexes on the surfaces of antigen-preventing cells (APCs) for presentation to CD4$^+$ T cells (31, 40). Hence, polysaccharides, by themselves, do not appear to be capable of recruiting cognate CD4$^+$ T-cell help, which may explain their relative T-cell and CD40-CD40 ligand independence and their inability to induce classic memory responses (61). Although antipolysaccharide responses occur in the absence of T cells, noncognate forms of T-cell help and suppression have been described to play a modulating role (61). Polysaccharides, unlike proteins, however, display repeating, identical antigenic epitopes capable of mediating sustained and powerful signaling through the BCR (11), which may help to explain the rapidity of the antipolysaccharide response and likely other functional outcomes in the responding B cells.

In contrast to purified antigens, intact S. pneumoniae, when considered as an immunogen, is a complex particle consisting of associated protein and polysaccharide moieties. In this regard, polysaccharide- and protein-specific Ig isotype responses to the intact bacterium could more closely resemble those observed using conjugate vaccines consisting of covalently linked polysaccharide and protein, as opposed to purified antigens (68). Additionally, the bacterium contains a number of adjuvanting moieties (pathogen-associated molecular patterns [59]) such as lipoteichoic acid and peptidoglycan, which signal via TLR2, and DNA containing unmethylated CpG which signals via TLR9 (84). Zwitterionic polysaccharide, such as the C polysaccharide (C-PS) (24), may demonstrate T-cell-stimulating properties (41). The potential ability of the serotype-specific capsular polysaccharide expressed by a given strain to further impact humoral immunity by influencing the rate of bacterial clearance, the relative ability to process associated protein for presentation to T cells, and the masking of subcapsular structures is largely unexplored (42, 50, 57). Finally, whereas protein antigens are typically mono- or paucivalent for any given epitope, many studies have utilized highly haptenated proteins which may deliver BCR signals that are not physiological. Similarly, hapten density on haptenated polysaccharides may not reflect the true epitope density of the repeating sugars comprising various natural polysaccharides. Thus, the conceptual paradigms established for the nature of antipolysaccharide and antiprotein Ig isotype responses using purified soluble antigens, with or without haptenation and/or inclusion of an adjuvant, may fall short of adequately explaining physiological Ig responses to the intact bacteria. Recent studies on the humoral immune response to intact S. pneumoniae, described below, lend support to this notion.

MARGINAL-ZONE AND B-1 B CELLS AND THE TI RESPONSE

B cells are divided into distinct subpopulations on the basis of defining cell surface phenotypes, associated with characteristic functional abilities, migratory patterns, and tissue localization. In particular, both B-1 cells (comprising the majority of B cells in the peritoneum and pleural cavity) and splenic marginal-zone B cells (MZB) (subpopulation of B-2) have been implicated in the early, T-cell-independent (TI) IgM response to polysaccharide antigens (54, 55). B-1 cells are also implicated in producing a significant percentage of the circulating natural Ig in the naïve host, which is germ line encoded, low affinity, polyreactive, and specific for conserved microbial structures, including polysaccharides. Indeed, a significant fraction of this natural Ig shows specificity for phosphorylcholine (PC), expressed on the C-PS (teichoic acid), in the *S. pneumoniae* cell wall. PC is also expressed by a number of other bacterial and nematode pathogens and is a self-antigen, being found on oxidized low-density lipoprotein and apoptotic cells (13, 18, 32, 71). The more abundant follicular B cells (FB) (subpopulation of B-2) appear to play the major role in T-cell-dependent (TD) responses to protein antigens, participating in the germinal center reaction and giving rise to isotype-switched, high-affinity memory B cells and long-lived plasma cells.

B-1 cells and MZB subserve their early host protective function on the basis of their strategic locations at initial portals of pathogen entry, a germ line BCR repertoire encoding conserved microbial structures, and a level of preimmune, heightened activation leading to rapid responses to microbial stimuli, with a propensity for IgM plasmablast differentiation (48, 54, 55, 64, 67, 80). In this regard, recent work using BCR transgenic mice directly demonstrated a concerted action of both B-1 cells and MZB, but not FB, in the early IgM anti-PC response to systemic challenge with heat-killed intact *S. pneumoniae* (56). In particular, splenic MZB leave the marginal zone and within 4 h after immunization are located at the B cell-T cell interface, where by 24 h they have undergone extensive proliferation. By 48 h a large number of IgM plasmablasts are observed in the splenic red pulp. Plasmablast differentiation peaks by day 3 to 4 and then declines over the subsequent 2 weeks. This rapid IgM response occurred in the absence of T cells. Adoptive-transfer studies using B cells expressing distinct PC-specific idiotypes (T15 for B-1 cells and M167 for MZB) demonstrated that the IgM anti-PC response to *S. pneumoniae* arose from both B-cell subpopulations. As alluded to above, it remains to be determined which B-cell subpopulation(s) plays the major role in IgG antipolysaccharide responses to intact extracellular bacteria, which, as is discussed below, may depend on cognate CD4$^+$-T-cell help.

"SECOND SIGNALS" FOR COSTIMULATION OF AN ANTIGEN-DRIVEN TI RESPONSE

BCR cross-linking stimuli by themselves can be mitogenic for FB, although much less so for B-1 cells or MZB (55, 56), but in the absence of costimuli fail to induce differentiation or isotype switching (77). CD4$^+$ T-cell receptor-α/β^+ (TCR-α/β^+) T cells engaging in cognate interactions with FB during responses to protein antigens provide these additional differentiative signals, such as membrane CD40 ligand in concert with secreted cytokines (63). However, the second signals required for antibacterial TI IgM antipolysaccharide responses by B-1 cells and MZB are poorly defined. Although BlyS/April released rapidly by activated macrophages and dendritic cells have been implicated in TI antipolysaccharide responses (53), including those to *S. pneumoniae* (3), their role appears to be largely to promote B-cell survival. In contrast, various bacterial TLR ligands, either alone or synergistically with a BCR cross-linking signal, can promote IgM as well as IgG secretion (77). Thus, highly purified resting B cells induced to proliferate by anti-IgD antibodies conjugated to dextran, an in vitro polyclonal model for B-cell activation by polysaccharide antigens, secrete IgM and undergo Ig isotype switching in response to bacterial TLR ligands such as lipopolysaccharide (LPS), lipoproteins, porins, and unmethylated CpG DNA. However, the contribution of bacterial TLR ligands to IgM antibacterial TI responses in vivo is largely unknown, although initial studies suggest that they may not be critical. Thus, mice lacking the major TLR adaptor protein, MyD88, have normal or even elevated IgM responses either to a soluble TI antigen (70) or to intact *S. pneumoniae* (C. M. Snapper, unpublished data). However, selective TD IgG2a responses, including those to *S. pneumoniae*, associated with decreased gamma interferon (IFN-γ) production are observed in these mice, whereas TD IgG1 responses are normal. Thus, TLR ligands may be particularly important in mediating the Th1 TD component of an antibacterial response, perhaps by inducing APCs to secrete interleukin-12 (IL-12) (62), but do not appear to be critical for the Th2 or TI component. As described below, primary immunization with intact *S. pneumoniae* elicits a cytokine response, comprising both type 1 and type 2 cytokines, that can regulate the induction of Ig in either a positive or negative fashion (44).

NK cells are non-antigen-specific lymphoid cells that become activated early after bacterial infection and may play a role in delivering second signals for TI antipolysaccharide responses. Thus, in vitro studies have demonstrated that NK cells, in part through release of IFN-γ, can directly stimulate murine B cells to secrete Ig (78, 79, 87, 95). Of interest, in vitro-activated NK cells, in the absence of exogenous stimuli, preferentially induced Ig secretion from MZB, but not FB, suggesting a selective role for NK cells in antipolysaccharide responses (78). NK cells also were shown to induce Ig production by human CD5[+] or CD27[+] (memory) B cells, a process that required cell contact and CD40-CD40 ligand interaction (7). NK cells may further play an indirect role in stimulating TI responses through their ability to augment dendritic cell (DC) function, especially in the presence of limited stimulation by bacterial pathogens (27). DCs in turn can directly stimulate activated B cells to secrete Ig in vitro (22) and, as discussed below, can play an active role

in stimulating in vivo TI and TD antipolysaccharide, as well as antiprotein, responses to intact *S. pneumoniae* (16). In vivo studies so far suggest only a limited role for endogenous NK cells in augmenting a humoral immune response, by demonstrating their ability to selectively stimulate trinitrophenyl (TNP)-specific IgG2a in response to the soluble TI antigen TNP-LPS (46, 89). NK cell induction of IgG2a may be mediated through its release of IFN-γ (46, 81) or perhaps through an IFN-γ-independent mechanism (26, 69). Despite these observations, a role for NK cells in regulating an in vivo antibacterial humoral immune response remains to be shown.

T CELLS

In addition to NK cells, NK T cells and TCR-γ/δ[+] T cells could potentially stimulate early humoral responses to bacterial pathogens, before recruitment of conventional CD4[+] TCR-α/β[+] T cells (34, 85). These unconventional T cells are present at strategic portals of pathogen entry, can recognize conserved nonprotein (carbohydrate and lipid) microbial structures, and may account for the early release of cytokines during an immune response to a pathogen. Although earlier studies demonstrated that Ig responses to polysaccharides expressed by intact bacteria were reduced in the absence of T cells (9, 10), no detailed studies had been published that described the parameters that regulated polysaccharide versus protein-specific Ig isotype responses to intact extracellular bacteria.

To further explore a potential role for T cells in regulating an in vivo, antigen-specific Ig response to intact extracellular bacteria, more recent initial studies with mice involved the use of heat-killed, intact *S. pneumoniae* (the nonencapsulated variant of *S. pneumoniae* capsular type 2 [strain R36A]), with later studies using the encapsulated *S. pneumoniae* serotype 14 (Pn14) (43, 75, 91–93). Both the primary and secondary murine Ig isotype responses specific for the PC determinant of the cell wall C-PS (teichoic acid), the cell wall pneumococcal surface protein A (PspA), and, when using Pn14, the capsular polysaccharide (Cps14) were determined through measurement of titers in serum by antigen-specific enzyme-linked immunosorbent assay.

A Major Inductive Role for CD4$^+$ TCR-α/β$^+$ T Cells

Initial kinetic studies indicated that the titers of both primary IgM and IgG anti-PC and anti-Cps14 antibodies in serum were maximal by day 5 to 6, whereas IgG anti-PspA titers peaked by day 10 to 11 (43, 92, 93). Upon secondary immunization, both the IgM and IgG antipolysaccharide responses showed little to moderate boosting (≤5-fold), whereas the anti-PspA response was consistently boosted up to 10- to 20-fold. Kinetics and boosting of the IgG response specific for the pneumococcal cell membrane lipoprotein, pneumococcal surface adhesin A (PsaA), were similar to those observed for PspA. The more rapid kinetics and more limited boosting of the antipolysaccharide relative to the antiprotein response to intact S. pneumoniae thus resembled that observed after immunization with analogous purified soluble antigens.

Athymic nude mice, which have a congenital absence of T cells, immunized with S. pneumoniae had IgM anti-PC and IgM anti-Cps14 responses similar to those of euthymic controls, whereas the IgG responses specific for PC, Cps14, and PspA were strikingly reduced (43, 91). Similarly, depletion of CD4$^+$ T cells with a monoclonal anti-CD4 monoclonal antibody (MAb) (GK1.5) had little, if any, effect on IgM responses but strongly inhibited all IgG responses. Adoptive transfer of CD4$^+$ T cells to athymic nude mice restored the defective IgG responses to S. pneumoniae and had no effect on the normal IgM response (91). Stimulated T cells are induced to express CD40 ligand, which mediates activation of B cells, DCs, and macrophages expressing CD40 (29). CD40-CD40 ligand interactions are critical for TD responses to soluble proteins but do not appear to play a role in TI responses to soluble polysaccharide antigens (25). Injection of a blocking anti-CD40 ligand MAb (MR1) to

wild-type mice immunized with S. pneumoniae inhibited IgG responses specific for PC, Cps14, and PspA, with no significant effect on IgM responses (43). To determine whether the CD4$^+$-T-cell help for the IgG responses was mediated by TCR-α/β$^+$ and/or TCR-γ/δ$^+$ T cells, mice genetically deficient in TCR-β, TCR-δ, or doubly deficient in TCR-β and -δ were immunized with S. pneumoniae. TCR-β$^{-/-}$ and TCR-β$^{-/-}$ × TCR-δ$^{-/-}$, but not TCR-δ$^{-/-}$, mice showed strong reductions in both IgG anti-PC and IgG anti-PspA responses, with no significant effect on the IgM anti-PC response (93). Collectively, these data demonstrate that both the IgG antipolysaccharide and IgG antiprotein responses to intact S. pneumoniae depend on CD4$^+$ TCR-α/β$^+$ T cells and CD40-CD40 ligand interactions, whereas the IgM antipolysaccharide response is TI. In this regard, the IgG antipolysaccharide response to the intact bacteria differs from IgG responses to purified soluble polysaccharide antigens in that the latter are TI and independent of CD40-CD40 ligand interactions. However, both types of responses still show rapid kinetics and limited boosting.

A Requirement for T-Cell Costimulation

Naïve CD4$^+$ T cells require at least two signals in order to be recruited into the primary immune response. The first is TCR-mediated signaling consequent to binding of MHC class II peptide complexes on the surface of APCs, and the second is via interaction of CD28, expressed constitutively on the T cell, with either B7-1 and/or B7-2 on the APCs (28). In this regard, injection of CTLA4Ig (which blocks B7-1 and B7-2 interactions with CD28 and CTLA4) into naïve mice immunized with R36A inhibited the primary IgG, but not IgM, anti-PC response, as well as the primary IgG anti-PspA response (92, 93). Using blocking anti-B7-1 and anti-B7-2 MAbs, it was demonstrated that B7-2, but not B7-1, was critical (92). Blocking B7-dependent costimulation at the time of primary R36A immunization also prevented the generation of PspA-specific memory, since secondary immunization (performed after CTLA4Ig was cleared) failed to show a boosted response. Further, injection of R36A-primed mice (PspA memory established) with CTLA4Ig at the time of secondary immunization with R36A also inhibited the secondary IgG anti-PspA response but had no effect on the anti-PC response. Of interest, primed mice receiving CTLA4Ig at the time of secondary challenge with *S. pneumoniae* exhibited a complete abrogation of PspA-specific memory, as evidenced by an absence of boosting upon tertiary immunization, at a time when the CTLA4Ig had already cleared. An additional challenge with *S. pneumoniae* demonstrated that memory was once again restored by this tertiary immunization in the absence of CTLA4Ig. Thus, B7-dependent costimulation is critical for the primary IgG (not IgM) anti-PC and IgG anti-PspA responses, as well as for the generation of PspA-specific memory and the elicitation of the anti-PspA secondary responses in mice already primed. These data are consistent with the requirement of CD4$^+$ T cells for these responses (91).

4-1BB (CD137) is another T-cell costimulatory molecule induced on activated CD4$^+$ and CD8$^+$ T cells that delivers a costimulatory signal upon binding to 4-1BB ligand expressed on APCs (86, 88). Induction of 4-1BB on T cells is dependent on activation via the TCR and possibly CD28. Mice genetically deficient in 4-1BB ligand showed a markedly reduced IgM and IgG anti-PC response but normal primary and secondary IgG anti-PspA responses to *S. pneumoniae* relative to that of wild-type mice (90). However, injection of an agonistic anti-4-1BB MAb, while having no significant effect on the anti-PC response, strongly inhibited the primary anti-PspA response, the generation of PspA-specific memory, and germinal center formation, but it did not induce a lasting state of tolerance. This inhibition was independent of CD8$^+$ T cells and was associated with expansion of CD4$^+$ T cells with an activated phenotype, which was partly dependent on B7-dependent costimulation. In con-

trast, anti-4-1BB MAb had no effect on the anti-PspA response when injected only at the time of secondary immunization (90). These data suggest a stimulatory role for endogenous 4-1BB–4-1BB ligand interactions during the anti-PC, but not anti-PspA, response to *S. pneumoniae*, and they further demonstrate a strong selective inhibitory effect of excessive 4-1BB activation on the anti-PspA, but not anti-PC, response. The effect of 4-1BB–4-1BB ligand interactions on anti-capsular PS responses has not been determined.

Delivery of Cognate versus Noncognate CD4$^+$ T-Cell Help

The above observations on kinetics, generation of memory, and T-cell costimulation for induction of antipolysaccharide and antiprotein responses suggested both shared and distinct roles for CD4$^+$ T cells in these two types of responses. To determine the time during which CD4$^+$ T cells were required for delivery of help for the IgG responses, CD4$^+$ T cells were acutely depleted (using anti-CD4 MAb) at different times after *S. pneumoniae* immunization (43, 91). These studies indicated that the T-cell help for optimal IgG anti-PC and IgG anti-Cps14 responses were required only during the first 2 to 3 days following immunization. In contrast, CD4$^+$ T cells were required for up to 5 days after challenge with *S. pneumoniae* in order to obtain an optimal IgG anti-PspA response. These more rapid kinetics of delivery of CD4$^+$-T-cell help for the IgG antipolysaccharide responses were mirrored by the observation that B7-dependent costimulation for IgG anti-PC responses was required only up to 2 days following immunization, whereas optimal IgG anti-PspA responses required costimulation for up to 6 days postimmunization (92). Thus, the relatively rapid kinetics of the IgG antipolysaccharide response is mirrored by a more rapid delivery of CD4$^+$-T-cell help relative to that observed for the antiprotein response.

In light of the requirement for CD4$^+$ T cells and CD40-CD40 ligand interactions for IgG antipolysaccharide, as well as IgG antiprotein, responses to intact *S. pneumoniae*, coupled with the fact that bacteria coexpress polysaccharide and protein antigens within a single particulate structure, it was of interest to determine whether or not the T-cell help for the IgG antipolysaccharide response was cognate (i.e., requiring specific T cells). As envisioned for soluble covalent conjugates of proteins and polysaccharides (conjugate vaccines) (68), the specific uptake of intact bacteria or bacterial fragments by a polysaccharide-specific B cell would allow such a B cell to process and present the associated bacterial protein as MHC class II peptides to specific T cells for recruitment of T-cell helper function. To determine this, H-Y $\alpha^{-/-}$ versus wild-type mice were used in *S. pneumoniae* immunization studies (43, 91). Specifically, H-Y $\alpha^{-/-}$ mice were created by crossing TCR transgenic mice (specific for the male, H-Y antigen in association with MHC class I) with mice genetically deficient in TCR-α (33). These mice expressed no endogenous TCR, possessed relatively normal numbers of CD4$^+$ and CD8$^+$ T cells, and had normal numbers of B cells. CD8$^+$ T cells in H-Y $\alpha^{-/-}$ mice expressed specificity for H-Y, whereas CD4$^+$ T cells were idiotype negative but did express TCR-β, perhaps in association with another, undetermined protein (33). Importantly, although both the idiotype-positive and -negative T cells from H-Y $\alpha^{-/-}$ mice responded to concanavalin A, they failed to respond to immunization with a conventional protein antigen in adjuvant. This observation is consistent with other studies using nontransgenic TCR-$\alpha^{-/-}$ mice which also demonstrate complete abrogation of Ig responses to conventional protein antigens in the absence of TCR-α/β^+ T cells.

H-Y $\alpha^{-/-}$ mice immunized with intact *S. pneumoniae* exhibited IgM anti-PC and IgM anti-Cps14 responses similar to those observed in wild-type controls, consistent with the TI nature of these responses (43, 91). As expected, the IgG anti-PspA response in H-Y $\alpha^{-/-}$ mice was completely inhibited, associated with a complete lack of splenic germinal center formation in response to *S. pneumoniae* immu-

nization (91). Of interest, the IgG anti-Cps14 response was also markedly reduced in H-Y $\alpha^{-/-}$ mice, strongly suggesting that this response also required specific CD4$^+$-T-cell help as hypothesized above (43). In contrast, the IgG anti-PC response in H-Y $\alpha^{-/-}$ mice was similar to that observed in wild-type controls (43, 91). Nevertheless, injection of *S. pneumoniae*-immunized H-Y $\alpha^{-/-}$ mice with either anti-CD4 MAb or CTLA4Ig resulted in a strong reduction in the IgG anti-PC response without effecting induction of PC-specific IgM (91). These data suggested that the IgG anti-PC response is mediated by a non-TCR-specific form of CD4$^+$-T-cell help. This is consistent with studies with cathepsin S$^{-/-}$ mice, which are defective in formation of new MHC class II peptide complexes (72), in which IgG anti-PspA, but not IgG anti-PC, responses are reduced relative to those in wild-type controls (91). Thus, three categories of CD4$^+$-T-cell help can be recognized during the humoral response to intact *S. pneumoniae*: (i) TCR-specific, prolonged kinetics (antiprotein); (ii) TCR-specific, rapid kinetics (anticapsular PS); and (iii) TCR non-specific, rapid kinetics (anti-PC). The generality of these findings to other *S. pneumoniae* serotypes and to other bacterial species is of active interest.

DCs

DCs are APCs with the unique property of inducing primary immune responses, leading to the establishment of immunological memory (4). Immature DCs in peripheral tissues internalize (37, 49) and process bacteria and other particulate antigens for MHC class I and MHC class II antigen presentation to T cells (83) migrating to secondary lymphoid organs, where DCs prime naïve T cells (38, 39) and naïve and memory B cells (6, 94) and activate NK cells (23) to induce specific immune responses. A role for DCs in conferring antimicrobial immunity is suggested by studies in which DCs, pulsed with soluble microbial antigen and transferred into naïve hosts, elicited an immune response that conferred protection against a further challenge with the pathogen

(58, 82). These and other observations indicate that DCs play a key role in linking innate and adaptive immunity, critical for host protection against microbial pathogens.

As mentioned, TI antigens, such as bacterial polysaccharides, are unable to bind to MHC molecules (31, 40) and hence cannot directly trigger cognate interactions of APCs and T cells. These observations suggested a minimal, if any, role for APCs and T cells in humoral immune responses to polysaccharide antigens. Nevertheless, until recently, little was known regarding the potential role of APCs in these responses. Only in vitro studies describing a direct role of macrophages in B-cell activation in response to TNP-haptenized antigens, such as TNP-Ficoll, TNP-LPS, and TNP-*Brucella abortus*, had been reported (8, 73).

S. pneumoniae Activates DCs

As mentioned above, *S. pneumoniae* expresses a number of structures that can mediate DC activation through TLRs, such as peptidoglycan (TLR2), lipoteichoic acid (TLR2), and unmethylated CpG-containing DNA (TLR9). Upon exposure to *S. pneumoniae*, murine DCs (CD11c$^+$ CD11b$^+$ CD8α^- cells obtained from bone marrow cells cultured in vitro for 6 days in granulocyte-macrophage colony-stimulating factor) up-regulate mRNA specific for, and secrete, both proinflammatory (IL-6, tumor necrosis factor α [TNF-α], and IL-12) and anti-inflammatory (IL-10) cytokines, with kinetics that vary for the individual cytokine (16). In particular, TNF-α mRNA expression and secretion are both induced (within 1 h) and subsequently down-regulated rapidly, whereas the majority of IL-10 mRNA and protein is expressed relatively late (>5 h). No detectable IL-2, IL-3, IL-4, or IFN-γ was observed after *S. pneumoniae* activation of DCs. Cytokine secretion was followed, 24 h after *S. pneumoniae* exposure, by phenotypic maturation as evidenced by increased cell surface expression of MHC class II, CD40, CD86 (B7-2), and CD25 (IL-2Rα) and down-regulation of CD16/CD32 (FcγRII/-III). These effects could be mimicked by purified peptidoglycan.

DC uptake of unopsonized *S. pneumoniae* occurred rapidly and was not dependent on the degree of encapsulation (15, 16). However, phenotypic maturation and the secretion of IL-6 were unaffected by treatment with an inhibitor of phagocytosis (cytochalasin), although IL-10 secretion was partially inhibited and IL-12 and TNF-α were completely inhibited (15). In the latter instance it is not clear whether this effect of cytochalasin was due to inhibition of bacterial uptake or selective effects of cytochalasin on DC signaling pathways independent of uptake (60). DCs obtained from MyD88$^{-/-}$ mice failed to secrete any cytokine in response to *S. pneumoniae*, but phenotypic maturation was only partially inhibited (15). These data suggest that DCs become activated, at least in part, through contact of bacterial cell wall structures (e.g., peptidoglycan and lipoteichoic acid) with DC TLR. Indeed, heavy bacterial encapsulation partially inhibits DC activation, ostensibly by the masking of cell wall structures by capsular polysaccharide (15).

S. pneumoniae-Pulsed DCs Induce In Vivo Antipolysaccharide and Antiprotein Ig Isotype Responses

As mentioned, bone marrow-derived CD8α^- DCs avidly take up *S. pneumoniae* in vitro. Adoptive transfer of *S. pneumoniae* (Pn14)-pulsed DCs into naïve wild-type mice elicited a primary in vivo IgM and IgG isotype response specific for Cps14 and PC, as well as for PspA (16). DC-mediated induction of antipolysaccharide and antiprotein responses required uptake of *S. pneumoniae* and adoptive transfer of *viable* DCs, indicating an active role for DC in Ig induction in vivo. Thus, DC exposure to *S. pneumoniae* followed by induction of either necrosis or apoptosis of the pulsed DCs, before adoptive transfer, abrogated the in vivo Ig response (14, 16). *S. pneumoniae*-pulsed viable DCs also induced the generation of memory, as evidenced by strong boosting of all anti-PspA IgG isotype responses (17- to 19-fold) after secondary immunization with low doses of free *S. pneumoniae* (16). In contrast, the IgM, IgG3, and

IgG2b anti-PC responses in mice receiving pulsed DCs alone were equivalent to those observed after secondary immunization of DC-primed mice with free bacteria, with the exception of IgG1 anti-PC, which was ~10-fold higher in boosted mice. Anti-Cps14 responses also showed no boosting for IgM and IgG2b, although IgG1 and IgG3 were boosted.

Depletion of T cells using anti-CD4 plus anti-CD8 MAbs prior to adoptive transfer of *S. pneumoniae*-pulsed DCs resulted in a striking reduction in IgG anti-PC, anti-Cps14, and anti-PspA responses, with a smaller but significant effect on IgM anti-PC and IgM anti-Cps14 (16). A similar effect was observed using CTLA4Ig to block B7-dependent T-cell costimulation in the recipient host. These data are similar, though not identical, to what was observed using free bacteria, as described above, and support the notion that T cells, as well as DCs, play a significant role in in vivo IgG antipolysaccharide, as well as antiprotein, responses to intact *S. pneumoniae*. Of interest, the mechanisms by which DCs induce antipolysaccharide versus antiprotein Ig responses show clear-cut differences. Thus, CD40$^{-/-}$, MHC class II$^{-/-}$, or B7-1 × B7-2$^{-/-}$ DCs were markedly deficient in eliciting IgG anti-PspA responses but, with the exception of IgG1, were able to elicit normal IgM and IgG anti-PC and anti-Cps14 responses relative to wild-type DCs (16). These data suggest that whereas cognate DC-T cell interactions are critical for induction of anti-PspA responses, a mechanistically different role for DC in TD antipolysaccharide responses is possible. The observation of a requirement for TCR-specific help for the anti-Cps14 response (43), described above, may instead be at the level of the B cell-T cell interaction.

A recent study also supports a role for *endogenous* DCs in stimulating in vivo TI Ig responses to intact *S. pneumoniae* (3). Murine peripheral blood contains a population of highly phagocytic, immature, CD11c^{low} DCs. These DCs, as well as Gr-1$^+$ neutrophils, rapidly take up systemically administered intact *S. pneumoniae* and enter the spleen in large numbers, where CD11c^{low} DCs undergo phenotypic maturation. Using mice transgenic for the M167 (anti-PC) idiotype, it was shown that these DCs can directly stimulate PC-specific MZB to undergo plasma cell differentiation through delivery of B-cell survival signals mediated by DC release of BlyS/April (3). This process occurred in a TI manner and likely has general significance for induction of TI IgM antipolysaccharide responses to intact bacteria.

DC Apoptosis

Despite the critical importance of the DC in the response to primary infections, very little is known regarding the ability of bacteria to induce apoptosis in DCs. Only *Listeria monocytogenes* has been demonstrated to induce apoptosis in DCs via the action of hemolysin (30). DCs, however, appear to exhibit mechanisms that counterbalance apoptotic stimuli that otherwise efficiently induce apoptosis in macrophages. Mature DCs are relatively resistant to the proapoptotic action of TNF-α (51, 52) and CD95-mediated apoptosis (2), and this resistance is associated with the up-regulation of FLIP (2) and Bcl-X$_L$ (52). A relative resistance of DCs for apoptosis induction in response to bacteria would potentially play an important role in optimizing their ability to act as the key APCs during a primary immune response.

Recently, immature CD11c$^+$ CD11b$^+$ CD8α$^-$ DCs obtained from granulocyte-macrophage colony-stimulating factor-supplemented bone marrow cell cultures were shown to undergo apoptosis in response to intact *S. pneumoniae* through two distinct mechanisms (15). The first mechanism, in response to live *S. pneumoniae*, resulted in a rapid induction of DC apoptosis (>15% apoptotic cells by 3 h and >85% by 24 h), was caspase independent, and was critically dependent on bacterial expression of pneumolysin. The rapid kinetics of apoptosis induction in response to live bacteria or to fresh culture supernatants from live bacteria (containing active pneumolysin) was roughly similar to that observed using direct inducers of apoptosis, such as actinomycin D or camptothecin, consistent with the toxic nature of this response. The second mechanism of apoptosis occurred in response to intact, heat-killed *S. pneumoniae* (which contained no active pneumolysin) or a live pneumolysin-deficient mutant strain (15). This mechanism of apoptosis induction was delayed in onset (only beginning after 20 h, with ~30% apoptotic cells by 48 h and ~80% by 72 h), caspase dependent, and closely associated with terminal DC maturation. Delayed-onset apoptosis did not require bacterial internalization but, as described above for *S. pneumoniae*-induced DC cytokine induction and phenotypic maturation, was triggered by the interaction of bacterial subcapsular components and DC (likely Toll-like) receptors acting in a MyD88-dependent manner. Again, heavy polysaccharide encapsulation interfered with both DC maturation and apoptosis induction (15). In contrast, neither CD95-CD95 ligand interactions nor TNF-α appeared to play a role in the delayed onset of apoptosis. Thus, the mechanisms by which *S. pneumoniae* induces DC apoptosis differ from that observed in macrophages, in that the latter is associated with intracellular killing of bacteria and is relatively resistant to the action of pneumolysin (21).

Apoptotic, in addition to necrotic, DCs appear to be ineffective at inducing antibacterial humoral immunity in vivo. Thus, *S. pneumoniae*-pulsed DCs poised to undergo massive apoptosis (beginning at 20 h following bacterial stimulation) exhibit a strong reduction in their ability to induce antipolysaccharide and antiprotein Ig isotype responses when transferred into naïve hosts, relative to DCs transferred only 6 h after bacterial pulsing (14). Inhibition of delayed-onset apoptosis of DCs by in vitro treatment with a caspase-10 inhibitor during an extended period of pulsing with *S. pneumoniae* (when substantial apoptosis would normally occur) restores the in vivo Ig-inducing capacity of DCs when they are transferred into a naïve host (14). This observation is consistent with a recent report showing that neither necrotic nor apoptotic DCs loaded with a peptide are able to induce T-cell responses when transferred in vivo (45).

Naïve CD4$^+$ T cells require prolonged (~20 h) contact with mature DCs in order to reach a cumulative signaling threshold for effective recruitment into the adaptive response (35). Thus, factors that regulate DC longevity could significantly impact the generation of T-cell effector and memory cells and the subsequent humoral and/or cell-mediated immune response. One such factor may be IL-10. IL-10 is widely regarded as an inhibitor of immunity in part through its ability to inhibit DC function (17, 20, 36). However, a recent study evaluating the ability of *S. pneumoniae*-pulsed DC to induce in vivo Ig responses suggests a modification of this view (14). Thus, it is demonstrated instead that a critical balance exists between signals mediated by *S. pneumoniae*-derived pathogen-associated molecular patterns and DC-derived IL-10 for optimization of DC induction of an in vivo humoral immune response. Specifically, following bacterial activation, DCs are shown to have a limited time during which they can function as effective APCs in vivo due to the onset of maturation-associated apoptosis, as described above. Autocrine IL-10, by limiting the time during which DCs are responsive to widely varying levels of bacterial stimulation, delays the onset of DC apoptosis and thus prolongs the time during which DCs are able to elicit in vivo humoral immunity (14). These data demonstrate a requirement for properly balanced positive and negative signaling in DCs in order to optimize an in vivo immune response to *S. pneumoniae*.

THE ROLE OF CYTOKINES

S. pneumoniae induces the early release of proinflammatory (type 1) cytokines (TNF-α, IL-6, IL-12, and IFN-γ), the anti-inflammatory cytokine IL-10, and the type 2 cytokine IL-4 (44). These cytokines, in addition to being implicated in regulating innate immune responses to extracellular bacteria, also act in balance to regulate both polysaccharide- and protein-specific Ig isotype responses to an in vivo challenge with intact *S. pneumoniae* (44). In particular, using cytokine knockout mice

TABLE 1 Regulation of antiprotein and antipolysaccharide responses

Parameter	Antiprotein	Anti-PC	Anti-capsular PS
Peak primary	10 days	6 days	6 days
Boosting	>10-fold	<3-fold	Minimal
IgM response	ND^a	TI	TI
IgG response	TD	TD	TD
T-cell subset	CD4$^+$ TCR-α/β^+	CD4$^+$ TCR-α/β^+	CD4$^+$
TCR specificity	Specific	Nonspecific	Specific
T-cell kinetics	>4 days	<3 days	<3 days
Costimulation	B7-2, prolonged	B7-2, rapid	Not tested
CD40 dependence	IgG	IgG	IgG
DC induction	+	+	+

^{a} ND, not determined.

and neutralizing anticytokine MAbs, it was demonstrated that endogenous proinflammatory cytokines are inductive for humoral immunity, whereas IL-10 is inhibitory during primary challenge with bacteria. Endogenous IL-4 promotes class switching to IgG1. TNF-α, released within the first 48 h as part of the innate response, was shown to play an inductive role in the subsequent adaptive (Ig) response (44). Analysis of *S. pneumoniae*-pulsed DCs, obtained from various cytokine knockout mice and transferred into naïve wild-type mice, indicated a particular role for DC-derived IL-6 in stimulating antipolysaccharide and antiprotein Ig responses (16). IL-6 may function both to induce local macrophage release of complement, important for augmenting the immune response, and to directly promote plasma cell differentiation (47).

SUMMARY

The use of intact *S. pneumoniae* as an immunogen to study the regulation of in vivo antipolysaccharide and antiprotein responses has revealed levels of complexity and distinct mechanistic pathways not anticipated from studies using isolated soluble antigens (Table 1). Most notably, an important role for DCs and T cells in mediating a polysaccharide-specific antibacterial humoral response is strongly suggested. Despite the TD nature of the IgG antipolysaccharide response to an intact pathogen, the rapid kinetics and limited generation of

memory of this response, in contrast to those observed for antiprotein responses, further underscore the intrinsic importance of the antigen itself in how the induction of an Ig response may be manifested during bacterial infections.

REFERENCES

1. **AlonsoDeVelasco, E., A. F. M. Verheul, J. Verhoef, and H. Snippe.** 1995. *Streptococcus pneumoniae*: virulence factors, pathogenesis, and vaccines. *Microbiol. Rev.* **59:**591–603.
2. **Ashany, D., A. Savir, N. Bhardwaj, and K. B. Elkon.** 1999. Dendritic cells are resistant to apoptosis through the Fas (CD95/APO-1) pathway. *J. Immunol.* **163:**5303–5311.
3. **Balazs, M., F. Martin, T. Zhou, and J. Kearney.** 2002. Blood dendritic cells interact with splenic marginal zone B cells to initiate T-independent immune responses. *Immunity* **17:**341–352.
4. **Banchereau, J., F. Briere, C. Caux, J. Davoust, S. Lebecque, Y. Liu, B. Pulendran, and K. Palucka.** 2000. Immunobiology of dendritic cells. *Annu. Rev. Immunol.* **18:**767–811.
5. **Barrett, D. J., and E. M. Ayoub.** 1986. IgG2 subclass restriction of antibody to pneumococcal polysaccharides. *Clin. Exp. Immunol.* **63:**127–134.
6. **Berney, C., S. Herren, C. A. Power, S. Gordon, L. Martinez-Pomares, and M. H. Kosco-Vilbois.** 1999. A member of the dendritic cell family that enters B cell follicles and stimulates primary antibody responses identified by a mannose receptor fusion protein. *J. Exp. Med.* **190:**851–860.
7. **Blanca, I. R., E. W. Bere, H. A. Young, and J. R. Ortaldo.** 2001. Human B cell activation by autologous NK cells is regulated by CD40-CD40 ligand interaction: role of memory B cells and CD5+ B cells. *J. Immunol.* **167:**6132–6139.

8. **Boswell, H. S., S. O. Sharrow, and A. Singer.** 1980. Role of accessory cells in B cell activation. I. Macrophage presentation of TNP-Ficoll: evidence for macrophage-B cell interaction. *J. Immunol.* **124**:989–996.

9. **Braun, D. G., B. Kindred, and E. B. Jacobson.** 1972. Streptococcal group A carbohydrate antibodies in mice: evidence for strain differences in magnitude and restriction of the response, and for thymus dependence. *Eur. J. Immunol.* **2**:138–143.

10. **Briles, D. E., M. Nahm, N. Marion, R. M. Perlmutter, and J. M. Davie.** 1982. Streptococcal group-A carbohydrate has properties of both a thymus-independent (TI-2) and a thymus-dependent antigen. *J. Immunol.* **128**:2032–2035.

11. **Brunswick, M., F. D. Finkelman, P. F. Highet, J. K. Inman, H. M. Dintzis, and J. J. Mond.** 1988. Picogram quantities of anti-Ig antibodies coupled to dextran induce B cell proliferation. *J. Immunol.* **140**:3364–3372.

12. **Carroll, M. C.** 1998. The role of complement and complement receptors in induction and regulation of immunity. *Annu. Rev. Immunol.* **16**:545–568.

13. **Claflin, J. L., R. Lieberman, and J. M. Davie.** 1974. Clonal nature of the immune response to phosphorylcholine. *J. Exp. Med.* **139**:58–73.

14. **Colino, J., and C. M. Snapper.** 2003. Opposing signals from pathogen-associated molecular patterns and IL-10 are critical for optimal dendritic cell induction of in vivo humoral immunity to *Streptococcus pneumoniae. J. Immunol.* **171**:3508–3519.

15. **Colino, J., and C. M. Snapper.** 2003. Two distinct mechanisms for induction of dendritic cell apoptosis in response to intact *Streptococcus pneumoniae. J. Immunol.* **171**:2354–2365.

16. **Colino, J., Y. Shen, and C. M. Snapper.** 2002. Dendritic cells pulsed with intact *Streptococcus pneumoniae* elicit both protein- and polysaccharide-specific immunoglobulin isotype responses in vivo through distinct mechanisms. *J. Exp. Med.* **195**:1–13.

17. **Corinti, S., C. Albanesi, A. la Sala, S. Pastore, and G. Girolomoni.** 2001. Regulatory activity of autocrine IL-10 on dendritic cell functions. *J. Immunol.* **166**:4312–4318.

18. **Cosenza, H., J. Quintans, and I. Lefkovits.** 1975. Antibody response to phosphorylcholine in vitro. I. Studies on the frequency of precursor cells, average clone size, and cellular cooperation. *Eur. J. Immunol.* **5**:343–349.

19. **der Balian, G. P., J. Slack, B. L. Clevinger, H. Bazin, and J. M. Davie.** 1980. Subclass restriction of murine antibodies. III. Antigens that stimulate IgG3 in mice stimulate IgG2c in rats. *J. Exp. Med.* **152**:209–218.

20. **De Smedt, T., M. Van Mechelen, G. De Becker, J. Urbain, O. Leo, and M. Moser.** 1997. Effect of interleukin-10 on dendritic cell maturation and function. *Eur. J. Immunol.* **27**:1229–1235.

21. **Dockrell, D. H., M. Lee, D. H. Lynch, and R. C. Read.** 2001. Immune-mediated phagocytosis and killing of *Streptococcus pneumoniae* are associated with direct and bystander macrophage apoptosis. *J. Infect. Dis.* **184**:713–722.

22. **Dubois, B., B. Vanbervliet, J. Fayette, C. Massacrier, C. V. Kooten, F. Briere, J. Banchereau, and C. Caux.** 1997. Dendritic cells enhance growth and differentiation of CD40-activated B lymphocytes. *J. Exp. Med.* **185**:941–951.

23. **Fernandez, N. C., A. Lozier, C. Flament, P. Ricciardi-Castagnoli, D. Bellet, M. Suter, M. Perricaudet, T. Tursz, E. Maraskovsky, and L. Zitvogel.** 1999. Dendritic cells directly trigger NK cell functions: cross-talk relevant in innate anti-tumor immune responses in vivo. *Nat. Med.* **5**:405–411.

24. **Fischer, W.** 2000. Pneumococcal lipoteichoic acid and teichoic acid, p. 155–177. *In* A. Tomasz (ed.), Streptococcus pneumoniae. *Molecular Biology and Mechanisms of Disease.* Mary Ann Liebert, Inc., Larchmont, N.Y.

25. **Foy, T. M., D. M. Shepherd, F. H. Durie, A. Aruffo, J. A. Ledbetter, and R. J. Noelle.** 1993. In vivo CD40-gp39 interactions are essential for thymus-dependent humoral immunity. II. Prolonged suppression of the humoral immune response by an antibody to the ligand for CD40, gp39. *J. Exp. Med.* **178**:1567–1575.

26. **Gao, N., T. Dang, and D. Yuan.** 2001. IFN-gamma-dependent and -independent initiation of switch recombination by NK cells. *J. Immunol.* **167**:2011–2018.

27. **Gerosa, F., B. Baldani-Guerra, C. Nisii, V. Marchesini, G. Carra, and G. Trinchieri.** 2002. Reciprocal activating interaction between natural killer cells and dendritic cells. *J. Exp. Med.* **195**:327–333.

28. **Greenfield, E., K. Nguyen, and V. Kuchroo.** 1998. CD28/B7 costimulation: a review. *Crit. Rev. Immunol.* **18**:389–418.

29. **Grewal, I. S., and R. A. Flavell.** 1998. CD40 and CD154 in cell-mediated immunity. *Annu. Rev. Immunol.* **16**:111–135.

30. **Guzman, C. A., E. Domann, M. Rohde, D. Bruder, A. Darji, S. Weiss, J. Wehland, T. Chakraborty, and K. N. Timmis.** 1996. Apoptosis of mouse dendritic cells is triggered by listeriolysin, the major virulence determinant of *Listeria monocytogenes. Mol. Microbiol.* **20**:119–126.

31. **Harding, C. V., R. W. Roof, P. M. Allen, and E. R. Unanue.** 1991. Effects of pH and

polysaccharides on peptide binding to class II major histocompatibility complex molecules. *Proc. Natl. Acad. Sci. USA* **88:**2740–2744.

32. **Harnett, W., and M. M. Harnett.** 2000. Phosphorylcholine: an immunomodulator present on glycoproteins secreted by filarial nematodes. *Mod. Asp. Immunobiol.* **1:**40–42.

33. **Hayashi, R. J., and O. Kanagawa.** 1999. Unique CD4(+) T cells in TCR alpha chain-deficient class I MHC-restricted TCR transgenic mice: role in a superantigen-mediated disease process. *Int. Immunol.* **11:**1581–1590.

34. **Hayday, A. C.** 2000. Ag cells: a right time and a right place for a conserved third way of protection. *Annu. Rev. Immunol.* **18:**975–1026.

35. **Iezzi, G., K. Karjalainen, and A. Lanzavecchia.** 1998. The duration of antigenic stimulation determines the fate of naive and effector T cells. *Immunity* **8:**89–95.

36. **Igietseme, J. U., G. A. Ananaba, J. Bolier, S. Bowers, T. Moore, T. Belay, F. O. Eko, D. Lyn, and C. M. Black.** 2000. Suppression of endogenous IL-10 gene expression in dendritic cells enhances antigen presentation for specific Th1 induction: potential for cellular vaccine development. *J. Immunol.* **164:**4212–4219.

37. **Inaba, K., M. Inaba, M. Naito, and R. M. Steinman.** 1993. Dendritic cell progenitors phagocytose particulates, including bacillus Calmette-Guerin organisms, and sensitize mice to mycobacterial antigens in vivo. *J. Exp. Med.* **178:**479–488.

38. **Inaba, K., J. P. Metlay, M. T. Crowley, and R. M. Steinman.** 1990. Dendritic cells pulsed with protein antigens in vitro can prime antigen-specific, MHC-restricted T cells in situ. *J. Exp. Med.* **172:**631–640.

39. **Ingulli, E., A. Mondino, A. Khoruts, and M. K. Jenkins.** 1997. In vivo detection of dendritic cell antigen presentation to CD4(+) T cells. *J. Exp. Med.* **185:**2133–2141.

40. **Ishioka, G. Y., A. G. Lamont, D. Thomson, N. Bulbow, F. C. A. Gaeta, A. Sette, and H. M. Grey.** 1992. MHC interaction and T cell recognition of carbohydrates and glycopeptides. *J. Immunol.* **148:**2446–2451.

41. **Kalka-Moll, W. M., A. O Tzianabos, P. W. Bryant, M. Niemeyer, H. L. Ploegh, and D. L. Kasper.** 2002. Zwitterionic polysaccharides stimulate T cells by MHC class II-dependent interactions. *J. Immunol.* **169:**6149–6153.

42. **Kamboj, K., H. L. Kirchner, R. Kimmel, N. Greenspan, and J. R. Schreiber.** 2003. Significant variation in serotype-specific immunogenicity of the seven valent Streptococcus pneumoniae-CRM$_{197}$ conjugate vaccine occurs despite vigorous T cell help induced by the carrier protein. *J. Infect. Dis.* **187:**1629–1638.

43. **Khan, A. Q., A. Lees, and C. M. Snapper.** 2003. Differential regulation of IgG anti-capsular polysaccharide and anti-protein responses to intact *Streptococcus pneumoniae* in the presence of cognate CD4[+] T cell help. *J. Immunol.* **172:**532–539.

44. **Khan, A. Q., Y. Shen, Z. Q. Wu, T. A. Wynn, and C. M. Snapper.** 2002. Endogenous pro- and anti-inflammatory cytokines differentially regulate an in vivo humoral response to *Streptococcus pneumoniae*. *Infect. Immun.* **70:**749–761.

45. **Kleindienst, P., and T. Brocker.** 2003. Endogenous dendritic cells are required for amplification of T cell responses induced by dendritic cell vaccines in vivo. *J. Immunol.* **170:**2817–2823.

46. **Koh, C. Y., and D. Yuan.** 1997. The effect of NK cell activation by tumor cells on antigen-specific antibody responses. *J. Immunol.* **159:**4745–4752.

47. **Kopf, M., S. Herren, M. V. Wiles, M. B. Pepys, and M. H. Kosco-Vilbois.** 1998. Interleukin 6 influences germinal center development and antibody production via a contribution of C3 complement component. *J. Exp. Med.* **188:**1895.

48. **Lam, K. P., and K. Rajewsky.** 1999. B cell antigen receptor specificity and surface density together determine B-1 versus B-2 cell development. *J. Exp. Med.* **190:**471–477.

49. **Leenen, P. J., K. Radosevic, J. S. Voerman, B. Salomon, N. van Rooijen, D. Klatzmann, and W. van Ewijk.** 1998. Heterogeneity of mouse spleen dendritic cells: in vivo phagocytic activity, expression of macrophage markers, and subpopulation turnover. *J. Immunol.* **160:**2166–2173.

50. **Leonard, E. G., D. H. Canaday, C. V. Harding, and J. R. Schreiber.** 2003 Antigen processing of the heptavalent pneumococcal conjugate vaccine carrier protein CRM197 differs depending on the serotype of the attached polysaccharide. *Infect. Immun.* **71:**4186–4189.

51. **Leverkus, M., H. Walczak, A. McLellan, H. W. Fries, G. Terbeck, E. B. Brocker, and E. Kampgen.** 2000. Maturation of dendritic cells leads to up-regulation of cellular FLICE-inhibitory protein and concomitant down-regulation of death ligand-mediated apoptosis. *Blood* **96:**2628–2631.

52. **Lundqvist, A., T. Nagata, R. Kiessling, and P. Pisa.** 2002. Mature dendritic cells are protected from Fas/CD95-mediated apoptosis by up-regulation of Bcl-X(L). *Cancer Immunol. Immunother.* **51:**139–144.

53. **MacLennan, I. C. M., and C. G. Vinuesa.** 2002. Dendritic cells, BAFF, and APRIL: innate players in adaptive antibody responses. *Immunity* **17:**235 238.

54. **Martin, F., and J. F. Kearney.** 2002. Marginal-zone B cells. *Nat. Rev. Immunol.* **2:**323–335.

55. **Martin, F., and J. F. Kearney.** 2001. B1 cells: similarities and differences with other B cell subsets. *Curr. Opin. Immunol.* **13**:195–201.

56. **Martin, F., A. M. Oliver, and J. F. Kearney.** 2001. Marginal zone and B1 B cells unite in the early response against T-independent blood-borne particulate antigens. *Immunity* **14**:617–629.

57. **Mawas, F., I. M. Feavers, and M. J. Corbel.** 2000. Serotype of Streptococcus pneumoniae capsular polysaccharide can modify the Th1/Th2 cytokine profile and IgG subclass response to pneumococcal-CRM(197) conjugate vaccines in a murine model. *Vaccine* **19**:1159–1166.

58. **Mbow, M. L., N. Zeidner, N. Panella, R. G. Titus, and J. Piesman.** 1997. *Borrelia burgdorferi*-pulsed dendritic cells induce a protective immune response against tick-transmitted spirochetes. *Infect. Immun.* **65**:3386–3390.

59. **Medzhitov, R., and C. A. J. Janeway.** 1997. Innate immunity: the virtues of a nonclonal system of recognition. *Cell* **91**:295–298.

60. **Melamed, I., G. P. Downey, K. Aktories, and C. M. Roifman.** 1991. Microfilament assembly is required for antigen-receptor-mediated activation of human B lymphocytes. *J. Immunol.* **147**:1139–1146.

61. **Mond, J. J., A. Lees, and C. M. Snapper.** 1995. T cell independent antigens type 2. *Annu. Rev. Immunol.* **13**:655–692.

62. **Murphy, K. M., and S. L. Reiner.** 2002. The lineage decisions of helper T cells. *Nat. Rev. Immunol.* **2**:933–944.

63. **Noelle, R. J., M. Roy, D. M. Shepherd, I. Stamenkovic, J. A. Ledbetter, and A. Aruffo.** 1992. A 39-kDa protein on activated helper T cells binds CD40 and transduces the signal for cognate activation of B cells. *Proc. Natl. Acad. Sci. USA* **89**:6550–6554.

64. **Oliver, A. M., F. Martin, and J. F. Kearney.** 1999. IgM^{high}CD21^{high} lymphocytes enriched in the splenic marginal zone generate effector cells more rapidly than the bulk of follicular B cells. *J. Immunol.* **162**:7198–7207.

65. **Pepys, M. B.** 1974. Role of complement in induction of antibody production in vivo. Effect of cobra venom factor and other C3-reactive agents on thymus-dependent and thymus-independent antibody responses. *J. Exp. Med.* **140**:126–145.

66. **Pryjma, J., J. H. Humphrey, and G. G. B. Klaus.** 1974. C3 activation and T-independent B cell stimulation. *Nature* **252**:505–506.

67. **Qian, Y., C. Santiago, M. Borrero, T. F. Tedder, and S. H. Clarke.** 2001. Lupus-specific antiribonucleoprotein B cell tolerance in nonautoimmune mice is maintained by differentiation to B-1 and governed by B cell receptor signaling thresholds. *J. Immunol.* **166**:2412–2419.

68. **Robbins, J. B., and R. Schneerson.** 1990. Polysaccharide-protein conjugates: a new generation of vaccines. *J. Infect. Dis.* **161**:821–832.

69. **Satoskar, A., L. Stamm, X. Zhang, M. Okano, J. David, C. Terhorst, and B. Wang.** 1999. NK cell-deficient mice develop a Th1-like response but fail to mount an efficient antigen-specific IgG2a antibody response. *J. Immunol.* **163**:5298–5302.

70. **Schnare, M., G. M. Barton, A. C. Holt, K. Takeda, S. Akira, and R. Medzhitov.** 2001. Toll-like receptors control activation of adaptive immune responses. *Nat. Immunol.* **2**:947–950.

71. **Shaw, P. X., S. Horkko, M. K. Chang, L. K. Curtiss, W. Palinski, G. J. Silverman, and J. L. Witztum.** 2000. Natural antibodies with the T15 idiotype may act in atherosclerosis, apoptotic clearance, and protective immunity. *J. Clin. Investig.* **105**:1731–1740.

72. **Shi, G.-P., J. A. Villadangos, G. Dranoff, C. Small, L. Gu, K. J. Haley, R. Riese, H. L. Ploegh, and H. A. Chapman.** 1999. Cathepsin S required for normal MHC class II peptide loading and germinal center development. *Immunity* **10**:197–206.

73. **Sinha, A. A., C. Guidos, K. C. Lee, and E. Diener.** 1987. Functions of accessory cells in B cell responses to thymus-independent antigens. *J. Immunol.* **138**:4143–4149.

74. **Slack, J., G. P. Der-Balian, M. Nahm, and J. M. Davie.** 1980. Subclass restriction of murine antibodies. II. The IgG plaque-forming cell response to thymus-independent type I and type 2 antigens in normal mice and mice expressing an X-linked immunodeficiency. *J. Exp. Med.* **151**:853–862.

75. **Snapper, C. M., Y. Shen, A. Q. Khan, J. Colino, P. Zelazowski, J. J. Mond, W. C. Gause, and Z.-Q. Wu.** 2001. Distinct types of T-cell help for the induction of a humoral immune response to *Streptococcus pneumoniae*. *Trends Immunol.* **22**:308–311.

76. **Snapper, C. M., and F. D. Finkelman.** 1997. Immunoglobulin class switching, p. 831–861. *In* W. E. Paul (ed.), *Fundamental Immunology*, 4th ed. Raven Press, New York, N.Y.

77. **Snapper, C. M., and J. J. Mond.** 1996. A model for induction of T cell-independent humoral immunity in response to polysaccharide antigens. *J. Immunol.* **157**:2229–2233.

78. **Snapper, C. M., H. Yamaguchi, M. A. Moorman, and J. J. Mond.** 1994. An in vitro model for T cell-independent induction of humoral immunity. A requirement for NK cells. *J. Immunol.* **152**:4884–4892.

79. **Snapper, C. M., H. Yamaguchi, M. A. Moorman, R. Sneed, D. Smoot, and J. J.**

Mond. 1993. Natural killer cells induce activated murine B cells to secrete Ig. *J. Immunol.* **151:** 5251–5260.

80. **Snapper, C. M., H. Yamada, D. Smoot, R. Sneed, A. Lees, and J. J. Mond.** 1993. Comparative in vitro analysis of proliferation, Ig secretion, and Ig class switching by murine marginal zone and follicular B cells. *J. Immunol.* **150:**2737–2745.

81. **Snapper, C. M., and W. E. Paul.** 1987. Interferon-γ and B cell stimulatory factor-1 reciprocally regulate Ig isotype production. *Science* **236:**944–947.

82. **Su, H., R. Messer, W. Whitmire, E. Fischer, J. C. Portis, and H. D. Caldwell.** 1998. Vaccination against chlamydial genital tract infection after immunization with dendritic cells pulsed ex vivo with nonviable Chlamydiae. *J. Exp. Med.* **188:**809–818.

83. **Svensson, M., B. Stockinger, and M. J. Wick.** 1997. Bone marrow-derived dendritic cells can process bacteria for MHC-I and MHC-II presentation to T cells. *J. Immunol.* **158:**4229–4236.

84. **Takeda, K., T. Kaisho, and S. Akira.** 2003. Toll-like receptors. *Annu. Rev. Immunol.* **21:**335–376.

85. **Taniguchi, M., M. Harada, S. Kojo, T. Nakayama, and H. Wakao.** 2003. The regulatory role of Vα14 NKT cells in innate and acquired immune response. *Annu. Rev. Immunol.* **21:**483–513.

86. **Vinay, D. S., and B. S. Kwon.** 1998. Role of 4-1BB in immune responses. *Semin. Immunol.* **10:**481–489.

87. **Vos, Q., C. M. Snapper, and J. J. Mond.** 1999. Heterogeneity in the ability of cytotoxic murine NK cell clones to enhance Ig secretion in vitro. *Int. Immunol.* **11:**159–168.

88. **Watts, T. H., and M. A. DeBenedette.** 1999. T cell costimulatory molecules other than CD28. *Curr. Opin. Immunol.* **11:**286–293.

89. **Wilder, J. A., C. Y. Koh, and D. Yuan.** 1996. The role of NK cells during in vivo antigen-specific antibody responses. *J. Immunol.* **156:**146–152.

90. **Wu, Z. Q., A. Q. Khan, Y. Shen, K. M. Wolcott, W. Dawicki, T. H. Watts, R. S. Mittler, and C. M. Snapper.** 2003. 4-1BB (CD137) differentially regulates murine in vivo protein- and polysaccharide-specific immunoglobulin isotype responses to Streptococcus pneumoniae. *Infect. Immun.* **71:**196–204.

91. **Wu, Z. Q., Y. Shen, A. Q. Khan, C. L. Chu, R. Riese, H. A. Chapman, O. Kanagawa, and C. M. Snapper.** 2002. The mechanism underlying T cell help for induction of an antigen-specific in vivo humoral immune response to intact *Streptococcus pneumoniae* is dependent on the type of antigen. *J. Immunol.* **168:**5551–5557.

92. **Wu, Z. Q., A. Q. Khan, Y. Shen, J. Schartman, R. Peach, A. Lees, J. J. Mond, W. C. Gause, and C. M. Snapper.** 2000. B7 requirements for primary and secondary protein- and polysaccharide-specific Ig isotype responses to *Streptococcus pneumoniae.* *J. Immunol.* **165:**6840–6848.

93. **Wu, Z.-Q., Q. Vos, Y. Shen, A. Lees, S. R. Wilson, D. E. Briles, W. C. Gause, J. J. Mond, and C. M. Snapper.** 1999. In vivo polysaccharide-specific IgG isotype responses to intact *Streptococcus pneumoniae* are T cell dependent and require CD40- and B7-ligand interactions. *J. Immunol.* **163:**659–667.

94. **Wykes, M., A. Pombo, C. Jenkins, and G. G. MacPherson.** 1998. Dendritic cells interact directly with naive B lymphocytes to transfer antigen and initiate class switching in a primary T-dependent response. *J. Immunol.* **161:**1313–1319.

95. **Yuan, D., J. Wilder, T. Dang, M. Bennett, and V. Kumar.** 1992. Activation of B lymphocytes by NK cells. *Int. Immunol.* **4:**1373–1380.

NEW PNEUMOCOCCAL VACCINES: BASIC SCIENCE DEVELOPMENTS

James C. Paton

24

THE NEED FOR NEW PNEUMOCOCCAL VACCINES

The potential contribution of vaccines to the control of pneumococcal disease has been recognized for nearly a century. The first experimental vaccines, comprising killed whole cells, were tested in the early 1900s, albeit with inconclusive results. Although interest fluctuated in the decades that followed, the steady increase in knowledge of the immunobiology of pneumococcal disease enabled a rational approach to vaccine design. This has resulted in the licensing of polyvalent capsular polysaccharide (PS) vaccines and, more recently, the PS-protein conjugates. These vaccines are aimed at preventing invasive diseases such as pneumonia, meningitis, and bacteremia, as well as less serious but highly prevalent infections such as otitis media. They are being targeted principally at specific groups at high risk of pneumococcal disease, particularly children under 2 years and adults over 65 years of age.

Research aimed at the development of more effective vaccines against *Streptococcus pneumoniae* is being driven by several factors,

some of which have been addressed in detail elsewhere in this volume. First among these is the fact that the pneumococcus continues to cause high morbidity and mortality throughout the world, even in regions where antibiotics are readily available. *S. pneumoniae* is the single most common cause of community-acquired pneumonia and has become the most common cause of meningitis in many regions. Although determining the true burden of pneumococcal disease is complicated by difficulties in establishing an etiological diagnosis, particularly in cases of nonbacteremic pneumonia, the pneumococcus is conservatively estimated to kill more than a million children under the age of 5 years each year in developing countries. This accounts for 20 to 25% of all deaths in this age group (21, 55). Even in developed countries, where effective antimicrobial drugs are readily available, morbidity and mortality from pneumococcal disease are significant. For example, in the United States there are approximately 500,000 cases of pneumococcal pneumonia, 50,000 cases of bacteremia, and 3,000 cases of pneumococcal meningitis each year, collectively resulting in an estimated 40,000 deaths (4, 24, 55). *S. pneumoniae* is also the single most common cause of otitis media, which in the United States results in over 24 million visits to pediatricians each

James C. Paton, School of Molecular and Biomedical Science, University of Adelaide, Adelaide, S.A. 5005, Australia.

The Pneumococcus. Volume Editor, Elaine I. Tuomanen,
© 2004 ASM Press, Washington, D.C.

year and more prescriptions for antibiotics than any other infectious disease (24, 56). Thus, otitis media has a significant impact on health care costs in developed countries, and estimates for the United States exceed $5 billion per annum (56).

The second major driver for vaccine development has been the increasing threat posed by antibiotic-resistant pneumococci. In the first two decades after introduction of penicillin, clinical isolates of *S. pneumoniae* were universally and exquisitely sensitive to this drug. However, strains with reduced susceptibility to penicillin (MIC > 0.1 μg/ml) were detected in the late 1960s and since then have steadily increased in prevalence throughout the world. The problem is greatest in areas in which the use of antibiotics has been poorly regulated, and rates of resistance to beta-lactams may exceed 50% of isolates (57). Moreover, the degree of resistance has been increasing as a consequence of accumulation of multiple mutations in penicillin-binding protein genes, resulting in strains with high-level penicillin resistance (MIC ≥ 2 μg/ml). Penicillin-resistant pneumococci are also often resistant to one or more other classes of antibiotics, and multiply resistant clones of *S. pneumoniae* have spread globally (58, 71). The increasing prevalence of penicillin-resistant and multiply resistant pneumococci is complicating management of patients with suspected pneumococcal disease, particularly those with meningitis. In developed countries this is necessitating the use of more expensive alternative antimicrobials, but this option is not available in poorer parts of the world.

PS-BASED VACCINES

Purified PS Vaccines

Another major driver for development of new pneumococcal vaccines is the real and perceived deficiencies of the currently licensed polysaccharide-based formulations. The protection imparted by pneumococcal PS vaccines is largely a result of binding of specific antibody to the capsule, resulting in opsonization and rapid clearance of invading pneumococci. A well-constructed trial of a tetravalent PS vaccine in U.S. military recruits during World War II demonstrated a high degree of protection against pneumonia caused by types included in the vaccine (65). However, the resultant commercial production of two hexavalent PS vaccines coincided with the introduction of penicillin and other antibiotics, which at the time appeared to be spectacularly effective against the pneumococcus. As a result, the vaccines were not utilized to any great extent and were eventually withdrawn from the market (5). Nevertheless, it soon became clear that prompt and appropriate antibiotic therapy could not be relied upon to prevent death from invasive pneumococcal disease in certain high-risk patient groups. The continued high morbidity and mortality rekindled interest in PS vaccines, and further trials were conducted with healthy young adults (South African miners) who had high attack rates of pneumococcal pneumonia and bacteremia. Multivalent formulations were shown to be approximately 80% effective in preventing invasive pneumococcal disease caused by serotypes contained in the vaccine (5). A 14-valent PS vaccine was licensed in 1977, and coverage was expanded to 23 types in 1983. The latter formulation includes types 1, 2, 3, 4, 5, 6B, 7F, 8, 9N, 9V, 10A, 11A, 12F, 14, 15B, 17F, 18C, 19A, 19F, 20, 22F, 23F, and 33F (24).

A number of additional trials have been conducted with older adults since the introduction of the PS vaccines (reviewed in reference 19). Comparisons of the results of the different studies are complicated by differences in study design and clinical criteria, but they have tended to show somewhat lower efficacy rates (usually on the order of 60%), particularly in elderly recipients, the immunocompromised, and those with underlying chronic diseases. Nevertheless, controversy surrounding the efficacy of PS vaccines for prevention of pneumococcal pneumonia has continued, particularly since some prospective randomized trials in older adults failed to demonstrate any protection whatso-ever (84, 103). In spite of this, the vaccine is currently recommended for all persons aged over 65 years and for those under 65 belong-ing to other high-risk groups (24). PS vac-cines are not recommended for children aged <2 years, even though they are a particularly high-risk group, because efficacy has not been demonstrated in clinical trials.

The likely explanation for the failure of the vaccine in young children is the poor immunogenicity of many of the component PS antigens. Children <2 years of age can mount an adequate antibody response to some types (e.g., type 3), but responses are particularly poor for the PS types which most commonly cause invasive disease in children, namely, 6A/B, 14, 18C, 19F, and 23F (32). Indeed, responses to these types are weak up to the age of 5 years and do not reach adult levels until 8 to 10 years of age (93). Elderly adults also exhibit antibody responses to some of these PS serotypes that are weaker and more transient than those of younger adults, and this undoubtedly accounts for the poorer clinical efficacy in this age group re-ferred to above (19). PSs are referred to as "thymus-independent type 2" antigens and activate B lymphocytes independently of $CD4^+$ cells by directly binding and cross-linking antigen receptors on the B-cell sur-face. This process is distinct from that in-duced by protein antigens and involves costimulation by CD21 (type 2 complement receptor) after binding of C3d (generated by activation of the alternative complement pathway by PS). Neonatal B lymphocytes express low levels of CD21, and this may be one explanation for the hyporesponsiveness to PS during infancy (95). PS antigens do not induce immunological memory, and anti-bodies produced are mainly of the im-munoglobulin G2 (IgG2) subclass. Even in healthy adults, antibody levels begin to de-cline about 1 year postvaccination and for many types return to preimmunization levels after about 5 years (19). Interestingly, Musher et al. (77) have reported that healthy individuals varied markedly in the capacity to mount an antibody response to various pneu-mococcal PS serotypes and that this was con-trolled genetically and inherited in a codom-inant pattern. Thus, a subset of the adult population may be refractory to immuniza-tion with pneumococcal PS vaccines.

The other principal weakness of PS vac-cines is that the protection they elicit is strictly serotype dependent. As mentioned above, the current vaccine formulation in-cludes PS purified from 23 of the 90 known serotypes of S. pneumoniae, and clinical trials have confirmed that it provides negligible cover against serotypes not included in the formulation. Fortunately, not all types are equally prevalent, and the formulation was determined with reference to available data on the distribution of types causing invasive disease in adults and children (96). Most of these data emanated from the United States or Europe, and the 23 included serotypes currently account for about 90% of invasive pneumococcal infections in these regions. However, there are geographic and temporal differences in the serotype distribution of dis-ease-causing pneumococci, and the existing formulation may cover as little as 60% of strains in parts of Asia (60). Moreover, serotype prevalence data are scanty for many developing countries, and vaccine coverage in these regions is uncertain.

Some PS types included in the vaccine (for example, type 6B) are known to elicit anti-bodies that cross-react with structurally related PS types that are not included (in this example, type 6A). Type 6A and 6B pneumococci are both important causes of invasive disease in children, but the vaccine was formulated with the expectation that the cross-reacting anti-bodies would provide cross-protection (96). However, this assumption may be incorrect, as the cross-reacting antibodies appear to be of low avidity and function poorly in in vitro op-sonophagocytic assays against the heterologous type. More recent clinical trials are also strongly suggestive of weaker-than-expected cross-protection (19).

The inevitable conclusion that must be drawn from the above is that notwithstanding the high protective efficacy in healthy adults, existing PS vaccines have suboptimal efficacy in groups who are most at risk from life-threatening invasive pneumococcal disease. The combination of incomplete protection against included serotypes and variation in serotype distribution affecting vaccine coverage will undoubtedly attenuate the overall global impact of PS vaccines. Nevertheless, even a 50% effective vaccine will prevent countless deaths in groups for whom the PS vaccine is currently recommended, and so until a better alternative is available, its continued use should be vigorously encouraged. Urgent efforts are required, however, to develop vaccines that are efficacious in young children, for whom the existing PS vaccine provides little demonstrable clinical benefit.

PS-Protein Conjugate Vaccines

A possible solution to the poor immunogenicity of PS in young children emanated from the seminal work of Avery and Goebel (6), who reported in 1931 that chemical conjugation of type 3 pneumococcal PS to a protein carrier massively increased its immunogenicity in rabbits. This approach was subsequently used against *Haemophilus influenzae* type b (Hib), and the Hib PS-protein conjugate vaccine has been spectacularly successful (97). This has encouraged development of multivalent pneumococcal PS-protein conjugate vaccines, the first generation of which are now licensed in several countries. Conjugation to a protein carrier converts the PS into a T-cell-dependent antigen. The PS component is thought to react with receptors on B cells, which then internalize the conjugate, process it, and present peptide fragments in association with class II major histocompatibility complex (MHC) molecules to peptide-specific T cells. Memory responses are generated, and primed B cells can be boosted either with conjugate or free PS (19).

Development of pneumococcal PS-protein conjugate vaccines has been considerably more complex than was the case with Hib, owing to the multiplicity of disease-causing serotypes. A number of parameters which influence immunogenicity of conjugate antigens need to be optimized for each type, including the molecular size of the PS component, the carrier protein, the PS/carrier ratio, and the method used to covalently link the two components. In view of this developmental complexity, the number of serotypes that can be included is by necessity less than in the PS vaccine. However, the conjugate vaccines are principally designed to prevent invasive disease and otitis media in young children, for whom the range of disease-causing serotypes is more restricted than in adults.

Conjugate vaccines developed to date by various manufacturers are either 7-, 9-, or 11-valent, use different cross-linking chemistries, and employ a range of carriers such as tetanus or diphtheria toxoids, the diphtheria toxin derivative CRM_{197}, or outer membrane proteins from *Neisseria meningitidis* group B or nontypeable *H. influenzae*. The 7-valent formulation includes types 4, 6B, 9V, 14, 18C, 19F, and 23F, and it is estimated that this would cover 60 to 90% of pediatric infections based on North American and European seroprevalence studies (34). The 9-valent vaccine includes these same types with the addition of types 1 and 5, which although uncommon in Europe and North America are important causes of invasive pediatric disease in other geographic regions. Indeed, a 10-year study of the seroprevalence of pneumococci causing invasive disease in children in Southern Israel indicated that inclusion of these two types would increase coverage from 41 to 67% in Jewish children and from 22 to 63% in Bedouin children (37). Types 3 and 7F were also included in the 11-valent formulation.

These conjugate vaccines are typically administered as a course of three injections at 2, 4, and 6 months of age, followed by a booster of either conjugate or PS vaccine at 12 to 15 months. Several clinical studies have demonstrated that they are well tolerated by infants and elicit strong, boostable antibody responses (reviewed in reference 115). A large study of the 7-valent vaccine in northern California demonstrated 97% protection against invasive (bacteremic) disease caused by vaccine types (14). A Finnish study designed to test the protective efficacy of the same vaccine against otitis media, which included microbiological analysis of middle ear fluid from all suspected cases, demonstrated a 57% reduction in infections caused by vaccine types (35). This figure is similar to the 67% reduction in otitis caused by vaccine types reported in the Californian study, although microbiological analysis had been confined to spontaneously draining ears (14).

Although the degree of type-specific protection imparted by the conjugate vaccine was less spectacular against otitis media than against invasive disease, this outcome was not unexpected, as higher antibody concentrations are probably required for prevention of the former. Nevertheless, the prevalence of pneumococcal otitis media is such that even a partially protective vaccine would prevent a very large number of cases. Interestingly, in the Finnish study the vaccine also reduced the number of otitis episodes caused by pneumococci belonging to nonincluded types such as 6A and 19A, which cross-react with vaccine types (6B and 19F, respectively), by 51% (35). This occurred even though antibodies to type 6B PS elicited by the conjugate have been shown to have weaker in vitro opsonophagocytic activity against type 6A pneumococci than against type 6B strains (110).

A major concern emanating from the Finnish study, however, was the finding that otitis media caused by all other nonincluded S. pneumoniae serotypes increased by 33% (35). This finding was also not unexpected. In previous trials conducted in The Gambia (80), South Africa (67), and Israel (28), the conjugate vaccine significantly reduced nasopharyngeal carriage of vaccine types in children, but this was offset at least partially by an increase in carriage of nonvaccine types, many of which were known to be capable of causing disease.

Nasopharyngeal carriage of S. pneumoniae is generally accepted as a prerequisite for pneumococcal disease, and serotypes being carried usually correlate with those causing disease in a community. Carriers are the major source for transmission of pneumococci, and communities with high rates of carriage also have high attack rates of pneumococcal disease. Carriage rates are high in young children, particularly in developing countries, where infants acquire pneumococci (presumably from their colonized mothers) in the first few days of life. Individuals may be colonized by multiple strains or serotypes of S. pneumoniae, and these presumably compete (with each other and perhaps with other microflora) for occupation of the nasopharyngeal niche (63). Detection of multiple S. pneumoniae serotypes in nasopharyngeal cultures is technically difficult if one strain is significantly outnumbered. Thus, it is hard to determine the extent to which "replacement carriage" observed in a conjugate vaccine recipient is a result of acquisition of new S. pneumoniae types not previously present or of facilitated detection of a preexisting nonvaccine serotype whose numbers have increased after elimination of interference from vaccine types. However, regardless of the mechanism, replacement carriage does appear to translate into increased disease caused by nonvaccine serotypes (35).

S. pneumoniae strains undoubtedly differ in their capacity to colonize the nasopharynx, as well as in their capacity to cause either otitis media or invasive disease once carriage has been established. These differences have a multifactorial basis and depend upon capsular serotype as well as upon other ill-defined virulence traits. This accounts for the nonuniform distribution and relative prevalence of the 90 serotypes of S. pneumoniae, as well as for the existence of highly successful, widely distributed clones, some of which are resistant to multiple antibiotics. Molecular analysis of one such highly transmissible, multiply resistant strain (the so-called Spanish type 23F clone) demonstrated that pneumococci are capable of switching serotype by recombinational exchange of capsule biosynthesis loci in vivo (26). It is easy to see how antibiotic therapy would facilitate such exchanges between cocolonizing sensitive and resistant strains; DNA released from the sensitive strain would directly transform the resistant type, enabling it to assume the serotype of the donor.

Most, but by no means all, serotype exchanges in resistant pneumococci detected to date have been from one vaccine type to another vaccine type, presumably because the other vaccine types are also commonly carried (104). However, increased colonization by nonvaccine types due to use of the conjugate vaccines will increase the likelihood of in vivo transformation of multiply resistant pneumococci to nonvaccine serotypes. Introduction of the conjugate vaccines will also provide direct selective pressure for acquisition of nonvaccine serotype capsule loci by highly virulent S. pneumoniae clones which had hitherto expressed a vaccine-type capsule.

The impact of widespread use of conjugate vaccines on the complex biology of pneumococcal disease is difficult to predict. The full effect of the vaccine on serotype prevalence may take many years to become apparent, and it will vary from region to region depending upon levels of endemic carriage and rates of vaccine utilization. In well-vaccinated populations, disease caused by vaccine types will be markedly reduced, and reduction of carriage of vaccine types will undoubtedly provide a degree of homotypic herd immunity. In the short term, the vaccine may also help to control the spread of antibiotic-resistant pneumococci, because they are more prevalent in children (the principal target population) than in adults, and the majority of such isolates belong to vaccine serotypes.

However, these major clinical benefits may be ephemeral. As vaccine use grows, so too will the rate of nasopharyngeal carriage of nonvaccine types. This, in turn, will increase the likelihood of transmission of nonvaccine serotype pneumococci to others in the community, including vaccinees, further increasing overall carriage rates of such strains. Capsule

type switching may also facilitate vaccine escape, as well as enabling spread of antibiotic resistance to a broader range of serotypes than is currently the case. Continued surveillance of the serotype distribution of pneumococci causing disease or being carried will be essential. Inclusion of additional conjugated polysaccharides in the formulation may be required if nonvaccine types become too prevalent.

There are limits, however, on just how many capsular types can be accommodated. Polyvalent PS-protein conjugate vaccines are very expensive to produce, and addition of further PS types or periodic reformulation to take account of altered serotype prevalence will add further to this cost. This may place the vaccine even further out of the reach of many developing countries, whose need for effective pneumococcal vaccines is greatest. In countries that can afford them, conjugate vaccines are likely to have a major impact upon the burden of pneumococcal disease in the short term, but their overall efficacy is likely to diminish with time, necessitating ongoing investment in development of alternative vaccination strategies capable of eliciting more broad-based and affordable protection.

PURIFIED PROTEIN VACCINES

The known and potential shortcomings of the PS and PS-conjugate vaccines referred to above have prompted extensive research aimed at development of vaccines based on pneumococcal proteins that contribute to virulence and are common to all serotypes. Such vaccines should be highly immunogenic and elicit immunological memory in infants and young children, who respond well to T-cell-dependent protein antigens. High-level expression of proteins can also be engineered in recombinant bacteria, enabling large-scale production at relatively low cost, resulting in vaccines that are more affordable, particularly for developing countries. A number of candidate pneumococcal protein antigens have been examined for vaccine potential, and these are discussed below.

Pneumolysin

Pneumolysin was the first pneumococcal protein to be proposed as a vaccine antigen (89), and its structure, function, and role in virulence are discussed in detail elsewhere in this volume. Briefly, it is a potent 53-kDa thiol-activated pore-forming cytolysin produced by virtually all strains of *S. pneumoniae*. It is a bifunctional toxin, and in addition to its cytotoxic properties, it is capable of directly activating the classical complement pathway (with a concomitant reduction in serum opsonic activity) by binding to the Fc region of human IgG. In vitro studies using purified toxin have demonstrated that pneumolysin has a variety of detrimental effects on cells and tissues, which undoubtedly contribute to the pathogenesis of disease (reviewed in reference 86). These properties include inhibition of the bactericidal activity of leukocytes, blockade of proliferative responses and Ig production by lymphocytes, reduction of ciliary beating of human respiratory epithelium, and direct cytotoxicity for respiratory endothelial and epithelial cells. Thus, pneumolysin may function in pathogenesis by interfering with both phagocytic and ciliary clearance of pneumococci, by blocking humoral immune responses, and by aiding penetration of host tissues (86).

Pneumolysin is also capable of direct induction of inflammatory responses (46), and injection of purified pneumolysin into rat lungs induces severe lobar pneumonia indistinguishable histologically from that seen when virulent pneumococci are injected (36). Additional insights into the role of pneumolysin in pathogenesis have been gained by studies of the behavior of defined pneumolysin-negative mutants of *S. pneumoniae* in a number of animal models. Such strains have significantly reduced virulence in mouse models of sepsis and pneumonia (13). Intranasal challenge with these mutants results in a less severe inflammatory response, a reduced rate of multiplication within the lung, a reduced capacity to injure the alveolar-capillary barrier, and a delayed onset of bacteremia compared with the wild-type strain (86). Additional site-directed mutagene-

sis studies have shown that both the cytotoxic and complement activation properties of the toxin contribute to the pathogenesis of pneumococcal pneumonia (9, 99).

Although native pneumolysin is a protective immunogen in mice, it is not suitable as a human vaccine antigen, because of its toxicity. To overcome this, mutations have been introduced into the pneumolysin gene in regions essential for its cytotoxic and/or complement activation properties, resulting in expression of nontoxic but immunogenic "pneumolysoids" which are easily purified from recombinant *Escherichia coli* expression systems (90). Pneumolysin is a highly conserved protein, and extensive analysis of genes from a wide range of *S. pneumoniae* serotypes has detected negligible variation in deduced amino acid sequence, auguring well for broad coverage. Indeed, immunization of mice with a pneumolysoid carrying a $Trp_{433} \rightarrow Phe$ mutation resulting in >99% reduction in cytotoxicity (designated PdB) provided a significant degree of protection against all nine serotypes of *S. pneumoniae* that were tested (2).

Humans are known to mount an antibody response to pneumolysin as a result of natural exposure to *S. pneumoniae*, and a recent study has shown that purified human antipneumolysin IgG also passively protects mice from challenge with virulent pneumococci (76). Thus, it is anticipated that the various pneumolysoids will be immunogenic in humans. However, pneumolysoid may not provide a sufficient degree of protection to be an effective stand-alone human vaccine antigen. Pneumolysin is not displayed on the surface of the pneumococcus. Rather, it is located in the cytoplasm and is released into the external milieu when pneumococci undergo spontaneous autolysis in some strains (10), as well as by an as-yet-uncharacterized export mechanism in others (8). Antibodies to pneumolysin are presumed to impart protection by neutralization of the biological properties of the toxin, thereby impeding the kinetics of infection, rather than by stimulating opsonophagocytic clearance of the invading bacteria. Thus, protein-based vaccines combining pneumolysoid with pneumococcal surface proteins capable of eliciting opsonic antibodies would be expected to be more effective.

PspA

A second pneumococcal protein with strong vaccine credentials is pneumococcal surface protein A (PspA), which is found on the surface of all pneumococci (27). It is a member of a family of choline-binding surface proteins, which are discussed in detail elsewhere in this volume. These proteins bind to the pneumococcal surface via phosphorylcholine moieties on cell wall teichoic acid and membrane lipoteichoic acid, a property mediated by a domain comprising 9 or 10 highly conserved 20-amino-acid repeats near their respective C termini (117, 118). PspA is known to be important in the pathogenesis of disease, as evidenced by significantly reduced virulence of defined PspA-negative pneumococci in animal models (70, 109). Its principal function appears to be inhibition of complement-dependent host defenses mediated by factor B. This results in reduced deposition of C3b on the pneumococcal surface and concomitant impairment of complement receptor-mediated clearance (109). PspA also binds lactoferrin (41, 42), and although the significance of this is uncertain, it may aid colonization of host mucosae, perhaps by facilitating acquisition of iron or adherence to host cells which express lactoferrin receptors (41).

The biological properties of PspA reside in the N-terminal portion of the molecule, which forms a highly charged, largely alpha-helical anti-parallel coiled-coil structure (45, 52). Immunization of mice with a soluble 43-kDa N-terminal PspA fragment has been shown to be highly protective against systemic challenge (107). This region of the molecule is highly variable in terms of its amino acid sequence, although this does not appear to impact upon biological function (94). PspA proteins produced by various S. pneumoniae strains have been grouped into three families, with 95% of isolates producing PspA belonging to family 1 or 2 (45, 111). In spite of the significant variation in amino acid sequence, the helical domain of PspA contains epitopes that elicit antibodies that are highly protective against challenge with S. pneumoniae strains producing heterologous PspA types (69).

Studies on the vaccine potential of PspA have extended to human trials, and immune sera from volunteers immunized with a family 1 PspA fragment reacted with 37 different S. pneumoniae strains belonging to diverse capsular and PspA types (78). Moreover, the sera passively protected mice against challenge with S. pneumoniae strains of three different capsular types expressing either family 1 or family 2 PspAs (17). Thus, it appears likely that a human vaccine may only need to include two or three different PspA types in order to provide near-species-wide protection.

In addition to its promise as a parenteral vaccine antigen capable of preventing systemic disease, PspA exhibits considerable promise as a mucosal vaccine antigen for prevention of nasopharyngeal carriage. As described previously, a vaccine capable of preventing carriage is likely to impart substantial herd immunity. Intranasal immunization with full-length native PspA, using cholera toxin B subunit (CTB) as an adjuvant, has been shown to elicit significant mucosal and serum antibody responses and to protect mice against both nasal carriage of S. pneumoniae and systemic disease (114, 116). However, a recombinant N-terminal PspA fragment (also administered with CTB) appeared to be less effective, reducing the level of colonization after intranasal challenge but not preventing it altogether (16). On the other hand, parenteral immunization with PspA elicits negligible levels of mucosal antibody and no detectable protection against carriage (114). The potential efficacy of mucosal immunization with PspA for prevention of carriage in humans is also supported by the findings of a human volunteer study, which demonstrated that preexisting (naturally acquired) antibody to PspA prevented colonization after intranasal administration of S. pneumoniae (68).

Other Choline-Binding Proteins

Several other members of the choline-binding protein family have been proposed as vaccine antigens. The best characterized of these was isolated independently in three laboratories and is referred to as either pneumococcal surface protein C (PspC) (A. Brooks-Walter, R. C. Tart, D. E. Briles, and S. K. Hollingshead, Abstr. 97th Gen. Meet. Am. Soc. Microbiol., abstr. 35, 1997), choline-binding protein A (CbpA) (98), or SpsA (43) (the foremost terminology is used here). Its choline-binding repeat region is approximately 95% identical to that of PspA, while its N-terminal helical portion, like PspA, is highly variable. The N-terminal half mediates binding to the secretory component of IgA (43, 44), as well as to C3 (25) and factor H (29). PspC has been shown to be involved in adherence of pneumococci to cytokine-activated lung epithelial cells in vitro as well as to glycoconjugates previously identified as pneumococcal binding ligands. Furthermore, PspC-deficient pneumococci have a reduced capacity to colonize the nasopharynges of infant rats and mice (7, 98). Interestingly, PspC has been shown to be expressed at greater levels in transparent-phase pneumococci, which are favored in the nasopharynx over opaque-phase variants (98). PspC may also be directly involved in invasion of nasopharyngeal cells, through interaction with the secretory component associated with the polymeric Ig receptor (pIgR) (119). The fact that PspC can interact with C3 and factor H is also strongly suggestive of a role in systemic disease, and significant differences in virulence between pspC-negative and otherwise isogenic wild-type pneumococci have been demonstrated in mouse models of lung infection and bacteremia (7). An earlier study did not detect differences in virulence between wild-type and pspC-negative pneumococci in an intraperitoneal challenge model, but mutation of both pspC and the pneumolysin gene had an additive attenuating effect (12).

Immunization of mice with PspC has been shown to be highly protective against intravenous or intraperitoneal challenge with S. pneumoniae (20, 83). Theoretically, the suitability of PspC as a vaccine antigen might be diminished to some extent by the fact that it is present on only about 75% of S. pneumoniae strains (20, 43). However, immunization with PspC was shown to provide significant protection against a strain that did not produce the protein (20). This can be explained by the fact that polyclonal antibodies to PspC cross-react with PspA as well as other protein species (20). In those strains that lack pspC, the locus is occupied by an allele hic, which encodes a protein with a high degree of similarity to PspC in the N-terminal half (48, 50), and this might also be recognized by PspC antibodies. Interestingly, the C-terminal portion of Hic does not have a choline-binding domain but instead has a sortase-dependent cell wall anchorage domain typical of surface proteins of gram-positive organisms, including an LPXTG motif (50). Like PspC, Hic can bind to factor H and thereby interfere with complement activation (51). However, there are differences in the precise nature of this interaction, as Hic binds to short consensus repeats 8 to 11 of factor H, whereas PspC binds to repeats 13 to 15 (33).

The release of the S. pneumoniae genome sequence by The Institute for Genomic Research in 1997 was a watershed event in the quest for additional pneumococcal vaccine antigens. Whereas previous studies had identified individual proteins as virulence factors, usually by studying the behavior of gene knockout mutants in animal models, access to the genome sequence permitted targeting of entire families of genes encoding proteins with recognizable structural features (88). The choline-binding proteins are good examples of this approach. Although several members of this family were previously identified by conventional techniques, such as elution from the cell surface with choline, a search of the genome sequence identified a total of 12 functional genes encoding proteins with choline-binding motifs. Site-specific mutagenesis was then used to demonstrate that five of the novel choline-binding proteins (CbpD, CbpE, CbpG, LytB, and LytC) were involved in in vitro adherence to epithelial cells, nasopharyngeal colonization, or sepsis, thereby identifying them as vaccine candidates (40). LytB and LytC are unusual in that their choline-binding domains are located in the N-terminal part of the molecule, while the C-terminal portions have murein hydrolase activity (64). Purified recombinant LytB and LytC were subsequently tested for protective efficacy as part of another large scale study. Immunization with these proteins conferred significant protection against intraperitoneal challenge in mice, although the degree of protection observed was marginally less than that observed using PspA, which was used as a control antigen (112).

PsaA and Other Metal-Binding Lipoproteins

Another candidate vaccine antigen is pneumococcal surface antigen A (PsaA), a highly conserved 37-kDa surface protein produced by all pneumococci. It was initially thought to be an adhesin based on sequence homology with putative lipoprotein adhesins of oral streptococci, but it is actually the metal-binding lipoprotein component of an ATP-binding cassette (ABC) transport system with specificity for Mn^{2+} (31). Defined *psaA*-negative mutants of *S. pneumoniae* are virtually avirulent for mice and exhibit markedly reduced adherence in vitro to human type II pneumocytes (11). This is presumed to be a consequence of a requirement for Mn as a cofactor or for regulation of expression of other virulence factors (e.g., adhesins), and/or growth retardation due to an inability to scavenge Mn in vivo. *psaA*-negative pneumococci also exhibit hypersensitivity to superoxide and hydrogen peroxide, suggesting that PsaA plays an important role in regulation of expression of oxidative stress response enzymes and intracellular redox homeostasis (108).

One study has shown that parenteral immunization of mice with purified PsaA in the presence of strong adjuvants elicits significant protection against systemic challenge with *S. pneumoniae* (106). However, in other studies immunization with PsaA elicited only marginal protection and was less efficacious

than pneumolysoid in an intraperitoneal challenge model (81). The dimensions of PsaA (approximately 7 nm at its longest axis) (59) are such that if it is anchored to the outer face of the cell membrane via its N-terminal lipid moiety, it is unlikely to be exposed on the outer surface of the pneumococcus. This is consistent with the fact that whereas the known surface-exposed domains of PspA and PspC are variable, the amino acid sequence of PsaA is highly conserved (100). Moreover, flow cytometric analysis failed to demonstrate surface accessibility of PsaA to antibody in exponentially growing bacteria (39). Thus, any protection elicited by immunization with PsaA is unlikely to be a consequence of enhanced opsonophagocytic clearance. Rather, it is presumably due to in vivo blockade of ion transport, which necessitates diffusion of antibody through the capsule and cell wall layers. Such penetration of antibody is likely to be concentration dependent, so high anti-PsaA titers may be required for protection. Moreover, accessibility of PsaA to exogenous antibody may well be influenced by the thickness of the capsule, which may vary from strain to strain. Expression of pneumococcal capsule biosynthesis genes has also been shown to be up-regulated during invasive infection (82). In contrast, pneumococci colonizing the nasopharynx are thought to down-regulate capsule expression, thereby facilitating interaction between surface adhesins and the host mucosa. Consistent with this hypothesis, several studies have shown that intranasal immunization of mice with PsaA in the presence of strong mucosal adjuvants such as CTB significantly reduces the level of nasopharyngeal carriage of *S. pneumoniae* (16, 30). A lesser, but still significant, reduction in susceptibility to carriage was also achieved by subcutaneous immunization of mice with synthetic lipidated multiantigenic PsaA peptides (53). Thus, at least in the nasopharynx, PsaA appears to be accessible to exogenous antibody.

Two other metal-binding lipoproteins have been proposed as pneumococcal vaccine antigens. These proteins, designated PiuA and

PiaA, are components of two separate ABC iron transport systems. At least one of these proteins is required for optimal growth of pneumococci in iron-depleted media, and they are capable of acquiring iron from hemoglobin (22). Indeed, PiuA has recently been shown to be capable of directly binding both hemin and hemoglobin (105). PiuA and PiaA are produced by all pneumococci and have been shown to contribute to virulence in mice in both lung and intraperitoneal models of infection (22). They are immunologically cross-reactive, and immunization of mice with either protein conferred a degree of protection against intraperitoneal challenge similar to that elicited by the pneumolysoid PdB. Moreover, immunization with a combination of PiuA and PiaA resulted in additive protection (23). Although a direct comparison has not been conducted, immunization with either PiuA or PiaA provided a higher degree of protection against systemic disease than that previously found for PsaA, using the same mouse model and *S. pneumoniae* challenge strain (81). Like PsaA, PiuA and PiaA are located on the outer face of the plasma membrane (105), and so the superior protective efficacy of the latter proteins is unlikely to be due to a difference in accessibility to exogenous antibody. These findings imply that blockade of pneumococcal iron scavenging mechanisms may have a more significant impact on the course of a systemic infection than blockade of Mn uptake. Nevertheless, the significant protection against nasopharyngeal carriage afforded by mucosal immunization with PsaA indicates that Mn uptake is very important for survival in that niche. Whether mucosal immunization with PiuA and PiaA has a similar protective effect against nasopharyngeal carriage remains to be determined.

Other Pneumococcal Proteins

As mentioned above, access to the pneumococcal genome sequence has permitted targeting of entire families of genes encoding proteins with recognizable structural features consistent with surface exposure (88). Apart from choline-binding motifs, targeted features have included signal peptides and sortase-dependent cell wall anchorage domains, as well as proteins with similarity to known virulence factors in other bacteria. One study (112) identified over 100 candidate genes which were then expressed in recombinant *E. coli*, enabling purification of the respective proteins (or fragments thereof). Five of these were demonstrated to be protective immunogens in a mouse model. The protective antigens were the choline-binding proteins LytB and LytC (discussed above), the cell wall-associated serine protease PrtA, and two proteins of unknown function designated PvaA and PhtA. Immunization with LytC, PvaA, and PhtA elicited a degree of protection against systemic challenge similar, but not superior, to that achieved using PspA (112). Interestingly, PhtA contains five copies of an unusual histidine triad motif (HXXHXH). Further examination of the genome sequence revealed three additional related open reading frames each encoding five or six copies of the motif, and immunization with two of these (PhtB and PhtD) also protected mice against some, but not all, *S. pneumoniae* challenge strains tested (1). A separate study (85) used two-dimensional electrophoresis to separate hydrophobic surface-associated pneumococcal proteins and identified three that elicited opsonophagocytic antibody responses in mice. These were subsequently identified as PspA, the oligopeptide-binding lipoprotein AmiA, and the putative proteinase maturation protein PpmA. In addition to these, several other pneumococcal virulence proteins have been examined for vaccine potential, including the neuraminidases NanA and NanB, the major autolysin LytA (another choline-binding protein), and hyaluronidase. Some of these imparted modest protection against systemic challenge, but they were not as efficacious as pneumolysin derivatives (reviewed in reference 87).

From the above it can be seen that there are a number of pneumococcal proteins that exhibit potential as vaccine antigens. However, assessment of their protective efficacy has been carried out in different laboratories, using a variety of animal models and challenge strains. Clearly, a comprehensive series of direct comparative protection studies needs to be performed in order to determine which of these proteins provides the strongest protection against the widest variety of *S. pneumoniae* strains.

Combination Protein Vaccines

Virtually all of the pneumococcal proteins under consideration as vaccine antigens are directly or indirectly involved in the pathogenesis of pneumococcal disease. Mutagenesis of some combinations of virulence factor genes, for example, those encoding pneumolysin and either PspA or PspC, or both PspA and PspC, has been shown to synergistically attenuate pneumococcal virulence in animal models, implying that the respective proteins function independently in the pathogenic process (7, 12). This strongly suggests that immunization with combinations of these antigens might provide additive protection. Moreover, there may be differences in the relative protective capacities of the individual antigens against particular *S. pneumoniae* strains, particularly for surface-exposed antigens that exhibit some degree of sequence variation. Thus, a combined pneumococcal protein vaccine may elicit a higher degree of protection against a wider variety of strains than any single antigen. To date, only a limited number of combination experiments have been performed. Immunization of mice with a combination of the pneumolysoid PdB and PspA provided significantly more protection against intraperitoneal challenge than immunization with either protein alone. However, combining either protein with PsaA did not result in enhanced protection (81). The potential benefits of combination protein vaccines are also well illustrated using a recently developed mouse model of nonbacteremic pneumonia, which closely reflects the most common form of pneumococcal respiratory disease in humans (18). In this system, subcutaneous immunization (using alum adjuvant) with either PdB or PspA, but not PsaA, significantly reduced numbers of *S. pneumoniae* bacteria in the lungs 7 days after challenge (Fig. 1). A significant additional reduction in bacterial load was achieved by immunization with a combination of PdB and PspA, but not when either protein was combined with PsaA (Fig. 1) (18). These findings contrast with those obtained using mucosal (intranasal) immunization with the same proteins with CTB as adjuvant. As discussed above, immunization with either PspA or PsaA, but not PdB, reduced the level of carriage of *S. pneumoniae* after intranasal challenge, and the combination of PspA and PsaA was more effective than either antigen alone (16). These findings imply that PsaA is either more important for survival of *S. pneumoniae* in the nasopharynx than in the lung or that it is more accessible to exogenous antibody in the former niche. On the other hand, pneumolysin appears to play only a minor role during the colonization phase but is clearly important once the organism has been aspirated into the lungs. Thus, optimum vaccine formulation will be dependent upon the mode of vaccine delivery and the stage of the pathogenic process being targeted for immunoprophylaxis.

To date, only a limited number of other pneumococcal protein combinations have been tested for additive protective immunogenicity. Immunization with both the iron transporters PiuA and PiaA was more effective than either antigen alone (23). PdB and PspC have also been tested in combination, but in that study, PspC imparted such strong protection against intraperitoneal challenge on its own that additional protection was not evident when combined with PdB (83). Clearly, additional comparative studies of the protective efficacy of the better-characterized proteins, as well as the more recently identified vaccine candidates (both singly and in combination), are required to enable informed decisions on the formulation of a protein-based pneumococcal vaccine.

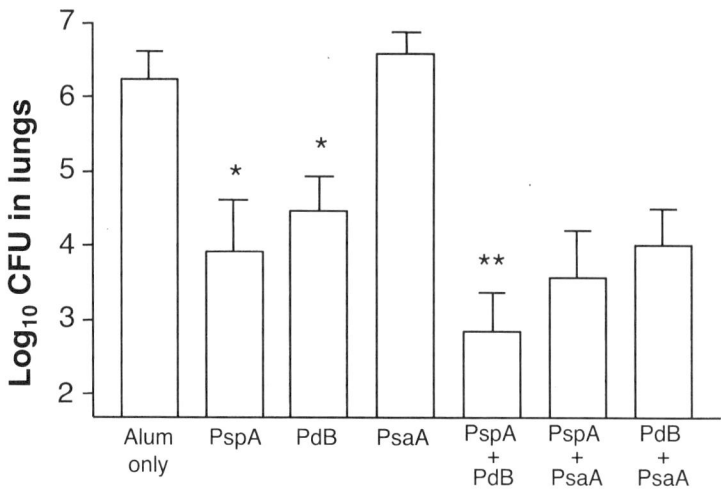

Subcutaneous immunogens

FIGURE 1 Protection against pulmonary infection with *S. pneumoniae* elicited by immunization with PspA, PdB (genetically toxoided pneumolysin), PsaA, or combinations thereof. CBA/N mice were immunized with the indicated proteins on alum or with alum alone, challenged with $10^{6.89}$ CFU of *S. pneumoniae* strain EF3030 (capsular group 19), and sacrificed 7 days later to determine numbers of CFU in their lungs. Significance of difference relative to control mice (alum only): ★, $P < 0.04$; ★★, $P < 0.001$. (Reproduced with permission from the *Journal of Infectious Diseases* [18].)

Consideration should also be given to using protein antigens as supplements to PS-protein conjugate vaccines. Incorporation of one or more proteins common to all *S. pneumoniae* serotypes may significantly reduce the problems associated with limited serotype cover and replacement carriage associated with the conjugate vaccines, although the problem of high cost remains. Pneumolysoid has also been proposed as an alternative carrier in PS-protein conjugate vaccines, and conjugates of PdB with type 19F PS have been shown to be highly immunogenic and protective in mice (61, 90). In a more recent study, a similar detoxified pneumolysin derivative was shown to be a very effective carrier protein in a tetravalent conjugate vaccine formulation including PS types 6B, 14, 19F, and 23F (72). Use of pneumolysoid, or other suitable pneumococcal proteins, as carriers for PS in conjugate vaccines may also minimize any problems associated with overuse of existing carrier proteins.

MUCOSAL VACCINATION STRATEGIES

Given the pivotal role of nasopharyngeal colonization in the transmission of *S. pneumoniae* and as a precursor of pneumococcal disease, vaccination strategies specifically designed to elicit mucosal immune responses may be more efficacious than parenteral immunization for certain antigens, particularly those implicated in colonization. To date this has been examined in animal models using direct intranasal administration of vaccine formulations (killed whole cells or purified antigens) with a strong mucosal adjuvant such as cholera toxin (CT), the related *E. coli* heat-labile enterotoxin (LT), or cytokines such as interleukin-1, interleukin-12, or granulocyte macrophage colony-stimulating factor (3, 113). Use of CT and LT holotoxins as adjuvants in human vaccine formulations is somewhat controversial, owing to their reactogenicity. However, significant mucosal adjuvant activity resides in the B subunits of CT and LT, and these are much less reactogenic, although there are residual concerns because of their capacity to bind to GM_1 receptors on olfactory nerve endings.

Intranasal administration of heat-killed type 4 pneumococci resulted in strong humoral and mucosal responses to type 4 PS and protection from homologous challenge (47). Similar anti-PS responses in mice have been achieved using purified PS conjugated to either CTB or an LT derivative (49, 102). On the other hand, use of killed nonencapsulated pneumococci has been shown to prevent nasopharyngeal carriage of type 6B pneumococci in a mouse model and to protect rats from intrathoracic challenge with virulent type 3 pneumococci (66). This study is a further demonstration of non-serotype-dependent protection achieved using non-PS antigens. The protective efficacy against carriage of mucosal immunization with purified PspA and/or PsaA in the presence of CTB has already been discussed (16). In a more recent study, coencapsulation of PsaA and CTB within alginate microspheres was shown to elicit higher serum and mucosal immune responses after oral immunization of mice than did nonencapsulated antigens, and this translated into stronger protection against intranasal challenge (101).

An alternative means of eliciting mucosal immune responses involves oral administration of live recombinant carrier bacteria expressing pneumococcal antigens. Recombinant attenuated salmonellae expressing PdB (92), PspA (54, 79) or both of these antigens as well as PsaA (E. M. Barry, A. R. Santiago, J. Sampson, E. Ades, G. Carlone, D. Briles, J. Paton, and M. Levine, Abstr. 3rd Int. Symp. Pneumococci Pneumococcal Dis., p. 102, 2002) have been constructed and shown to elicit mucosal and humoral antibody responses in mice after oral immunization. Furthermore, for strains expressing PspA, protection against systemic challenge was also demonstrated (54, 79). Expression of type 3 PS has also been achieved in *Lactococcus lactis*, which has been proposed as an alternative carrier for vaccine antigens (38). However, the mechanism of biosynthesis of

this PS serotype is much simpler than that of all the other clinically significant pneumococcal PS types and requires expression of only a small number of genes (91). Expression of the other much larger PS biosynthesis loci in heterologous bacteria may be extremely difficult, and any such live vaccines would also suffer from the disadvantages of serotype-dependent protection and poor immunogenicity of PS antigens in high-risk groups.

DNA VACCINES

A further strategy under consideration for prevention of pneumococcal disease is the use of DNA vaccines. This involves introduction of naked plasmid DNA carrying genes encoding protective antigens under the control of a eukaryotic promoter, either by intramuscular injection or transdermally using a gene gun. The naked DNA is taken up by host cells and the antigens are expressed in vivo. This approach has been used for a variety of pathogens and is attractive because DNA vaccines are potentially cheap to produce on a large scale. They usually elicit both humoral and cell-mediated immune responses, and although protection against *S. pneumoniae* is generally considered to be antibody dependent, the role (if any) of cell-mediated immune responses has not been investigated to any significant extent. One study reported construction of a DNA vaccine plasmid encoding the alpha-helical N-terminal half of PspA (the region which contains the cross-protective epitopes). This induced strong antibody responses in mice and conferred long-lasting protection against both homologous and heterologous *S. pneumoniae* challenge strains (15). However, in another study, immunization with constructs directing expression of similar regions of PspA from different *S. pneumoniae* strains resulted in production of cross-reacting antibodies, but protection against challenge was confined to strains expressing related PspA types (75). The same group also examined DNA vaccine constructs directing expression of either the C-terminal two-thirds of PspA or full-length PsaA, which elicited significant antibody responses to the respective protein in mice (73). However, protection against challenge was not elicited by the C-terminal PspA-expressing construct in a subsequent study (74), while that for the PsaA-expressing construct has not been examined.

Use of DNA vaccine delivery systems for PS antigens is extremely problematic, not only because of the multiplicity of serotypes but also because the genetic loci encoding PS biosynthesis are very large, comprising up to 20 or more genes for each PS type (91). An innovative solution to the latter problem has been achieved using an anti-idiotype approach. Firstly, a monoclonal antibody to type 4 PS was used to screen a phage display library, and this identified a peptide mimic capable of eliciting an anti-type 4 PS response. An oligonucleotide encoding this peptide was then inserted into a DNA vaccine vector and this elicited an anti-type 4 antibody response in mice (62). It remains to be seen whether such antibodies are protective against challenge with type 4 pneumococci, and whether peptide mimics can be developed for a sufficient number of the other PS serotypes.

CONCLUSIONS

The ongoing high global morbidity and mortality associated with pneumococcal disease, and the complications caused by increasing rates of resistance to antimicrobials, have underpinned extensive efforts in recent years to develop more effective vaccination strategies against *S. pneumoniae*. These efforts have benefited from a better understanding of the mechanisms of pathogenesis of pneumococcal disease and the advances made possible by the advent of recombinant DNA technology and access to genome sequence data. The polyvalent PS vaccines have doubtless prevented a large number of deaths from invasive disease in recipients belonging to those patient groups for whom this vaccine is currently recommended. The newer PS-protein conjugate formulations will also confer a very high degree of protection on young children against included serotypes, and they may also have an impact on prevalence of drug-resistant strains. However, there is now general acceptance that this vaccination approach is not without its drawbacks, and as explained above, the initially substantial clinical benefits that are expected to be derived from widespread use of conjugate vaccines may diminish with time. It will take many years for the overall impact of conjugate vaccines on disease burden and the population biology of *S. pneumoniae* to become apparent. At the very least, use of the conjugate vaccines will buy time for development of cheaper, non-serotype-specific vaccines based on combinations of protein antigens. It must be emphasized, however, that the success of these protein vaccines is not dependent upon real or perceived failure of the conjugates. Rather, the two approaches should be viewed as complementary, each having an important role to play in global prevention of pneumococcal disease. Neither should development of parenteral protein vaccines impede future research on mucosal or DNA-based delivery systems, which may further improve presentation of protective antigens to the immune system, thereby optimizing host responses.

REFERENCES

1. Adamou, J. E., J. H. Heinrichs, A. L. Erwin, W. Walsh, T. Gayle, M. Dormitzer, R. Dagan, Y. A. Brewah, P. Barren, R. Lathigra, S. Langermann, S. Koenig, and S. Johnson. 2001. Identification and characterization of a novel family of pneumococcal proteins that are protective against sepsis. *Infect. Immun.* **69:**949–958.

2. Alexander, J. E., R. A. Lock, C. C. A. M. Peeters, J. T. Poolman, P. W. Andrew, T. J. Mitchell, D. Hansman, and J. C. Paton. 1994. Immunization of mice with pneumolysin toxoid confers a significant degree of protection against at least nine serotypes of *Streptococcus pneumoniae*. *Infect. Immun.* **62:**5683–5688.

3. Arulanandam, B. P., J. M. Lynch, D. E. Briles, S. Hollingshead, and D. W. Metzger. 2001. Intranasal vaccination with pneumococcal surface protein A and interleukin-12 augments

antibody-mediated opsonization and protective immunity against *Streptococcus pneumoniae* infection. *Infect. Immun.* **69:**6718–6724.

4. **Austrian, R.** 1981. Some observations on the pneumococcus and on the current status of pneumococcal disease and its prevention. *Rev. Infect. Dis.* **3**(Suppl.):S1–S17.

5. **Austrian, R.** 2001. Pneumococcal otitis media and pneumococcal vaccines, a historical perspective. *Vaccine* **19:**S71–S77.

6. **Avery, O. T., and W. F. Goebel.** 1931. Chemo-immunological studies on conjugated carbohydrate-proteins. V. The immunological specificity of an antigen prepared by combining the capsular polysaccharide of type III pneumococcus with foreign protein. *J. Exp. Med.* **54:**437–447.

7. **Balachandran, P., A. Brooks-Walter, A. Virolainen-Julkunen, S. K. Hollingshead, and D. E. Briles.** 2002. Role of pneumococcal surface protein C in nasopharyngeal carriage and pneumonia and its ability to elicit protection against carriage of *Streptococcus pneumoniae*. *Infect. Immun.* **70:**2526–2534.

8. **Balachandran, P., S. K. Hollingshead, J. C. Paton, and D. E. Briles.** 2001. The autolytic enzyme LytA of *Streptococcus pneumoniae* is not responsible for releasing pneumolysin. *J. Bacteriol.* **183:**3108–3116.

9. **Berry, A. M., J. E. Alexander, T. J. Mitchell, P. W. Andrew, D. Hansman, and J. C. Paton.** 1995. Effect of defined point mutations in the pneumolysin gene on the virulence of *Streptococcus pneumoniae*. *Infect. Immun.* **63:**1969–1974.

10. **Berry, A. M., R. A. Lock, D. Hansman, and J. C. Paton.** 1989. Contribution of autolysin to the virulence of *Streptococcus pneumoniae*. *Infect. Immun.* **57:**2324–2330.

11. **Berry, A. M., and J. C. Paton.** 1996. Sequence heterogeneity of PsaA, a 37-kilodalton putative adhesin essential for virulence of *Streptococcus pneumoniae*. *Infect. Immun.* **64:**5255–5262.

12. **Berry, A. M., and J. C. Paton.** 2000. Additive attenuation of virulence of *Streptococcus pneumoniae* by mutation of the genes encoding pneumolysin and other putative pneumococcal virulence proteins. *Infect. Immun.* **68:**133–140.

13. **Berry, A. M., J. Yother, D. E. Briles, D. Hansman, and J. C. Paton.** 1989. Reduced virulence of a defined pneumolysin-negative mutant of *Streptococcus pneumoniae*. *Infect. Immun.* **57:**2037–2042.

14. **Black, S., H. Shinefield, B. Fireman, E. Lewis, P. Ray, J. R. Hansen, L. Elvin, K. M. Ensor, J. Hackell, G. Siber, F. Malinoski, D. Madore, I. Chang, R. Kohberger, W. Watson, R. Austrian, K. Edwards, and**

Northern California Kaiser Permanente Vaccine Study Center Group. 2000. Efficacy, safety and immunogenicity of heptavalent pneumococcal conjugate vaccine in children. *Pediatr. Infect. Dis. J.* **19:**187–195.

15. **Bosarge, J. R., J. M. Watt, D. O. McDaniel, E. Swiatlo, and L. S. McDaniel.** 2001. Genetic immunization with the region encoding the alpha-helical domain of PspA elicits protective immunity against *Streptococcus pneumoniae*. *Infect. Immun.* **69:**5456–5463.

16. **Briles, D. E., E. Ades, J. C. Paton, J. S. Sampson, G. M. Carlone, R. C. Huebner, A. Virolainen, E. Swiatlo, and S. K. Hollingshead.** 2000. Intranasal immunization of mice with a mixture of the pneumococcal proteins PsaA and PspA is highly protective against nasopharyngeal carriage of *Streptococcus pneumoniae*. *Infect. Immun.* **68:**796–800.

17. **Briles, D. E., S. K. Hollingshead, J. King, A. Swift, P. A. Braun, M. K. Park, L. M. Ferguson, M. H. Nahm, and G. S. Nabors.** 2000. Immunization of humans with recombinant pneumococcal surface protein A (rPspA) elicits antibodies that passively protect mice from fatal infection with *Streptococcus pneumoniae* bearing heterologous PspA. *J. Infect. Dis.* **182:**1694–1701.

18. **Briles, D. E., S. K. Hollingshead, J. C. Paton, E. W. Ades, L. Novak, F. W. van Ginkel, and W. H. Benjamin, Jr.** 2003. Immunizations with pneumococcal surface protein A and pneumolysin are protective against pneumonia in a murine model of pulmonary infection with *Streptococcus pneumoniae*. *J. Infect. Dis.* **188:**339–348.

19. **Briles, D. E., J. C. Paton, M. H. Nahm, and E. Swiatlo.** 1999. Immunity to *Streptococcus pneumoniae*, p. 263–280. *In* M. W. Cunningham and R. S. Fujinami (ed.), *Effects of Microbes on the Immune System*. Lippincott, Williams and Wilkins, Philadelphia, Pa.

20. **Brooks-Walter, A., D. E. Briles, and S. K. Hollingshead.** 1999. The *pspC* gene of *Streptococcus pneumoniae* encodes a polymorphic protein PspC, which elicits cross-reactive antibodies to PspA and provides immunity to pneumococcal bacteremia. *Infect. Immun.* **67:**6533–6542.

21. **Broome, C.** 1996. Meningococcal and pneumococcal disease vaccines, p. 28–32. *In* Progress of Vaccine Research and Development—1996. Document WHO/VRD/GEN/96.02. World Health Organization, Geneva, Switzerland.

22. **Brown, J. S., S. M. Gilliland, and D. W. Holden.** 2001. A *Streptococcus pneumoniae* pathogenicity island encoding an ABC transporter involved in iron uptake and virulence. *Mol. Microbiol.* **40:**572–585.

23. **Brown, J. S., A. D. Ogunniyi, M. C. Woodrow, D. W. Holden, and J. C. Paton.** 2001. Immunization with components of two iron-uptake ABC transporters protects mice against systemic *Streptococcus pneumoniae* infection. *Infect. Immun.* **69:**6702–6706.

24. **Centers for Disease Control and Prevention.** 1997. Prevention of pneumococcal disease: recommendations of the Advisory Committee on Immunization Practices (ACIP). *Morb. Mortal. Wkly. Rep.* **46**(RR-8):1–24.

25. **Cheng, Q., D. Finkel, and M. K. Hostetter.** 2000. Novel purification scheme and functions for a C3-binding protein from *Streptococcus pneumoniae. Biochemistry* **39:**5450–5457.

26. **Coffey, T. J., M. C. Enright, M. Daniels, J. K. Morona, R. Morona, W. Hryniewicz, J. C. Paton, and B. G. Spratt.** 1998. Recombinational exchanges at the capsular polysaccharide biosynthetic locus lead to frequent serotype changes among natural isolates of *Streptococcus pneumoniae. Mol. Microbiol.* **27:**73–84.

27. **Crain, M. J., W. D. Waltman II, J. S. Turner, J. Yother, D. F. Talkington, L. S. McDaniel, B. M. Gray, and D. E. Briles.** 1990. Pneumococcal surface protein A (PspA) is serologically highly variable and is expressed by all clinically important capsular serotypes of *Streptococcus pneumoniae. Infect. Immun.* **58:**3293–3299.

28. **Dagan, R., N. Givon-Lavi, O. Zamir, and D. Fraser.** 2003. Effect of a nonavalent conjugate vaccine on carriage of antibiotic-resistant *Streptococcus pneumoniae* in day-care centers. *Pediatr. Infect. Dis. J.* **22:**532–540.

29. **Dave, S., A. Brooks-Walter, M. K. Pangburn, and L. S. McDaniel.** 2001. PspC, a pneumococcal surface protein, binds human factor H. *Infect. Immun.* **69:**3435–3437.

30. **De, B. K., J. S. Sampson, E. W. Ades, R. C. Huebner, D. L. Jue, S. E. Johnson, M. Espina, A. R. Stinson, D. E. Briles, and G. M. Carlone.** 2000. Purification and characterization of *Streptococcus pneumoniae* palmitoylated pneumococcal surface adhesin A expressed in *Escherichia coli. Vaccine* **18:**1811–1821.

31. **Dintilhac, A., G. Alloing, C. Granadel, and J. P. Claverys.** 1997. Competence and virulence of *S. pneumoniae*: Adc and PsaA mutants exhibit a requirement for Zn and Mn resulting from inactivation of metal permeases. *Mol. Microbiol.* **25:**727–739.

32. **Douglas, R. M., J. C. Paton, S. J. Duncan, and D. Hansman.** 1983. Antibody response to pneumococcal vaccination in children younger than five years of age. *J. Infect. Dis.* **148:**131–137.

33. **Duthy, T. G., R. J. Ormsby, E. Giannakis, A. D. Ogunniyi, U. H. Stroeher, J. C. Paton, and D. L. Gordon.** 2002. The human complement regulator factor H binds pneumococcal surface protein PspC via short consensus repeat domains 13 to 15. *Infect. Immun.* **70:**5604–5611.

34. **Eskola, J.** 2001. Polysaccharide-based pneumococcal vaccines in the prevention of acute otitis media. *Vaccine* **19:**S78–S82.

35. **Eskola, J., T. Kilpi, A. Palmu, J. Jokinen, J. Haapakoski, E. Herva, A. Takala, H. Kayhty, P. Karma, R. Kohberger, G. Siber, P. H. Makela, and the Finnish Otitis Media Study Group.** 2001. Efficacy of a pneumococcal conjugate vaccine against acute otitis media. *N. Engl. J. Med.* **344:**403–409.

36. **Feldman, C., N. C. Munro, P. K. Jeffery, T. J. Mitchell, P. W. Andrew, G. J. Boulnois, D. Guerreiro, J. A. Rohde, H. C. Todd, P. J. Cole, and R. Wilson.** 1991. Pneumolysin induces the salient histological features of pneumococcal infection in the rat lung *in vivo. Am. J. Respir. Cell Mol. Biol.* **5:**416–423.

37. **Fraser, D., N. Givon-Lavi, N. Bilenko, and R. Dagan.** 2001. A decade (1989–1998) of pediatric invasive pneumococcal disease in 2 populations residing in 1 geographical location: implications for vaccine choice. *Clin. Infect. Dis.* **33:**421–427.

38. **Gilbert, C., K. Robinson, R. W. Le Page, and J. M. Wells.** 2000. Heterologous expression of an immunogenic pneumococcal type 3 capsular polysaccharide in *Lactococcus lactis. Infect. Immun.* **68:**3251–3260.

39. **Gor, D. O., X. Ding, Q. Li, J. R. Schreiber, M. Dubinsky, and N. S. Greenspan.** 2002. Enhanced immunogenicity of pneumococcal surface adhesin A by genetic fusion to cytokines and evaluation of protective immunity in mice. *Infect. Immun.* **70:**5589–5595.

40. **Gosink, K. K., E. R. Mann, C. Guglielmo, E. I. Tuomanen, and H. R. Masure.** 2000. Role of novel choline binding proteins in virulence of *Streptococcus pneumoniae. Infect. Immun.* **68:**5690–5695.

41. **Hakansson, A., H. Roche, S. Mirza, L. S. McDaniel, A. Brooks-Walter, and D. E. Briles.** 2001. Characterization of the binding of human lactoferrin to pneumococcal surface protein A (PspA). *Infect. Immun.* **69:**3372–3381.

42. **Hammerschmidt, S., G. Bethe, P. Remanen, and G. S. Chhatwal.** 1999. Identification of pneumococcal surface protein A as a lactoferrin-binding protein of *Streptococcus pneumoniae. Infect. Immun.* **67:**1683–1687.

43. **Hammerschmidt, S., S. Talay, P. Brandtzaeg, and G. S. Chhatwal.** 1997. SpsA,

a novel pneumococcal surface protein with specific binding to secretory immunoglobulin A and secretory component. *Mol. Microbiol.* **25:**1113–1124.

44. **Hammerschmidt, S., M. P. Tillig, S. Wolff, J. P. Vaerman, and G. S. Chhatwal.** 2000. Species-specific binding of human secretory component to SpsA protein of *Streptococcus pneumoniae* via a hexapeptide motif. *Mol. Microbiol.* **36:**726–736.

45. **Hollingshead, S. K., R. S. Becker, and D. E. Briles.** 2000. Diversity of PspA: mosaic genes and evidence for past recombination in *Streptococcus pneumoniae*. *Infect. Immun.* **68:**5889–5900.

46. **Houldsworth, S., P. W. Andrew, and T. J. Mitchell.** 1994. Pneumolysin stimulates production of tumor necrosis factor α and interleukin-1β by human mononuclear phagocytes. *Infect. Immun.* **62:**1501–1503.

47. **Hvalbye, B. K., I. S. Aaberge, M. Lovik, and B. Haneberg.** 1999. Intranasal immunization with heat-inactivated *Streptococcus pneumoniae* protects mice against systemic pneumococcal infection. *Infect. Immun.* **67:**4320–4325.

48. **Iannelli, F., M. R. Oggioni, and G. Pozzi.** 2002. Allelic variation in the highly polymorphic locus *pspC* of *Streptococcus pneumoniae*. *Gene* **284:**63–71.

49. **Jakobsen, H., D. Schulz, M. Pizza, R. Rappuoli, and I. Jonsdottir.** 1999. Intranasal immunization with pneumococcal polysaccharide conjugate vaccines with non-toxic mutants of *Escherichia coli* heat-labile enterotoxins as adjuvants protects mice against invasive pneumococcal infections. *Infect. Immun.* **67:**5892–5897.

50. **Janulczyk, R., F. Iannelli, A. G. Sjoholm, G. Pozzi, and L. Bjorck.** 2000. Hic, a novel surface protein of *Streptococcus pneumoniae* that interferes with complement function. *J. Biol. Chem.* **275:**37257–37263.

51. **Jarva, H., R. Janulczyk, J. Hellwage, P. F. Zipfel, L. Bjorck, and S. Meri.** 2002. *Streptococcus pneumoniae* evades complement attack and opsonophagocytosis by expressing the *pspC* locus-encoded Hic protein that binds to short consensus repeats 8–11 of factor H. *J. Immunol.* **168:**1886–1894.

52. **Jedrzejas, M. J., S. K. Hollingshead, J. Lebowitz, L. Chantalat, D. E. Briles, and E. J. Lamani.** 2000. Production and characterization of the functional fragment of pneumococcal surface protein A. *Arch. Biochem. Biophys.* **373:**116–125.

53. **Johnson, S. E., J. K. Dykes, D. L. Jue, J. S. Sampson, G. M. Carlone, and E. W. Ades.** 2002. Inhibition of pneumococcal carriage in mice by subcutaneous immunization with peptides from the common surface protein pneumococcal surface adhesin A. *J. Infect. Dis.* **185:**489–496.

54. **Kang, H. Y., J. Srinivasan, and R. Curtiss III.** 2002. Immune responses to recombinant pneumococcal PspA antigen delivered by live attenuated *Salmonella enterica* serovar Typhimurium vaccine. *Infect. Immun.* **70:**1739–1749.

55. **Klein, D. L.** 2000. Pneumococcal disease and the role of conjugate vaccines, p. 467–477. *In* A. Tomasz (ed.), Streptococcus pneumoniae. *Molecular Biology and Mechanisms of Disease.* Mary Ann Liebert, Inc., Larchmont, N.Y.

56. **Klein, J. O.** 2001. The burden of otitis media. *Vaccine* **19:**S2–S8.

57. **Klugman, K. P.** 1990. Pneumococcal resistance to antibiotics. *Clin. Microbiol. Rev.* **3:**171–196.

58. **Klugman, K. P.** 1996. Epidemiology, control and treatment of multiresistant pneumococci. *Drugs* **52**(Suppl. 2):42–46.

59. **Lawrence, M. C., P. A. Pilling, A. D. Ogunniyi, A. M. Berry, and J. C. Paton.** 1998. The crystal structure of pneumococcal surface antigen PsaA reveals a metal-binding site and a novel structure for a putative ABC-type binding protein. *Structure* **6:**1553–1561.

60. **Lee, C. J., S. D. Banks, and J. P. Li.** 1991. Virulence, immunity and vaccine related to *S. pneumoniae*. *Crit. Rev. Microbiol.* **18:**89–114.

61. **Lee, C.-J., R. A. Lock, T. J. Mitchell, P. W. Andrew, G. J. Boulnois, and J. C. Paton.** 1994. Protection of infant mice from challenge with *Streptococcus pneumoniae* type 19F by immunization with a type 19F polysaccharide-pneumolysoid conjugate. *Vaccine* **12:**875–878.

62. **Lesinski, G. B., S. L. Smithson, N. Srivastava, D. Chen, G. Widera, and M. A. Westerink.** 2001. A DNA vaccine encoding a peptide mimic of *Streptococcus pneumoniae* serotype 4 capsular polysaccharide induces specific anti-carbohydrate antibodies in Balb/c mice. *Vaccine* **19:**1717–1726.

63. **Lipsitch, M., J. K. Dykes, S. E. Johnson, E. W. Ades, J. King, D. E. Briles, and G. M. Carlone.** 2000. Competition among *Streptococcus pneumoniae* for intranasal colonization in a mouse model. *Vaccine* **18:**2895–2901.

64. **Lopez, R., M. P. Gonzalez, E. Garcia, J. L. Garcia, and P. Garcia.** 2000. Biological roles of two new murein hydrolases of *Streptococcus pneumoniae* representing examples of module shuffling. *Res. Microbiol.* **151:**437–443.

65. **MacLeod, C. M., R. G. Hodges, M. Heidelberger, and W. G. Bernhard.** 1945. Prevention of pneumococcal pneumonia by vaccination. *J. Exp. Med.* **82:**445–465.

66. Malley, R., M. Lipsitch, A. Stack, R. Saladino, G. Fleisher, S. Pelton, C. Thompson, D. Briles, and P. Anderson. 2001. Intranasal immunization with killed unencapsulated whole cells prevents colonization and invasive disease by capsulated pneumococci. *Infect. Immun.* **69:**4870–4873.

67. Mbelle, N., R. E. Huebner, A. D. Wasas, A. Kimura, I. Chang, and K. P. Klugman. 1999. Immunogenicity and impact on nasopharyngeal carriage of a nonavalent pneumococcal conjugate vaccine. *J. Infect. Dis.* **180:**1171–1176.

68. McCool, T. L., T. R. Cate, G. Moy, and J. N. Weiser. 2002. The immune response to pneumococcal proteins during experimental human carriage. *J. Exp. Med.* **195:**359–365.

69. McDaniel, L. S., B. A. Ralph, D. O. McDaniel, and D. E. Briles. 1994. Localization of protection-eliciting epitopes on PspA of *Streptococcus pneumoniae* between amino acid residues 192 and 260. *Microb. Pathog.* **17:**323–337.

70. McDaniel, L. S., J. Yother, M. Vijayakumar, L. McGarry, W. R. Guild, and D. E. Briles. 1987. Use of insertional inactivation to facilitate studies of biological properties of pneumococcal surface protein A (PspA). *J. Exp. Med.* **165:**381–394.

71. McGee, L., K. P. Klugman, and A. Tomasz. 2000. Serotypes and clones of antibiotic-resistant pneumococci, p. 375–379. *In* A. Tomasz (ed.), Streptococcus pneumoniae. *Molecular Biology and Mechanisms of Disease.* Mary Ann Liebert, Inc., Larchmont, N.Y.

72. Michon, F., P. C. Fusco, C. A. Minetti, M. Laude-Sharp, C. Uitz, C. H. Huang, A. J. D'Ambra, S. Moore, D. P. Remeta, I. Heron, and M. S. Blake. 1998. Multivalent pneumococcal capsular polysaccharide conjugate vaccines employing genetically detoxified pneumolysin as a carrier protein. *Vaccine* **16:**1732–1741.

73. Miyaji, E. N., W. O. Dias, M. Gamberini, V. C. Gebara, R. P. Schenkman, J. Wild, P. Riedl, J. Reimann, R. Schirmbeck, and L. C. Leite. 2001. PsaA (pneumococcal surface adhesin A) and PspA (pneumococcal surface protein A) DNA vaccines induce humoral and cellular immune responses against *Streptococcus pneumoniae.* *Vaccine* **20:**805–812.

74. Miyaji, E. N., W. O. Dias, M. M. Tanizaki, and L. C. Leite. 2003. Protective efficacy of PspA (pneumococcal surface protein A)-based DNA vaccines: contribution of both humoral and cellular immune responses. *FEMS Immunol. Med. Microbiol.* **37:**53–57.

75. Miyaji, E. N., D. M. Ferreira, A. P. Lopes, M. C. Brandileone, W. O. Dias, and L. C. Leite. 2002. Analysis of serum cross-reactivity and cross-protection elicited by immunization with DNA vaccines against *Streptococcus pneumoniae* expressing PspA fragments from different clades. *Infect. Immun.* **70:**5086–5090.

76. Musher, D. M., H. M. Phan, and R. E. Baughn. 2001. Protection against bacteremic pneumococcal infection by antibody to pneumolysin. *J. Infect. Dis.* **183:**827–830.

77. Musher, D. M., D. A. Watson, and R. E. Baughn. 2001. Genetic control of the immunological response to pneumococcal capsular polysaccharides. *Vaccine* **19:**623–627.

78. Nabors, G. S., P. A. Braun, D. J. Herrmann, M. L. Heise, D. J. Pyle, S. Gravenstein, M. Schilling, L. M. Ferguson, S. K. Hollingshead, D. E. Briles, and R. S. Becker. 2000. Immunization of healthy adults with a single recombinant pneumococcal surface protein A (PspA) variant stimulates broadly cross-reactive antibodies. *Vaccine* **18:**1743–1754.

79. Nayak, A. R., S. A. Tinge, R. C. Tart, L. S. McDaniel, D. E. Briles, and R. Curtiss III. 1998. A live recombinant oral *Salmonella* vaccine expressing pneumococcal surface protein A induces protective responses against *Streptococcus pneumoniae.* *Infect. Immun.* **66:**3744–3751.

80. Obaro, S. K., R. A. Adegbola, W. A. Banya, and B. M. Greenwood. 1996. Carriage of pneumococci after pneumococcal vaccination. *Lancet* **348:**271–272.

81. Ogunniyi, A. D., R. L. Folland, S. Hollingshead, D. E. Briles, and J. C. Paton. 2000. Immunization of mice with combinations of pneumococcal virulence proteins elicits enhanced protection against challenge with *Streptococcus pneumoniae.* *Infect. Immun.* **68:**3028–3033.

82. Ogunniyi, A. D., P. Giammarinaro, and J. C. Paton. 2002. The genes encoding virulence-associated proteins and the capsule of *Streptococcus pneumoniae* are upregulated and differentially expressed *in vivo.* *Microbiology* **148:**2045–2053.

83. Ogunniyi, A. D., M. C. Woodrow, J. T. Poolman, and J. C. Paton. 2001. Protection against *Streptococcus pneumoniae* elicited by immunization with pneumolysin and CbpA. *Infect. Immun.* **69:**5997–6003.

84. Ortqvist, A., J. Hedlund, L. A. Burman, E. Elbel, M. Hofer, M. Leinonen, I. Lindblad, B. Sundelof, M. Kalin, and Swedish Pneumococcal Vaccine Study Group. 1998. Randomised trial of 23-valent pneumococcal capsular polysaccharide vaccine in prevention of pneumonia in middle-aged and elderly people. *Lancet* **351:**399–403.

85. Overweg, K., A. Kerr, M. Sluijter, M. H. Jackson, T. J. Mitchell, A. P. de Jong, R. de

Groot, and P. W. Hermans. 2000. The putative proteinase maturation protein A of Streptococcus pneumoniae is a conserved surface protein with potential to elicit protective immune responses. *Infect. Immun.* **68:**4180–4188.

86. Paton, J. C. 1996. The contribution of pneumolysin to the pathogenicity of *Streptococcus pneumoniae. Trends Microbiol.* **4:**103–106.

87. Paton, J. C., A. M. Berry, and R. A. Lock. 1997. Molecular analysis of putative pneumococcal virulence proteins. *Microb. Drug Resist.* **3:**1–10.

88. Paton, J. C., and P. Giammarinaro. 2001. Genome-based analysis of pneumococcal virulence factors: the quest for novel vaccine antigens and drug targets. *Trends Microbiol.* **9:**515–518.

89. Paton, J. C., R. A. Lock, and D. J. Hansman. 1983. Effect of immunization with pneumolysin on survival time of mice challenged with *Streptococcus pneumoniae. Infect. Immun.* **40:**548–552.

90. Paton, J. C., R. A. Lock, C. J. Lee, J. P. Li, A. M. Berry, T. J. Mitchell, P. W. Andrew, D. Hansman, and G. J. Boulnois. 1991. Purification and immunogenicity of genetically obtained pneumolysin toxoids and their conjugation to *Streptococcus pneumoniae* type 19F polysaccharide. *Infect. Immun.* **59:**2297–2304.

91. Paton, J. C., and J. K. Morona. 2000. *Streptococcus pneumoniae* capsular polysaccharide, p. 201–213. *In* V. Fischetti, R. Novick, J. Ferretti, D. Portnoy, and J. Rood (ed.), *Gram-Positive Pathogens.* ASM Press, Washington, D.C.

92. Paton, J. C., J. K. Morona, S. Harrer, D. Hansman, and R. Morona. 1993. Immunization of mice with *Salmonella typhimurium* C5 *aroA* expressing a genetically toxoided derivative of the pneumococcal toxin pneumolysin. *Microb. Pathog.* **14:**95–102.

93. Paton, J. C., I. R. Toogood, R. Cockington, and D. Hansman. 1986. Antibody response to pneumococcal vaccine in children aged 5 to 15 years. *Am. J. Dis. Child.* **140:**135–138.

94. Ren, B., A. J. Szalai, O. Thomas, S. K. Hollingshead, and D. E. Briles. 2003. Both family 1 and family 2 PspA proteins can inhibit complement deposition and confer virulence to a capsular serotype 3 strain of *Streptococcus pneumoniae. Infect. Immun.* **71:**75–85.

95. Rijkers, G. T., E. A. Sanders, M. A. Breukels, and B. J. Zegers. 1998. Infant B cell responses to polysaccharide determinants. *Vaccine* **16:**1396–1400.

96. Robbins, J. B., R. Austrian, C. J. Lee, S. C. Rastogi, G. Schiffman, J. Henrichsen, P. H. Makela, C. V. Broome, R. R. Facklam, R. H. Tiesjema, and J. C. Parke, Jr. 1983. Considerations for formulating the second-generation pneumococcal capsular polysaccharide vaccine with emphasis on the cross-reactive types within groups. *J. Infect. Dis.* **148:**1136–1159.

97. Robbins, J. B., R. Schneerson, P. Anderson, and D. H. Smith. 1996. Prevention of systemic infections, especially meningitis, caused by *Haemophilus influenzae* type b. Impact on public health and implications for other polysaccharide-based vaccines. *JAMA* **276:**1181–1185.

98. Rosenow, C., P. Ryan, J. N. Weiser, S. Johnson, P. Fontan, A. Ortqvist, and H. R. Masure. 1997. Contribution of novel choline-binding proteins to adherence, colonization and immunogenicity of *Streptococcus pneumoniae. Mol. Microbiol.* **25:**819–829.

99. Rubins, J. B., D. Charboneau, C. Fasching, A. M. Berry, J. C. Paton, J. E. Alexander, P. W. Andrew, T. J. Mitchell, and E. N. Janoff. 1996. Distinct roles for pneumolysin's cytotoxic and complement activities in the pathogenesis of pneumococcal pneumonia. *Am. J. Respir. Crit. Care Med.* **153:**1339–1346.

100. Sampson, J. S., Z. Furlow, A. M. Whitney, D. Williams, R. Facklam, and G. M. Carlone. 1997. Limited diversity of *Streptococcus pneumoniae psaA* among pneumococcal vaccine serotypes. *Infect. Immun.* **65:**1967–1971.

101. Seo, J. Y., S. Y. Seong, B. Y. Ahn, I. C. Kwon, H. Chung, and S. Y. Jeong. 2002. Cross-protective immunity of mice induced by oral immunization with pneumococcal surface adhesin a encapsulated in microspheres. *Infect. Immun.* **70:**1143–1149.

102. Seong, S. Y., N. H. Cho, I. C. Kwon, and S. Y. Jeong. 1999. Protective immunity of microsphere-based mucosal vaccines against lethal intranasal challenge with *Streptococcus pneumoniae. Infect. Immun.* **67:**3587–3592.

103. Simberkoff, M. S., A. P. Cross, M. Al-Ibrahim, A. L. Baltch, P. J. Geiseler, J. Nadler, A. S. Richmond, R. P. Smith, G. Schiffman, D. S. Shepard, et al. 1986. Efficacy of pneumococcal vaccine in high-risk patients: results of a Veterans Administration Cooperative Study. *N. Engl. J. Med.* **315:**1318–1327.

104. Spratt, B. G., and B. M. Greenwood. 2000. Prevention of pneumococcal disease by vaccination: does serotype replacement matter? *Lancet* **356:**1210–1211.

105. Tai, S. S., C. Yu, and J. K. Lee. 2003. A solute binding protein of *Streptococcus pneumoniae* iron transport. *FEMS Microbiol. Lett.* **220:**303–308.

106. Talkington, D. F., B. G. Brown, J. A, Tharpe, A. Koenig, and H. Russell. 1996.

Protection of mice against fatal pneumococcal challenge by immunization with pneumococcal surface adhesin A (PsaA). *Microb. Pathog.* **21:**17–22.

107. **Talkington, D. F., D. L. Crimmins, D. C. Voellinger, J. Yother, and D. E. Briles.** 1991. A 43-kilodalton pneumococcal surface protein, PspA: isolation, protective abilities, and structural analysis of the amino-terminal sequence. *Infect. Immun.* **59:**1285–1289.

108. **Tseng, H. J., A. G. McEwan, J. C. Paton, and M. P. Jennings.** 2002. Virulence of *Streptococcus pneumoniae:* PsaA mutants are hypersensitive to oxidative stress. *Infect. Immun.* **70:**1635–1639.

109. **Tu, A.-H. T., R. L. Fulgham, M. A. McCory, D. E. Briles, and A. J. Szalai.** 1999. Pneumococcal surface protein A (PspA) inhibits complement activation by *Streptococcus pneumoniae. Infect. Immun.* **67:**4720–4724.

110. **Vakevainen, M., C. Eklund, J. Eskola, and H. Kayhty.** 2001. Cross-reactivity of antibodies to type 6B and 6A polysaccharides of *Streptococcus pneumoniae* evoked by pneumococcal conjugate vaccine in infants. *J. Infect. Dis.* **184:**789–793.

111. **Vela Coral, M. C., N. Fonseca, E. Castaneda, J. L. Di Fabio, S. K. Hollingshead, and D. E. Briles.** 2001. Families of pneumococcal surface protein A (PspA) of *Streptococcus pneumoniae* invasive isolates recovered from Colombian children. *Emerg. Infect. Dis.* **7:**832–836.

112. **Wizemann, T. M., J. H. Heinrichs, J. E. Adamou, A. L. Erwin, C. Kunsch, G. H. Choi, S. C. Barash, C. A. Rosen, H. R. Masure, E. Tuomanen, A. Gayle, Y. A. Brewah, W. Walsh, P. Barren, R. Lathigra, M. Hanson, S. Langermann, S. Johnson, and S. Koenig.** 2001. Use of a whole genome approach to identify vaccine molecules affording protection against *Streptococcus pneumoniae* infection. *Infect. Immun.* **69:**1593–1598.

113. **Wortham, C., L. Grinberg, D. C. Kaslow, D. E. Briles, L. S. McDaniel, A. Lees, M. Flora, C. M. Snapper, and J. J. Mond.** 1998. Enhanced protective antibody responses to PspA after intranasal or subcutaneous injections of PspA genetically fused to granulocyte-macrophage colony-stimulating factor or interleukin-2. *Infect. Immun.* **66:**1513–1520.

114. **Wu, H. Y., M. H. Nahm, Y. Guo, M. W. Russell, and D. E. Briles.** 1997. Intranasal immunization of mice with PspA (pneumococcal surface protein A) can prevent intranasal carriage and infection with *Streptococcus pneumoniae. J. Infect. Dis.* **175:**839–846.

115. **Wuorimaa, T., and H. Kayhty.** 2002. Current state of pneumococcal vaccines. *Scand. J. Immunol.* **56:**111–129.

116. **Yamamoto, M., L. S. McDaniel, K. Kawabata, D. E. Briles, R. J. Jackson, J. R. McGhee, and H. Kiyono.** 1997. Oral immunization with PspA elicits protective humoral immunity against *Streptococcus pneumoniae* infection. *Infect. Immun.* **65:**640–644.

117. **Yother, J., and D. E. Briles.** 1992. Structural properties and evolutionary relationships of PspA, a surface protein of *Streptococcus pneumoniae*, as revealed by sequence analysis. *J. Bacteriol.* **174:**601–609.

118. **Yother, J., and J. M. White.** 1994. Novel surface attachment mechanism for the *Streptococcus pneumoniae* protein PspA. *J. Bacteriol.* **176:**2976–2985.

119. **Zhang, J. R., K. E. Mostov, M. E. Lamm, M. Nanno, S. Shimida, M. Ohwaki, and E. Tuomanen.** 2000. The polymeric immunoglobulin receptor translocates pneumococci across human nasopharyngeal epithelial cells. *Cell* **102:**827–837.

VACCINE-INDUCED IMMUNITY TO PNEUMOCOCCAL INFECTION

P. Helena Mäkelä and Helena Käyhty

25

INTRODUCTION

Pneumococcal vaccines have a history of nearly 100 years. They have had their ups and downs, but research on them is still strong—in fact more active than ever—and an ideal vaccine is still not in hand. It therefore seems pertinent to evaluate the present state of the art and try to identify the problems and prospects in sight. The overriding questions deal with vaccine efficacy in an environment used to demanding very high efficacy of vaccines intended for wide use in the public health context. We therefore first focus on the evidence of efficacy of the pneumococcal vaccines and problems associated with them, and then we go on to examine the antibody response as a determinant of protection and as a possible surrogate measure of the protection elicited by them. Neither aspect is trivial since with the pneumococcal vaccines we have to deal with differences according to the clinical manifestation of the infection, the age and possible risk factors of the vaccinated individual, and both the quantity and the quality of the immune response.

P. Helena Mäkelä, Department of Vaccines, National Public Health Institute, 00300 Helsinki, Finland. Helena Käyhty, Vaccine Immunology Laboratory, National Public Health Institute, 00300 Helsinki, Finland.

A consensus of the importance of the bacterial capsule for protective immunity has prevailed from very early on, based on the serotype specificity of the first whole-cell vaccines and soon after that on the protective immunity observed after vaccination with the purified polysaccharide (4, 25, 56, 74). The importance of the serotypes in pneumococcal immunology is further supported by the now historic therapeutic use of pneumococcal antisera as well as by experience in experimental animal studies.

There are at least 90 different pneumococcal polysaccharide antigens, but luckily only one-fifth of them cause the majority of pneumococcal diseases (34, 35). Polysaccharide antigens have their unique immunological properties as described in detail in chapter 23. They are T-cell-independent antigens and thus do not evoke immunological memory, one of the cornerstones of long-term protection evoked by vaccination. Further, they are not immunogenic in young children. The immunological properties of the pneumococcal polysaccharide vaccines are changed dramatically when the polysaccharides are conjugated to protein carriers (5, 23, 94). This finding, made already in the 1930s, created a totally new possibility to protect children and high-risk groups that do not respond to the polysaccharide vaccine, culminating in the licensure of the conjugate vaccine in 2000.

Thus, the vaccines that we have now are based on the capsular polysaccharide, either the polysaccharide as such or conjugated to a carrier protein (Table 1). In each case, the vaccines consist of a mixture of various numbers of individual serotypes, selected to cover as well as possible the majority of serotypes found in pneumococcal disease.

Nevertheless, the reliance on this one molecule—the capsular polysaccharide—in vaccine design is a double-edged sword: it is very attractive because of proven efficacy and a wealth of clinical experience, but it has the drawbacks of strict serotype specificity and the special immunological properties of polysaccharide antigens. Then, on the other hand, the alternative vaccines based on pneumococcal protein antigens are at a very early stage of exploration too early to allow predictions of their potential as vaccines especially as long as there are no data on their protective ability in humans. The alternative approaches are discussed in chapter 23.

PROTECTIVE IMMUNITY INDUCED BY PNEUMOCOCCAL POLYSACCHARIDE VACCINES

Invasive Infections

The evidence of serotype-specific protection against bacteremia in immunocompetent adults, starting approximately 2 weeks after a single dose of the polysaccharide vaccine, is unequivocal based on both clinical trials before vaccine licensure and the subsequent experience with the wide use of the polysaccharide vaccine.

In the placebo-controlled blinded clinical studies in the 1940s among military recruits, using 3- or 4-valent vaccines, and in the 1970s among South African mineworkers or Papua New Guinea highlanders, using 6- to 14-valent ones, the focus was on pneumonia, but in a population in which a high proportion of

TABLE 1 Pneumococcal vaccines at the end of 2003

Polysaccharides	Carrier	Efficacy trials	Market situation
Polysaccharide vaccines			
3-, 4-, 6-, 12-valent		Extensive trials in young adults in different epidemic situations	A 6-valent vaccine marketed in the 1950s, then withdrawn
14-valent		Several trials	Licensed in 1977, later replaced by the 23-valent vaccine
23-valent		No prelicensure trials	Licensed in 1983
Conjugate vaccines			
7-valent	CRM$_{197}$[a]	Three trials	Licensed in 2000 in the United States and in 2001 in the European Union
7-valent	OMPC[b]	One trial	
9-valent	CRM$_{197}$	One trial completed, another ongoing	
11-valent	D[c] and T[d]	One trial ongoing	
11-valent	Hi protein D[e]	One trial ongoing	

[a] Nontoxic variant of diphtheria toxin.
[b] OMPC, outer membrane protein complex from *Neisseria meningitidis*.
[c] D, diphtheria toxoid.
[d] T, tetanus protein.
[e] Lipoprotein D from *H. influenzae* (Hi).

the pneumonia patients were bacteremic (4, 46, 56, 86, 101). Considering the definitive, blood culture-positive cases only, the aggregate serotype-specific vaccine efficacy (VE) was 85%, based on a total of 82 cases among the controls and 12 among the vaccinees (4, 46, 86).

The licensing of the 14-valent (and later, the 23-valent) vaccine in 1977 was soon followed by general recommendations in the United States for broad use of the vaccine to all those considered at increased risk of pneumococcal pneumonia or its complications; in the present recommendations this includes all those 65 years of age or older, plus a number of younger ones in medically defined risk groups (15). It is to be noted that these vaccine target populations are very different from those with which the prelicensure clinical studies had been carried out—typically young men in military or labor force camps with a very high incidence and epidemic spread of pneumococcal infection. To make this concrete, consider the incidence of pneumococcal bacteremia in 16/1,000 in South African miners, compared to that of 0.5/1,000 in elderly U.S. citizens (3, 4, 28). Against this difference it is in fact remarkable that postmarketing epidemiological studies in the United States, carried out by either case control or indirect serotype-specific cohort methodologies, show consistently an overall 60 to 75% efficacy of the polysaccharide vaccine for bacteremia and meningitis in immunocompetent adult target groups (10, 12, 14, 96).

Together these data stress the robustness of the protection from invasive infections, characterized by pneumococci cultured from either the blood or the cerebrospinal fluid, over a broad range of ages and underlying diseases. The protection also seems to apply over a wide range of serotypes, although many in the 23-valent vaccines are too rare for a significant effect to be demonstrated. On the other hand, the most common serotypes in bacteremia in the South African studies showing protection were 1, 2, 12, and 25, serotypes that are now rare in the United States.

The epidemiological studies have not been able to demonstrate effectiveness of the polysaccharide vaccine in various immunocompromising conditions, including sickle cell disease (14). Nowadays human immunodeficiency virus (HIV) infection is the most important of the immunocompromised states and carries a very high risk of pneumococcal infection (45, 67). One case control study in HIV-infected adults has shown 76% protection from invasive disease in white Americans, while the effect was not significant in African-Americans; however, the HIV infection in the latter was not equally controlled (11). There are no data as of yet on the efficacy of the vaccine in HIV-positive adults receiving modern antiretroviral therapy.

An unexpected effect of pneumococcal vaccine was seen in a randomized double-blind trial in HIV-infected adults in Uganda (32). Not only was there no reduction of invasive disease due to pneumococci (vaccine type or not) but also there was a significant increase in pneumonia of any cause. The apparent untoward effect was most clear-cut in the first 6 months after vaccination—also an unexplained finding. It should be noted that the rates of both pneumonia and invasive pneumococcal disease were very high during the trial: in the 2 years of its duration there were 76 episodes of pneumonia and 34 episodes of invasive disease among the 1,392 participants.

Immunogenicity studies have given data mostly concordant with those of the epidemiological and vaccine efficacy studies. They have shown that high-risk groups, e.g., patients with HIV, sickle cell disease, or bone marrow transplantation or the elderly, have in general lower antibody concentrations and lower and shorter lasting antibody responses after vaccination than normal healthy adults (29, 31, 54, 77, 88, 90, 91, 93, 108). Even if antibodies are found by standard enzyme immunoassay (EIA), they can have low functional activity suggesting inefficient protection (78, 88). Recent data have shown that the choice of the method for determination of the antibody concentrations seems to be critical, especially when adult sera are used (17, 102). The standard EIA is not specific enough, but an extra neutralization step with 22F polysaccharide in addition to the standard cell wall polysaccharide neutralization has improved the determination of specific responses to the polysaccharides. However, there are still very few studies using this methodology in adult high-risk groups.

Pneumonia in Adults

Pneumococcal pneumonia was the major target when developing the polysaccharide vaccine and also the primary endpoint in the clinical trials in the 1940s and 1970s referred to above. They all attested to its efficacy for this indication, when the endpoint was defined as pneumonia with the serotype-specific pneumococci cultured from blood, lung aspirate, sputum, or, in one case, a nasopharyngeal swab (4, 46, 56, 86, 101). The trial in Papua New Guinea furthermore showed a statistically significant 42% reduction of mortality for pneumonia (determined by verbal autopsy) and 22% reduction of overall mortality (86).

Against this background it has been a disappointing surprise that protection from pneumococcal pneumonia has been an elusive target in both the United States and northern Europe. Already in the 1970s two placebo-controlled trials were initiated in the United States, aiming at demonstrating the efficacy of the 14-valent vaccine for pneumococcal pneumonia in elderly populations, one involving 1,300 inpatients of a psychiatric hospital and the other involving 13,600 members of a health care plan (3). Pneumococcal pneumonia was believed to be a major cause of illness and even death in these populations. However, a vaccine effect could not be seen in either of the trials with respect to overall or serotype-specific pneumococcal pneumonia or pneumonia-associated deaths. Pneumococcal bacteremia was infrequent, with altogether 4 cases among the controls (corresponding to an incidence of 0.5/1,000) and none among the vaccinees, thus

consistent with protection against invasive disease but not statistically significant.

Over the following 20 or so years several controlled trials have been carried out among elderly populations, either unselected or selected because of increased risk for pneumonia. In four of these there was no evidence of a protective effect of the vaccine on "pneumococcal pneumonia" identified by various diagnostic criteria (37, 39, 75, 99). In the other three an effect was observed for a selected high risk population for the incidence of hospitalization or death due to pneumonia (33, 53, 65). In none of these studies did the polysaccharide vaccine show efficacy towards overall pneumonia; whether or not it protects from pneumococcal pneumonia cannot be determined from the studies because of uncertainties in the etiological diagnosis. Why did one not have similar problems of diagnosis in the early clinical trials that had so convincingly shown a protective effect? We do not know for sure, but several features of them suggest an answer. Very likely the epidemic spread had selected for highly virulent clones of the bacteria—and indeed, the serotypes met in the early trials were different from the ones today. The disease was more fulminant during the outbreaks in the previously healthy young men than in the present elderly population with many underlying diseases. The more fulminant disease was associated with production of more copious purulent sputum in which the bacteria were trapped and easier to find in the laboratory. However, a less effective immune response to the polysaccharide vaccine in the elderly population (88, 91, 92) is also a possible explanation and is further discussed later in this chapter. Finally, it is possible that pneumococci are not an important cause of pneumonia in these populations (in which viruses and other, less pathogenic bacteria are sufficient to cause the clinical picture of pneumonia).

Whether or not the polysaccharide vaccine protects from pneumonia is important for acceptance of the vaccine in nationwide programs: if the vaccine only protects the elderly from rare invasive disease with an incidence of

0.5/1,000, its benefits will be much less than if it also protects from the more common pneumococcal pneumonia with an estimated incidence of 10/1,000—although the figure must be taken with caution because of the diagnostic difficulties. Uncertainty about this may be the real reason for the large gap existing between recommendations and actual use of the vaccine. In the United States less than 50% of those 65 or older use it, and in Europe it is used very little indeed (27, 112).

Pneumonia in Children

Pneumonia is also an important illness in young children and a major cause of death in developing countries, where 2 million deaths annually are estimated to be due to pneumonia, half of these to pneumococcal pneumonia (7). The possibility of the pneumococcal polysaccharide vaccine offering a means to prevent this huge disease burden was the impetus for a series of double-blind placebo-controlled trials in Papua New Guinea, starting in 1974 and continuing a decade later (85). Altogether 7,220 children from 6 to 59 months of age participated in them and were monitored by monthly household survey and verbal autopsy methods to determine mortality for all causes and specifically for pneumonia. The results indicated 59% vaccine efficacy for death because of acute lower respiratory tract infection (ALRI, most of it pneumonia) as the sole cause in the whole group, somewhat less in those vaccinated when younger than 2 years. More detailed follow-up in a subset of these children indicated a 37% reduction of morbidity for ALRI (84). Thus, the vaccine seemed as protective in the children as in adults in the same region (86), a result at odds with the growing realization of the very low immune responses in children to most of the vaccine polysaccharides. Unfortunately, there was no identification of pneumococci, let alone their serotypes, in the disease that might give us a clue about the unexpectedly high protection rate. In view of the importance of pneumonia for children in developing countries, it is regrettable that further studies to confirm or refute these findings have not been conducted in the face of a lack of funding and an expectation of an improved vaccine (70, 95).

Otitis Media in Children

Otitis media is a very common respiratory infection in which pneumococci have a large role. In recent studies in Finland almost every child has been shown to have had at least one episode, and many several, of otitis media before the age of 2 years, with pneumococci present in the purulent middle ear fluid in 30% of these cases (50). A vaccine that could prevent this painful condition would be highly welcomed. Encouraged by the demonstration of protective activity in children of another experimental polysaccharide vaccine, that for group A meningococcal infection, clinical trials were initiated in the early 1980s in both the United States and Finland to directly probe the potential of the then-new 14-valent vaccine in preventing otitis media (44, 58, 104). The setup of the trials varied to some extent: vaccine was given after an attack of otitis media to prevent recurrences, or early in infancy with the hope that preventing the first attack early in life would block the chain of developing an otitis-prone state. The results were disappointing in general; however, protection observed in selected subgroups suggested that parenteral vaccination could have an effect on the infection in spite of its mucosal localization. The protection was limited to serogroups that were most immunogenic and to children older than 2 years, speaking to a protective role of antibodies.

Nasopharyngeal Carriage of Pneumococci

Carriage of pneumococci by healthy people is an important step in their transmission. Thus, a vaccine that would reduce carriage might have secondary—herd immunity—effects on pneumococcal disease also among those not vaccinated. Does the polysaccharide vaccine reduce carriage? This question was addressed in two of the early trials, with contradictory findings. In the study with military recruits, nasopharyngeal swabs were analyzed extensively by direct culture, enrichment culture followed by type-specific microscopy (the capsular swelling reaction), and mouse passage—methods that together had given 60% carriage rates in a recent large survey (56). The carriage rates of the four vaccine types (1, 2, 5 and 7) were low, from 0.3 to 1.7% (aggregate, 3.26%) among the 1,800 nonimmunized men sampled and still lower (aggregate, 1.79%) among those vaccinated, indicating a 45% reduction of carriage. The authors argue by comparison to a nonvaccine type that while the vaccine reduced carriage among the vaccinated men, this indirectly led to reduced carriage among their nonvaccinated mates and herd immunity seen as reduced pneumonia rates among the nonvaccinated.

In the Papua New Guinea study, the overall carrier rate was also high, 68%; 7% of the placebo recipients and 10% of those vaccinated carried vaccine-type pneumococci, indicating no reduction of carriage (86). There are many differences between the studies that may explain the different outcomes. The military study was carried out in a closed situation in which epidemic spread of respiratory infection, including pneumococcal infection, was the rule, while the Papua New Guinea study participants lived in their homes. In the first study the vaccine consisted of only four serotypes, all rare in carriage (5% of all carriage isolates) while important in disease, while in the second study the 14-valent vaccine covered 21% of the pneumococci isolated from the noses of the placebo recipients. Further studies on carriage among children older than 2 years have also given equivocal results, ranging from no effect to a weak protective effect (24, 36). The consensus seems to be that the pneumococcal polysaccharide vaccine does not protect from carriage.

PROTECTIVE IMMUNITY INDUCED BY PNEUMOCOCCAL CONJUGATE VACCINES

The experience with another polysaccharide vaccine, the one against *Haemophilus influenzae* type b (Hib), showed the way to solve the problem of lacking immune responses in infancy: covalent conjugation to a protein carrier antigen. Experience with several different Hib conjugates showed a number of other properties usually associated with a T-cell-dependent response: induction of immunological memory allowing for higher antibody concentrations sufficient to affect mucosal carriage, maturation of the antibody response, and lifelong immunity (57, 59). Eventually, conjugation was applied to pneumococcal polysaccharides, and the conjugates were shown to have gained similar T-cell-dependent properties (21, 22, 48, 61, 68, 69, 81–83, 98; A. Nurkka, J. Joensu, M. Malm, A. Holm, J. Poolman, C. Laferriere, P. Peeters, H. Käyhty, and T. Kilpi, 3rd Int. Symp. Pneumococci Pneumococcal Dis., 2002).

Efficacy against Invasive Disease

The major study to test the protective efficacy of the 7-valent conjugate vaccine was carried out in California, within the framework of the large health insurance plan of Kaiser Permanente (8). The vaccine (or placebo) was given as four injections at 2, 4, 6, and 12 to 15 months of age to nearly 40,000 participants. The cases were detected through the records of Kaiser Permanente, and the primary endpoint was invasive pneumococcal disease proven by culturing a vaccine-type pneumococcus from a normally sterile site (blood or cerebrospinal fluid). At the conclusion of the trial there were 39 endpoint cases in the placebo group and 1 in the vaccine group, corresponding to the excellent efficacy of 97.4% against invasive disease. This result was met with enthusiasm and led to both a rapid licensure of the vaccine and recommendations for its inclusion in the infant immunization program in the United States (1). The vaccine was well received by the public, and invasive pneumococcal disease in the primary target age group, those younger than 2 years, started a dramatic decline (111).

Would the vaccine be equally effective in other circumstances? Two recent studies, both with the same endpoint of serotype-specific invasive disease, suggest that this is true. A trial was carried out with the same 7-valent vaccine, enrolling 8,292 children younger than 2 years of the Navajo and White Mountain Apache tribes in the United States (73). In the intention-to-treat analysis there were 11 cases among the controls and 2 among the recipients of the 7-valent vaccine (vaccine efficacy, 83%); in the per-protocol analysis the corresponding numbers were 8 and 2.

In a trial in Soweto, South Africa, the vaccine was a 9-valent one, including the serotypes of the 7-valent vaccine plus types 1 and 5, considered especially important causes of invasive disease in developing countries (52). The vaccine was given with the South African infant immunization schedule at 6, 10, and 14 weeks of age to a total of 19,922 infants. A complication to the interpretation of the results of the study is the high prevalence of HIV infection among the cases, consistent with the high risk of pneumococcal disease associated with HIV infection. The intention-to-treat analysis showed 83% efficacy among the children without HIV infection and 65% efficacy among HIV-infected children. The latter is especially good news to African settings where HIV infection and the associated mortality for pneumococcal infection are very high.

Efficacy against Pneumonia

The difficulties of obtaining a bacteriological confirmation of pneumococcal pneumonia and identification of its serotype, as already referred to, apply to an even greater extent to pneumonia of children because of their generally high carriage rates. Thus, trials have to rely on an endpoint of probable pneumococcal pneumonia defined by a combination of clinical and radiological signs. This affects both the sensitivity and the specificity of the diagnosis and thus the power of the study to identify a vaccine effect. In the past several years the World Health Organization (WHO), together with the investigators in the different trials, has made an effort to standardize the endpoint definition in order to improve the evaluation of trial results and facilitate comparisons between

trials (79). The working group has developed a hopefully commonly acceptable and implementable algorithm for the diagnosis of "endpoint pneumonia" that would have a high likelihood of representing true pneumococcal pneumonia. Serotype specificity is still beyond the ambition of the group, and indeed the concept of the endpoint pneumonia would badly need validation through a vaccine probe analysis, i.e., showing high efficacy of a vaccine toward this endpoint (63, 64).

In the Kaiser Permanente trial pneumonia was a secondary endpoint (9). Chest X rays were taken if the physician treating the patient considered it clinically relevant. The readout of the films was done carefully but not fully to the algorithm of the WHO group. The result indicated 21% protective efficacy toward the endpoint pneumonia used. In the South African trial, considering only those not HIV infected, the vaccine efficacy was 20% for the endpoint pneumonia defined according to the WHO algorithm. The confidence intervals were, however, wide, from 2 to 35%. No protection (13% [confidence interval, −7 to 29%]) was seen for pneumonia in the HIV-positive patients (52). The trial among the Navajo and White Mountain Apache tribes had also planned to assess vaccine efficacy for pneumonia, but unexpected problems with the study design were encountered. This leaves us with only two more, ongoing trials in developing-country populations, one in Africa (The Gambia) with the 9-valent vaccine and the other in Asia (the Philippines) with another, experimental 11-valent vaccine (40, 82). The results of these are expected in 2005.

Because of the unspecificity of the endpoint, none of these studies will give exact information on protection against pneumonia caused by pneumococci in general or vaccine-type pneumococci. It is likely that the endpoint pneumonia, however defined, will include other bacterial and viral pneumonias, leading to an underestimation of the efficacy of the vaccine. In very simple terms, a 20% protection could indicate a true vaccine efficacy of 100% if only one-fifth of the endpoint cases were due to pneumococcal serotypes covered by the vaccine; conversely, it could indicate a true vaccine efficacy of 20% if all endpoint cases were true ones. What, then, is the value of these studies for an assessment of the vaccine for the public health goal of preventing pneumonia or mortality associated with it? The answer is "limited, unfortunately." First of all, in the real-life situation—not under the close surveillance of a trial—most pneumonia deaths occur in the home (40). The trials are likely to give an underestimation of the burden of the severe pneumococcal disease that nevertheless would be prevented by the vaccine. The question looking for an answer still is what fraction of child deaths the vaccine could prevent. To answer it properly we would need a large controlled study in the target population with minimal interference with the health services of the area—very difficult to achieve in a trial situation. Alternatively, the information could be obtained by careful surveillance after stepwise introduction of the vaccine, ethically well justified because of its proven efficacy against invasive disease. In these direct approaches it would be furthermore important to have death for all causes as an additional endpoint because of the possible misclassification of death due to pneumonia when identified by verbal autopsy (40).

Efficacy against Otitis Media

When the pneumococcal conjugate vaccine had been shown to be able to immunize young infants, its effect on otitis media was among the first targets for study. The Kaiser Permanente trial included otitis media as an endpoint but with limitations similar to those described for pneumonia endpoints: the causative agent of the disease was not identified because obtaining a sample from the middle ear fluid was not part of the routine diagnosis. Therefore, various clinical endpoints were considered without precise knowledge of the role in them of the pneumococcus (or the serotypes covered by the vaccine). When the endpoint was "otitis media episodes," the vaccine appeared to have an efficacy of only 7.8%; when "ventila-

tory tube placement" (usually done because of frequently repeating otitis) was taken as an endpoint, the efficacy was 24% (30). More definitive information of the efficacy of the vaccine was obtained in two trials carried out in Finland, in which the pneumococci were cultured from the middle ear fluid, yielding serotype-specific data (26, 49). Two conjugate vaccines, PncCRM and PncOMPC, were used, both with the same seven serotypes but with different carrier proteins; the results of the two were very similar. The serotype-specific efficacy against otitis media episodes was about 60%, but at the same time episodes caused by other, nonvaccine types of pneumococci increased, as did episodes due to other causes so that the total efficacies against otitis media episodes in the two trials were 6 and 0%—not significantly different from the 7.8% in the Kaiser Permanente trial. A follow-up of one arm of the Finnish trials showed a 44% efficacy in preventing ventilatory tube placement at the age of 2 to 4 years, also consistent with the Kaiser Permanente findings and now shown to last for several years (A. Palmu, T. Kaijalainen, E. Herva, R. Syrjänen, P. H. Mäkelä, and T. Kilpi, 21st Annu. Meet. Eur. Soc. Paediatr. Infect. Dis., 2003).

Important new knowledge had been gained. First, it was learned that parenteral immunization with pneumococcal vaccine can prevent otitis media in a vaccine-specific way, indicating protection on the mucosal site in the middle ear by antibodies and confirming the tentative conclusions from the trials with the polysaccharide vaccine. The protective antibodies derive most probably from the serum, since the local immunoglobulin A (IgA) production has been found to be rare and short-lived (16; A. Nurkka, M. Lahdenkari, T. Kilpi, H. Käyhty, and the FinOM Study Group, 41st Intersci. Conf. Antimicrob. Agents Chemother., 2001; A. Nurkka, M. Lahdenkari, A. Palmu, T. Kilpi, H. Käyhty, and the FinOM Study Group, 3rd Int. Symp. Pneumococci Pneumococcal Dis., 2002). Second, it was learned that the vaccine efficacy toward otitis was only about 60% with a vaccine that had over 95% efficacy toward invasive disease, suggesting that a higher concentration of antibodies is needed for protection at the mucosal site than in the blood. This is consistent with observations with the Hib vaccines: a lower serum antibody concentration is needed to prevent invasive disease than to prevent nasopharyngeal carriage as seen in both vaccinated children and experimental studies in the infant rat (6, 47). Third, the pneumococcal types not covered by the vaccine as well as other microbes may take the place of vaccine-type pneumococci in otitis media, although there are indications that the pneumococcus may lead to a more severe clinical picture (87; A. Palmu, J. Verho, P. H. Mäkelä, P. Karma, and T. Kilpi, 8th Int. Symp. Recent Adv. Otitis Media, 2003).

Effect on Carriage

Early studies on the immunogenicity of the pneumococcal conjugate vaccines showed an effect on pneumococcal carriage unlike anything seen with the polysaccharide vaccine (20, 61, 71). The effect was specific to the serotypes in the vaccine studied and was often accompanied by an increase of other, nonvaccine types. Vaccine efficacy in preventing carriage of vaccine types calculated in the Finnish otitis trials (T. Kilpi, R. Syrjänen, A. Palmu, E. Herva, J. Eskola, P. H. Mäkelä, and the FinOM Study Group, 19th Annu. Meet. Eur. Soc. Paediatr. Infect. Dis., 2001) was approximately 40% less than the efficacy against invasive disease and even less than the efficacy against otitis media (26, 49), but it was highly reproducible and lasted at least up to the age of 4 to 5 years (Palmu et al., 21st AMESPID).

The reduction of the serotypes in the vaccine also means a reduction of the chance of transmission of the infection and is expected to result in a true herd immunity and overall reduction of the diseases in the population, even among those not vaccinated (72). The experience in the United States after the introduction of the 7-valent conjugate in the infant immunization schedule gives support to the theoretical chain of events: the rate of invasive pneu-

mococcal disease decreased considerably not only in the vaccinated age group but also among adults, and the bulk of the decrease was in the seven serotypes covered by the vaccine (111).

The replacement and change of the pneumococcal populations colonizing humans in vaccinated regions is a fact that must be accepted. Does the arrival of the new serotype suggest that there is a niche assigned for pneumococci, a site where they have a competitive advantage against other bacterial species? Or does the vaccine-type pneumococcal colony just decrease in size, leaving space for a second type to grow into the majority of the pneumococcal colony? And what would the consequences of these alternatives be (55)? We do not have answers to these questions at the moment. Then there is the more practical question of whether the nonvaccine types, when they become abundant in the human micro flora, also cause the same diseases for which the vaccine serotypes have been responsible. So far this has not been seen in invasive disease, suggesting that the nonvaccine types are in general less pathogenic. However, they do have some pathogenic potential as attested by their increase in otitis media of vaccinated children (26, 49, 55). Furthermore, knowing the capability of pneumococci to adapt to new situations through genetic exchange and mutation, it seems foolish to believe that a nonvaccine type might not gain invasive properties. More likely (because there is already evidence of such capsule switching) a virulent vaccine-type pneumococcus would acquire a new capsular type and thus escape the protection afforded by the capsule-specific antibodies (13).

MECHANISMS OF PROTECTION

From the times of early studies with polysaccharide vaccines on, it has been clear that antibodies to capsular antigens are important for protection. In blood, together with the complement cascades they augment phagocytosis by opsonizing the bacteria. Pneumococci have phase variation; the transparent colonies are isolated more often from a mucosal site, while invasive isolates grow often in opaque colonies (19, 109). The invasive isolates have armed themselves so that they survive better in blood; they have thicker capsules and the expression of some proteins (like PspA or CbpA) important for invasion or mucosal adhesion has changed (51, 76, 89). The larger amount of capsule renders the pneumococci resistant to phagocytosis and the functional action of antibodies to subcapsular antigens. It has also been suggested that some capsules can interfere with the complement activation needed for optimal opsonophagocytosis (38). Even small amounts of preformed antibody can be sufficient to prevent the growth of pneumococci in blood; the number of invasive bacteria in blood is low to begin with, and if those bacteria can be eliminated, the invasion can be stopped at an early phase. However, in most cases the pneumococcus is colonizing the nasopharynx before invasion, and this event can cause an antibody response that prevents the actual invasion. Vaccination with a conjugate vaccine induces B-cell memory, and thus the antibody production upon contact is vigorous and protection against invasive disease can be achieved even if the antibody concentration at the time of the infection is low. Conjugate vaccines seem to induce high-avidity antibodies (2, 113). The high avidity is important especially when the antibody concentration has decreased to a low level.

In pneumonia, the inflammation process happens in the epithelia of the alveoli. The bacteria bind to the epithelia via their adhesive surface molecules, such as choline, CbpA, and PspA (18, 89). The transparent pneumococci are better able to bind to the N-acetylglucosamine and platelet-activating factor receptors on cytokine-activated lung epithelial cells than are opaque pneumococci (19). The pneumococci inhabiting mucosal surfaces seem to have a thin capsule, which is optimal for adhesion (51, 103). The bacteria may also invade and start an invasive disease via a tissue injury caused by, e.g., pneumolysin and cell wall polysaccharide. The key questions for the protection induced by vaccination are how to induce antibodies at the local sites, and how they would function there. It can be assumed that serum antibodies, mainly IgG, transude to the mucosa of the alveoli and the nasopharynx. On the other hand, vaccination with polysaccharide or conjugate vaccines seems to induce, though in small amounts, secretory IgA antibodies on the mucosal surfaces (16, 66, 69). The protection could rely on both of these. IgG but also IgA can function as an opsonin for phagocytes (106). Further, the binding of antibodies to the capsule can function as steric hindrance for adhesion, and in the case of Hib antibodies have even been suggested to prevent the growth of bacteria by blocking nutrient uptake (105). It has also been suggested that the Fab fragments of IgA1 antibodies, cleaved by IgA1 protease, might even enhance adherence by neutralizing the inhibitory effect of the negatively charged capsule on the adhesive interaction with the host cell (110).

The protection against otitis media can be speculated to happen very much via the same mechanism as described above. The bacteria attach to the epithelia and the inflammation process brings in phagocytosing cells. Because the fluid accumulating into the middle ear cavity is believed to derive mostly from serum (62), IgG and IgM antibodies can be found from the middle ear fluid in addition to locally produced secretory IgA (43, 100, 107). Secretory IgA has been shown to be beneficial for the resolution of the otitis media (43, 100), but the long-term protection against pneumococcal otitis media afforded by vaccination is most probably due to existing serum antibody.

Carriage is the least invasive infection and occurs at mucosal surfaces, which have small amounts of phagocytosing cells or complement components. Despite that, vaccination with pneumococcal conjugate vaccines seems to prevent carriage acquisition effectively, but it has less of an effect on existing carriage. The effect can be thought to be mediated by antibodies via steric hindrance and a change in the electrostatic forces. In the case of Hib, the prevention of carriage has been most notable when using vaccination schedules that induce high concentrations of antibody (6). In animal colonization models for Hib or pneumococci, passive immunization both intraperitoneally (IgG) and via the nose (IgG and IgA) induced protection from carriage acquisition, suggesting that both serum IgG and mucosal IgA play a role (47, 60).

The clinical trials and animal studies suggest that many fewer antibodies are needed for protection against invasive infection than against pneumonia or other less severe mucosal infections. This implies also that the protection against invasive diseases will probably last longer than against local infections. However, in the Finnish studies the carriage of vaccine serotypes even 3 to 4 years after vaccination was less among the vaccinated than control children (Palmu et al., 21st AMESPID). On the other hand, continuing carriage after vaccination can cause natural boosting of immunity and thus in fact help in inducing long-lasting protection against invasive infection. The persistence of immunity after pneumococcal vaccination should in fact be lifelong, since pneumococcal diseases occur through adulthood and especially among the elderly. It is too early to say how long the immunity lasts; the antibody measurements 3 to 4 years after vaccination speak for persistent immunity at least in early childhood. Further, there can be serotype-specific differences so that immunity to those serotypes that are carried most often even after vaccination seems to persist best (H. Åhman, A. Palmu, T. Kilpi, S. Grönholm, T. Wuorimaa, H. Käyhty, and FinOM Study Group, 3rd Int. Symp. Pneumococci Pneumoccal Dis., 2002). The kinetic studies (N. Ekström et al., submitted for publication) in connection with the Finnish efficacy trial suggest that there might be differences in the persistence of antibodies after vaccination with vaccines having different carrier proteins.

Correlates or Surrogates of Protection

After the licensure of the pneumococcal conjugate vaccine, the need to find correlates or surrogates of protection has been high. Placebo-controlled efficacy trials with novel vaccines or with vaccines containing extra serotypes in addition to the seven serotypes in the licensed vaccine would as a rule not be feasible for ethical and practical reasons, and the licensure of new vaccines would need to rely on serologic criteria. The simplest way would be to establish a threshold concentration that would predict protection at the population level (41). To this end, WHO consulted in June 2003 a group of experts on pneumococcal epidemiology and vaccine evaluation and representatives of regulatory agencies. A pooled analysis of efficacy and immunogenicity data from the three efficacy trials with pneumococcal conjugate vaccine with an invasive disease endpoint was used as a basis of the recommendations for the criteria for use as a surrogate to establish noninferiority to a licensed vaccine in head-to-head comparisons. It was pointed out that these criteria should not be used to evaluate vaccines against clinical endpoints other than invasive disease and that this threshold does not necessarily predict protection in an individual subject. The primary endpoint was recommended to be a threshold IgG antibody concentration of 0.35 μg/ml measured 4 weeks after the three-dose vaccine series. Since the licensure was based on aggregate efficacy, the recommendation included an option that protective concentrations are similar for each serotype. It was recommended that while noninferiority for each of the serotypes in the registered vaccine was desirable, it should not be an absolute requirement. Additional supportive criteria were suggested to be the demonstration of functional activity of the antibodies after the third dose and the evidence of the development of immunological memory shown either by giving a booster dose of pneumococcal polysaccharide vaccine to trigger the memory B cells or by showing the avidity maturation of antibodies.

While there seem to be no clear serotype specific differences in the efficacy against invasive disease, the efficacy against acute otitis media varied clearly between serotypes. For example, the efficacy against acute otitis media caused by serotype 6B was high (approximately 80%), while there was no significant (approximately 30%, with negative lower 95% confidence intervals) protection against serotype 19F (26, 49). The comparison of the serotype-specific mean antibody concentrations either after the third or fourth dose did not correlate with the serotype-specific efficacy. Further, an association model, using the immunogenicity and efficacy data from the Finnish efficacy trials, developed for predicting vaccine efficacy against otitis media at different antibody concentrations showed that higher mean anti-19F concentrations than, e.g., anti-6B or -23F concentrations are needed for protection (42).

At this moment it is too early to determine the details of the optimal immune response for prevention of mucosal infections, including carriage acquisition. The mucosal secretory IgA response to the present pneumococcal conjugate vaccines seems to be only modest, and thus the long-lasting efficacy after systemic vaccination against mucosal infections would rely on persistent systemic immunity.

WHERE TO NOW?

The conjugate vaccines seem to have filled the main gap remaining with the polysaccharide vaccines, i.e., protection of infants and young children from invasive pneumococcal infec-

tion (Table 2). However, we are very far from being able to recommend general worldwide use of the vaccine for this purpose because of economic problems: the licensed vaccine is too expensive for the majority of the world's children, and the technology of its production is demanding enough to preclude, for a considerable time at least, less expensive manufacture in a developing country. If the vaccine would also prevent a substantial amount of pneumonia in the developing countries, its use would become much more cost-effective but probably still not feasible.

The price is indeed proving a problem also for the introduction of the vaccine in industrialized countries outside the United States. Efficacy against otitis media is an important consideration in many of these, and in this context the data on long-term reduction of tube replacement are promising. On the other hand, only a fairly small proportion of the disease would be prevented.

Would the conjugate also solve the problems of the polysaccharide vaccine in the elderly? An improved antibody response with better-quality antibodies might make the difference needed for prevention of pneumonia; on the other hand, available data on the immunology of the conjugate (80, 97) in this age group are not encouraging.

TABLE 2 Summary compilation on the efficacy of the pneumococcal, polysaccharide, and conjugate vaccines in different age groups and clinical entities[a]

Pneumococcal infection	Efficacy of vaccine in:					
	Elderly		Young adults and children >2 years		Infants	
	PPV	PCV	PPV	PCV	PPV	PCV
Invasive disease	60–75%	NT	85%	NT	ND	80–95%, no replacement (yet)
Pneumonia	ND	NT	65%[b]	NT .	ND	20% of radiological endpoint pneumonia
Otitis media	NT	NT	NT	NT	ND	60%, vaccine serotype specific, 0–7.8% overall
Carriage	NT	NT	ND	NT	ND	40%, vaccine serotype specific, replacement common

[a] Immunocompromised individuals were not included. PPV, pneumococcal polysaccharide vaccine; PCV, pneumococcal conjugate vaccine; NT, not tested; ND, not demonstrated.
[b] Also, a 42 to 59% reduction of pneumonia mortality and a 22% reduction of overall mortality were seen in studies in Papua New Guinea.

What about serotype replacement after extensive use of the vaccine? Replacement certainly will occur among the pneumococcal populations in healthy carriers, but to what extent the new types will cause disease is still open. Emergence of one or a few new serotypes of highly virulent clones seems a possibility in the near future, demanding new formulations of the vaccine—but who should then be revaccinated? The present selection of serotypes in the conjugate vaccines has been based on the type distribution in industrialized countries but has lower coverage of serotypes in many other parts of the world; should new formulations be developed for such areas? The cost of such enterprises would be very high in view of the strict requirements of licensing authorities. The alternative approach with protein-based vaccines would sound very attractive, but so far we do not have any demonstration of a protective effect of such vaccines in humans and no assurance that corresponding problems due to continuing evolution of the very versatile pneumococci would not emerge.

REFERENCES

1. **American Academy of Pediatrics, Committee on Infectious Diseases.** 2000. Policy statement: recommendations for the prevention of pneumococcal infections, including the use of pneumococcal conjugate vaccine (Prevnar), pneumococcal polysaccharide vaccine, and antibiotic prophylaxis. *Pediatrics* **106:**362–326.

2. **Anttila, M., J. Eskola, H. Åhman, and H. Käyhty.** 1999. Differences in the avidity of antibodies evoked by four different pneumococcal conjugate vaccines in early childhood. *Vaccine* **17:**1970–1977.

3. **Austrian, R.** 1981. Some observations on the pneumococcus and on a current status of pneumococcal disease and its prevention. *Rev. Infect. Dis.* **3**(Suppl.)**:**1–17.

4. **Austrian, R., R. M. Douglas, G. Schiffman, A. M. Coetzee, H. J. Koornhof, S. D. Hayden-Smith, and R. D. W. Reid.** 1976. Prevention of pneumococcal pneumonia by vaccination. *Trans. Assoc. Am. Phys.* **89:**184–194.

5. **Avery, O. T., and W. F. Goebel.** 1931. Chemo-immunological studies on conjugated carbohydrate-proteins. V. The immunological specificity of an antigen prepared by combining the capsular polysaccharide of type III pneumococcus with foreign protein. *J. Exp. Med.* **54:**437–447.

6. **Barbour, M. L.** 1996. Conjugate vaccines and the carriage of Haemophilus influenzae type b. *Emerg. Infect. Dis.* **2:**176–182.

7. **Black, R. E., S. S. Morris, and J. Bryce.** 2003. Where and why are 10 million children dying every year? *Lancet* **361:**2226–2234.

8. **Black, S., H. Shinefield, B. Fireman, E. Lewis, P. Ray, J. R. Hansen, L. Elvin, K. M. Ensor, J. Hackell, G. Siber, F. Malinoski, D. Madore, I. Chang, R. Kohberger, W. Watson, R. Austrian, K. Edwards, and Northern California Kaiser Permanente Vaccine Study Center Group.** 2000. Efficacy, safety and immunogenicity of heptavalent pneumococcal conjugate vaccine in children. *Pediatr. Infect. Dis. J.* **19:**187–195.

9. **Black, S. B., H. R. Shinefield, S. Ling, J. Hansen, B. Fireman, D. Spring, J. Noyes, E. Lewis, P. Ray, J. Lee, and J. Hackell.** 2002. Effectiveness of heptavalent pneumococcal conjugate vaccine in children younger than five years of age for prevention of pneumonia. *Pediatr. Infect. Dis. J.* **21:**810–815.

10. **Bolan, G., C. V. Broome, R. R. Facklam, B. D. Plikaytis, D. W. Fraser, and W. F. Schlech III.** 1986. Pneumococcal vaccine efficacy in selected populations in the United States. *Ann. Intern. Med.* **104:**1–6.

11. **Breiman, R. F., D. W. Keller, M. A. Phelan, D. H. Sniadack, D. S. Stephens, D. Rimland, M. M. Farley, A. Schuchat, and A. L. Reingold.** 2000. Evaluation of effectiveness of the 23-valent pneumococcal capsular polysaccharide vaccine for HIV-infected patients. *Arch. Intern. Med.* **160:**2633–2638.

12. **Broome, C. V., R. R. Facklam, and D. W. Fraser.** 1980. Pneumococcal disease after pneumococcal vaccination: an alternative method to estimate the efficacy of pneumococcal vaccine. *N. Engl. J. Med.* **303:**549–552.

13. **Brueggemann, A. B., D. T. Griffiths, E. Meats, T. Peto, D. W. Crook, and B. G. Spratt.** 2003. Clonal relationships between invasive and carriage Streptococcus pneumoniae and serotype- and clone-specific differences in invasive disease potential. *J. Infect. Dis.* **187:**1424–1432.

14. **Butler, J. C., R. F. Breiman, J. F. Campbell, H. B. Lipman, C. V. Broome, and R. R. Facklam.** 1993. Pneumococcal polysaccharide vaccine efficacy. An evaluation of current recommendations. *JAMA* **270:**1826–1831.

15. **Centers for Disease Control and Prevention.** 1997. Prevention of pneumococcal disease: recommendations of the Advisory Committee on

Immunization Practices (ACIP). *Morb. Mortal. Wkly. Rep.* **46**:1–24.

16. **Choo, S., Q. Zhang, L. Seymour, S. Akhtar, and A. Finn.** 2000. Primary and booster salivary antibody responses to a 7-valent pneumococcal conjugate vaccine in infants. *J. Infect. Dis.* **182**:1260–1263.

17. **Concepcion, N. F., and C. E. Frasch.** 2001. Pneumococcal type 22f polysaccharide absorption improves the specificity of a pneumococcal-polysaccharide enzyme-linked immunosorbent assay. *Clin. Diagn. Lab. Immunol.* **8**:266–272.

18. **Cundell, D. R., N. P. Gerard, C. Gerard, I. Idänpään-Heikkila, and E. I. Tuomanen.** 1995. Streptococcus pneumoniae anchor to activated human cells by the receptor for platelet-activating factor. *Nature* **377**:435–438.

19. **Cundell, D. R., J. N. Weiser, J. Shen, A. Young, and E. I. Tuomanen.** 1995. Relationship between colonial morphology and adherence of *Streptococcus pneumoniae. Infect. Immun.* **63**:757–761.

20. **Dagan, R., R. Melamed, M. Muallem, L. Piglansky, D. Greenberg, O. Abramson, P. M. Mendelman, N. Bohidar, and P. Yagupsky.** 1996. Reduction of nasopharyngeal carriage of pneumococci during the second year of life by a heptavalent conjugate pneumococcal vaccine. *J. Infect. Dis.* **174**:1271–1278.

21. **Dagan, R., R. Melamed, O. Zamir, and O. Leroy.** 1997. Safety and immunogenicity of tetravalent pneumococcal vaccines containing 6B, 14, 19F and 23F polysaccharides conjugated to either tetanus toxoid or diphtheria toxoid in young infants and their boosterability by native polysaccharide antigens. *Pediatr. Infect. Dis. J.* **16**:1053–1059.

22. **Daum, R. S., D. Hogerman, M. B. Rennels, K. Bewley, F. Malinoski, E. Rothstein, K. Reisinger, S. Block, H. Keyserling, and M. Steinhoff.** 1997. Infant immunization with pneumococcal CRM197 vaccines: effect of saccharide size on immunogenicity and interactions with simultaneously administered vaccines. *J. Infect. Dis.* **176**:445–455.

23. **Dick, W. E., and M. Beurret.** 1989. *Glycoconjugates of Bacterial Carbohydrate Antigens,* vol. 10. S. Karger Ab, Basel, Switzerland.

24. **Douglas, R. M., D. Hansman, H. B. Miles, and J. C. Paton.** 1986. Pneumococcal carriage and type-specific antibody. Failure of a 14-valent vaccine to reduce carriage in healthy children. *Am. J. Dis. Child.* **140**:1183–1185.

25. **Ekwurzel, G. M., J. S. Simmons, L. I. Dublin, and L. D. Felton.** 1938. Studies on immunizing substances in pneumococci. VIII. Report on field tests to determine the prophylac-

tic value of *pneumococcus* antigen. *Public Health Rep.* **53**:1877–1893.

26. **Eskola, J., T. Kilpi, A. Palmu, J. Jokinen, J. Haapakoski, E. Herva, A. Takala, H. Käyhty, P. Karma, R. Kohberger, G. Siber, and P. H. Mäkelä.** 2001. Efficacy of a pneumococcal conjugate vaccine against acute otitis media. *N. Engl. J. Med.* **344**:403–409.

27. **Fedson, D. S.** 1997. Pneumococcal vaccination: four issues for Western Europe. *Biologicals* **25**:215–219.

28. **Fedson, D. S., J. A. Scott, and G. Scott.** 1999. The burden of pneumococcal disease among adults in developed and developing countries: what is and is not known. *Vaccine* **17**(Suppl. 1):S11–S18.

29. **Feikin, D. R., C. M. Elie, M. B. Goetz, J. L. Lennox, G. M. Carlone, S. Romero-Steiner, P. Holder, W. A. O'Brien, C. G. Whitney, J. Butler, and R. F. Breiman.** 2002. Randomized trial of the quantitative and functional antibody responses to a 7-valent pneumococcal conjugate vaccine among HIV-infected adults. *Vaccine* **20**:545–553.

30. **Fireman, B., S. B. Black, H. R. Shinefield, J. Lee, E. Lewis, and P. Ray.** 2003. Impact of the pneumococcal conjugate vaccine on otitis media. *Pediatr. Infect. Dis. J.* **22**:10–16.

31. **French, N., M. Moore, R. Haikala, H. Käyhty, and C. F. Gilks.** A case control study to investigate serological correlates of clinical failure of 23-valent pneumococcal vaccine in HIV-1-infected Ugandan adults. *J. Infect. Dis.,* in press.

32. **French, N., J. Nakiyingi, L. M. Carpenter, E. Lugada, C. Watera, K. Moi, M. Moore, D. Antvelink, D. Mulder, E. N. Janoff, J. Whitworth, and C. F. Gilks.** 2000. 23-valent pneumococcal polysaccharide vaccine in HIV-1-infected Ugandan adults: double-blind, randomised and placebo controlled trial. *Lancet* **355**:2106–2111.

33. **Gaillat, J., D. Zmirou, M. R. Mallaret, D. Rouhan, J. P. Bru, J. P. Stahl, P. Delormas, and M. Micoud.** 1985. Clinical trial of an antipneumococcal vaccine in elderly subjects living in institutions. *Rev. Epidemiol. Sante Publique* **33**:437–444. (In French.)

34. **Hausdorff, W. P., J. Bryant, C. Kloek, P. R. Paradiso, and G. R. Siber.** 2000. The contribution of specific pneumococcal serogroups to different disease manifestations: implications for conjugate vaccine formulation and use, part II. *Clin. Infect. Dis.* **30**:122–140.

35. **Hausdorff, W. P., J. Bryant, P. R. Paradiso, and G. R. Siber.** 2000. Which pneumococcal serogroups cause the most invasive disease: implications for conjugate vaccine formulation and use, part I. *Clin. Infect. Dis.* **30**:100–121.

36. Herva, E., J. Luotonen, M. Timonen, M. Sibakov, P. Karma, and P. H. Mäkelä. 1980. The effect of polyvalent pneumococcal polysaccharide vaccine on nasopharyngeal and nasal carriage of Streptococcus pneumoniae. *Scand. J. Infect. Dis.* **12:**97–100.

37. Honkanen, P. O., T. Keistinen, L. Miettinen, E. Herva, U. Sankilampi, E. Läärä, M. Leinonen, S. L. Kivelä, and P. H. Mäkelä. 1999. Incremental effectiveness of pneumococcal vaccine on simultaneously administered influenza vaccine in preventing pneumonia and pneumococcal pneumonia among persons aged 65 years or older. *Vaccine* **17:**2493–2500.

38. Hostetter, M. K. 1986. Serotypic variations among virulent pneumococci in deposition and degradation of covalently bound C3b: implications for phagocytosis and antibody production. *J. Infect. Dis.* **153:**682–693.

39. Jackson, L. A., K. M. Neuzil, O. Yu, P. Benson, W. E. Barlow, A. L. Adams, C. A. Hanson, L. D. Mahoney, D. K. Shay, and W. W. Thompson. 2003. Effectiveness of pneumococcal polysaccharide vaccine in older adults. *N. Engl. J. Med.* **348:**1747–1755.

40. Jaffar, S., A. Leach, P. G. Smith, F. Cutts, and B. Greenwood. 2003. Effects of misclassification of causes of death on the power of a trial to assess the efficacy of a pneumococcal conjugate vaccine in The Gambia. *Int. J. Epidemiol.* **32:**430–436.

41. Jodar, L., J. Butler, G. Carlone, R. Dagan, D. Goldblatt, H. Käyhty, K. Klugman, B. Plikaytis, G. Siber, R. Kohberger, I. Chang, and T. Cherian. 2003. Serological criteria for evaluation and licensure of new pneumococcal conjugate vaccine formulations for use in infants. *Vaccine* **21:**3265–3272.

42. Jokinen, J., H. Åhman, T. Kilpi, P. H. Mäkelä, and H. Käyhty. The concentration of anti-pneumococcal antibodies as a correlate of protection: an application to acute otitis media. *J. Infect. Dis.,* in press.

43. Karjalainen, H., M. Koskela, J. Luotonen, E. Herva, and P. Sipilä. 1990. Antibodies against Streptococcus pneumoniae, Haemophilus influenzae and Branhamella catarrhalis in middle ear effusion during early phase of acute otitis media. *Acta Oto-Laryngol.* **109:**111–118.

44. Karma, P., J. Pukander, M. Sipilä, M. Timonen, S. Pöntynen, E. Herva, P. Grönroos, and H. Mäkelä. 1985. Prevention of otitis media in children by pneumococcal vaccination. *Am. J. Otolaryngol.* **6:**173–184.

45. Karstaedt, A. S., M. Khoosal, and H. H. Crewe-Brown. 2001. Pneumococcal bacteremia in adults in Soweto, South Africa, during the course of a decade. *Clin. Infect. Dis.* **33:** 610–614.

46. Kaufman, P. 1947. Pneumonia in old age. Active immunization against pneumonia with pneumococcus polysaccharide, results of a six year study. *Arch. Intern. Med.* **79:**518–531.

47. Kauppi, M., L. Saarinen, and H. Käyhty. 1993. Anti-capsular polysaccharide antibodies reduce nasopharyngeal colonization by Haemophilus influenzae type b in infant rats. *J. Infect. Dis.* **167:**365–371.

48. Käyhty, H., H. Åhman, P. R. Rönnberg, R. Tillikainen, and J. Eskola. 1995. Pneumococcal polysaccharide-meningococcal outer membrane protein complex conjugate vaccine is immunogenic in infants and children. *J. Infect. Dis.* **172:**1273–1278.

49. Kilpi, T., H. Åhman, J. Jokinen, K. S. Lankinen, A. Palmu, H. Savolainen, M. Grönholm, M. Leinonen, T. Hovi, J. Eskola, H. Käyhty, N. Bohidar, J. C. Sadoff, and P. H. Mäkelä. 2003. Protective efficacy of a second pneumococcal conjugate vaccine against pneumococcal acute otitis media in infants and children: randomized, controlled trial of a 7-valent pneumococcal polysaccharide-meningococcal outer membrane protein complex conjugate vaccine in 1666 children. *Clin. Infect. Dis.* **37:**1155–1164.

50. Kilpi, T., E. Herva, T. Kaijalainen, R. Syrjänen, and A. K. Takala. 2001. Bacteriology of acute otitis media in a cohort of Finnish children followed for the first two years of life. *Pediatr. Infect. Dis. J.* **20:**654–662.

51. Kim, J. O., S. Romero-Steiner, U. B. Sorensen, J. Blom, M. Carvalho, S. Barnard, G. Carlone, and J. N. Weiser. 1999. Relationship between cell surface carbohydrates and intrastrain variation on opsonophagocytosis of Streptococcus pneumoniae. *Infect. Immun.* **67:**2327–2333.

52. Klugman, K. P., S. A. Madhi, R. E. Huebner, R. Kohberger, N. Mbelle, and N. Pierce. 2003. A trial of a 9-valent pneumococcal conjugate vaccine in children with and those without HIV infection. *N. Engl. J. Med.* **349:** 1341–1348.

53. Koivula, I., M. Sten, M. Leinonen, and P. H. Mäkelä. 1997. Clinical efficacy of pneumococcal vaccine in the elderly: a randomized, single-blind population-based trial. *Am. J. Med.* **103:** 281–290.

54. Kroon, F. P., J. T. van Dissel, E. Ravensbergen, P. H. Nibbering, and R. van Furth. 1999. Antibodies against pneumococcal polysaccharides after vaccination in HIV-infected individuals: 5-year follow-up of antibody concentrations. *Vaccine* **18:**524–530.

55. **Lipsitch, M.** 2001. Interpreting results from trials of pneumococcal conjugate vaccines: a statistical test for detecting vaccine-induced increases in carriage of nonvaccine serotypes. *Am. J. Epidemiol.* **154:**85–92.

56. **MacLeod, M., R. G. Hodges, M. Heidelberger, and W. G. Bernhard.** 1945. Prevention of pneumococcal pneumonia by immunization with specific capsular polysaccharides. *J. Exp. Med.* **82:**445–465.

57. **Mäkelä, P. H., H. Käyhty, T. Leino, K. Auranen, H. Peltola, N. Ekström, and J. Eskola.** 2003. Long-term persistence of immunity after immunisation with Haemophilus influenzae type b conjugate vaccine. *Vaccine* **22:**287–292.

58. **Mäkelä, P. H., M. Leinonen, J. Pukander, and P. Karma.** 1981. A study of the pneumococcal vaccine in prevention of clinically acute attacks of recurrent otitis media. *Rev. Infect. Dis.* **3**(Suppl.)**:**S124–S132.

59. **Mäkelä, P. H., and H. Käyhty.** 2002. Evolution of conjugate vaccines. *Expert Rev. Vaccines* **1:**399–410.

60. **Malley, R., A. M. Stack, M. L. Ferretti, C. M. Thompson, and R. A. Saladino.** 1998. Anticapsular polysaccharide antibodies and nasopharyngeal colonization with Streptococcus pneumoniae in infant rats. *J. Infect. Dis.* **178:**878–882.

61. **Mbelle, N., R. E. Huebner, A. D. Wasas, A. Kimura, I. Chang, and K. P. Klugman.** 1999. Immunogenicity and impact on nasopharyngeal carriage of a nonavalent pneumococcal conjugate vaccine. *J. Infect. Dis.* **180:**1171–1176.

62. **Mogi, G., S. Maeda, T. Yoshida, and N. Watanabe.** 1976. Immunochemistry of otitis media with effusion. *J. Infect. Dis.* **133:**126–136.

63. **Mulholland, K., S. Hilton, R. Adegbola, S. Usen, A. Oparaugo, C. Omosigho, M. Weber, A. Palmer, G. Schneider, K. Jobe, G. Lahai, S. Jaffar, O. Secka, K. Lin, C. Ethevenaux, and B. Greenwood.** 1997. Randomised trial of Haemophilus influenzae type-b tetanus protein conjugate vaccine [corrected] for prevention of pneumonia and meningitis in Gambian infants. *Lancet* **349:**1191–1197.

64. **Mulholland, K., O. Levine, H. Nohynek, and B. M. Greenwood.** 1999. Evaluation of vaccines for the prevention of pneumonia in children in developing countries. *Epidemiol. Rev.* **21:**43–55.

65. **Nichol, K. L., L. Baken, J. Wuorenma, and A. Nelson.** 1999. The health and economic benefits associated with pneumococcal vaccination of elderly persons with chronic lung disease. *Arch. Intern. Med.* **159:**2437–2442.

66. **Nieminen, T., H. Käyhty, O. Leroy, and J. Eskola.** 1999. Pneumococcal conjugate vaccination in toddlers: mucosal antibody response measured as circulating antibody-secreting cells and as salivary antibodies. *Pediatr. Infect. Dis. J.* **18:**764–772.

67. **Nuorti, J. P., J. C. Butler, L. Gelling, J. L. Kool, A. L. Reingold, and D. J. Vugia.** 2000. Epidemiologic relation between HIV and invasive pneumococcal disease in San Francisco County, California. *Ann. Intern. Med.* **132:**182–190.

68. **Nurkka, A., H. Åhman, M. Korkeila, V. Jäntti, H. Käyhty, and J. Eskola.** 2001. Serum and salivary anti-capsular antibodies in infants and children immunized with the heptavalent pneumococcal conjugate vaccine. *Pediatr. Infect. Dis. J.* **20:**25–33.

69. **Nurkka, A., H. Åhman, M. Yaich, J. Eskola, and H. Käyhty.** 2001. Serum and salivary anti-capsular antibodies in infants and children vaccinated with octavalent pneumococcal conjugate vaccines, PncD and PncT. *Vaccine* **20:**194–201.

70. **Obaro, S., A. Leach, and K. W. McAdam.** 1998. Use of pneumococcal polysaccharide vaccine in children. *Lancet* **352:**575.

71. **Obaro, S. K.** 2002. The new pneumococcal vaccine. *Clin. Microbiol. Infect.* **8:**623–633.

72. **O'Brien, K. L., and R. Dagan.** 2003. The potential indirect effect of conjugate pneumococcal vaccines. *Vaccine* **21:**1815–1825.

73. **O'Brien, K. L., L. H. Moulton, R. Reid, R. Weatherholtz, J. Oski, L. Brown, G. Kumar, A. Parkinson, D. Hu, J. Hackell, I. Chang, R. Kohberger, G. Siber, and M. Santosham.** 2003. Efficacy and safety of seven-valent conjugate pneumococcal vaccine in American Indian children: group randomised trial. *Lancet* **362:**355–361.

74. **Ordman, D.** 1938. Pneumococcus types in South Africa. A study of their occurrence and distribution in the population and the effect thereon of prophylactic inoculation. South African Institute of Medical Research **9:**1–28.

75. **Örtqvist, Å., J. Hedlund, L. A. Burman, E. Elbel, M. Hofer, M. Leinonen, I. Lindblad, B. Sundelöf, M. Kalin, and Swedish Pneumococcal Vaccination Study Group.** 1998. Randomised trial of 23-valent pneumococcal capsular polysaccharide vaccine in prevention of pneumonia in middle-aged and elderly people. *Lancet* **351:**399–403.

76. **Overweg, K., C. D. Pericone, G. G. Verhoef, J. N. Weiser, H. D. Meiring, A. P. De Jong, R. De Groot, and P. W. Hermans.** 2000. Differential protein expression in phenotypic variants of Streptococcus pneumoniae. *Infect. Immun.* **68:** 4604–4610.

77. **Parkkali, T., H. Käyhty, T. Ruutu, L. Volin, J. Eskola, and P. Ruutu.** 1996. A comparison of early and late vaccination with Haemophilus influenzae type b conjugate and pneumococcal polysaccharide vaccines after allogeneic BMT. *Bone Marrow Transplant.* **18:**961–967.

78. **Parkkali, T., M. Väkeväinen, H. Käyhty, T. Ruutu, and P. Ruutu.** 2001. Opsonophagocytic activity against Streptococcus pneumoniae type 19F in allogeneic BMT recipients before and after vaccination with pneumococcal polysaccharide vaccine. *Bone Marrow Transplant.* **27:**207–211.

79. **Pneumonia Vaccine Trial Investigators' Group.** 2001. Standardization of interpretation of chest radiographs for the diagnosis of pneumonia in children. WHO/V&B/01.35. World Health Organization, Geneva, Switzerland.

80. **Powers, D. C., E. L. Anderson, K. Lottenbach, and C. M. Mink.** 1996. Reactogenicity and immunogenicity of a protein-conjugated pneumococcal oligosaccharide vaccine in older adults. *J. Infect. Dis.* **173:**1014–1018.

81. **Puumalainen, T., R. Dagan, T. Wuorimaa, R. Zeta-Capeding, M. Lucero, J. Ollgren, H. Käyhty, and H. Nohynek.** 2003. Greater antibody responses to an eleven valent mixed carrier diphtheria- or tetanus-conjugated pneumococcal vaccine in Filipino than in Finnish or Israeli infants. *Pediatr. Infect. Dis. J.* **22:**141–149.

82. **Puumalainen, T., M. R. Zeta-Capeding, H. Kayhty, M. G. Lucero, K. Auranen, O. Leroy, and H. Nohynek.** 2002. Antibody response to an eleven valent diphtheria- and tetanus-conjugated pneumococcal conjugate vaccine in Filipino infants. *Pediatr. Infect. Dis. J.* **21:**309–314.

83. **Rennels, M. B., K. M. Edwards, H. L. Keyserling, K. S. Reisinger, D. A. Hogerman, D. V. Madore, I. Chang, P. R. Paradiso, F. J. Malinoski, and A. Kimura.** 1998. Safety and immunogenicity of heptavalent pneumococcal vaccine conjugated to CRM197 in United States infants. *Pediatrics* **101:**604–611.

84. **Riley, I. D., F. A. Everingham, D. E. Smith, and R. M. Douglas.** 1981. Immunisation with a polyvalent pneumococcal vaccine. Effect on respiratory mortality in children living in the New Guinea highlands. *Arch. Dis. Child.* **56:**354–357.

85. **Riley, I. D., D. Lehmann, M. P. Alpers, T. F. Marshall, H. Gratten, and D. Smith.** 1986. Pneumococcal vaccine prevents death from acute lower-respiratory-tract infections in Papua New Guinean children. *Lancet* **ii:**877–881.

86. **Riley, I. D., P. I. Tarr, M. Andrews, M. Pfeiffer, R. Howard, P. Challands, and G. Jennison.** 1977. Immunisation with a polyvalent pneumococcal vaccine. Reduction of adult respiratory mortality in a New Guinea Highlands community. *Lancet* **i:**1338–1341.

87. **Rodriguez, W. J., and R. H. Schwartz.** 1999. Streptococcus pneumoniae causes otitis media with higher fever and more redness of tympanic membranes than Haemophilus influenzae or Moraxella catarrhalis. *Pediatr. Infect. Dis. J.* **18:**942–944.

88. **Romero-Steiner, S., D. M. Musher, M. S. Cetron, L. B. Pais, J. E. Groover, A. E. Fiore, B. D. Plikaytis, and G. M. Carlone.** 1999. Reduction in functional antibody activity against Streptococcus pneumoniae in vaccinated elderly individuals highly correlates with decreased IgG antibody avidity. *Clin. Infect. Dis.* **29:**281–288.

89. **Rosenow, C., P. Ryan, J. N. Weiser, S. Johnson, P. Fontan, A. Ortqvist, and H. R. Masure.** 1997. Contribution of novel choline-binding proteins to adherence, colonization and immunogenicity of Streptococcus pneumoniae. *Mol. Microbiol.* **25:**819–829.

90. **Rubins, J. B., M. Alter, J. Loch, and E. N. Janoff.** 1999. Determination of antibody responses of elderly adults to all 23 capsular polysaccharides after pneumococcal vaccination. *Infect. Immun.* **67:**5979–5984.

91. **Rubins, J. B., A. K. Puri, J. Loch, D. Charboneau, R. MacDonald, N. Opstad, and E. N. Janoff.** 1998. Magnitude, duration, quality, and function of pneumococcal vaccine responses in elderly adults. *J. Infect. Dis.* **178:**431–440.

92. **Sankilampi, U., P. O. Honkanen, A. Bloigu, E. Herva, and M. Leinonen.** 1996. Antibody response to pneumococcal capsular polysaccharide vaccine in the elderly. *J. Infect. Dis.* **173:**387–393.

93. **Sankilampi, U., P. O. Honkanen, A. Bloigu, and M. Leinonen.** 1997. Persistence of antibodies to pneumococcal capsular polysaccharide vaccine in the elderly. *J. Infect. Dis.* **176:**1100–1104.

94. **Schneerson, R., J. B. Robbins, C. Chu, A. Sutton, W. Vann, J. C. Vickers, W. T. London, B. Curfman, M. C. Hardegree, J. Shiloach, and S. C. Rastogi.** 1984. Serum antibody responses of juvenile and infant rhesus monkeys injected with *Haemophilus influenzae* type b and pneumococcus type 6A capsular polysaccharide-protein conjugates. *Infect. Immun.* **45:**582–591.

95. **Shann, F.** 1998. Pneumococcal vaccine: time for another controlled trial. *Lancet* **351:**1600–1601.

96. **Shapiro, E. D., A. T. Berg, R. Austrian, D. Schroeder, V. Parcells, A. Margolis, R. K.**

Adair, and J. D. Clemens. 1991. The protective efficacy of polyvalent pneumococcal polysaccharide vaccine. *N. Engl. J. Med.* **325:** 1453–1460.

97. Shelly, M. A., H. Jacoby, G. J. Riley, B. T. Graves, M. Pichichero, and J. J. Treanor. 1997. Comparison of pneumococcal polysaccharide and CRM197-conjugated pneumococcal oligosaccharide vaccines in young and elderly adults. *Infect. Immun.* **65:**242–247.

98. Sigurdardottir, S. T., G. Ingolfsdottir, K. Davidsdottir, T. Gudnason, S. Kjartansson, K. G. Kristinsson, F. Bailleux, O. Leroy, and I. Jonsdottir. 2002. Immune response to octavalent diphtheria- and tetanus-conjugated pneumococcal vaccines is serotype- and carrier-specific: the choice for a mixed carrier vaccine. *Pediatr. Infect. Dis. J.* **21:**548–554.

99. Simberkoff, M. S., A. P. Cross, M. Al-Ibrahamin, A. L., Baltch, P. J. Geiseler, J. Nadler, A. S. Richmond, R. P. Smith, G. Schiffman, D. S. Shepard, et al. 1986. Efficacy of pneumococcal vaccine in high-risk patients. Results of a Veterans Administration Cooperative Study. *N. Engl. J. Med.* **315:**1318–1327.

100. Sloyer, J. L., Jr., V. M. Howie, J. H. Ploussard, G. Schiffman, and R. B. Johnston, Jr. 1976. Immune response to acute otitis media: association between middle ear fluid antibody and the clearing of clinical infection. *J. Clin. Microbiol.* **4:**306–308.

101. Smit, P., D. Oberholzer, S. Hayden-Smith, H. J. Koornhof, and M. R. Hilleman. 1977. Protective efficacy of pneumococcal polysaccharide vaccines. *JAMA* **238:** 2613–2616.

102. Soininen, A., G. van den Dobbelsteen, L. Oomen, and H. Käyhty. 2000. Are the enzyme immunoassays for antibodies to pneumococcal capsular polysaccharides serotype specific? *Clin. Diagn. Lab. Immunol.* **7:**468–476.

103. Talbot, U. M., A. W. Paton, and J. C. Paton. 1996. Uptake of *Streptococcus pneumoniae* by respiratory epithelial cells. *Infect. Immun.* **64:** 3772–3777.

104. Teele, D. W., J. O. Klein, L. Bratton, G. R. Fisch, O. R. Mathieu, P. J. Porter, S. G. Starobin, L. D. Tarlin, R. P. Younes, and The Greater Boston Collaborative Otitis Media Study Group. 1981. Use of pneumococcal vaccine for prevention of recurrent acute otitis media in infants in Boston. *Rev. Infect. Dis.* **3**(Suppl.):S113–S118.

105. van Alphen, L., P. Eijk, H. Käyhty, J. van Marle, and J. Dankert. 1996. Antibodies to Haemophilus influenzae type b polysaccharide affect bacterial adherence and multiplication. *Infect. Immun.* **64:**995–1001.

106. van Egmond, M., C. A. Damen, A. B. van Spriel, G. Vidarsson, E. van Garderen, and J. G. van de Winkel. 2001. IgA and the IgA Fc receptor. *Trends Immunol.* **22:**205–211.

107. Watanabe, N., H. Yoshimura, and G. Mogi. 1988. Induction of antigen-specific IgA-forming cells in the middle ear mucosa. *Arch. Otolaryngol. Head Neck Surg.* **114:**758–762.

108. Weintrub, P. S., G. Schiffman, J. E. Addiego, Jr., K. K. Matthay, E. Vichinsky, R. Johnson, B. Lubin, W. C. Mentzer, and A. J. Ammann. 1984. Long-term follow-up and booster immunization with polyvalent pneumococcal polysaccharide in patients with sickle cell anemia. *J. Pediatr.* **105:**261–263.

109. Weiser, J. N., R. Austrian, P. K. Sreenivasan, and H. R. Masure. 1994. Phase variation in pneumococcal opacity: relationship between colonial morphology and nasopharyngeal colonization. *Infect. Immun.* **62:**2582–2589.

110. Weiser, J. N., D. Bae, C. Fasching, R. W. Scamurra, A. J. Ratner, and E. N. Janoff. 2003. Antibody-enhanced pneumococcal adherence requires IgA1 protease. *Proc. Natl. Acad. Sci. USA* **100:**4215–4220.

111. Whitney, C. G., M. M. Farley, J. Hadler, L. H. Harrison, N. M. Bennett, R. Lynfield, A. Reingold, P. R. Cieslak, T. Pilishvili, D. Jackson, R. R. Facklam, J. H. Jorgensen, and A. Schuchat. 2003. Decline in invasive pneumococcal disease after the introduction of protein-polysaccharide conjugate vaccine. *N. Engl. J. Med.* **348:**1737–1746.

112. Whitney, C. G., W. Schaffner, and J. C. Butler. 2001. Rethinking recommendations for use of pneumococcal vaccines in adults. *Clin. Infect. Dis.* **33:**662–675.

113. Wuorimaa, T., R. Dagan, M. Väkeväinen, F. Bailleux, R. Haikala, M. Yaich, J. Eskola, and H. Käyhty. 2001. Avidity and subclasses of IgG after immunization of infants with an 11-valent pneumococcal conjugate vaccine with or without aluminum adjuvant. *J. Infect. Dis.* **184:**1211–1215.

SUBJECT INDEX